014393304
KV-429-908

Moulds, Toxins and Food

Moulds, Toxins and Food

Claude Moreau
Laboratoire de Biologie Végétale
Faculté des Sciences
Brest, France

Translated with additional material by
Maurice Moss
Department of Microbiology
University of Surrey
Guildford

A Wiley–Interscience Publication

JOHN WILEY & SONS
Chichester · New York · Brisbane · Toronto

First published 1974 © Masson & Cie, Éditeurs, Paris
under the title Moisissures Toxiques dans L'Alimentation
2nd Edition by Claude Moreau

Library of Congress Cataloging in Publication Data:

Moreau, Claude.
Moulds, toxins and food.

'A Wiley–Interscience publication.'
Translation of Moisissures toxiques dans
l'alimentation.
Bibliography: p.
Includes index.
1. Food poisoning. 2. Mycotoxins–Toxicology.
3. Moulds (Botany) 4. Veterinary mycology.
I. Title. [DNLM: 1. Food poisoning. 2. Mycotoxins–Poisoning. QZ65 M837m]
QR201.F6M6713 615.9'52'92 78-8715
ISBN 0 471 99681 5

Typeset in IBM Press Roman by Preface Ltd, Salisbury, Wilts
Printed in Great Britain by Unwin Brothers, The Gresham Press, Old Woking, Surrey

CONTENTS

PREFACE TO THE FRENCH EDITION

The interest shown in France and abroad in the first edition of this work (which was published in 1968 at the request of Le Comité de Technologie Agricole de la Délégation générale à la Recherche Scientifique et Technique) and the rapid advances made in the study of toxic moulds in foodstuffs, have led us to consider that it would be useful to enlarge upon the information given in our first edition. We intended to produce a 'standard work' for the many research workers and technologists interested in the subject. Here they will find references to original works more detailed than this study, the only aim of which is to assist, not only mycologists, doctors, veterinarians and biochemists, but also toxicologists, nutritionists, dieticians, hygienists, and all those concerned with food technology.

We should like to thank all our colleagues and friends who have been kind enough to stimulate interest in this book by their reviews in various publications, correcting errors and informing us of their recent work on the subject.

May we express out gratitude to all the various collaborators who have contributed in different ways to the completion of the present edition.

Claude Moreau

PREFACE TO THE ENGLISH EDITION

It is the translator's primary desire to make available in the English language the remarkable synthesis by Dr. Moreau of his very complete bibliography covering the many aspects of the study of mycotoxins up to 1972.

The author himself is very aware of the many advances which have been made in several areas of study since 1972 and has encouraged the addition of new material to the translation. I have attempted to illustrate these advances by reviewing some of the literature which appeared during the period from 1976 to the beginning of 1978, and this review appears as an additional chapter (Chapter 13) at the end of the book. This approach has meant that the period 1973–75 has not been covered with the thoroughness with which the author himself dealt with the earlier literature.

The translation is not always a literal one for I have occasionally taken the liberty of imposing my own interpretation of the literature where it significantly differed from that of the author. For this reason, blame for any errors should be placed upon the translator rather than upon the original author.

I would like to thank Dr. Moreau for his encouragement and help and hope that others will find his multidisciplinary approach to the study of mycotoxins as inspiring as I have.

Maurice Moss
Guildford, 1978.

INTRODUCTION

A TOPICAL PROBLEM

Through their use in sorcery and medicine ancient civilizations already had some idea of the toxicity of fungi. However it required the methods of modern science before the full implications of these properties could be appreciated.

Although studies on macrofungi had advanced at a relatively early stage, work on the microfungi, especially the moulds, proceeded particularly rapidly during the last few decades especially since the discovery of the antibiotics.

Agricultural and food industrial practices have undergone considerable changes recently and these changes have brought to the forefront problems hardly appreciated before. The stakes are high: the agricultural and food industries have reached second place in the economy of many nations. In France, for example, the turnover in 1964 reached of the order of 40 thousand million francs (1505).* However in the world at large food production has increased at a rate of 1.5% per annum, whereas, in the same time, population growth has been 2.5%. Thus the increase in food production is insufficient and every effort at improvement, however small, should be considered.

Moreover, there are some parts of the world where surpluses in production of some food commodities lead to large-scale storage, and other parts of the world, associated with significant demographic expansion, are threatened by a shortage of nourishment leading to the consideration of new sources of food, especially of protein. This represents one of the great paradoxes of our age stressed in a recent editorial by Linières (1505).

Now, wherever there is a concentration of stored food products the danger of mould damage is high; another aspect of the spoilage of fresh foods, use of which is not yet widespread, is the risk of production of highly toxic compounds which could be disastrous to those for whom the foods are destined.

If human foods pose grave problems, those associated with animal feeds, perhaps also affected by new methods, cause considerable anxiety. The gradual introduction of pelleted feeds containing cereal products, oil-meal cake, milk powder, and various waste products has considerably changed the conditions under which meat is produced. These changes have not occurred without introducing new problems, particularly arising from the growth of moulds (830). The role of soil fungi, capable of producing diffusible metabolites in the rhizosphere which may be taken up by

* Numbers in brackets refer to references in the bibliography.

the plants themselves, and the adverse effects of moulds so often present on dry forage, are topics that have not been tackled sufficiently.

Even the most recent accounts of food microbiology concentrate particularly on problems associated with the growth of bacteria in foods, paying little attention to the fungi (52, 345, 576, 836, 1100, 1220, 1453, 1514, 2147, 2160, 2550, 2580, 2739, 2797). Much of the work stressing the importance of mould toxins in foods is fragmented and scattered. In addition to the socio-economic aspects, this is a question which doctors, veterinarians, food hygienists, animal technicians, biochemists, those concerned with the storage and transport of food, and consumers should not be disinterested in. For their part mycologists have not been particularly interested until recently.

Since 1948, we have ourselves been preoccupied with the problems of changes in stored fruit, but it was in 1960, in establishing the accidental death of cattle following the consumption of forage heavily contaminated with the mould *Aspergillus clavatus,* that we appreciated the gravity of the problem (1738, 1739). The abundance of moulds on food products (1724, 1737), particularly on stored grain (1740), and their importance in manufactured products (1741), is a cause for anxiety shared by doctors, veterinarians, and food manufacturers who have drawn our attention to the danger which the fungi present to the consumer.

From mycological studies, of which several are original, we have brought together a large number of references. Nevertheless, the magnitude of the problem is such that it can only be touched on here in this single review of work which is constantly being added to by new research.

CHAPTER 1

FOOD-BORNE MOULDS

The growth of microfungi on the surface of, and within products to be incorporated into foods has frequently been recorded, especially in stored commodities. Delays in the consumption of foods, and increased transport over long distances, make food storage more and more necessary. Rapid change continually poses new problems which, if not appreciated, may result in an increase in the spread of the moulds.

CONDITIONS FOR THE GROWTH OF MOULDS

The conditions of harvesting, and to a greater extent those of storage, of agricultural commodities and their products have a considerable influence on the growth of moulds. We only mention here some of the essential facts and, for further details, suggest that general treatises on fungal physiology be consulted (532, 1045, 1498).

Temperature

Temperature plays a dominant role in the growth of mycelium as well as in the formation and germination of spores. The majority of moulds grow between 15–30 °C with optimum growth at about 20–25 °C. However, *Cladosporium herbarum* is capable of slow growth at –6 °C on refrigerated meat (317); we have been able to isolate several species of *Penicillium* from refrigerated warehouses used for keeping fish at –20 °C. They certainly could not grow at this temperature but the spores remained viable. Thus the survival as well as the growth of fungi may be important in those commodities destined as foods. It is worth recalling, out of curiosity, the observations of Becquerel (177); the viability of spores of *Rhizopus nigricans, Mucor mucedo,* and *Aspergillus niger* was not reduced after 77 hours in liquid hydrogen (–253 °C) or 492 hours in liquid air (–190 °C).

This retention of viability by fungi at low temperatures makes it possible to preserve many of them in the mycological laboratory by freeze-drying.

Conversely certain fungi can grow at relatively high temperatures, thus *Aspergillus flavus* may dominate the microbial flora at 35 °C in drying tunnels used in the food industry (1737), *Aspergillus fumigatus* may tolerate temperatures of 50 °C, and *Humicola lanuginosa* is able to grow at 60 °C (545). Some fungi are able to survive, but not grow, at even higher temperatures. Ascospores of *Byssochlamys fulva,* held at 85 °C for 10 minutes or at 70 °C for 2 hours, subsequently showed up to 20% germination. In a dry atmosphere, the ascospores of *Neurospora* are able to survive a period of 15 to 20 minutes at 130 °C (2611), which explains their survival in bakery ovens.

Some fungi have a wide temperature range for growth whereas for others the minimum and maximum cardinal temperatures are fairly close (Figure 1). *Botrytis cinerea,* an undesirable pathogen of glasshouse crops as well as an agent in the spoilage of fruit in refrigerated warehouses, grows very well at 20 °C and still quite vigorously at 5 °C. On the other hand *Aspergillus clavatus,* which is found in profusion on germinated cereals at 22 °C is not able to grow at temperatures below 10 °C.

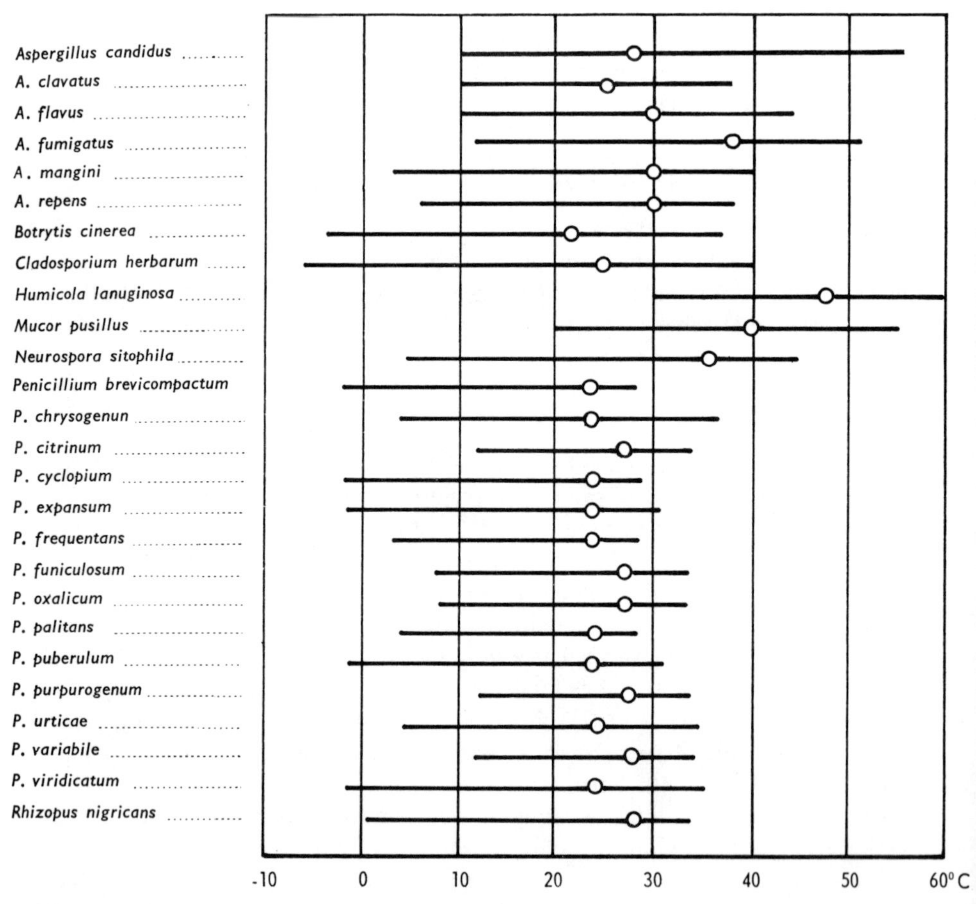

Figure 1. Minimum and maximum temperatures for the growth of a number of moulds from food commodities, based on the results of several authors (545, 1498, 1688)

It is thus possible to recognize the following groups (545):

1. *Thermophilic fungi* which grow well at 50 °C but are unable to grow below 20 °C. Within this group some may be distinguished as microthermophiles (optimum temperatures for growth 25–35 °C, maximum 40–48 °C, e.g. *Byssochlamys nivea, Thielavia sepedonium,* others as psychrotolerant thermophiles (growing over a wide range of temperature but well at 48 °C, e.g. *Absidia ramosa, Aspergillus fumigatus*), and others as true thermophiles (optimum temperature for growth 40–50 °C, maximum may reach 60 °C, but the minimum temperature is always greater than 20 °C, e.g. *Mucor pusillus, Thermomyces lanuginosus*) (65).
2. *Thermotolerant fungi* which may have a maximum temperature for growth close to 50 °C but are able to grow at temperatures well below 20 °C (e.g. *Aspergillus niger*).
3. *Mesophilic fungi* growing in the range 10–40 °C with optima generally close to 25 °C (e.g. *Penicillium chrysogenum, Aspergillus versicolor*).
4. *Psychrophilic fungi* showing optimum growth at temperatures between 5–10 °C.
5. *Cryophilic fungi* particularly able to grow at very low temperatures (996).

It is essential to note that, although often very similar, the cardinal temperatures for growth and for sporulation may be different; thus the observed optimum for sporulation may be at either a lower temperature than that for growth (e.g. *Fusarium conglutinans, Aspergillus versicolor, Penicillium cyclopium*) (2632), or at a higher temperature. It is worth recalling that for some species (notably members of the *Aspergillus glaucus* group) the optimum temperature for sporulation may vary according to the type of spores formed. The production of cleistothecia is restricted to a narrow range around the optimum temperature for growth (18–27 °C) whereas conidia are produced over the whole range at which growth occurs (10–33 °C) (1746).

Temperature may also influence the morphology of sporophores. A classical example is that of *Aspergillus giganteus* (a species of the *clavatus* group) which, at 20 °C, produces very long conidiophores reaching up to 10 cm in height whereas cultures incubated at 30 °C have conidiophores less than 1 cm in length (2105). Such morphological variation is often observed at extreme temperatures not very favourable for growth.

As well as playing an important role in the growth and morphogenesis of moulds, temperature may also influence the quantity and nature of the metabolites which they produce. Neither is the optimum temperature for growth always the best temperature for the production of all metabolites. Thus *Pithomyces chartarum,* for which the optimum growth temperature is 24 °C (311), produces sporidesmin optimally at 20 °C (602). *Penicillium chrysogenum* produces an increased yield of penicillin if, after incubation at 30 °C for 42 hours (which is the optimum for growth), it is then cultured at 20 °C (1903).

Humidity

It is well established that, among the factors involved in the growth of fungi, humidity plays an important role (2326, 2433, 2434). It influences, not only the growth of mycelium and the production of fruiting bodies, but also the germination of spores. The role of humidity in the growth of moulds is particularly apparent among those attacking stored cereals (1498). Table 1 and Figures 2 and 3 give several examples showing the minimum relative humidity required for the growth of a range of fungi.

It is possible to classify fungi according to the range of relative humidity to which they respond (1058, 1949):

1. *Xerophilic fungi* in which spores can germinate at less than 80% RH and for which optimum growth is observed at about 95% RH (e.g. *Aspergillus repens, A. restrictus, A. versicolor, Hemispora stellata*).
2. *Mesophilic fungi* in which spores require between 80 and 90% RH to germinate and optimum growth occurs between 95 and 100% RH. (e.g. *Alternaria tenuissima, Cladosporium cladosporioides, Penicillium cyclopium*).
3. *Hydrophilic fungi* in which spores will only germinate at RH values greater than 90% and for which the optimum for growth is close to 100% (e.g. *Epicoccum nigrum, Mucor circinelloides, Trichothecium roseum*).

Relative humidity should not be confused with the concentration of water in a substrate. Of course, at a given temperature, for each concentration of water in a substrate there is a corresponding relative humidity representing the ease with which the substrate loses water to the atmosphere (982). At low relative humidities (about 25–30%), water is bound to the substrate by a significant bonding energy but, as the relative humidity increases, the availability of water also increases, the

Table 1. The minimum relative humidity for the spore germination, growth, and sporulation of moulds (2891)

Species	*% Relative humidity*		
	Germination	*Growth*	*Sporulation*
Aspergillus echinulatus	71	62	–
A. chevalieri	65–73	65	–
A. candidus	72–75	72–75	80
A. versicolor	76–78	75	–
A. repens	71–80	–	–
A. flavus	80	80	85
Penicillium expansum	82–86	82	85
Aspergillus niger	80	88–89	92–95
Mucor racemosus	88	92	95
Rhizopus nigricans	90–92	92–94	96
Alternaria tenuis	94	–	–
Cladosporium herbarum	94	–	–

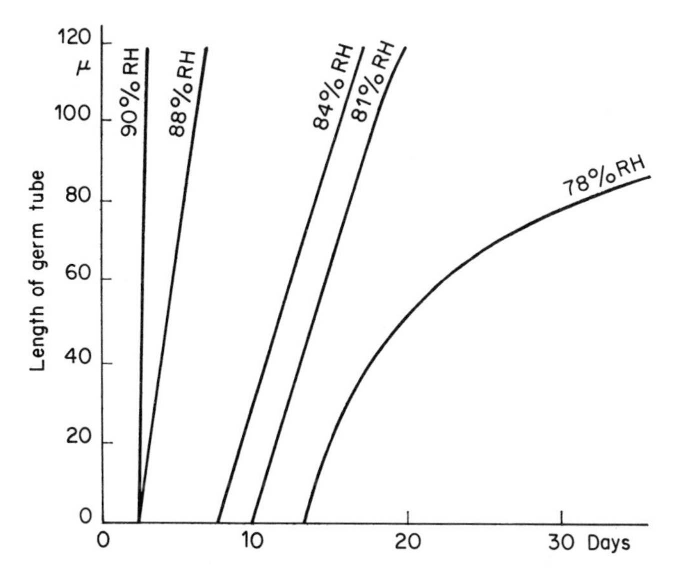

Figure 2. The influence of relative humidity on the growth of *Aspergillus repens* on agar plates at 20 °C (2434)

bonding becoming more and more feeble. It is the degree of mobility of water which makes it possible for moulds to grow on solid substrates. For each substrate, at a particular temperature, it is possible to draw a curve, representing the equilibrium relationship between the relative humidity of the atmosphere and the water content of the substrate, referred to as a *water sorption isotherm*.

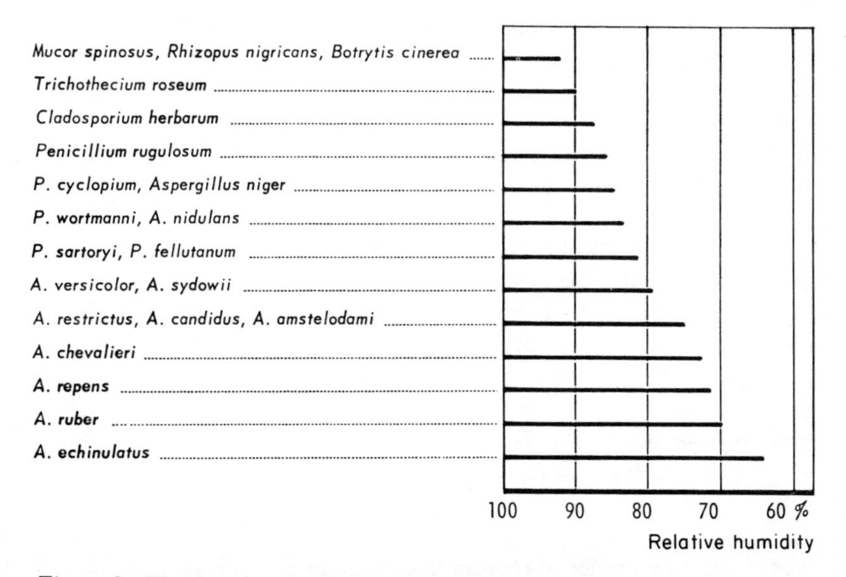

Figure 3. The range of relative humidity over which the spores of a number of fungi will germinate on nutrient agar at 25 °C (2434)

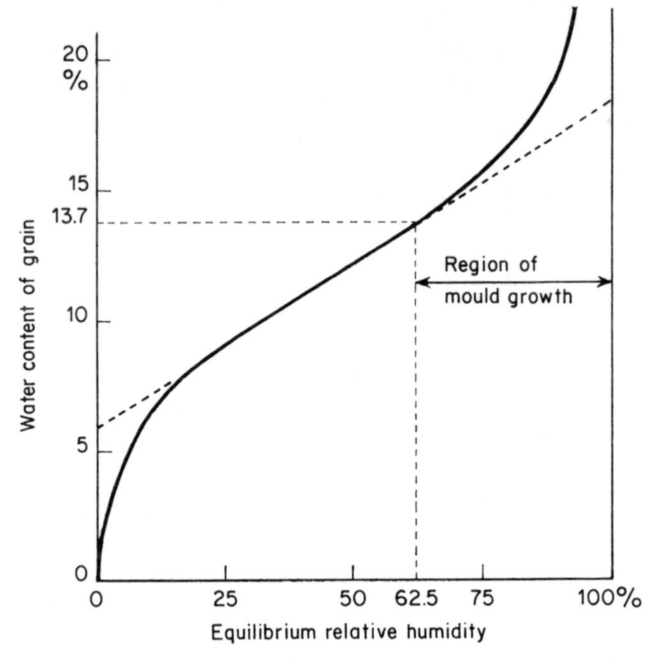

Figure 4. Water sorption isotherm for wheat at 25 °C showing the region in which fungi may grow (2007)

For example, in the case of wheat, Poisson and Guilbot (2007) have shown that, at 25 °C, the growth of moulds will not occur at relative humidities below 62.5% corresponding to a water content in the grain of 13.7% (Figure 4). It is thus possible to understand why, for substrates which may seem to have a high water content, the fungal flora is nevertheless represented only by xerophilic species (2847).

Moulds behave in the same way on media of high osmotic pressure (substrates containing high concentrations of sodium chloride or sugars for example). Thus the *glaucus* group of the genus *Aspergillus* contains members which do not grow well unless media contain 20–40% sucrose or an equimolar concentration of sodium chloride. In a liquid medium *Aspergillus halophilicus* requires even higher concentrations of osmotic agents for optimum growth (70% sucrose or 20% sodium chloride which, at 25 °C, corresponds to an equilibrium relative humidity of approximately 73%). Depending on the concentration of sugars in the medium, members of the *Aspergillus glaucus* group may aquire different morphological characteristics reflecting a variation in their metabolism (Figure 5) (454, 1746). Selective media are essential for a realistic assessment of the osmosphilic mycoflora (1742).

The equilibrium relative humidity, or water activity, is thus a very important parameter and a knowledge of changes in this parameter during storage often allows us to understand the evolution of the fungal flora (2445).

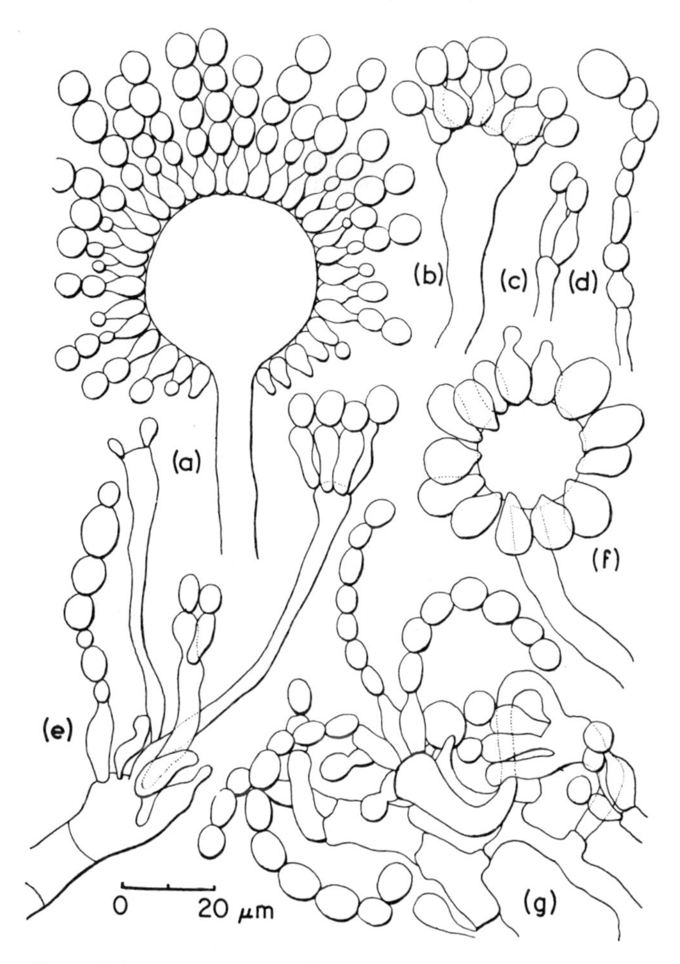

Figure 5. Morphological variations in the conidial structures of *Aspergillus mangini*, a member of the *glaucus* group, as a function of temperature and sugar concentration.

(a) Normal conidial head on Czapek medium containing 1200 g glucose per litre, 33 °C.
(b) Reduction of the number of phialides on a medium containing 2600 g sucrose, 33 °C.
(c) Conidial head not swollen with 2 phialides, from a medium containing 2000 g sucrose, 27 °C.
(d) Mycelium terminating in a single phialide with irregular conidia, from a medium containing 800 g glucose, 33 °C.
(e) Anarchic proliferation of phialides on a medium containing 400 g glucose, 33 °C.
(f) Conidial head with swollen phialides and few spores typical of growth on a concentrated medium, 1200 g glucose, and growth at 18 °C.
(g) Formation of aberrant conidia on media poor in sugar, 200 g sucrose, and growth at 18 °C. (1746)

Nutritional and other factors

A number of other parameters influence the growth of fungi and affect the range of metabolites produced. The exact nutritional requirements for each species differ from all others and thus, for a given substrate, only those species occur for the growth of which the substrate is suitable. In the laboratory it is possible to prepare culture media corresponding to the optimum production of biomass or of certain metabolites by a particular mould.

Most moulds grow at a pH in the range 4–8 although some are capable of growing on very acid, or even very alkaline media; some only grow over a very narrow range others tolerate a wide range (532, 1045, 1498, 2546, 2632).

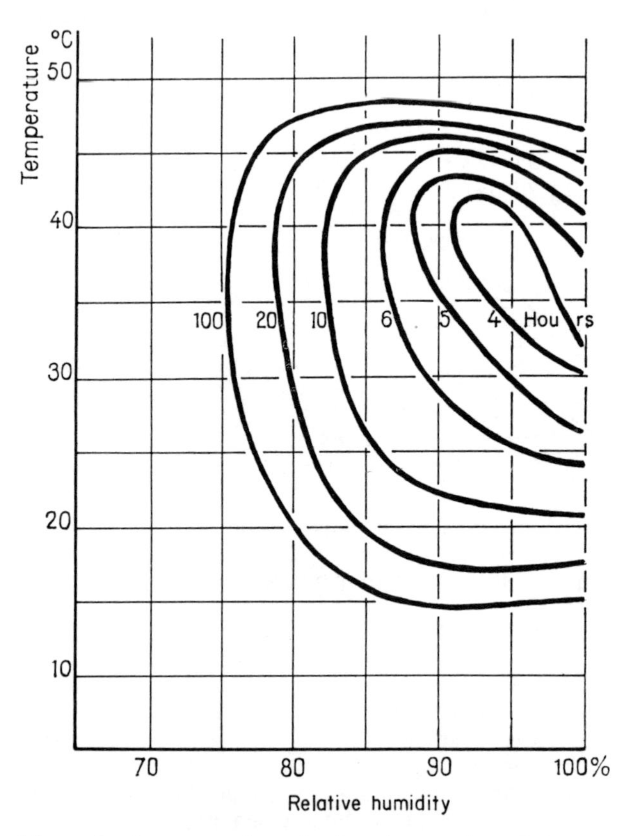

Figure 6. An example of the interaction between humidity and temperature. The diagram shows the conditions for the growth and germination of *Aspergillus niger*. The numbers by each curve indicate the number of hours required for germination at each set of conditions. It can be seen, for example, that the optimum temperature for germination at 90% relative humidity is close to 40 °C (1498)

The concentration of oxygen available to fungi for respiration is an important factor affecting growth. The majority of moulds are aerobic; thus *Mucor* and *Trichoderma,* which have a high oxygen requirement for complete development, grow on the surface of substrates, whereas *Periconia* and *Stachybotrys,* which have a reduced requirement, may grow deeper in the substrate. The growth and sporulation of such genera as *Aspergillus* and *Penicillium* vary according to whether they are cultured in shake flasks or stationary culture (2634). Some fungi are capable of growing very well in an atmosphere of nitrogen others show a very reduced growth (2537).

Carbon dioxide concentration can be a significant factor affecting the growth and morphogenesis of fungi. Low concentrations of carbon dioxide are required for the germination of the spores of *Aspergillus niger* whereas high concentrations, on the other hand, inhibit growth (2912). *Aspergillus flavus* behaves in a similar manner (1074).

It is as well to remember that all the physicochemical parameters interact with each other (see Figure 6), and that added to these parameters there is the possibility of interaction between several species of mould present at the same time on the same substrate. Depending on the situation there may be mutualism or antagonism between fungi (1736); there may be simple competition for available substrate, or the production of antibiotic substances by one of the antagonists, or the transformation of the substrate by one of the fungi leading to an unfavourable medium for other species (for example by the formation of alcohols or acids). The study of competition in the ecology of fungi merits further attention.

It can be seen how complex may be the conditions for growth of moulds on food commodities. However, owing to their large number, they appear to have fantastic plasticity and a remarkable facility for adaptation to the very varied conditions for life.

SOME EXAMPLES OF THE DETERIORATION OF FOOD COMMODITIES

A recent survey estimates that, of the world losses in food commodities amounting to about 20%, more than half is due to the growth of moulds. FAO have calculated the annual losses as many thousands of millions of dollars (thus in 1960, in thousands of millions of dollars, rice 8, potatoes 3, sugar 3, coffee 2.5, millet and sorghum 2, maize 2, wheat 1.5, groundnuts 1, etc.) (933, 1098).

Considerable efforts have been made to increase yields: in France, for example, the yield of wheat has almost doubled since 1939 due essentially to three factors:

- the selection of improved varieties;
- the improvement of cultural techniques (ploughing, sowing, weeding, harvesting);
- the use of good fertilizers adapted to the physiology of each species or variety.

But this splendid work, resulting from the collaboration of agronomists, biologists, geneticists, biochemists, phytopathologists, and soil scientists, for increasing the yield and improving the quality is often short lived. Their efforts are partially extinguished by post-harvest deterioration and losses.

Most stored food products are susceptible to mould spoilage (519). An investigation carried out in German supermarkets led to the isolation of 191 different species of mould from 185 separate samples; these consisted of 26 *Aspergillus*, 118 *Penicillium*, 19 members of the Mucorales, and 28 other species (249). From 385 food samples examined in Japan it was possible to isolate 2,940 cultures of fungi, mainly of the genera *Aspergillus* and *Penicillium* (1375).

The following sections briefly indicate the fungal species most frequently associated with the moulding of different food commodities, giving with each commodity a useful bibliography for further reference.

Cereals and cereal products

Annual losses are high and estimated to be about 55 million tonnes (482, 673). There are many records of the abundant mould flora of rice (625, 634, 1148, 1308, 1370, 1704, 2645, 2647, 2651, 2662, 2663). Among the fungi isolated from rice some are associated with the corn in the field (*Fusarium, Epicoccum, Nigrospora, Chaetomium, Trichoderma, Alternaria, Curvularia,* and *Helminthosporium*) whereas others are associated with the grain after harvest (notably *Cladosporium, Oospora, Rhizopus, Absidia, Mucor, Penicillium, Aspergillus,* as well as prokaryotes such as *Streptomyces*). The moulds identified from rice in storage can be listed as a function of the concentration of water in the substrate (Table 2) (473, 2645). In Japan, the water content of rice is often 15–16% allowing the growth of many aspergilli and penicillia (1372, 2643, 2652, 2909). For this reason a reduction to the level of 14.5% has been recommended for long-term storage (1805).

Although ventilation of stored grain reduces the level of mould attack (2299),

Table 2. The principal moulds on rice in storage as a function of water content of the substrate

Water content (%)	*Moulds*
14–15	*Apergillus chevalieri*
15–16	*Aspergillus candidus* *A. nidulans* *Penicillium citreoviride*
16–17	*Aspergillus oryzae* *A. fumigatus* *A. niger* *Penicillium notatum* *P. islandicum* *P. urticae*
17–18	*Fusarium* *Rhizopus*

the introduction of new mechanical techniques for harvesting (threshing in the field), prestorage, bulk drying, and storage in silos of huge dimensions have all given rise to mould problems of cereals in temperate regions affecting wheat, rye, barley, and oats. These changes have been recognized by phytopathologists in the United States such as in Illinois (1327), Iowa (2333, 2334, 2335, 2338), and especially in Minnesota (469, 470, 471, 472, 480, 481, 482, 492, 1919, 2653, 2654, 2655).

It is possible to recognize a succession of three distinct mycofloras during the storage of cereals (481):

1. *Field fungi* growing and established before harvesting (*Alternaria, Fusarium, Helminthosporium, Cladosporium*);
2. *Storage fungi* taking over and dominating in the silo (*Aspergillus* and *Penicillium*);
3. Advanced decay fungi (*Papulospora, Sordaria, Fusarium graminearum* and members of the Mucorales).

Observations made in France have led to conclusions which are in general agreement with this classification (1743, 1945, 1946, 1947, 1953).

There is thus a succession of species starting with strongly cellulolytic, hydrophilic fungi which tend to cause an increase in the pH of the substrate and originate from the haulms and glumes (*Chaetomium, Gliocladium, Trichoderma, Trichothecium*). These make way, before harvest, for species which, though still cellulolytic, are indifferent to water activity and colonize the integument and embryo (*Cladosporium*). In storage these are all gradually replaced by moulds such as *Aspergillus* and *Penicillium,* with a low cellulolytic activity, which tolerate a reduced water activity and generally cause a reduction in the pH of the substrate. Finally, after deterioration of the grain and an increase in water activity, the last group consisting mainly of members of the Mucorales appears.

Such an evolution of the mycoflora is essentially controlled by ecological factors: temperature plays an important role (1333, 1748) but a high humidity, during storage, brings about a rapid invasion by moulds and a subsequent heating of the grain.

It has been possible to demonstrate a relationship between the minimum water content of wheat and the species of *Aspergillus* able to invade it (Table 3) (481).

Table 3. Various species of *Aspergillus* invading wheat at different values of water content (481)

Minimum water content (%)	*Fungi*
13–13.2	*Aspergillus halophilicus*
13.2–13.5	*A. restrictus*
14–14.2	*A. amstelodami, A. repens, A. ruber*
15–15.2	*A. candidus, A. ochraceus*
17.5–18	*A. flavus*

These results confirm laboratory observations concerning the minimum relative humidity required for germination and growth of moulds.

Now, in Europe, cereals on harvesting may easily have a water content of 16% and even 18% in wet seasons, whereas in the United States, where farming conditions are very different, 14% is expected for soft wheats and 14.5% for hard wheats. According to Parkin (1922), it is advisable not to exceed a water content of 13.5% for storage over several months and it is necessary to reduce it to 12% for long-term storage.

Interactions between moulds might explain certain aspects of the development of the grain flora (1948) but it should not be forgotten that a grain of wheat has a heterogeneous structure (1145) and that, for example, the water content of the embryo will be much higher than that of the rest of the grain and could be selective for certain fungal species (1952).

The following species (in order of decreasing frequency and abundance) may be isolated from wheat and are characteristic of the storage flora in France (1945): *Aspergillus repens, Penicillium cyclopium, A. versicolor, A. echinulatus, A. amstelodami, A. candidus, P. spinulosum, P. stoloniferum, P. frequentans, A. flavus.* Essentially the same species are found on wheat in other countries (1213, 1774, 1942, 2542). Several studies have been carried out on barley parallel to those on wheat (492, 787, 1337, 1389, 1779, 1780, 1781, 1957, 1990). Whereas at harvest *Alternaria, Fusarium, Cladosporium,* and *Mucor* are dominant, *Aspergillus, Penicillium,* and *Absidia* are more frequent after storage. Three months after the harvest of barley the most common species are *Penicillium cyclopium, P. roquefortii, Absidia corymbifera, Aspergillus candidus,* and *A. terreus.* In moist stored barley, if temperatures reach 37–45 °C, thermophilic species such as *Aspergillus fumigatus, A. terreus, Absidia corymbifera, Mucor pusillus,* and *Dactylomyces crustaceus* may be isolated (1780).

Analogous problems have been reported with rye (462) as well as with maize, although in the latter case at a much higher water content (approximately 30%). Numerous species have been isolated from maize (306, 1477, 2064, 2251, 2335). Members of the *Aspergillus glaucus* group often predominate (2338). Within the genus *Penicillium,* whereas *P. oxalicum* and *P. funiculosum* are common before storage, *P. cyclopium, P. brevicompactum,* and *P. viridicatum* ultimately become more frequent (1688). The development of the mycoflora varies according to the conditions of storage (1410) and even with the variety of maize (1751).

Other cereals, sorghum for example, are equally susceptible to invasion by moulds during storage (474, 1518, 1774).

It is evident that flour, resulting from the grinding of grain, which assures the uniform dispersal of spores, may yield as many, if not more moulds as, for example with rice flour (1370, 1372) or wheat flour (51, 302, 464, 465, 466, 477, 535, 667, 834, 881, 891, 953, 954, 955, 1072, 1073, 1075, 1092, 1114, 1155, 1212, 1213, 1363, 1727, 1803, 1984, 1985, 1986, 2005, 2245, 2304, 2442, 2457, 2458, 2581, 2591, 2660, 2752, 2753, 2754). However, qualitatively the fungal flora of flour is not the same as that of grain because the component corresponding to the field flora has practically disappeared, or been removed with the bran, whereas the storage flora is well developed (1076). Moreover there may be further contaminants resulting from poor handling conditions (2581).

Although the temperature of an oven is usually sufficient to kill any spores present in dough, bread may be contaminated after baking by the flour which is inevitably present in the bakery atmosphere (835, 1741, 2457). However, some moulds are found on bread (*Neurospora, Rhizopus*) and only occasionally in flour (466).

Moulds from flour often turn up in animal feeds in the compounding of which cereal flours and milling wastes play a significant role. In France 60% of the food rations for pigs and poultry, and 30% of those for young cattle, are based on cereal products (536, 1944, 2432, 2433, 2435). These compounded feeds have become more and more important in new feeding techniques applied to the raising of animals for they give a better nutritional balance and increased yields (770). In 1966, French farming had absorbed 4,500,000 tonnes and Dutch farming 17,000,000!

Three examples will suffice to show the origin of contamination in these foods (1944). *Aspergillus flavus* was isolated from a pig feed (corresponding to about 6.5 $\times 10^3$ spores per gram food) and the same organism was isolated from 100% of the grains of wheat used in the preparation of this particular feed; in another case it was *Mucor circinelloides* which was most abundant (about 4.5×10^3 spores per gram) and this species could be isolated from 44% of the grains of barley used in the feed; in the third case, a feed based on oats gave 8×10^5 spores of *Aspergillus versicolor*, and this same organism infected 43% of the grains of the oats used.

Furthermore, the contamination of the substrate in cattle or poultry feeds simultaneously results in the dispersion of microbial propagules and their subsequent growth. Humidification involved in certain preparations, notably the production of granules, also favours the spreading of the dominant moulds.

Milk and Milk Products

It is mainly milk powders which, despite their low water content (3–4%), become contaminated with moulds if they are prepared and stored in poor conditions (*Aspergillus glaucus* group, *A. versicolor, Penicillium cyclopium, P. oxalicum, P. stoloniferum, P. viridicatum, Scopulariopsis brevicaulis*, etc.) (541). If hygienic precautions are not taken when sealing containers of milk concentrate *Aspergillus repens* may frequently occur (2187).

The deterioration of cream may be due to the proteolytic and lipolytic activities of a variety of moulds. Several moulds may be responsible for spoilage during the preparation of butter, from which about fifty species have been identified (2247). *Aspergillus repens* is essentially lipolytic, *Cladosporium herbarum* and *Oospora suaveolens* decompose the milk. In the summer *Alternaria tenuis* and *Fusarium* frequently occur and in some regions *Aspergillus flavus* is significant.

Attempts have been made to control these moulds in a number of ways:

– by cold storage, but *Stemphylium botryosum, Cladosporium herbarum,* and *Alternaria tenuis* are all capable of growing at 5 °C;

– by the addition of salt, but members of the *Aspergillus glaucus* group comfortably tolerate a medium containing 25% sodium chloride;

– by the addition of lactic acid, but several penicillia grow even in the presence of 5% lactic acid (*P. spinulosum, P. fellutanum, P. citreoviride, P. terlikowskii, P. sartoryi, P. roquefortii,* and *Citromyces pfefferianus*);

– by pasteurization of the cream before churning (1562), but contamination may take place during churning rendering any previous treatment ineffective.

In the cheese industry it is well known that certain moulds are essential for the preparation of both soft and hard cheeses (*Penicillium caseicolum* in Camembert for example), but there are numerous spoilage species which give rise to problems in manufacture: *Mucor racemosus, M. globosus, Penicillium brevicompactum* (288, 633) *P. expansum, P. funiculosum* (1286), *P. aurantio-virens* (1188), *Scopulariopsis brevicaulis* (1209), *Cladosporium herbarum* (1208), etc. Some cause blemishes on the cheese whereas others give rise to disagreeable flavours. Pasteurized soft cheese may also be spoiled by moulds (*Penicillium commune, P. expansum, Aspergillus candidus, Cladosporium herbarum,* etc.). The pH of the curd may play an important role in the growth of some spoilage fungi (2449).

Meats

The list of moulds associated with meat has been growing (993, 1231, 1452, 2550), especially from studies in abattoirs of Chicago (2922). It includes *Penicillium expansum, Aspergillus clavatus, A. niger, Mucor racemosus, Monascus purpureus, Neurospora sitophila, Fusarium* spp, *Mortierella* spp, *Oidium lactis,* and many others. The Mucorales are particularly frequent during the period of storage between slaughter and consumption.

On refrigerated meats it is not uncommon to see white patches of *Sporotrichum carnis* or black patches of *Cladosporium herbarum.*

Dried and cured meats are often contaminated and there are many problems, for example, in preventing the growth of *Penicillium expansum* or members of the *Aspergillus glaucus* group during the manufacture of dried sausages, because they interfere with a balanced, useful, surface flora of yeasts (1737). On Yugoslavian hams species of *Penicillium* and *Aspergillus* as well as 36 other moulds such as *Cladosporium* and *Alternaria* have been identified (2527).

Eggs

Mallman and Michael (1584) have isolated a whole series of moulds both from the surface of the shell and from the interior of the egg. The include *Mucor racemosus, Aspergillus niger, A. flavus, Penicillium oxalicum, P. puberulum, P. citrinum, P. viridicatum,* and *P. verruculosum.* None would grow if the relative humidity of the storage room were less than 85%.

Egg powder is also associated with a particular mycoflora.

Oilseeds and derivatives

Oilseeds and their derivatives (719) have increased in importance over these last few years, especially in the compounding of animal feeds (703, 772).

During the processes of drying, threshing, decorticating, and grinding there are many opportunities for the mould damage of commodities such as groundnuts (267, 654, 655, 656, 756, 876, 877, 1055, 1172, 1248, 1872, 2036, 2843). Black discoloration may be caused by *Lasiodiplodia theobromae* or *Macrophomina phaseoli* (270, 271) or blue discoloration caused by *Sclerotium rolfsii* (92).

Some pods are already contaminated before touching the soil during the growth of the plant (1017), but numerous species, associated with the soil, infect the shells once the pods become buried (1244, 1245, 1343, 1540).

During the storage of groundnuts it is species of *Aspergillus,* especially *Aspergillus flavus* and members of the *glaucus* group, which are most frequently observed followed by *Fusarium, Penicillium, Pythium, Rhizoctonia,* and *Trichoderma viride* (874). Nearly 200 different species capable of damaging husks and kernels have been identified (1174, 1182, 1245, 1248, 1537, 1539, 1540). Among them 12 species of *Penicillium* and 25 of *Aspergillus. Aspergillus flavus*

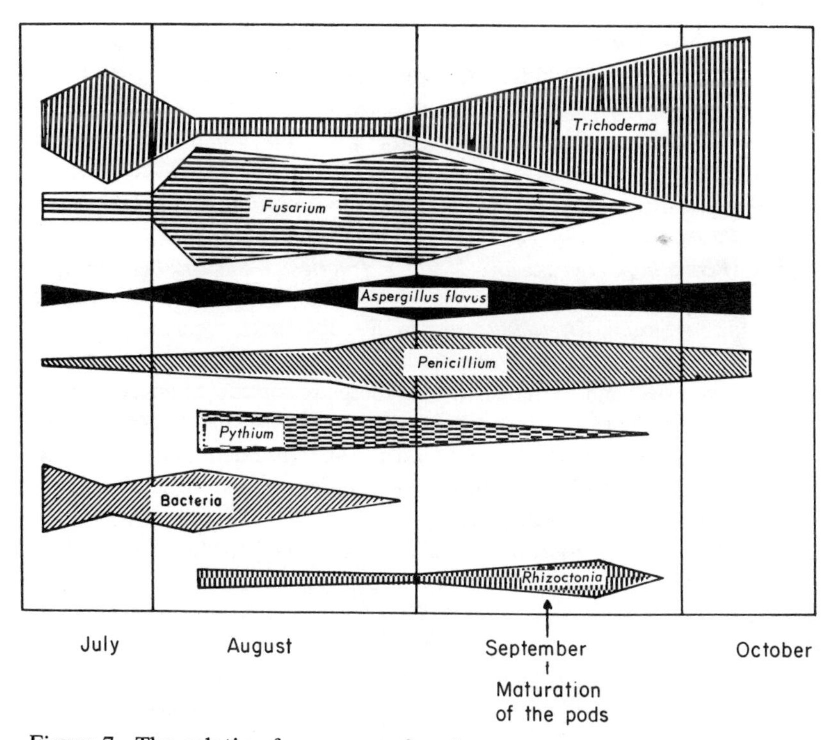

Figure 7. The relative frequency of various species of fungi isolated from groundnuts (in Virginia) as a function of the maturation of the pods and the times of storage (874)

appears to be especially common in certain African and South American batches, it being frequently possible to isolate it from more than half the kernels (105).

A succession of species can be demonstrated depending on the state of maturation of the pods as well as the duration and condition of storage (Figure 7) (874, 2800).

Oilcake from groundnuts contains little water (5–6%) and if any outside source of water is excluded moulds are not able to grow on this substrate.

Cottonseed, which has been recommended for use in human food especially because of an interest in its high protein content (289), is equally susceptible to attack by *Aspergillus flavus* (1515, 2023) and many other moulds (490). The same is true of sunflower seed (35, 476), linseed (2436), and soya beans. The latter are also invaded by *Aspergillus halophilicus, A. restrictus, A. repens* and *A. amstelodami* even at a water content of less than 13% (478, 1290, 1671, 2575). Oilseed cake from soya may also be infected by a number of moulds (1729). On olives, the growth of *Aspergillus versicolor* may cause spoilage by lipolysis (937), and margarine is sometimes contaminated by fungi such as *Phialophora bubakii* (=*Margarinomyces bubakii*) (1427).

Fruits and vegetables

Because of the diversity of storage conditions there is a wide range of fungi associated with these commodities. As examples it is sufficient to recall those that occur on ripening bananas (*Colletotrichum musae, Thielaviopsis paradoxa, Penicillium expansum, Rhizopus nigricans, Fusarium semitectum, Aspergillus oryzae, Cladosporium cladosporioides,* etc.). (1258), associated with the storage of citrus fruit (*Penicillium italicum, P. digitatum, Phytophthora parasitica, Phomopsis citri, Diplodia natalensis, Botrytis cinerea, Geotrichum candidum,* etc.) (1725), or of apples (*Penicillium expansum, Monilia fructigena, Trichothecium roseum, Botrytis cinerea, Cylindrocarpon mali, Fusarium avenaceum, F. lateritium, Cladosporium herbarum, Alternaria tenuis*) (259, 2749).

Others cause spoilage during transport, for example *Botrytis cinerea* on peaches, cherries, strawberries, and currants. Few of these are considered dangerous to man.

On dry fruits (prunes, figs etc.) or on preserves it is mainly members of the *Aspergillus glaucus* group which are found (1746) although *Xeromyces bisporus* sometimes takes over (579). Nuts have a very varied mycoflora some members of which are reputed to be toxigenic (2789).

Regarding the deterioration of vegetables, some fungi appear early in the field, others only after harvest (1653): thus carrots stored in silos or cellars may be infected with *Stemphylium radicinum* and *Sclerotinia sclerotiorum,* as they are in the field, but additionally with *Aspergillus niger, Mucor* spp, *Botrytis cinerea,* and various species of *Penicillium.* Onions and garlic in storage may be infected by three species of *Botrytis* (*B. alli, B. byssoidea,* and *B. squamosa*) as well as *Penicillium corymbiferum* and *P. cyclopium.* Stored potato tubers are often attacked by *Fusarium coeruleum, Phoma foveata,* and *Oospora pustulans.* It is particularly during the transport of French beans, peas, and lettuce that they are susceptible to attack by *Botrytis cinerea,* but even dry peas may become mouldy (779).

These scattered examples will suffice to demonstrate the variety of moulds able to spoil fruits and vegetables between harvest, storage, and consumption.

Other food commodities

Among the beverages, fruit juices are sometimes affected by *Byssochlamys fulva*. Moulds enter particularly when the bottles are opened (2122, 2341).

Coffee beans are infected by numerous species: *Cunninghamella, Trichoderma, Aspergillus, Penicillium, Fusarium* (1576, 1810), and particularly by *Aspergillus tamarii* (1056, 1851). During the fermentation of cacao pods, in the preparation of cocoa, they usually become moulded and some contaminants may cause problems (*Aspergillus glaucus, A. tamarii, A. flavus, A. fumigatus, Penicillium* spp, etc.) (344, 512, 577, 1927, 2139, 2140).

Spices (pepper, nutmeg, etc.). are frequently infected by moulds (479, 488) and their use in foods prepared in advance of cooking is equivalent to inoculating the foods with moulds some of which may be toxigenic (*Aspergillus flavus, A. fumigatus*, etc.) (1731).

It is worth mentioning the potential danger of certain moulds associated with lucerne, hay, and dried straw. A number of cases of allergies in the respiratory tract have been reported among English farmers inhaling spores which may be abundant in hay (962, 963, 964); such forage may also be toxic for cattle because of the presence of moulds (1432, 1707, 1758, 1894, 2327).

Despite the relatively anaerobic conditions achieved in the wet silage of forage grasses, this material is not entirely free from moulds, especially on the exposed areas from which a number of toxic species have been isolated (748, 749, 750). This is true of silage prepared from other materials especially if made under poor conditions (1732).

Finally we would like to draw attention to the possibility that certain tobaccos may contain mould-derived toxins due to the presence of species from such genera as *Alternaria, Epicoccum, Penicillium,* and *Aspergillus* (1013).

THE CONTAMINATION OF FOOD PRODUCTS AND THE SPREAD OF MOULDS

Sources of contamination

For moulds to grow on a food commodity it is obviously necessary for it to become contaminated in the first place and the sources of contamination are very varied.

Sometimes contamination occurs in the field but infection is deferred and spoilage of the product only manifests itself after several months of storage. A good example is the lenticular spot disease of the variety of apple known as Golden Delicious. The fungus *Phlyctaena vagabunda* penetrates the lenticels of apples during their growth (in July) and at the time of storage there is nothing in the appearance of the apples which makes it possible to distinguish a sound one from one infected with *Phlyctaena* (in October). It is only after several months storage, when the fruit reaches maturity, following rupture of the suberin layer at the bottom of the lenticel cavity, that the mycelium of the fungus penetrates the underlaying parenchyma and causes deterioration (258).

Many field infections hardly develop any further after harvest but give way to a characteristic storage mycoflora; this is particularly so with cereals and oilseeds but may also occur in fruits and vegetables.

Frequently the origin of fungi in food commodities is the very buildings used for storage and ripening. The common moulds capable of spoiling food products may proliferate on the walls and ceilings, and even on the floors, of these buildings, or on the wood of packing cases, wrappings, jute or cotton sacks, paint or varnish used in the building (857). Contamination is particularly easy for the spores may be dispersed in the atmosphere or carried by water of condensation. It was easy to recognize, for example, the drying rooms as the source of infection of Camembert spoiled by *Penicillium brevicompactum* (633). On the other hand it was necessary to search as far as the mildewed ceilings of the paper manufacturer to find the origin of moulds on butter which were thus contaminated from the wrapping paper! In a ripening store for plums problems arose from an abnormal contamination of the containers. The workshop where they were manufactured was nearby and was contaminated from a pile of refuse on which was thrown all the mouldy fruit!

Most spoilage contamination can be avoided by hygienic precautions, requiring logic and common sense.

Rate of growth

When ambient conditions of temperature, humidity, etc. are favourable for them, most of the moulds can grow very rapidly (Table 4). Some, such as *Rhizopus* and *Trichoderma viride,* have a very high rate of mycelial growth and can cover all the available surface in a few days. Others, such as most of the penicillia and aspergilli, have a restricted mycelial growth but spore abundantly, even when the mycelium is young, the spores dispersing to give new colonies and in this way the substrate is rapidly invaded. A single spore culture of *Aspergillus clavatus,* for example, may produce 50 million spores in four days and five times as many in six days (1747). It is thus easy to understand how it is possible to count nearly 120 million spores on a single grain of barley one week after contamination!

In a particular silo of wheat destined for consumption, 29 million spores per grain were counted consisting mainly of *Penicillium cyclopium* but in addition *Alternaria tenuissima, Aspergillus flavus, Rhizopus nigricans, Actinomucor repens* and *Cladosporium cladosporioides.* A cleaning and drying process using fans reduced this spore load to 71,000 per grain but the storage conditions were not modified and, after several weeks, a new count yielded results comparable to those originally obtained.

The sporulation of a mould on a given medium first shows an exponential phase followed by a more or less clearly defined attenuation phase. Autolysis, common in some moulds, is often compensated by a new wave of spore production, especially

Table 4. Daily growth rate of a number of moulds on malt agar (2%) at 22 °C and 32 °C (1749)

Species	*Growth rate (mm/day)*	
	22 °C	*32 °C*
Absidia lichtheimii	6.2	13.5
Alternaria tenuissima	5.6	2.9
Aspergillus candidus	1.3	0
A. flavus	5.2	7.8
A. fumigatus	4.9	11.6
A. niger	2.0	4.6
Cladosporium cladosporioides	1.9	0
Fusarium culmorum	8.5	0
Mucor racemosus	13.5	0
Penicillium brevicompactum	1.1	0
P. cyclopium	0.7	0
P. expansum	2.1	0
P. frequentans	1.8	0
P. funiculosum	2.1	0
P. stoloniferum	2.8	0
Rhizopus nigricans	54.9	1.4
Trichoderma viride	23.2	11.6
Tricothecium roseum	4.6	0

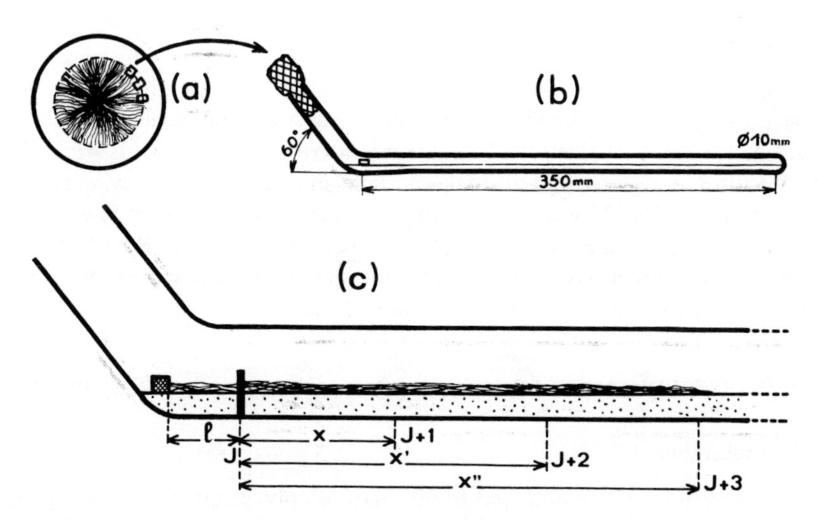

Figure 8. A method for measuring the rate of mycelial growth:
(a) Inoculum taken from the edge of a young colony growing in a Petri dish.
(b) Transferred to the open end of a long tube containing nutrient agar medium.
(c) After a lag period (l), linear growth is measured at daily intervals (1749)

in species with aerial dissemination such as *Aspergillus candidus, Penicillium cyclopium* and *P. spinulosum* (1749).

A useful method of evaluating mycelial growth is illustrated in Figure 8.

Dispersal

The least breath of air allows the dispersal of dry spores and it is easy to explain the contamination by moulds of food commodities in warehouses (1737). By experimentally exposing 20 g of mouldy hay to a wind of 1.2 m s^{-1} for 3 minutes it has been shown that the number of spores thus dispersed may reach 112 million per gram (964). They were mainly spores of *Aspergillus flavus, A. nidulans, A. niger, A. ochraceus, A. terreus, A. versicolor, A. fumigatus* and members of the *A. glaucus* group which were recovered. It is thus understandable that, in a cattle shed, counts of 12–21 million spores of these species per cubic metre of air may be counted (150, 1390).

Water of condensation may serve to disperse spores in the locality of mouldy walls and ceilings. When a drop of water 5 mm in diameter falls to the ground from a ceiling 7 m high, more than 5,000 droplets of variable diameter (5–2,400 μm) spread out some 10–20 cm from the point of impact. Thousands of spores of *Fusarium solani* may be spread in this way (961).

Insects, rats, and various animals may also play a significant role in the spreading of spores thus contributing to the contamination of healthy foodstuffs.

Survival of spores

The dispersal of mould spores being assured by many processes, it is not in fact necessary for them to immediately regain a substrate favourable for their germination, for many of them have remarkable survival ability (2525). When the spores of *Rhizopus nigricans* were dried and kept in a sealed tube they were still found to be viable after 22 years! Under the same conditions, spores of *Penicillium brevicaule, P. luteum,* and *P. camemberti* have remained viable for at least 12 years.

Spores having thick walls, such as chlamydospores and ascospores, are more suited than others to survive extreme conditions.

THE CONSEQUENCES OF MOULD CONTAMINATION OF FOODS

Unsightly appearance

The proliferation of moulds on foods may simply result in it having an unacceptable appearance. This is so in the case of the blackening of flour by *Cladosporium* (120), the blue coloration of macaroni by *Monilia albo-violacea* (764) or the orange coloration due to the dust of spores of *Neurospora sitophila* (1918).

Sometimes simply the smell of an infected product is sufficient to make it unacceptable as a food (1984).

Chemical changes caused by moulds

The fungi modify the substrates on which they are growing; they remove those elements required for the synthesis of their cell material and for the production of the energy necessary for growth processes. Many produce powerful enzymes capable of breaking down food constituents (811, 1750).

Some compounds, such as vitamins and amino acids, may be taken up by fungi without modification but high molecular weight compounds, such as cellulose, starch or proteins, have first to be broken down by hydrolases (cellulases, amylases, proteinases, etc.). Other reactions which are mediated by enzymes include oxidations and reductions.

The study of fungal metabolism demonstrates that these organisms have remarkable biosynthetic activity (958) as well as being able to carry out bioconversions. However, even the smallest change in the physical and chemical environment in which they grow may qualitatively or quantitatively modify the products of their metabolism.

The moulds are thus well suited to remove or change most of the constituents of food materials:

Carbon – Generally most of the sources of carbon, such as sugars, amino acids, organic acids, various polycyclic compounds, and alkaloids may be utilized by fungi. However, many species show a marked preference, thus, for example, when *Penicillium chrysogenum* is grown on a medium containing glucose, lactose, acetate, and lactate, it is the acetate that is first metabolized, then the lactate, and finally glucose and lactose together (1218). Strains, such as mutants of *Aspergillus niger* and of *Penicillium chrysogenum,* may be very selective in the carbon compounds metabolized.

The process by which a group of closely related compounds, such as sugars, is metabolized varies from one species to another as shown in a study of members of the *Aspergillus glaucus* group (453).

The carbohydrates of a foodstuff, cereals for example, may be completely removed by mould activity (1529, 1671, 2781).

Nitrogen – Depending on the species moulds may utilize a wide range of sources of nitrogen from simple inorganic compounds such as nitrate or ammonium ions to complex organic nitrogen derivatives. Nitrate may be either assimilated or used for respiration but the oxidation of ammonium ions to nitrite and nitrate, which is carried out by some bacteria, is of rare occurrence among fungi. The biosynthesis by amino acids and proteins is carried out by well-known classical pathways (1849).

Others – *Sulphur, phosphorus, potassium, magnesium,* and other oligonutrients are usually readily available to moulds from the substrates invaded.

Changes in substrates brought about by moulds have sometimes been made use of in industry, for example in the synthesis of organic acids (citric, lactic, gluconic,

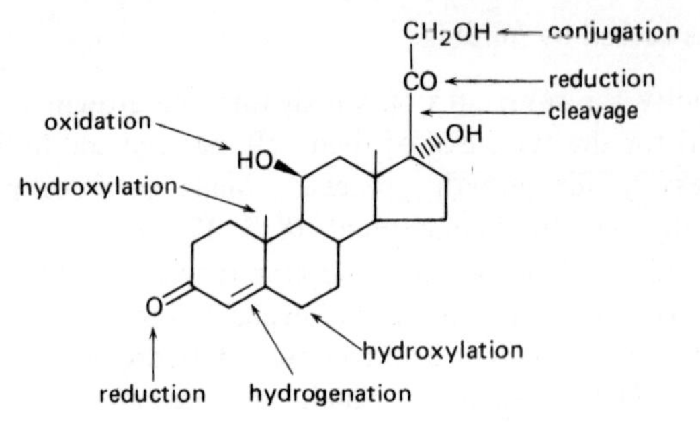

Figure 9. Some of the modifications which micro-organisms may carry out on a typical steroid molecule (hydrocortisone)

fumaric, itaconic acids), fats, and proteins. Several commercial antibiotics are produced by moulds and a number of useful transformations of steroids may be carried out by the action of moulds (see Figure 9) (408).

In practical terms the chemical changes brought about in foods manifest themselves as changes in the nutritive value, organoleptic quality, difficulties in preservation, or occupational diseases and toxicoses.

Changes in nutritional value

Rats fed on grain infected with *Fusarium moniliforme* or *Diplodia* (*Phaeostagonosporopsis*) *zeae* show a reduced body weight when compared with rats fed on an equal quantity of healthy grain (1691). Clearly some nutrient constituents have been removed by the moulds. In the case of poultry fed with mouldy grain, Pomeranz (2014) noticed that the delay in growth could be compensated by the addition of lysine and methionine, amino acids which were probably removed by the growth of the moulds. Similar changes in the quality and quantity of amino acids have been recorded for rice infected with *Fusarium chlamydosporum* (2289).

The breakdown of protein by proteolytic enzymes produced by some moulds sometimes renders flour unfit for the manufacture of bread (149, 1306, 2016, 2445).

Mouldy forage is usually not as rich in vitamin A as sound material and it is possible that some of the metabolites produced by fungi act as antivitamins. Inversely it should be considered that sometimes a moulded food material has a higher level of vitamins than the same food uncontaminated by moulds: this may be because of the high vitamin content of the mycelium of many fungi (943).

Lipolysis in various seeds (1552), in groundnuts, and in copra may be brought about by moulds, especially *Aspergillus flavus, A. fumigatus, A. awamori, A. niger, Syncephalastrum racemosum,* and to a lesser extent *Aspergillus nidulans, A. tamarii, A. sulphureus, A. chevalieri, Penicillium steckii, Paecilomyces varioti* (1098). These same moulds cause rancidity of oil, usually following lipolysis (2781); they impair the manufacture of emulsified fats (margarine for example) by oxidation or decarboxylation giving rise to ketonic rancidity (136, 2699).

Table 5. Analysis of straw before and after its use as a substrate for the growth of *Aspergillus*

Component	*% before*	*% after*
Ash	9.1	14.0
Ash (water soluble)	3.9	4.9
Protein	0.9	8.0
Ammonium sulphate	–	0.1
Cellulose	50.3	50.7
Pentosans	26.8	10.2
Lignin	12.9	17.0
Fibre	55.0	49.5
Available nitrogen	35.0	28.4

The suppression of growth of animals by particular metabolites, such as oxalates produced by numerous species of fungi, is not negligible (101).

The growth of moulds on food need not always be associated with a reduction of nutritional value; it is sometimes beneficial. Thus, during the First World War, Pringsheim and Lichtenstein attempted to increase the protein level of feed for cows by inoculating straw, previously sprayed with a solution of ammonium sulphate, with a species of *Aspergillus.* After a period of incubation during which the fungus synthesized proteins, the straw was shown to have an increased nutritive value as shown in Table 5.

An analogous technique has made it possible to improve the protein content of rice (959) and, for the same reasons, cattle foods have been supplemented with the mycelium of a *Penicillium* (676). It is known that the preruminant calf digests glucose better than flour starch (1626) and, because some moulds degrade starch to glucose, flour infected with such moulds may be more easily digested by calves. On the one hand, an enzymic degradation of proteins may improve the nutritional value of a food, on the other hand the degradation of macromolecules may result in spoilage (1264). Moulds may be involved, depending on the situation, in improvement or in spoilage!

Modification of organoleptic quality

The organoleptic qualities of a food may be altered by the presence of a mould, in the majority of cases for the worst:

- the growth of *Geotrichum candidum* on cheese makes it unpalatable;
- a species of *Aspergillus* is responsible for a particularly disagreeable bitterness in coffee (1098);
- *Aspergillus tamarii,* members of the *A. glaucus* group, and *Penicillium citrinum* cause rancidity of groundnut oil (2781);
- Aspergilli and penicillia cause rancidity of margarines by the formation of long chain methyl ketones (2699).

Sometimes, however, the presence of moulds improves flavour. Thus *Aspergillus repens* may impart to salted, dry ham the delicate aroma of Parma ham (1737): it is moulds which give the piquancy and spicy flavour to Roquefort cheese (brought about by the hydrolysis of fats and the liberation of free caproic, caprylic, and capric acids); similarly the strong and fine flavours of soft cheeses such as Brie and Camembert are due to mould activities.

It is worth noting a rather unexpected effect of a mould on food flavour: Weigmann and Wolff (2791) have established that the presence of *Penicillium brevicaule* (= *Scopulariopsis stercoraria*) on cattle food imparted a flavour of mustard to the milk and butter.

Difficulties in preservation

On a commercial and economic scale the major repercussion of the growth of moulds is in the reduction of shelf life of food products.

Just as the housewife knows that jams containing insufficient sugar preserve badly (because penicillia and aspergilli will grow on the surface) so also do importers and retail agents know to what extent *Penicillium italicum* and *P. digitatum* are an obstacle to the long distance transport and sound preservation of oranges. Industrial bakers well know the problems resulting from badly contaminated flour.

Occupational hazards

More injurious yet are those moulds capable of causing occupational hazard to those handling them. Two examples from past history are those involved with the force-feeding of pigeons and those who used to scour wigs with flour, both groups of people being susceptible to pulmonary aspergillosis (2143). Today cutaneous mycoses caused by *Candida* are not rare among people working dough or carrying flour from the mill. There is often a build up of thermophilic fungi, some of which may be pathogenic, in the drying tunnels used in the manufacture of pasta (1737).

Mycoses may thus result from the ingestion of mouldy foods or contact with moulded products. The frequent occurrence of *Aspergillus fumigatus* on straw and hay in farmyards may give rise to pulmonary mycoses in poultry (1739) and eventually in the farmer. Some allergies associated with moulds should also be considered as occupational hazards.

Toxicoses

However the gravest consequences of the moulding of food are undoubtedly the dangers of poisoning due to toxic metabolites produced by fungi, liberated into the food, and hence entering the body of the consumer. Such compounds are known as 'mycotoxins' and the subsequent 'mycotoxicoses' may affect man or other animals.

The importance of the problem is indicated by the observation that official organizations of the United States of America have not hesitated to set aside more than two million dollars annually for the study of mycotoxins (189, 1158).

CHAPTER 2

MYCOTOXICOSES AND MYCOTOXINS

THE DISEASES CAUSED BY FUNGI

In human or veterinary medicine the fungi may be implicated in three types of illness (102):

1. Infections (*mycoses*) corresponding to an invasion of the living tissue by a fungus which may penetrate directly into a healthy organ (primary infection), or only develop following a preliminary lesion of some other kind (secondary infection);
2. Allergies which are specific reactions of an individual following inhalation of the spores (respiratory allergy), or any contact with a particular fungus;
3. Toxicoses, poisoning resulting from the ingestion of food containing toxic fungal metabolites.

Mycoses

Mycoses are illnesses in which a living fungus is involved and are infectious, often contagious. They may take the form of simply inflammation of a particular organ, thus infection of the ear is referred to as otitis, of the nails as keratitis, of the cardiac valve as endocarditis, etc., but in each case it is advisable to specify the fungus incriminated.

They may also affect the bronchia and the lungs, indeed bronchiomycoses and pneumomycoses caused by aspergilli are among the most serious of these diseases. When granulomatous lesions develop the term mycetoma is frequently used. Infections of the skin, or dermatomycoses, are fairly common.

Some thirty human mycoses are known and well defined involving about sixty fungi. Those which may be lethal involve *Coccidioides immitis, Histoplasma capsulatum, Cryptococcus neoformans, Aspergillus fumigatus, Absidia corymbifera, Rhizopus oryzae,* and the actinomycete *Nocardia asteroides* (731) and more rarely *Candida albicans, Blastomyces dermatitidis, B. brasiliensis, Sporothrix schenckii,* and *Cladosporium trichoides* (701).

Allergies

The allergies caused by fungi may take many forms such as rhinitis, conjunctivitis, dermatitis, bronchial asthma, etc. (443, 2587). Those fungi, such as

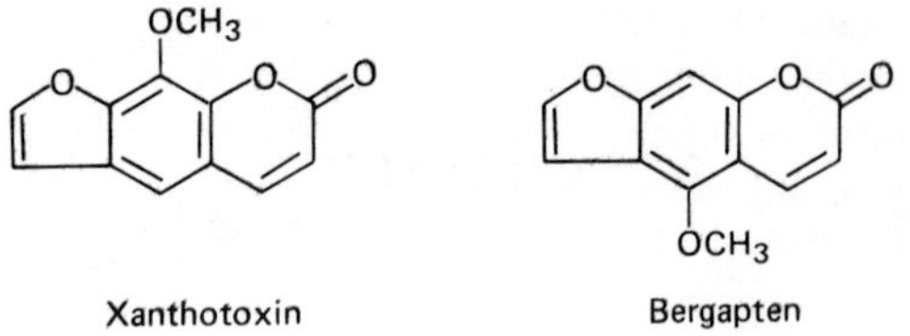

OCH_3

Xanthotoxin

OCH_3

Bergapten

Figure 10. Photoactive furocoumarins of plant origin

Cladosporium and *Alternaria,* with wind borne spores which are frequently found in the air flora, have been considered as one cause of those asthmas which occur seasonally (261, 1011). A respiratory disorder found among lumbermen cutting maples has already been described and associated with *Cryptostroma corticale* (1735).

The disorder known as farmers' lung is caused by the inhalation of spores of moulds and actinomycetes present in hay. It is sometimes very severe and complicated by cyanosis of the lungs and the production of purulent phlegm. Gregory and Lacey (962, 963) demonstrated an association between water content of hay and the possibility of farmers' lung. They were able to distinguish good hay with a water content of about 15%, not containing more than 5×10^5 spores per gram and not causing any allergy; mouldy hay with 25% water content containing $5–100 \times 10^5$ spores per gram, especially species of the *Aspergillus glaucus* group; very mouldy hay containing more than 35% water, able to undergo the process of self-heating to temperatures of up to 65 °C, and containing large amounts of moulds and thermophilic actinomycetes such as *Thermoactinomyces vulgaris* (= *Micromonospora vulgaris*) and *Micropolyspora faeni* (= *Thermopolyspora polyspora*). It is this last type of hay which is responsible for respiratory allergies in (farmers' lung). Although particularly associated with British farmers some isolated cases have been recorded in France. Other pneumopathies have been described in industrial and craft workers as well as agricultural workers (1388, 1712).

The skin lesions found on market gardeners handling celery may also be considered in this section on allergies. They have been shown to be associated with a *Sclerotinia* rot of celery and the dermatitic reaction is caused by photoactive furocoumarins such as xanthotoxin (8-methoxy psoralene), bergapten (5-methoxy psoralene), and 4,5,8-trimethyl psoralene (Figure 10) (1962, 2884).

Mycotoxicoses

The fundamental characteristic distinguishing mycoses and toxicoses is that the fungi associated with the latter are not contagious or infectious.

According to the principle upheld by Gaumann (884), and which he suggests is absolutely applicable in human and veterinary medicine as well as plant pathology, a micro-organism is only pathogenic if it produces toxins. Put another way, the agents of disease can only harm their hosts if they form toxins which penetrate the host tissues. Without being so dogmatic it must be admitted that there are connections between mycoses and mycotoxicoses. It is essentially by the presence of the fungus responsible *in situ,* rather than as an indirect result of its activity, that mycoses may be distinguished from mycotoxicoses. Moreover, several fungi, notably *Aspergillus fumigatus, A. flavus, A. versicolor, A. sydowii,* and *A. terreus,* may be equally responsible for mycoses as well as outbreaks of poisoning (102, 1954, 2229). Likewise, *Penicillium rubrum,* known to produce mycotoxins, has been shown to induce allergy in people sensitive to asthma and hay fever (1020, 1956).

In most cases toxic moulds grow and proliferate on food products, their toxins diffusing into the food so that the consumer suffers the results, possibly in the absence of the fungus responsible. However, thermotolerant species and those resistant to the acids of the gastric juices may be able to grow in the digestive tract, elaborating toxic metabolites in sufficient quantities to have a marked effect; it is plain that very small quantities may not have any serious effects. An extension of the retention time of the food (in ruminants for example) may increase the possibility of toxin production although it is doubtful whether any moulds could grow in the anaerobic conditions of the rumen. Although many fungi are killed by passage through the digestive tract, others which have resistant propagules, such as thick walled chlamydospores or ascospores, may pass straight through the alimentary canal and are often recovered as the dung fungi (1734).

THE FUNGI RESPONSIBLE FOR MYCOTOXICOSES

The variety of toxins elaborated by the fungi

Fungi elaborate numerous toxic substances. Although originally the term toxin was used to describe substances of biological origin with a high molecular weight, which were both toxic and antigenic (446), there is the tendency today to use the term for all toxic substances of biological origin.

Some fungal toxins are poisonous to animals (zootoxic), others are toxic to plants (phytotoxic) and yet others may act against micro-organisms, both fungi and bacteria, and are known as antibiotics.

Definition of mycotoxins

Although it may appear logical to call all those toxins elaborated by fungi mycotoxins (and hence mycotoxicoses the effects for which they are responsible), common usage has restricted the term to a more precise category: extracellular zootoxic metabolites (exotoxins) produced by moulds in food consumed by man or animals. Mirocha (1678), for example, has clearly defined the mycotoxicoses as those illnesses of man and of animals caused by the toxins of fungi, but, like Austwick (104), he restricts the mycotoxicoses to illness resulting from the consumption of food borne toxins.

Other zootoxins

Among compounds toxic to animals the mycotoxins may be distinguished from the intracellular toxins (endotoxins) elaborated by poisonous macrofungi and several microfungi parasitic on plants. Poisoning resulting from the ingestion of the toxic fruitbodies of certain toadstools is among the best known. A recent review by Roger Heim (1057) describes the subject well including many observations and personal experiences of the author.

Numerous works report the problems of ergotism caused by the consumption of a fungus which is usually parasitic on the ovaries of rye but may also occur on other members of the Gramineae. The fungus, *Claviceps purpurea* (Fr.) Tul., produces sclerotia from which several alkaloids have been isolated. As well as the now classical work of Stoll, Hofmann, and their colleagues, it is worth mentioning the many contributions from the Institute of Fermentation of Osaka, Japan.

The rust of wheat, *Puccinia graminis* Pers., has occasionally been accused of being responsible for poisoning (24), but much more serious are the diseases caused by smuts which have been referred to as 'ustilaginism'. Debré and Névot (608) have described an infantile disease similar to erythroedema, which may occur frequently, especially in central Europe, in those infants fed on gruel made with maize parasitized by *Ustilago maydis* (DC) Tul. It is thought that consumption of *Ustilago* may inhibit the action of adrenaline. It has been reported elsewhere that it provokes abortions in cattle and may cause the contraction of uterine muscle and the stimulation of smooth muscle. In fact, a number of alkaloids and various other compounds have been isolated from maize smut the nature and activity of which deserve further study. For example, Niccolini (1848) has isolated a compound similar to acetylcholine which, by acting on the parasympathetic nerve system, initiates low blood pressure in the rabbit causing a cyanosis of the face. *Ustilago avenae* (Pers.) Rostr. and *U. hordei* (Pers.) Lagerh. have similar activities (1586, 1860).

Tilletia caries (DC) Tul. and *T. foetida* (Wallr.) Liro, which belong to the same order as the Ustilaginales and are the causative agents of bunt or stinking smut of wheat, have also been considered as toxic. The spore mass of *Tilletia* has a characteristic smell of trimethylamine. Nielsen (1856) has recently isolated ergothioneine, choline, betaine, hercynine, and two unidentified derivatives of trimethylammonia; one or more of these compounds may act as the precursors of trimethylamine itself. When animals ingest a certain number of these spores they suffer digestive disorders accompanied by nervous upsets (2798). However, there are reports that raise some doubts about these observations (161, 675).

Vicia villosa, one of the vetches cultivated as a forage, has been implicated in a toxicosis (520) associated with convulsions in cattle consuming it. The plant contains cyanogenic glycosides. It is possible that vetch parasitized by the mould *Ovularia viviae* is involved in these disorders (1630).

In a similar manner lupins such as *Lupinus albus* L. parasitized by *Phomopsis leptostromiformis* (Kühn) Bubák are responsible for a disorder known as lupinosis. This is a hepatotoxic disease quite distinct from the nervous disorder arising from the consumption of alkaloids present in bitter lupins. It is particularly associated with sheep and has been reported in Europe, North America, New Zealand, Australia, and South Africa (1364, 2729, 2730).

Sclerotium rolfsii Sacc., a widespread plant parasite in warm temperate climates, is said to produce a compound toxic to sheep, ducks, and chickens. In South Africa it has been associated with a reduction of growth rate, anorexia, and nervous disorders (2075, 2571).

From the mycelium and culture fluid of *Cercospora beticola* Sacc. it is possible to extract a complex with haemolytic activity and a compound of the coumarin group which causes internal haemorrhages. It is possible that a problem of toxicity of beet leaves for cattle may be associated with the presence on them of *Cercospora* (583).

Fungi producing zootoxic exotoxins

Although several yeasts are involved in mycoses in man and animals, they do not seem to have been implicated as yet in any cases of toxicoses (unless one includes alcohol poisoning as a mycotoxicosis!).

The filamentous fungi, on the other hand are frequently responsible for outbreaks of poisoning. A certain number among them are those which grow on senescent plant material, stems or dead leaves, but the majority are those hyaline moulds which occur on, either cereals and their products, or on oilseeds and seed cake, and on all sorts of food commodities.

Among toxigenic moulds there are a few in the Mucorales, but there are many representatives from the genera *Aspergillus* and *Penicillium*, imperfect stages of ascomycetes of the Plectascales; *Fusarium*, the imperfect stage of ascomycetes of the Hypocreales; *Neurospora*, an ascomycete of the Sphaeriales. The discomycetes are represented by *Gloeotinia temulenta* and the dematiaceous hyphomycetes by such genera as *Pithomyces*, *Stachybotrys*, *Trichothecium*, and *Epicoccum*.

Toxigenic moulds thus come from a wide range of systematic groupings.

List of fungi responsible for mycotoxicoses

Table 6 is taken from numerous references. It gives the names of moulds which have been associated with toxicoses, although the gravity of the cases of poisoning is very variable depending on the species involved.

It has been estimated that 30–40% of all the moulds recognized may be capable of producing toxic metabolites in more or less dangerous quantities. Included in the table, where it is possible, are the names of the principal toxins produced. It should be noted that the same toxin may be produced by a range of different fungi (patulin, for example, is produced by *Aspergillus clavatus*, *A. giganteus*, *A. terreus* as well as by *Penicillium urticae*, *P. griseofulvum*, *P. expansum*, *P. claviforme*, *P. divergens*, *P. melinii*, *P. novae zelandiae*, and *Byssochlamys nivea*). On the other hand, some fungi may produce several different toxins, thus *Aspergillus fumigatus* produces fumigatin, spinulosin, helvolic acid, fumagillin, gliotoxin, etc.

It is essential to note that, among several species, only certain isolates or strains

Table 6. List of moulds responsible for toxicoses and the toxins associated with them (some remain to be confirmed)

Absidia lichtheimii (Lucet and Cost.) Lendn. = *A. corymbifera* (Cohn) Sacc. and Trott.
Absidia ramosa (Lindt) Lendn.
Alternaria humicola Oud. (*)
A. longipes (Ell. and Ev.) Tisdale and Wadkins.
A. tenuis Neés (*)
Aspergillus alliaceus Thom and Church . . . ochratoxins.
A. amstelodami (Mangin) Thom and Church . . . anthraquinones?
A. avenaceus G. Smith (*) . . . avenaciolide.
A. candidus Link . . . candidulin, kojic acid.
A. carneus (v. Tiegh) Blochw. (*) . . . flavipin?
A. chevalieri (Mangin) Thom and Church . . . anthraquinones? gliotoxin, xanthocillin X.
A. clavato-flavus Raper and Fennell.
A. clavatus Desm. . . . patulin, ascladiol, cytochalasin E, tryptoquivaline.
A. flavipes (Bain. and Sart.) Thom and Church . . . flavipin.
A. flavus Link . . . aflatoxins.
A. foetidus (Naka) Thom and Raper (*)
A. fumigatus Fres. . . . gliotoxin, helvolic acid, fumagillin, fumitremorgin.
A. giganteus Wehm. . . . patulin.
A. janus Raper and Thom.
A. luchuensis Inui . . . oxalic acid?
A. melleus Yukawa. . . . ochratoxins.
A. nidulans (Eidam) Wint. . . . nidulin, nornidulin, kojic acid, asperthecin, nidulotoxin.
A. niger v. Tiegh . . . oxalic acid, malformin C.
A. niveus Blochw. . . . citrinin.
A. ochraceus Wilh. . . . ochratoxins.
A. oryzae (Ahlb.) Cohn. . . . kojic acid, oryzacidin.
A. oryzae (Ahlb.) Cohn var. *effusus* (Tir.) Ohara . . . kojic acid.
A. oryzae (Ahlb.) Cohn var. *microsporus* Sakaguchi . . . maltoryzine.
A. ostianus Wehmer . . . aflatoxins, ochratoxins.
A. parasiticus Spear . . . aflatoxins.
A. petrakii Vorös (*) . . . ochratoxins.
A. phoenicis (Cda) Thom (*)
A. restrictus G. Smith (?)
A. ruber (Spieck and Brem.) Thom and Church . . . aflatoxins, anthraquinones ?
A. sclerotiorum Huber . . . ochratoxins.
A. sulphureus (Fres.) Thom and Church . . . ochratoxins.
A. sydowii (Bain. and Sart.) Thom and Church . . . sterigmatocystin.
A. tamarii Kita . . . kojic acid.
A. terreus Thom . . . terrein, patulin, citrinin.
A. terricola Marchal . . . kojic acid.
A. thomii Smith . . . kojic acid.
A. umbrosus Bain. and Sart. (*)
A. ustus (Bain.) Thom and Church . . . austocystins, austamide, austdiol.
A. versicolor (Vuil.) Tir. . . . sterigmatocystin, aversin, cyclopiazonic acid.
A. viride nutans . . . viriditoxin.
A. wentii Wehmer . . . kojic acid, aflatoxin.
Byssochlamys fulva Olliver and Smith . . . byssochlamic acid.

Table 6. (continued)

B. nivea Westl. . . . byssochlamic acid, patulin.
Cephalosporium acremonium Cda. . . . cephalosporin P_1
Chaetomium cochliodes Palliser (*)
C. globosum Kunze . . . oosporein, chaetomin, chaetocin.
Cladosporium exoasci Link (*)
C. fagi Oud. (*) . . . fagicladosporic acid.
C. fuligineum Bon. (*)
C. gracile Cda. (*)
C. herbarum (Pers.) Link . . . epicladosporic acid.
C. molle Cke. (*)
C. penicillioides Preuss (*)
C. sphaerospermum Penz. (*)
Curvularia sp.
Dendrodochium toxicum Pidoplichko and Bilai . . . dendrodochin.
Diplodia zeae (Schw.) Lév.
Epicoccum nigrum Link . . . flavipin.
Fusarium avenaceum (Fr.) Sacc.
F. culmorum (W. G. Sm.) Sacc. (= *F. roseum* (Link) Sn. and H.).
F. diversisporum Sherb. (*) . . . diacetoxyscirpenol.
F. equiseti (Cda.) Sacc. . . . diacetoxyscirpenol.
F. graminearum Schw. (= *Gibberella zeae* (Schw.) Petch.).
F. graminum Cda (*)
F. lateritium Neés (*)
F. moniliforme Sheld. (= *Gibberella fujikuroi* (Saw.) Wr.).
F. nivale (Fr.) Ces. . . . nivalenol, fusarenone.
F. oxysporum Schl. (*)
F. poae (Peck.) Wr. (= *F. tricinctum* (Cda.) Sn. and H. f. *poae*).
F. redolens Wr. (*)
F. roseum (Link) Sn. and H. . . . diacetoxyscirpenol.
F. sambucinum Fuck. (= *Gibberella pulicaris* (Fr.) Sacc.).
F. scirpi Lamb. and Fautr. (*)
F. semitectum Berk. and Rev. (*) (= *F. roseum* (Link) Sn. and H.).
F. gr. *sporotrichiella* Wr. and Reink. (= *F. tricinctum* (Cda.) Sn. and H.).
F. sporotrichioides Sherb. (= *F. tricinctum* (Cda.) Sn. and H.).
F. tricinctum (Cda.) Sn. and H. . . . sporofusarin, T_2 toxin.
F. tricinctum (Cda.) Sn. and H. f. *poae* . . . poaefusarin, poïn.
Gibberella fujikuroi (Saw.) Wr.
G. pulcaris (Fr.) Sacc.
G. zeae (Schw.) Petch. . . . zearalenone.
Gliocladium virens Miller, Giddens, and Foster . . . gliotoxin, viridin.
G. roseum Bain. . . . paraquinones.
Gloeotinia temulenta (Prill. and Del.) Wilson, Noble and Gray.
Hemispora stellata Vuil. (= *Wallemia ichtyophaga* Johan-Olsen).
Mucor albo-ater Naum.
M. circinelloides v. Tiegh (*)
M. corticolus Hag. (*)
M. fumosus Naum. (*)
M. globosus Naum. (*)
M. hiemalis Wehm.
M. humicola Raillo (*)
M. pusillus Lindt.

Table 6. (continued)

M. racemosus Fres (*)
Myrothecium verrucaria (Alb. and Schw.) Ditmar . . . verrucarol, verrucarin, muconomycin.
Neurospora sitophila Shear and Dodge.
Oospora colorans v. Beyma . . . oosporein.
Paecilomyces varioti Bain. ?
Penicillium atrovenetum G. Smith . . . β-nitropropionic acid.
P. aurantio-violaceum Biourge . . . citrinin.
P. brefeldianum Dodge . . . decumbin.
P. brevicompactum Dierckx . . . mycophenolic acid.
P. brunneum Udagawa . . . rugulosin, emodin, skyrin.
P. charlesii Smith . . . carolic acid.
P. chermesinum Biourge . . . costaclavin.
P. chrzaszszi Zaleski . . . citrinin.
P. citreo-viride Biourge . . . citreoviridin, citrinin.
P. citrinum Thom . . . citrinin, aflatoxin.
P. claviforme Bain. . . . patulin.
P. commune Thom.
P. concavorugulosum Abe.
P. corylophilum Dierckx . . . citrinin, gliotoxin.
P. crustosum Thom (*)
P. cyaneum (B. and S.) Biourge . . . decumbin.
P. cyclopium Westl. . . . penicillic acid, emodic acid, cyclopiazonic acid.
P. decumbens Thom . . . decumbin.
P. divergens Bain. and Sart. . . . patulin.
P. duclauxii Delacr.
P. expansum Link. . . . patulin.
P. fellutanum Biourge . . . carolic acid.
P. fenelliae Stolk . . . penicillic acid.
P. frequentans Westl. . . . frequentic acid, aflatoxin.
P. gilmanii Thom.
P. griseofulvum Dierckx . . . patulin.
P. herquei Bain. and Sart.
P. implicatum Biourge . . . citrinin.
P. islandicum Sopp. . . . luteoskyrin, islanditoxin, cyclochlorotin.
P. italicum Wehmer (*)
P. janthinellum Biourge.
P. jenseni Zal. (*)
P. lanosum Westl. (*)
P. lilacinum Thom (*)
P. lividum Westl. . . . citrinin.
P. martensii Biourge . . . puberulic acid, penicillic acid.
P. melinii Thom . . . patulin.
P. nigricans Bain (*)
P. notatum Westl. (*) . . . notatin, xanthocillin X.
P. novae zelandicae v. Beyma . . . patulin.
P. obscurum Biourge (= *P. corylophilum* Dierckx).
P. ochrosalmoneum Udagawa . . . citreoviridin.
P. olivino-viride Biourge . . . penicillic acid.
P. oxalicum Currie and Thom (*) . . . secalonic acid D.
P. palitans Westl. . . . palitantin, penicillic acid
P. patulum Bain. (= *P. urticae* Bain).

Table 6. (continued)

P. phoeniceum v. Beyma . . . phoenicin.
P. piceum Raper and Fennell (*) . . . helenin.
P. puberulum Bain. . . . penicillic acid. aflatoxin, puberulic acid.
P. pulvillorum Turfitt . . . citreoviridin.
P. purpurogenum Stoll . . . glaucanic acid, glauconic acid, rubratoxins.
P. roqueforti Thom.
P. roseo-purpureum Dierckx . . . frequentic acid.
P. rubrum Stoll . . . phoenicin, rubratoxins.
P. rugulosum Thom . . . rugulosin.
P. sartoryi Thom . . . citrinin.
P. spinulosum Thom . . . spinulosin.
P. steckii Zal. . . . citrinin.
P. stoloniferum Thom . . . mycophenolic acid.
P. tardum Thom . . . rugulosin.
P. terlikowskii Zal. . . . gliotoxin.
P. terrestre Jensen . . . patulin, terrestric acid.
P. toxicarium Miyake (= *P. citreoviride* Biourge).
P. umbonatum Sopp. (*)
P. urticae Bain. . . . patulin.
P. variabile Sopp. . . . aflatoxin.
P. verruculosum . . . verruculogen.
P. viridicatum Westl. . . . viridicatin, ochratoxins, citrinin, oxalic acid, viridicatic acid.
P. waksmani Zal. (*)
P. westlingi Zal. (= *P. waksmani* Zal.).
P. wortmanni Klocker . . . rugulosin.
Periconia minutissima Cda.
Piptocephalis freseniana de Bary (*)
Pithomyces chartarum (Berk. and Curt.) M. B. Ellis . . . sporidesmins.
Rhizoctonia leguminicola Gough and Elliot . . . slaframine.
Rhizopus nigricans Ehr. (*)
Sclerotium rolfsii Sacc.
Scopulariopsis brevicaulis (Sacc.) Bain.
S. candida (Guéguen) Vuil.
Sporidesmium bakeri Syd. (= *Pithomyces chartarum* (Berk. and Curt.) M. B. Ellis).
Stachybotrys alternans Bon (= *Stachybotrys atra* Cda.).
S. atra Cda.
Stemphylium sarcinaeforme (Cav.) Wiltshire . . . stemphone.
Thamnidium elegans Link (*)
Trichoderma lignorum (Tode) Harz.
T. viride (Pers.) Fr. . . . trichodermin.
Trichothecium roseum Link. . . . trichothecolone, trichothecin.
Verticillium psalliotae Treschow. . . . oosporein.
Wallemia ichtyophaga Johan-Olsen.

(*) Toxic in experimental tests but not yet found from natural toxicoses.

may produce toxic metabolites, on the other hand the production of some toxins may only occur on particular substrates (2838).

Table 6 includes fungi from which the production of toxins has not yet been confirmed, but which have been found to be frequently associated with foods connected with outbreaks of poisoning.

A systematic search for toxins has become an undertaking for the experimental laboratory (1033, 1235, 1951, 1954, 2315) and it is surprising to find that certain species which are capable of producing toxic metabolites in the laboratory are rarely, if at all, incriminated in natural outbreaks of poisoning.

This list in Table 6 will probably increase as future laboratory studies demonstrate the production of toxic metabolites by species previously considered as safe.

When confronted with so impressive a list of moulds reputed to be toxic, it is tempting to conclude, with Steyn (2490), that in practice all mouldy foods should be considered as suspect until proven to be safe. One can then, for example, only be pleasantly surprised to find so little consumer complaint when one thinks of the large number of species of mould concealed by flour (1727).

It is essential to recognize that errors in the diagnosis of the aetiology of a toxicosis are possible; thus Biester *et al.* (199) have indicated that, having incriminated *Trichoderma viride* in a toxicosis of horses, they were not able to reproduce the symptoms experimentally. There have also been errors in identification, the systematics of fungi being a study for specialists!

BRIEF HISTORY OF MOULD TOXICOSES

Past history

Ancient civilizations already recognized the malicious nature of some fungi. From fragmentary information it would seem that knowledge of fungal poisoning was present in ancient Greco-Roman times, although it was not until the Renaissance that more precise descriptions are found. These early accounts, however, are only concerned with intoxication due to the consumption of macromycetes (1057). It is not really until the end of the nineteenth century that reference is made to the toxicity of moulds and such early references usually contain a number of confusing observations which give them anecdotal rather than scientific value.

In 1862 (894), and subsequently in 1882 (1661), a disease of horses, ascribed to the consumption of mouldy feed, was described but not given much attention. In 1891, Woronin (2876) had established a relationship between dizziness and headaches with the consumption of bread made from grain contaminated with a *Fusarium.* Comparable observations are reported in older books about fungi, such as that of Cordier (546) published in 1876, who states on page 166: 'Mouldy bread may not be eaten without some danger to man and animals.'

Doctor Vesteroff reported having seen two children who, after eating bread made from rye moulded by *Mucor mucedo* Lin., had red, swollen faces, a startled animated stare and complained of a dry tongue, a weak, fast pulse, giddiness, an unquenchable thirst, followed by a desire to sleep, depression, despondency, etc. These symptoms disappeared following the use of an emetic.

The following example demonstrates that moulds growing on biscuits may be lethal to poultry:

> M. Simon, a solicitor of Arlon (Luxembourg, Belgium), had received for several years a case of Champagne containing some biscuits. The case was stored in a cave, for it was not known that the biscuits were there, and on being opened several months later, the biscuits were found to be entirely covered with moulds. They were thrown to the chickens in the farmyard which eagerly consumed them, became ill, and died.

A little further in the book, Cordier extrapolates by writing on page 172:

> Undoubtedly many other illnesses, such as dental caries and syphilis, may be attributed, by analogy, to the activity of cryptogams. Could not cancer, atonal ulcers, for example, be regarded as the products of a plant which, multiplying without respite, eats slowly away at the organ with which it is associated?

As Gibbons recalls (902) numerous intoxications of cattle have been recognized by veterinarians since 1850 (670, 951) under the name 'forage poisoning' and according to Pearson (1940) a form of forage poisoning of horses characterized by dizziness, assumed to result from cerebrospinal meningitis, was clearly related to the growth of moulds on the feed. This fact is confirmed by Pammel (1915). In the same way a variety of nervous diseases of horses consuming mouldy grain have been reported in the United States from 1896–1934 (338, 363, 717, 951, 1708, 1719) and may have been due to the presence of moulds. They have been variously called meningitis, leucoencephalitis, staggers, choking, and haemorrhagic encephalitis depending on the symptoms reported. In some cases cervical necrosis or necrosis of a large part of the spine marrow has been described, in other cases changes in the liver or kidneys were reported (1637). According to veterinary reports more than 5,000 horses died in Illinois after having consumed mouldy grain during the winter of 1933–34 (952). It must however be accepted, as Ronk and Carrick (2191) emphasize, that incorrect diagnosis could occur, especially with encephalitis of the horse which has been shown since 1930 (2307) to be due to a virus.

In 1916, an important step was taken with the work of Turesson (2659) on 'the presence and significance of moulds in the digestive tract of man and higher animals'. He recalls that Hammerl (1015) in 1897, Moro (1752) in 1900, Cao (407) in 1900, and Kohlbrugge (1331) in 1901 had demonstrated the presence of fungi, usually without further identification, in the digestive tract of humans. But Turesson established experimentally a relation between the chance accumulation of spores and the production of harmful toxins. He considered that *Aspergillus fumigatus, A. niger, A. nidulans, A. umbrinus, A. terreus, Penicillium avellaneum*, and *P. divaricatum* were particularly toxic and that their ingestion invoked muscular convulsions resembling tetanus, a weakening of the organism, and paralysis often followed by death.

Recent studies

Although the report by Steyn (2490) in 1933 and the review by Sarkisov (2241) in 1954 already provide a large number of references to earlier work, it was not until after 1955 and more particularly 1960, that there was a renewed interest in mycotoxins and increased precision in research on the topic. The scientific literature of the whole world ranges from original articles, bibliographic reviews to popular articles, and several important works have been published recently (212, 506, 927, 1067, 1492, 1726, 2048).

Numerous contributions from the United States treat the subject in its entirety but give a dominant place to the toxicity of *Aspergillus flavus.* The papers of Borker (265), Christensen (487, 488, 1518, 1519, 1648), Davis and Diener (595, 596, 598, 599, 654–659, 664, 665, 666, 1417, 1418), Forgacs and Carll (799, 800–807), Garren (874–877), Goldblatt (924–931), Hesseltine (1071, 1074, 1077, 1078, 1079, 1081, 2337, 2379), Mirocha (487, 489, 754, 1679–1685), Pomeranz (2015), Smalley (2417), and of Wogan (2854, 2856), as well as their collaborators are among the best known. Some publications particularly concern the poisoning of pigs and chickens and the dangerous moulds present in forages and litter (262, 801, 802, 805). Several reviews have been published (265, 315, 936, 1520, 2080, 2834).

Papers have also appeared in many other parts of the world. Thus from Canada there are those of Van Walbeek, Scott, and Thatcher (2316, 2318, 2320, 2323, 2324, 2325, 2727, 2765). Work has been carried out in South America (232, 690, 796, 935, 1865) and in Japan research has particularly concentrated on the toxic moulds of rice. The history of these studies has been described by Miyaki (1701) in a joint colloquium with the Americans held in Honolulu (1067). Several publications come from Formosa (461, 2658), India (458, 585, 940, 1583, 1796, 1807, 2092, 2656, 2657), Malaysia (455, 456, 457), and Australia (870, 1214, 2043, 2880).

In New Zealand work was concentrated on facial eczema of sheep (311–315, 557, 602, 669, 687, 688, 793, 845, 1124, 1598, 1958, 1959, 1964, 2232, 2393, 2420, 2421, 2535, 2596–2600, 2810, 2811, 2880, 2881). Some important work has come out of South Africa (10, 555, 1346, 1875, 1876, 1943, 2046, 2047, 2048, 2074, 2076, 2077, 2079, 2332, 2572, 2585, 2708, 2728) and East Africa (1534, 1536, 1542, 1543, 1594, 1899). A range of studies has been made in Israel by Joffe and his colleagues (267, 1234–1250, 2685).

A specific organization, the VNIL was established in the USSR for the study of mycotoxicoses, especially stachybotryotoxicosis of horses and alimentary toxic aleukia in man. Papers by Bilai (206, 209, 212) in the Ukraine, of Sarkisov (2241), and Spesivtseva (2456) in the region of Moscow, as well as numerous other publications testify to this work (64, 123, 253, 627, 674, 700, 725–727, 795, 810, 862–865, 917, 945, 979, 1159, 1162, 1163, 1279, 1298, 1335, 1337, 1338, 1340, 1377, 1378, 1466–1468, 1473, 1580, 1631, 1651, 1677, 1766, 1793, 1802, 1860, 1882, 1894, 2001, 2011, 2198, 2207, 2226, 2229, 2239, 2240, 2242, 2243, 2250–2252, 2364, 2385, 2456, 2741–2743, 2931, 2934).

Although in central Europe there is an interest in mycotoxins in general (115,

202, 698, 1216, 1275, 1693, 1912, 1914, 2026, 2541, 2612, 2697), in Scandinavian countries research is particularly concerned with porcine nephrotoxicosis, especially that of Krogh and his colleagues (849, 998, 1041, 1267, 1351–1359, 1501, 1531, 1568, 1569, 1570, 1855).

Since the large number of deaths among turkey poults and ducklings in 1960, work in Great Britain has centred around the toxins of *Aspergillus flavus,* however numerous agents of toxicoses are mentioned by Ainsworth and Austwick (24) in their book on the fungal diseases of animals and there are numerous publications about a wide range of organisms (23–25, 33–45, 81–85, 93, 101–107, 140, 141, 224–229, 240–243, 366, 370–376, 421–423, 522, 526–529, 531, 776–778, 902, 903, 1515, 1516, 1592, 1593, 1623, 1624, 1928, 2119, 2191, 2234–2238, 2386, 2453, 2455, 2486, 2487, 2532, 2776–2778, 2822) as well as some interesting biochemical work, notably that of Moss and his colleagues (1767–1773, 2623).

From Italy there is the work of Cantini, Scurti, and their collaborators (404–406, 2327), Ceni and Besta (430–435), and several others (450, 533, 631, 897–899, 1524, 2309, 2310, 2472), and from Spain that of Hernandez (1063).

In Germany research has proceeded in a methodical manner at Kulmbach, Munich, Karlsruhe, Hanover, Detmold, and Berlin (165, 246, 247–250, 352, 353, 814, 815, 817, 818, 888, 889, 1023, 1024, 1104, 1215, 1216, 1320, 1656, 1658, 2122, 2126–2136, 2267, 2271, 2272, 2285, 2302–2304, 2339–2341, 2457–2461, 2559, 2560, 2588, 2613, 2614, 2885). At least two bibliographies have been published (1655, 1657).

The work carried out in France is fairly scattered among scientists in various laboratories: Institut de la Sante et de la Recherche Médicale (194, 1398–1409), Institut National de la Recherche Agronomique (31, 59, 617–619, 748–750, 767, 768, 983, 984, 1223–1228, 1431, 1432, 1460), Centre National de la Recherche Scientifique (13–15, 280, 388, 822–833, 1393, 1394, 1435–1437, 1715, 1726, 1738, 1776, 1777, 2112–2116), and universities (272–278, 409, 1189–1207, 1944, 1951, 1954), but colloquia regularly bring workers together.

THE NATURE OF MYCOTOXICOSES ASSOCIATED WITH MOULDS

One factor holding back work on the mycotoxins produced by moulds is the difficulty in establishing a definitive classification of the syndromes associated with them. After a few remarks on the notion of toxicity a brief description is given on the principal manifestations currently recognized.

Toxicity

The very notion of toxicity is relative. According to Fabre and Truhaut (760, 761), one should consider as toxic: 'all substances which, after penetrating the organism in a transitory or permanent manner, either rapidly in a single relatively high dose, or over a long period of time in repeated small doses, provoke disturbances in one or more cell functions, disturbances which may lead to complete destruction and death.'

It is evident that when referring to a toxic dose, it is necessary to specify the route of ingestion of the toxin. When considering toxic moulds in foods it is usually the oral route which is relevant, nevertheless the pulmonary route should sometimes be considered and even absorption through the skin may be relevant on occasions.

The quantitative expression of toxicity (usually the LD_{50}, the dose required to kill 50% of a statistically valid group of animals, frequently expressed per unit of body weight) is only a convenient criterion for defining the acute toxicity of a compound. It does not take into account effects less severe than those leading to death, but which nevertheless constitute toxic symptoms, neither does it indicate chronic or cumulative toxicity which may result from the repeated ingestion of small doses. Furthermore the LD_{50} is determined from experiments on laboratory animals and can only be extrapolated to man with extreme caution.

Mycotoxicoses are frequently long-term phenomena in which chronic toxicity oftens plays a more important role than acute toxicity. For this reason the LD_{50} values of compounds incriminated, when these are known, are only of relative importance compared with a knowledge of the symptoms and aetiology following ingestion of mouldy food or mycotoxins.

Classification and nomenclature of mycotoxicoses

Classification based on aetiology

The most rational manner of classifying mycotoxicoses is undoubtedly on the basis of the causative agent. This is not always easy, first because earlier authors have often described toxicoses without being precise about the causative agent, and second because several species of fungi may be isolated from some types of mycotoxicoses, each playing a part but belonging to quite different taxonomic groups.

Terms such as aspergillotoxicosis, stachybotryotoxicosis, fusariotoxicosis, etc., corresponding respectively to toxicoses due to *Aspergillus, Stachybotrys,* and *Fusarium*, etc., fit such an aetiological classification.

A very vague nomenclature, also corresponding to the aetiology of illness, is to name the disease after the substrate on which the toxic moulds were growing. Thus, for example, the literature contains such terms as mouldy corn toxicosis of swine, forage poisoning, mouldy feed toxicosis of poultry, mouldy corn toxicosis of horses, mouldy rice poisoning, etc.

Classification based on the symptoms of disease

A number of mycotoxicoses are described, more or less precisely, by the symptoms observed. Thus in the literature there is mention of facial eczema of ruminants, alimentary toxic aleukia, cardiac ulceration in pigs, forage tetany, haemorrhagic syndrome of poultry, canine hepatitis, hepatoma of trout, etc.

Sometimes provisional names, such as turkey 'X disease', are given to certain conditions. More rarely it is the toxin itself which gives a name to the disease, e.g.

Table 7. The main characteristics of the principal mycotoxicoses

Predominant syndrome	*Fungi responsible*	*Substrates*	*Toxin*	*Sensitive animals*
	Aspergillus flavus	Oilseed and oil cake, grain, flour, etc.	Aflatoxins	Pigs, cattle, horses, sheep; turkey, duckling, pheasant, chicken, mink, dog, monkey, man ?, trout
Liver Damage	*Aspergillus versicolor*	Grain and oilseed products	Sterigmatocystin Aversin	
	Aspergillus ochraceus	Grain, rice, groundnuts	Ochratoxins	Chickens, lambs, heifers, pigs
	Penicillium islandicum	Rice, sorghum, millet, barley	Rugulosin, luteoskyrin, Islanditoxin	Man (?)
Kidney Damage	*Penicillium citrinum* *Penicillium viridicatum*	Groundnuts, rice, cereals, maize, barley, wheat, rye-grass	Citrinin	Pigs
Cardio-toxins	*Penicillium charlesii*	Cereals	Carolic acid	
	Penicillium terrestre		Terrestric acid	
	Penicillium viridicatum		Viridicatic acid	
	Penicillium cyclopium		Penicillic acid	
Gastro-enteric toxins	*Fusarium nivale* *Fusarium tricinctum* *Fusarium roseum*	Seeds	Scirpenes	Pigs, horses, sheep
	Trichothecium roseum *Trichoderma viride*	Various substrates		

Table 7. (Continued)

Predominant syndrome	*Fungi responsible*	*Substrates*	*Toxin*	*Sensitive animals*
Haemorrhagic symptoms	*Aspergillus fumigatus*	Seeds, flour	Quinones	All
	Aspergillus gr. *glaucus*	Seeds, flour, dried fruit	Anthraquinones	Cattle, poultry, rabbits
	Penicillium rubrum	Seeds	Rubratoxins	Pigs, horses, geese
	Penicillium purpurogenum	Cereal products	Glaucanic, glauconic acids	Chickens
	Byssochlamys fulva	Fruit juices	Byssochlamic acid	
	Stachybotrys atra	Straw	Stachybotryotoxin	Horses (Ukraine)
	Fusarium sporotrichioides	Cereals, especially millet, (overwintered)	Fusariogenin	Man (Ukraine, Siberia)
Neurotoxins	*Aspergillus clavatus*	Germinating seed	Patulin	Cattle
	Aspergillus oryzae	Malt	Maltoryzine	Cattle
	Gloeotinia temulenta	Grasses (esp. *Lolium*)	? Butenolide	Man, domestic animals Horses, man
	Fusarium nivale	Cereals (esp. oats)	Citreoviridin	
	Penicillium citreo-viride	Rice	?	Cattle, sheep (South Africa)
	Diplodia zeae	Maize		
Oestrogenic abortive	*Fusarium graminearum*	Stored maize	Zearalenone	Pigs, poultry
	Mucor, Absidia		?	Laying chickens
Facial eczema	*Pithomyces chartarum*	Pasture plants	Sporidesmins	Sheep (New Zealand)
	Periconia minutissima	*Cynodon dactylon*	?	Cattle (United States)

aflatoxicosis, patulin toxicosis, luteoskyrin toxicosis, etc., but such a terminology would be ambiguous when the disease resulted from the action of several toxins.

Table 7 indicates briefly the characteristics of the principal mycotoxicoses based on the major syndromes observed.

Minor toxicoses

In the presence of some fungi, or of low doses of some mycotoxins, symptoms which are not particularly serious may be observed: the irritation of gastric membranes provoking vomiting, diarrhoea, and sometimes dizziness, but not followed by any further symptoms unless the quantity of toxin ingested increases. *Gibberella zeae* (= *Fusarium graminearum*) produces an emetic compound (492, 691), whereas an irritant substance has been isolated from *Trichothecium roseum* which may cause the formation of stomach ulcers. The smoke of cigarettes made from tobacco infected with *Alternaria* and other dematiaceous fungi may cause pulmonary emphysema (804, 1013).

When poisoning is more serious a whole series of symptoms may appear and vomiting and diarrhoea may only be present as secondary phenomena depending on the trauma suffered by a particular tissue or organ (761).

Liver and kidney lesions

Mycotoxicoses are frequently associated with effects on the liver and the kidneys, and metabolites elaborated by moulds are thus frequently hepatotoxic and nephrotoxic. The gravity of such toxicoses should be stressed (1352, 1353).

The role of the liver in detoxification is well recognized, the hepatic tissues being capable of absorbing, and thus removing from circulation, toxic substances. Such compounds are often chemically transformed into less toxic metabolites in which form they may then be secreted. Or else the liquid secretion, bile, may be the route by which metabolites are excreted; certainly this is so in the case of the products of haemoglobin catabolism.

The kidney is an essential part of the excretory process and is concerned above all with the elimination of waste products and of foreign substances not utilized by the body. It can be seen then why, in the case of food poisoning, the first organs affected may be the liver and the kidney, followed by the gall bladder and bile ducts.

Doctors know well the disadvantages arising from the misuse of antibiotics of microbial origin, particularly erythromycin (1044) and triacetyloleandomycin (2602), in patients with a weakened liver. A recent colloqium organized by the Academy of Sciences of New York (588, 589) has emphasized the fact that a number of hepatic disorders may be due to fungi (1550) especially of the genera *Asperillus* and *Penicillium.* Indeed the appearance of adenomas followed by hepatomas has often been reported as resulting from the action of toxins produced by moulds. Some provoke the degeneration of cells of the hepatic parenchyma, fibrosis, and proliferation of the bile ducts in the portal spaces. Or else uraemia, albuminurea, and glomerulonephritis have been associated with such mycotoxins as citrinin, produced by *Penicillium citrinum* (1041, 1352, 1421, 1799, 2222, 2790), and the toxin of *P. viridicatum* (1358). It is important to know what role mycotoxins play in liver cancer in man (912, 1386, 1666, 1876, 2866).

Effect on the heart

Some mycotoxins with similar chemical formulae are cardiotoxic. These include the series of tetronic acids, such as carolic acid, produced by *Penicillium charlesii* and *P. fellutanum,* terrestric acid, produced by *P. terrestre,* viridicatic acid, produced by *P. viridicatum,* and penicillic acid, isolated from *P. cyclopium,* several other penicillia, and strains of *Aspergillus ochraceus* (2832).

Effect on the blood system

Once it has arrived in the blood, by whatever route, a toxic substance is transported in as little as 23 seconds throughout the entire organism (761). Many of the lesions associated with the activity of mycotoxins are accompanied by haemorrhages. The haemorrhagic syndrome is one of the most frequently described both in animals (862) and in man (732), especially in connection with the mycotoxicoses of *Stachybotrys atra*, *Fusarium tricinctum*, *Dendrodochium toxicum,* and *Penicillium citreoviride.* These haemorrhages are associated with considerable fragility of the blood capillaries and may be produced on the surface of the skin or in various organs such as the gastrointestinal tract, liver, kidneys, adrenal glands, lungs, brain, etc. Species of *Aspergillus* and *Penicillium,* as well as members of the Mucorales may be responsible for similar internal haemorrhages in chickens fed on mouldy feeds.

Simple haemorrhages usually result from acute intoxications whereas chronic forms frequently give rise to more subtle effects (801). There may be a progressive disappearance of the haematopoietic function of the bone marrow and leucopoenia (reduction of leucocytes). There may often be a lymphocytosis (destruction of lymphocytes) which may be dramatic in fusariotoxicosis or following the action of mycophenolic acid produced by members of the *Penicillium brevicompactum* series.

A particularly serious toxic effect has been demonstrated for a compound produced by *Aspergillus ochraceus* which causes an agglutination of the red blood cells in the liver leading inevitably to a fatal outcome (2483).

The haemolytic activity of several fungi should also be considered. *Aspergillus fumigatus, Alternaria tenuis,* certain mucorales, and several dermatophytes may all give a similar reaction with erythrocytes (853, 916).

Effect on the nervous system

Several products of mould metabolism are neurotoxic and such activity on the nervous system may have repercussions on the activity of the muscular system. In some cases there is a weakening, apathy, vertigo, and headache; in others there is by contrast a considerable agitation with fast breathing and hyperactivity, then nausea associated with convulsions or numbness, prostration, muscular paralysis and incoordination of movement.

These symptoms may be caused by the toxins of *Aspergillus clavatus, A. oryzae,* and *A. ochraceus.* They correspond to those seen in animals consuming grasses invaded by *Gloeotinia temulenta.*

Dermatoses resulting from photosensitization

Some moulds cause a classical type of photosensitization reaction (514, 1342). The mechanism of these reactions is well understood following the work of Blum (244). Some plants, such as *Senecio,* trefoil, lucerne, *Hypericum,* and *Agave lechuguilla* contain photodynamic substances which, after ingestion, sensitize those regions in the skin of the animal with little or no pigment to sunlight. It is known that, if the mechanism of biliary excretion is upset, photodynamic agents such as phylloerythrin and bilirubin are produced in excess, passing into the blood circulation and reaching the peripheral blood capillaries. This results in photosensitization and jaundice.

Photosensitizing agents, having reached the capillaries under the skin are activated by the sun causing disturbances of the cell structure and an alteration in permeability. This leads to a release of compounds triggering off local inflammation.

The toxins of *Pithomyces chartarum,* and possibly of *Periconia minutissima,* act in just this way causing the release of photosensitizing agents into the blood stream and giving rise to eczema and dermatoses.

Hyperkeratosis

Hyperkeratosis in cattle is a skin disease characterized by epithelial excrescence resulting in a thickening of the skin and loss of hair especially in the region of the neck and shoulders (1887). It is known that such lesions may be caused by a reduction of available vitamin A, or by the action of chloronaphthalenes used as wood preservatives. Hyperkeratosis has also been associated with the presence of certain fungi (382, 412, 413, 414, 805, 807). *Aspergillus chevalieri, A. clavatus,* and *A. fumigatus* have been particularly incriminated in this syndrome but all three are equally involved in haemorrhagic syndromes.

Oestrogenic activity

Oestrogenic compounds of plant origin are not generally classed as toxins, although Moule, Braden, and Lamond (1775), in their study of these compounds, consider that their effects may be compared with those of true toxins. For example, the high concentration of coumoestrol in the leaves of lucerne parasitized by *Pseudopeziza medicaginis* (Lib.) Sacc. or *Leptosphaerulina briosiana* (Poll.) Graham and Luttrel causes grave disorders in animals consuming the forage (1022). Among the fungi, *Gibberella zeae* (Schw.) Petch, (*Fusarium graminearum* Schw.), *Trichothecium roseum* Link, and various mucorales, especially *Rhizopus nigricans* Ehrh., can either elaborate such compounds or carry out bioconversions leading to oestrogenic substances. Such moulds have been implicated in some problems of the genital organs such as vulvovaginitis in the sow and swelling of the foreskin in the pig.

Abortive activity

Abortions may result from the consumption of certain compounds produced by fungi or they may be one aspect of mycoses due to such species as *Absidia ramosa, Mucor pusillus,* and *Aspergillus fumigatus* (188, 914, 1355).

Indirect activity

Morbid symptoms may arise in which fungi play an indirect role. For example, metabolic products of moulds may act as antivitamins (431–435) or the presence of fungi may cause a loss of appetite, with repercussions on the health of the animals, by conferring an evil taste or smell to the food.

Serious problems may be brought about by moulds which, though not directly producing toxic metabolites, are able to transform previously harmless materials in feeds into toxic compounds. For example, the oxidation of lipids, which may be influenced by moulds, may give rise to some problems. Thus Mme Mercier (1264) has recorded a distinct increase in the degradation of lipids in a very mouldy consignment of maize and Pomeranz (2015) has noted that the oxidation of unsaturated free fatty acids in infected grain may be incriminated in the appearance of muscular dystrophy of pigs.

The growth of *Ceratocystis fimbriata,* and other fungi, on the tubers of sweet potato triggers off in the plant the production of compounds toxic to man. Both terpenes such as ipomeamarone and ipomeanine, and coumarin derivatives such as umbelliferone, scopoletin, esculetin, scopolin, and skimmine may be produced (2695, 2841).

It is known that an *Aspergillus* growing on melilot may transform the non-toxic courmarin into dicoumarol which is a well-known anticoagulant and antivitamin K agent. Further work may possibly reveal other examples of toxicoses due directly or indirectly to moulds.

THE CHEMICAL NATURE OF MYCOTOXINS

A mycotoxicosis is essentially characterized by being neither contagious nor infectious but by the fact that toxins are responsible. Ceni and Besta (431–435) were among the first to have recognized the involvement of toxins elaborated by moulds in the diseases of animals. They showed that extracts of spores and mycelium of *Aspergillus fumigatus, A. flavus, A. niger, A. ochraceus,* and several strains of *'Penicillium glaucum'* when injected into animals by the intraperitoneal route, subcutaneously or intravenously, often caused death 'following spasms, tetanic and epileptic convulsions'. These authors clearly demonstrated an association between the symptoms resulting from the consumption of mouldy feed, especially in the case of *Aspergillus fumigatus,* and those elicited by injection of toxic compounds extracted from the fungus.

Several fungi are toxic simply because of the production of oxalic acid (*Aspergillus niger, A. luchuensis*), but the majority of mycotoxins have a very much more complex structure (1623, 1644–1646).

The range of mycotoxins produced by microfungi

The mycotoxins responsible for the poisoning of warm blooded animals, the subjects of this review, often have their analogies among the toxins produced by plant parasites, or with toxins active against micro-organisms, known as phytotoxins and antibiotics respectively.

Phytotoxins

Many microfungi parasitic on higher plants have been shown to produce toxins. Those which have been particularly well studied include the toxins of *Fusarium,* which produce more than ten such compounds (134, 1294, 1863, 1864), *Phytophthora* (1361), *Pythium irregulare* (1607), *Helminthosporium victoriae* (2037, 2260), *Periconia circinata* (2038), and many other parasites such as *Rhizoctonia, Verticillium, Phialophora, Endothia, Dothichiza, Deuterophoma, Fusicoccum,* and *Colletotrichum.* A number of relevant reviews have recently been produced (285, 639, 2039, 2805).

Many phytotoxins do not seem to have any relationship with the mycotoxins active against animals. However, it is worth noting that endothianine, isolated from *Endothia parasitica,* the agent of chestnut blight, is a dianthraquinone similar to skyrin, a mycotoxin produced by *Penicillium.* Diaporthin, another metabolite of *E. parasitica,* has a similar structure to that of the ochratoxins of *Aspergillus ochraceus.* Helminthosporin, isolated from several species of *Helminthosporium,* is isomeric with islandicin produced by *Penicillium islandicum,* and lycomarasmin and other toxins associated with vascular wilts caused by *Fusarium* have structures similar to the aspergillomarasmins of *Aspergillus.*

Inversely, several compounds toxic to animals, such as patulin produced by *Aspergillus clavatus,* and aflatoxin produced by *A. flavus,* are equally toxic to plants in laboratory tests (293, 1317, 1711, 2882).

Antibiotics

The fungi produce several antibiotic compounds (305). By definition an antibiotic is a compound produced by one living organism (usually a micro-organism) which inhibits the growth of micro-organisms. The term phytoalexin is used specifically to describe antimicrobial compounds produced by vascular plants in response to attack by micro-organisms.

A significant number of mycotoxins are also antibiotics, on the other hand not all antibiotics produced by moulds are implicated in mycotoxicoses. As emphasized by Cantini and colleagues (406), there are many reports of moulds with both toxigenic and antibiotic activities (for example, *Penicillium citrinum, P. corylophilum, P. urticae, Aspergillus chevalieri, A. flavus, A. nidulans, A. niger, A. repens,* and *Fusarium moniliforme*). Nevertheless, certain moulds are very toxic to man and animals but only have feeble antibiotic activity. Thus, because some antibiotics, those of *Aspergillus clavatus* for example (1189, 1202), are highly toxic, and others not so, it is not safe to assume any correlation between mammalian toxicity and antibiotic activity.

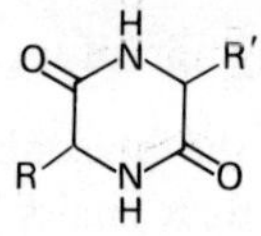

Figure 11. The general structure of diketopiperazines

Compounds derived from peptides

Peptide derivatives, especially polypeptides, are among the most dangerous compounds produced by macrofungi. Several microfungi, parasitic on plants, and saprophytic moulds produce such compounds.

Diketopiperazine derivatives

A number of mycotoxins contain the diketopiperazine nucleus (2924), and they are all derived by a combination of different amino acids. A generalized formula (Figure 11) indicates how the amino acids are built into these molecules.

An example is gliotoxin which can be derived by the dehydration of a dipeptide formed from metatyrosine and serine (2848). The gliotoxin is a result of diketopiperazine dehydrogenation, methylation of one of the nitrogen atoms, and

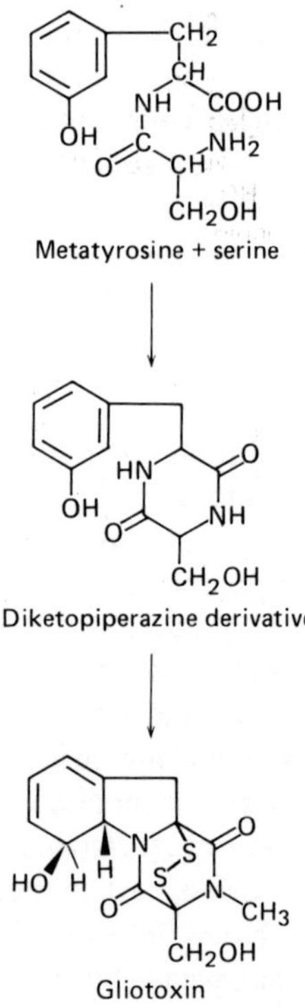

Metatyrosine + serine

Diketopiperazine derivative

Gliotoxin

Figure 12. A possible route to the biosynthesis of gliotoxin

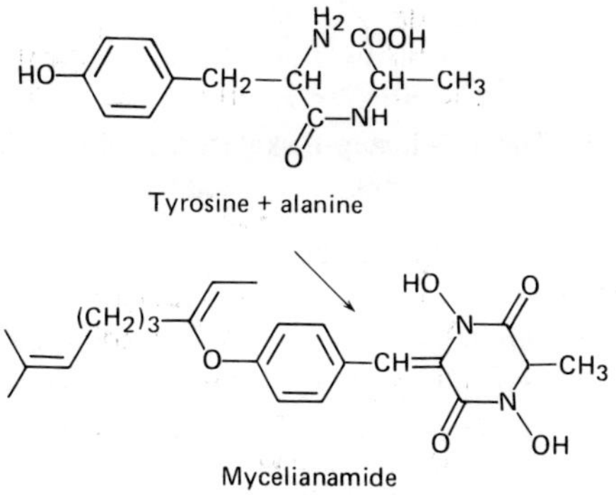

Tyrosine + alanine

Mycelianamide

Figure 13. A possible route to the biosynthesis of mycelianamide

the introduction of a disulphide bridge across the piperazine ring (Figure 12) (1645). It should be noted that such an introduction of sulphur occurs in other mycotoxins derived from peptides such as sporidesmin produced by *Pithomyces chartarum.* This biosynthetic pathway to gliotoxin has been confirmed using precursors, such as tyrosine and serine, labelled with ^{14}C.

In an analogous manner it has been shown (Figure 13) that mycelianamide, produced by *Penicillium griseofulvum,* is derived from tyrosine and alanine (218).

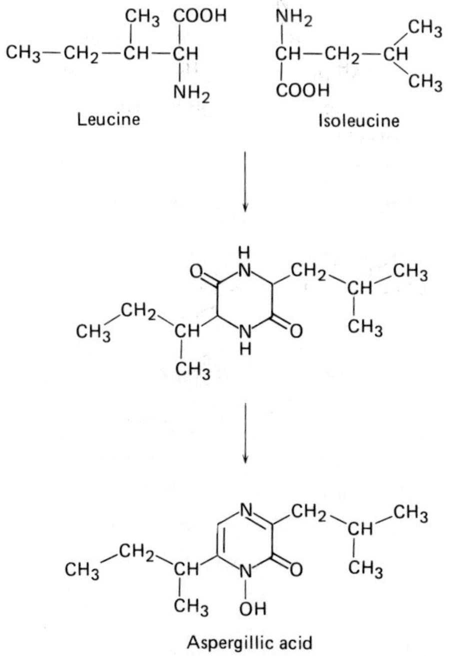

Leucine

Isoleucine

Aspergillic acid

Figure 14. Biosynthesis of aspergillic acid

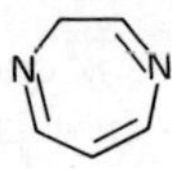

Figure 15. The 1,4-diazepin skeleton of the cyclopenins

Aspergillic acid produced by *Aspergillus flavus* is derived from the precursors leucine and isoleucine (Figure 14) (1549).

The metabolites aspergillomarasmine B and anhydro-aspergillomarasmine B, produced by *Aspergillus oryzae,* although not strictly diketopiperazines, do give rise to aspartic acid and glycine on prolonged acid hydrolysis (2162).

Cyclopenins

The cyclopenins, isolated from *Penicillium cyclopium,* are fairly similar to the diketopiperazines, their formation involving the aromatic ring of phenylalanine or metatyrosine (Figure 15). The resulting heterocyclic ring is the 1,4-diazepin nucleus which is a seven membered ring (1644).

$HOOC-CH(NH_2).(CH_2)_3-COOH$ — L-α-Aminoadipic acid

$H_2N-CH(COOH)-CH_2SH$ — L-Cysteine

$(CH_3)_2CH-CH(NH_2)-COOH$ — L-Valine

↓

$HOOC-CH(NH_2).(CH_2)_3.CO.NH-CH(CH_2SH)-CO-NH-CH(CH(CH_3)_2)-COOH$

L-Aminoadipylcysteinylvaline

↓

$HOOC-CH(NH_2).(CH_2)_3.CO.NH$– (penam nucleus: CH–CH, S, $C(CH_3)_2$, CH–COOH, N, C=O)

Isopenicillin N

$HOOC-CH(NH_2).(CH_2)_3CO.NH$– (cephem nucleus: CH–CH, S, CH_2, $C-CH_2OAc$, C–COOH, N, C=O)

Cephalosporin C

Figure 16. The biosynthesis of penicillins and cephalosporins

Penicillins and related compounds

The tripeptide α-aminoadipylcysteinylvaline may cyclize to give either cephalosporin N, which is in fact a penicillin with the characteristic β-lactam ring fused to the five membered thiazolidene ring (1645), and should be called penicillin N, or cephalosporin C in which the β-lactam ring is fused to the six membered dihydrothiazine ring (8). The relationships between these compounds are indicated in Figure 16 for they are good examples of metabolites with very little toxicity but considerable antibiotic activity.

Other peptide derivatives

Fungal polypeptides include such compounds as the depsipeptides, an example of which is sporidesmolide II extracted from *Pithomyces chartarum* along with the sporidesmins (195, 2651); the echinulins, produced by *Aspergillus echinulatus* and other members of the *glaucus* group, for which tryptophan and isoprene derivatives are probably precursors (1668).

Islanditoxin is an interesting chlorine-containing toxic metabolite produced by *Penicillium islandicum,* containing in its structure α-aminobutyric acid, β-phenyl-β-aminopropionic acid, serine, and dichloroproline.

Although the indole based alkaloids are best known as metabolites of *Claviceps,* a number of moulds, notably *Aspergillus fumigatus,* also produce them.

Quinone derivatives

Benzoquinones

These compounds are based on the structure indicated in Figure 17. A number of toxins produced by *Aspergillus fumigatus*, such as fumigatin and spinulosin, belong to this group. The pigments phoenicin, produced by *Penicillium rubrum* and *P. phoeniceum* and oosporein (= chaetomidin), produced by *Chaetomium, Oospora, Verticillium,* and *Acremonium,* are metabolites in which two benzoquinone groups are linked directly by a carbon–carbon bond. They correspond in their substitution pattern to fumigatin and spinulosin respectively.

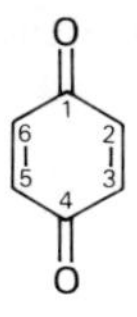

Figure 17. The benzoquinone nucleus

Anthraquinone derivatives

The structure and numbering system of the anthraquinones are given in Figure 18.

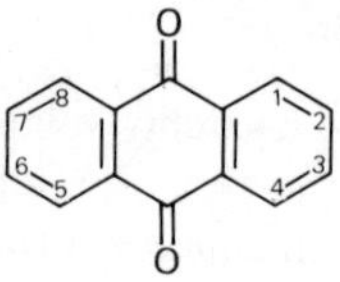

Figure 18. The anthraquinone nucleus

Most of these compounds are pigments, some possibly having a role in respiratory activity. Their absorption spectra and biological activity differ according to the position of substituents. There are two groups of positions for substitution, 1, 4, 5, 8 may be grouped together as α-substituents and 2, 3, 6, 7 as β-substituents (1764, 1971). Of the simple substituted anthraquinones the following have been recorded as mould metabolites:

Chrysophanol	1,8-OH ; 3-CH_3	*Penicillium*
Islandicin	1,4,8-OH ; 3-CH_3	*P. islandicum*
Helminthosporin	1,5,8-OH ; 3-CH_3	*Helminthosporium*
Emodin	1,6,8-OH ; 3-CH_3	*Cladosporium, Penicillium*
Methylemodin	1,6-OH ; 3-CH_3 ; 8-OCH_3	*P. frequentans*
Emodic acid	1,6,8-OH ; 3-COOH	*P. cyclopium*
Physcion	1,8-OH ; 3-CH_3 ; 6-OCH_3	*Penicillium, Aspergillus*
Endocrocin	1,6,8-OH ; 2-COOH ; 3-CH_3	*Penicillium, Aspergillus*
Cyanodontin	1,4,5,8-OH ; 3-CH_3	*Helminthosporium*
Catenarin	1,4,6,8-OH ; 3-CH_3	*Helminthosporium*
Tritisporin	1,4,6,8-OH ; 3-CH_2OH	*Helminthosporium, Penicillium*
Erythroglaucin	1,4,8-OH ; 3-CH_3 ; 6-OCH_3	*Aspergillus glaucus* group
Asperthecin	1,2,5,6,8,-OH ; 3-CH_2OH	*A. nidulans*

A number of metabolites which are important from a toxicological point of view are formed by joining two anthraquinone molecules together through position five with or without other links (Figure 19). For example, among the toxins of *Penicillium rugulosum* and *P. islandicum,* there are skyrin, iridoskyrin, and luteoskyrin, and from a species of *Fusarium,* fusaroskyrin.

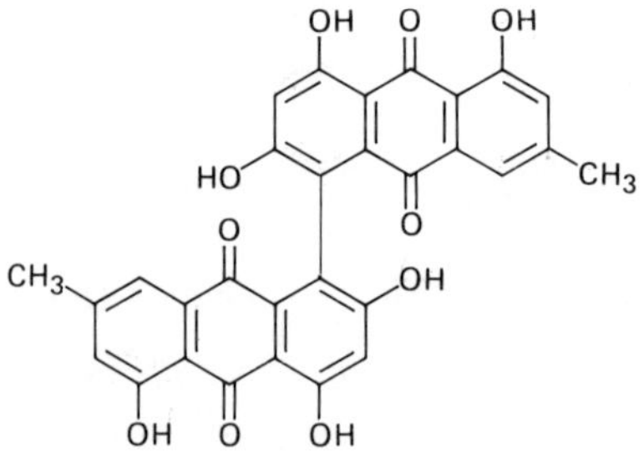

Figure 19. Skyrin, an example of a dianthraquinone

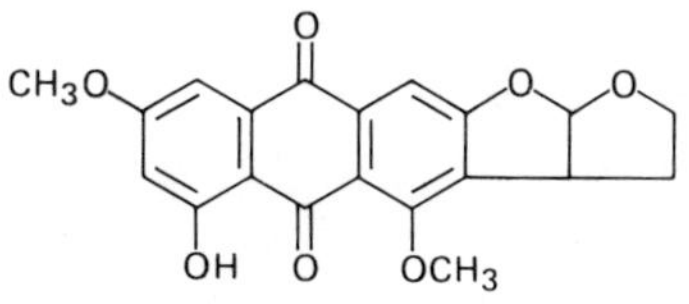

Figure 20. Aversin, an anthraquinone pigment from *Aspergillus versicolor*

Aspergillus versicolor produces a number of anthraquinone metabolites with more complex substituents such as the difuran ring of aversin and the versicolorins (Figure 20), the heterocyclic system of averufin, and the linear chain of six carbon atoms of averantin.

Compounds with a pyrone nucleus and precursors

Gentisaldehyde and its products

It is considered that a diphenol such as gentisaldehyde may, after fission, give rise to a pyran nucleus attached to a γ-lactone ring (Figure 21) (2871), a process which probably occurs in the biosynthesis of patulin by *Aspergillus clavatus* and several *Penicillium* species. Indeed, gentisyl alcohol has been isolated from strains of *Penicillium griseofulvum* able to produce patulin.

Maltoryzine, a metabolite of *Aspergillus oryzae,* is a triphenol with a pentenone side chain giving it some features in common with gentisaldehyde.

Flavoglaucin and auroglaucin, two pigments produced by members of the *Aspergillus glaucus* group, both contain a gentisaldehyde nucleus, whereas citreoviridin, a toxic pigment of *Penicillium citreo-viride,* contains a hydrofuran nucleus and an α-pyrone nucleus linked by a long unsaturated chain.

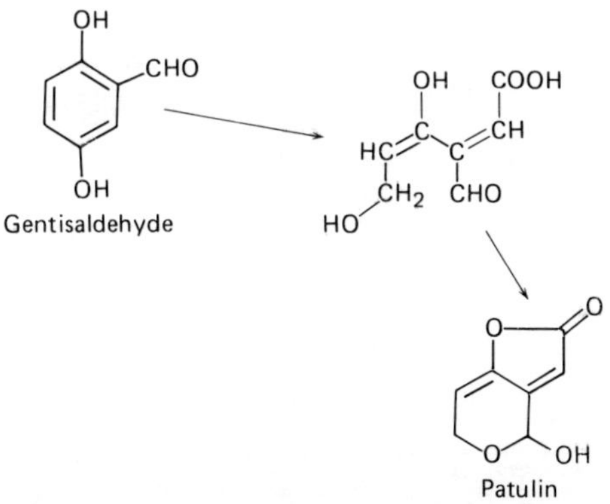

Figure 21. Biosynthesis of patulin

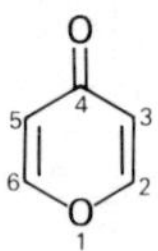

Figure 22. The γ-pyrone nucleus

Kojic acid and derivatives

The biosynthesis of the γ-pyrone nucleus of kojic acid may occur directly by oxidation and dehydration of glucose without breaking the carbon chain (73). Kojic acid and several derivatives, all with low toxicity, occur among the metabolites of a number of moulds and they all contain the nucleus shown in Figure 22.

Xanthones

There are relatively few members of this group of metabolities (Figure 23) obtained from the fungi (214). Sterigmatocystin and its monomethoxy derivative, isolated from *Aspergillus versicolor,* were among the first to be recognized.

Sterigmatocystin (Figure 24) contains the same difuran ring system as that found in aversin, one of the anthraquinone pigments of *A. versicolor,* and the aflatoxins produced by *A. flavus,* indeed, this latter organism also produces *O*-methylsterigmatocystin and aspertoxin, a hydroxy derivative of *O*-methylsterigmatocystin.

Closely related to the xanthones are toxins such as rubrofusarin produced by *Fusarium culmorum,* and flavasperone produced by *Aspergillus niger,* which contain the nucleus shown in Figure 25.

Figure 23. The xanthone nucleus

OH O OCH$_3$

Figure 24. Sterigmatocystin

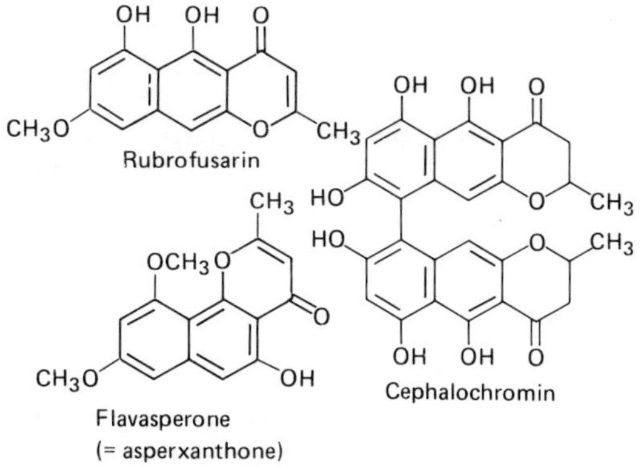

Figure 25. Chromone metabolites produced by moulds

If two such molecules are joined through position five, in a manner analogous to the skyrins, another group of compounds is obtained. An example is cephalochromin isolated from a *Cephalosporium* species; now *Cephalosporium* may be a microconidial form of *Fusarium* and the chemical relationship between the metabolites may reflect a systematic relationship between the moulds.

Another similar structure is that shown in Figure 26 and found in the metabolite produced by *Penicillium frequentans* known as frequentic acid.

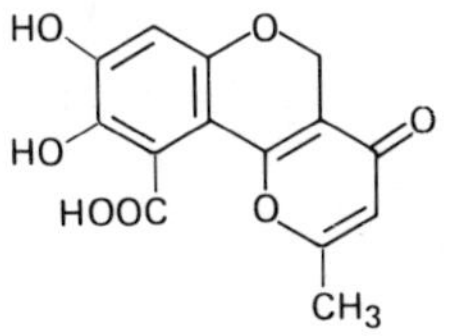

Figure 26. Citromycetin (= frequentic acid)

Coumarins and derivatives

Coumarin itself (Figure 27) is not toxic to animals and yet there are metabolites containing this nucleus which are among the most toxic of small molecules. Examples are ochratoxin produced by *Aspergillus ochraceus,* melitoxin, formed in melilot by the activity of *Aspergillus fumigatus,* and aflatoxin.

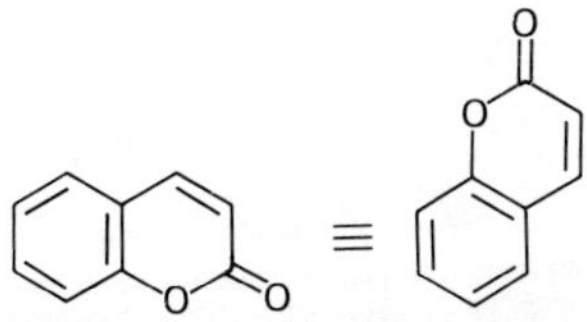

Figure 27. The coumarin nucleus

Terpene derivatives

Triterpenes

Two biologically active triterpenes may be isolated from moulds in the form of their diacetates: helvolic acid (which is present in the fumigacin described by Waksman) from *Aspergillus fumigatus*, and cephalosporin P from *Cephalosporium* (254).

Sesquiterpenes

A group of closely related sesquiterpenes have been isolated from several phialidic moulds. They form the classes of scirpenes or derivatives of trichothecin (127), a number of which are collected together in Figures 28–31.

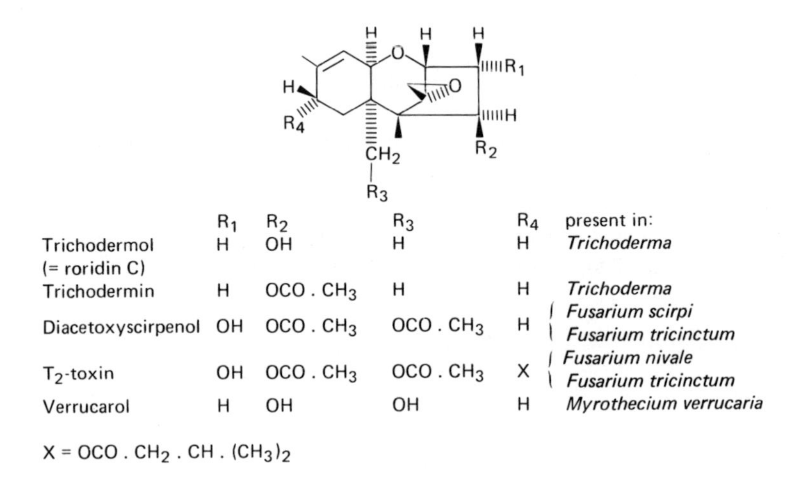

	R_1	R_2	R_3	R_4	present in:
Trichodermol (= roridin C)	H	OH	H	H	*Trichoderma*
Trichodermin	H	OCO . CH_3	H	H	*Trichoderma*
Diacetoxyscirpenol	OH	OCO . CH_3	OCO . CH_3	H	*Fusarium scirpi*, *Fusarium tricinctum*
T_2-toxin	OH	OCO . CH_3	OCO . CH_3	X	*Fusarium nivale*, *Fusarium tricinctum*
Verrucarol	H	OH	OH	H	*Myrothecium verrucaria*

X = OCO . CH_2 . CH . $(CH_3)_2$

Figure 28. 12,13-Epoxytrichothecene derivatives

	R_1	R_2	R_3	R_4	present in:
Trichothecin	H	OCO . CH=CH . CH_3	H	H	*Trichothecium roseum*
Trichothecolone	H	OH	H	H	*Trichothecium roseum*
Toxic diacetate	OH	OCO . CH_3	OCO . CH_3	OH	*Fusarium equiseti*, *Fusarium scirpi*
Nivalenol	OH	OH	OH	OH	*Fusarium nivale*
Fusarenone	OH	OCO . CH_3	OH	OH	*Fusarium nivale*

Figure 29. 8-Oxo-12,13-epoxytrichothecene derivatives

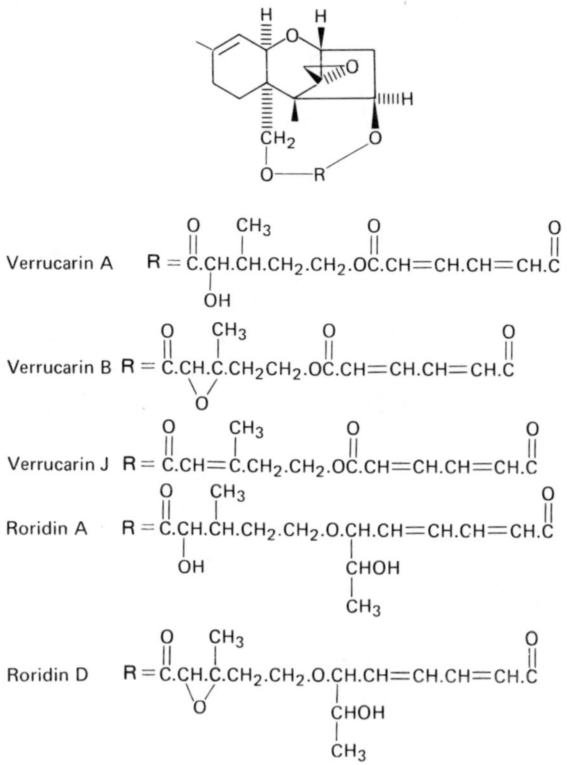

Verrucarin A R = C(=O).CH(OH).CH(CH$_3$).CH$_2$.CH$_2$.OC(=O).CH=CH.CH=CH.C(=O)

Verrucarin B R = C(=O).CH.C(CH$_3$).CH$_2$CH$_2$.OC(=O).CH=CH.CH=CH.C(=O) (CH–C epoxide, O)

Verrucarin J R = C(=O).CH=C(CH$_3$).CH$_2$.CH$_2$.OC(=O).CH=CH.CH=CH.C(=O)

Roridin A R = C(=O).CH(OH).CH(CH$_3$).CH$_2$.CH$_2$.O.CH(CHOH.CH$_3$).CH=CH.CH=CH.C(=O)

Roridin D R = C(=O).CH.C(CH$_3$).CH$_2$.CH$_2$.O.CH(CHOH.CH$_3$).CH=CH.CH=CH.C(=O) (CH–C epoxide, O)

Figure 30. Toxic metabolites of *Myrothecium roridum* and *M. verrucaria*

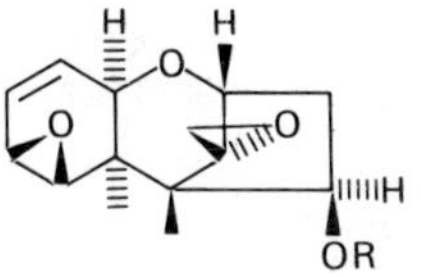

Crotocin R = CO.CH=CH.CH$_3$
Crotocol R = H

Figure 31. Toxic trichothecenes from *Cephalosporium*

Nonadrides

This is a group of compounds with complex formulae based on the presence of anhydride groups attached to a nine membered carbocyclic ring, including glaucanic and glauconic acids from *Penicillium purpurogenum,* byssochlamic acid from *Byssochlamys fulva,* and the rubratoxins of *Penicillium rubrum.*

It can thus be seen that the mycotoxins form a chemically diverse group of compounds ranging from polypeptides, alkaloids, benzoquinones, anthraquinones, xanthones, coumarin derivatives, terpene derivatives, etc. Further investigations using modern biochemical and chemical techniques will continue to increase our knowledge of these compounds.

PRODUCTION, MODE OF ACTION, AND DETECTION OF MYCOTOXINS

There has been a great deal of study on the mode of action of some mycotoxins such as the aflatoxins, but considerably less on the biosynthesis and conditions of production. In these areas a number of hypotheses emerge and are subjected to further study using new techniques. The pressure of the practical requirement to detect the presence of some specific mycotoxins in foods has ensured that sensitive techniques have been developed but there are some toxins for which such techniques have not yet been developed.

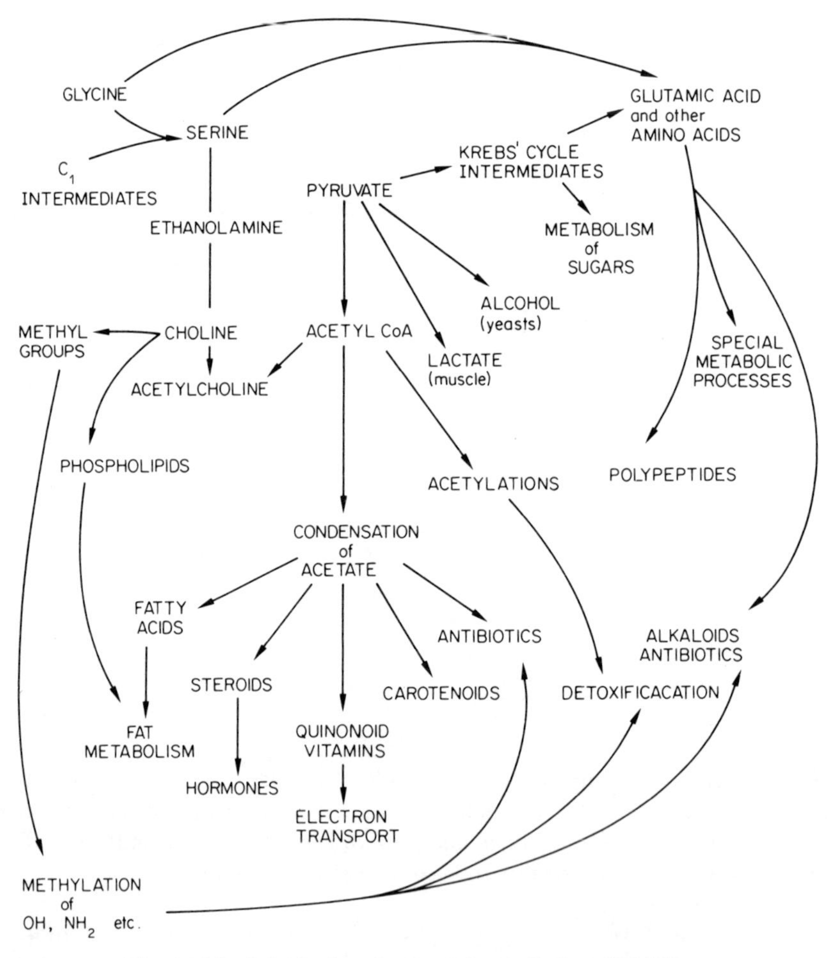

Figure 32. Principal pathways of metabolism (2345)

Production

The very idea of a toxin dates back to the classical experiments of Loeffler in 1884: having injected subcutaneously diphtheria bacteria into guinea-pigs he noticed, after they died several days later, that they showed internal lesions from which bacteria could not be isolated. He concluded that the bacteria grew close to the site of infection and produced a 'poison' which was disseminated by the blood.

These conclusions were confirmed five years later when Roux and Yersin proved that the injection of bacteria-free filtrates could produce the specific symptoms of the disease resembling those which would have been produced by injecting bacteria.

The toxic compounds produced within the mycelium or spores of fungi are called endotoxins and are only active if sufficient quantity of the fungus itself is consumed. This is the situation with most of the mycotoxins of the macrofungi. The exotoxins are released into the ambient medium by the mycelium. However, such a distinction may often have only a theoretical value, thus the aflatoxins may be detected in the mycelium and spores of *Aspergillus flavus,* but may equally well be secreted into foods such as groundnut meal invaded by this fungus.

Some mycotoxins are destroyed by heat, others are thermostable. The latter may be the more dangerous in food for sterilization may remove the viable moulds present but may not destroy any thermostable toxins already produced.

As soon as a mould spore germinates biosynthetic processes begin, first of RNA and DNA which are detectable during the first few hours. Toxic metabolites are usually produced a little later when the mycelium has become established in the substrate. Many factors are involved in the biosynthesis of mycotoxins by moulds. Some of the biosynthetic processes which may be involved are indicated in Figure 32.

Environmental factors

Environmental factors certainly play a major role (597) and one of the most important is temperature. It is known, for example, that *Fusarium sporotrichioides* only produces its toxin at a low temperature, between –4 and 1.5 °C (2243). *Stachybotrys atra* produces toxin at an optimum rate when it starts to produce spores. As an example of a nutritional factor the requirement of *Aspergillus clavatus* for an amyloid compound in its production of patulin may be cited (1202). On the other hand environmental factors may be equally important in processes by which toxins are degraded (1228).

Genetic factors

Otto (1900) was one of the first to recognize that differences in the production of toxins by *Aspergillus fumigatus* not only depended on the season but also on the strain. Thus Italian strains which he possessed were more toxic than German strains. Toxin production by such moulds as *A. fumigatus* and *A. flavus* is subject to genetic variability.

As well as variability in the production of toxins another aspect of mycotoxicoses is the variation in the sensitivity of the consumer. Certain toxins only affect a particular species other toxins affect many, and even within a single species individual resistance may vary. It must also be recognized that an animal already weakened for some other reason may be more sensitive to the action of a toxin.

Young ducklings are very sensitive to low doses of aflatoxin, the calf is much less so; it thus requires a much higher dose of toxin to produce symptoms in the calf which are recognized in ducklings at very much lower doses.

Mode of action

As would be expected for such a chemically diverse group of compounds the mode of action varies according to the compound. Although at one time it was considered (1900) that the toxins alone had no activity but that it was necessary for the fungi themselves to invade the body, this view was already questioned when it was proposed (1897, 2286) and would certainly not be accepted today.

The tissues of animals may respond to the presence of a foreign compound by carrying out one or more of a number of chemical reactions such as oxidations or reductions, condensations or hydrolyses, alkylation or dealkylation (2345). If a compound is going to be toxic it will perturb the metabolism of the poisoned animal perhaps triggering off a cascade of secondary effects.

An examination of any recently published map of biochemical pathways in living organisms will be sufficient to indicate the complexity of interacting processes which take place in the animal. It may suffice to inhibit or alter any one of these processes, by acting for example on one of the enzymes involved, to produce disorders the consequence of which may be more or less severe (2815).

Reviews such as that of Newton (1845) and work such as that of Gottlieb and Shaw (948) demonstrate that considerable information is available about the detailed mode of action of antibiotics and some of this work may be relevant to the studies of mycotoxins.

Some antibiotics act at the surface of cells either by blocking cell wall synthesis, or by disrupting the integrity of membranes, or by acting on the permeases which control the active transport of, for example, amino acids and sugars.

Other compounds act by decoupling oxidative phosphorylation and inhibiting some respiratory pathways. The important role of quinone compounds in respiratory processes is well known (1660), for, by losing an electron, a hydroquinone becomes a semiquinone which may either revert back to the

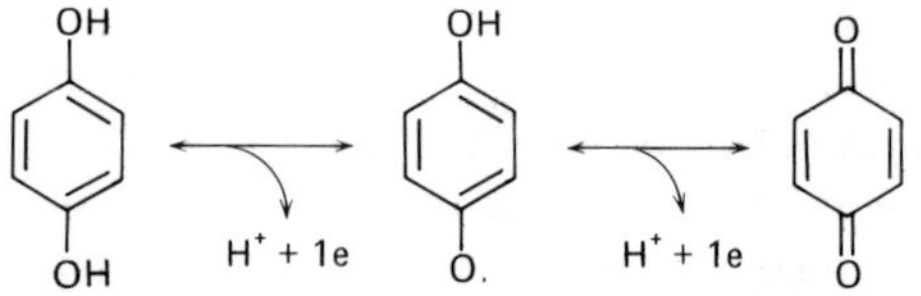

Figure 33. Interactions between quinones and hydroquinones

hydroquinone, or lose a second electron to form a quinone (Figure 33). A number of fungal toxins are quinones and their presence may upset normal respiratory processes.

Several compounds have strong chelating activity, for example aspergillic acid has an affinity for metal cations (947), and novobiocin forms a strong complex with magnesium causing a magnesium deficiency in the poisoned organism. Many metal ions play an important role as enzyme activators or in maintaining the integrity of the cell structure. Organic compounds, such as hexachlorophene and 8-hydroxyquinoline, which are capable of forming complexes with metal ions are potential growth inhibitors.

A number of toxins affect the transcription of genetic information or inhibit the synthesis of proteins. For example, fumagillin produced by *Aspergillus fumigatus*, poïn and sporofusarin, produced by *Fusarium*, are said to inhibit the synthesis of DNA. Some toxins may have a multiple effect but it would seem that any resulting inhibitation of protein biosynthesis would be one of the more serious consequences.

Detection of mycotoxins

Confirmation that a particular mycotoxin is responsible for a toxicosis can only result from the collaboration of specialists working in several disciplines.

From doctors and veterinarians

The doctor or the vet is usually the first to be called to diagnose a mycotoxicosis and the diversity of symptoms makes theirs a thankless task. Poisons which give very characteristic symptoms are rare, the majority having a complex activity. As well as any primary activity there are frequently secondary reactions which will depend on the individual, the more so if the primary reaction is less specific (761). Usually moulds are only incriminated when all other possible causes of the illness have been eliminated. Often it is only after autopsy, especially if the liver and kidneys have been examined, that a mycotoxin may be suspected.

The medical practitioner may then pass the problem on to other specialists, chemists, and mycologists, but ought nevertheless to become involved again during aetiological studies of the disease. It may be necessary to resort to experimental toxicity testing once a suspected toxin has been detected. This may involve the administering of food deliberately infected with moulds or to which toxins have been added; alternatively the intravenous or intraperitoneal injection of culture extracts or purified toxins may be carried out using a range of experimental animals. The recent work of Scott (2315), on the feeding of one-day-old ducklings with cereals experimentally infected with isolates of *Aspergillus* and *Penicillium* may be taken as a model in the verification of Koch's postulates. Information can sometimes be obtained using special techniques, thus the skin test on rabbits allows the detection of fusariotoxins and the production of corneal opacity in the eye of the rabbit may show the presence of sporidesmin.

From the biochemist

The toxins responsible for illness are often present in trace amounts and the biochemist will need to have sensitive analytical techniques (394, 1147). The use of biological indicators is often recommended but it is advisable to take special precautions if any one such test is going to have any value (327, 671).

Spectroscopy may be particularly valuable and the ultraviolet spectra of many of the quinone derivatives are now well known (859, 1764, 1971) and the analysis of these particular compounds is relatively easy. An examination for fluorescence has been used to distinguish aflatoxin-producing cultures of *Aspergillus flavus* from those which do not produce aflatoxin. Fluorodensitometry can be used to provide quantitative data (928). Chromatographic techniques have been developed for the estimation of those mycotoxins which are considered a particular health hazard (784, 2098, 2494).

From the mycologist

The mycologist, whose help is often not sought at all, may play an important role if a sample of the food, suspected of being toxic, is still available. A simple enumeration of spores is totally inadequate and often misleading. It may be necessary to isolate the moulds in pure culture and study their toxicity individually. For the determination of the species present in a particular food it is usually necessary to use several selective media (1743, 1945). The mycologist may help in assessing the quality of foods, in determining the presence of known toxigenic species, and in unravelling the aetiology of suspected mycotoxicoses.

CHAPTER 3

AFLATOXICOSIS

INTRODUCTION

A Formidable Mycotoxicosis

Aspergillus flavus Link, the toxicity of which is referred to as aflatoxicosis, is without doubt the mould which has aroused the most interest, especially in Great Britain. It is perhaps sufficient to note the considerable interest of the popular media in the fight against cancer to understand why this fungus, which for many years has only commanded interest because of its use with *A. oryzae* in the production of koji, has excited so much concern. It is known to have been responsible for the death of a large number of domesticated animals, poultry, cattle, and fish and it is possible that man himself may be a victim of this organism.

The number of scientific papers treating diverse aspects of aflatoxicosis continues to grow: according to Meyer (1655) there were 15 publications on aflatoxin in 1960, 106 in 1963, 216 in 1966, and more than 400 in 1970!

Formidable in its effects, *A. flavus* has also proved a useful agent to science, for it has made it possible to study certain aspects of carcinogenesis which are of interest in the search for suitable therapy.

Historical

For some time, it had been established that cattle cake based on oilseeds used as feed supplements may become toxic to animals (1289, 1507, 1862). However, as the incidents were isolated, little attention was given to them.

It was not until 1960, after considerable losses of turkey poults, aged three to six weeks, in the south and east of England, that particular disquiet was expressed in the British veterinary literature in the form of a series of brief notes in *The Veterinary Record* (38, 93, 240, 241, 903, 905, 1592, 1593, 2234, 2237, 2386, 2402, 2426, 2486, 2487, 2532, 2776, 2777, 2822). The animals usually died a week after the appearance of the first symptoms which consisted of a loss of appetite, feeble fluttering, and lethargy. Autopsy revealed haemorrhages and necroses of the liver frequently accompanied by kidney lesions (2386).

The cause of this disease being quite unknown at the time it was called turkey ‘X disease’ (241, 242).

Meanwhile a similar disease was recognized in young ducks in Great Britain, Austria, and Hungary (632), as well as Uganda and Kenya (14,000 ducklings died in four weeks on a single farm!) (997). Autopsy revealed symptoms very similar to those described in the turkey poults, even the remarkable destruction of the hepatic parenchyma as well as the proliferation of bile duct epithelial cells (93, 2778).

A similar disease in chickens was also reported in Spain (420, 1513).

Rapidly, a link between these various toxicoses and the consumption of groundnut meal, a by-product from the extraction of groundnut oil, was established. Not only poultry (turkeys, ducks, pheasants, and rarely chickens) but also other livestock such as pigs, calves, and sheep may be affected (1516, 1673, 2108, 2893).

Some of the symptoms resemble those caused by poisoning from the consumption of various plants such as *Senecio, Crotalaria,* and *Heliotropium* (1089, 1090). Because the toxicity of these plants is due to the presence of pyrrolizidine alkaloids (400), a search was made for these compounds in toxic groundnut meal without any success (38).

Although it was groundnut meal originating from Brazil which was first implicated, certain batches from other countries, notably Nigeria, West Africa, Gambia, East Africa, and India, also showed similar toxic properties (2234). Such a distribution of the toxic principle led to the consideration that it was really produced by a microbial contaminant and in September 1960, Austwick and Ayerst (105) had observed the presence of fungal mycelium in the incriminated feeds. Following this observation, the common mould *Aspergillus flavus* was isolated from Ugandan groundnuts (2238), and the conclusion was reached that all such cases of poisoning were associated with this species.

In retrospect a number of diseases described much earlier have the same aetiology:

– those relating to hepatoma in trout raised in fish farms first recorded in 1933 (992, 2309) and intensively studied more recently (1464, 1465, 2864, 2867);
– the observations of Shilo (2364) who, as early as 1940, emphasized the toxic nature of *Aspergillus flavus*;
– the observations of Ninard and Hintermann (1862) in 1944, of Burnside and his colleagues (357) in 1957, of Zinehenko (2934) in 1959, and of Raynaud (2109) in 1960, all establishing that the toxicity of groundnuts or maize to pigs was associated with the presence of *Aspergillus flavus* and *Penicillium rubrum*;
– the observations of Bailey and Groth (117) who, in 1959, had demonstrated that canine hepatitis was due to the same cause.

Similarly it would seem that the sporadic intoxications of turkeys noted since 1957 (2276), of ducks (985) and outbreaks of illness among laboratory animals such as rats (1414, 1435, 1436) or guinea-pigs (1909, 1928, 2457) were also aflatoxicoses. In fact Kulik (1366) had, in 1954, shown that an extract of peas moulded with *Aspergillus flavus* was toxic to cats and rats.

Work continued on a study of the toxic metabolites produced by *A. flavus* for which the name aflatoxin was agreed in the report of an Interdepartmental Working

Party which appeared in 1962 (2488). It was subsequently demonstrated that there was not one, but several aflatoxins, all closely related chemically.

Numerous reviews have appeared in many countries (13–15, 209, 385, 690, 923, 926, 1216, 1368, 1415, 1459, 1531, 1535, 1547, 1703, 1800, 1832, 1847, 1865, 1943, 2318, 2372, 2656, 2658, 2836). There have been major works edited by Goldblatt (927), Ciegler, Kadis, and Ajl (506), and Purchase (2048, 2939) as well as a number of significant bibliographies (1655, 2900). A brief review of the toxicology of aflatoxin was presented by Butler (368) at a symposium in Pretoria. Several committees, both at national and international level, regularly meet and report on the aflatoxin problem, for example the joint AOAC–AOCS aflatoxin committee in the United States (781).

Fungi which produce aflatoxins

Aspergillus flavus Link is the mould most studied and some strains are known to produce large amounts of aflatoxin. *Aspergillus parasiticus* Speare, usually isolated as an insect pathogen, also gives high yields of aflatoxins in culture (106, 534, 640, 722) however, according to several authors (2339, 2833), it may only be a variety, or even a particular strain, of *A. flavus.* A very toxigenic isolate, ATCC 15517, has been referred to as *A. parasiticus* var. *globosus* (1788).

Penicillium puberulum Bain. has been reported to produce aflatoxins (1101), but in low yield (1365) and it is possible that another fluorescent metabolite, chromatographically similar to aflatoxin, was mistaken for it (638). A number of other organisms have been reported to produce aflatoxin including; *A. tamarii* Kita (1413), *A. niger* v. Tiegh., *A. ostianus* Wehmer, *A. ruber* (Spiek and Brem.) T. and C., *A. wentii* Wehmer (2300), as well as *Penicillium citrinum* Thom, *P. variabile* Sopp. and *P. frequentans* Westl. all of which have been reported to produce it in trace amounts (1365, 1925, 2324). Aflatoxin has also been reported among the metabolites of *Aspergillus versicolor* (Vuil.) Tirab., but it is possible that, in some cases at least, there has been confusion with sterigmatocystin, aversin, and other metabolites with some of the structural and physicochemical properties of the aflatoxins (341). Other reports have suggested that *A. ochraceus,* which produces other mycotoxins anyway, and a species of *Rhizopus* has also been reported to produce aflatoxins (818, 2727).

Aspergillus flavus is frequently associated with other moulds, especially *Penicillium rubrum* Stoll, and it may sometimes be difficult to determine the part played by each species in a particular toxicosis.

Reports of aflatoxin production by species other than *A. flavus* and *A. parasiticus* should be treated with caution (3022).

ASPERGILLUS FLAVUS

Morphology

Aspergillus flavus Link (= *Monilia flava* Pers.) is readily recognized by the greenish yellow colour, and the more or less floccose nature of the thallus.

At the end of the erect, rough-walled conidiophore there is a subglobose conidial head, from which, either the phialides emerge directly (uniseriate, see Figure 34 b, c), or are attached to metulae (biseriate, see Figure 34 a, d); sometimes the two types coexist.

The conidiospores are relatively large (5–7 μm diameter), globose, greenish ochre, and slightly roughened.

It has been suggested that only those isolates with rough conidiophores and biseriate phialides should be placed in the species *A. flavus,* whereas those with smooth conidiophores and uniseriate phialides should be placed in *A. parasiticus.* These characters have been used to separate the varieties *globusus*, described by Murakami (1788) and *columnaris* of Van Walbeek (2725).

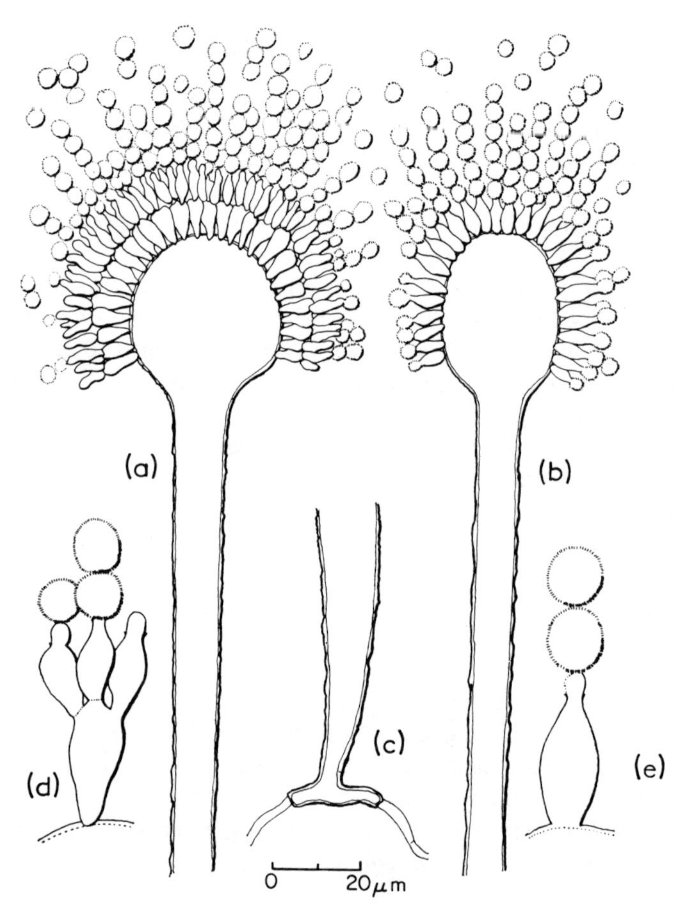

Figure 34. *Aspergillus flavus* Link
(a) Sporing head with biseriate phialides.
(b) Sporing head with uniseriate phialides.
(c) Base of conidiophore.
(d) Detail of biseriate phialides.
(e) Detail of isolated phialide

Ecology

Generalities

A. flavus is considered to be a cosmopolitan organism occurring in the soil, on organic matter and on seeds, especially oilseeds. Its occurrence in soil has been known for a long time (2), and it has been isolated from that of forests (2528), peat bogs (2738), the deserts of the Sahara (1307), as well as from cultivated soil, humus (2258), the rhizosphere of the tomato plant (2577), or of wheat (2779), etc. It is considered to be capable of rapidly recolonizing the soil immediately after steam sterilization (1499). The soils of tropical regions seem to yield higher numbers of *A. flavus* than those of temperate regions (1200).

It is common on cereal grain (484, 1215, 1919, 1945, 1946, 2233, 2401) and can be isolated from flour (477, 1371, 1727, 1742, 2581), from pasta (483, 1737, 2727), and even from bread (249, 818, 1207, 2339, 2459, 2460). It is frequently isolated from maize (362, 1326, 2065, 2375, 2620, 2783) and from rice (256, 384, 1146, 1371, 1629, 2293, 2296, 2299, 2620) and their products. It is abundant on the fibres and especially on the seed of the cotton plant as well as meal derived from it (86–89, 354, 733, 856, 1515, 1559, 1599, 1635, 1636, 2023, 2275, 2736, 2737, 2817). The mould penetrates into the grain through the basal region of the ovule where the stalk is attached (chalaza) (1600) or through regions of the grain damaged by insects (90).

Aspergillus flavus has also been isolated from the following substrates:

soya beans and soya meal (1672)
copra (1807, 1926, 2780)
cassava (1809, 1976)
cacao pods (511, 1428, 2139)
coffee beans (1471, 2316)
Brazil nuts (231, 2452, 2727)
pecan nuts (694, 1487)
tobacco (1036, 1121, 1122, 1277)
sorghum (152, 1463, 2620, 2783)
sunflower seed (232, 830, 1398)
jute seed (1153)
peas (779)
pine seeds (868)
carob beans (361)
haricot pods (2620)
pistachios (1207)
millet (2617)
red peppers (*Capsicum frutescens*)(2262)
beet pulp (2566)
pears (737)
hams, mould ripened sausages, and numerous other food products (339, 1207, 1400, 2302).

Its presence on cattle food, even in the absence of groundnut meal, is not surprising (1398, 1399) and it is readily isolated from forage (1079, 2541).

It has been looked for on cheeses (1303, 1905) but it seems that the presence of other moulds inhibits the growth of *A. flavus*, even though it is capable of growing on this substrate if artificially inoculated (1478).

If conditions are favourable it sometimes grows profusely; Olafson (1886) reported that the spores of *A. flavus* represented 50–100% of all mould spores present on maize in a silo at 15.2–17% water content. It can be so abundant that it may grow on the surface of the grain to form a crust some 0.6 m deep (2334). It becomes common on cobs of maize as soon as the water content is greater than 15.5% (1519, 2250, 2576).

A. flavus and the groundnut

The groundnut and its products undoubtedly present a particularly favourable environment for *Aspergillus flavus*. It is usually accompanied by numerous other fungi (cf. Chapter 1) a number of which are reputed to be toxic to animals (459, 1173):

some *Fusarium* (449), especially *F. moniliforme* (2708)
Rhizopus (2852)
Penicillium citrinum (803), *P. purpurogenum* (628), *P. rubrum* (2852), *Aspergillus amstelodami* (2076), *A. chevalieri* (803), *A. clavatus* (803), *A. fumigatus* (628), *A. glaucus* (628), *A. nidulans* (628), *A. oryzae* (2852), *A. restrictus* (628), *A. carneus* (2708), *A. ochraceus* (2708), *A. ruber* (2708), *Trichothecium roseum* (2708), and *Paecilomyces varioti* (2708).

Competition occurs between these many micro-organisms (1239, 1509, 1538) and the growth of *A. flavus* may sometimes be inhibited by moulds such as *Aspergillus niger* (267, 1240, 1243, 1244), *Rhizoctonia solani* (92) or species of *Penicillium* (2821). Inversely *A. flavus* itself may slow down the growth of species such as *Sclerotium bataticola* (1176).

Infection of groundnuts during cultivation

The majority of moulds which attack groundnuts, particularly *A. flavus*, are capable of growing and certainly persisting in the soil (1248). These fungi may then easily infect the surface of the pods as they come into contact with the soil (666, 1872, 2099, 2843). It is difficult, however, for *A. flavus* to penetrate the pod and invade the kernel of the living groundnut (1546). As well as the natural barriers of the healthy pod, the normal endogeocarpic mycoflora also forms an inhibitory barrier (1503). The colonization of the kernel seems to be easier when the fruit is mature, and especially after harvest (1018).

Studying peanuts taken from apparently intact pods, Jackson (1172) noticed that their teguments had an abundant mycoflora and that *A. flavus* was almost always present. The mycelium, scattered and often visible between the periderm and the surface of the cotyledon, is supported, it seems, on the internal face of the periderm. Now, it is said that the periderm begins to wither at about the seventh week in the growth of the fruit (2261) and it is probably then dead tissue. When the pod is mature it therefore constitutes, not only a protection for the cotyledons, but also it is normally a physical barrier against infection of the seeds in good physiological condition, while nevertheless being a permanent reservoir of infection. When some form of disequilibrium occurs then the cotyledons and the embryo may also be invaded. If these observations are confirmed then one has here a truly latent infection.

The penetration of *A. flavus* into the groundnut has been studied using the electron microscope (1442).

Although Norton *et al.* (1872) have established in Texas the infection of seeds in the soil which were in perfect condition, healthy and able to germinate, this is rather rare. On the contrary damaged pods, previously attacked by *Macrophomina phaseoli* or *Sclerotium rolfsii*, are easily infected (85, 91, 622, 657, 878, 1369, 1534, 2294). Mechanical damage of the shell of the pod caused by insects also favours infection, and termites are a particular nuisance in this respect in South Africa (2332). Nematodes, especially *Meloidogyne arenaria*, may also have an important role (1674). But the pods may also be spoilt by drought causing a cracking of their surface (1548). Very rapid growth, or over-advanced maturity may also lead to such lesions (657, 1179, 1545, 2294, 2332).

Agricultural practices may themselves be the origin of injury to the pods thus making it possible for fungal infection to occur (1534, 2294). It can be seen then that numerous factors may predispose the groundnut to infection by *A. flavus* in the field before harvest (85, 137).

A. flavus occasionally occurs on young plants affecting their growth and causing necrosis of the hypocotyl axis and the cotyledons (183, 530, 621, 1710, 1711). This usually occurs with young plants produced from infected seed, which have a lower germination success than healthy seed (844, 1929). The cotyledons may rot soon after emerging (904) or the fungus may persist in the senescent cotyledons for up to a month after planting (182).

Infection of pods and kernels; growth of A. flavus during harvest, drying and storage

The atmospheric conditions at the time of harvest are very important for the subsequent growth of *A. flavus* (692). In northern Nigeria where harvesting may occur either during the dry season or during the rainy season, a large difference in post-harvest infection is noted. Pods harvested during the dry season show little infection, whereas those harvested during the rains are badly infected (1534).

It is a frequent practice to leave the withering haulm of the plant in place until harvest but the pods break off from the stem carrying them and remain in the soil. Pods thus separated from the parent plant dry slowly in the soil and it is these which may be consumed by the natives of some groundnut growing regions. Such pods are particularly susceptible to the growth and survival of *A. flavus* (1854) and for this reason a very late harvest is usually forbidden (1547).

Injury associated with mechanical harvesting predisposes the pod to penetration by *A. flavus,* but this may not be a major factor (1980). It is said that when the peanuts are in the pod they are less susceptible to *A. flavus* than after decorticating (662).

The water content of groundnuts has a major influence on infection by *A. flavus* and its subsequent growth. At the time of harvesting the groundnuts contain 30–60% water and drying reduces this to 10% or even less. During natural drying it is essential to avoid accidental rehydration (1085, 1534, 1542, 1543, 1545, 1548). For this reason, in Nigeria, the pods are protected during the night or during periods of rain to avoid them becoming damp again (358). If drying is too slow this seems to be favourable for the growth of *A. flavus* (1180, 1181).

The optimum water content for attack by *A. flavus* seems to be in the region of 15–25%, which occurs during the initial period of drying (621), but growth is possible in the range from 9–35% water content (91, 648, 1029). At higher humidity it is perhaps the turgor of the tissues, or their active metabolism, which gives the kernels some resistance, whereas below 9% water content the fungus is unable to grow (105, 656); indeed, a minimum of 15.8% is required for the germination of the spores (2333).

The fact that the literature on the influence of water content of kernels on the growth of *A. flavus* is sometimes contradictory, probably reflects the method of study. Some authors have studied the growth of *A. flavus* on groundnuts dried naturally after harvest, whereas others have made their observation on groundnuts which have been artificially rehydrated after first completely drying. In the latter situation infection may take place at lower water content because kernels previously dehydrated are much more susceptible (1169, 1171). Analogous results have been obtained during studies on the moulds of corn (1519, 1945).

A fairly low temperature during drying seems to be unfavourable for good growth of *A. flavus,* and hence for the production of aflatoxin. This was demonstrated by Dickens and Pattee (648) in Virginia and Carolina to be so despite the need for a longer period of drying. It is possible that at certain temperatures there is interference and antagonism between several fungi (1176).

Several experiments on artificial drying have been carried out by Dickens and Pattee (648) who found that:

– at 32 °C and 85% RH *A. flavus* grows with the production of aflatoxin;
– at 32 °C and 50% RH or at 21 °C and 5% RH groundnuts remain safe.

Lowering the relative humidity is thus more important than raising the temperature for drying groundnuts in order to avoid the production of aflatoxin. Other workers have confirmed these results (663). During storage it is best to avoid the presence of mites which are capable of carrying the spores of *A. flavus* and hence increase the chances of infection (96).

Strain differences in the susceptibility of groundnuts

Only little information is available on the relative susceptibility of different strains of groundnuts. None appears to be completely immune but Spanish 205,

Natal Common, Samara 38, and 48 – 14 appear to be the least susceptible (1534). Florigiant, Argentine, and Early Runner have a relatively low susceptibility but *A. flavus* is nevertheless nearly always present on the surface of the pods. Some varieties of groundnut are reputed to only harbour non-toxigenic strains of *A. flavus.* As Doupnik (693) indicates it is advisable to be somewhat circumspect about such reports.

A comparison of the toxicity of groundnuts and products derived from them

From a report by Hiscocks (1098), based on over a thousand samples, it is noted that:

Whereas, in groundnuts themselves, 3.3% of the pods were very toxic, containing more than 0.25 mg aflatoxin B_1 per kilogram, and 21.7% of the pods were moderately toxic, 75% were non-toxic, in groundnut meal, on the other hand, 42% of the samples were very toxic, 49.3% moderately toxic, and only 8.7 non-toxic.

It can be seen that there is an accumulation of toxins in the groundnut meal which may arise, either from its preparation, or from further proliferation of *Aspergillus flavus.* Aflatoxin may also appear in many samples of peanut butter prepared from contaminated groundnuts whereas, in contrast, groundnut oil appears to be free from contamination.

Pathogenicity of *A. flavus*

Dangerous though the toxic metabolites of *A. flavus* are, the same mould can also behave as a true pathogen (102, 1190). It is especially associated with pulmonary infections arising from the inhalation of spores. Birds such as pigeons and turkeys (23), chickens (2502), geese (1275), and other wild and domesticated species (25, 986, 2209), seem to be most susceptible, but horses may also be affected (2639), and mice, especially after treatment with cortisone (2383). In man, *A. flavus* has been shown to be associated with bronchial and pulmonary infections (1806, 2382), an intracavitary lesion (66), and mycotic endocarditis (1315), as well as a mycotic infection of the bladder (320, 2799), and dermatitic lesions of the ear (2863). It was, in fact, from the human ear that this species was described for the first time by Link in 1806. It has also been associated with infections of the sinuses (117, 2799). In the cow, *A. flavus* has been isolated from the placenta (2000).

This fungus has also been recognized as a possible endoparasite of insects. It has been observed on the internal integuments of the maize pyralid moth (*Pyrausta nubilalis*) and on *Platysamia cecropia* (2524, 2615), as well as on the house-fly (*Musca domestica*) (54), but it does not seem able to infect the bee or the grass-hopper by this particular route (356). On the other hand, it is able, as is the closely related species *A. parasiticus*, to penetrate the tracheal tubes of some of the Orthoptera (1458, 1877). The germ tubes of *A. flavus* may be able to penetrate the chitin layer in the coecal region of the larvae of *Platysamia cecropia* (2524). It has been reported from *Proceras indicus,* an insect pest of the sugar cane in India (587) and may also be associated with the scolytid beetle *Xyloterinus politus* (1558).

It is still necessary to determine the role of the toxic metabolites of *A. flavus* in its pathogenicity for such insects as the larvae of the cockchafer, silkworm, cricket, etc. (2744).

Conditions for growth and sporulation

Aspergillus flavus is capable of growth over the temperature range from 6–54 °C (1179, 2547). Its optimum growth rate on such diverse substrates as groundnuts (105), dry lucerne (1034), and laboratory culture media (284, 1748, 2075) occurs between 30–35 °C (Table 8). However, the optimum growth yield was obtained at 20 °C (2547).

A medium with a high sucrose concentration (30–200 g l^{-1}), and with nitrate as the source of nitrogen, is particularly favourable for growth. *A. flavus* requires at least 80–85% RH for growth which explains why it is rarely found on groundnuts in competition with aspergilli of the *glaucus* group for which the requirement for water is lower (Figure 35a).

The appearance of conidia coincides with a marked decrease in nitrogen content of the mycelium (1989) and, whereas darkness seems to favour mycelial growth, a low light intensity stimulates the formation of conidia (2552). Sporulation is also dependent on temperature and the composition of the culture medium, being better at 32 °C than at 22 °C and reduced by the addition of sucrose to a malt medium. The production of spores increases rapidly during the first eight days of culture and is subsequently much slower (Figure 35b).

The spores of *A. flavus* retain their viability for a long time, several years under favourable conditions. If the spores are maintained at –70 °C, only about 7% germinate if they are allowed to thaw slowly (0.9 °C per minute), whereas about 75% will germinate if the temperature is raised rapidly (1640). As well as being resistant to fairly low temperatures, the spores are not killed at temperatures as high as 53 °C (2401). They are, however, very sensitive to changes in relative humidity and germinate rapidly as soon as the relative humidity reaches 79–81% (70). One author reports that they do not survive a month at 75% RH and yet there is 40% survival after six months at 85% RH (2564).

A reduction in the concentration of atmospheric oxygen causes a decrease in the mycelial growth rate (1183, 1417), as also does an increase in the concentration of carbon dioxide from 20–100% (1417). In submerged culture, in 14 l fermenters containing 10 l of medium, an aeration rate of 9 l min^{-1} gave optimum growth, and hence optimum yields of aflatoxin (1046).

Table 8. Linear growth (expressed in mm/day) of *Aspergillus flavus* (strain isolated from wheat) at various temperatures (1748)

Culture medium	22 °C	32 °C	37 °C
2% Malt	5.23	7.78	6.24
2% Malt + 10% NaCl	6.44	12	5.95

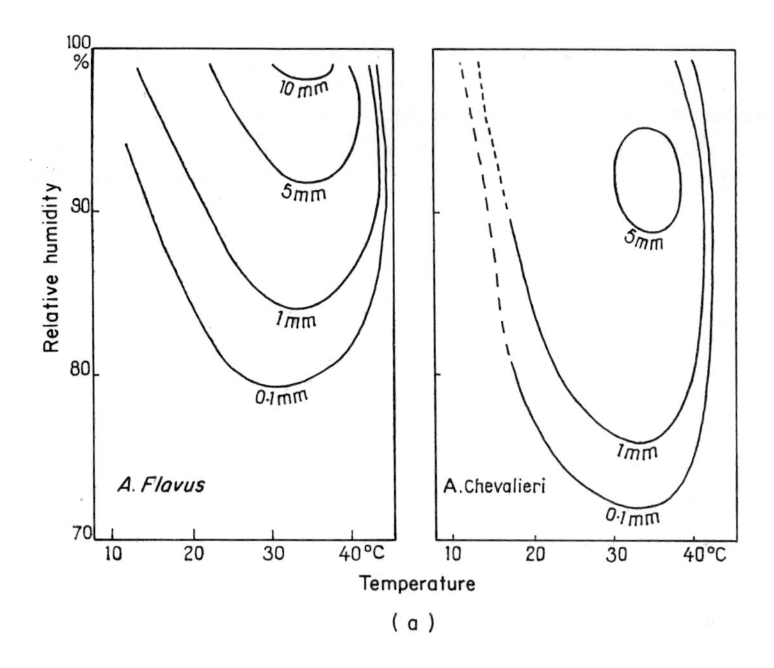

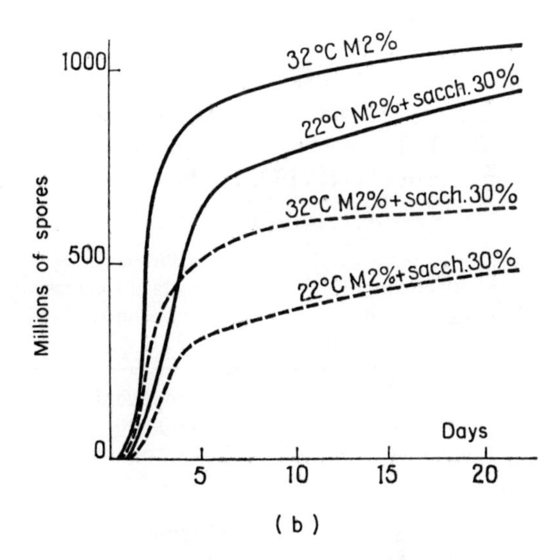

Figure 35. (a) Growth in mm per day, as a function of temperature and humidity, of *Aspergillus flavus* and *Aspergillus chevalieri* (105). (80% RH corresponds to a water content in groundnuts of 9% of the dry weight, 90% RH corresponding to 14% and 99% RH to 37% water content.) (b) Sporulation of *Aspergillus flavus* (strain isolated from wheat) grown at 22 °C and 32 °C on 2% malt or 2% malt + 30% sucrose (1960)

Growth of *A. flavus* is negligible at pH 3.9 and below, increasing to a maximum at pH 5.5 and decreasing to a negligible level again at pH 9.1 (284).

As well as a wide range of simple carbohydrates and polysaccharides such as starch, *A. flavus* is able to utilize carboxymethylcellulose as the sole source of carbon (1334, 2121). The growth of *A. flavus,* as with other micro-organisms, is profoundly influenced by the nature of the source of nitrogen and Table 9 indicates the amount of protein produced from a range of different sources of nitrogen (2481). Another aspect of nitrogen metabolism is the ability of *A. flavus* to produce nitrites and nitrates from organic nitrogen (1165, 1581, 2268–2270) by a process of nitrification (1603, 1604).

There are many papers on the culture of this, now famous, mould ranging from a report of the favourable influence of tomato juice on growth (1524), the isolation of proteolytic non-sporing mutants (1278), and the use of botran (2,6-dichloro-4-nitroaniline) in culture media as a selective agent in the isolation of *A. flavus.*

Compounds produced by *A. flavus*

The name aflatoxin had been given to the toxic metabolites of *A. flavus* before it was appreciated that it was a complex mixture of compounds and before their structures had been elucidated. In fact most strains of *A. flavus* which are toxigenic produce mainly aflatoxin B_1 with smaller amounts of aflatoxin G_1, B_2, and G_2. B_2 and G_2 are dihydro derivatives of B_1 and G_1 respectively. Such compounds as aflatoxins M_1, M_2 and P_1 (578) are usually found as animal transformation products of the earlier described aflatoxins. Other compounds met with in the literature include aflatoxins GM_1 and B_3 (1051), aflatoxin G_3 (2925) and even 3B (1401) or toxin B3 (831). The names flavatoxins or flavacoumarins have been proposed for these compounds (1191, 1199, 1200).

Table 9. Yield of protein synthesized by *Aspergillus flavus* as a function of the nitrogen source in the culture medium; the results are expressed as percentages, 100% representing the yield from the nitrogen source most favourable for growth, thus allowing maximum synthesis of protein (2481)

$NaNO_3$	85	Phenylalanine	0
$(NH_4)_2SO_4$	95	Tyrosine	45
Urea	85	Tryptophan	0
Glycine	75	Arginine	80
Alanine	100	Lysine	70
Serine	80	Valine	70
Threonine	80	Leucine	50
Aspartic acid	100	Isoleucine	55
Glutamic acid	100	Purines	100
Proline	100	P.A.B.O.A.B.	0
Histidine	75	Nicotinic acid	15
Cysteine	55	Guanidine	80
Methionine	50	Pyrimidines	80

Moreover *A. flavus* produces numerous other metabolites, some of which have been studied because of their antibiotic properties. The production of all these metabolites is very strain dependent.

Several metabolites of *A. flavus* belong to the sterigmatocystin family of compounds (394) which are of particular interest because of their possible relationship to the biosynthesis of the aflatoxins themselves (Figure 36). Sterigmatocystin ($C_{18}H_{12}O_6$) itself and 6-methoxysterigmatocystin ($C_{19}H_{14}O_7$) are both toxins produced by *Aspergillus versicolor*, whereas *o*-methylsterigmatocystin ($C_{19}H_{14}O_6$) has been isolated from a strain of *A. flavus* also producing aflatoxin (347). Aspertoxin ($C_{19}H_{14}O_7$), or 3-hydroxy-6,7-dimethoxy-difuroxanthone, which, like the aflatoxins, is fluorescent (2429), has been isolated from two different strains of *A. flavus.* Its chemical structure was established by two groups at the same time but independently (2761, 2182). Aspertoxin is toxic for both the chick embryo (2181) and the larvae of the 'zebra' fish (394).

	R_1	R_2	R_3
Sterigmatocystin	H	H	H
6-Methoxysterigmatocystin	H	H	OCH_3
o-Methylsterigmatocystin	H	CH_3	H
Aspertoxin	OH	CH_3	H

Figure 36. Derivatives of sterigmatocystin

Aspergillus flavus also produces aspergillic acid and a number of related compounds. Aspergillic acid itself ($C_{12}H_{20}N_2O_2$) (1261, 2807–2809) has activity against both Gram positive and Gram negative bacteria and is resistant to acid, alkali, and heat but, unfortunately, it is toxic to animals, having an intraperitoneal LD_{50} for mice of 150 mg/kg (2450). Biosynthetically it is derived from leucine and isoleucine (1549) (see Chapter 2). Neoaspergillic acid, with the same empirical formula, shows both antibacterial and antiviral activity and has been isolated from the closely related organism *A. sclerotiorum.* It is interesting that hydroxyaspergillic acid has a different spectrum of antibacterial activity than does the parent compound (2246). A compound, referred to as granegillin, is similar to aspergillic acid, having a wide range of antibacterial activity, but is very much less toxic (207).

The monocyclic pyrone, kojic acid ($C_6H_6O_4$) (Figure 37) is produced, not only by *A. flavus*, but by a number of other aspergilli (*A. tamarii, A. oryzae* and its variant *effusus, A. parasiticus, A. candidus*, and members of the *wentii* group), four species of *Penicillium*, and *Verticillium dahliae* (274, 1924, 2829, 2901).

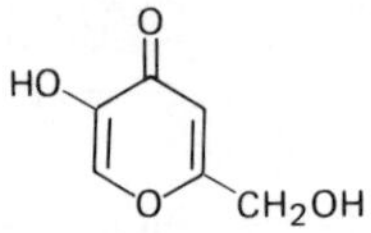

HO, O, O, CH_2OH

Figure 37. Kojic acid

Although the contrary opinion is often expressed, kojic acid has some toxicity to animals (2801). A single dose of 250 mg/kg is toxic to mice when given intravenously, subcutaneously or intraperitoneally, the animal becoming prostrate, stretching its limbs and tail, respiration becoming slow and difficult; convulsions occur followed by death (1229). Injection of 150 mg/kg is fatal to the dog and during *in vitro* experiments the metabolite is toxic to leucocytes at a low concentration (207). Kojic acid has been shown to inhibit the development of insects and it has been suggested that it could be used as an insecticide at high concentrations (173).

Other metabolites which have been isolated from *A. flavus* include flavicin and flavicidin, which have antibiotic properties similar to those of the penicillins to which they may be related (11, 359, 1553, 1554), flavicidic acid, considered primarily as a phytotoxin (1144), nitropropionic acid (2829), oxalic acid (2829), and orizazin, an uncharacterized antibiotic active against both Gram positive and Gram negative bacteria.

A. flavus may also produce a tremorgenic compound which is more potent than tremorine (dipyrolidine-1,4-but-2-yne) and is inhibited by compounds used in the control of Parkinson's disease (757, 2828, 2829, 2837).

It is possible to extract a growth factor from this same fungus (773), which, at a dose of 10 mg per day per animal in the food of baby mice, or of 200 mg in the feed of pigs, gives increased growth.

Among its enzymic activities *Aspergillus flavus* produces polygalacturonase and protopectinase but has negligible cellulase activity (2747). Both proteolytic (1271) and lipolytic (1460) activities have been studied.

THE AFLATOXINS

Production

Whereas nearly all strains of *Aspergillus parasiticus* are toxigenic, the production of aflatoxins by *A. flavus* varies considerably from strain to strain. In both species it is also a function of the environmental conditions (1217). Thus, as Schroeder and Ashworth (1295) have written, the production of aflaxtoxin is the result of the interaction of the genotype of a strain and the environment in which is it growing.

Studies on the production of aflatoxin stem from three areas of interest; research into conditions which inhibit its production, information derived may help in preventing aflatoxin occurrence in foods; research into conditions for optimizing yields so that pure aflatoxins may be obtained for fundamental studies; research into the biosynthesis of aflatoxins.

Toxigenic strains

A large number of observations have been made on the more or less toxic nature of various isolates and strains of fungi either on natural substrates, or in laboratory culture (Table 10) (1407, 1924, 2289, 2302).

Significant variations in toxigenesis have been noted depending on the substrate from which *Aspergillus flavus* has been isolated and its geographic origin. Thus, for example, it has been reported that 94% of the strains isolated from rice in the United States were toxigenic (256), whereas 86% of those isolated from groundnuts in the same country were so (597), and only 71% of isolates from stored groundnut kernels in Israel (267). Strains of tropical origin seem to be more toxigenic than strains from temperate regions, 46% on the one hand and 15% on the other of 427 isolates studied by Jacquet and Boutibonnes (1200). However, in India, from several different surveys, only 3–26% of isolates produced aflatoxin. Statistics based on 1,390 isolates obtained from six different countries indicate that 60% of the isolates were toxigenic (664).

Aflatoxin production also varies quantitatively from culture to culture and toxigenic isolates grown on the same substrate and under the same conditions may give yields ranging from 100–2,000 mg/kg of substrate (1077, 1788), indeed, Arseculeratne *et al.* (76) reported production of up to 8 g/kg from coconut flesh! More recent analytical data give figures for the relative yields of the different aflatoxins as well as for total aflatoxin production. In general, aflatoxin B_1 is formed in the largest amount both in nature (658, 1200) and in culture (1624), then aflatoxin G_1 followed by B_2, G_2, and other aflatoxins in very much lower

Table 10. The influence of the strain and culture conditions on the production of aflatoxins by *Aspergillus flavus* (1624)

Strain	*Culture medium*	*Total aflatoxins* mg/l, mg/kg	*Relative amounts* (%)				*Reference*
			B_1	B_2	G_1	G_2	
ATCC 15517	Synthetic	45	87	4	9	<1	1621
?	Groundnuts	265	44	1	54	1	534
MRE 1	Groundnuts	14	98	2	0	0	534
NRRL 2999	Wheat	870	35	9	48	7	1077
NRRL 2999	Wheat + methionine	1,700	44	11	38	7	1077
NRRL 2999	Rice	?	23.8	6.3	6.8	0.9	351
NRRL 3000	Sucrose + amino acids (submerged culture) 72 h at 20°C	86	26	0	74	0	510
NRRL 3000	Sucrose + amino acids (submerged culture) 72 h at 25 °C	154	70	0	30	0	510

yields. However certain strains produce only one aflatoxin in the absence of any of the others (742, 2622).

An attempt has been made to distinguish toxigenic from non-toxigenic isolates on the basis of morphological characters. It has been suggested that toxigenic strains always have green conidial heads in old cultures, biseriate sterigmata, and conidiophores with echinulate walls (1789). Jaquet and Boutibonnes (1194, 1195, 1200) noticed a hypertrophy of certain parts of the mycelium of toxigenic strains. The spherocysts and the anomalous appearance were considered as characteristic of clones producing aflatoxin. Nevertheless, it often appears to be difficult to detect any biological or chemical differences between those cultures which do, and those which do not produce aflatoxin (2701). Neither should it be forgotten that a toxigenic strain may lose its ability to produce aflatoxin after successive subculturing on synthetic media and, also, that it is sometimes possible to increase the toxigenicity of a strain by successive subculture on the appropriate natural substrate (68, 640, 1365). Both natural and induced mutants may also be obtained.

Substrates and environmental conditions

The situations in which aflatoxin is produced on naturally contaminated media and on natural or synthetic substrates deliberately infected with known strains should perhaps be distinguished. Cultures growing on oilseeds, and especially on groundnuts and their products, seem to be more toxigenic than those isolated from other cereal products (2375). Strains isolated from spoiled meat (994), bread (2458–2460), pasta (483) or naturally contaminated cheeses are usually only slightly toxigenic or not at all. On the other hand nearly a third of the cultures isolated from spices produce aflatoxins (994). It is not possible to generalize about this topic. It is said, for example, that *Aspergillus flavus* growing on certain varieties of groundnuts (such as US26) does not produce aflatoxins, but there is no explanation for this phenomenon (2522).

The toxigenic properties of a culture may be masked if the toxic metabolites are subsequently transformed into non-toxic derivatives by other micro-organisms. It was surprising to find in Texas (91) groundnuts which had a low toxicity despite being heavily contaminated with *A. flavus.* Further examination of these samples revealed the presence of bacteria and fungi capable, either of inhibiting the production of aflatoxin, or of converting it to products with lower toxicity. Such observations provide some grounds for the active research into biological methods of detoxifying infected products.

The production of aflatoxin is usually proportional to the weight of mycelium formed in culture, being a maximum when the biomass reaches its optimal value and rapidly declining from the moment that the mycelium starts to autolyse (2290). At the onset of lysis, degradation of aflatoxin occurs, both being favoured by high aeration and strong agitation of the culture (508, 1228).

In general, under usual culture conditions, aflatoxin production by *A. flavus* begins at the same time as the formation of conidia (68), increasing up to the period of intense sporulation (2096), which is usually after six days, and then decreases. Some strains show a second increase in production several days later (1403). Yields are usually higher in shake flasks than in stationary culture (534, 1621), but this is not always the case (1404).

Many factors, both physical and nutritional, have an effect on the production of aflatoxins in culture and, by analogy, under natural conditions. On groundnuts the optimum temperature for aflatoxin production by *A. flavus* is 25 °C after seven to nine days (660, 2265), whereas on rice 28 °C seems to be more favourable (2446). Maintaining a constant temperature of 45 °C inhibits both growth and toxin production (358), and little or no aflatoxin is formed below 12–15 °C (597, 659, 663). However certain strains do seem able to produce aflatoxin at 7.5 °C, a temperature at which many domestic refrigerators operate (2725). Very precise studies seem to indicate that it is, in fact, the production of aflatoxin B_1 which is optimal at 24–28 °C, whereas at 30 °C aflatoxin G_1 may be produced in the highest yield (2075, 2077, 2297, 2446).

Fluctuations in temperature, which may of course be found in nature, with a mean of 25 °C but reaching 40–50 °C are less favourable for aflatoxin production than a constant temperature of 25 °C. On the other hand fluctuations down to 10 °C seem to be without effect (2298).

The water content of the substrate plays an important role in aflatoxin production (105) undoubtedly associated in its effect on the growth of *A. flavus.* Dickens and Pattee (649) have shown, for example, that at 32 °C aflatoxin is formed in groundnuts after two days when the kernels have a water content between 15–30%. This fits in with the observations of Bampton (132) that, under tropical conditions, if *A. flavus* grows on groundnuts previously free of aflatoxins, the latter become detectable only 48 hours later. Aflatoxins are rapidly formed, if the temperature is warm enough, on rice with a water content of 24–26% (384) or on maize with a water content of 19–20% (1141, 2731). Rehumidified grain deteriorates much faster than grain harvested damp and this is reflected in the increased levels of aflatoxin in the former (2630).

The refined assays achieved by Jemmali, Poisson, and Guilbot (1228) further emphasize that the production of aflatoxin may be more important in products which have involved a drying stage (such as semolina and flour products) than those which do not (such as food starch). It is possible that residues of the sulphite used in the manufacture of food starch may inhibit fungal activity and, in particular, the production of aflatoxin.

The initial pH of the medium has only a small influence on the formation of aflatoxins; whatever it is there is a tendency for it to drift to a value of 4–5 as a result of fungal activity (599, 1046, 1479, 1621). An increase in the concentration of atmospheric carbon dioxide reduces the growth of *A. flavus* and consequently the production of aflatoxins (597, 932, 1417, 1418, 2231). The same is true for both a decrease in atmospheric oxygen (1418, 1665) and an increase in nitrogen concentration (746).

Considered as sensitive to light (1486), the aflatoxins are, in fact, sensitive to ultraviolet radiation especially at 360 nm (1200).

Numerous studies have been made on the significance of various nutritional factors on aflatoxin production, either in static culture fermenters, or in shake flasks (31, 510, 984, 1899, 2572).

Carbon sources

The influence of different carbohydrates in the medium on the production of aflatoxins by *A. flavus* has been studied (Table 11) (616).

It can be seen that glucose, mannose, and fructose all favour aflatoxin production, as does glyceraldehyde. It has been pointed out that the three hexoses all have the same configuration at positions 3 and 4 in the molecule but it is probably more important that they are all readily assimilated. Further work has confirmed the results shown in Table 11 and shown that sucrose is equally favourable (598, 599, 1081, 1621).

Table 11. The influence of carbohydrates on the production of aflatoxins. The number of crosses indicates their relative abundance, 0 their absence, and – not measured (616)

Carbohydrate	*Concentration*	
	1%	3%
D-glucose	+++	+++
D-mannose	+++	+++
D-fructose	+++	+++
D-galactose	0	++
D-gulose	0	–
D-arabinose	0	0
D-xylose	+	++
D-ribose	0	+
D-erythrose	0	–
D-glyceraldehyde	+++	+++

Nitrogen source

On media in which nitrate is the sole source of nitrogen little aflatoxin is produced by *A. flavus,* whereas it may be produced in large quantities on media containing ammoniacal nitrogen, uric acid, or glutamic acid (9). It has therefore been suggested that the sodium nitrate of the classical recipe of Czapek be replaced by ammonium chloride (616) or another ammonium salt.

Improved yields of aflatoxin are obtained on media containing yeast extract or peptone (598) or certain amino acids (720), especially glycine and glutamate, with alanine and aspartic acid being a little less favourable (265, 598). Thiamine and other vitamins of the B group stimulate the synthesis of aflatoxins, although riboflavin and pyridoxine have no effect (151, 720).

Metal ions

The production of aflatoxin is stimulated by the presence of zinc (68, 598, 1439, 1621, 1624, 1825), cadmium (1439), magnesium and iron (598), whereas cobalt, chromium, calcium, and manganese are said to have little effect (1439). The addition of 3.9 μmoles of barium acetate inhibits the formation of aflatoxins (1439).

Other nutritional factors

When *A. flavus* is grown on wheat grain, the quantity of aflatoxin produced is very much higher in the embryo than in the testa (378, 984). It has also been reported that the addition of lipids, in the form of a pentane extract of wheat germ, has a favourable effect on aflatoxin production on a medium composed of lipid-free wheat germ (1228). The beneficial effect of fatty acids on the biosynthesis of the toxins, recognized by several authors (720, 1931), indicates that they play an important role. It is possible that their degradation gives rise to the formation of precursors, such as acetyl coenzyme A, which enter the metabolic pathway to aflatoxin biosynthesis (1460).

The addition of dimethylsulphoxide to culture media provokes either a slight increase in aflatoxin production, or a considerable reduction which may reflect some metabolic interaction rather than a chemical reaction between aflatoxin and DMSO (168, 169).

Large-scale production of toxins

It is not only possible to produce aflatoxin on a preparative scale in the laboratory, but it is now available from several chemical companies.

Very toxigenic strains maintained in culture collections include NRRL 2999 (2375), NRRL 3145 (1077, 2517), NRRL A-13794 (2821), and ATCC 15517 (designated by Murakami *et al.* (1788) as *A. parasiticus* Speare var. *globosus* Murakami, but it seems to be a strain of *A. flavus*) (664).

Natural substrates giving the highest yields are rice (2378), pulverized wheat (2170, 2821), grated coconut (75), papaya (155), and milk cream (1200). High concentrations of water in the substrate and a high relative humidity of the atmosphere are recommended. Incubation is usually at about 28 °C for one week. The extraction, separation, and purification of the aflatoxins are carried out by fairly routine methods.

Biological detection

It is important to be able to detect the presence of the aflatoxins in a wide range of food commodities (818). Simply the presence of *Aspergillus flavus* in commodities such as cereals (2804), groundnuts (651), cottonseed (2507) or their products (450, 744), does not automatically imply the presence of aflatoxins in these commodities. There may be non-toxigenic strains, growth of the fungus may not actively have occurred following contamination by spores, or growth may be at too early a stage for toxin production to have started. Conversely, aflatoxins may be detected without it being possible to isolate the mould. This may be because the fungus grew long ago and has since disappeared having produced the toxin, or the food may be treated by a process which kills the fungus without destroying the toxin, or there is even the possibility that moulds other than *A. flavus* were responsible for the production of aflatoxin. The visual detection of *A. flavus,* when harvesting groundnuts for example, may thus only be indicative of the presence of aflatoxin in the batch (650, 651).

Numerous procedures for detecting aflatoxins have been proposed, and they all have some interest and indeed some of them are complementary. Several are based on the recognition of fairly specific biological changes in animals or plants. The diverse methods advocated have been reviewed by Brown (327) and Legator (1448), and certain criteria of validity have been defined by Di Palma (671).

Test using one-day-old ducklings

One of the first tests used as evidence of the toxic properties of the aflatoxins involved one-day-old ducklings of about 50 g body weight. It was considered satisfactory to extract groundnuts, or groundnut products, with chloroform, remove this solvent and extract the residue with methanol followed by partition with petroleum ether (methanol 10 vol., water 1 vol., petroleum ether 10 vol.). The aflatoxin remained in the aqueous methanol phase and, after removal of the solvent, could be made into an emulsion in water. This was usually done so that the final concentration per millilitre was equivalent to 40 g of the original sample. The emulsion was then introduced into the gizzard of the ducklings with a plastic tube. On the first day a dose corresponding to 10 g of the original sample was given to each duckling and the dose gradually increased until each animal had received the equivalent of 80 g. A sample would be considered very toxic if the ducklings died during the seven days following the start of the test. The surviving ducklings are examined for hepatotoxic lesions, the severity of which allows a further assessment of the toxicity of the original sample. The duckling is particularly sensitive to aflatoxin and it is thus possible to detect minute quantities of these toxins (243, 2237, 2583).

The ‘one-day-old duckling test’ is thus widely used its value depending, in part, on a knowledge of the acute LD_{50} (422):

Aflatoxin B_1 0.36 mg/kg B_2 1.69 mg/kg
Aflatoxin G_1 0.78 mg/kg G_2 2.45 mg/kg

and in part on a knowledge of the chronic and sublethal effect on the proliferation of the bile ducts when ducklings have received as little as 0.04 mg/kg (2853).

There are nevertheless some difficulties such as the necessity for continuous breeding of the animals, and problems arising from the animals regurgitating the toxic ration administered.

Chick embryo test

After the one-day-old duckling test, this is the next most frequently used biological test (365, 1222, 1397, 1556). It consists of injecting the extract to be tested, either into the air sac, or into the yolk sac of the fertile egg of a White Leghorn, incubated for five days. Some workers make the injection before incubation, others after nine days!

The chicken embryo, being very sensitive to aflatoxin, dies within two days of being injected with a sample containing at least 0.3 μg of aflatoxin. This test is thus 200 times more sensitive than the duckling test. It is necessary to carry out a histological examination of the lesions arising from aflatoxin, examination of the liver being particularly instructive (1676). The LD_{50} values following injection into the air sac have been determined to be:

Aflatoxin B_1	0.025 μg/egg	B_2	0.125 μg/egg
Aflatoxin G_1	1.1 μg/egg	G_2	2.7 μg/egg
Aflatoxin M_1	0.2 μg/egg		

Brine shrimp larva test

The brine shrimp, *Artemia salina*, is a phyllopod crustacean of brackish water which has proved to be particularly easy to use as an indication of the presence of mycotoxins. The test is complete in 24 hours, it doesn't require expensive apparatus, the test organisms are easily maintained and bred; in fact, the eggs of this shrimp are cheap and remain viable for many years if kept under sufficiently dry conditions. The eggs hatch in less than 24 hours at 27 °C when placed in saline water (sp. gr. 1.02). It is relatively easy to select larvae of comparable vitality by allowing them to swim from the dark into an illuminated zone.

The suspected sample is extracted with chloroform which is subsequently removed on a water bath. Using a graduated pipette, 30–50 larvae are removed in 0.5 ml water and transferred to a watch glass containing the residue from the extract to be tested. After a period of 24 hours at 37.5 °C the percentage of dead larvae is calculated from a count. It has been demonstrated that, under these conditions, a concentration of 0.5 μg of aflatoxin B_1 per millilitre will cause a 60% kill of the larvae, and that double this concentration will cause a 90% kill (328).

Harwig and Scott (1038) have described a technique using paper discs impregnated with the substance to be tested and have shown that 1.3 μg of aflatoxin G_1 per millilitre is lethal to the larvae of *Artemia* after 16 hours.

Jacquet and Boutibonnes (1199, 1201) have used the larvae of freshwater crustaceans in a similar way; copepods of the genus *Cyclops,* ostracods of the genus *Cypris,* and brachiopods of the genus *Daphnia. Daphnia pulex* appears to be the most sensitive of those tested and, at 20 °C in 24 hours, 1 μg of aflatoxins per millilitre will give 100% mortality of the larvae. It is sufficient to use a 5 g sample, if it contains at least 0.1 mg of aflatoxin per kilogram, to reach the lethal concentration for the daphnid larvae.

'Zebra' fish test

The 'zebra' fish, *Brachydanio rerio,* has transparent eggs making their development easy to follow. The young fish, or larvae, are produced within three to four days at ambient temperature and have a large embryo yolk sac, so they do not require any nutrients or special care. Acetone, at sufficiently low concentrations, is not toxic to them. It has been shown that a concentration of 1 μg aflatoxin B_1 per millilitre, prepared using 0.8% acetone, causes the larvae to show abnormal movements after 30 minutes, and to become moribund after five hours. There is a darkening of the yolk and a twisting of the tail after 20 hours, rupture of the yolk sac and death after 24–36 hours.

Tests on insects

It has been shown that several insects are sensitive to aflatoxins so they have been considered as possible biological agents for the detection of mycotoxins. Using the yellow fever mosquito, *Aedes aegypti,* the house-fly, *Musca domestica,* and the fruit-fly, *Drosophila melanogaster,* it is possible to demonstrate a reduction in the number of eggs layed, as well as a reduction in the percentage which hatch, after the adults have consumed food contaminated with 0.005–0.03% aflatoxins (171, 172, 1628). The larvae of the rice moth, *Corcyra cephalonica* (1053) and those of *Heliothis virescens* (980) are equally sensitive.

It has also been reported (2744) that the aflatoxins cause lesions in the adipose tissue or haemocytic granulomas in the visceral cavity, and even paralysis followed by rapid death in silk worm larvae and cockchafer larvae.

Boutibonnes and Jaquet (276) have recorded the early death in the pupal stage and a distinct slowing down of the metamorphosis of the meal worm, the larva of the flour beetle *Tenebrio molitor,* when the flour in which it is growing contains 25–200 μg aflatoxin/g.

Tests using molusc eggs

The eggs of the marine borer, *Bankia setacea,* are incubated in sea water at a temperature between 10–20 °C. Once fertilized they normally pass through a two-cell stage within two hours, a multicellular stage within three to four hours and, within 18 hours, the trochophore stage is reached, this being a free-swimming larval stage. The addition of aflatoxin B_1 (in propylene glycol), at a concentration of 0.05–40 μg/ml, inhibits the formation of separate cells but not the division of the nuclei. After five hours incubation a single multinucleate cell results. At concentrations less than 0.05 μg aflatoxin/ml, some of the eggs divide but their development ceases before the trochophore stage.

Similar phenomena are not shown by the common garden snail, *Helix aspersa,* because it seems that the aflatoxins are transformed into a new metabolite (276). In contrast, the South American snail, *Biomphalaria tenagophila,* is killed by aflatoxin B_1 at a concentration of 0.5 mg/kg (2061).

Tests using the tadpoles of amphibia

Whereas the frog, *Rana temporaria,* only shows a congestion of the liver after absorption of aflatoxins (276), its tadpoles are killed by concentrations greater than 2 μg/ml and, below that level, are not able to complete the metamorphosis to the adult (277). The tadpoles of *Bufo melanostrictus, Racophorus leucomystax maculatus,* and *Uperodon* spp are all killed by aflatoxin and it has been shown that changes occur in both the liver and the kidney (75).

Tests using isolated living cells and tissue culture

Although the toxicity of aflatoxin to various laboratory mammals is well documented, they could not be used as a simple, rapid screen for the detection of aflatoxin. Local irritant activity (276) and the occurrence of skin lesions (863, 1250, 2685) are sometimes used. By contrast, tests founded on the effects on mitosis in tissue cultures of embryonic lung (858, 1447, 1450), of liver (858, 2427, 2428, 2569, 2936–2938), or kidney (22, 1265), as well as with fibroblasts (581) have all been proposed.

The polynuclear leucocytes of guinea-pigs show, under the influence of aflatoxins, a reduction of their resazurin reductase activity which is totally inhibited at 10 μg/ml (276).

The spermatozoa of the bull (used for artificial insemination) are inhibited in their motility within 30 minutes in the presence of 25 μg/ml and in three hours in the presence of 5 μg/ml (276).

The amoeba (165) and some paramecia (2133) are also sensitive to aflatoxin B_1.

Tests using bacteria

The classical methods used in the assay of antibiotics have also been tried with aflatoxins, for they offer the advantages of rapidity and simplicity. Among 329 micro-organisms studied by Burmeister and Hesseltine (351), *Bacillus megaterium* gave the best response in a growth inhibition test. The test is carried out by placing known quantities of aflatoxin onto small paper discs using chloroform as the solvent. After complete evaporation of the chloroform, the discs are placed on a nutrient agar culture medium seeded with a suspension of 10^{10} spores of *B. megaterium* per millilitre. After incubating for 15–18 hours at 35–37 °C, the zones of inhibition of growth of the bacteria are measured. Lillehoj and Ciegler (1490) have studied the mechanism of this inhibition.

Using this technique it is possible to detect as little as 1 μg of aflatoxin B_1 in as little as seven hours of incubation (524).

Bacillus brevis, and other Gram positive bacteria (351), B. *stearothermophilus* var. *calidolactis* (1198, 1206), *Clostridium sporogenes,* some *Streptomyces* and *Nocardia* (351), and *Escherichia coli* (2428, 2877) may equally well be used.

It has been reported that the addition of 10 μg aflatoxin B_1 per millilitre of nutrient broth produces a significant reduction in the oxygen consumption of bacteria (1846).

Other tests are based on the inhibition by aflatoxins of the DNA polymerase activity of *Escherichia coli* (2428, 2877) and other bacteria such as *Staphylococcus* and *Serratia marcescens* (1225). There is also a report on the activity of aflatoxin against mycobacteria (1408).

Tests using moulds

It has been reported that the growth of several fungi, such as species of *Aspergillus* and *Penicillium*, is inhibited by aflatoxin B_1 at a concentration of 20 μg/ml (1493), although it is possible to repress this activity by replacing the nitrate, used as the source of nitrogen in the Czapek medium used in the assay, by yeast extract. Reiss (2129) has also shown that aflatoxin B_1 induces atypical germination of the spores of *Mucor hiemalis* causing a reduction in the diameter of the hyphae, and in the activity of several enzymes of this mould. Aflatoxin will also inhibit the sporulation of *Aspergillus niger, Cladosporium herbarum, Thamnidium elegans*, and *Rhizopus nigricans*, whereas it only has a weak activity on *Penicillium expansum* (2132).

Tests using algae and higher plants

Species of the unicellular alga *Chlorella* are easily cultured and their growth is affected by compounds containing a coumarin nucleus, such as the aflatoxins. Ikawa *et al.* (1150) have reported that an aflatoxin B_1 concentration of 4 μg/ml is active against these algae.

Aflatoxin can inhibit the synthesis of chlorophyll by many plant chloroplasts causing virescence, or albinism, in young plants of corn, water cress (*Nasturtium officinale*), citrus and duckweed (*Lemna minor*) (706, 1203, 1328, 1329, 2282). The research of Jacquet *et al.* (1203) on the biological effects of aflatoxins on higher plants suggests that it should be possible to exploit these phenomena in the detection and assay of aflatoxins, and perhaps other mycotoxins. A clear inhibition of *Nigella sativa*, the wild cabbage (*Brassica sylvestris*), and particularly of canary grass (*Phalaris canariensis*) has been observed. Furthermore, it has been shown that aflatoxin inhibits the growth of canary grass, millet (*Panicum miliaceum*), water cress, lentil (*Ervum lens* = *Lens culinaris*), onion (*Allium cepa*), and shallots (*A. ascalonicum* = *A. cepa*). In the case of the cress the growth of the hypocotyl is more affected than that of the radical (2136).

The stems of *Vicia faba, Arachis hypogaea, Pisum sativum, Melilotus albus, Trifolium alexandrinum, Glycine hispida, Phaseolis vulgaris*, and *Lycopersicum esculentum* wither and fade in a few days when soaked in a medium inoculated with a toxigenic strain of *Aspergillus flavus*, whereas they remain healthy when treated in the same way with a non-toxigenic strain (1242).

The ultrastructure of plants is profoundly altered by aflatoxins (1203). Undoubtedly, as with most other biological tests for detecting toxins, these observations do not provide a specific test for aflatoxins, but biological tests such as these have the merit of generally being rapid and simple and may be used to complement physical and chemical tests (823).

Physicochemical detection

The physicochemical methods for determining the presence and quantity of aflatoxins in a sample have been the subjects of numerous publications and have been reviewed by Pons and Goldblatt (2025). Their judicious use is essential but, if not carried out by experienced specialists, there are risks of errors (1972).

Extraction and purification of the extract

Most methods of extraction and purification (931, 395, 567, 742, 823, 1206, 1826, 2294, 2470, 2626) are based on the solubility of the aflatoxins in such organic solvents as chloroform, methanol, ethanol, acetone, and benzene, and their insolubility in the lipid solvents such as hexane, petroleum ether, and diethyl ether. As early as 1961, Sargeant *et al.* (2237) extracted aflatoxins from groundnut meal with methanol after defatting with diethyl ether; the extract, diluted with distilled water, was further extracted with chloroform and the chloroform layer separated. After removal of the solvent the dark residue was dissolved in a mixture of petroleum ether, methanol, and water and decanted after vigorous shaking. The lower methanolic phase was rinsed twice with petroleum ether and then evaporated under reduced pressure. Further purification was by paper chromatography, a blue fluorescent spot seen under long wave ultraviolet light being indicative of aflatoxin.

This method has since had numerous improvements. The extraction of lipids from the sample, which is essential when there is more than 2% lipid, is carried out, either with petroleum ether, or with pentane in a Soxhlet extractor, or with hexane using mechanical shaking.

The extraction of the aflatoxins may be by one of the following methods:

- methanol for four hours (307, 1228) or six hours (544),
- chloroform (1798, 2626)
- aqueous acetone (2020)
- by mixtures such as:

  petroleum ether/methanol/water (544, 614),
  chloroform/methanol/water (307, 543, 893, 2636),
  chloroform/acetone (99:1 v/v) (1205)
  chloroform/acetone + water (88:12+1.5) (2515),

(this last formula favouring subsequently a separation of aflatoxins B_2 and G_1)

  acetonitrile/water (9:1) (2923)
  methanol/water/hexane/chloroform (1826, 2773).

The scheme suggested by Jemmali and Lafont (1227) is useful:

Extraction with chloroform 50 g of the material to be analysed is dispersed by shaking for 30 minutes in 250 ml of chloroform and 25 ml of water. The chloroform phase is filtered through Whatman paper and 50 ml of this extract passed through a silica gel column. The column is then treated successively with

150 ml hexane to remove lipids followed by 100 ml anhydrous ether to remove pigments. Aflatoxin is eluted with 250 ml of a mixture of chloroform/methanol (97:2 v/v) and the eluate evaporated to dryness under dry nitrogen.

Extraction with acetone – the material to be analysed is dispersed in five times its own weight of aqueous acetone (80:20 v/v). The extraction is carried out for an hour in the dark at 20 °C, after which the aqueous acetone phase is filtered through Whatman paper and the filtrate immediately evaporated under reduced pressure to just completely remove the acetone. The residual aqueous phase is then extracted three times successively with chloroform, the chloroform extracts combined and concentrated under reduced pressure.

To isolate the aflatoxins formed by strains of *Aspergillus flavus* growing on maize, Petit (1976) used a simple chloroform extraction for six hours in a Soxhlet apparatus. The solvent was evaporated under reduced pressure in a rotary evaporator and the residue dissolved in a mixture of methanol and petroleum ether in equal parts. The mixture was thoroughly shaken, with 5 ml of distilled water. The extract was decanted into a separating funnel, the methanolic layer recovered, and the methanol removed by evaporation under reduced pressure.

For studying the distribution of aflatoxin in damaged and undamaged groundnuts workers of the United States Department of Agriculture (2294) used the following technique: kernels (50 g) or pods (20 g) were pulverized in a Waring blender with 200 ml methanol for six minutes. After adding a further 300 ml of methanol the mixture was refluxed for two hours, filtered through a Whatman No. 42 filter paper and the methanol removed leaving a thick syrup. The residue was redissolved in a mixture (100 ml) of acetone/hexane/water (60:35:5) and extracted three times with 10 ml of water removing the aqueous phase each time. The combined aqueous phases were extracted with petroleum ether to ensure complete removal of fats and the acetone removed by evaporation. The aflatoxins were removed from the aqueous phase into chloroform (4 x 25 ml) and the chloroform solution dried by passing through a bed of anhydrous sodium sulphate, concentrated to a small known volume in preparation for thin layer chromatography.

Other methods using organic solvents have been described (31) and a process involving differential extraction into aqueous solutions of a variety of inorganic salts has been investigated (2470).

Chromatographic separation

The extracts obtained by the methods described above may be further analysed by chromatography using known standards as markers.

Paper chromatography

The first successful separations were on paper using a mixture of benzene/toluene/cyclohexane/ethanol/water for developing the chromatogram (544).

The recommended methods are standard, for example Petit *et al.* (1976) used two solvent systems; butanol/water/acetic acid (20:1:19) and chloroform/methanol (95:5). The first solvent mixture gives two phases and the tank is equilibrated with the bottom phase and the chromatogram developed with the upper phase at 24 °C. Development takes 20 hours on Whatman No. 1 paper (550 x 30 mm strip), the sample being spotted 60 mm from the end of the strip and 30 mm from the level of the solvent. The second mixture gives a single phase, equilibration being for one hour and development for three and a half hours. This solvent has been used by several authors (741, 1365).

Whatever the solvent the chromatogram is dried at 100 °C for 20 minutes.

Column chromatography

This is a frequently used form of chromatography and, depending on the situation, it may involve the use of:

– columns of alumina (2179),

– columns of silica gel, eluting with chloroform and a chloroform/methanol mixture (95:5). It is possible to obtain pure aflatoxin B_1 although B_2, G_1, and G_2 are not well separated from each other. Aflatoxin M_1 is cleanly separated from M_2 (743, 2516, 2824). Other solvents which have been suggested for use with silica gel columns include diethyl ether/hexane (3:1), chloroform/acetone (8:2), and ethyl acetate.

– columns of silicic acid using chloroform/ethanol (99:1) as the eluant. This gives a good separation of aflatoxin B_1 and cleanly separates B_1 and B_2 from M_1 and M_2 (560, 2516, 2518).

– columns of cellulose; after absorption onto cellulose the pigments are first eluted with hexane and the aflatoxins with a mixture of hexane and chloroform (1:1) (2504, 2507).

– columns of Sephadex G_{10} eluting with water which separates the aflatoxins in the order G_{2a}, B_{2a}, G_2, B_2, G_1, and B_1 (1587). Sephadex G_{25} and LH 20 may also be used (1588).

Thin layer chromatography

Without a doubt this is the method most widely used. As long ago as 1962 de Iongh (613) described the use of plates coated with a layer of silica gel (Kieselgel G) for separating the four aflatoxins B_1, B_2, G_1, and G_2 using a mixture of chloroform and methanol (98:2). Several types of silica gel have been used usually in layers 0.25 or 0.5 mm thick. Among the wide range of solvent mixtures used the following are most frequent:

– chloroform/methanol (from 93:7 to 99:1)

– chloroform/acetone (9:10) (85:15)

– chloroform/acetone/ethanol (97.25:2.0:0.75) (89:10:1)

– chloroform/acetone/propan-2-ol (850:125:25) (825:150:25)

– methanol/water/ether (3:1:96)

– benzene/ethanol/water (46:35:19)
– toluene/isoamyl alcohol/methanol (90:32:3).

The time of development varies, according to the solvent system used and the temperature at which the tank is maintained, from 45 minutes to three hours (250, 614, 733, 738, 742, 1019, 1084, 1193, 1383, 1486, 1588, 1589, 1827, 1829, 2023–2025, 2117, 2362, 2493, 2502, 2565).

Thin layer chromatography is sometimes carried out using alumina, kieselguhr, cellulose powder or polyacrylamide gel (1441).

An important aspect of the successful separation of aflatoxins by thin layer chromatography seems to be the preliminary equilibration of the plates before developing. Plates should be left long enough for equilibration to occur but not so long that artefacts arise (1976).

The chromatograms are usually prepared on glass plates of 200 x 220 mm but a simple procedure using glass slides (26 x 76 mm) has also been described (1206) and pliable supports such as plastic film have been used (1975).

Many compounds behave chromatographically like the aflatoxins (818, 2380) and to decrease the possibility of error two dimensional chromatography may be used. Such a system also favours the more complete resolution of aflatoxins B_2 and G_1 which usually run very close to each other (1972). Using silica gel G-HR as the absorbent, the solvent for the first dimension may be acetone/chloroform (1:9) and for the second ethyl acetate/isopropanol/water (10:2:1).

Reiss (2130) claimed an improved resolution of the aflatoxins by allowing the solvent to migrate radially from a point in the centre of a horizontal plate using methanol/chloroform (3:97).

Quantitative estimation

Most methods are based on the intense fluorescence of the aflatoxins when exposed to ultraviolet light. In 1962 Schmidt and Bankole (2273) demonstrated the use of the fluorescent antibody technique for detecting the mycelium of *Aspergillus flavus* on slides buried in the soil, applying the methods reviewed by Beutner (197). The mycelia of aflatoxin-producing strains fluoresce without the use of specific fluorescent antibody staining; this phenomenon has been used successfully in the examination of cottonseed (86, 2816).

The earliest methods of quantitatively measuring aflatoxin B_1 were by the dilution of extracts to extinction, making use of the knowledge of the concentration of aflatoxin which was just detectable from its fluorescence under u.v. light (543). This method was later replaced by the direct comparison of the intensity of spots on chromatograms examined under the u.v. lamp with those of standards of known concentration (742, 1826, 2020, 2021, 2168–2170). Such methods may involve errors of as much as 50%.

Greater precision is obtained with the use of a fluorodensitometer, the TLC plate being examined under u.v. light (320–390 nm) and scanned with a photometer which allows the exact location of the position of fluorescent spots as well as a precise measure of the intensity of their fluorescence. Suggested by Ayres and

Sinnhuber (111) this method is now widely used (176, 279, 928, 1974, 2019, 2021, 2024, 2025, 2169, 2362, 2363, 2517, 2613, 2614).

The use of fluorescence for the detection of aflatoxins is not without the risk of error. It appears, for example, that under certain conditions more than 20% of damp but sound groundnuts themselves contain a fluorescent substance, insoluble in water, but soluble in methanol and chloroform as are the aflatoxins. This compound appears during storage, especially at 30 °C, first near the surface then progressing into the interior of the kernel. Chromatography of the extract gives spots of blue-violet fluorescence at three positions, the centre one having almost the same R_f as aflatoxin. The formation of these fluorophores is inhibited by heat (above 50 °C), KCN, NaF, As_2O_3, as well as by classical inhibitors of germination such as coumarin; *p*-chloromercuribenzoate and streptomycin do not inhibit this process. During germination of the seed, these compounds concentrate particularly in the cotyledons and not the germ (2470). Confirmation of this phenomenon comes from as far afield as Senegal (623) and the United States (58).

Several fungi produce fluorescent metabolites which may be confused with aflatoxins. Examples include scopoletin produced by *Ceratocystis fimbriata* (2620), a metabolite of *Macrophomina phaseoli* (2292), and several metabolites of aspergilli and penicillia (1687, 2010). Among the moulds used industrially in Japan, especially for koji manufacture, a number of strains of *Aspergillus* produced compounds of the pyrazine group some of which are fluorescent (1791, 1792, 2925–2927) and others not so (2246). The fluorescent compounds include 2-hydroxy-3,6-di-sec-butylpyrazine ($C_{12}H_{20}ON_2$), deoxyaspergillic acid ($C_{12}H_{20}ON_2$), flavacol ($C_{12}H_{20}ON_2$), deoxyhydroaspergillic acid ($C_{12}H_{20}O_2N_2$), and deoxyhydroxymutaspergillic acid ($C_{11}H_{18}O_2N_2$). The related compounds, aspergillic acid and hydroxyaspergillic acid, in which the >N–H is replaced by >N–OH, do not fluoresce (2927).

So the detection of fluorescent compounds in groundnuts or rice may not necessarily imply the presence of aflatoxins. This is true for various other commodities such as tobacco, coffee, Sauterne wine, and cheeses some of which may contain other derivatives of coumarin (1206).

Spectrophotometric methods can also be used for the detection and assay of aflatoxins (18, 991, 1445, 1798, 2472).

Chemical detection

A number of chemical properties may be used to confirm the presence of aflatoxins (31):

– the reduction of ammoniacal silver nitrate, sodium tungstate or molybdate;
– the formation of a characteristic violet colour with carbazole in perchloric acid and the weak reaction with vanillin in 30% sodium hydroxide;
– the condensation of the furan nucleus with phenol in sulphuric acid;
– sensitivity to alkali.

Derivatization directly on the chromatogram using, for example, mixtures of acetic acid or formic acid with thionyl chloride, or trifluoroacetic acid can be used to differentiate aflatoxins from artefacts (58, 2378, 2504, 2823, 2824). Jacquet *et al.* (1206) have similarly used the action of a number of oxidizing agents, acids, and other active agents which destroy or modify spots of aflatoxin. Nitric acid specifically destroys aflatoxins B_1 and G_1 (1639).

It has been reported that aflatoxins react with di-*o*-anisidine tetrazolium chloride to form blue-violet derivatives with B_1 and B_2 and brown derivatives with G_1 and G_2.

Application of methods

These various methods have been adapted and applied to the examination of a variety of commodities, e.g. groundnuts and their products (450, 614, 744, 1109, 1356, 1639, 2771, 2773, etc.), cottonseed (568, 2506, 2507, 2737), cereals and cereal products (1227, 1256, 1511, 2150, 2375, 2459, 2460, 2804), cocoa beans (2317, 2322), coffee (1470), meat (340, 994, 2560), milk products (278, 1206, 1304, 1305, 1479, 2363), and various other foods (818, 1320, 2929).

Recommended methods include those described by the Tropical Products Institute, London (Reports G6 and G13) which include the following stages:

1. Defatting with petroleum ether or hexane.
2. Extraction of aflatoxins with methanol and (or) chloroform.
3. Partial purification of the aflatoxin extract by partition between chloroform and aqueous methanol.
4. Thin layer chromatography on silica gel G using chloroform/methanol (95:5).
5. Estimation of intensity of fluorescence of spots by comparison with those of marker spots, or better using a scanning fluorodensitometer.

Chemical structure

The results of the purification and characterization of the toxic principle have demonstrated that the 'aflatoxin' defined in 1962 was, in fact, a mixture of closely related chemical entities.

Aflatoxins B_1, G_1, B_2, G_2

Early studies led to the isolation of two closely related compounds (613, 1825). TLC on alumina using chloroform/carbon tetrachloride/water/methanol (2:2.5:1:3) revealed two u.v. fluorescent spots. One compound fluoresced blue-violet and hence was called aflatoxin B, the other, with a lower R_f fluoresced green and was called aflatoxin G.

These two compounds were crystallized and shown to have the empirical formulae $C_{17}H_{12}O_6$ and $C_{17}H_{12}O_7$ and molecular weights 312 and 328 respectively. Work carried out in the laboratories of Unilever resulted in the isolation of a compound which was called aflatoxin FB_1 which was identical with aflatoxin B_1, the subscripts becoming necessary when it was appreciated that

Aspergillus flavus synthesized two dihydroaflatoxins subsequently called B_2 and G_2.

The presence of lactone and methoxyl and the absence of free hydroxyl groups was quickly established. Aflatoxin B_1 is an unsaturated compound which, on catalytic hydrogenation in ethanol, gives a tetrahydrodesoxy derivative the spectral properties of which are more easily interpreted, being similar to those of 5,7-dimethoxycoumarin (Figure 38). Of all the derivatives of 5,7-dialkoxycoumarins studied, the spectrum of the 5,7-dimethoxy-cyclopentene derivative (Figure 39) was closest to that of tetrahydrodesoxyaflatoxin B_1.

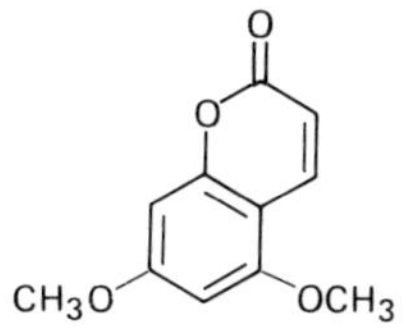

Figure 38. 5,7-Dimethoxycoumarin

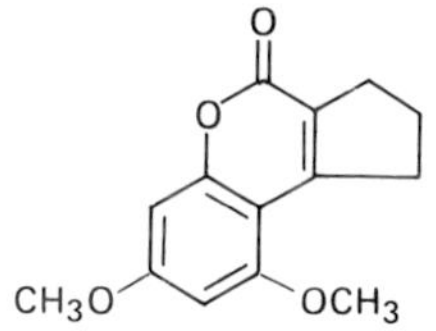

Figure 39. 5,7-Dimethoxy-cyclopent (c) coumarin

It was thus possible to establish a part formula for aflatoxin B_1. Studies of nuclear magnetic resonance spectra of aflatoxin B_1 showed that four atoms of hydrogen were attached to two adjacent carbon atoms in a cyclopentenone ring. Comparison of the spectrum with that of a previously encountered dihydrofuran (1168) and those obtained in earlier studies of substituted benzene derivatives (653) suggested that the formula of aflatoxin had something in common with that of sterigmatocystin established by Bullock *et al.* (341, 342). Thus Asao *et al.* (79) proposed what is now accepted as the structure of aflatoxin B_1 (Figure 40).

Aflatoxin G_1 (green fluorescence) has a structure very similar to that of B_1 but there are two lactone functions rather than one (Figure 41). The work of several

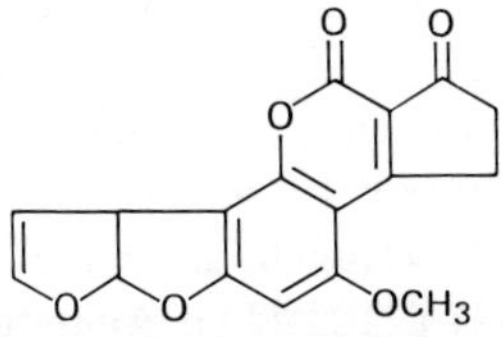

Figure 40. Aflatoxin B_1

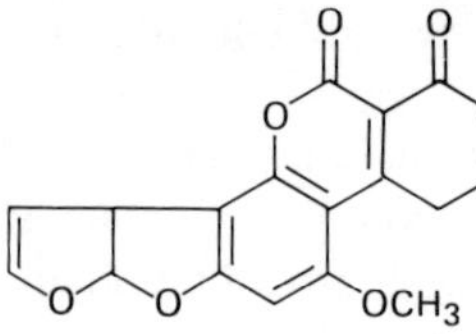

Figure 41. Aflatoxin G_1

laboratories (2715) confirmed these results. A report from the Tropical Products Institutes, London (1037) summarized studies confirming this structure and rejecting a structure which had been proposed by van der Merwe *et al.* (2705). An X-ray analysis not only confirmed the proposed structure but revealed that the two dihydrofuran rings are fused in a *cis* configuration (445).

The two other toxic metabolites produced by strains of *A. flavus* known as aflatoxins B_2 and G_2 are respectively dihydroaflatoxins B_1 and G_1 (441, 1037, 2715). They may be synthesized by reduction of the isolated double bond in the terminal dihydrofuran ring (Figures 42 and 43) (422).

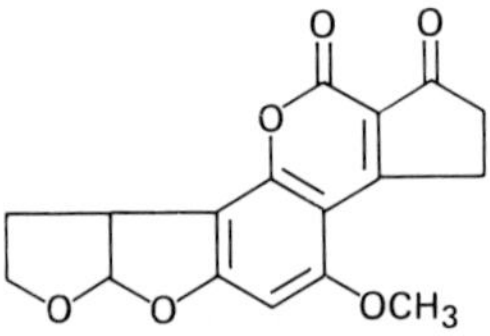

Figure 42. Aflatoxin B_2

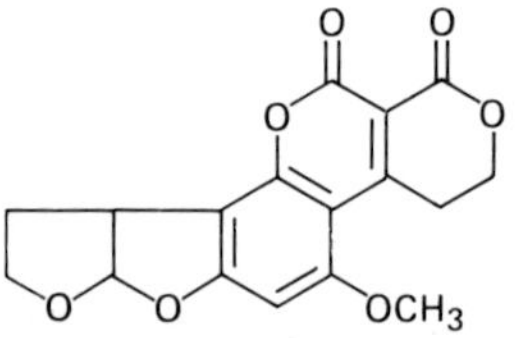

Figure 43. Aflatoxin G_2

One report (2429) indicated the existence of as many as 12 fluorescent compounds of which five showed toxicity when administered to ducklings. However it is possible that artefacts appear during the period of extraction and purification.

X-ray crystallography has been used to confirm the structures of aflatoxin G_1 (445) as well as B_1 and B_2 (287, 2718, 2719).

Hydroxyaflatoxins

In 1966, Dutton and Heathcote (709) described two new aflatoxins in cultures of *Aspergillus flavus,* one with a blue fluorescence the other with a green fluorescence. Studies of their u.v., i.r., mass, and nuclear magnetic resonance spectra, as well as a number of chemical reactions allowed their identification as the 2-hydroxy derivatives of aflatoxin B_2 and G_2 respectively. They were shown to have the empirical formulae $C_{17}H_{14}O_7$ and $C_{17}H_{14}O_8$ and called aflatoxins B_{2a} and G_{2a} (Figures 44 and 45) (710, 711). Andrellos and Reid (58) had used the name aflatoxin W for aflatoxin B_{2a}. These compounds are 60–100 times less toxic

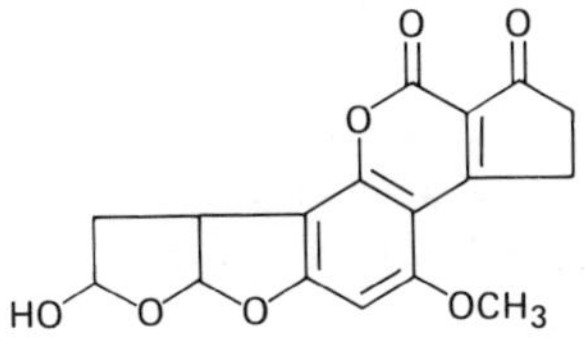

Figure 44. Aflatoxin B_{2a}

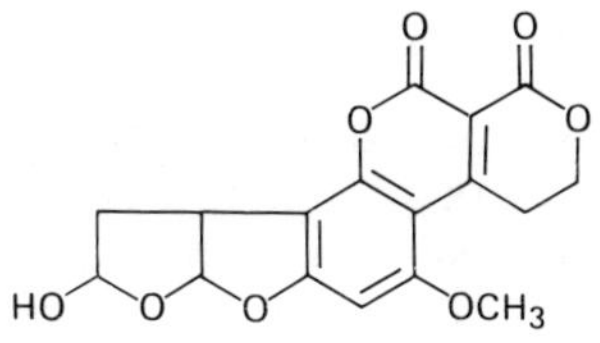

Figure 45. Aflatoxin G_{2a}

to the duckling than aflatoxin B_1 and are therefore not so dangerous unless there is the possibility of them being dehydrated to aflatoxins B_1 and G_1.

During their studies on the possibility of toxic compounds being present in the milk and flesh of cows which have eaten food contaminated with aflatoxins, Allcroft and Carnaghan (35) demonstrated the presence of compounds capable of giving symptoms in ducklings resembling those caused by aflatoxins. At first called 'milk toxin' it was shown to be a mixture of two compounds, one with a blue-violet fluorescence called aflatoxin M_1, the other with a lower R_f having a violet fluorescence and called aflatoxin M_2 (Figures 46 and 47) (615, 1118). Identical compounds have been recovered from the liver, kidneys, and the urine of sheep (45, 1118, 2352), as well as from groundnuts themselves (1118). These compounds are in fact 4-hydroxyaflatoxins B_1 and B_2.

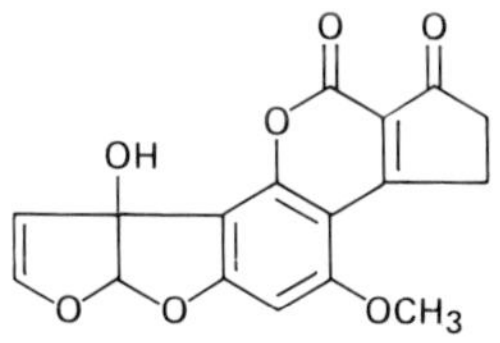

Figure 46. Aflatoxin M_1

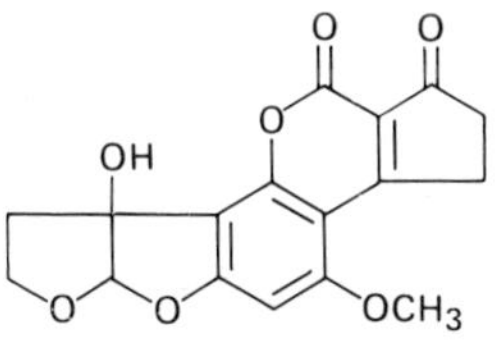

Figure 47. Aflatoxin M_2

Aflatoxin P_1

Recently, Dalezios, Wogan, and Weinreb (578), while analysing the urine of monkeys which had consumed aflatoxin B_1, identified a phenolic derivative, identified as a metabolite of aflatoxin B_1, which they called aflatoxin P_1. It was isolated on an amberlite XAD-2 (Rohm and Haas) column and shown, using mass spectroscopy, to have a molecular weight of 298. It results from demethylation of aflatoxin B_1 (Figure 48) and may represent at least 60% of the metabolites of aflatoxin in the urine.

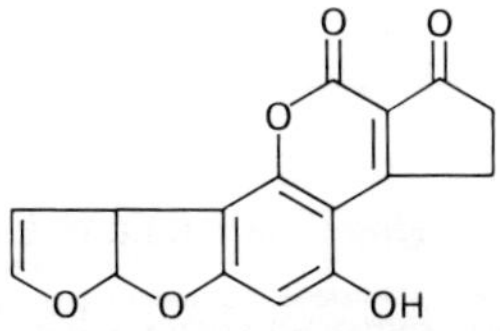

Figure 48. Aflatoxin P_1

Other aflatoxin derivatives

Among the compounds which have been called aflatoxins are aflatoxin 3B (or toxin B_3) (831, 1401), a compound with a low R_f isolated from cultures of *Aspergillus flavus,* crystallizing as yellow needles and having a toxicity of 40–50 times less than that of aflatoxin B_1 by the egg embryo test.

A compound considered to be a precursor of the aflatoxins, and named aflatoxin B_3 was isolated by Heathcote and Dutton (1051). The same compound was isolated by Stubblefield *et al.* (2515, 2516) who called it parasiticol (Figure 49). It was shown to be aflatoxin B_1 in which the cyclopentenone ring had been replaced by an open ethanol chain (Figure 49) and is thus 6-methoxy-7-(2′hydroxyethyl) difurocoumarin.

Louria (1521) isolated from cigarette tobacco from New Jersey a compound which was called T-2, which seemed to be very similar to aflatoxin B_2, having the same R_f the same long wave emission maximum in the fluorescence spectrum, but differing slightly in the long wave absorption in the u.v. spectrum; it differed also in its toxicity to the chick embryo (LD_{50} 134 μg/egg instead of 92 μg/egg).

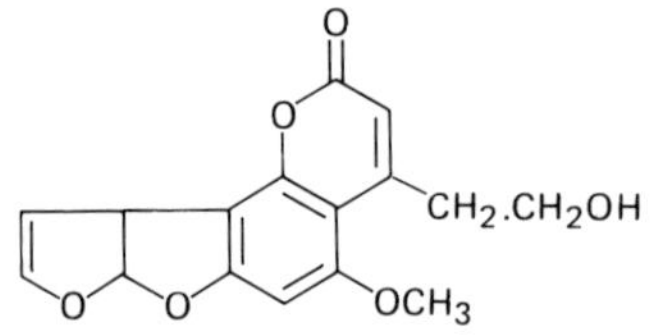

Figure 49. Parasiticol (= aflatoxin B_3)

Related compounds

It has sometimes been emphasized that aflatoxin has a structure similar to that of the isocoumarone of Kuč (1362) and pisatin described by Perrin and Bottomley (1965), and Cruickshank (563) $C_{17}H_{14}O_6$ (Figure 50). These highly oxygenated compounds are sometimes referred to as phytoalexins (1784) being produced by plants in response to fungal attack, or other forms of damage.

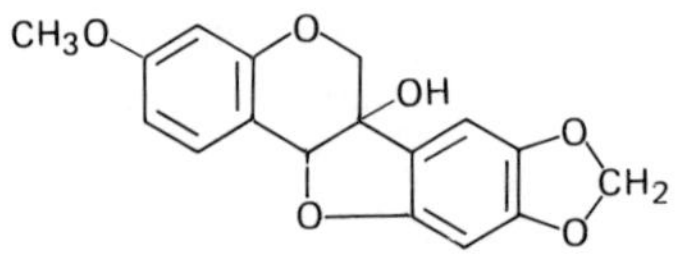

Figure 50. Pisatin

Physicochemical characteristics

These have been reported by many authors who do not always agree. In Table 12a we show the information provided by Beljaars (181) on the absorption and

Table 12a. The major peak in the fluorescence emission spectra of the aflatoxins (181)

State of sample	*Absorption maximum λ(nm)*	*Fluorescence emission maximum λ(nm)*			
		B_1	B_2	G_1	G_2
Solution in methanol	365	430	430	450	450
Solution in ethanol	365	430	430	450	450
Solution in chloroform	365	413	413	430	430
Solution in acetonitrile	365	415	412	440	437
Absorbed on silica gel (solid)	368–369	432	427	455	450

fluorescence characteristics of the aflatoxins (see also Figure 51). In Table 12b are collected together a number of physicochemical properties, on most of which there is general agreement (79, 265, 422, 1118, 1618, 2236) although, where there are some differences we have used the results of Townsend (2622), Stubblefield *et al.* (2518), and Beljaars (181).

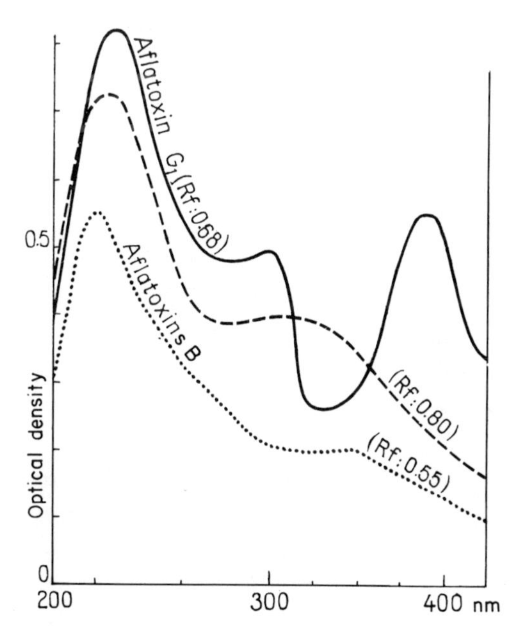

Figure 51. Ultraviolet absorption curves for aflatoxins B_1, B_2, and G_1 (31)

Synthesis

Once the structure of the aflatoxins had been established chemists set out to synthesize these compounds. Büchi *et al.* (344) demonstrated an elegant synthesis

Table 12b. Principal physicochemical properties of the aflatoxins, results from (*) Townsend (2622), (**) Stubblefield *et al.* (2518), and (***) Beljaars (181)

Aflatoxin	Formula	Mol. wt.	Melting point			Fluorescence	Optical rotation $[\alpha]_D^{CHCl_3}$	Ultraviolet absorption in ethanol		Infrared absorption $\nu_{max}^{CHCl_3}$ (cm^{-1})
				(*)	(**)			λ(nm)	ϵ	
B_1	$C_{17}H_{12}O_6$	312	268–269	265–270	265.2– 266.2	blue	–558°	223 265 362	25,600 13,400 21,800	1,760 (intense) 1,684 (weak) 1,632, 1,598 1,562
B_2	$C_{17}H_{14}O_6$	314	286–289	305–309	280–283	blue	–492° (***) –430°	222 265 362	19,600 9,200 (***) 11,000 14,700 20,800	
G_1	$C_{17}H_{12}O_7$	328	244–246	247–250	246.7– 247.3	green	–556°	243 257 264 362	11,500 9,900 10,000 16,100	1,760 1,695 1,630 1,595
G_2	$C_{17}H_{14}O_7$	330	229–231	237–240		green	–473°	221 245 265 365	28,000 12,900 11,200 19,000	
M_1	$C_{17}H_{12}O_7$	328	299			blue-violet	–280°	226 265 357	23,100 11,600 19,000	3,425 1,760 1,690
M_2	$C_{17}H_{14}O_7$	330	293			violet		221 264 357	20,000 10,900 21,000	3,350 1,760 1,690

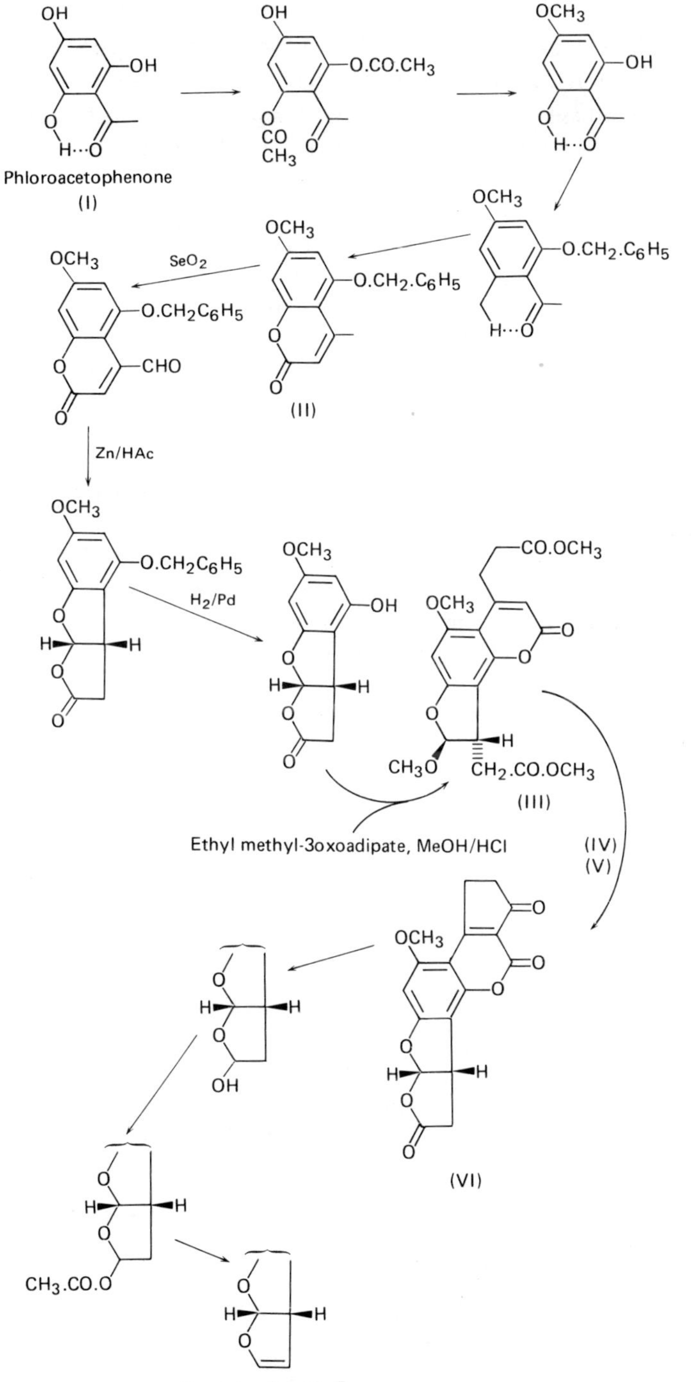
OH
OH
Phloroacetophenone
(I)
O.CO.CH3
CO
CH3
OCH3
O.CH2.C6H5
SeO2
O.CH2C6H5
CHO
(II)
Zn/HAc
H2/Pd
CO.OCH3
CH3O
CH2.CO.OCH3
(III)
Ethyl methyl-3oxoadipate, MeOH/HCl
(IV)
(V)
(VI)
CH3.CO.O
Racemic aflatoxin B1
(VII)

Figure 52. Some stages in the total synthesis of the aflatoxins (334) starting from phloroacetophenone (I) from which was prepared 5-benzoyloxy-7-methoxy-4-methylcoumarin (II) followed by the benzofuran derivative (III). The synthesis continued via the compound (IV) and the tetracyclic lactone (V) to the pentacyclic structure (VI) related to aflatoxin B_2 and hence to the racemic form of aflatoxin B_1 (VIII) (Figure 52).

Roberts *et al.* (2165) were able to synthesize the aflatoxins B_1 and B_2 by condensing tetrahydro-4-hydroxy-6-methoxyfuro (2,3-b)-benzofuran (I) with ethyl cyclopentanone-2-carboxylate (II) (Figure 53).

In an attempt to synthesize aflatoxin G_2, Bycroft, Hatton, and Roberts (380) described a synthesis of the coumarino-lactone system but not of aflatoxin itself.

Speculations on the biosynthesis of the aflatoxins have given rise to many hypotheses and the use of radioactively labelled substrates has made it possible to test some aspects of them (1485). It has been possible to produce aflatoxin labelled with both ^{14}C and ^{3}H (109). The origin of 13 of the 17 carbon atoms in the molecule of aflatoxin B_1 has been determined (16). It was long considered that phenylalanine may be a precursor of the coumarin moiety in the biosynthesis of the aflatoxins by *Aspergillus flavus.* However, using ^{14}C labelled amino acids, it has been demonstrated that either tyrosine or phenylalanine must first be degraded to acetate before incorporation into aflatoxin. This is also the case in the biosynthesis of novobiocin, an antibiotic produced by an actinomycete (219). There is no doubt

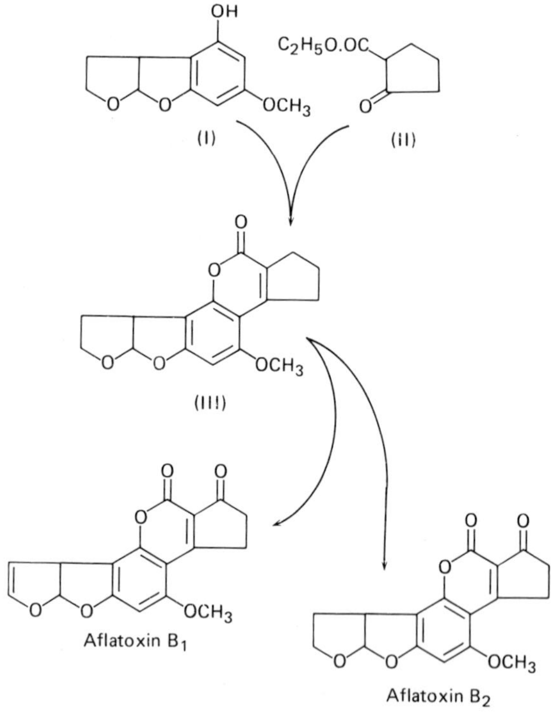

OH

$C_2H_5O.OC$

OCH_3

(I)

(II)

OCH_3

(III)

OCH_3

Aflatoxin B_1

OCH_3

Aflatoxin B_2

Figure 53. Alternative route to the synthesis of the aflatoxins (2165)

now that aflatoxins are essentially polyketide metabolites and are probably synthesized outside the mitochondria (1135). Papers by Holker and Underwood (1110), Thomas (2593), Biollaz, Büchi, and Milne (217), and Donkersloot *et al.* (689) all suggest a pathway in which acetate is incorporated into an anthraquinone pigment, perhaps related to averufin, which is converted via sterigmatocystin, or a related xanthone, to aflatoxin B_1 which is further metabolized to aflatoxin G_1. However the addition of ^{14}C labelled sterigmatocystin into a growing culture of *A. flavus* does not lead to its incorporation into aflatoxin (1110) giving some doubt about its role as an intermediate in the pathway. The pigment versicolorin A may also be an intermediate between a linear polycyclic compound and aflatoxin (216). It has been possible to convert demethylsterigmatocystin, another metabolite of *Aspergillus versicolor,* into 5-hydroxydihydrosterigmatocystin which is a presursor of aflatoxins B_2 and G_2 (729).

Moody (1716) suggested that mevalonate may be a precursor of aflatoxin whereas Heathcote *et al.* (1050) considered kojic acid to be the precursor of the non-furan part of the molecule. Jaquet and Boutibonnes (1200) have indeed shown that strains of *A. flavus* which are most toxigenic usually also produce the most kojic acid, but it is now considered unlikely that it is a precursor.

The effect of cycloheximide on the biosynthesis of aflatoxins has been studied. This antibiotic blocks the synthesis of proteins and, if added at a concentration of 150 μg/ml before the appearance of the B_1 synthetase, no toxin is produced. If it is added during the production of the synthetase, biosynthetic activity remains constant for, once the synthetase has been formed, the synthesis of aflatoxin is not inhibited, even though the synthesis of protein is blocked (637).

The importance of fatty acids has already been indicated and this probably arises from the fact that their biodegradation gives rise to acetyl coenzyme A, a precursor of aflatoxin (1460).

In contrast to observations with ^{14}C labelled acetate, there are only small differences in the incorporation of ^{32}P labelled orthophosphate by toxigenic and non-toxigenic strains of *A. flavus* (989).

If the biosynthesis of the aflatoxins has given rise to a large number of studies, so also have metabolic transformations and degradation by chemical means, especially in order to modify these toxic molecules to non-toxic compounds (1625, 2733).

Metabolic transformations and excretion

The question arises as to what happens to aflatoxins in the bodies of animals which have consumed them. As well as its intrinsic interest, this problem is in fact important in the context of the foods of animal origin consumed by man.

It was Allcroft and Carnaghan (35, 36) who first demonstrated that if lactating cows were fed cattle cake rich in aflatoxin, their milk became toxic to one-day-old ducklings. Analysis of the milk revealed only traces of aflatoxin B_1 but demonstrated the presence of a related compound to which the name 'milk toxin' or aflatoxin M was given (614, 616).

It has since been demonstrated that two hydroxylated derivatives of aflatoxin

B_1 and B_2 were in fact produced and they have been called aflatoxin M_1 and M_2 respectively (336, 1118, 1618). Small quantities of these aflatoxins have also been detected in cultures of *A. flavus* on groundnuts and a number of other substrates (615, 616, 1078, 1118, 1618, 2053). In order to confirm that the milk toxins really are transformation products of ingested aflatoxins, aflatoxin B_1 has been administered orally to lactating rats; aflatoxin M_1 was subsequently recovered from their milk (615, 616). This compound has also been detected in the milk of nanny-goats (619), in the bile of rats (373), in the faeces and urine of cows and sheep (44, 45, 1260, 1617, 1796, 1797), in the blood, urine, and bile as well as the kidneys of young rabbits (768), and in the urine of the goose, and the guinea-pig (1933).

Recent studies have confirmed the presence of aflatoxin M_1 in the urine of human beings who have consumed peanut butter contaminated with aflatoxins as well as in the milk of an Indian mother under the same circumstances (402). Some milk sold in South African markets has been shown to be contaminated (2060).

The metabolism of ingested aflatoxins is fairly rapid (461). If a cow receives a single dose (0.5 mg/kg) of material containing a mixture of aflatoxins (B_1, 44% G_1, 44%; B_2, 2%) and the milk, urine, and faeces regularly analysed, it has been demonstrated that 85% of the aflatoxins detected in the milk and urine are excreted within the first 48 hours. None could be detected in the milk after four days or in the urine and faeces after six days. Only aflatoxin M_1 could be detected in the milk and the quantity recovered represented 0.35% of the total aflatoxin B_1 ingested (43, 44, 2163). Elsewhere it has been shown that, if 67–200 mg of aflatoxin B_1 is added to the weekly ration of a cow than 0.07–0.15 mg of aflatoxin M_1 per kilogram of freeze-dried milk could be detected (1295).

The ewe behaves a little differently from the cow. Excretion of aflatoxin M_1 in the urine and faeces is much earlier and in much smaller amounts; in contrast much more aflatoxin B_1 is excreted unchanged. Another fluorescent compound, referred to as 'U', was also detected in the urine and may have been the hydroxy derivative related to G_1 in the same way that aflatoxin M_1 is related to B_1 (45, 1476, 1796).

What is the toxicity of these aflatoxins? The oral LD_{50} of aflatoxin M_1 in the one-day-old duckling is 0.32 mg/kg compared with 0.24 mg/kg for aflatoxin B_1 under the same conditions (2047). Thus the acute toxicities of the two are very similar. The lesions produced by aflatoxin M_1 in the liver of the duckling are very similar to those caused by aflatoxin B_1 and low doses of M_1 may give rise to a progressive degeneration of the liver cells and proliferation of the bile duct (2047). In fact, it is considered that, in the trout for example, aflatoxin M_1 is incontestably as potent a liver carcinogen as is B_1 (2405). Aflatoxin M_2 is considerably less toxic than M_1, the LD_{50} for the one-day-old duckling being 1.248 mg/kg (2047).

Several authors (619, 2046, 2052, 2053) have pointed out that, if we consider that a mammal may secrete as much as 1% of the quantity of aflatoxin B_1 ingested in the form of M_1, and that this is nearly as toxic as B_1, it is possible for there to be as much as 0.8–3.5 μg/l in the milk given the average contamination of cattle feeds. Now this milk will be consumed particularly by the young and the frail, both groups being more susceptible to the toxic effects of aflatoxins, and of course

Table 13. The distribution of ^{14}C in male Fischer rats 24 hours after the intraperitoneal injection of ^{14}C labelled aflatoxin

Locality	*% of the total ^{14}C recovered when the label was present in:*	
	Methoxyl group	Nucleus
CO_2	32.6	0.3
Urine	26.1	14.8
Faeces	14.1	69.8
Intestine	11.8	3.3
Stomach	0.3	0.3
Carcass	7.5	0.8
Kidneys	0.4	0.4
Blood	0.8	2.3
Liver	5.9	7.7

consumption will be daily. It is considered, therefore, that rapid methods for the estimation of aflatoxin M_1 in milk are required (1187, 1619, 2052).

The unstable hemiacetals, such as aflatoxin B_{2a} and G_{2a} have been identified in the liver of certain animals. These compounds are very much less toxic than aflatoxin B_1 when given orally to animals (710, 711, 1934).

In the mammal, therefore, the metabolism of aflatoxin B_1 is usually by hydroxylation, although demethylation has also been demonstrated giving rise to compounds such as aflatoxin P_1 recently detected in urine (578, 2859). If aflatoxins B_1 and G_1 are given to rats, analysis of the urine shows that the coumarin nucleus remains intact; the molecule is degraded by opening the furan ring with the formation of an aliphatic side chain terminating in an aldehyde group which is oxidized on excretion (4).

The most precise observations have been made using aflatoxin B_1 labelled with ^{14}C, either in the methoxy side chain, or in the nucleus (see Table 13) (2347, 2352, 2859).

The testicles, pancreas, spleen, and brain contained less than 0.1% of ^{14}C whether the label was in the nucleus or the methoxyl group. It can be seen that when the label is in the methoxyl group a significant amount of label appears as CO_2 indicating that demethylation and oxidation of the methyl group is a significant route of metabolism. When the label is in the nucleus, on the other hand, very little of it appears as CO_2 indicating that the nucleus is not metabolized very far.

Considering the distribution of label following metabolism of aflatoxin B_1 labelled with ^{14}C in the nucleus, it can be seen that the intestinal route is the most important, especially biliary excretion (2311). The urinary route is secondary in importance. Faulk, Thompson, and Kotin (763) obtained similar results by studying the biliary excretion of aflatoxin B_1 using fluorescence. Bassir and Osiyemi also confirm this result (159). The liver retains aflatoxin B_1, or metabolites derived from it, in much larger quantities, and for much longer periods, than do other tissues.

In summary, one of the routes of metabolism of aflatoxin B_1 involves hydroxylation usually to the monohydroxy derivative M_1, which is readily demonstrated in the milk, faeces, and liver. The other important route involves demethylation leading to a phenolic compound, aflatoxin P_1, recently isolated from urine. Information about the metabolism of aflatoxins B_2, G_1, and G_2 is lacking at the moment.

MYCOTOXICOSIS OF *ASPERGILLUS FLAVUS*

The name aflatoxicosis is given to the intoxication caused by the aflatoxins. Only the symptoms recognized following experimental ingestion of isolated purified aflatoxins are relevant to aflatoxicosis *sensu stricto*. Toxicoses caused by the mould *Aspergillus flavus* may be more complex due to the interaction of other toxic compounds produced by the mould and it is often difficult to recognize the parts played by each of them.

Characteristics of the toxicity of the aflatoxins

Lethal dose for the duckling

Although the LD_{50} is only an indication of the acute toxicity, nevertheless its determination gives a valuable piece of comparative information.

The results of Asao *et al.* (78) were based on the use of one-day-old Pekin white ducklings (body weight 51 ± 4 g) receiving a range of doses of aflatoxins dissolved in propylene glycol. Each animal was given 0.1 ml delivered directly into the gizzard and mortality was recorded 48 hours after administration. Under these conditions:

– LD_{50} of aflatoxin B_1 is 0.564 mg/kg (95% confidence values being 0.494–0.644 mg/kg).

– LD_{50} for aflatoxin G_1 is 1.8 mg/kg.

Most determinations of the LD_{50} using ducklings are recorded six days after administration of the aflatoxins either in propylene glycol or, more often than not, in dimethylformamide. A range of values from different authors is given in Table 14.

From these figures it can be seen that reduction of the double bond in the difuran ring system lowers the toxicity by about four and a half times (cf. B_1 to B_2; G_1 to G_2). There is also a reduction in toxicity following the expansion of the cyclopentenone ring to the second lactone (cf. B_1 to G_1; B_2 to G_2).

It is possible to detect subacute toxicity from the induction of hyperplasias in the epithelium of the bile ducts. Ingestion of 0.4 μg of aflatoxin B_1 per day gives rise to detectable lesions; the production of similar effects requires 1.56 μg per day of aflatoxin G_1 (1826).

Table 14. Values of the LD_{50} of the aflatoxins for the duckling as determined by a number of authors

Aflatoxin	LD_{50} (mg/kg)	*Reference*
B_1	<0.4	2453
	0.364 (95% confidence limits 0.28–0.476)	422
	0.4	1825
B_2	1.696 (95% confidence limits 1.3–2.2)	422
G_1	1.2	1825
	1.2	2453
	0.784 (95% confidence limits 0.542–1.134)	422
G_2	3.450 (95% confidence limits 3.16–3.76)	422
M_1	0.332	1345 2047
M_2	1.24	1345 2047

The lethal dose for a variety of laboratory mammals

For the weanling male Fischer rat the oral LD_{50} of aflatoxin B_1 is 5.55 mg/kg (95% confidence limits 4.92–6.27 mg/kg), that is to say about ten times more than the comparable value for the duckling. It is of interest to note that the female rat is less sensitive than the male (LD_{50} = 7.44 mg/kg, 95% confidence limits 6.56–8.44 mg/kg) and one-day-old animals are more sensitive, having an LD_{50} of about 1.0 mg/kg (2853).

The Wistar rat shows an even greater difference between the male and female; oral LD_{50} of aflatoxin B_1 for the male is 7.2 mg/kg and for the female 17.9 mg/kg (364). This difference may be due to the more rapid metabolism in the female with the formation of aflatoxin M_1 in the milk resulting, at least in part, from hormonal interactions (2054).

The mouse is particularly resistant to aflatoxins (2030) whereas the rabbit has a sensitivity comparable to that of the duckling (617), the oral LD_{50} for the young rabbit being between 0.2–0.4 mg/kg (768). In the guinea-pig, the maximum dose which does not provoke observable histopathological changes is of the order of 6.25 μg of aflatoxin B_1, 100 μg of B_2, 25 μg of G_1, and 800 μg of G_2 (889). The optimum teratogenic dose for the eight day foetus of the hamster is 4 mg/kg by injection (721).

Toxicity to tissue cultures

In vitro experiments using cultured animal cells have shown that mortality is produced by concentrations of the order of 1–5 μg/ml of medium (858). On the other hand inhibition of growth and nuclear division can be observed at 0.03 μg/ml (581, 1450). It has been noted that aflatoxins B_2 and G_2 are much less toxic to embryonic human liver cells than aflatoxin B_1, the lethal doses being 35 and 10 μg/ml respectively after 48 hours exposure to the toxins (2520).

At a concentration of 1.0 μg aflatoxin B_1 per millilitre cultures of the primary kidney epithelial cells of the monkey show symptoms of degeneration including karyorrhexis, pycnosis, and increased cytoplasmic vacuolation (736).

Toxicity to fish

One experiment using nine-month-old trout of 60 g body weight indicated that the LD_{50} by the intraperitoneal route is 0.81 mg/kg for aflatoxin B_1 and 1.9 mg/kg for aflatoxin G_1. It is not possible to determine the oral LD_{50} because the trout regurgitates the stomach contents as soon as it ingests aflatoxin (163).

The sensitivity of various animals

Although farm animals are generally sensitive to the aflatoxins ruminants are among the most resistant. Young animals are always more sensitive than adults, and whereas a calf of one to six months is very sensitive, cows of eight to ten years are practically insensitive.

Farm mammals

In order of decreasing susceptibility livestock may be classified as follows: pigs > cows > horses > sheep. Whereas the young pig of three to twelve weeks is very sensitive to aflatoxin, this sensitivity decreases in the adult and it is necessary to have considerable quantities of toxin in the food (a ration containing 17.5–20% toxic feed for a period of 22–182 days) in order to observe a noticeable reduction in growth and the appearance of changes in the liver (69, 1025, 1093, 2898).

Calves fed on a ration containing 5–8% of a toxic feed (corresponding to 0.22–0.44 mg aflatoxin B_1 per kilogram) every day for 16 weeks do not show any clinical symptoms, except a slight reduction in weight during the first three months. If the toxic feed is added at the rate of 18% to the daily ration for 16 weeks, the reduction in weight is more evident and, if feeding is prolonged, death may follow (40).

In two-year-old milking cows concentrations corresponding with 0.22, 0.44, or 0.66 mg aflatoxin B_1 per kilogram of the daily feed have little effect after 20 weeks except for slight hepatic lesions (522). When the feed contains 2.4 mg aflatoxin B_1 per kilogram (corresponding to about 20% addition of toxic feed to the diet) and is fed for 13 months to eight to ten-year-old cows, effects are noticed after the seventh month. A significant reduction in lactation results from ingestion of a diet containing 2.7 mg aflatoxin B_1 for a period of 20–53 days (40). The most important effect in the lactating cow is the transfer of toxic substances to the milk (37). It appears that aflatoxin ingested by gestating cows has no effect on the health of the calf for in the few experimental observations available calves have been born in perfect health (40).

Similar observations have been made from experimental trials in pregnant rats from which foetal malformations have not been recorded (828) although it has been possible to detect aflatoxin in the placenta (376).

Sheep are remarkably resistant. The lamb is practically immune from aflatoxicosis from the age of three months; at this age absorption of a daily ration containing 20% toxic feed has no effect, even if continued for three years. Biopsy of the liver does not reveal any changes (33). An explanation of this resistance may come from an examination of the faeces and urine of those animals ingesting aflatoxin. When fed on the same ration containing 2.4 mg aflatoxin B_1 per kilogram for three weeks, the faeces and urine of the sheep are toxic to ducklings, whereas those of calves are not. It is possible then that the considerable resistance of the sheep is explained by their ability to excrete aflatoxins very efficiently (2238).

Poultry

Poultry are classified in decreasing order of susceptibility duckling > turkey, goose, and pheasant > chicken (1785). The chicken is relatively resistant but, even though mortality is low, hepatic changes nevertheless do occur (93, 870). There is no significant loss of weight in ten-week-old chickens fed on diets containing up to 15% toxic feed. Turkeys and pheasants are somewhat more sensitive and ducklings fed on a diet containing 2.5% toxic feed (corresponding to 0.3 mg aflatoxin B_1 per kilogram) each day for six weeks have a mortality of 30% and the survivors show a considerable loss in weight compared with controls.

When 500 μg of aflatoxin is given to one-day-old ducklings as five daily doses of 100 μg each, hepatic lesions are produced; 2,000 μg over five days caused proliferations of bile duct tissue in all cases. A diet containing only 6.25% contaminated groundnut may kill ducklings within 13 days if the feed is started within the first six days of their life (420). Prolonged exposure to a diet containing 0.5% of badly contaminated feed gave rise to malignant liver tumours in ducks.

There is also a variation in susceptibility from one species to another at the embryonic stage. The embryo of the chicken is 200 times more sensitive than that of the duck. Addition of vitamins A and E give some protection to the embryo (388).

Other animals

A variety of animals are susceptible to aflatoxin poisoning. Both mink and a wart-hog fed on a diet containing contaminated groundnut meal showed symptoms of aflatoxicosis (1387, 2066). The dog is also sensitive to aflatoxins (117, 705, 1835).

A wide range of hepatic lesions (hyperplasia of the epithelium of the bile duct, cholangitis, hepatoma, cholangioma) have been observed in trout fed on toxic groundnut flour, cottonseed, and soya (2443).

It should be noted that some forms of stress, such as a protein deficiency (1563), or a previous history of cirrhosis (1842), may increase the sensitivity of the liver to the effects of the aflatoxins.

General symptoms of toxicity

Conditions of study

The symptoms of aflatoxin poisoning have been studied, either following natural intoxication, or following feeding trials with diets experimentally contaminated with *Aspergillus flavus* or, more recently, with diets to which pure aflatoxins have been added. The first toxicity tests were carried out on animals long before the toxic principle had been isolated and characterized.

In 1954, Paget described an illness, exudative hepatitis of guinea-pigs, which was not infectious and was characterized by considerable oedema of the subcutaneous tissue and unusual changes in the liver. This disease always appeared when the animals were fed on a diet called 'pelleted 18' (329) which contained 15% groundnut meal. In 1956 this disease was again recorded (2475) and it was confirmed that it was associated with the diet. It was only after 1962 that it was appreciated that toxicity was in fact due to the groundnut meal (1928).

In a similar manner Mlle le Breton and Boy (1435) had described the appearance of hepatomas in laboratory raised rats receiving a particular diet in 1960, but it was only in the following year (1436) that the relationship with the presence of aflatoxins was established.

Most of the first rational tests for toxicity were carried out using ducklings but, although the duckling is still widely used (1565), the guinea-pig or rat, which are widely available, are often preferred (2588).

More recently the monkey has been used as an experimental animal in order to compare the symptoms with those possibly associated with man.

Acute and chronic toxicity

The aflatoxins give rise to a number of symptoms associated with the acute toxicity and changes associated with chronic intoxication. The latter include three types of activity: carcinogenic, teratogenic, and mutagenic (1448).

The acute toxicity is expressed as the death of the animal in a time dependent on the particular sensitivity. In general the liver appears pale, decolorized, and increases in size. Necroses of the hepatic parenchyma and haemorrhages are often recorded in young pigs, ducklings, and turkeys. If death has not occurred within a week there is a characteristic proliferation of undifferentiated cells in the portal spaces (830). The kidney may show lesions such as glomerular nephritis whereas in the lung there may be congestion.

The changes associated with chronic poisoning are of a quite different order. The first visible symptoms appear as a loss of appetite and a reduction of growth, but it is the liver which is affected most rapidly by the effects of the toxin (1988). It is congested and shows haemorrhagic and necrotic zones (771). Proliferation of the epithelial cells of the bile ducts is characteristic in cattle, ducklings, and turkeys; in chickens destruction of the cells of the hepatic parenchyma and proliferation of the epithelial cells occur during the first two to three weeks followed by an infiltration of lymph cells (93). Such proliferation may alleviate hepatic inadequacy (7) but it may equally result in obstruction of the periportal bile ducts associated with peripheral necrosis (it is known that in such cases of obstruction or cirrhosis the proliferation of the bile ducts is not uncommon) (366). When exposure is prolonged, cancer of the liver may be observed.

Kidneys are congested (1566) and occasionally haemorrhagic enteritis may be observed.

The most characteristic indications of toxicity often appear only a few days before death. The animal is melancholy, staggers, some show nervous symptoms, muscular spasms, incoordination of movement. At the moment of death, the animals are in extension (771).

Clinical and histopathological examination

Pigs

The characteristic symptoms of aflatoxicosis in the pig are ataxia and severe tenesmus with prolapse of the muscle of the rectum; jaundice is frequently observed, although it has not been possible to produce this experimentally (1023); the liver shows a white and yellowish marbling. Anorexia and loss of weight accompany these symptoms when the dose of aflatoxin is greater than 0.1 mg/kg (2536).

In an incident occurring in Madagascar (2108) the following were recorded:

– cirrhosis in patches with zones of necrosis, haemorrhages or dilation of capillaries and yellow nodules along the tracts of the bile ducts;

– hypertrophic cirrhosis; the kidney being enormous, marbled or totally necrosed.

Ulcers of the upper opening of the stomach may be associated with the consumption of food contaminated with *Aspergillus flavus* (1944).

In experimental toxicosis changes of the liver can be seen progressing from fatty degeneration, proliferation of the bile ducts and fibrous tissue leading to karyomegaly, and finally nodular hyperplasia (64, 202, 575, 861, 1023, 1093, 1568, 2109, 2734).

Cows

In the cow, hepatic lesions consist mainly of cirrhosis with ascites accompanying visceral oedema. In experimental intoxication of the cow (879, 1090, 1530) slight hyperplasia of the hepatic cells from the end of the first month is noted; this phenomenon increases during the second and third months until centrolobular degeneration of the hepatic cells becomes apparent. During the fourth month central necrosis of the hepatic cells marked by a proliferation of the bile ducts and occlusion of the centrolobular veins is recorded. Such occlusion is similar to that observed in cattle poisoned by the pyrrolizidine alkaloids of *Senecio jacobaea* (1089). It is worth noting that only cattle respond to aflatoxin poisoning by vascular occlusion and this explains the confusion which has existed between aflatoxicosis and Senecio poisoning. There are no noticeable changes in the clinical pattern in milking cows (940). Adult ruminants are among the most resistant animals to aflatoxin and the milk production is only affected by high levels of contamination, of the order of 2.5 mg/kg of food. For the state of health of the animals to be noticeably disturbed, and for cases of death to occur, it is necessary for the diet to contain exceptionally high levels of aflatoxin B_1, in the region of 60 mg/kg (767).

Poultry

In poultry anorexia accompanied by a reduction in growth are frequently recorded with a tendency for pulling out down and feathers (2424). Subcutaneous haemorrhages in the foot give rise to a red coloration (771). Contrary to the situation in mammals, the liver of poultry suffering aflatoxicosis does not show fibrosis. Degeneration of the cells of the hepatic parenchyma is particularly noticeable accompanied by active regeneration by proliferation of the epithelium of the bile ducts from the periportal regions. After three weeks on a diet containing 10% toxic feed, only small islands of normal parenchymatous tissue remain surrounded by dense masses of epithelial cells of the bile ducts as well as fibrous tissue. The kidneys become congested; they show multiple haemorrhages; a membranous glomerular nephritis occurs in the young turkey but not in the duckling. To these hepatic and renal lesions may be added duodenal lesions in the form of catarrhal enteritis (93, 2386).

In the duckling, a single dose of aflatoxin is sufficient to obtain rapid proliferation which reaches its maximum after three days. Associated with this is a degeneration of peripheral parenchymatous cells. This lesion rapidly regresses after ten days, some accumulation of basophilic cells is observed and, in some regions, it is very difficult to distinguish regenerating basophils from parenchymatous cells. Such lesions may also be produced after a single dose of cycasine.

In the duckling, after 30 days of treatment with aflatoxin, the liver is enlarged and shows a brown or greenish colour; the surface has irregular granulations sometimes forming into large nodules. As early as the end of the first week of treatment, the parenchymatous cells are inflated and the cytoplasm is granular and eosinophilic; some cells are already dead; the nuclei are inflated and hyperchromatic; hyperplasia of the bile ducts is already clear at this stage and the 'egg-shaped' cells appear radially dispersed from the periportal zone. After two weeks there is a very clear cytoplasmic vacuolation and a considerable hyperplasia of the bile ducts. After three weeks hepatic nodules, with hyperplastic cells, are separated by bands of bile duct epithelium (2584).

Renal disorders are visible after three weeks; the kidneys are swollen and pale brown. Protein debris is present in abnormal quantities in the dilated tubules and large nuclei with a strange morphology are visible in some of the degenerating epithelial cells (1834).

It may be noted that the enlargement of the hepatic cells (megalocytosis) and swelling of the nuclei (karyomegaly) are seen in cattle, ducks, and turkeys, whereas in chickens, swelling of the nucleus without megalocytosis may be seen.

In the chicken, a minimum dose of 0.8 mg aflatoxin per kilogram is required to produce primary hepatic disorders (1344).

Rabbits

The rabbit, after ingesting aflatoxin, shows a particularly extensive fatty degeneration of the hepatic parenchyma; the cells are swollen, very vacuolated, and show nuclear changes. Sometimes there appears to be some regeneration of cells. Proliferation of cells of the periportal tissue and bile ducts is observed (1246).

The lactation of rabbits is affected by consumption of groundnut meal contaminated with aflatoxin (619).

Guinea-pigs

The guinea-pig responds to acute toxicosis in a manner analogous to the duckling. A diet containing 1.4–1.6 mg aflatoxin per kilogram is lethal within three weeks. Lower doses allow much longer survival but cause severe liver lesions (370): centrolobular necrosis and massive bile duct proliferation. The hamster shows similar changes accompanied by megalocytosis (1066).

Rats

Preliminary experience with the rat indicated that it may tolerate a much higher dose of toxic food than the duckling. The rat is less sensitive to the acute effects of the aflatoxins than many other animals; the heart, spleen, intestinal tract, and urogenital tracts appear to be unaffected. The kidneys are apparently normal, nevertheless changes in the glomerules and necrosis of the tubule epithelium of the inner layer of the renal cortex have been noted (375, 745). The lungs show macroscopic grey lesions with areas of haemorrhage. Many of the rats ingesting aflatoxin become obese, in fact, their livers grow abnormally becoming nearly double their normal volume; irregular, brown-yellow nodules appear on the surface as well as numerous yellowish lesions (1414), and a diffuse necrosis in the periportal region is noted (369, 375).

As oral administration of aflatoxin produces hepatoma in the rat, so subcutaneous injection provokes local sarcomas (645). Aflatoxins upset the gestation of the rat; the lethal dose is very much reduced; the psychological behaviour of the female rat is modified (she does not make a nest, etc.); in the foetus, acute toxicity is expressed by the death of all or part of the litter, but foetal malformations are never observed (828) which confirms observations made in cattle (40).

When a single LD_{50} dose of aflatoxin B_1 is given to the rat, the histological changes which appear in the liver develop progressively and fairly slowly (366). After 24 hours, there is a destruction of the basophilic properties of the peripheral cytoplasm with some pycnotic cells. After 36–48 hours, all the lobules are well marked by a peripheral region of necrosis (note that the lesions are peripheral in the rat and central in the guinea-pig). After four days in the areas of necrosis there is a clear bile duct proliferation. There is no cirrhosis, but fibrous cholangiomas occur and haemorrhages are frequent (1664, 1834).

Lacassagne *et al.* (1387) have followed the progession of hepatic lesions in the rat fed exclusively on toxic groundnut meal:

at 74 days there is an absence of lipids in the hepatocytes, the cell nuclei are pleomorphic, often very large, and multinucleolar;

– at 89 days, the bile ducts increased in number and size; as well as a proliferation of egg-shaped cells, which may be of biliary origin (969) and are well described by Farber (765), there is also a lymphohistocytic reaction around the ducts;

– at 164 days, a dislocation of the lobules and formation of nodules of regeneration were noted; these consisted of genuine bile tubules and had a tendency to group in small adenomas, some showing the beginnings of cholangiectasis:

– at 254 days, these have become common and voluminous.

It requires less than 7 mg of aflatoxin per kilogram to provoke hepatitis which may be followed by death from necrosis of the parenchyma (if sufficient toxin has been administered), or to the chronic lesions of cirrhosis. A daily dose of 0.15 μg aflatoxin per rat leads to cancer of the liver which may arise after a more or less lengthy period of time (825, 826).

With 0.94 mg aflatoxin B_1 per day for 23 days, giant hepatocytes of irregular shape and with large nuclei are found in the region of the central vein (1664). Moreover, in chronic poisoning, testicular lesions accompanying anastomoses and fusion of spermatozoids may be observed in the rat within the first month followed by a severe necrosis of the germinal epithelium by the end of six months (1221, 1714). Numerous other observations have been made using the rat as a laboratory animal (30, 1448, 1839, 2529, 2610).

Mice

Allcroft and Carnaghan (36) did not report an acute toxicity of aflatoxin for the mouse, neither did Platinow (1996) have any success in observing toxicity after feeding mice of 20–25 g for three months with groundnut meal containing 4.5 mg aflatoxin B_1 or G_1 per kilogram. Weight loss and slight hyperaemia of the renal glomerules have sometimes been reported (2717). Newberne (1834), however, has in some cases obtained hepatomas.

Minks

When an epidemic occurred in mink, Lacassagne *et al.* (1387) noted initial symptoms of prostration, loss of appetite, and faecel material showed a change in the characteristic colour (from brown to green or dark grey, or even appearing like white putty with traces of blood being frequently present). The animals drink more than usual. Their eyes appear oval rather than circular. Death follows between the third and sixth day, rarely after 15 days.

Autopsy of mink poisoned by groundnut meal contaminated by *Aspergillus flavus* revealed, in addition to congestion of the lungs and the presence of black blood in the intestine and golden yellow fat around the kidneys, a clear yellow liver with small disseminated granules. Microscopic examination of the liver lesions (1387) showed a lot of congestion and, above all, considerable proliferation of dilated bile ducts in the portal spaces. Nodules of incomplete regeneration surrounded by thin layers of connective tissue to newly formed bile ducts were observed, sometimes giving rise to a proliferation of oval cells. Furthermore, numerous centres of hyperplastic reactions have been reported with masses of endothelial cells in the sinuses of the liver.

Dogs

Mouldy grain, causing a toxicosis in pigs, known as 'mouldy corn toxicosis', has been fed in experimental doses to dogs. The symptoms recorded were in many ways comparable to those observed in the pig, and corresponded to a disease of dogs which, at the time, was referred to as 'hepatitis X' (117).

When young dogs were given a mixture of aflatoxins (B_1 37.5%; B_2 5.4%; G_1 17.1%; G_2 1%) by oral administration in daily doses of 1, 5, or 20 μg/kg for five days a week over a period of 10 weeks, it was noted that, whereas those receiving 1 or 5 μg/kg showed no symptoms of aflatoxicosis, those receiving 20 μg/kg showed numerous symptoms. These included icterus, loss of appetite, yellow-orange urine, increased concentration of prothrombin, moderate proliferation of bile ducts, accumulation of bile pigments in the portal regions, multiplication of vascular vessels around the central and portal veins (436).

Monkeys

Acute toxicity tests have recently been made on primates (2657). When a preparation of 1 mg of aflatoxin per kilogram was administered daily to monkeys aged two to three months, there was a high mortality after four weeks. Post-mortem examination revealed severe hepatic changes characterized by fatty infiltrations in the region of the portal vein. Furthermore, changes in the concentrations of transaminases and of bilirubin were recorded.

In young monkeys receiving daily doses of 0.15–0.25 mg of aflatoxins per kilogram it was recorded that hepatic necroses (resembling those of viral origin) occurred with some regeneration of the periportal liver parenchyma. In a variety of cases no authentic adenomas were recorded, but nodules of cells with an irregular disposition and very large nuclei (2154, 2621), or giant multinucleate cells (2530), have been noted.

In *Papio cyanocephalus* testicular atrophy with undeniable fibrosis has also been recorded (1713). Hepatitis with similar characteristics has been induced in *Cercopithecus aethiops* by daily doses of 0.01–1.0 mg/kg (48). *Macaca irus* showed no visible symptoms following daily ingestion of 0.3 mg aflatoxin B_1 per kilogram over a period of three years; however many animals died after absorbing 1.8 mg/kg daily for about a month of two (574).

It has been found that experiments using *Papio cyanocephalus* are frequently upset by infections of these animals by eelworms.

Trout

The problem of hepatoma in trout farms merits some detailed attention.

Historical

It was in California, in the spring of 1960, that large numbers of liver tumours in rainbow trout were observed (2864). On histological examination they were shown to involve primary carcinoma of the liver cells (2867). Similar but isolated cases had been noted previously throughout America (1858, 1859, 2201), in Europe (569, 992, 1464, 1465, 2309), and in the Far East (1119).

Distribution

As emphasized during a symposium dedicated to diseases of fish (2896), this is a widespread phenomenon and not just a chance peculiarity. It would seen that wild trout do not show the same hepatomas which have only been recorded in farmed trout. Trout used in the fish farms of two continents are almost all descended from American stock taken from one river in the State of Washington and exported to Europe in 1925.

The rainbow trout (*Salmo irideus* = *S. gairdnerii*) is the most frequently used species in fish farms and it was long considered that it alone was susceptible. Lately, heptatic tumours have also been described in the American brook trout (*Salvelinus fontinalis*) grown in Italy (897–899) and in France (1464). Furthermore a liver adenoma has, in fact, been reported in a char (*Umbla salvelinus*) taken from Lake Leman. The problem is thus associated with all salmonids raised in fish farms (533, 2885).

Cancerous nature of the disease

Robertson has reviewed the surgical techniques available for the study of fish (2173). They were adapted to obtain an improved examination of the evolution of the nodules. It was then a simple process to effect explants, implants, and transplants, and small fragments of hepatoma tissue have been inserted carefully in various parts of the animal making use of rapid surgical methods. In general, these transplants developed well in the liver, the stomach musculature, and in the coecum, less well in the ovaries, testicles, heart, and fatty subcutaneous tissue. Transfer from one individual to another is easy when they are of the same genetic stock, otherwise only 10% of transplants take (1003). It has been confirmed histologically that the carcinomas are derived from liver cells. Metastases have been observed and investigated; they may join the myocardium at the tip or the base of the ventricle (1464) and others may reach the kidney (1137).

Aetiology

Based on other precedents and the fact that no geographical link exists between the diverse localities where the disease has been recorded, it was deduced, using the ideas of Mlle le Breton (1436), that nutritional factors played an important role in the genesis of hepatomas and their spontaneous appearance. However, some investigators have suggested that genetic predisposition and associated phenomena may be a factor. Sniesko (2431), for example, pointed out that other species of trout, placed with the rainbow trout and submitted to the same diet, were not affected and Nielson (1857) has observed large variations in the rate of hepatoma formation depending on the origin of the trout. It would seem that in a farm where 75–100% of the oldest fish are affected, only about 25% of younger fish are ill. An hepatic adenoma, suspected of malignancy, occurring in trout of two to three years usually precedes the hepatomas found in the older generation of three or more years (1464). It is also possible that the quality and temperature of the water may play a significant part (1348).

The possibility of a viral or bacterial origin was at one time considered but the non-transmission of the disease from ill trout to healthy trout placed in contact with them led to the abandonment on this idea (569).

The food of trout

Hepatomas are not found in wild trout so it is worth comparing them with those raised on fish farms and the major difference lies in their diet. If a comparative analysis is made of fish farm and wild trout the differences shown in Table 15 are observed (2869).

Table 15. Comparative global analysis of wild and fish farm trout (2869)

	Fish farm trout		*Wild trout*	
	content (%)	mean (%)	content (%)	mean (%)
Protein	56–80	68.1	64–81	75.8
Lipids	12–38	22.5	7–25	13.4
Ash	7–14	9.6	8–16	12.3
Carbohydrates	0–8	0.6	0–13	0.4

Table 16. Global analysis of the food of fish farm trout and wild trout (2869)

	Fish farm (%)	*Wild trout* (%)
Proteins	62.4	64.6
Lipids	23.0	12.4
Ashes	10.0	9.6
Carbohydrates	4.8	13.5

The normal food of fish farm trout is always very rich in lipids and poor in carbohydrates compared with estimates made of the composition of the food of wild trout (Table 16).

On the other hand it was apparent that, in the United States, the appearance of hepatomas in trout was correlated with the use of dry pelleted feed in place of moist foods. These new feeds contain fishmeal, and groundnut meal as a source of protein, with a cottonseed base. Such feeds are frequently stored and sometimes undergo self-heating, a phenomenon suspected of being associated with the production of hepatomas (682, 1003). It has been considered that carcinogenic degradation products may be formed, especially by peroxidation of fats (1420). Among the oxidized fats, those of waxy constituency may appear in a sporadic manner in the parenchymatous cells of the liver of fish depending on their age and nourishment. They especially appear after continuous ingestion of partially oxidized unsaturated fatty acids in the absence of suitable intracellular antioxidants such as α-tocopherol.

Table 17. Frequency of hepatomas in trout fed for 20 months on a diet containing added carcinogen (1003, 1004, 1005, 1008, 1009, 1420)

Added carcinogen	*Dose in feed* (mg/100 g)	*Frequency of hepatomas* (%)
Control	–	0
Dimethylnitrosamine	480	82
Aminoazotoluene	120	52
Aminoazotoluene	30	6
Dichlorodiphenyltrichloroethane	8	6
2-Acetyl aminofluorene	30	20
Thiourea	480	18
p-Dimethylaminoazobenzene	30	13
Tannic acid	120	13
Urethane	480	11
Carbon tetrachloride	120	10
Carbarsone	480	10

Experimental carcinogenesis in the trout

Several recognized carcinogenic agents, which may occur accidentally in the food of fish farm trout, have been given to trout and the frequency of hepatomas recorded after 20 months on each diet (Table 17).

In most cases initiation of tumours is already visible after four to nine months of such feeds (1008). It has been noted that many of these substances are lipid soluble, and it is thus understandable why the problem of lipid metabolism in trout has received attention.

Inversely, it has been reported that a feed containing a high protein concentration, a low concentration of carbohydrates, few fats, and rich in riboflavin, α-tocopherol, choline, nicotinic acid, and methionine, when compared with a standard diet, causes a regression in the frequency of tumours (1004).

Experimental aflatoxicosis in the trout

Since 1963, it has become apparent that, among the components of the feed of fish farm trout, cottonseed meal may be particularly responsible for hepatomas (1184, 1348, 2403). Because it may be infected with *Aspergillus flavus,* the possibility that aflatoxins could be incriminated in these disorders was considered. Studies continue to confirm this hypothesis and there is now experimental evidence that aflatoxins induce hepatomas similar to those observed in nature (84, 1005, 2404, 2406, 2407).

The LD_{50} for both aflatoxins B_1 and G_1 is between 0.5–1.0 mg of crystalline material per kilogram fresh weight of rainbow trout. A dose of 0.2 mg (per kilogram fresh weight) of crystalline aflatoxin produces significant disorders in the liver of trout.

To determine the chronic toxicity, crystalline aflatoxin was absorbed onto cellulose powder and incorporated at various concentrations into the feed. Histological examination was conducted after either six months or a year of such treatment. A variety of symptoms have been described depending on the extent of toxic activity:

- haemorrhages were particularly frequent;
- spots in more or less large numbers were visible on the liver;
- the liver was swollen and discrete nodules appeared;
- tumours of various dimensions appeared.

Proliferation of bile ducts was noted, as well as typical hyperplasias (1005). All these symptoms recall those observed in other animals following the action of aflatoxins.

Conclusions

Allcroft (34) has condensed in tabular form a comparison of the principal symptoms of aflatoxicosis in a variety of domestic animals (Table 18).

Table 18. The principal symptoms of aflatoxicosis among a variety of animals (34). (+ = presence of a symptom, – = absence)

	Cow	*Pig*	*Sheep*	*Duck-ling*	*Turkey*	*Chicken*
Lesions of liver:						
acute necrosis and haemorrhage	–	–	–	+	+	–
chronic fibrosis	+	+	–	–	–	–
regenerative nodules	+	+	–	–	+	–
bile duct proliferation	+	+	–	+	+	+
vascular occlusions	+	–	–	–	–	–
Hepatic cells:						
megalocytosis	+	+	–	+	+	–
giant nuclei	+	+	–	+	+	+
infiltration of mast cells	–	–	–	–	+	+

Cellular alterations

Electron microscopic examination of liver cells from rats exposed to the action of aflatoxin has been carried out by Frayssinet and Lafarge (826), and they have made the following observations:

– The cell nucleus is first affected; a nucleolar segregation, said to be a condensation and redistribution of the constituents, is noted in the nucleolus; the chromatin is ejected to the outside, fibrils and particles of ribonucleoproteins coming together in compact and distinct zones, while interchromatin particles form voluminous aggregates (1715).

– Lesions are much later in the cytoplasm but also longer lasting; they consist of dilation of ergastoplasmic vesicles, swelling of the mitochondria, the internal cristae of which become indistinct, and in particular a degradation of the polysomes to form monomers (124, 2017) with a significant reduction in the number of ribosomes.

Similar observations have been made on the liver cells of trout with aflatoxicosis (2254, 2255) and the embryo cells of chickens (2567).

The formation of nucleolar protuberances in the epithelial cells of the foot of the freshwater molusc, *Planorbarius corneus,* after receiving a toxic dose of aflatoxin B_1 has also been noted (166).

The action of aflatoxin B_1 on *Bacillus thuringiensis* and *B. megaterium* has been interpreted as an intense fragmentation of the nuclear apparatus, followed by cell elongation and random division (277).

Profound changes in cell structure have also been reported in plants after contact with aflatoxins (1203); attack of the membranes and intracellular binding, disappearance of ribosomes, multiplication of dictyosomes and golgi apparatus, the folding of the endoplasmic reticulum with the formation of cytolysosomes, the disappearance of lamellae and internal grana, followed by deformation, of the chloroplasts. Such changes in the chloroplasts have previously been described in the leaves of maize (2415), as also have chromosomal aberrations in the root cells of *Vicia faba* (1499) similar to those found in animals (680).

Carcinogenic properties of the aflatoxins

The action of the aflatoxins forms two groups of phenomena:

1. a series of rapid phenomena associated with toxicity;
2. the slow phenomenon of carcinogenesis.

The suppression of the first does not necessarily obviate the occurrence of the second (824).

Historical

As early as 1933, Haddow and Blake (992) had described the appearance of liver nodules in fish but, of course, in this period there was no knowledge of the aflatoxins.

In 1944, Ninard and Hintermann (1862), in Morocco, drew attention to the frequency of liver lesions in pigs examined in the abattoir. Hypertrophy with multiple nodules, accompanied by more or less marked cirrhosis, were especially noted, and histological studies of the nodules indicated that they were either adenomas or carcinomas.

In 1961, Guillon and Renault (985) also reported several cases of liver tumours in ducks of nine months or more, from several regions of France. Sometimes they were associated with adenomas, sometimes with malignant tumours. An hypothesis of viral origin had been suggested.

Similarly, in 1960, Mlle le Breton and Boy (1435) recognized the existence of hepatomas in rats fed on a particular diet. In the following year, in collaboration with Frayssinet (1436), they defined the role of a toxin produced by *Aspergillus flavus*: carcinogenesis of the liver was obtained in rats only six months from weaning, having ingested a contaminated feed (containing a quantity of aflatoxins of the order of one part in ten million).

Lancaster *et al.* (1414) described the carcinogenic properties of groundnut meal contaminated with aflatoxins in 1961.

Hepatomas in the rat

After feeding rats material containing aflatoxin, it was observed that multicentric hepatocellular carcinomas appear in most cases. The tumorous nodules are greyish yellow or haemorrhagic and necrotic (1834). Earlier observations, such as those reviewed by Wogan (2853), on the conditions giving rise to these hepatomas are summarized in Table 19.

Table 19. Cases of hepatoma in the rat fed on material containing toxic groundnut meal

Author	*% Ground-nut meal in feed*	*Conc. of aflatoxins* (mg/kg)	*Time of feeding* (weeks)	*Presence of tumours*
Lancaster *et al.* (1961)	20	?	24	9/11
Schoental (1961)	15	?	52	2/5
le Breton *et al.* (1962)	16	?	72	26/48
Salmon and Newberne (1963)	33.3	0.1–3.5	48–73	64/73
Butler and Barnes (1963)	40	2.8–4.0	35–38	5/6
		2.8–4.0	16	4/6
		1.4–1.6	12	5/7*
		1.4–1.6	26	9/10**
		0.35–0.40	81	2/7

*38–73 weeks after stopping toxic feed
** 40–58 weeks after stopping toxic feed

If a fairly large quantity of groundnut meal infected by *Aspergillus flavus* is added to the feed of rats, the incidence of hepatic tumours is proportional to the concentration of aflatoxin in the food (1834, 1835) cf. Table 20 and Figure 54.

It has been shown that the composition of the diet may influence the relationship between the presence of aflatoxin and the appearance of cancerous lesions (1564). Similar results are obtained if pure aflatoxins are added to the feed instead of groundnut meal infected with *A. flavus* (141). When rats, which had received 1.75 mg pure aflatoxin B_1 per kilogram in their food over a period of 89 days, were examined 316 days later, it was noted that one in three had tumours. After 485 days this proportion had reached two in three and those animals not having tumours had characteristic precancerous lesions.

The younger the animal, the more sensitive it is to the carcinogenic activity. Induction in weanlings requires 0.2–0.5 mg of aflatoxin and the effect is

Table 20. The relation between the concentration of aflatoxin, determined by fluorescence (1826), and the incidence of liver tumours in the rat (1835)

Concentration of aflatoxin in food (mg/kg)	*Duration of experiment* (days)	*Incidence of liver tumours*
5.0	370	14/15
3.5	340	11/15
3.5	335	7/10
1.0	294	5/9
1.0	323	8/15
0.2	360	2/10
0.2	361	1/10
0.005	384	0/10

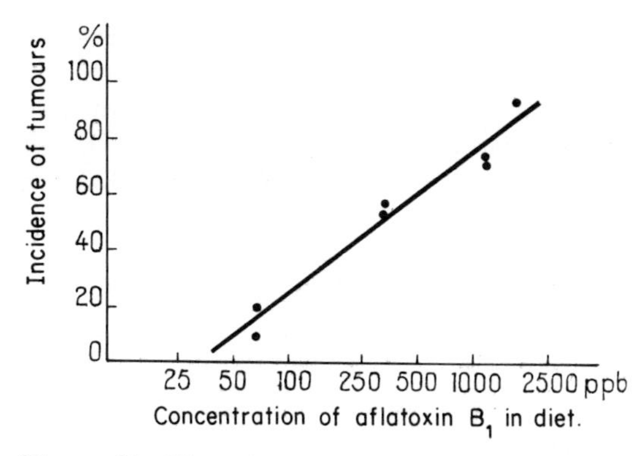

Figure 54. The relation between the concentration of aflatoxin B_1 in food given to rats and the incidence of liver tumours (1834)

irreversible. It is only necessary for there to be a brief period of induction (ingestion over two to three weeks) but the latent period before the appearance of tumorous nodules is long (12–18 months) (372, 374, 830, 1837, 2155–2157, 2899).

It has been reported that ingestion of 5 μg per day of aflatoxins does not give rise to any precocious lesions, even after a year, and animals seem to be in excellent general health, and yet 50–60% of rats subjected to this treatment eventually show tumours. In contrast ingestion of 20 μg per day, while also not causing any early lesions, does give rise to a state of ill health after a year with 60–70% of individuals having tumours. Thus an increase in the dose of aflatoxins causes a change in the general state of health without causing the appearance of very many more tumours (824).

Moreover, in one case, cancer of the glandular stomach of the rat was observed (371) leading one to think that aflatoxins may be the cause of a variety of carcinomas in organs other than the liver, for example the lungs (2819).

It is incontestable that the diet plays an important part in the induction of carcinomas, thus a diet deficient in lipids favours the development of carcinomas (2184–2186).

Recently, hepatic nodules have been obtained in culture *in vitro* by the action of aflatoxin B_1. They have the same appearance as those obtained using another hepatocarcinogen, 2-fluorenylacetamide (2414).

Localization of the carcinogenic fraction

If a rat is fed on a low protein diet containing groundnut meal, contaminated with *A. flavus,* from which the chloroform-soluble material has been removed, growth is arrested, steatosis occurs, but there is no periportal proliferation and no hepatomas. If the chloroform extract is further fractionated into a fraction soluble in petroleum ether and a fraction soluble in aqueous methanol and each is added separately to the previously extracted diet the following is observed:

1. Addition of the fraction soluble in petroleum ether causes an arrest of growth, steatosis but no periportal proliferation or hepatomas.
2. Addition of the fraction soluble in aqueous methanol causes an arrest of growth, no steatosis but periportal proliferations and some hepatomas.

Thus the carcinogenic agent of groundnut meal is strictly localized in the latter fraction. The lipotropic agents, which reduce or eliminate steatosis, are also in this fraction.

It is considered that the dihydrodifuran nucleus alone does not confer carcinogenicity on the aflatoxin molecule (2774). It would appear that the carcinogenicity of aflatoxin B_1 depends on the presence of both the dihydrodifuran system and the unsaturated δ-lactone moiety (110). It has been suggested (1994) that aflatoxin B_1 is only a precarcinogen and that it needs to be converted, probably by microsomal enzymes, to a carcinogenically active compound. It has been further suggested that it is necessary for epoxide formation at the double bond terminating the difuran ring system in aflatoxins B_1, G_1, and M_1 which is lacking in B_2, G_2, and M_2, thus explaining the lower activity of the latter (2278).

The cancerous nature of hepatomas associated with aflatoxins

The question has been raised as to whether the hepatomas caused by aflatoxin are of a malignant nature.

General characters

By definition (630), the property of a cancer is the possession of undifferentiated, invasive, and metastasic characters.

An abnormal proliferation of cells may either give rise to a benign tumour, formed of identical non-invasive cells, or a malignant tumour of undifferentiated cells, some growing more rapidly than others, separating from each other and infiltrating into neighbouring tissue. The principle focus breaks up moreover and, by way of the blood and lymph vessels, moves to invade the ganglions and to create new centres, called metastases, in other organs such as the liver, bone, etc.

In the case of a liver carcinoma the diagnosis is easy if the tumour has metastases or can be transplanted (366). In the absence of these two criteria it is possible to use a combination of macroscopic and microscopic criteria. In the first place, for sectioning it has been proposed (2027) that a minimum diameter of 1 cm is required for a nodule, it being difficult to confirm a carcinoma when the liver nodules are all very small. Then one should look for haemorrhage and necrosis. Histologically, localized invasion, pleomorphism of the cells, loss of polarity, and numerous mitotic figures should be observed. This type of lesion has been called a hepatoma, although there is some confusion for, whereas some authors reserve this term for malignant liver tumours, others use it equally for benign hepatic lesions or hyperplastic nodules.

Changes in the parenchymatous cells are also noted in liver carcinomas. There are very large cells (which have been called megalocytes) with hypertrophied nuclei, although the term megalocyte may lead to confusion as it has already been used for abnormal erythrocytes. What is the significance of these cells? Two hypotheses have been advanced: on the one hand they may be precursors of malignant cells; on the other hand they may lead to a type of degenerate cell. If they are precursors of malignant cells, they should have, like the precursors of meristomatic cells, the potential for diverse activity but only rarely is mitosis seen in these large cells and, when observed, it is usually abnormal. In those experiments in which toxic feed is discontinued, it has been observed that the number of these cells decreases and the recent publications of Lippincott *et al.* (1508), on the behaviour of the giant cells of tumours, indicate that they may not be viable. Butler (366) has suggested that the large parenchymatous cells are similar, in this respect, to giant tumour cells.

Comparison with the action of other hepatocarcinogens

Lesions caused by the aflatoxins are similar, in many respects, at doses of 3–4 mg/kg, to those caused by other known hepatocarcinogens. In addition to the proliferation of large parenchymatous cells, fibrosis observed as a result of aflatoxin activity is similar to that seen in other cancers.

The activity of 4-dimethylaminoazobenzene, described by Glinos, Butcher, and Aub (919), gives rise to, as the first stage of malignancy, proliferation of the biliary system, nodules of regeneration of the biliary system, and nodules of regeneration of some large parenchymatous cells. Compare these changes with those observed following exposure to other hepatotoxic agents (873, 1484, 2573). For example, carbon tetrachloride produces rapid histological changes and necroses are visible after 18 hours. They are centrolobular, and not peripheral, and they do not disturb the overall structure of the liver (391). Dimethylnitrosamine (DMN) also produces centrolobular lesions characterized by haemorrhagic necrosis occurring a few hours after exposure to the toxic agent (142, 1573). It has been demonstrated (1574) that DMN acts as an alkylating agent, is very carcinogenic as a chronic toxin, but, from a single dose, it gives rise to renal tumours while the liver appears to be histologically normal. Extracts of the nuts of *Cycas circinalis* containing cycasine are also hepatocarcinogenic following chronic intoxication (1419). Acute poisoning causes centrolobular lesions very similar to those of DMN as well as renal tumours. The alkaloids of *Senecio,* such as lasiocarpin and retrorsin, produce an acute poisoning in the rat with centrolobular necrosis (590) but at slightly different doses the liver is not able to recover (2280).

Thus, all these hepatotoxic agents produce centrolobular changes which may, or may not be reversible. However, remarkably few compounds produce the periportal necrosis observed with the aflatoxins. Allyl alcohol, manganese, phosphorus do so, but it is not proven that any of these are carcinogens.

There is an increase in the capillary permeability in liver lesions produced by sporidesmin in sheep (2413); aflatoxin has a similar effect (at least in the guinea-pig) but sporidesmin has not so far been considered as carcinogenic.

Following a single dose of CCl_4, DMN, allyl alcohol, and most other hepatotoxic agents, the liver may recover its structure. This is not so in the case of either the aflatoxins or the *Senecio* alkaloids (2281). This irreversibility and the slow, progressive, irremediable growth of these lesions is considered by Butler (366) to be one of the most fundamental characters of these intoxications. A month after the application of aflatoxin there is a biliary proliferation which spreads between and into the lobules.

The presence of urethane (1836) and especially of diethylstilboestrol (1841) appears to reduce the incidence of tumours. Hypophysectomy, in the rat, also inhibits the formation of liver tumours. It has been considered that, as recognized for other carcinogenic agents, aflatoxin may have to be metabolically activated before becoming an effective carcinogen of the liver (938).

Conclusions on the carcinogenicity of the aflatoxins

All the hepatocarcinogens produce similar tumours with variations in the type of cells involved. With aflatoxin the predominant type is the more or less differentiated liver cell which produces metastases with, occasionally, cholangiocarcinomas which may also give metastases (366). These metastases may, in some cases, reach the lungs of affected animals (1834).

It is possible to see in a single tumour all types of cells from well differentiated parenchymatous cells to anaplastic carcinomas and cholangiocarcinomas.

Aflatoxin is one of the most powerful oral hepatocarcinogens known. The absorption of a total of 2.5 mg aflatoxin over a period of 89 days has resulted in cancer in the liver more than a year later (141). If doses capable of inducing cancer are compared under identical conditions it is found that 1,000 times more of the dye butter yellow than aflatoxin is required.

Being such active liver carcinogens, the aflatoxins have proved to be useful in the study of the nature of cancer cells. In fact, the induction phase of carcinogenesis is very brief (when applied to young animals or growing liver tissue) and the quantity necessary for initiation is very small (*c.* 5 μg per day for several months in the rat).

It is possible to obtain hepatomas which develop very slowly in an animal having a normal diet once induction with the toxin has occurred. These hepatomas are at that time not very undifferentiated and, in particular, the parenchymatous cells are analogous to normal hepatocytes thus facilitating their study (822, 828).

Biochemical effects of aflatoxins

Numerous studies have been undertaken in order to elucidate the mechanism of action of the aflatoxins at the biochemical level. They have been particularly concerned with interactions with cell constituents especially nucleic acids and the intermediates of protein metabolism. These studies have, in many cases, benefited from those on actinomycin D, on the action of which there is already a great deal of information (194, 261, 1393, 1394, 2466, 2568).

Frayssinet and Lafarge (826) have emphasized that aflatoxin may be considered as a biosynthetic inhibitor, large doses causing total inhibition and lower doses progressively affecting different systems. Clifford and Rees (526) have pointed out that it is possible to tabulate the successive stages in the biological activity of aflatoxins on the liver cell, each step being a consequence of the previous one.

1. Interaction with DNA and inhibition of the polymerases responsible for DNA and RNA synthesis.
2. Suppression of DNA synthesis.
3. Reduction of RNA synthesis and inhibition of messenger RNA.
4. Alterations of nucleolar morphology.
5. Reduction in protein biosynthesis.

Interaction with DNA

It has been demonstrated that the aflatoxins may bind with DNA causing a change in its characteristic spectrum. Aflatoxin B_1, biologically the most active, binds more strongly than G_1 or G_2 (529). Based on the interaction of aflatoxins with purines and purine derivatives detected spectroscopically, Clifford and Rees (526) have suggested a preferential interaction with the purine ring. A major difference between aflatoxins and most other compounds binding with DNA is that aflatoxin may also interact with single stranded DNA. Binding with DNA is nevertheless so weak that simply passage through a Sephadex column dissociates the complex. These interactions may explain the inhibition by aflatoxin of the synthesis of DNA and RNA at one and the same time (126).

Suppression of DNA synthesis

The inhibition of nucleic acid biosynthesis may be by inactivation of the synthetic enzyme systems or because the DNA molecules present in the injured cell no longer act as suitable models for their duplication (1434, 2115). This failure of DNA duplication due to aflatoxin is analogous with the action of actinomycin (especially as the structures of the two molecules have some features in common). Actinomycin inserts itself into the double helix of DNA at a site containing guanine whereas adenine and thymine remain unchanged (2113).

A precise study of the action of aflatoxin B_1 on DNA metabolism in the liver has been made by Frayssinet *et al.* (827, 2115, 2116) using the stimulation of synthesis obtained by hepatectomy. If a partial hepatectomy is made on a healthy individual, compensatory hypertrophy results, a process which is inhibited by aflatoxin. It is not so much action of mitosis as such, but an inhibition of biosynthetic processes required for cell division, especially of nucleic acids themselves. If toxin (30–60 μg/day) is injected during the five days previous to surgical removal of two-thirds of the liver, the animals being sacrificed 24 hours after surgery, there is a significant inhibition of DNA synthesis. If toxin is injected 24 hours after partial removal of the liver, when the many biosynthetic processes for cell regeneration have already started, all biosynthetic processes are blocked. These results have been confirmed using ^{14}C labelled orotic acid (828, 2197).

Using tritiated thymidine, it is possible to follow the synthesis of DNA in rats hepatectomized either 1, 2, 4, or 12 hours before sacrifice, which is 36 hours after receiving aflatoxin. Considerable inhibition of DNA synthesis is found from 65% in 1 hour, 80% in 2 hours, 90% in 4 hours and 95% in 12 hours (1776, 2113, 2115).

Autoradiographic studies show that the nucleus, which is profoundly changed, may have begun to synthesize DNA but that it is subsequently inhibited (2114). These results have been confirmed by the work of Rogers and Newberne (2184).

Several authors emphasize the possibility that aflatoxin suppresses the synthesis of DNA by inactivating the enzyme systems involved (322, 867, 2585). Thus, it is stated that aflatoxin inhibits the DNA polymerase of *Escherichia coli* (2877) in a manner analogous to mitomycin C. The polymerase is reduced to 60% in its activity and, significantly, filamentous forms appear in the culture.

Aflatoxin B_1 inhibits the growth of *Flavobacterium aurantiacum*, acting on cell partition, and giving rise to aberrant morphological forms (1494). Aflatoxin G_1 has a similar effect but is less toxic (1495). Cell division is affected in *Bacillus megaterium* (1490), *Bacillus cereus*, and *Bacillus licheniformis* (1491). It is known that cell division, or cytokinesis, is associated with an interaction between the nucleus and the cell membrane as shown by Jacob, Brenner, and Cuzin (1185). These observations provide evidence that aflatoxin certainly affects the biosynthesis of nucleic acids.

Aflatoxins B_1, B_2, and M_2 all promote the activity of pancreatic desoxyribonuclease at concentrations of the order of 80 μM (2256). For their part, Sporn *et al.* (2466) have shown that the absorption spectrum of aflatoxin B_1 is modified by the presence of denatured DNA; the maximum in the region of 362–364 nm is displaced to 366–368 nm and, furthermore, the extinction coefficient is reduced by about 30%. Jemmali (1223) had the original idea of studying the effects of aflatoxin B_1 on a form of DNA surrounded by a protein envelope as it is found in a virulent bacteriophage. He showed that, in the presence of aflatoxin B_1, either directly or absorbed on discs of Whatman filter paper, a virulent bacteriophage, h9, is no longer able to lyse a culture of *Streptococcus lactis*, normally sensitive to this phage.

Reduction in the synthesis of RNA

By interacting in the first place with DNA, aflatoxins may also affect the biosynthesis of RNA by preventing the transcription of DNA by RNA polymerase (528, 892, 1031, 1311). The activity of cytoplasmic RNA is totally inhibited whereas nuclear RNA is less so (823, 824, 828, 1393, 1437, 2116, 2466). The biosynthesis of nucleolar RNA is inhibited at an early stage (791, 1399, 1777, 2759). After 15 minutes contact with aflatoxin biosynthesis is inhibited by 90–95%.

In vivo tests on rat liver cells give more rapid results than *in vitro* tests on human tissue culture cells (e.g. kidney cells T, HeLa S_3, Chang liver) or with mouse 3T3 fibroblasts (1251, 2195, 2253, 2928, 2937). Activity on RNA polymerase is not direct but through an interaction with components of chromatin (2857).

At low doses (0.5–1.0 mg/kg) these phenomena are reversible and the various biosynthetic processes restart in the following order:

after 24 hours total synthesis of nuclear RNA;
then synthesis of nucleolar RNA;
after 48 hours synthesis of DNA (1392).

Changes in nucleolar morphology

Treatment of male Fischer rats of 100 g body weight with one intraperitoneal dose of 1 mg aflatoxin B_1 per kilogram causes an ultrastructural change in the nucleolus of the liver cell after 36 hours which correlates with an inhibition of enzyme activity. However, this nucleolar disorganization is only observable following a dose of at least 0.2 mg aflatoxin B_1 per kilogram. It is not detectable with 0.1 mg/kg despite the observation that this concentration inhibits enzyme activity by 5% (2018). Analogous observations have been made on the nucleolus of the embryonic chicken liver cell in tissue culture (2570).

Morphologically these changes are interpreted as 'nucleolar segregation' described much earlier.

Reduction of protein biosynthesis

The reduction in protein biosynthesis (Table 21) is the final consequence of the inhibitions described earlier (2750). Inhibition of protein biosynthesis appears to be independent of the processes affecting fat metabolism (2120) and is widespread amongst animals (2101).

Pretreatment of rats with phenobarbital (1 mg/ml in their drinking water) confers some protection against the effects of the aflatoxins such as the *in vivo* inhibition of nuclear and cellular RNA in the liver (988). Phenobarbital induces enzymes involved in hydroxylation reactions in the liver leading to the metabolism of aflatoxins to non-carcinogenic products (1557, 1935). Similar results have been obtained with the mouse kidney (27).

Table 21. Effect of the concentration of aflatoxin B_1 on the synthesis of DNA, RNA, and protein by *Flavobacterium aurantiacum* (1488). The values given represent the means of three observations on the increase in the amount of DNA, RNA, and protein in 10 ml of culture medium after 4 hours incubation compared with that present in the inoculum

Aflatoxin B_1 (μg/ml)	*DNA* (μg)	*RNA* (μg)	*Protein* (μg)
0	33	70	190
10	27	60	190
25	21	60	180
50	0	60	170
100	0	50	165

A single lethal dose of aflatoxin B_1 causes a 50% suppression of the incorporation of precursors into total kidney RNA within three hours. The activity of RNA polymerase is inhibited and the concentration of DNA decreases in the nuclei of kidney cells two hours after treatment.

A comparison of the inhibitory effects of the different aflatoxins demonstrates that $B_1 > G_1 > B_2$ (2857).

Other biochemical changes in poisoned animals

Pig – Several enzymic activities increase in the pig after ingestion of aflatoxin: alkaline phosphatase in serum, glutamic-oxaloacetic transaminase, isocitric dehydrogenase (135, 987). The concentration of lipids increases but that of vitamin A decreases (1025) and the latter is sometimes totally absent (37).

Cattle – The most notable changes (40) are an almost complete absence of vitamin A in the liver and an increase in the activity of serum alkaline phosphatase during the first 12 weeks, followed by a reduction in this activity in spite of the continued proliferation of the bile ducts. There is no marked change in the concentration of liver lipids or of the level of serum glutamate-oxaloacetate transaminase.

In the rumen of cattle aflatoxin causes a decrease in cellulolysis, as well as in the total concentration of volatile fatty acids and ammoniogenesis. It also causes an increase in the percentage of propionic acid, significantly reducing the ratio of acetic/propionic acid during the fermentation of hay in rumen juice *in vitro.* It seems that aflatoxin affects the activity of micro-organisms of the rumen (767).

Poultry – The aflatoxins play a role in the degeneration of the liver of the laying hen. The concentration of liver lipids increases from 1.65–2.06 times depending on the diet containing from 2.4–4 mg aflatoxin kg. The increase of fat in the liver is accompanied by a reduction in the number of eggs laid (59).

In the egg, aflatoxin provokes a reduction in the concentration of lipids in the vitellus without a change in protein concentration. In contrast there is a change in both the quantity and the distribution of proteins in the white, the concentration of the albumin decreases whereas that of the globulins increases (59).

In the one-day-old chick all the tissues show a reduction in glycogen, lipids, and proteins following the action of aflatoxins (423, 2092). Moreover, it has been reported that the concentration of selenium increases from the normal level of 0.5–3 mg/kg to 10–30 mg/kg (10).

The influence of aflatoxin on the biochemical activity of the duckling has been examined and perturbations in carbohydrate metabolism were demonstrated by the following experiment (2853):

Animals of 100 g body weight received a dose of 6 μg aflatoxin B_1 each day for five days. Such a dose causes changes in the liver but there is a low mortality. On the seventh day each animal was injected with 2 μg of ^{14}C labelled glucose and sacrificed one hour later. The livers were then removed and analysed showing a net reduction in glycogen (from 4.3 to 1.3%), a change in lipids (from 7.7 to 10.3%), but hardly any change in protein concentration. It is in the protein fraction from which most of the labelled carbon from the glucose is recovered. In parallel with this work, there have been studies in the United States on the variations in concentrations of plasma proteins in ducklings receiving aflatoxins (584, 1824).

As in the calf, histochemical studies reveal an increase in the activity of acid phosphatase and a reduction in the activities of other enzymes (2584), especially adenosine triphosphatase, inosine diphosphatase, and thiamine pyrophosphatase (2585).

Rat – Similar changes in metabolism to those already described in the duckling have been found in the rat (2853). Aflatoxin B_1 induces tryptophan pyrrolase in the liver, as does actinomycin D (526, 2860) as well as thymidine kinase (468). There is an increase in the activities of at least five lysosomal enzymes in the liver (2009) and all the acid hydrolases during the first hours after administration. This is probably related to an increase in the number of lysosomes (2008).

Aflatoxin has a better anticoagulant activity than the 4-hydroxycoumarins (112, 2422, 2423).

Monkey – *Anti-mortem* changes in the monkey, *Macaca irus*, include an increase in the serum concentration of glutamic-pyruvic transaminase, glutamic-oxaloacetic transaminase, alkaline phosphatase, bilirubin, and isocitric dehydrogenase (2102).

Biochemical activity in plants

Several physiological effects of aflatoxins on higher plants have already been described as tests for the detection of these mycotoxins; inhibition of growth, inhibition of chlorophyll biosynthesis, etc.

There are cases where aflatoxin acts synergistically with indolylacetic acid just as do derivatives of coumarin and other unsaturated lactones. For example, the addition of 0.02–200 μg aflatoxin B_1 per litre accelerates the activity of 100 μg IAA per litre on the pea, *Pisum sativum*, (2131), as well as tobacco and tomato (119).

Although inhibiting the replication of mitochondrial DNA, aflatoxin B_1 has very little effect on protein biosynthesis in the tissues of higher plants (2695). It does not affect the activity of peroxidase but acts in a manner similar to mitomycin C, 5-iododeoxyuridine, and triethylene triphosphamide.

Aflatoxin B_1 does inhibit synthesis of α-amylase and the lipase induced by gibberellic acid in germinating seeds of barley and cotton (234).

AFLATOXICOSIS AND MAN

Do aflatoxins cause food poisoning, or even cancer, in man? Are the foods which we eat, and which are contaminated with *Aspergillus flavus*, or the meat of animals, which frequently ingest aflatoxins, dangerous to our health?

Such are the questions which are being asked and which some studies are trying to answer.

Aflatoxins and cancer in man

The epidemiologist asks whether a link has been established between the frequency of hepatoma and the consumption of groundnuts in certain regions of black Africa. Such an hypothesis has been questioned (31, 822, 912, 1315, 1345, 1386, 1666, 1865, 2866). Nevertheless, in the Bantus of the Transvaal and Swaziland and in all of Southern Africa, where a lot of food is contaminated with *Aspergillus flavus,* the number of liver cancers is significant (556, 1071, 1282–1284, 1610, 1875, 1876, 2062). Observations made in Mozambique (1594) and in Uganda (1336, 1517, 2342) are also very worrying and it has been shown, for example, that cases of liver cancer are 58 times more frequent in Mozambique than in the United States (1594). Each year there is one case of liver cancer per 100,000 inhabitants of countries such as Holland, Norway or Canada, 103.8 among the Bantu, 19.2 in South Africa, 9.8 in Nigeria, and 9.7 in Hawaii (2049).

In 1970, a 15-year-old African child died in Uganda. Autopsy revealed pulmonary oedema, a dilated heart, and diffuse centrolobular necrosis, all symptoms frequently associated with aflatoxicosis. Other members of the family had, at the same time, suffered violent abdominal pain and the major part of the family's diet was made up of cassava contaminated with 1.7 mg aflatoxin/kg (2343).

In India, where cirrhosis is frequent among children, it has been considered that *Aspergillus flavus* may play a dominant role, but it does not yet seem to have been proven (2902).

In Thailand and in all South East Asia, is has been suggested that there is a link between the frequency of liver cancer in man and the abundance of *A. flavus* contamination of food (2346, 2349).

In the Philippines, the excretion of aflatoxin M_1 in urine has been measured, all the children having eaten peanut butter contaminated with aflatoxin B_1 (at the rate of about 0.5 mg/kg). Aflatoxin M_1 was detected in the urine of the children but neither M_1 nor B_1 was detected in the milk of wet nurses consuming the same groundnut butter (403). The faeces of seven children who had consumed 11.2–15.0 μg of aflatoxin per day were shown to contain no toxins.

Malnutrition, cirrhosis, and liver cancer coexist in many parts of the world (1838). Primary liver cancer is mainly a disease of young people in tropical and semitropical regions and is relatively rare in Europe and North America. May this not be due to the action of aflatoxins?

Several possibilities have been suggested:

1. Malnutrition undoubtedly plays a part and, in fact, most of the hepatotoxins act as antagonists to pyridoxin and carcinogenicity is highest in those individuals whose diet is deficient in pyridoxin (53, 813). However, straightforward nutritional cirrhosis does not lead to cancer in the rat (1838).

2. It is only a simple evolution of cirrhosis derived from the consumption of alcohol. Now, although most liver cancers do follow cirrhosis, those areas where the incidence of cancer is the highest do not correspond to those areas of highest incidence of cirrhosis.

3. It has been suggested that haemosiderosis, a disease common among the Bantu, may be the origin of these cancers but Higginson (1086) has shown that the siderotic cirrhosis rarely becomes malignant.

4. It has been considered that plants consumed by the people of those countries where liver cancer is high may play an important role. The alkaloids extracted from *Crotalaria,* legumes scattered widely in hot countries, are recognized as carcinogenic as are those of the nut of *Cycas circinalis,* common in Indonesia and Polynesia (1419). The latter is often covered with potentially dangerous moulds (392).

There are unquestioned cases of hepatic fibrosis in man following prolonged consumption of flour contaminated with *Aspergillus flavus* (1939).

Whatever their origin, it seems that this type of liver cancer evolves in two successive stages (1086).

1. A primary lesion first appearing during youth (this may be the result of protein malnutrition but may also be the consequence of ingestion of milk contaminated with carcinogens such as aflatoxins).

2. Much later, after a long latent period, the cancerous process evolves (and it has been suggested that viral hepatitis may be involved).

It has been reported that the witchdoctors of a tribe in British Guinea (1717) know of the practice of adding mouldy groundnuts to the drink of people judged to be undesirable in order to poison them. A missionary father was killed in this manner in 1964 after suffering violent pains in the liver (631). On the other hand these same tribal doctors knew of the haemostatic activity of much lower doses (recently confirmed by Boudreaux and Frampton (269) and that such material may cause abortions probably resulting from action on the smooth muscles of the uterus (268).

Do the experiments recently carried out on the monkey (48, 574, 1813, 2154, 2621, 2657) throw any light on the risks incurred by man? It is known that in the rat there is a large difference in the quantity of aflatoxin required to provoke either hepatitis or cancer. On the other hand the difference is much less in the monkey. Does man behave like the rat or the monkey in his response to aflatoxins? (825).

The presence of aflatoxins in foods of vegetable origin

The problem of aflatoxin poisoning from our food, and the possible consequences, disturbs and even alarms some nutritionists (29, 814, 816, 1024). It is right that these fears be put into proper perspective without denying the existence of the problem. The most grave situation with the aflatoxins is chronic poisoning. It may therefore be useful to consider the possible toxicity of constituents most frequently present in our diet.

Cereals, bread, and derivatives – Some batches of wheat and other seed are contaminated with aflatoxins (2475, 2377) and up to half of the samples of stored grain analysed by Richard and Cysewski (2150, 2153) contained them. This is to be expected when one considers the frequency of *Aspergillus flavus* on wheat (1945) and the fact that, of the strains of *A. flavus* isolated from wheat, the percentage which are toxigenic is high (1407). These considerations lead to the suspicion that consignments of wheat and of flour may be polluted with aflatoxins and as much as 0.2 mg/kg has been reported in some samples (1406).

What happens to aflatoxins during bread making (2903)? Jemmali and Lafont (1227) have reported that oxidation occurs during kneading although the fermentation and hydrothermal process during cooking cause the degradation of most of the aflatoxin B_1. Thus, from a flour containing 0.2 mg aflatoxin/kg, no more than 0.044 mg/kg could be recovered from the bread. Several studies of the possible occurrences of aflatoxin in commercial bread have been made in Germany (249, 814, 816, 1023, 2339, 2458–2460):

– from 1964–66, of 120 loaves examined, 15 (12.5%) yielded *A. flavus* or *A. parasiticus,* but no toxins were found.
– in 1968, aflatoxin was indisputably found in one type of bread but others had none,
– in 1969, from 42 loaves from a supermarket three isolates of *A. flavus* were obtained all able to produce both aflatoxin B and G,
– in 1970, from 91 more or less moulded loaves from 24 different bakeries, aflatoxin was found in 16% of the samples.

If bread is infected with toxigenic strains of *A. flavus,* aflatoxins may be recovered several days later (814), although the mould is not able to grow freely in wrapped bread because of lack of oxygen (2134).

From food pies, on which *A. flavus* often grows (1737), Christensen and Kennedy (483) were not able to obtain any evidence of the presence of aflatoxins. Nevertheless five out of nine isolates were shown to be toxigenic to ducklings.

Oilseed products – In those countries where the groundnut has an important place in the human diet, there is a serious risk of aflatoxicosis.

In fact, in the markets of Uganda, for example, it has been reported that groundnuts for human consumption were badly contaminated, 15% containing more than 1 mg of aflatoxin B_1 per kilogram and 2.5% having more than 10 mg/kg (626, 1517). Analogous results were obtained in the Philippines (403), in the Transvaal, and in Swaziland (1609).

It is of some concern to know whether groundnut oil obtained for our consumption has any risk of contamination. Jacquet *et al.* (1204) have reported that, in the oil first pressed, from nine samples three contained 13–100 μg aflatoxin per kilogram whereas from refined oil, two samples out of 18 contained 10–15 μg/kg.

Likewise, under very unusual conditions, Chong and Beng (455, 456) were able to detect 3–16 μg of aflatoxins per kilogram in the groundnut oil imported into Hong Kong and Singapore. But practically all authors are unanimous in recognizing that, even if groundnuts are contaminated, the oil obtained from them is not toxic, all traces of aflatoxins being removed by the solvents or destroyed by the alkaline washing which is part of the refining process (396, 1504, 1921, 2366, 2453, 2735, 2756, 2894).

Cigarettes – Commercially manufactured cigarettes have been treated with 100–300 μg aflatoxin B_1, then smoked in a smoking machine, and no trace of aflatoxin was detected either in the gas phase or in the ashes (1273).

The presence of aflatoxins in foods of animal origin

Because aflatoxins occur in the foods consumed by domestic animals, attention has been given to whether the flesh of these animals, the milk, eggs, etc. have any toxicity to man (1204). On the other hand *A. flavus* may also grow on some of these commodities.

Meat – Hadlok (994) has not detected any toxigenic strains among those isolated from damaged meat.

Some aflatoxin B_1 and its metabolities have been detected in the liver and the flesh of chickens which had consumed feeds rich in aflatoxins (2732). If toxigenic strains of *A. flavus* are inoculated onto cured sausages it is possible to recover large quantities of aflatoxins after 30 days at 25 °C; G_1 is predominant and B_2 practically absent (2560). In Naples, no aflatoxins have been detected in salami-type sausages, even after artificial infection. On the other hand, a fluorescent substance with an R_f comparable to that of aflatoxin G_2 has been found to be produced by a culture of *Aspergillus sclerotiorum* present on the skin of the sausages (2464).

Aflatoxin has apparently never been isolated from fish meal, unless it was inoculated with *A. flavus* and incubated for 35 days at 28 °C at a water concentration of 18% (1522).

Milk and milk products – It has already been noted that milking cows ingesting feeds contaminated with aflatoxins secrete M_1 and M_2 in their milk. Certainly the quantity of toxin recovered in the milk is not more than 1% of that consumed in the animals' food. Thus Jacquet *et al.* (1207) have detected several micrograms per litre in the milk of cows ingesting feed containing 0.3 mg of aflatoxin per kilogram, and 5–20 μg/l in the milk of cows consuming 1.3 mg of aflatoxin per kilogram feed. It will only be the milk of certain farms which will be incriminated and, because milk is usually bulked by dairy cooperatives before being pasteurized for distribution, the concentration of aflatoxin is usually not detectable.

Children are more sensitive than adults to the effects of aflatoxin. However, as emphasized by Abrams (10), a child of 5 kg would need to consume 27 l of milk per day from cows ingesting 2 kg of groundnut meal containing 4 mg of aflatoxin

B_1 per kilogram each day, in order to show the same symptoms as ducklings, supposing that man and ducklings are equally sensitive. Opinions are therefore divided as to the gravity of this problem.

If a freshly cut cheese is inoculated with a toxigenic strain of *A. flavus* it is possible to follow the penetration of aflatoxin into the cheese (1478), however, under natural conditions, cheese is generally unaffected by aflatoxins (1304, 1305, 2363).

Eggs – There is no evidence that aflatoxin passes into the eggs of laying hens even if they are ingesting food contaminated with aflatoxin (10, 1531).

Although they play a significant part in the disease of farm animals, the aflatoxins do not, at the moment, seem to be particularly important in the food of man, at least in temperate regions. However, the possibility is not excluded that they may be involved in the aetiology of cancer in tropical regions. Vigilant supervision is essential to avoid the growth of toxigenic strains on foods for the presence of several milligrams of aflatoxin per kilogram in food would constitute a hazard, especially of chronic poisoning. Competent agencies should be functioning in order to protect the consumer.

THERAPY

An economic problem of world importance

Animal feeds based on oilseed products (cattle cake, meal or flour), especially groundnuts, are increasingly used because they are appetizing, rich in protein, and have a high nutritive value as well as being economically acceptable. Thus, for example, in 1970, 1,600,000 tonnes of cattle cake meal were used for animal feed in France, 400,000 tonnes of which were groundnut meal. In the European Community, the consumption of groundnut meal is between 700,000 and 800,000 tonnes per year.

Now, most of these materials are contaminated by *Aspergillus flavus,* or metabolites produced by it, and often contain significant quantities of aflatoxins. This has resulted in an increasing number of incidents among farm animals (1205).

Recent analyses by Krogh and Hald (1356) of animal feed constituents from several different countries have shown that 86.5% of the batches analysed were contaminated with aflatoxins and 82.7% had more than 0.1 mg/kg. Cattle cake from Norway often contained as much as 1.5–1.9 mg/kg (788). One of the highest concentrations recorded was a sample from Brazil containing 3,465 mg/kg, of which 2,520 mg/kg was aflatoxin B_1 (1354). Another sample imported into Finland contained 4,056 mg/kg of which 2,222 mg/kg was aflatoxin B_1 (1357).

As has already been shown the sensitivity to aflatoxins varies from one animal species to another. Thus pigs may ingest feed containing 0.233 mg aflatoxin per kilogram without any noticeable effect (1297), and even a diet containing as much as 0.45 mg aflatoxin B_1 per kilogram has hardly any effect on the growth or appetite of young bacon pigs (1093, 1094).

The health regulations in preparation within the European Community (540) may contain very strict limits for the maximum concentration of aflatoxins based on dry weight, varying from 0.02–0.05 mg/kg depending on the destination of the feed. It is required that a negligible quantity, that is to say less than the lowest limit of sensitivity of the analytical method, be present in the feed of calves, piglets, lambs, chickens, and trout and, obviously, for confectionary peanuts.

The Tropical Products Institute, London, has classed as unsuitable for consumption groundnuts and products containing more than 0.05 mg of aflatoxins per kilogram (631). For its part, the Ministry of Agriculture of Denmark considers that all batches of meal containing more than 0.1 mg/kg should be eliminated from animal feeds (1354). An FAO/WHO group of experts also provisionally proposed a stringent safe threshold of 0.03 mg/kg, and the threshold recognized in the United States is even lower. All the cattle cake containing more than these limits is supposed then to be directed towards the fertilizer market (684).

These are evidently wise health precautions. But it is not an economic proposition to utilize groundnut meal, or castor meal, as a fertilizer and so the downgrading of contaminated groundnut meal, which may be as much as half that produced, would represent a considerable economic loss. It is thus urgently necessary to obtain groundnut meal free from aflatoxins.

Particular attention should be given to aflatoxins by organizations charged with surveillance of public health because of the possibility that products contaminated with *A. flavus,* or its metabolites, enter the human food chain.

Treatment of aflatoxicosis

Faced with an established case of aflatoxicosis, the very first measure must obviously be to change the feed so as to remove the primary source of illness. Partial regression of hepatomas in the rat has been obtained with the help of cortisone acetate and hydrocortisone (280), but such treatment has no effect on carcinomas in trout.

A mixture of choline, inositol, vitamin B_{12}, and vitamin E has been found to be ineffective in preventing the development of fatty degeneration in liver induced in hens by the aflatoxins (1012). It is worth recalling that it has been possible to inhibit the action of aflatoxin by pretreatment with phenobarbital (988, 1557, 1935).

Destruction of *A. flavus* during the cultivation of groundnuts

The easiest way to prevent the formation of aflatoxins is to eliminate contaminated material (13, 776). Efforts ought therefore to be primarily aimed at preventing contamination of the pods and groundnuts on the one hand, and the destruction of *A. flavus* as soon as detected on infected grain, even though abundant growth of the mould on groundnuts does not necessarily indicate the production of toxin (139). *Aspergillus flavus* is a common soil organism in groundnut growing areas and conditions limiting its growth should be studied.

Several fungicides have been tried both for disinfection of the soil and for treatment of the base of the groundnut plants or the pods. The following compounds have been used for the treatment of the soil in trials reported by Howell (1133):

- botran (2,6-dichloro-4-nitroaniline)
- D-D (a mixture of 1,3-dichloropropene and 1,2-dichloropropane)
- dexon (sodium *p*-dimethylaminobenzene diazo sulphonate)
- difolatan (N-(1,1,2,2-tetrachloroethylthio)-4-cyclohexane-1,2-dicarboximide)
- triphenyltin hydroxide
- PCNB (pentachloronitrobenzene)
- polyram (polyethylene polymer with zinc carbamate)
- vapam (sodium N-methyldithiocarbamate)
- potassium sorbate and acetic acid (5%)

Such treatments may play a significant role in the evolution of the soil mycoflora (1249). In another approach, by treatment of the pods, the best results seem to have been obtained with dichlofluanid and difolatan (2765), captan (1710), oxinate and phaltane (1854).

TMTD is not effective at concentrations of up to 100 mg/kg (1710). Thiabendazole inhibits *A. flavus* at a dose of 30 mg/ml (corresponding to 10 mg of fungicide per 100 g of dried food treated) (31, 2481). In laboratory trials, sodium propionate and potassium sorbate have been shown to be effective (1178).

A. flavus survives surface disinfection of groundnut kernels by 0.1% mercuric chloride, whereas *A. versicolor, Fusarium solani,* and *Penicillium rubrum* were destroyed under these conditions (1240). At the present time it does not seem that any fungicide has been used with sufficient success to form a certain means of preventing the invasion of groundnuts by these moulds.

A useful area of research is into the development of cultivars of groundnut resistant, or with reduced susceptibility, to attack by *A. flavus* (909, 2890). In India the varieties Asiriya mwitunde (PI 268893 and PI 295170) and US 26 (PI 246388) have been developed (1367, 2104).

In all respects, the improvement of agricultural practices would certainly help in the struggle against *A. flavus.* Damage of pods during growth and at the time of harvest must be avoided, over-late harvesting must be forbidden (1547), etc.

Modern methods of mechanical harvesting do not necessarily damage the pods more than manual harvesting (139). Several studies are being carried out in groundnut growing areas and it is hoped that teams, such as those at the Centre of Agronomic Research of Bambey (Senegal) (623) and in other countries (1976), are able to establish the best conditions for the growth and harvest of groundnuts.

Detoxification of foods contaminated with aflatoxins

Preventive measures

Goldblatt (929) has suggested that preventive measures should be foremost in the battle against aflatoxins.

Sorting pods and kernels – On an industrial scale, before considering detoxification (this term, which describes the physicochemical processes of removing toxicity from a product, should not be confused with detoxication, which describes the elimination of toxic substances from the body), by physical, chemical or biological means, it is best to establish a number of preventive measures. Problems arise from the considerable heterogeneity of batches of groundnuts. During drying, grinding, and the extraction of oil, batches of different origin are mixed and the resulting cake may well contain material from both sound and contaminated batches. It is certainly possible, and at the same time desirable, to sort out the kernels on arrival at the drying and decortication stations.

Separation of contaminated from healthy groundnuts has been attempted using differences in their density. But kernels with a high fatty acid content have a low density like those badly contaminated with aflatoxins and the process has been adandoned (684).

An original technique consists of selective decortication under pressure; at low pressures only previously damaged seed, considered to be the most contaminated, is decorticated; at increased pressures, 4–7 atmospheres depending on the concentration of water in the product, the undamaged seed, presumed to be healthy, is decorticated (Creusot–Loirs process).

In countries where groundnut flour plays a large part in the human diet (in India, for example, where it is a major source of protein) the risk is reduced by hand sorting. But the elimination of all wrinkled grain, immature kernels or apparently mouldy kernels is difficult to achieve and costly (1000, 1979).

Some careful analyses (2294) have shown that even apparently healthy kernels may contain 0.5–1.0 mg of aflatoxin per kilogram and damaged pods, especially those with cracks from mechanical damage or insect attack, may be very much more contaminated. However, it is possible that certain analytical techniques, especially fluorescence, may have led to confusion between aflatoxins and other quite unrelated metabolites (620, 2470).

Drying – Although groundnut kernels contain about 40% water at the moment of harvesting, they should contain only 4% after drying before oil extraction. Groundnut meal is normally a dry product with a water content of 5–6% (which may rise to 9–10% during transport). So long as there is no external source of water, *Aspergillus flavus* cannot grow under these conditions (germination of spores and growth of mycelium requires at least 80–85% relative humidity, corresponding to 16–17% water content). It is thus best to maintain the cake in a dry state and prevent any dampness.

The drying process is very important and various types have been tried (1547, 2029). Rapid artificial drying is preferable to slow natural drying (1181, 1542, 1979).

Storage conditions – It has already been pointed out that a controlled atmosphere gives improved preservation of groundnuts compared with free exposure to air. Growth of *A. flavus,* and hence aflatoxin production, is completely inhibited in an atmosphere of nitrogen. However, such an atmosphere seems to favour the growth of *Fusarium,* giving a disagreeable smell to groundnuts. Arnold and Petitt (72) have therefore recommended storage in an atmosphere containing 87% carbon dioxide and 13% nitrogen. Other authors have arrived at similar conclusions (138, 1183, 1932).

Ambient humidity and temperature influence the effect of carbon dioxide and it is reported that aflatoxin biosynthesis is inhibited at 86% RH at 17 °C by 20% CO_2 or at 25 °C by 40% CO_2. At constant temperature an increase in CO_2 concentration reduces the production of aflatoxins (2231).

Physical treatment

Heat – The first trials in the destruction of aflatoxin by heat treatment were not very successful (242, 776, 2015). Pure aflatoxins were found to be stable up to their melting point, close to 250 °C.

There has been disagreement in the literature about the effect of heat on aflatoxins and this reflects the fact that there is a relationship between the stability of the aflatoxins and the water activity of the substrate; the damper the substrate the greater the destruction of aflatoxin (1590). For example, a meal containing 0.144 mg of aflatoxin B_1 per kilogram kept at 100 °C for two and a half hours shows no loss if the water content is 6.6%, but the concentration of aflatoxin is reduced to about a quarter if the water content is 15–30% (542, 684).

The opening of the lactone ring of the molecule is, first, a hydrolytic process followed by decarboxylation (1590, 2183). In practice it is thus necessary to add water to contaminated meals before heat treatment in order to destroy aflatoxins. In fact, the reduction in the concentration of aflatoxins is generally insufficient to be acceptable under the more rigorous legislation. Another disadvantage is that heat treatment alters the quality of the protein, especially the availability of lysine (55, 2870). If heating is insufficient (60 °C for 30–60 minutes) there is a danger, not only of not destroying the aflatoxins, but also of increasing the toxigenic ability of *A. flavus* if it is present (1225, 1226).

Roasting groundnuts at 150 °C for half an hour gives an average reduction of 80% in the concentration of aflatoxin B_1 and 60% in that of aflatoxin B_2 (1443, 1444, 2772).

It has been reported that tortillas contain practically no aflatoxin even when they are made from flour contaminated with the toxin and this is almost certainly due to heat inactivation (2680).

There is some doubt as to whether aflatoxins are produced at low temperatures, such as those of a domestic refrigerator; some reports (2725) indicate that they are and others that they are not (1889).

Irradiation – Gamma radiation has been used with some success for destroying moulds and, at a water content of about 22% in the substrate, a dose of 1 Mrad is required. However this appears to have no effect on the aflatoxins (819, 1702, 2786).

At doses of less than 200 krad of γ-irradiation spores of *A. flavus* may be induced to produce significantly more aflatoxin, although very high doses considerably reduce the ability of synthesizing these metabolites (1224, 1226). Experiments carried out by Frank, Muentzer, and Diehl (820) gave very variable results which were fairly readily explained by the occurrence of mutation and the heterocaryotic nature of *A. flavus.* Thus, depending on the situation, the toxigenic nature of some strains may be enhanced and others reduced.

Ultraviolet light will often give degradation products of the aflatoxins (21). Dimethylsulphoxide provokes a pigmentation of the spores and the latter, when exposed to u.v. light, are easily destroyed (170). However, exposure of groundnut cake to u.v. for eight hours did not reduce its toxicity, previously formed aflatoxins being unaffected (776).

Chemical treatment

Solvent extraction – The oil industry uses saturated hydrocarbons in which the aflatoxins are insoluble. On the other hand, the use of polar solvents, in which aflatoxins are soluble, causes difficulties in the subsequent refining of oils. In 1965, Robertson (2168) proposed the use of a mixture of acetone/hexane/water in the proportions of 50:48.5:1.5 (v/v) for extracting aflatoxins from groundnuts as the first stage in their analysis.

A number of solvents, or mixtures of solvents, have been proposed for use on semi-industrial scale, either simultaneously with extraction of the oil, or before the oil is extracted by conventional means, or even after extraction of the oil (684). Among the solvents most frequently recommended are acetone, benzene, and chloroform, but they are not very efficient at removing all the aflatoxin from groundnut meal (776). Methanol gives excellent results so long as it is used in aqueous conditions (2227). Similarly, acetone, with the addition of 10% water (w/w) gives a significant reduction in the concentration of aflatoxin (from 0.113 to 0.008 mg/kg at 48 °C and from 0.180 to 0.011 mg/kg at 44 °C in the case of cottonseed cake) (684, 872). It has been suggested that as much as 25–30% water should be added to acetone (2022), thus giving a reduction from 0.535–0.026 mg of aflatoxin B_1 per kilogram of cattle cake.

Extraction with aqueous ethanol (95:6) at 24 °C gives a reduction of 93–96% of the aflatoxins in cottonseed cake and 96–98% in groundnut cake (2111). Aqueous isopropanol (80%) is also a promising solvent when used at 60 °C (2110).

If it is required to extract the aflatoxins and the oil simultaneously, it is necessary to use an azeotropic mixture of hydrocarbon and acetone or hydrocarbon and ethanol which allows the production of acceptable oils and relatively uncontaminated cake (2756). Hexane/ethanol (79:21), hexane/methanol (73:27), acetone/hexane (59:41), hexane/ethanol/water (85:12:3), acetone/hexane/water (42.1:56.5:1.4 or 54.5:44.4:1.1 or 54:44:2 or 50:48.5:1.5). This last mixture has already been recommended for the removal of gossypol which is toxic to monogastric animals (1310). In a pilot plant continuous extraction process (880) the results look very promising (924). With a ratio of solvent to cake of 2:1, a cake containing 14.7% oil and 0.160 mg aflatoxin B_1 per kilogram was extracted for 70 minutes at a solvent temperature of 43 °C. The resulting cake did not contain more than 0.002 mg aflatoxin per kilogram. In two other trials, the concentration of aflatoxin was reduced from 0.3 to 0.031 mg/kg and from 0.207 to 0.038 mg/kg (872).

A disadvantage of this technique is that, in some cases, the cake aquires an unacceptable odour of urine due to oxidation products from acetone and it is necessary to remove this. It is usually necessary to make several successive extractions with the solvent mixture in order to obtain the most efficient removal of aflatoxins.

In India attempts have been made to use solutions of inorganic salts both for the analysis of the aflatoxins and their removal from foods (2470). Unfortunately, this technique also removes the protein from the food, thus sodium bicarbonate removes as much as 33% and calcium chloride 6%.

Modification of the molecule – Instead of extraction of the aflatoxins with solvents an alternative approach is to inactivate them by modifying the structure to form a less toxic derivative. It is in this manner that animals which are most resistant to aflatoxins react to them (158).

Most of the chemical methods of detoxification involve oxidation or hydroxylation of the aflatoxin molecule opening the double bond of the terminal furan ring in aflatoxins B_1 and G_1 (2625). A systematic search has been made for oxidizing and reducing agents (684). Some compounds, such as choline chloride, triethylamine hydrochloride, 2-ethylaminoethanol, sodium sulphite, glucose, and fructose, are of little use, others, such as 3-aminopropanol, sodium glycate, 1-amino-2-propanol, trisodium phosphate, phosphoric acid, lime, and ammonium carbonate, are moderately useful. Most promising results have been obtained with methylamine, ethanolamine, trimethylamine hydrochloride with soda, choline, and soda alone.

The best results have been obtained with methylamine and sodium hydroxide (684). Thus, when a cake containing 4 mg of total aflatoxins per kilogram (of which 2.85 mg/kg was aflatoxin B_1) with an adjusted water content of 15%, was treated with warm 1.25% methylamine, the aflatoxin level was reduced to 0.065 mg/kg (of which 0.063 mg/kg was aflatoxin B_1). In a similar manner, a cake containing 0.113 mg of total aflatoxins per kilogram (of which 0.068 mg/kg was B_1) and an adjusted water content of 22%, contained 0.075 mg of aflatoxins per kilogram (0.047 mg/kg was aflatoxin B_1) after a simple heat treatment. When heated in the presence of 2% sodium hydroxide, a sample containing 0.068 mg of aflatoxin B_1 per kilogram was reduced to 0.011 mg/kg, although, if the water content was increased to 30%, treatment with 2% sodium hydroxide left no trace of aflatoxins.

At the experimental stage treatment with ammonia seemed very interesting (871, 2681). It gave a reduction of 99% in the concentration of aflatoxins in groundnuts without significantly altering the concentration of nitrogenous compounds except that some amino acids, such as cystine, were destroyed. American industry uses this process which has the advantage of taking place after extraction of the oil (684, 855), and new trials are under way in West Africa.

Among oxidizing agents, benzoyl peroxide and osmium tetroxide react with aflatoxins B_1 and G_1, but not with B_2 and G_2. Ceric ammonium sulphate,

$Ce(NH_4)_2(SO_4)_3$, NaOCl, $KMnO_4$, $NaBO_3$ and a mixture of 3% H_2O_2 and $NaBO_3$ (1+1) react with all four aflatoxins (2625).

Detoxification using oxygenated water at 80 °C for half an hour has been recommended in India (2471) where experience was obtained with an especially badly contaminated meal and the concentration of aflatoxin was reduced a hundred fold. It would be of interest to establish whether comparable results could be obtained with a less contaminated feed for it seems that such treatment has little effect on the proteins.

Dwarakanath *et al.* (713) have tried oxidizing the aflatoxins with ozone under different conditions of humidity, temperature, and time of exposure. Adjusting the water content of the meal to 22% for cotton and 30% for groundnut was strongly advised for, at 100 °C, detoxification was optimal after two hours. However, only aflatoxins B_1 and G_1 were affected and the concentration of B_2 and G_2 remained unchanged.

Many other trials have been carried out. Gaseous chlorine upsets the organoleptic qualities of the food by its action on fats often producing toxic products (776, 2470). Chlorinated hydrocarbons such as trichlorethylene, have sometimes been suggested (782, 2509) as well as hydrochloric acid (776) and 5% aqueous hypochlorite (509). Sulphur dioxide and propylene oxide do not seem to be of any use (776). Formalin, especially with 1% sodium carbonate, may be useful.

The possibility must not be lost sight of that some modifications of the aflatoxin molecule may give derivatives which are reconverted back into aflatoxins themselves in the acidic conditions of the animal stomach.

Biological control and bioconversions

Fungal competition – Numerous species of fungi are present in the rhizosphere of the groundnut plant and in the geocarposphere of the pods. Competition is the rule among such micro-organisms (91, 1239, 1509, 1538, 2294). It may be possible to encourage those species which inhibit the growth of *Aspergillus flavus.* This is the direction of work such as that carried out in Israel.

Confirmation of the interest in this type of work is given by the observations of Chung-Min Chang and Lynd (499) that the highest concentrations of aflatoxins are found in groundnuts growing in acid soils (pH 5.3–5.7) and the lowest in alkaline soils (pH 7.2–9.1). It was suggested that this may be due to the effect of pH on the antagonistic activity of micro-organisms against *A. flavus.* Not only fungi, but bacteria may also play a role in these competitive interactions (91, 2138).

Use of the antibiotic, aureofungin, has sometimes been recommended (2814).

Bioconversions – Fungi and bacteria are well known for their activity as agents of bioconversions. Thus, first *Dactylium dendroides,* then *Absidia repens,* and *Mucor griseo-cyanus* were recognized as able to hydroxylate steroids. Detroy and Hesseltine (635, 636) have been able to transform aflatoxin B_1 into a compound 18 times less toxic. Of numerous micro-organisms which Ciegler *et al.* (508, 509) have tested for this purpose, *Aspergillus niger, Penicillium raistrickii,* and especially *Flavobacterium aurantiacum* (strain NRRL B 184) have been shown to be capable

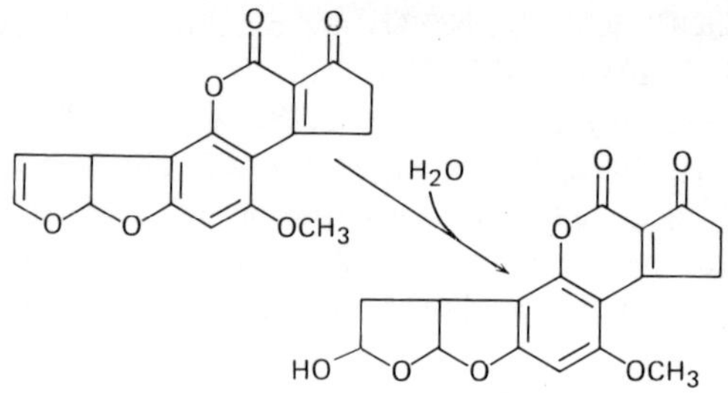

Figure 55. The formation of hydroxydihydroaflatoxin B_1 (= aflatoxin B_{2a})

of degrading aflatoxins. Forty-four hours after inoculating toxic feed with this organism most of the aflatoxin had been destroyed. These trials have been successful, not only with groundnut meal, but also with milk, unrefined oil, and wheat, but it would seem that soya may not respond to this treatment.

During this process, in acidic media, a compound is obtained which Ciegler and Peterson (509) believe to be hydroxydihydroaflatoxin B_1 (= B_{2a} Figure 55). It is indeed this compound which is obtained when citric acid is added to aflatoxin B_1. The structure is isomeric with aflatoxin M_2 and the derivative is less toxic than the parent compound. Another compound obtained in the same manner by Detroy and Hesseltine (637) has been shown to be aflatoxin R_0, which is also less toxic than B_1 (Figure 56). *Flavobacterium aurantiacum* metabolizes not only aflatoxin B_1 (508, 1494) but also G_1 (1495), and M_1 (1496).

In the same category of experimentation are those tests of the usefulness of ensiling corn with a high water content, which has been badly contaminated with aflatoxin B_1, in order to bring about a conversion to less toxic derivatives. Although the production of silage involves a lactic acid fermentation, there is insufficient acid to destroy all the aflatoxin B_1 (1502).

Using various species of *Rhizopus* (*R. stolonifer, R. arrhizus, R. oryzae*) Cole and Kirksey (537) have been able to convert aflatoxin G_1 to a compound already described as aflatoxin B_3 (a metabolite of *A. flavus*) and to parasiticol (a metabolite of *A. parasiticus*). A reduction of 50% in the concentration of aflatoxins in groundnut meal has been obtained by growing *Neurospora sitophila* on the meal. A product resembling tempeh, which is produced by growing *Rhizopus oligosporus* on soya beans, was obtained having good nutritional value and rich in riboflavin (2721).

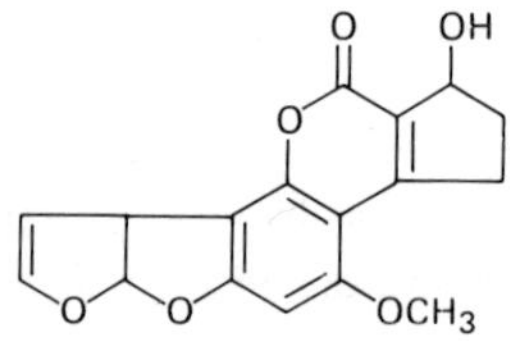

Figure 56. Aflatoxin R_0

Some animals are able to degrade aflatoxins, thus the protozoan *Tetrahymena pyriformis* will degrade up to 58% of aflatoxin B_1 in 24 hours to a compound with a bright blue fluorescence, but is unable to degrade aflatoxin G_1 (2578). In aflatoxin B_1 the carbonyl group of the cyclopentenone ring is reduced to a hydroxyl group with the formation of aflatoxin R_0 (2172). The insect *Trogium pulsatorium* is able to reduce aflatoxin G_1 to G_2 (796).

Useful conclusions

On an industrial scale, all physical, chemical or biological processes for treating groundnuts or meals need to be compatible with the following:

1. They should be simple to apply and not add more than a few per cent to the relatively low cost of such meals.
2. The process should not require rehydration of dry meals which will mean a second drying stage or problems during transport and storage.
3. The process should not alter the basic composition so that the food value, especially the concentration of proteins, is not changed.

Of the numerous trials reported many have been successful at the laboratory stage but their application on an industrial scale becomes difficult if these principles are to be adhered to. As pointed out by Dollear and Gardner (684), although there has been remarkable progress in research on the extraction and inactivation of aflatoxins in oilseed products, few processes have reached the industrial stage (925).

Perhaps treatment with ammonia, methylamine or sodium hydroxide is the easiest to apply, coupled with extraction using the solvent system acetone/hexane/water. However, the main disadvantage is that there is the risk of denaturing the protein, for it is in fact necessary to release the toxin bound to protein and other cell components. A preliminary stabilization of proteins by a tanning compound such as short chain aldehydes (acetaldehyde, glyoxal, glutaraldehyde) may perhaps be an elegant solution to this problem (Patents INRA No. 1,461, 364 and REY No. 1,453,261).

Techniques based on biological competition are probably the best long-term solution but they require numerous and precise observations as well as long-term experimentation.

CHAPTER 4

CLAVACITOXICOSIS

INTRODUCTION

An example of poisoning due to Aspergillus clavatus

The considerable drought of 1959 presented difficulties for many livestock farmers, who found that they urgently needed feed for their herds, and many went in for germination of malt. The hydroponic production of cereal sprouts has its defenders and detractors. The high humidity and elevated temperature associated with this particular type of hydroponic culture are favourable for the growth of certain fungi.

It was in this way that problems associated with *Aspergillus clavatus* first came to our attention. It did not seem to completely restrict the growth of the plants but it was nevertheless responsible for the deaths of many cows of herds in the Ile-de-France (1738). Subsequent investigations showed that similar accidents had occurred during the same period in other parts of France, notably Brittany, Nivernais, and Normandy (1202, 1739).

Patulin, a metabolite produced by this species of *Aspergillus,* is very toxic but it seems that patulin poisoning has not reached such dramatic proportions before, probably because the fungi which produce it rarely find conditions allowing sufficient growth.

However, patulin was found to be responsible for an outbreak of poisoning recently occurring in Japan following the consumption of malt infected with *Penicillium urticae* (2692–2694). Similar outbreaks have been reported from Bulgaria among cows on a diet of barley malt infected with *Aspergillus clavatus* (2612), and probably in Germany, close to Leipzig (1) where this mould was found in profusion on malt causing poisoning of cattle.

History of the discovery of patulin

The early history of patulin was reviewed by Lochhead, Chase, and Landerkin (1512). It was in March 1942 that Wiesner reported the antibacterial activity of a compound present in the filtrate of a culture of *Aspergillus clavatus.* The culture filtrate was capable of destroying staphylococci at dilutions of 1/100,000. A similar report appeared in August of the same year from the laboratory of Waksmann, Horning, and Spencer (2763) who gave the active compound the name clavacin.

During the same period, Chain, Florey, and Jennings (437) isolated a crystalline compound, active against both Gram positive and Gram negative bacteria, from the filtrates of *Penicillium claviforme* Bain. which they called claviformin.

A year later, Anslow, Raistrick, and Smith (63) isolated from *Penicillium patulum* Bain. (= *P. urticae* Bain.) and from *P. expansum* Link, a crystalline, antibiotic compound, the empirical formula of which was established and which they called clavatin. It was found to be identical with patulin (191, 192, 1120). Chain, Florey, and Jennings (438) also established that patulin and claviformin are identical.

Hooper *et al.* (1120) recognized the identity of clavacin and patulin thus confirming the important work of Katzman *et al.* (1280). Thus clavacin, claviformin, patulin, and clavatin as well as expansin, known since 1938 from cultures of *P. expansum* but only studied in 1944 and 1945 (712), are all one and the same compound. Although the name clavacin has historical precedence, patulin is the most widely used probably because, under this name, in 1943, Raistrick *et al.* (2083) investigated its use against the common cold.

The recent review of Ciegler, Detroy, and Lillehoj (504) points out that other synonyms for patulin include penicidin obtained by Atkinson in 1942 from a *Penicillium* sp (94, 95), mycoin described in 1943 (144, 702, 1291, 1455), tercinin and leucopin (942, 1319, 2355).

Fungi producing patulin

Many species of *Aspergillus* and *Penicillium* are able to produce patulin. Among the aspergilli it is mainly those of the *clavatus* group, that is to say *A. clavatus* Desm. and *A. giganteus* Wehmer, if the latter is accepted as distinct from the former (2812), which synthesize patulin (790). Filtrates of cultures of *A. terreus* Thom (1293) and of *Byssochlamys nivea* Westl. (2358) have also been reported to contain this metabolite. Among the penicillia, highest yields are obtained from *P. urticae* Bain. (= *P. patulum* Bain.) (1870), *P. griseofulvum* Dierckx (which is very close to *P. urticae*), *P. expansum* (Link) Thom (63), *P. claviforme* Bain. (789), *P. terrestre* Jensen. The following have also been reported as able to produce some patulin: *P. divergens* Bain. and Sart. (a species which is perhaps synonymous with *P. granulatum* Bain.), *P. melinii* Thom (1267), *P. novae zelandiae* v. Beyma, and an unidentified species of *Penicillium* (1014).

ASPERGILLUS CLAVATUS

Morphology (Figure 57)

A. clavatus is easily recognized because the vesicles terminating the conidiophores are particularly elongated and large, giving each conidial head the appearance of a minute pin 1–3 mm in length, 20–30 μm in diameter. It carries many tens of

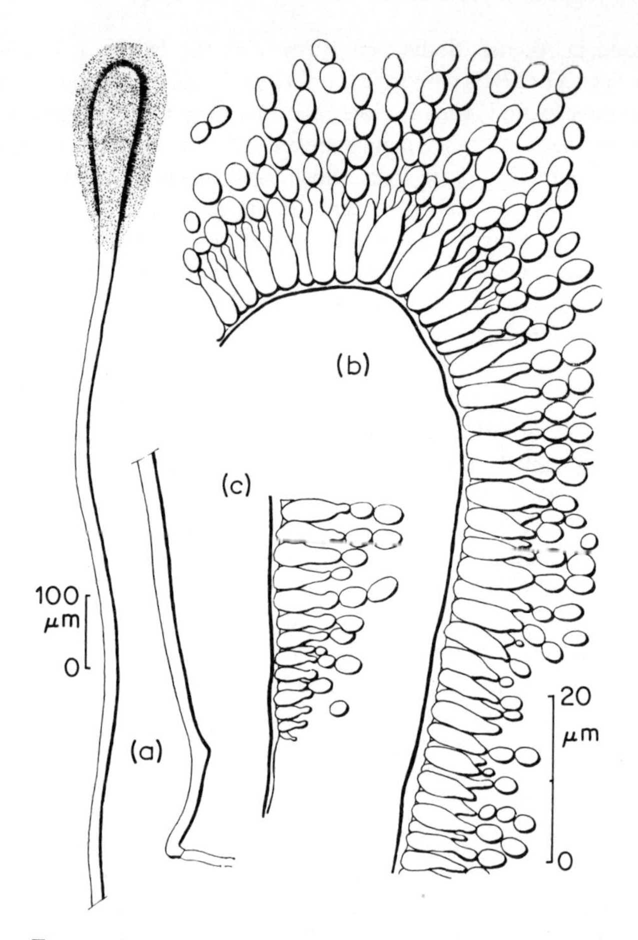

Figure 57. *Aspergillus clavatus* (a) A conidiophore. (b) Detail of the phialides and conidia in the region of the end of the claviform vesicle. (c) Detail of the phialides near the base of the fertile region of the vesicle

thousands of spores and is pale blue-green in colour. The whole surface of the claviform vesicle is fertile and the structure is uniseriate, bearing phialides directly on the swollen vesicle. The elliptical conidia are smooth and 3–4.5 x 2.5–3.5 μm having an especially thick, smooth wall.

Ecology

A. clavatus is a frequent inhabitant of the soil but its survival in an agricultural soil seems to depend on the crop grown (2827). Values of the mean arc sine percentages based on ten plate counts of soils associated with different crop plants were as follows; wheat was most favourable at 6.8, maize 2.3, lucerne 2.0, soya 1.4, and oats 0.6. Transmission by the atmosphere is common and the species has been isolated from ripening fruit such as bananas (1258). When it infects rice, *A. clavatus* causes the grain to break up like chalk.

The frequency of *A. clavatus* on livestock farms is perhaps related to its particular predilection for decomposing plant material and material characterized by a high nitrogen content, such as animal dung, especially that of chickens (2592).

A. clavatus may cause hyperkeratosis in cattle and pigs (1888, 2409), and it may also cause allergic reactions in the region of the pulmonary alveoli (2159).

Conditions for growth and sporulation

Aspergillus clavatus has a wide temperature range for growth and survival; it requires a period of at least six months at −12 °C to reduce the viability and it grows well at 37 °C, although its optimum temperature for growth is 23–26 °C. Its ability to grow at warm temperatures may account for it being isolated from the contaminated atmosphere of a tunnel for drying food pastes (1737).

Spore production by *A. clavatus* is extremely precocious and, in a favourable medium, the speed with which the spores mature allows many growth cycles to occur within a short time. Moreover the spores are particularly easily disseminated (1747). This species grows very well on a medium rich in nitrogen and is able to break down cellulose in a medium containing ammonium sulphate as the source of nitrogen.

Compounds produced by Aspergillus clavatus

This species is particularly well known for its ability to produce patulin, however, some strains may produce a penicillin-like compound active against staphylococci.

Tanabe and Suzuki (2526, 2544) have demonstrated that a strain of *A. clavatus* (WF-38-11), isolated from wheat flour, produces a second mycotoxin as well as patulin. The compound has been called ascladiol, $C_7H_8O_4$ (Figure 58) and is only about one-quarter as toxic as patulin from which it may be derived be reduction.

A number of compounds with antibiotic properties, and which may not be patulin, have been described from cultures of *A. clavatus* (191, 192, 409, 580, 971, 1765, 2820).

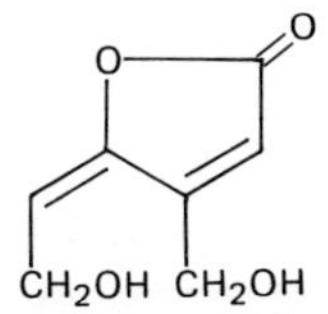

O O

CH_2OH CH_2OH

Figure 58. Ascladiol

Production

On an ordinary Czapek medium, it appears that the optimum production of patulin occurs after about nine days. Thus, using *P. griseofulvum,* Simonart and Lithouwer (2389) obtained the yields shown in Table 22.

Lochhead *et al.* (1512) reported optimum production in 11 days for a culture grown at 20 °C and, whereas at 25 °C production was not significantly reduced, at 30 °C it was very much less and levelled out after the sixth day.

The pH of the medium continues to increase as the concentration of patulin decreases and a pH between 4.5–5.0 appears to be the most favourable for its biosynthesis. It is possible to increase the yield of patulin by adding sodium phthalate to the medium (5 g/l) which may prevent the medium becoming too alkaline (582).

Production of patulin is not significantly better in shake flasks than in stationary culture, indeed it is usually higher, though reached more slowly, in the latter (582).

It seems that, in cultures of *A. clavatus,* the biosynthesis of patulin may only occur in the presence of carbohydrates (1202). Glucose is much better than sucrose, dextrin or lactose for the production of patulin (971, 1512). Norstadt and McCalla (1870) obtained yields of 1.2–1.7 g patulin per litre of medium from *P. urticae* grown on a potato dextrose medium at 25 °C for 14 days.

Simonart and Lithouwer (2389) obtained patulin by growing *Penicillium griseofulvum* on media containing fructose or galactose, but not xylose, lactic, malic, pyruvic or α-ketoglutaric acids.

If patulin is produced by moulds growing on meat substrates, it is readily inactivated by the sulphydryl groups present in proteins, especially at pH values close to 7 (1107).

Table 22. The relationship between age of culture and yield of patulin produced by *Penicillium griseofulvum* (2389)

Age of culture (days)	*Yield of patulin* (μg/100 ml)	*pH of medium*
4	10	3.6
6	52	3.6
7	104	–
8	168	5.0
9	182	–
10	174	5.7
12	152	–
16	96	5.6
19	54	6.4
25	24	–
35	0	6.9

Detection and extraction

Several methods of detection are based on colour reactions obtained with diazotized benzidine (1896), diazosulphanilic acid (2125), phenylhydrazine (yellow-orange colour) or alkali (brown-red colour) (2906). These reactions are not very sensitive. A procedure preferred by several authors (2002, 2003, 1870) is to extract with ethyl acetate, dry the extract over anhydrous magnesium sulphate, remove the solvent and redissolve in dry ether. The extract is then purified by passing through an alumina column adjusted to pH 4.5 and eluting with ether. After concentration of the eluate, the residue is further purified by TLC on silica gel plates using chloroform/methanol (95:5 v/v) and the plates sprayed with potassium permanganate to reveal patulin.

Chemical structure and biosynthesis

Patulin (Figure 59) is 4-hydroxy-4*H*-furo(3,2-*c*)pyran-2(6*H*)-one (63, 1293) with an empirical formula $C_7H_6O_4$ and molecular weight 154. A number of structures have been suggested for patulin, including the idea that it existed as an equilibrium mixture of the tautomeric structures shown in Figure 60 and suggested by Nauta *et al.* (1815). Woodward and Singh (2871) first proposed the correct structure which was rapidly confirmed by Dauben and Weisenborn (586). The presence of the hemiacetal was further confirmed by Lalau-Kéraly *et al.* (1412).

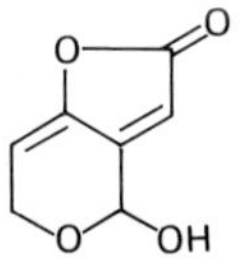

O O OH

Figure 59. Patulin

O O O O O O O O OH O O OH O O

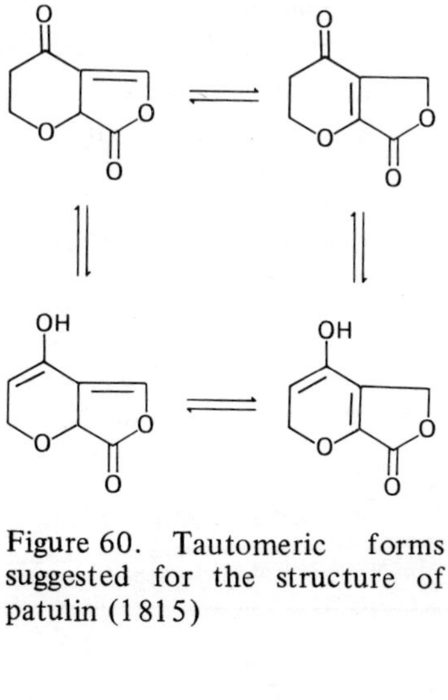

Figure 60. Tautomeric forms suggested for the structure of patulin (1815)

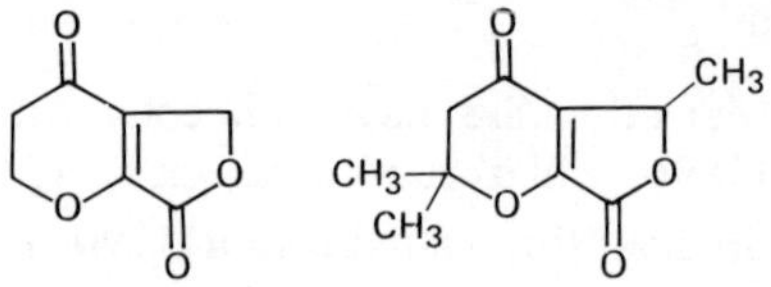

Figure 61. Isopatulin and trimethylisopatulin

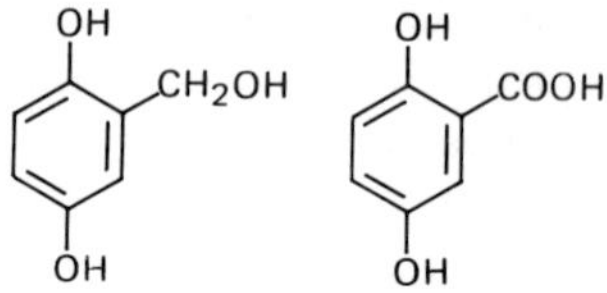

Figure 62. Gentisyl alcohol and gentisic acid

One of the γ-pyrone structures suggested by Nauta *et al.* for the structure of patulin has been synthesized and called isopatulin (Figure 61) (973) as has the trimethyl derivative of this isomer (57, 794, 2041, 2042). The former has a melting point of 87 °C, the latter 142 °C whereas patulin itself melts at 110 °C. The biosynthesis of patulin has already been discussed in Chapter 2 and probably involves the cleavage of an aromatic precursor such as gentisaldehyde (2871). In fact cultures of *Penicillium urticae, P. claviforme,* and especially *P. griseofulvum* produce both gentisyl alcohol and gentisic acid (Figure 62) as well as patulin. A correlation between the yields of gentisyl alcohol and patulin in Czapek medium has been demonstrated by Brack (281) whose results are shown in Table 23.

However, it has also been demonstrated that metal ions in the medium may influence the relative yields of these two metabolites (Table 24) (281). The production of gentisyl alcohol is especially enhanced in the presence of zinc ions, which also favour a high yield of mycelium.

It has been suggested by Bassett and Tannenbaum (153, 2548) that there are two mechanisms for the production of patulin by *P. urticae,* one route *via* gluconic and gallic acids, the other by conversion of glucose to methylsalicylic acid,

Table 23. Production of gentisyl alcohol and patulin in Czapek medium (281)

Age of culture (days)	*pH of medium*	*Yield of pure metabolite* (g/l)	
		gentisyl alcohol	*patulin*
7	4.6	0.3	0.6
9	4.4	0.5	0.9
13	5.1	0.2	0.7
16	6.5	0.2	0.5
19	7.4	0.02	0.05

Table 24. The influence of the presence of metal salts in the medium on the production of gentisyl alcohol and patulin

Culture medium	*Dry weight of mycelium* (mg/100 ml)	*Yield of metabolite* (g/l)	
		gentisyl alcohol	*patulin*
Without iron	115	0.92	0.80
$FeSO_4$ (1 mg/l)	250	1.28	0.20
+ $CuSO_4$ 10^{-6}M	142	0.60	0.10
+ $ZnSO_4$ 10^{-6}M	567	2.06	0.05
+ $MnSO_4$ 10^{-6}M	192	0.02	0.80

oxidation and cleavage to prepatulin followed by cyclization to patulin. It had been known since 1948 that *Penicillium claviforme* produces 6-methylsalicylic acid (972) and this metabolite has been isolated from *P. urticae* grown in Czapek medium containing 4% glucose in the presence of $CaCO_3$ (2548).

Prepatulin is the name given to a waxy compound produced by *P. urticae* in the presence of chalk. It does not contain aromatic phenolic groups and is converted to patulin on crystallizing from anhydrous solvents in the presence of traces of acid. It has been known that deutero *m*-cresol may be converted into patulin (2313) and may be considered as an intermediate in the biosynthetic pathway (Figure 63).

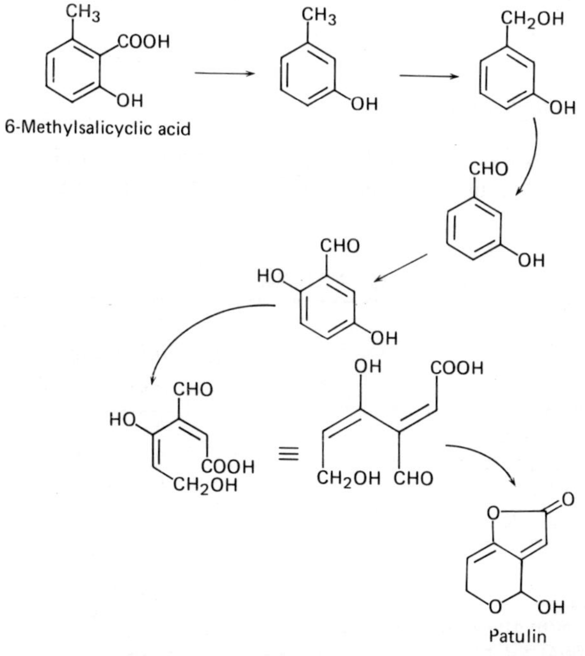

Figure 63. Biosynthesis of patulin from 6-methylsalicylic acid

Properties

Physiochemical properties

Patulin forms colourless rhombic or prismatic crystals (207), m.p. 109.3–110.5 °C. The ultraviolet spectrum has a single peak at 276 nm (2548) and fluorescence under u.v. is optimal at 360 nm (1868). Patulin is soluble in water and the usual organic solvents except petroleum ether (1314). It is stable in acidic media, but unstable, and its biological activity is destroyed, in alkaline media (437).

Antibiotic properties

Patulin is an antibiotic active against a wide range of both Gram positive and Gram negative bacteria as well as several fungi (191, 192, 207, 409, 1064, 2762, 2820) (Table 25). *Actinomyces scabies* is inhibited by a concentration of patulin of about 6 μg/ml. It is toxic to protozoa, thus *Tetrahymena pyriformis* Ehr., for example, is killed within 24 hours by a concentration of 2 μg/ml, or in one hour by 200 μg/ml. Species such as *Strigomonas* require a concentration of 200 μg/ml for 24 hours before they are killed whereas *Paramecium aurelia* requires 10 μg/ml (100).

Patulin inhibits the growth of isolated cells of tissue cultures (2755); it inhibits the motility of *Chlamydomonas* and inhibits the plasmolysis of *Spirogyra* (885, 1659). It shows some activity against viruses, inhibiting the replication of influenza virus (2200), inactivating some bacteriophage (999, 1259), but does not inactivate the phage of *Pseudomonas pyocyanea* (652).

Phytotoxicity to higher plants

Although *Aspergillus clavatus* did not seem to show any phytotoxicity to wheat or barley germinating in a malt house (1738), a number of cases of inhibition of

Table 25. Antibiotic activity of patulin

	Patulin concentration inhibiting growth μg/ml
BACTERIA	
Micrococcus lutea	2
Salmonella paratyphi	2.5–12.5
Bacillus subtilis	5.0–10
Escherichia coli	12.5–30
FUNGI	
Candida albicans	20–40
Microsporum audouini	6.25–25
Rhizoctonia solani	12.5–50
Verticillium dahliae	200

Table 26. Phytotoxicity for the tomato of patulin and other antibiotics (1317)

Antibiotic	*Minimum toxicity* (μg/ml)	*Lethal dose* (μg/ml)
Patulin	15	60
Penicillin	500	2,000
Streptomycin	250	500
Chloramphenicol	60	500

plant growth by patulin have been observed:

– inhibition of seed germination (207, 882, 2607, 2769);
– wilting of plants (886, 1166, 1317, 1850, 2702, 2775, 2882) especially pea seedlings (2128);

– reduction in growth rate (143, 1262, 1391, 1663, 2882);
– inhibition of protoplasmic streaming in *Elodia canadensis* (885);
– inhibition of scopoletin secretion by oat roots (1606).

The production of patulin in the soil by *Penicillium expansum* seems to play a role in a phenomenon referred to as 'soil fatigue' (294, 1532, 1533) which has been particularly studied in young plantations of apples by Borner (266).

According to Bilai (207) the application of solutions containing 50 μg patulin/ml is toxic to wheat seeds, whereas 1 μg/ml has an inhibitory effect on maize roots; 60 μg/ml is toxic to tomato plants. The phytotoxicity of patulin against the tomato is compared, under the same conditions, with other common antibiotics in Table 26.

Therapeutic use

The remarkable antibiotic properties of patulin led to the hope that it could be useful pharmacologically. Interesting results have been obtained in the treatment of bovine brucellosis (1457, 2196). However, after several unfortunate trials with British sailors, notably in the treatment of the common cold and bronchitis (224, 1123), its toxicity led to its being abandoned (1641).

Patulin has been used, with some success, as a local application in the treatment of infected wounds, even in human therapy (702, 1895), as well as for the healing of skin wounds (2755).

On the other hand, *in vitro* tests have shown that patulin has antineoplastic activity (207). A concentration of 20 μg/ml inhibits the growth of cancer cells and a concentration of 1 μg/ml destroys sarcomas and melanomas in mice.

Gerstner (895) has used the antifungal activity of patulin in the treatment of plant diseases finding it to be effective against smut and helminthosporium leaf spot of wheat. Timonin (2606, 2607) has also considered the use of patulin in the control of loose smut of wheat, however, its phytotoxicity precludes the widespread use of patulin in the treatment of plant diseases.

Toxic properties

Numerous investigations have been carried out on the toxicity of patulin to animals (1738, 1895).

1. Intravenous administration of 1 mg or more to mice causes the death of almost all the animals and the LD_{50} for the mouse has been established by Broom *et al.* (318) to be 0.5 mg. Ernst has reported that a dose of 10–15 mg/kg was fatal to cats, mice, and rabbits, an observation confirmed by Kinosita and Shikata (1314).
2. By intraperitoneal injection, Lochead (1512) has found that a dose as low as 0.1–0.2 mg was fatal to mice, whereas Stansfield reported that a dose of 0.25 mg was necessary.
3. The toxic dose is slightly less when the compound is administered subcutaneously compared with intravenous injection. The LD_{50} reported by Broom (318) is 0.3 mg, and by Katzman (1280) 0.2 mg.
4. Chain *et al.* (437) found that 2.5 mg caused the death of mice when given orally although Broom *et al.* reported an oral LD_{50} of 0.7 mg. Katzman and his colleagues (1280) reported the death of tropical fish in the presence of a concentration of 10 μg/ml. Increased concentration of patulin cause the death of all animals in a shorter time.
5. Applied in the form of a powder to the cerebral cortex, even quite large quantities of patulin did not seem to produce any clinical symptoms or electroencephalographic disturbances, however, a dose of 5–10 mg is solution in warm water or isotonic saline, when applied to the parietal cortex, did cause significant disturbances on the electroencephalogram as well as convulsions (1253).

When tested on egg embryos patulin has an LD_{50} of 0.7 mg (409). Toxic effects against leucocytes have been demonstrated by Chain, Florey, and Jennings (439) at concentrations of patulin in the region of 1.25 μg/ml whereas, using another method, Hopkins (1123) showed that phagocytosis of polynuclear leucocytes and macrophages was inhibited by 0.5 mg/ml. Jaquet *et al.* (1202) demonstrated a net decrease of red blood cells in mice receiving patulin by injection (mean erythrocyte count 8,600,000/mm^3 instead of 10,900,000 in healthy mice).

MYCOTOXICOSIS OF *ASPERGILLUS CLAVATUS*

Symptoms of poisoning

The toxic nature of patulin has been clearly demonstrated by the work of Hopkins (1123), Broom *et al.* (318), and Katzman *et al.* (1280). Very serious cases of poisoning have been especially recorded in cows: some in France after consumption of germinated cereal grain contaminated with *Aspergillus clavatus* (1738), others in Japan after ingesting malt contaminated with *Penicillium urticae* (2692).

Superficial symptoms

Whereas in mice, which have been injected with patulin, the first symptoms are increased agitation and rapid respiration (1314), in the case of naturally intoxicated cattle the main symptoms recorded are incoordination of movement, a curving of the back, and paralysis of the hind limbs resulting in the animal falling as soon as it reaches slippery ground. Badly affected cows show extreme nervousness associated with trembling, loss of appetite, they do not chew the cud, they show paralysis of the gastric system, and constipation all accompanied by a significant reduction in lactation (1738). Incontinence in urination is also noticed (1).

Internal symptoms

The pathological changes most clearly seen, both in the mouse in the laboratory (702, 1314) and in the cow on the farm (1738), are the following:

– the lungs are congested, either in patches or completely, with thickening of the alveolar walls; pulmonary oedema, accompanied by haemorrhage, is a characteristic symptom of poisoning;
– the liver shows areas of degeneration; the bile duct is large and distended with bile;
– the spleen is always congested;
– the kidneys are hypertrophied, congested with degeneration and lesions of the tubules (2717); the suprarenal glands doubling in volume; there is a reduction, sometimes even a suppression, of urine formation (this symptom is observed in the rat at much lower doses of patulin);
– the cardiac rhythm and blood pressure are disturbed (57);
– ocular troubles sometimes accompany these symptoms;
– after subcutaneous injection, the tissues show inflammation followed by oedema, necrosis, and finally an open sore. After repeated injections, Dickens and Jones (644) found the development of sarcomas analogous to those provoked by other unsaturated lactones.

Mode of action

The state of our knowledge in this area has been reviewed very well by Singh (2396, 2397).

Inhibition of cell division – Patulin does inhibit cell division and the appearance of giant cells in bacteria, following treatment with patulin, is associated with this property (113). It is also known that binucleate cells are formed in maize and onion roots exposed to patulin (2775). There appears to be inhibition of the formation of a cross-wall.

Some reports suggest that chromosomes are not affected (2775) but others suggest partial or complete fragmentation (2342).

Having confirmed that the growth of fibroblasts in tissue culture is arrested, several authors conclude that patulin has an antimitotic activity, mitosis being blocked at metaphase (318, 580, 624, 1287, 1288). Does this involve changes in the spindle? Steineger and Leupi (2484) do not believe this to be so from their studies of the nuclei of *Allium cepa* and *Lepidium sativum.* Rondanelli (2189), on the other hand, thinks that patulin reduces the elasticity of contractile fibres by reacting with sulphydryl groups; in fact, the addition of sulphydryl compounds such as cysteine or dimercaptopropanol inhibits the mitostatic effects of patulin.

However, although so many workers have reported inhibition of the toxicity of patulin by cysteine, sodium thioglycolate, and dimercaptopropanol (95, 360, 580, 624, 702, 890, 1663, 2161, 2189), Cavallito and Bailey (426) only obtained this inhibition with cysteine. On the other hand, cysteine increases the inhibitory activity of patulin on the formation of nitrate reductase by higher plants (17).

It is also reported that peptone, glycine, methionine, asparagine, and *para*-aminobenzoic acid on the one hand (702, 2762) and sodium sulphite, sodium thiosulphate, and sodium pyrosulphite on the other (1663) all reduce the antibiotic activity of patulin whereas tryptophan, urea, and thiourea enhance it (702).

Various authors envisage patulin acting at the level of cell metabolism by combining with sulphydryl groups of some enzymes or essential metabolites and disturbing mitosis by this interaction with thiol groups (426–428, 890, 2161), although Singh (2396) considers that such an hypothesis is not sufficiently justified.

Inhibition of respiration – Patulin inhibits the aerobic respiration of bacteria (437, 438, 1455, 1456), of fungi such as the mycelium of *Claviceps purpurea,* as well as cell-free extracts from this organism (949). Patulin also inhibits the respiration of phagocytes (624), brain cell tissue, and the kidney cell tissue cultures of guinea-pig (57).

It is possible that the inhibition of respiration by patulin may by the cause of the inhibition of growth. However, growth and respiration may both be perturbed by the action of patulin on the semipermeable membranes of cells (886). It has also been reported that patulin reduces the absorption of potassium ions by erythrocytes (1270) without affecting the haemoglobin (949). It inhibits the absorption of glucose by mycelium (2396) without affecting such metabolites as inorganic phosphate, sucrose, and amino acids. In short, patulin behaves as though it disturbs some transport mechanism. Absorption of glucose, or potassium, requires energy which would be obtained from respiration so inhibition of absorption may perhaps be interpreted as much as a result as a cause of the inhibition of respiration (2397).

THERAPY

The worst cases of poisoning associated with patulin seem to be those observed in cattle. Only rigorous hygiene in cattle stalls will reduce the risk of introducing *Aspergillus clavatus.*

Treatment of poisoned animals with vitamin preparations containing vitamin A and the B complex enriched with B_1, such as phenergan, seems to give some amelioration, whereas intramuscular and intravenous injections of acepromazine maleate can be used as a tranquillizer (1202).

There is evidence that patulin may be produced in large quantities in apples infected by *Penicillium expansum* and the consumption of such mouldy fruits should be avoided. The fruit juice industry should never use rotten apples (297), a practice which is happily decreasing. Even heating fruit juices to 80 °C is insufficient to destroy patulin, which is fairly thermostable at low pH (2323).

CHAPTER 5

OTHER ASPERGILLOTOXICOSES

INTRODUCTION

The two toxicoses associated with the genus *Aspergillus* described in the previous two chapters, namely aflatoxicosis associated with *A. flavus* Link and clavacitoxicosis associated, in the main, with *A. clavatus* Desm., are important but other representatives of the genus have been implicated in cases of poisoning. They are less well known, and their effects may be less grave, but their existence should not be neglected.

Thus, Frayssinet and Lafont (832) demonstrated that, of 110 cultures of *Aspergillus* isolated from cereals or foods made from them, 50 were toxigenic to ducklings. In another study, Semeniuk *et al.* (2336, 2337), investigating the toxicity of 392 isolates of *Aspergillus,* representing 132 species and 19 varieties, to chickens or guinea-pigs, have shown that 55% of them produced toxins.

It is noted that Jaquet and Boutibonnes (1197) recently proposed a classification of these moulds based essentially on spore characteristics. In this book the various aspergillotoxicoses are described following the systematic grouping of *Aspergillus* provided by Raper and Fennel (2105). Figure 64 shows the main outlines of this classification indicating those species known to be toxic.

INTOXICATIONS ASSOCIATED WITH THE *GLAUCUS* GROUP

A score of different species is now recognized among the moulds previously lumped together under the name *Aspergillus glaucus* Link, considered as conidial forms of *Eurotium herbariorum* Link. They are all remarkable in their abundant production of pigments (80, 452) and their nutritional requirements (1746).

Widespread in nature, they are particularly associated with media of high osmotic pressure or low water activity (jams, seeds, flour, dried fruits, leather, etc.) and are often difficult to obtain in culture on the usual isolation media. Members of the *glaucus* group have long been considered inoffensive (1592), perhaps because experiments have been of short duration or because of lack of concern (665).

Species incriminated

Depending on the conditions, it is usually either *Aspergillus chevalieri* (Mangin) Thom and Church or *A. amstelodami* (Mangin) Thom and Church which are associated with toxicity. They appear to be two closely related series of isolates which are distinguished particularly by the shape of the ascospores, in the case of *A. chevalieri* the lenticular ascospores have a well marked equatorial groove and are smooth whereas, in the case of *A. amstelodami,* the ascospores are covered with wart-like ridges (Figure 65).

On media rich in sugar or salt the mycelium of each fungus is a bright sulphur yellow with olive green conidial heads. On salt rich media, *A. chevalieri,* but not *A. amstelodami,* produces a red pigment as well. In both species, the phialides are short and dense, producing subglobose spores, of about 5 μm, which are finely echinulate. In *A. amstelodami* the phialides tend to cover the whole of the vesicle bearing them, whereas in *A. chevalieri* they are subradial.

	Groups and toxic species	Colour	Perithecia	Osmophilic	Conidiophore
1. Phialides strictly uniseriate	 *clavatus* *A. clavatus* Desm. *A. giganteus* Wehm.	blue - green to grey	–	–	smooth
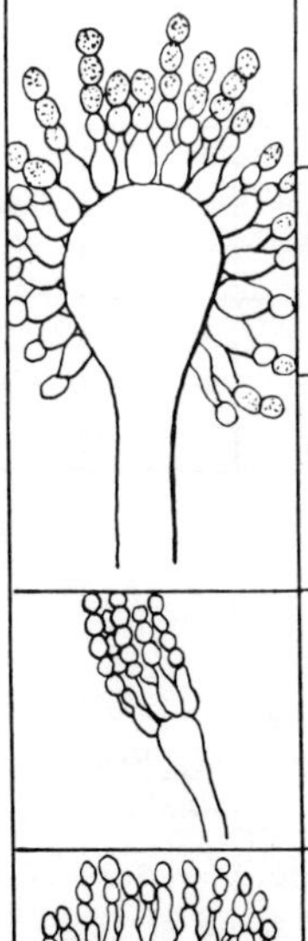	*glaucus* *A. chevalieri* (Mang.) T. et C. *A. amstelodami* (Mang.) T. et C. *A. ruber* (Spiek et Brem.) T. et C.	blue-green to olive green	+	+	smooth
	ornatus	grey to yellow-green	+	-	smooth
	cervinus	fawn to pinkish	–	–	smooth
	*restrictus* *A. restrictus* G. Smith	green	–	+	smooth
	*fumigatus* *A. fumigatus* Fres.	grey-green to blue-green	±	–	smooth

Figure 64. Key to the groups of the genus *Aspergillus*

	Groups and toxic species	Colour	Perithecia	Osmophilic	Conidiophore
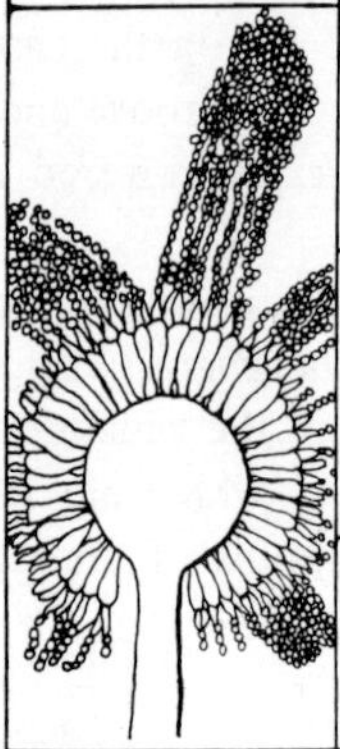 2. Phialides usually biseriate, sometimes uniseriate	*ochraceus* *A. ochraceus* Wilh.	yellow to ochre	+	–	rough
	niger *A. niger* v. Tiegh. *A. luchuensis* Inui	black	–	–	smooth
	candidus *A. candidus* Link	white to cream	–	–	smooth
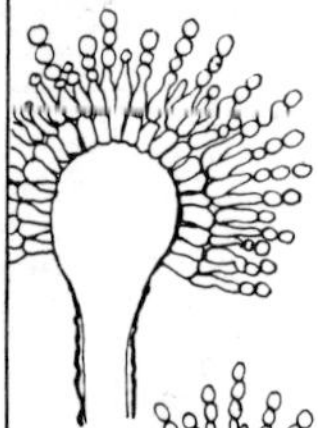	*flavus* *A. flavus* Link *A. parasiticus* Speare *A. oryzae* (Ahlb.) Cohn *A. oryzae* (Ahlb.) Cohn var. *microsporus* Sak. *A. oryzae* (Ahlb.) Cohn var. *effusus* (Tir.) Ohara *A. avenaceus* G. Smith *A. tamarii* Kita	yellow-green to olive brown	–	–	rough
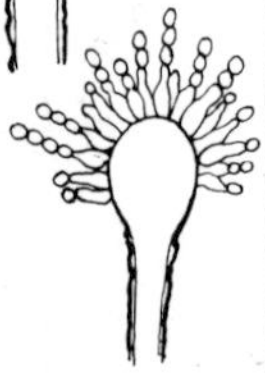	*wentii* *A. wentii* Wehm. *A. terricola* Marchal *A. thomii* Smith	brownish yellow	–	–	smooth
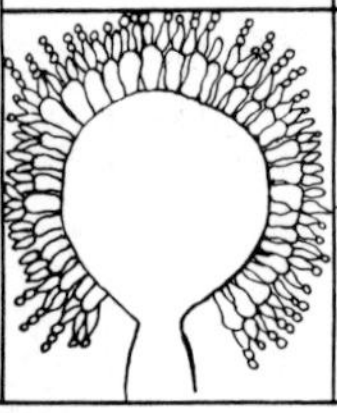	*cremeus*	pale yellow-green	+	+	smooth
	sparsus	grey, green to olive	–	–	smooth

Figure 64 (continued)

A. ruber (Spieck and Brem.). Thom and Church and *A. restrictus* G. Smith (now considered as not belonging to the *glaucus* group) frequently occur on groundnuts and have sometimes been considered to play a role in the toxicity of some meals (628, 2708). Kulik and Holaday (1365) have suggested that some strains of *A. ruber* may produce aflatoxin B_1. Notwithstanding the extent to which *A. restrictus* may infect wheat in storage (471), it does not seem to have been responsible for any serious toxicoses. Filtrates of cultures of *Aspergillus umbrosus* Bain. and Sartory have been shown to be toxic to mice (2216).

		Groups and toxic species	Colour	Perithecia	Osmophilic	Conidiophores
Phialides strictly biseriate		*versicolor* *A. versicolor* (vuil.) Tirab. *A. sydowii* (Bain et Sart.) T. et C.	greenish	scherotia	–	smooth
		ustus	olive to dull brown	–	–	smooth
		nidulans *A. nidulans* (Eidam) Wint.	yellowish green	+	–	smooth
		flavipes *A. flavipes* (Bain. et Sart.) T. et C. *A. niveus* Blochw. *A. carneus* (v. Tiegh.) Blochw.	white, hazel to pink	–	–	smooth
3.		*terreus* *A. terreus* Thom	red - orange	–	–	smooth

Figure 64 (continued)

Toxic compounds

The exact nature of the toxic metabolites produced by these fungi is still unclear but, because of their chemical relationship with known toxic fungal metabolites, the quinone pigments may play a significant role.

From 1934–39 Gould and Raistrick, with their collaborators, isolated from several species of the *glaucus* group three series of quinonoid pigments (950):

– a yellow pigment, flavoglaucin, which is formed in the perithecia,
– a red pigment, rubroglaucin, which was shown to be a mixture of physcion and erythroglaucin several years later,
– an orange pigment, auroglaucin (see Figure 66).

In 1939, Ashley, Raistrick, and Richards (80) isolated the anthranols A and B (Figure 67) thus bringing the number of characterized pigments of *A. glaucus* to six.

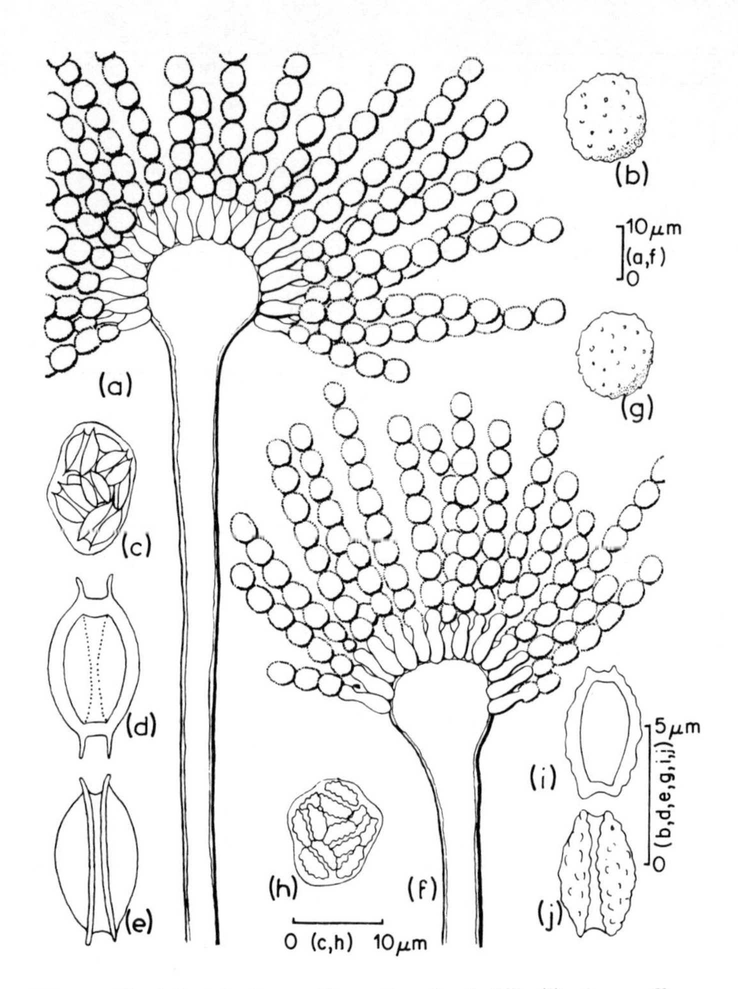

Figure 65. (a)–(e) *Aspergillus chevalieri.* (f)–(j) *Aspergillus amstelodami.* (a), (f) Conidial head. (b), (g) Conidia. (c), (h) Asci. (d), (i) Ascospores, in optical section. (e), (j) Ascospores, in profile

Flavoglaucin and auroglaucin do not seem to have any toxicity to animals, nevertheless the possibility cannot be ruled out that these pigments might be transformed into toxic metabolites. In the case of erythroglaucin, physcion, and the anthranols, these are related to the anthraquinone metabolities of *Penicillium islandicum,* one of the moulds involved in yellow rice toxicosis, some of which may be carcinogenic. In fact, erythroglaucin is none other than 7-methoxyislandicin (80).

The transformations which these pigments undergo, which are dependent on the pH and redox potential of the medium, indicate that these pigments may be involved in oxidation-reduction processes in the cytoplasm, a point emphasized by Mlle Chollet (452). It is thus possible to see that the formation of toxic metabolites

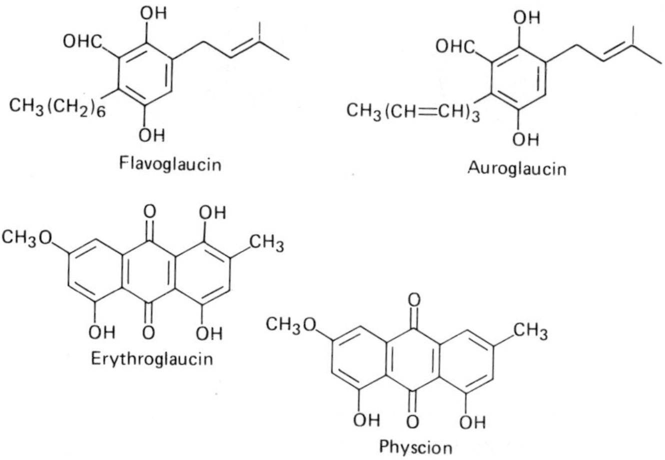

Flavoglaucin

Auroglaucin

Erythroglaucin

Physcion

Figure 66. Four pigments associated with the *glaucus* group

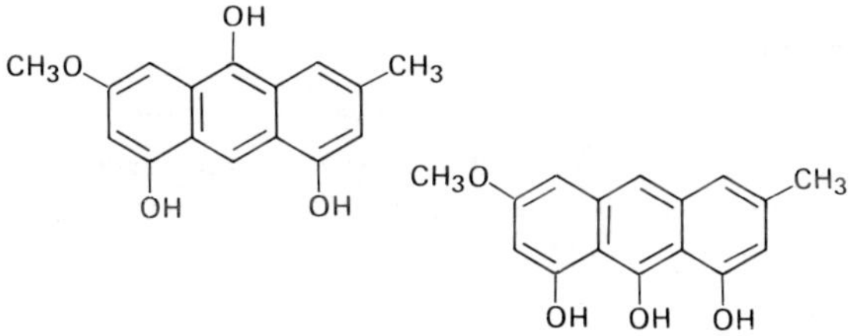

Figure 67. Two anthranols isolated from *Aspergillus glaucus* (80)

by fungi of the *glaucus* group is very sensitive to the precise factors to which the fungi are exposed during growth.

It has been shown that *A. chevalieri* is able to produce gliotoxin (2826), an antibiotic produced by *Gliocladium virens* and a number of other moulds. It has a relatively low toxicity (the LD_{50} for the mouse varies from 45–65 mg/kg depending on the route of administration) which, however, should not be ignored.

Xanthocillin X (Figure 68), produced by *A. chevalieri* and some strains of *Penicillium notatum,* is considered to be hepatotoxic (551). The LD_{50} for the mouse is 25 mg/kg by intramuscular injection, 35 mg/kg intraperitoneally and 40 mg/kg orally (2830).

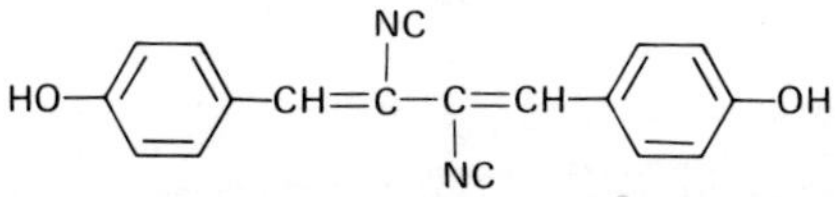

Figure 68. Xanthocillin

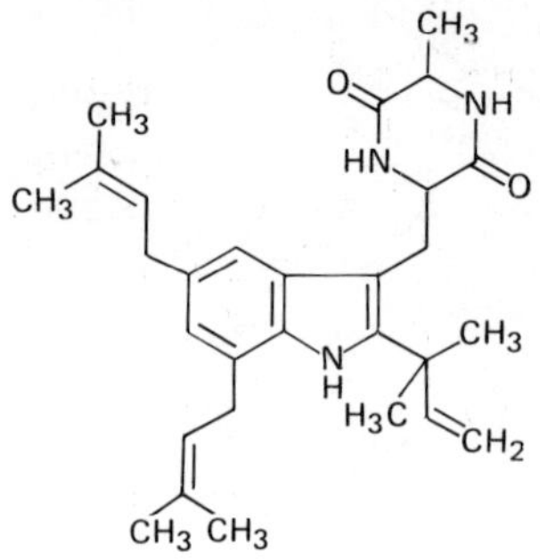

Figure 69. Echinulin

Finally, a compound with an unusual structure derived from indole, amino acids and terpene precursors (1668), has been isolated from *A. echinulatus, A. chevalieri, A. amstelodami,* and *A. repens.* It has been called echinulin (Figure 69), $C_{29}H_{39}O_2N_3$ (2068–2073) but its role in toxicoses has not been established.

Symptoms of poisoning

External symptoms – In cattle it is possible that cases of hyperkeratosis of the skin may be associated with *A. chevalieri* (1438). It has been demonstrated that sloughing and thickening of the skin, accompanied by loss of hair, are observed in the region of the neck and shoulders of the cow after application of toxic extracts of this same fungus (414, 415). Hyperkeratosis may also be provoked by substances produced by several other aspergilli, especially *A. clavatus* and *A. fumigatus* (412).

Systematic experiments have been carried out in South Africa (2076) by adding dried crushed maize, on which the fungus has been grown, to the food of animals, and it has been demonstrated that *A. amstelodami* has some toxicity to farm animals. Chickens and ducklings show a clear reduction in growth rate (Table 27) and White Leghorn and New Hampshire chickens show more significant lesions than do ducks (2075).

In the rabbit, nervous symptoms, such as hypersensitivity analogous to that caused by strychnine, are seen first followed by difficulty in walking. These symptoms are followed by a loss of appetite if the dose of toxin is sufficiently high.

Table 27. Rate of growth (g) of chickens and ducklings fed with material contaminated with *Aspergillus amstelodami* (2076)

CHICKENS	*1st Day*	*17th Day*	*28th Day*	
Control	38	177	320	
Feed with *A. amstelodami*	38	129	226	
DUCKLINGS	*1st Day*	*14th Day*	*28th Day*	*37th Day*
Control	44	452	1,031	1,091
Feed with *A. amstelodami*	44	178	482	598

A haemorrhagic syndrome, associated with *Aspergillus chevalieri,* has been reported in poultry (2306).

Histopathology – Histological examination of poisoned animals (2076) shows a clear proliferation of the bile ducts in the capsule of Glisson and the intralobular region, but their distribution is more diffuse than that observed in aflatoxicosis. When intoxication is particularly severe a marked fibrosis is noted.

In the rabbit, fatty infiltration and a slight lymphocytic encephalitis are also observed.

In the mouse affected by the toxins of *A. mangini,* renal lesions have been reported, with congestion of the cortex, the glomerules, and the medullary region, as well as degeneration of the proximal tubular epithelium (2717).

POISONING ASSOCIATED WITH *ASPERGILLUS FUMIGATUS*

The problems with *A. fumigatus*

Aspergillus fumigatus Fres. has been known for a long time as an agent of mycoses. Without a doubt, the observations of Gaucher and Sergent (883) in 1894 and those of Rénon (2142, 2143) in 1895 and 1897 now fall in the category of folklore. They concern the 'crammers' of pigeons who, professionally, masticate the grain and pass if from their mouth to the beak of the pigeon; also to those people involved in the cleaning of horsehair destined for the production of wigs.

Pulmonary aspergillosis is fortunately quite rare in mammals although it does seem to have reappeared again in recent years. In contrast, it is one of the principal causes of death among young birds, the fungus growing with such a profusion of spores (as many as 10–50 million spores have been counted in the lungs of young chickens) that they obstruct the airsacs and penetrate into neighbouring tissue (102, 1062). It thus produces many deaths on chicken farms. As a particular instance, aspergillosis is the manner in which many Emperor penguins have died throughout zoological gardens in the United States, despite the precautions made on their arrival (731).

However, the problems associated with *A. fumigatus* are not confined to pulmonary mycoses. The fungus may proliferate following a trauma in the eye giving rise to ulceration of the cornea (780); it may invade the sinuses (2141); *Aspergillus* otomycoses (967) are also known, infection of the bone (2353), generalized infections (429, 762, 960), and mycotic abortions (107).

Along with *A. flavus,* it may cause a very destructive disease of bees and its transmission to bee keepers has sometimes resulted in death.

Finally, the matter which particularly concerns us here, *A. fumigatus* may produce toxic effects through products of its metabolism.

Aspergillus fumigatus

Morphology – *Aspergillus fumigatus* Fres. forms close, velvety colonies which are slightly flocculose at the centre and are at first whitish, rapidly becoming green to

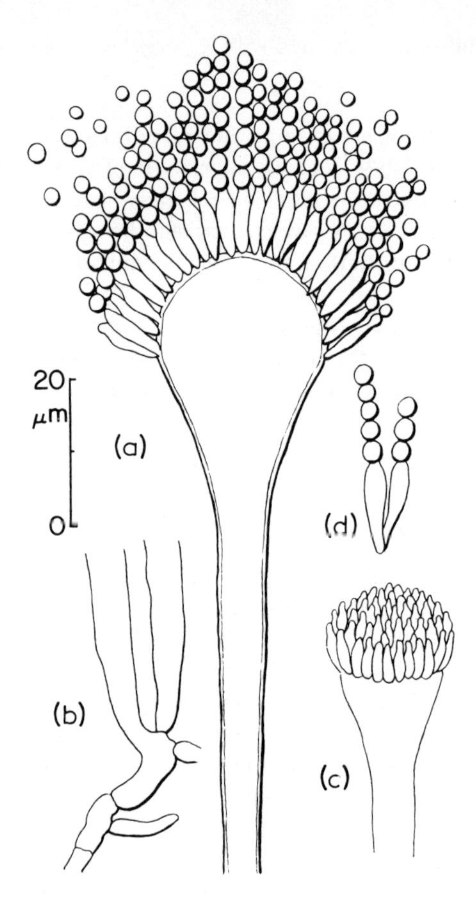

Figure 70. *Aspergillus fumigatus* (a) Conidial head. (b) Base of conidiophores. (c) Apical position of the phialides. (d) Phialides and spores (after Pelhâte)

greenish grey. The conidiophores are smooth, rather short (about 300 x 5–8 μm) and are terminated by a vesicle of 20–30 μm in diameter which is only fertile over the upper region (Figure 70). The phialides of 6–8 x 2–3 μm are uniseriate, erect, and parallel to the conidiophore axis and the conidia, which are dark green in mass, are globose to subglobose, finely echinulate and measure 2.5–3 μm in diameter. They form long adhering chains which are erect and parallel to one another to give a compact column of spores which may be up be to 400 μm long. In the electron microscope, the spores appear to be covered with an interlacing pattern of fibrous bundles orientated in all directions, with localized, prominent cones (1069).

Growth – This species is thermotolerant and is able to grow between 18–55 °C with an optimum for growth at 37 °C. At 55 °C most spores are still capable of germination but the mycelium grows very slowly (769).

Ecology – Although aspergilli are not frequently isolated from temperate soils (1983) *Aspergillus fumigatus* is recorded from time to time. Remacle (2137), for example, recorded it in the soil of oak copses in Belgium. According to Saksena (2225), who has isolated it in India, it is a pioneer species which is resistant to drying and may occur in denuded soils.

Its thermotolerant properties allow this species to grow in compost heaps or on organic matter undergoing decomposition (769, 1318). This explains, for example, its presence in peat with other thermophilic fungi (1159, 1382) as well as in stored wheat grain undergoing self-heating (1211).

It is frequently isolated as a pathogen of birds, domestic animals, and sometimes man, indeed, it was in a bustard in the zoological gardens of Frankfurt that Fresenius first described this species in 1863.

It has cellulolytic properties which allow it to grow on jute fibres, cotton, etc. and it is able to grow on media with a high concentration of arsenic (1341).

Toxic Metabolites

The toxicity of *A. fumigatus* has been known for some time (245, 309), but it has sometimes been difficult to know which metabolite is responsible, for it produces a number of metabolites, many of which are toxic. Among the antibiotics produced by this mould one called fumigacin was subsequently shown to consist of a mixture of gliotoxin and helvolic acid (1647).

Fumagillin (Figure 71) – This metabolite was first isolated from the filtrates of a strain of *A. fumigatus* in 1954 (1043), one particular strain, isolated later in Russia, producing enormous quantities of this substance (180). Its empirical formula is $C_{26}H_{34}O_7$ (265) and it is one of the most cytotoxic compounds produced by *A. fumigatus.* It inhibits the synthesis of deoxyribonucleic acid, thus acting like poïn, sporofusarin, barbituric acid, and streptomycin (175). The LD_{50} of fumagillin for the mouse is 800 mg/kg subcutaneously (1314), and 2,000 mg/kg by oral administration (207) and its toxicity is low enough for it to be used in medicine in the control of amoebic dysentry under the names fumidile, fumogilline, amebaciline and phagopedine sigma. It must, however, be used under strict control because of its toxicity.

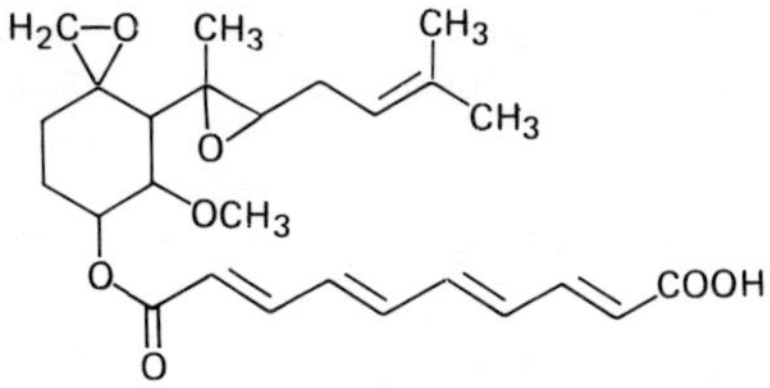

H₂C—O CH₃ CH₃
CH₃
O
OCH₃
O
COOH
O

Figure 71. Fumagillin

Helvolic acid (Figure 72) – Helvolic acid is a triterpene with an empirical formula $C_{33}H_{44}O_8$ which is known in the natural state as a diacetate (47, 207). It has the

Figure 72. Helvolic acid (see Iwasaki, S., Sair, M. I., Igarashi, H. and Okuda, S. *Chemical Communications*, 1970, p. 1119 and cf. Ref. 1885)

following physical characteristics: m.p. 208–211 °C; $[\alpha]^{25}_{D}$ −124 °C (chloroform); λ_{max} 322 nm (ϵ = 98), 231 nm (ϵ = 17,300).

The LD_{50} of helvolic acid for the mouse is 400 mg/kg by the intraperitoneal route and mice tolerate doses of up to 250 mg/kg intravenously or subcutaneously and up to 1,000 mg/kg by oral administration (1314). Repeated intravenous or subcutaneous injections of 1,000 mg/kg provoke a fatty degeneration of the liver (207).

There is a particular mutant referred to as *helvola* of *A. fumigatus* which produces a large yield of helvolic acid (439).

Gliotoxin – This compound, isolated in 1936 from *Gliocladium virens*, is also produced by *A. fumigatus* (1647). When first noticed by Stanley (2479) he referred to it as aspergillin and Plotnikov *et al.* (1999) called it 'antibiotic No. 13'.

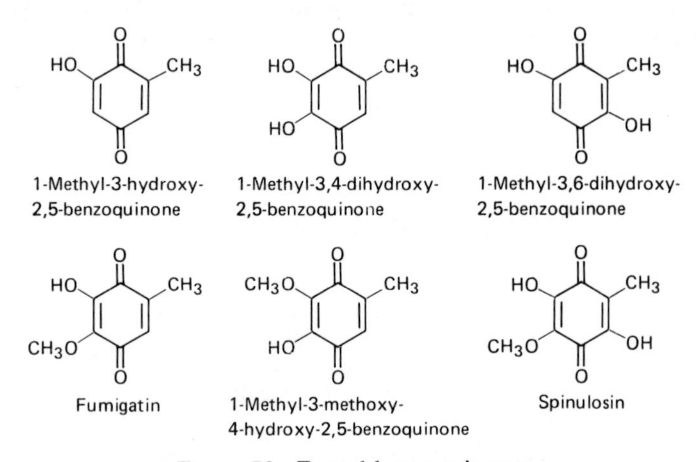

Figure 73. Fungal benzoquinones

Quinone derivatives – *Aspergillus fumigatus* produces a series of derivatives of 2,5-toluquinone (2907), the best known of which are fumigatin and spinulosin (Figure 73).

Fumigatin appears to destroy leucocytes and it has been suggested that the toxicity of the toluquinone derivatives is enhanced by the presence of the methoxyl group in the molecule (2355). Spinulosin has also been isolated from *Penicillium spinulosum* (1906).

Alkaloids and tremorgenic compounds – Some strains of *A. fumigatus* produce, in culture, alkaloids similar to those of *Claviceps purpurea,* the ergot of rye (Figure 74). They include such compounds as agroclavine, elymoclavine, secaclavine (= chanoclavine) $C_{16}H_{20}N_2O$, fumigaclavins A, B, and C ($C_{22}H_{30}N_2O_2$), and festuclavine $C_{16}H_{20}N_2$ (2462, 2908).

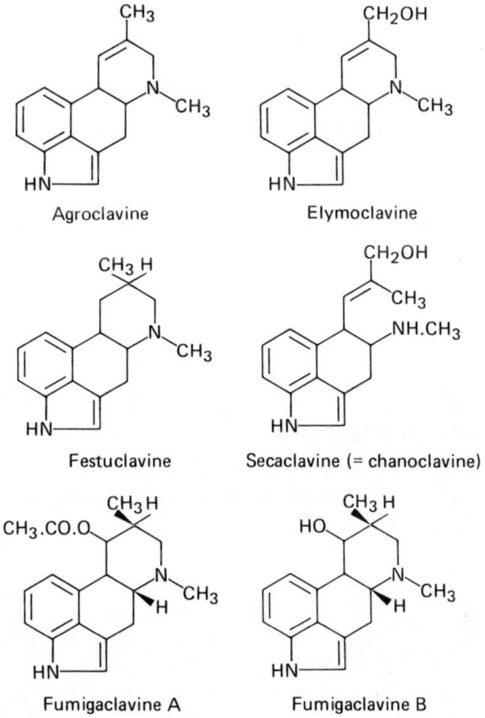

Agroclavine   Elymoclavine

Festuclavine   Secaclavine (= chanoclavine)

Fumigaclavine A   Fumigaclavine B

Figure 74. Alkaloids isolated from *Aspergillus fumigatus*

Two tremorgenic compounds have been isolated by chromatography on a silica gel column using n-hexane and benzene/acetone (4:1). They have been called fumitremorgin A ($C_{32}H_{41}O_7N_3$) and fumitremorgin B ($C_{27}H_{33}O_5N_3$) for which the structures given in Figure 75 have now been proposed (see Chapter 13).

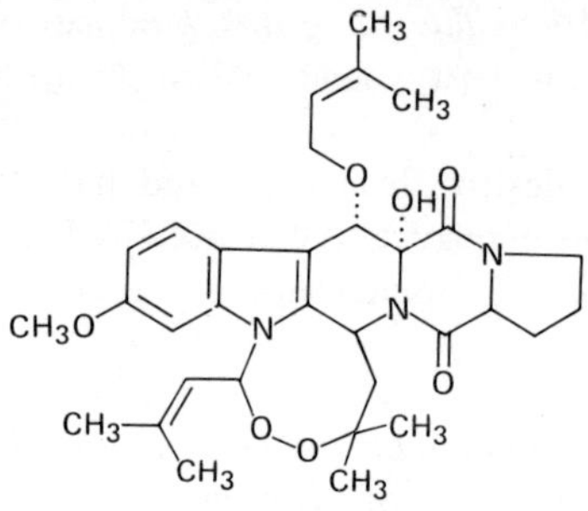

Fumitremorgin A

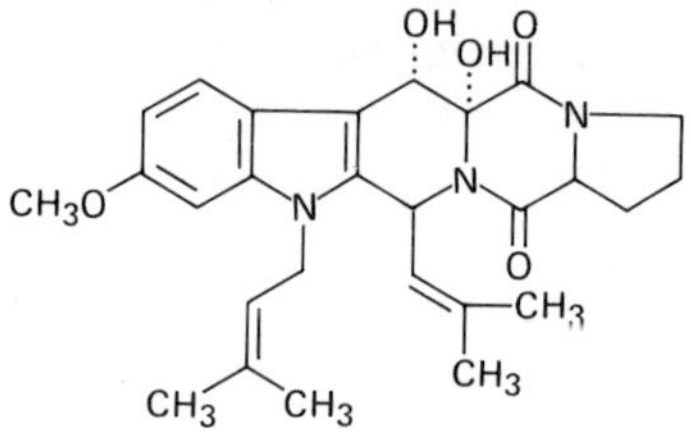

Fumitremorgin B

Figure 75. The fumitremorgins (2979, 3087)

Addition of L-tryptophan to the culture medium stimulates the production of these toxins and they are indeed indole derivatives (2910).

Other compounds which have been reported from *A. fumigatus* include monomethylsulochrin, $C_{18}H_{18}O_7$ (Figure 76) and trypacidin, $C_{18}H_{16}O_7$ (Figure 77) (1646).

Attempts have been made to localize the toxins of *A. fumigatus* and Ceni and Besta (431) have demonstrated that extracts of conidia using alcohol or ether have toxic effects on the nervous system of laboratory animals. As well as toxins liberated into the medium, it should be emphasized that *A. fumigatus* can produce endotoxins (2228, 2604, 2605).

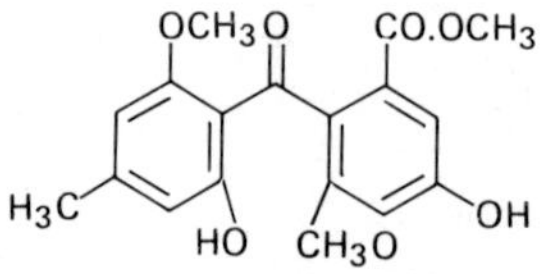

Figure 76. Monomethylsulochrin

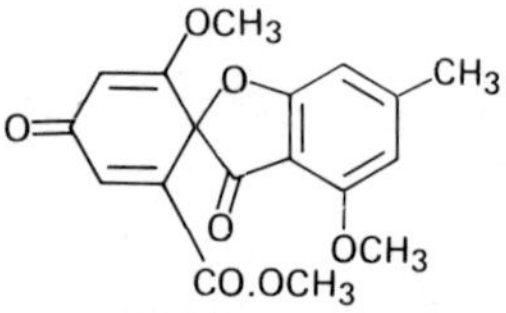

Figure 77. Trypacidin

Symptoms of poisoning

It is difficult to separate the symptoms of mycoses from those of toxicoses due to *Aspergillus fumigatus.* This is the case, for example, in mycotic abortions frequent in cattle, often following infection of the genital organs.

The symptoms observed will, of course, depend on which toxin predominates but activity on the muscular and nervous systems (431), with convulsions, tetany, and paralysis, followed often by death within several hours, are frequently described (245).

Pigs, having consumed cereal flour badly contaminated with *A. fumigatus,* showed signs of weakness, loss of appetite, and hypothermia. Some individuals had a staggering gait and showed nervous symptoms. Haemorrhagic enteritis was also observed and several animals died one or two days after the illness first appeared. Autopsy demonstrated significant haemorrhages in the thoracic and abdominal cavities, numerous infarctions of the spleen, congestion of the kidneys, acute hepatic toxicity, and hypertrophy of the mesenteric ganglia (1396).

The haemorrhagic symptoms, as well as lesions of the viscera, appeared to be reproducible in cattle fed with artificially infected cereals (415).

Hyperkeratosis of the skin of cows and horses has also been demonstrated by simply applying ether extracts of a culture (413).

Henrici (1061) has been able to induce oedema, haemorrhages, fatty necrosis of the liver, necrosis of the kidney tubules in rabbits and guinea-pigs by injecting extracts of the mycelium of *A. fumigatus.* Experimental kidney lesions have also been observed in mice, chickens, pigs (2206), and sheep (2597). Slight hydropic degeneration of the proximal tubules and occasional pycnosis of epithelial cells at the cortico-medullary junction were described (2717).

These metabolites have also been studied for their haemolytic properties (2107) and their dermal and nephrotoxicities. There is not always any clear relationship between the toxicity and the antibiotic properties of extracts from strains of *Aspergillus fumigatus* (2205, 2221).

Moulded melilot disease

A haemorrhagic syndrome in cattle, referred to as 'mouldy melilot disease', has been known since 1921 and associated with the consumption of *Melilotus officinalis* Lam. or, more rarely, of hay containing *Lespedeza.* This disease has been very significant in certain parts of the United States and Canada (310, 323, 397, 786, 821, 1021, 1136, 1309, 1638, 1938, 2176–2178, 2257, 2283, 2284).

The nature of the toxin responsible has been the subject of numerous studies but there is still doubt about the role of the moulds and opinions are contradictory.

As early as 1924, Schofield (2284) isolated an *Aspergillus* from the stems of melilot consumption of which had caused outbreaks of poisoning. He was able to reproduce a similar toxicosis by giving rabbits a feed infected with this *Aspergillus.* The mould grew abundantly on the hollow stems of the plant, especially if they were stored in a humid atmosphere and the material remained toxic when stored for three to four years.

There have been cases where toxicity has been associated with melilot which did not appear to be mouldy and, on the other hand, toxicity tests on moulds readily isolated from melilot have been negative. The species *Melilotus dentata* does not appear to cause any poisoning even when it is moulded (301, 2430). From these apparently contradictory facts it is tempting to propose the following explanation

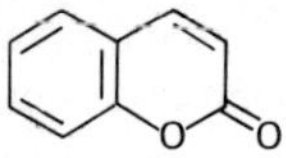

Coumarin

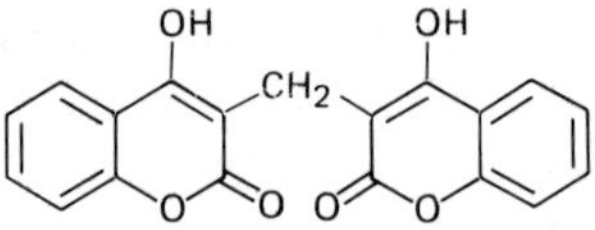

Dicoumarol (melitoxin)

Figure 78. Melitoxin and its relationship with coumarin

(397, 398, 1136. 2474). *Melilotus officinalis,* but not *M. dentata,* produces coumaric acid, which is converted into coumarin by the loss of a molecule of water. Certain fungi which grow on *M. officinalis* are apparently able to transform coumarin into methylene 3,3′ *bis* (4-hydroxycoumarin) (1506). This compound is usually called dicoumarol or melitoxin (Figure 78).

Coumarin is an astringent, healing compound considered to be non-toxic. On the other hand, dicoumarol is well known for its anticoagulant and anti vitamin K properties. The question remains as to which species of fungi are able to carry out this transformation of coumarin into dicoumarol.

Recently, Bye and King (381) have shown that a strain of *Aspergillus fumigatus,* isolated from mouldy hay, was able to convert melilotic acid (*o*-hydroxyphenylpropionic acid) and *o*-coumaric acid to 4-hydroxycoumarin and dicoumarol (Figure 79). The sequence of reactions is as follows:

MELILOTIC ACID $\xrightarrow{-2H}$ *O*-COUMARIC ACID $\xrightarrow{+H_2O}$

β-HYDROXYMELILOTIC ACID $\xrightarrow{-2H}$ OXOMELILOTIC ACID $\xrightarrow{-2H}$

4-HYDROXYCOUMARIN.

Melilot poisoning has been found almost exclusively in bovines. Some rare cases have been recorded in sheep and horses (1938) and sheep have been shown experimentally to be more resistant than cows (2178). The symptoms appear three to eight weeks after ingestion of the toxic feed and young animals are more sensitive than old (310). Typical symptoms have been observed in the new born calf when, with the same dose of toxic melilot, nothing was discernible in its mother. This sensitivity is partly due to a lack of reserves of vitamin K in the young animal.

Mouldy melilot poisoning is a haemorrhagic disease in which the animal bleeds to death either internally or externally. The fatal outcome is swift occurring one to three days after the first symptoms appear. In severe cases, sudden localized subcutaneous swellings are first noticed on some parts of the body, especially on either side of the vertebral column, on the thighs or near the shoulders. These

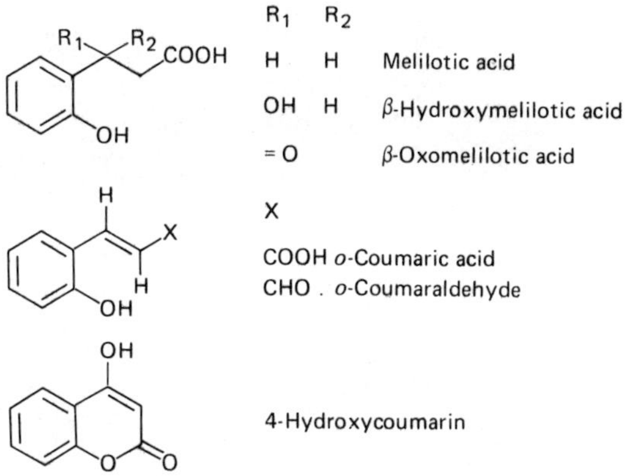

R_1	R_2	
H	H	Melilotic acid
OH	H	β-Hydroxymelilotic acid
= O		β-Oxomelilotic acid

X	
COOH	*o*-Coumaric acid
CHO .	*o*-Coumaraldehyde

4-Hydroxycoumarin

Figure 79. Intermediates in the metabolism of melilotic acid by *Aspergillus fumigatus* (381)

swellings may become quite large, up to several decimetres in circumference (1938). The mucous of the sick animal becomes pale, the pulse feeble, there is a general weakness and death occurs without any spasm.

The subcutaneous swellings are caused by hypodermic haemorrhages and are accompanied by internal haemorrhages in the muscles. If the haemorrhages are localized in particular tissues they may cause complications such as blindness, if haemorrhage occurs in the eyes, paralysis if it occurs in the nervous tissues. The presence of blood in the milk is frequent.

Clotting of the blood is much reduced and the slightest injury of the sick animal causes a considerable loss of blood.

Post-mortem examination shows subcutaneous and intermuscular haemorrhages which vary from petechiae to ecchymosis or haematomas. As well as parenchymatous organs, nearly all the tissues show general haemorrhages. On the heart, subendocardial haemorrhages occur especially in the left ventricle (2176).

The primary lesion consists of a reduction in the concentration of prothrombin in the blood, which explains the difficulties in coagulation (2178). Anaemia caused by mouldy melilot is not due to destruction of the blood cells, but results entirely from haemorrhages. This reduction in the concentration of prothrombin is progressive (2178), but it has not been demonstrated *in vitro* using the melilot toxin (397).

Administration of vitamin K counters the effects of dicoumarol (941, 1506) and oxalic acid and malonic acid have given satisfactory results in the treatment of haemorrhages in dogs poisoned by the coumarin derivative (222).

POISONING ASSOCIATED WITH *ASPERGILLIS OCHRACEUS*

Aspergillus ochraceus (Figure 80)

Aspergillus ochraceus Wilhelm, so-called because of the ochreous colour of its spores, is a common representative of the mycoflora of decomposing vegetable

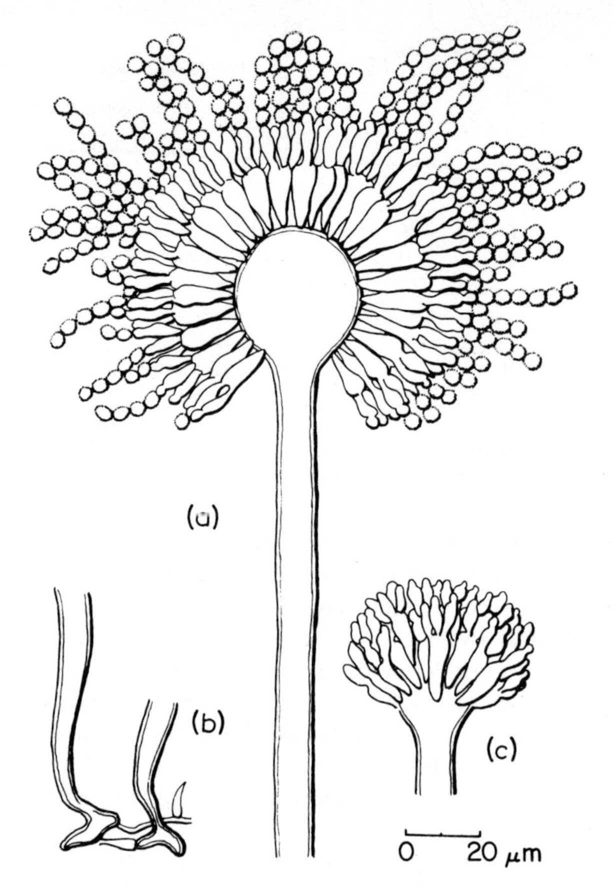

Figure 80. *Aspergillus ochraceus* (a) Conidial head. (b) Base of conidiophore. (c) Young conidial head in profile (after Pelhâte)

matter and, less frequently, it is isolated from mouldy seed (2376), rice (1704, 2909), groundnuts (696) or foods prepared from these materials. It plays a role in the deterioration of cottonseed and is also found in the soil.

Christensen (470) showed that its growth on wheat germ was rapid at 20–25 °C and a water concentration in the grain of more than 16%. Several toxic metabolites are produced by *A. ochraceus,* but not all strains are equally toxigenic (696, 1705), thus, for example, strain NRRL 3174 produces 29 μg ochratoxin per 100 ml nutrient medium containing 4% sucrose and 2% yeast extract (601) and strain M 298 only produces 2.5 μg on a potato dextrose medium (2263).

Toxic metabolites

Staron, Allard, Nguyen Dat Xuong and their colleagues (2483) purified three toxic metabolites from the mycelium and culture fluid of *A. ochraceus,* isolated from a feed which poisoned lambs or from mouldy flour which caused the death of heifers.

1. The major product had a molecular weight of 187 and contained C (80.51%), H (11.18%), and O (8.31%). (TRANSLATOR'S NOTE: a compound containing only carbon, hydrogen, and oxygen must have an even molecular weight!). The compound had a m.p. 160 °C and a single absorption maximum at 272 nm in the u.v. At low doses it caused a reduction of growth in mice and at higher doses led to death without any preliminary symptoms. It seemed to act by inducing the fatal agglutination of red blood cells in the liver.

2. The second compound, which is perhaps a secosteroid, had a molecular weight of 276 and gave an approximate analysis of C (69%), H (7%), O (25%). It had three maxima in the u.v. at 230, 300, and 392 nm. The LD_{50} by the intraperitoneal route in mice was 6 mg/kg, the animals showing incoordination of movement, followed by deep prostration with paralysis and loss of sensitivity. The compound appears to become mainly localized in the brain, the skin, and the kidneys.

3. The third compound, which is probably chemically related to the second judging from the similarity of their infrared spectra, had a molecular weight of 341 and analysis of C (68%), H (7.5%, O (24.5%). It showed two maxima in the u.v. at 225 and 322 nm and was only toxic at a relatively high dose (10 mg per mouse) causing erythema, petechiae of the head, reduction in skin sensitivity, and incoordination of movement.

The symptoms noted in the laboratory corresponded to those reported in animals affected by the feed from which the fungus was isolated.

Among the compounds known for some time as metabolites of *A. ochraceus* and related species are ochracin (= mellein), and 4-hydroxymellein which, although they have little toxicity, may possibly be precursors of the ochratoxins (538), penicillic acid (907, 1811), a metabolite produced in large quantities by *Penicillium cyclopium* and *P. puberulum* and which may be carcinogenic (643), hydroxyaspergillic acid, also produced by *Aspergillus flavus*, neohydroxyaspergillic acid, and 3-(1,2 epoxypropyl)-5,6-dihydro-5-hydroxy-6-methylpyran-2-one, as well as aspochracin (1811). However, the full toxic potential of *A. ochraceus* was only realized after the isolation of the ochratoxins from strain K-804 isolated originally from sorghum seed in South Africa (2707).

Ochratoxins – These are the most widely recognized toxins produced by *A. ochraceus* (2496, 2706, 2707). The isolation, purification, and characterization of these metabolites has been achieved, their degradation products known, and a complete synthesis accomplished (2167). A culture of *A. ochraceus* grown on crushed moist wheat for 20 days at 20 °C gives a high yield of ochratoxins (2264). Extraction of the toxins is by methods similar to those used in the extraction of aflatoxins. The ochratoxins can be extracted with warm chloroform, precipitated with hexane, and removed by filtration. The crude toxin is then redissolved in chloroform and purified by extraction with 0.5 M aqueous sodium bicarbonate

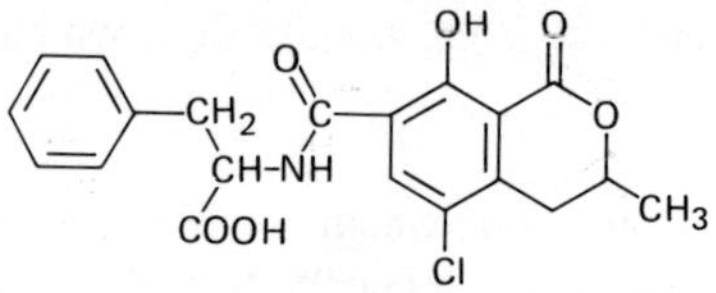

Ochratoxin A

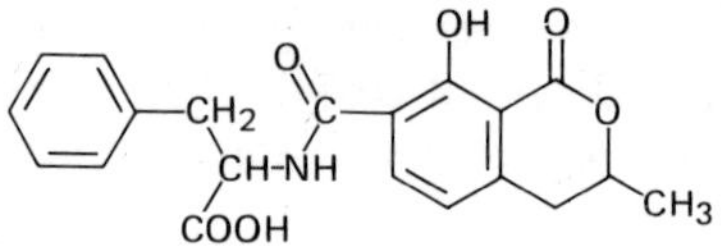

Ochratoxin B

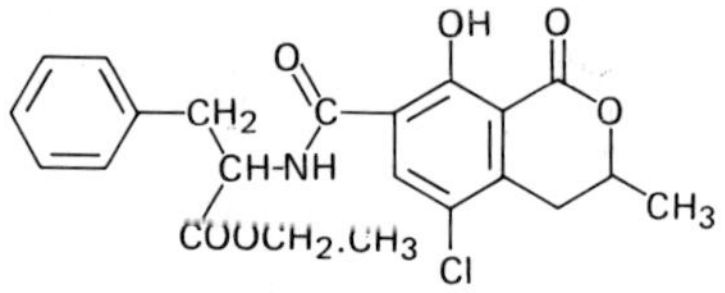

Ochratoxin C

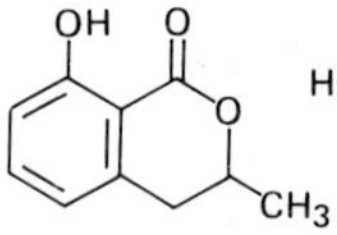

Ochracin (mellein)

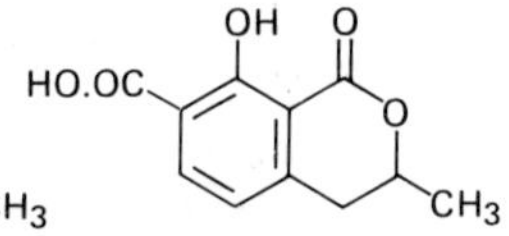

7-Carboxy-3,4-dihydro-8-hydroxy-3-methylisocoumarin

Figure 81. The ochratoxins and related metabolites

from which it can be re-extracted into solvent after acidification. Separation of the toxins may be achieved by TLC or on a column of silica gel using the following solvent mixtures for elution: benzene/acetic acid (4:1) or (9:1 v/v), benzene/methanol/acetic acid (12:2:1), toluene/ethyl acetate/90% formic acid (5:4:1) (1830, 2320, 2376). The intensity of fluorescence on chromatograms, or spectrofluorodensitometry compared with reference compounds allows the detection of the different ochratoxins (496, 2319, 2629). A bioassay using *Bacillus cereus* var. *mycoides* has also given good results (308).

The yield of ochratoxin varies with the culture medium and a good yield may be obtained in shake culture (both in Erlenmeyer flasks or in a fermenter) using glutamic acid (7.5–15 g/l) or proline (8 g/l) as the nitrogen source and sucrose (15–60 g/l) as the carbon source. The addition of 0.25% lactic acid is sometimes recommended (774, 775). Optimum production of ochratoxin A (by strain NRRL 3174 of *A. ochraceus*) on maize, rice or wheat occurs at 28 °C after 7–14 days depending on the substrate. The toxin is stable and withstands autoclaving for three hours (600, 2264, 2628). A pH of 6.0–6.3 is recommended for production media (1411). Ochratoxin seems to be produced with particular abundance on oats (2123).

The ochratoxins form a family of compounds of which the following have been recognized (Figure 81):

Ochratoxin A: $C_{20}H_{18}O_6NCl$ m.p. 94–96°C.
Ochratoxin B: $C_{20}H_{19}O_6N$ m.p. 221°C
Methyl ester of ochratoxin A: $C_{21}H_{20}O_6NCl$
Ochratoxin C = Ethyl ester of ochratoxin A: $C_{22}H_{22}O_6NCl$

The ochratoxins are essentially phenylalanine derivatives of an isocoumarin nucleus (2328). Degradation gives rise to the dihydroisocoumarin which is also detected in the urine of adult rats experimentally poisoned (2051). Hydrolysis of ochratoxin A with proteolytic enzymes gives L-phenylalanine and the compound referred to as ochratoxin α, 7-carboxy-5-chloro-8-hydroxy-3,4-dihydro-3R-methylisocoumarin, which has been found to be excreted in the bile (1993).

The structures of the ochratoxins are similar to those of diaporthin, produced by *Endothia parasitica,* and canescin, a metabolite of *Penicillium canescens* (Figure 82).

The toxicity of the various ochratoxins appears to be associated with the more or less facile dissociation of the phenolic hydroxyl group (498). The most toxic is ochratoxin A, which has an LD_{50} for one-day-old ducklings of 25 μg/duckling (2706). However, estimations of 150 μg/duckling have also been recorded (2051, 2497). For the albino rat, by oral administration, the LD_{50} has been reported to be 22 mg/kg for males and 20 mg/kg for females, being close to half that of aflatoxin B_1 (2056). For the larvae of the brine shrimp, *Artemia salina,* the LD_{50} is 10.1 μg/ml (1038).

The methyl ester of ochratoxin A is as toxic as ochratoxin A itself and, contrary to an earlier report (2705), ochratoxin C is also as toxic as ochratoxin A (2497), although its LD_{50} for one-day-old chicks is 216 μg compared with 166 μg in the case of ochratoxin A (497).

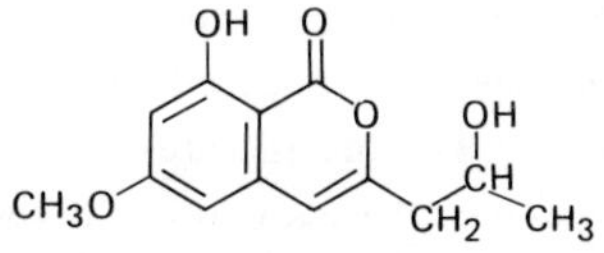

Diaporthin

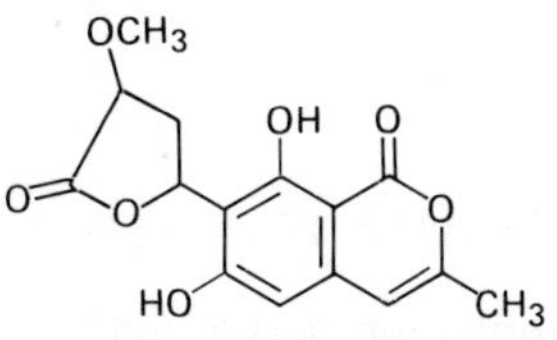

Canescin

Figure 82. Isocoumarin metabolites of fungi

Symptoms of toxicity

Ochratoxins affect many different animals and have caused the death of chickens (696), lambs, with paralysis and respiratory failure, heifers, with agglutination of red blood cells in the liver (2483), and of pigs, with cardiac ulcers (1944).

In 14-week-old White Leghorn chickens ochratoxin causes a retarding of sexual maturity and feeble egg production at low concentrations and, at higher concentrations, a very low egg production with hepatic and renal disorders and sometimes death (460).

The following symptoms have been observed in experimentally poisoned rats and ducklings (2056, 2586, 2622):

– enteritis;
– renal disorders and tubular necrosis of the kidneys;
– above all, hepatic disorders with necrosis of periportal cells in the liver which is generally associated with isolated cells and rarely with large groups. Just as with aflatoxicosis fatty infiltration into the liver cells is observed but there is not any proliferation of the bile ducts.

It is known that ochratoxin interferes with the enzyme phosphorylase causing an increase of glycogen in the liver (2056). This inhibitory effect is perhaps due to competition of the toxin with 3′,5′ cyclic AMP (1992).

Ochratoxin A, and one of its hydrolysis products (the dihydroisocoumarin) severely inhibit coupled respiration when applied to mitochondria isolated from rat liver at low doses (1718). Ochratoxin A is known to induce abortions in both laboratory animals as well as farm animals, especially bovines (2499).

Moulds other than *Aspergillus ochraceus* are known to be able to produce ochratoxins; the related species *A. sulphureus* (Fres.) Thom and Church, *A. sclerotiorum* Huber, *A. alliaceus* Thom and Church, *A. melleus* Yukawa (2337), as well as *Penicillium viridicatum* Westl. (2325, 2726). *A. sulphureus* produces a good yield of ochratoxin A in eight to ten days on a medium of glucose or sucrose, containing 100 mg of potassium per litre and 25 mg of phosphorus per litre at pH 6.0–6.3 (1411). *A. ostianus* has been reported to produce aflatoxins (2324) as well as ochratoxins A and B (1082). Culture filtrates of *A. petrakii* Vorös, isolated from rice, are toxic to mice (2216), probably due to the ochratoxin which may be produced (1082).

Aspergillus ochraceus is one of the moulds used in the bioconversion of steroids for it converts progesterone into 11α-hydroxyprogesterone and 6,11α-dihydroxyprogesterone (1622, 2266, 2398, 2745, 2785).

POISONING ASSOCIATED WITH *ASPERGILLUS NIGER*

There is relatively little information on the toxins produced by the *niger* group of aspergilli, and most of this concerns the production of oxalic acid in food commodities. This in itself is sufficient to cause a reduction in growth of animals

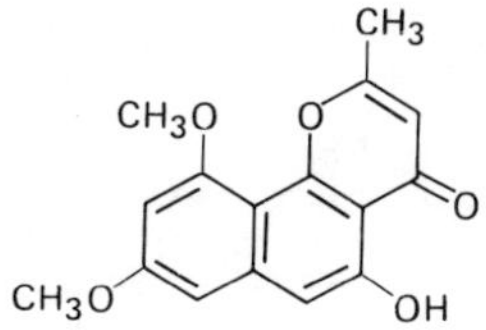

Figure 83. Flavasperone

(101). There is a record (850) of pigs having been experimentally poisoned with bread on which *A. niger* v. Tiegh was growing. This particular case is probably due to the production of oxalates (2837) unless it was due to aflatoxins which have also been reported, though not substantiated, as being produced by some strains of this species (1365).

When culture filtrates of the closely related species, *A. luchuensis* Inui, were injected into rabbits they caused pyrexia (1030) whereas filtrates from *A. foetidus* (Naka.) Thom and Raper have been shown to be poisonous to mice (2216).

Among the suspected substances isolated from *A. niger* is flavasperone $C_{16}H_{14}O_5$ (Figure 83) which has a m.p. 203–204 °C (379).

The species *A. microcephalus* Mosseray, which is related to *A. niger*, is abundant on poorly dried forage and has been shown to be toxic to young ruminants (1758).

POISONING DUE TO *ASPERGILLUS CANDIDUS*

Aspergillus candidus Link is widespread in the soil and is frequently recorded on grain in silos, being considered as an active agent in the process of self-heating (425). It has also been observed in necroses of birds and mammals (2210). This organism produces kojic acid, the role of which in certain toxicoses may not be negligible. It also produces an antibiotic, candidulin $C_{11}H_{15}NO_3$, which is active against bacteria (2480) but has little toxicity to animals, the subcutaneous LD_{50} for the mouse being 250 mg/kg. Timonin (2606) mistakenly suggested that this species produces citrinin, to which he attributed its toxic nature. Timonin's strain, studied by Rao and Thirumalachar (2100), was shown to be *A. niveus* Blochwitz which belongs to the *flavipes* group.

TOXICITY ASSOCIATED WITH *ASPERGILLUS ORYZAE*

Aspergillus oryzae (Ahlb.) Cohn belongs to the *flavus* group and is very similar to *A. flavus* itself, although it seems to be much less dangerous and is often used in the production of saké alcohol, miso, and soya sauce. It is used to prepare koji, which, once dried, is used in the manufacture of amazake or fermenting rice.

Rats and mice fed on some koji have shown focal necroses of the liver several days after the start of the diet, and it has been demonstrated that culture filtrates from the strain of *A. oryzae* used provoked serious lesions, with fatty degeneration of the liver, and incoordination of movement when injected into mice and guinea-pigs (1313, 1314).

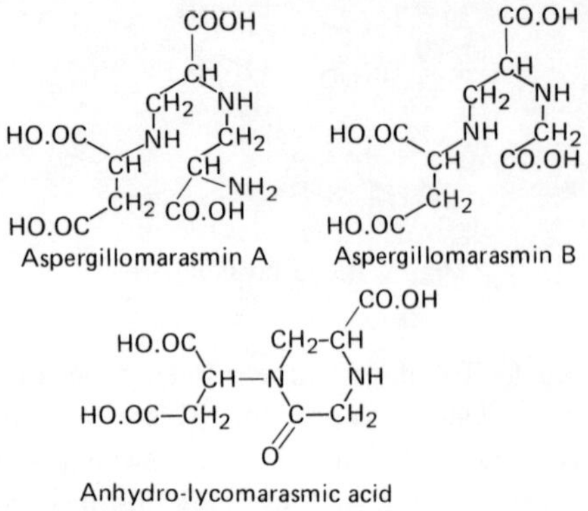

Aspergillomarasmin A

Aspergillomarasmin B

Anhydro-lycomarasmic acid

Figure 84. Amino acid derivatives produced by *Aspergillus oryzae*

Rigorous control over the strains of *Aspergillus* used in the industrial manufacture of saké (1792), koji, (20), or takadiastase (1873) has been established in Japan (1629) and none of these strains is known to be toxic.

The symptoms of the toxic strain are similar to those described by Malencon and Rieuf (1582) in Morocco where a herd of goats consumed some bait, prepared for the destruction of locusts, on which *A. oryzae* had grown.

As well as kojic acid, this species produces oryzacidin (2367, 2368, 2369) and a group of peptide derivatives similar to toxins associated with plant wilts. They include aspergillomarasmin A, $C_{10}H_{17}O_8N_3$ (812, 2162), aspergillomarasmin B, $C_9H_{14}O_8N_2$, and anhydrolycomarasmic acid, $C_9H_{12}O_7N_2$ (Figure 84).

Over a period of years, several milking cows died on experimental farms near Tokyo. It was eventually realized that mouldy malt, given as a feed to encourage growth of the animals, was probably responsible. The mould isolated was referred to as *Aspergillus oryzae* var. *microsporus* Sakaguchi, although Raper and Fennel (2105) consider this to be a synonym for *A. oryzae.* Iizuka and Iida (1149) demonstrated that the fungus could produce a potent toxin which they called maltoryzine (Figure 85), which is similar to hordeine, a toxin of less importance. Maltoryzine, $C_{11}H_{12}O_4$, which has a melting point of 68.5–69 °C, is only formed on a malt medium. The LD_{50} to mice is 3 mg/kg when injected by the intraperitoneal route and the compound causes a muscular paralysis. A detailed study of the symptoms has been made by Okubo *et al.* (1884).

Kojic acid has been isolated from a fungus closely related to *A. oryzae* which has

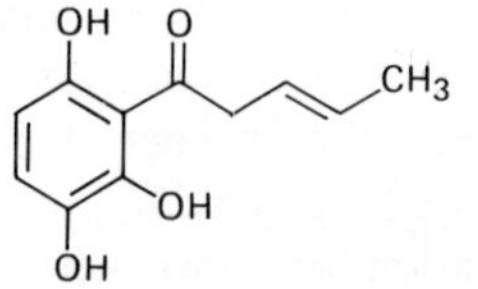

Figure 85. Maltoryzine

been referred to as *A. effusus* Tiraboschi (640), although Blochwitz (238) has compared it with *A. flavus* and Ohara considers it to be either a variety of *A. oryzae* (1878) or a related species (1879).

A toxic lactone, avenaciolide, which has been reported from *Corynocarpus leavigata* J. R. and G. Forst, is produced by *A. avenaceus* G. Smith growing on this plant (316). Culture filtrates of this species are toxic to ducklings (2315).

Apart from producing kojic acid, *A. tamarii* Kita, another member of the *flavus-oryzae* group, has not been incriminated in toxicosis (1398) except by Datta (585), who considers it to be responsible for cancer in India, perhaps in association with mycoplasmas. It is true that this species, with its tendency to be thermotolerant, is common in soil, but is not frequently isolated from food products. It has been suggested that it may produce aflatoxins (1413).

Muramaki and Suzuki (1790) insist on distinguishing two series of aspergilli within the group usually considered as the *flavus* group. They would call these the *flavus* series and the *oryzae* series, aflatoxin production being confined to the *flavus* series.

TOXICOSES ASSOCIATED WITH THE *ASPERGILLUS WENTII* GROUP

These fungi, which are common in the soil and on mouldy grain, produce kojic acid. This is particularly so with *A. wentii* Wehmer (333), *A. terricola* Marchal, often referred to as *A. lutescens* (Bainier) Thom or *A. luteo-virescens* Blochw. (908, 1605) and *A. thomii* Smith. *A. wentii* has been reported as producing aflatoxin B_1 (1365) as well as an antibiotic on which little information is available (237). Some of these species have been implicated in outbreaks of toxicosis. Thus liver lesions have been reported in ducklings, chickens, rabbits, and sheep in South Africa (2075, 2078, 2079).

TOXICOSES ASSOCIATED WITH THE *ASPERGILLUS VERSICOLOR* GROUP

Aspergillus versicolor and *A. sydowii* have both been incriminated in cases of poisoning. They are morphologically very similar and it seems that these two species have frequently been confused by toxicologists.

Toxicoses due to *Aspergillus versicolor* (Figure 86)

Aspergillus versicolor (Vuil.) Tiraboschi is particularly common on moulded seeds and their products (1945, 2233, 2633, 2634, 2653, 2767) and on oilseed products (937). It has the ability to produce a wide range of enzymes which cause the rapid deterioration of these commodities (2632).

Toxic metabolites – Among the metabolites suspected as toxic there are many anthraquinone derivatives:

– Averantin, $C_{20}H_{20}O_7$, and averythrin, $C_{20}H_{18}O_6$ (1646) (Figure 87), have m.p. 238–234 °C and 229–231 °C respectively.

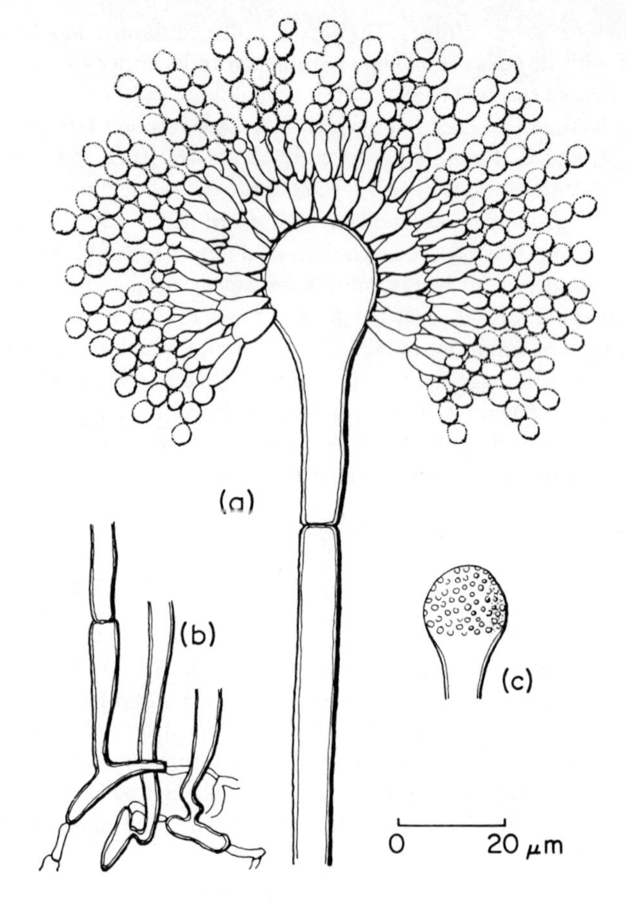

Figure 86. *Aspergillus versicolor* (a) Conidial head. (b) Base of conidiophores. (c) Top of conidial head before the development of phialides (after Pelhâte)

– Aversin, $C_{20}H_{16}O_7$ (Figure 88), m.p. 217 °C, $[\alpha]_D^{20}$ –222° (341).

– Pigments which crystallize as orange-yellow needles (1010), known collectively as the versicolorins (1646, 2094, 2095):

Versicolorin B, $C_{18}H_{12}O_7$ (Figure 89), m.p. 298 °C, $[\alpha]_D^{20}$ –223°.

Versicolorin C is a racemate of versicolorin B, m.p. >310 °C.

Versicolorin A, $C_{18}H_{10}O_7$, or dehydroversicolorin B, m.p. 289 °C, $[\alpha]_D^{18}$ –354 °.

– Averufin, $C_{20}H_{16}O_7$ (Figure 90), m.p. 280–282 °C.

These pigments, the toxicity of which deserves further study, have also been isolated from a related species, *A. crystallinus* Kwon and Fennel, isolated from the soil of Costa Rica. This species also produces pachybasin, $C_{15}H_{10}O_3$, and chrysophanic acid (776).

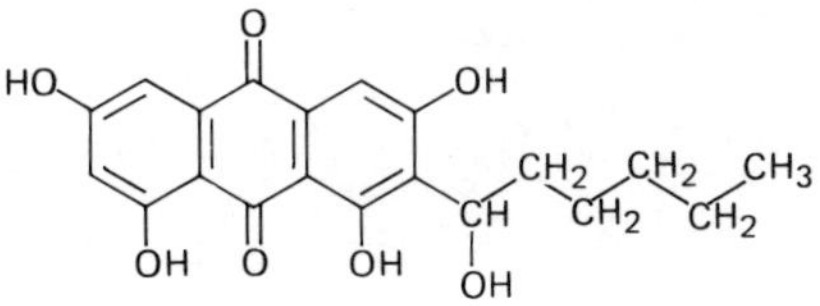

Averantin

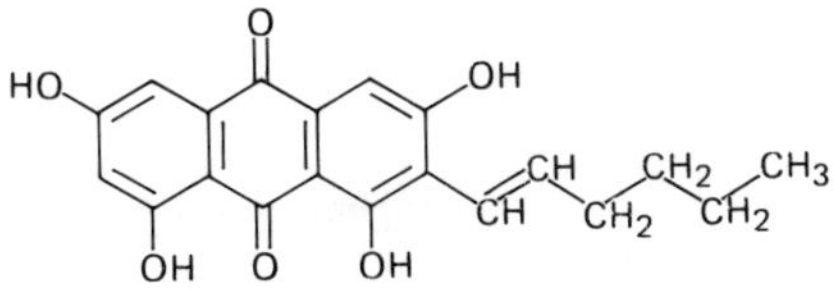

Averythrin

Figure 87. Anthraquinone derivatives of *Aspergillus versicolor* with a linear C_6 side chain

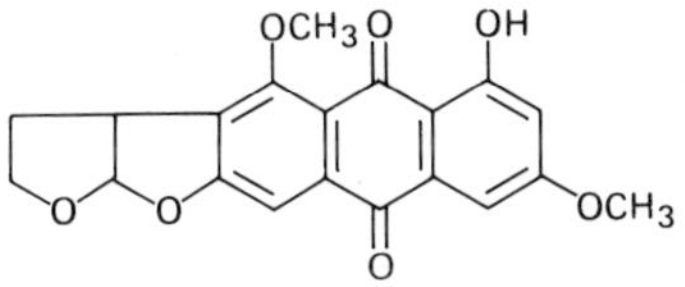

Figure 88. Aversin

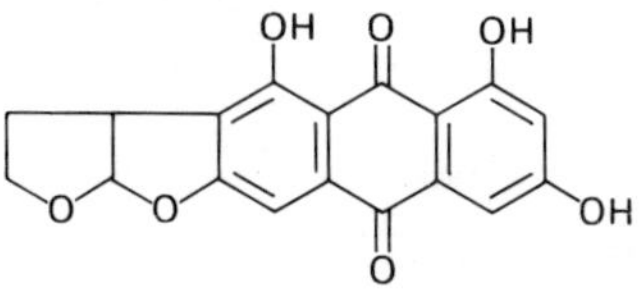

Figure 89. Versicolorin B

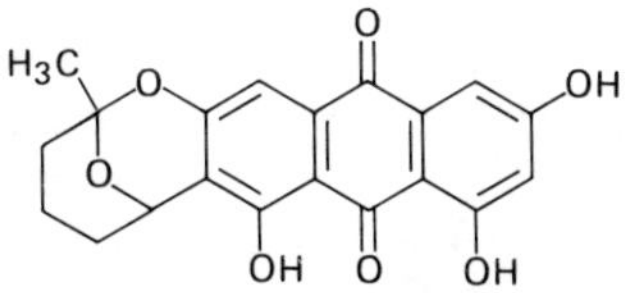

Figure 90. Averufin

Sterigmatocystin – Toxicoses which involve *A. versicolor* are probably due to sterigmatocystin, and related metabolites, which it produces. Sterigmatocystin itself, $C_{18}H_{12}O_6$ (Figure 91), is a pale yellow crystalline compound, m.p. 246 °C, which is insoluble in water. Its structure is based around a xanthone nucleus. Isolated and characterized by Bullock *et al.* (342), sterigmatocystin was the first natural product recognized as containing the dihydrofurobenzfuran nucleus (226, 341, 592, 593). Dickens and his colleagues (644, 647) on the one hand, and Purchase and Van der Watt (2057, 2058) on the other, have demonstrated that sterigmatocystin may be carcinogenic. Because the other carcinogenic metabolites, the aflatoxins, also contain this structure, it has been suggested that the carcinogenesis of this type of metabolite may be associated with this feature.

A practical method of analysing sterigmatocystin in grain and oilseeds has been described by Vorster and Purchase (2757). It is based on the conversion of this

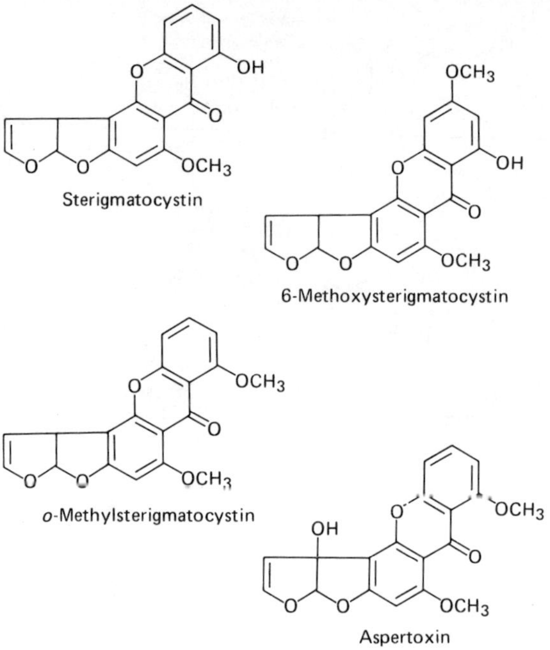

Sterigmatocystin

6-Methoxysterigmatocystin

o-Methylsterigmatocystin

Aspertoxin

Figure 91. Derivatives of sterigmatocystin produced by *Aspergillus versicolor*

metabolite into a monoacetate which is much more fluorescent. The u.v. absorption spectrum of sterigmatocystin is fairly characteristic and distinguishes it from the anthraquinones (Figure 92).

Sterigmatocystin is less toxic than the aflatoxins (1489), the LD_{50} for rats being:

oral 166 mg/kg (113–224) for males,
120 mg/kg (92–155) for females
intraperitoneal 60 mg/kg (46–77) for males (2057).

For the monkey, the LD_{50} has been shown to be 32 mg/kg (2711), and for the brine shrimp, *Artemia salina*, larvae the LD_{50} is 0.54 μg/ml.

Sterigmatocystin, at doses varying from 18–100 mg/kg, causes liver necrosis which may be either centrolobular (when administered orally) or periportal (when administered intraperitoneally). When the dose is very high (144 mg/kg) haemorrhages and severe necrosis of the tubules and glomerules of the kidney are seen (2057–2059, 2710, 2717).

Sterigmatocystin is metabolized and secreted within 12–24 hours in the urine, faeces, and especially in the gastrointestinal tract of the rat (1821).

Other mycotoxins similar to sterigmatocystin (394) include 6-methoxysterigmatocystin, $C_{19}H_{14}O_7$, m.p. 223 °C, $[\alpha]_D^{25}$ −360 °C, o-methylsterigmatocystin, which was isolated by Burkhardt and Forgacs (347) from a strain of *Aspergillus*

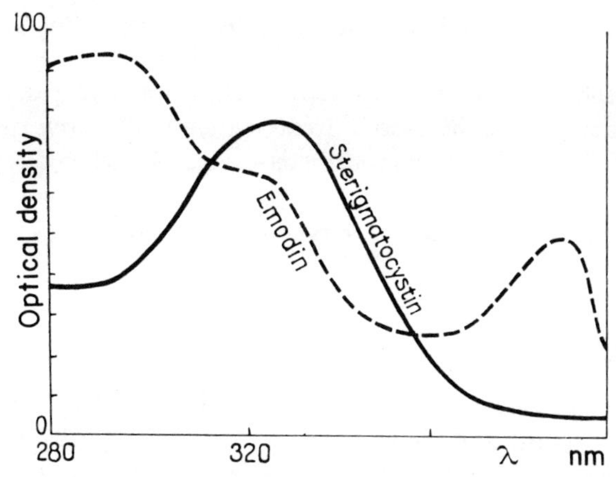

Figure 92. Ultraviolet absorption curve of sterigmatocystin. The spectrum of an anthraquinone derivative, emodin, is compared at the same concentration (2094)

(I)

(II)

(III)

(±) Dihydro *o*-methylsterigmatocystin

Figure 93. The synthesis of a derivative of sterigmatocystin (2097)

flavus which also produced aflatoxins, and aspertoxin, isolated from two different strains of *A. flavus*. Whereas the total synthesis of aflatoxins B_1 and B_2 has been achieved (334, 2165), that of sterigmatocystin has not as yet, although Rance and Roberts (2097) have synthesized (±) dihydro-O-methylsterigmatocystin by condensing methyl 2-bromo-6-methoxybenzoate (Figure 93, I) with tetrahydrofurobenzofuran (II), which, after hydrolysis, gives the corresponding carboxylic acid (III).

The recent isolation of two dihydrodifuroxanthones, dihydrosterigmatocystin and dihydrodemethylsterigmatocystin, from cultures of *A. versicolor* seems to indicate the possibility of a biological conversion of sterigmatocystin to the aflatoxins and aspertoxin (1042).

Finally, an antifungal antibiotic, versicolin, has also been isolated from *A. versicolor* (642).

Toxicoses associated with *Aspergillus sydowii*

A reduction in the growth of certain rats in a breeding unit was, at first, thought to be due to a nutritional deficiency. In fact, Wooley *et al.* (2873) have demonstrated that it was due to a toxicosis associated with the growth of *A. sydowii* (Bain. and Sart.) Thom and Church on some of the constituents of their diet. Simply autoclaving was sufficient to destroy the thermolabile toxin.

For several years, people in Japan who were treated with injections of glucose in Ringer's solution prepared in ampoules, developed rapid fever sometimes leading to their death. It was subsequently discovered that *A. sydowii* was growing on the glucose powder used for the preparation of these solutions (1322).

However, some authors have suggested that *A. sydowii*, growing on foods may sometimes be beneficial because the mycelium contains a high concentration of certain vitamins. *Aspergillus sydowii* will grow on a wide range of organic substrates and is sometimes isolated from the soil.

TOXICOSES ASSOCIATED WITH *ASPERGILLUS NIDULANS*

Aspergillus nidulans (Eidam) Wint. is very widespread in nature. It is essentially a soil micro-organism but may be isolated from a variety of substrates. This species has been the object of many genetic and physiological studies and produces a number of antibiotics.

Two antibiotics were first isolated from a culture which was misidentified as *A. ustus* (Bain.) Thom and Church (1381). One of these, which contains chlorine, was called ustin (679). The culture was subsequently identified as an anascosporic strain of *A. nidulans.* Another crystalline antibiotic was obtained from the same strain and called nidulin (605, 606). Nidulin, which was obtained by growing the culture at 20 °C, was then shown to be the monomethyl ether of ustin, which was formed at 30 °C. The name nidulin has been retained but ustin has been changed to nornidulin. Both nidulin, $C_{20}H_{17}O_5Cl_3$, and nornidulin, $C_{19}H_{15}O_5Cl_3$ (Figure 94) are chlorine-containing depsidones and a derivative lacking the chlorine, dechloronornidulin, has also been isolated by Dean, Erni, and Robertson (604).

These compounds are apparently not very toxic and mice can tolerate doses of 300–400 mg/kg of nornidulin. However, *A. nidulans* has been implicated in some problems with laying hens (1398). It is possible that kojic acid, which this species also produces, may be responsible or the symptoms may be due to an

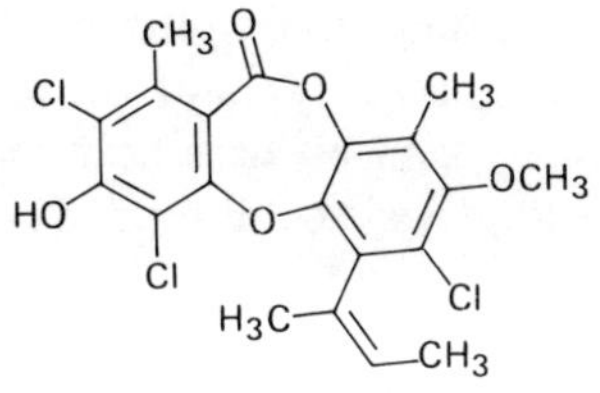

Nidulin

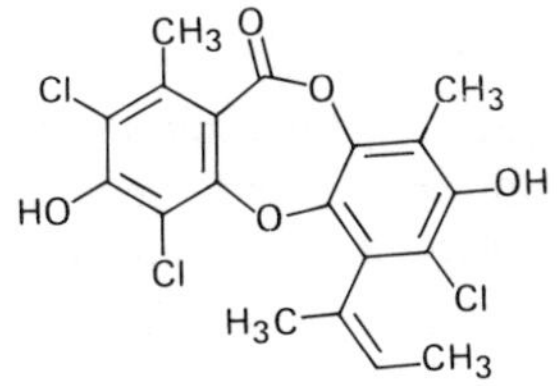

Nornidulin

Figure 94. Nidulin and nornidulin

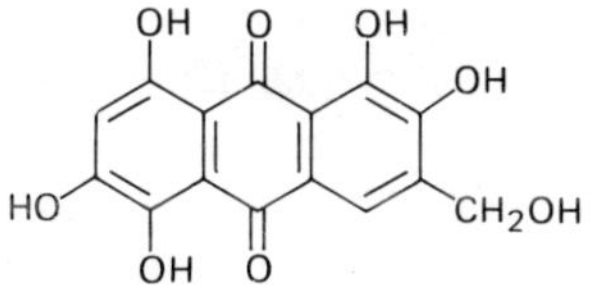

Figure 95. Asperthecin

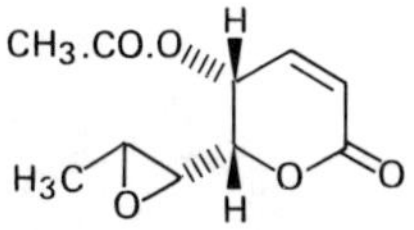

Figure 96. Asperline (antibiotic U-13, 933)

anthraquinone pigment, asperthecin (Figure 95) which is abundant in the cleistothecia and the biochemistry of which has been studied in detail (1132, 1817).

Some strains of *A. nidulans* produce sterigmatocystin (1118, 1821) and nidulol (2717), as well as an antibiotic called U-13,933 (1646). The antibiotic U-13,933, $C_{10}H_{12}O_5$ (Figure 96), is an α-pyrone, the toxicity of which deserves to be studied.

Lafont and his colleagues (1397, 1405, 1409) consider that the principal toxic compound produced by *A. nidulans* is nidulotoxin which is also produced by *A. versicolor* and *A. sydowii.* Nidulotoxin forms colourless, needle-like crystals, m.p. 240.5 °C with maxima in the u.v. at 320 and 248 nm, which are insoluble in water and hexane, slightly soluble in benzene and soluble in ethanol, methanol, acetone, dimethylsulphoxide, chloroform, and carbon tetrachloride. It gives a green colour with alcoholic ferric chloride and has an LD_{50} of 100 μg/animal in the duckling. In laboratory experiments in which hamsters have been inoculated with *A. nidulans,* the animals died after 32–35 days, the liver and kidneys showing spots of degenerate necrosis (1758).

TOXICOSES DUE TO ASPERGILLI OF THE *FLAVIPES* AND *TERREUS* GROUPS

Aspergillus flavipes (Bain. and Sart.) Thom and Church is a cosmopolitan soil organism associated with decaying vegetable matter and *A. terreus* Thom is also essentially a soil organism. Both species, and moulds related to them, produce flavipin which is 3,4,5-trihydroxy-6-methylphthalic aldehyde (2085). This compound, which is similar in structure to gladiolic and cyclopaldic acids, is apparently not toxic. However, it is also produced by those strains of *Epicoccum* which have been implicated in hepatic and renal disorders of chickens. *Aspergillus flavipes* has the capacity to convert progesterone into testolactone (846).

A. flavipes has been shown to produce, in mice, a degeneration of the kidney tubules. Pycnosis sporadically occurs in the region of the cells of the medullary tubules and sometimes even the cells of the cortex are affected (2717).

A. niveus Blochwitz, a member of the *flavipes* group, is known to produce citrinin.

A. terreus, although most frequently isolated from the soil, may also occur on vegetable matter, for example on grain in silos (625), on straw, forage, etc. As well as flavipin, it produces an antibiotic first called terreic acid, now known as terrein, $C_8H_{10}O_3$ (Figure 97), because it does not have a true acid function (1904, 2088).

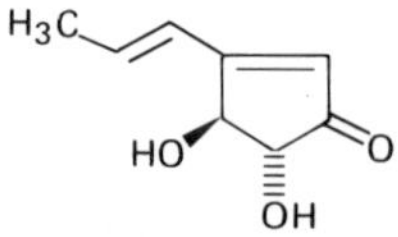

Figure 97. Terrein

The LD_{50} of terrein, for mice, varies from 71–119 mg/kg when given intravenously (2450).

Some strains of *A. terreus* produce citrinin at the same time as terrein, others produce succinic and oxalic acids. There are strains of *A. terreus* which synthesize the two chlorine-containing metabolites, geodin, $C_{17}H_{12}O_7Cl_2$, and erdin, $C_{16}H_{10}O_7Cl_2$, the former being the methyl ester of erdin (Figure 98).

Other metabolites of this species include geodoxin, a number of phenols (1039, 1040), asterric acid (2148), terrecin (1164) (Figure 99), patulin (971), and several imperfectly known metabolites. Many of these compounds have antibiotic properties and each has a different toxicity which explains the wide range of results obtained from toxicological studies of different strains of *A. terreus.*

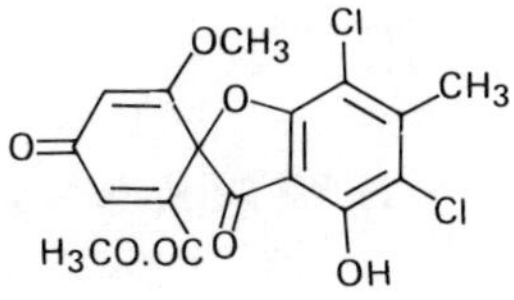

Figure 98. Geodin

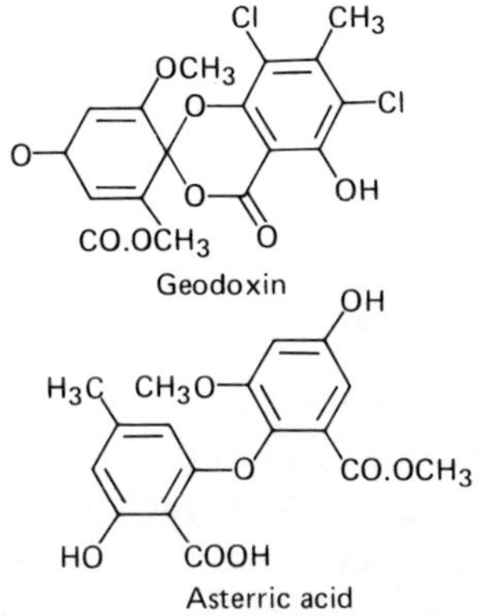

Geodoxin

Asterric acid

Figure 99. Metabolites of the *Aspergillus terreus* group

A. carneus (v. Teigh.) Blochwitz, which is closely related to *A. terreus,* has been isolated from groundnuts and incriminated in the toxicity of certain meals (2315, 2708), but its toxigenicity is still a matter of debate.

CHAPTER 6

ISLANDITOXICOSIS

INTRODUCTION

Moulds on rice in ancient times

Rice is a substrate for a rich flora of moulds and the fact that some members of this flora are able to produce antibiotics or toxins has been known for a long time. In one celebrated Buddhist temple a powder, formerly used for the external treatment of wounds, has been held sacred for thousands of years and seems to consist of mould spores from rice. Japanese folklore contains records of the use of these moulds in the protection of fish and meat against putrefaction for hundreds of years, long before the scientific recognition of antibiotics.

Several years before the discovery of penicillin as an antibiotic, the antibacterial properties of kojic acid, isolated from *Aspergillus oryzae,* had been recognized.

The early use of washed and polished rice was almost certainly aimed at conserving the rice grains from mould spoilage. As well as beneficial moulds, the existence of those causing illness of the consumer was recorded as early as 1891 (2224, 2687). They grow especially during warm and humid years, but there does seem to have been some confusion of disease due to these moulds and avitaminosis now known as beri-beri.

Toxic moulds from rice

Among the many moulds isolated from rice (see Chapter 1) it is those belonging to the genus *Penicillium* which are particularly associated with high toxicity (1523, 2641, 2643). Their growth produces a variety of colours in the grain, thus reference is made to:

– 'brown rice' associated with *P. brunneum* Udagawa;
– 'yellow rice' with *P. toxicarium* Miyake (now known to be a synonym of *P. citreo-viride* Biourge);
– 'yellow rice of Thailand' coloured by *P. citrinum* Thom.

The endosperm may also be coloured orange due to growth of *P. puberulum* Bain. (2287).

Members of the rice mycoflora which are reputed to be toxic include: *Penicillium tardum* Thom, *P. frequentans* Westl., *P. commune* Thom, *P. ochrosalmoneum* Udagawa, *P. viridicatum* Westl., *P. urticae* Bain., *Aspergillus clavatus* Desm., *A. chevalieri* (Mang.) Thom and Church, *A. fumigatus* Fres., *A. nidulans* (Eidam) Winter, *A. sydowii* (Bain. and Sart.). T. and C., *A. terreus* Thom, *A. candidus* Link, *Fusarium roseum* Link, etc., but the most dangerous of all is undoubtedly *Penicillium islandicum.*

In previous years, a shortage of food in Japan has necessitated importing rice from other regions in Asia and much of this arrives at its destination in a very mouldy state. An increasing frequency of both acute and chronic liver lesions was reported, including cirrhosis (2214) and even primary liver cancer.

The explanation of this phenomenon was at first considered to be due to a lack of protein or a choline deficiency in the food. But, considering that those regions where lesions were most frequent were also regions in which the largest quantities of moulded rice were consumed, the possibility was accepted that there may be a correlation between these two phenomena (1696) and that toxic mould metabolites, especially of *P. islandicum,* may be responsible.

PENICILLIUM ISLANDICUM

Toxic disturbances associated with the consumption of rice contaminated with *P. islandicum* are now known under the collective name of islanditoxicosis. *Penicillium islandicum* Sopp., responsible for the moulding of rice known as 'Island yellow rice', was isolated in 1953 by Tsunoda (2642) from rice imported into Japan. It has also been found on mouldy rice from Spain, Burma, Thailand, Pakistan, Vietnam, China, Turkey, Egypt, Peru, Argentina, and Ethiopia (2622). It has also been isolated from sorghum and millet in Ethiopia (531) and even from barley in Britain (2235). It is considered to be a widespread member of the soil flora.

P. islandicum belongs to the group referred to by Raper and Thom (2106) as the Biverticillata Symmetrica. The conidiophores, which are 2–3 μm in diameter, bear a verticil of 3–5 metulae, 6–10 x 2–3 μm, each of which carries 3–5 phialides, which are often fairly long, 6–11 x 2 μm, and give rise to long chains of elliptical spores, 3–4 x 2 μm (Figure 100). Very often the main axis continues to grow forming a second verticil of metulae, and a fairly dense conidial head may be formed. The long phialides, with frequently diverging neck, suggest some affinities with the genus *Paecilomyces.*

Colonies in culture show a mixture of orange, red, greenish, and dark yellow colours. Growth is slow but is optimal at 28–30 °C.

METABOLITES OF *PENICILLIUM ISLANDICUM*

The pigments of P. islandicum – Howard and Raistrick (1131) were the first to isolate a toxic substance from this mould and they called it islandicin. It was shown to be an anthraquinone isomeric with helminthosporin, a toxin isolated from a number of strains of *Helminthosporium* (Figure 101) (2418).

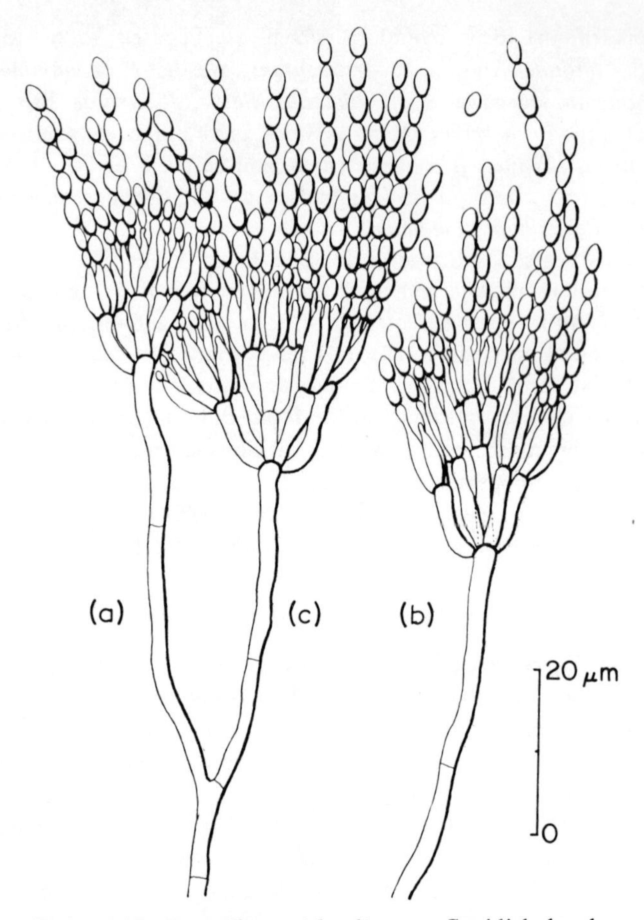

Figure 100. *Penicillium islandicum* – Conidial head. (a) A single verticil of metulae. (b) After having formed a primary verticil of metulae the axis has proliferated to form a second verticil of metulae, much shorter than the first. (c) The same as (b) but the verticil is more dense.

Since then very detailed chemical studies have provided a better understanding of the complex mixture of pigments produced by *P. islandicum*, among which are are rubroskyrin and the very toxic metabolite, luteoskyrin (2692, 2693, 2694), iridoskyrin, and erythroskyrin (2373). Erythroskyrin, $C_{26}H_{33}O_6N$ (Figure 102), has m.p. 130–133 °C, $[\alpha]_D$ +46.9 °. *P. islandicum* also produces gluconic acid which, by altering the pH of the medium, may affect the synthesis of the pigments by the mycelium (2904).

Toxic metabolites – Studies on the toxic metabolites of *P. islandicum* centre primarily around luteoskyrin and secondarily a chlorinated polypeptide (2215). During the studies on the structure of luteoskyrin, parallel studies were carried out

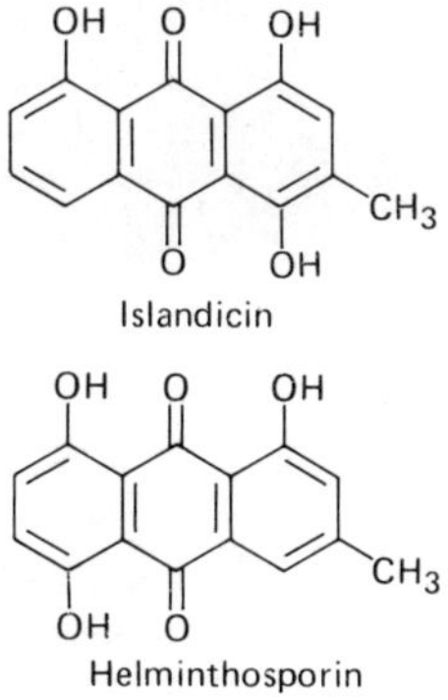

Islandicin

Helminthosporin

Figure 101. The relationship between islandicin and helminthosporin

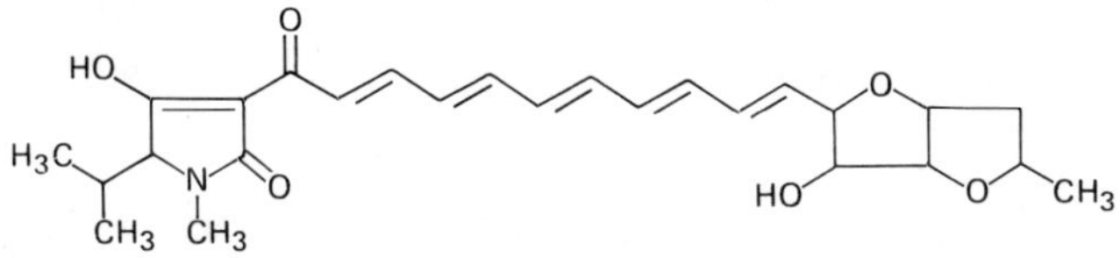

Figure 102. Erythroskyrine

on rugulosin, $C_{30}H_{22}O_{10}$ (a pigment produced by *P. rugulosum* which is related to *P. islandicum*), rubroskyrin, $C_{30}H_{22}O_{12}$, and luteoskyrin was eventually shown to have a complex cage-like structure (Figure 103).

The similarity of degradation products from these various metabolites, which are all hexahydrodianthraquinone derivatives, has demonstrated that luteoskyrin is isomeric with rubroskyrin and a homologue of rugulosin, which has two fewer hydroxyl groups (1995, 2356, 2357). In fact the following observations indicate the relationships between these compounds:

1. rugulosin yields chrysophanol and anhydroiridoskyrin on dehydration and emodin and skyrin after thermal decomposition;
2. rubroskyrin, blackish red in colour, readily dehydrates to give iridoskyrin; and on thermal decomposition yields islandicin, catenarin (7-hydroxyislandicin), and iridoskyrin;
3. luteoskyrin may be dehydrated to give islandicin and, by the action of heat, either islandicin and iridoskyrin, or iridoskyrin and catenarin, are obtained.

Luteoskyrin is soluble in water, is lipophilic, and may be isolated by solvent extraction of the dried mycelium when 100 g of mycelium may yield as much as 2 g of luteoskyrin.

It has been asked whether luteoskyrin is the most dangerous component among the metabolites of *P. islandicum.* Buu Hoi and Zajdela (377, 2911) have, in fact, demonstrated that purified luteoskyrin does not produce the same hepatotoxic

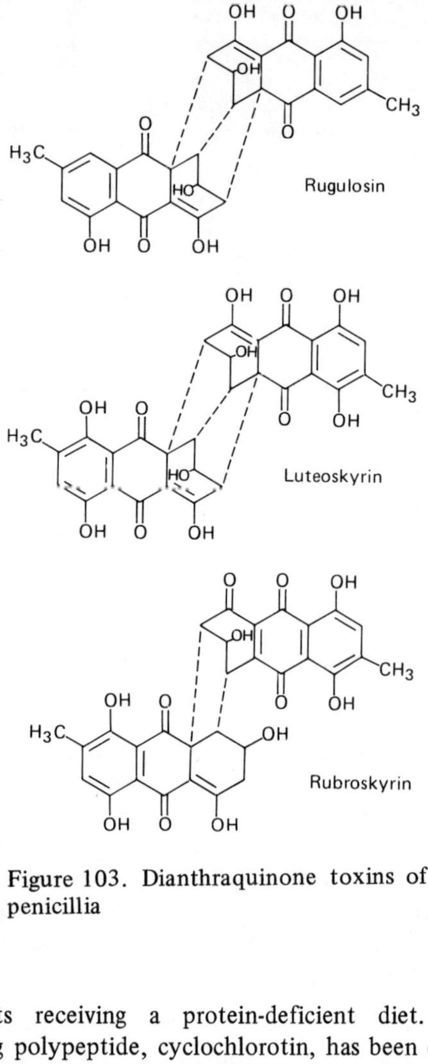

Rugulosin

Luteoskyrin

Rubroskyrin

Figure 103. Dianthraquinone toxins of penicillia

symptoms in rats receiving a protein-deficient diet. The presence of a chlorine-containing polypeptide, cyclochlorotin, has been observed (1611–1615). It has the empirical formula $C_{24}H_{31}O_7N_5Cl_2$ and contains the unusual amino acid residue dichloroproline.

Townsend (2622) described the extraction of culture filtrates with butanol followed by recrystallization of the product with methanol and then acetone to give colourless crystals of a cyclic peptide consisting of two molecules of L-serine linked on the one hand to a molecule of α-aminobutyric acid and on the other to D-β-phenyl-β-aminopropionic acid and L-dichloroproline.

Marumo (1611, 1612) named a compound, closely related to cyclochlorotin, islanditoxin. Only the infrared and nuclear magnetic resonance spectra of the amide obtained by partial hydrolysis of these two compounds allows them to be distinguished (2248). Islanditoxin (Figure 104) is the anhydride corresponding to L-seryl-L-seryl-L-dichloroprolyl-D-β-aminopropionyl-L-α-aminobutyric acid.

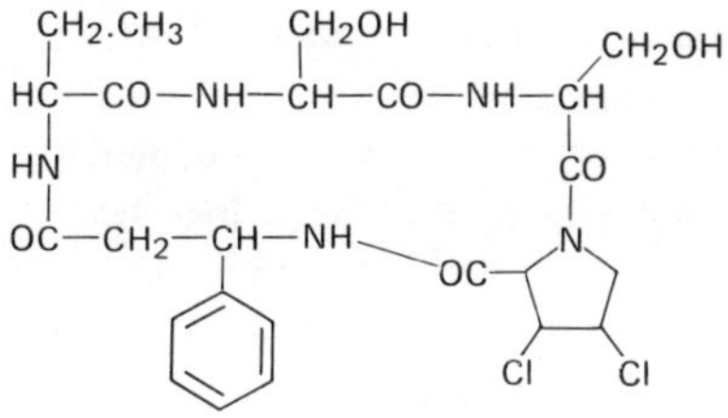

Figure 104. Islanditoxin

If the toxicities of luteoskyrin and islanditoxin are compared (Table 28), it can be seen that islanditoxin is the more toxic (1697, 2554). The toxicity of islanditoxin is considerably reduced if the chlorine atoms are removed from the molecule.

Table 28. A comparison of the toxicities for the mouse of luteoskyrin and islanditoxin, metabolites of *P. islandicum*

Route of administration	LD_{50} (mg/kg)	
	Luteoskyrin	*Islanditoxin*
Intravenous	6.65	0.338
Subcutaneous	147	0.475
Oral	221	6.550

MYCOTOXICOSIS OF *PENICILLIUM ISLANDICUM*

Symptoms of toxicosis – The symptoms associated with poisoning by *Penicillium islandicum* vary according to the quantity of infected rice consumed:

– a diet containing 100% moulded rice rapidly leads to the appearance of cirrhosis of the liver;

– with only 10–30% mouldy grain, a characteristic atrophy of the liver is first recorded and several forms of cirrhosis appear after 300 days.

– with rice containing only 1% moulded grain, a diffuse atrophy of the liver occurs after about 300 days associated with pleomorphism of the cells; after 360 days, adenomatous nodules appear with abnormalities in the arrangement of the cells and the dimensions of the nuclei (1160, 1698, 1699, 2214).

If the effects of luteoskyrin alone are noted, acute toxicity is manifested particularly by a centrolobular necrosis with fatty degeneration of the liver cells which do not show necrosis. Chronic toxicity is expressed as centrolobular pigmentation, a slow oedema with inflammation in the periportal region, and thickening of the hepatic vessels and formation of hepatomas (2681).

Acutely, islanditoxin provokes a rapid death associated with vacuolar degeneration of the liver cells with the formation of colourless droplets, followed by a progressive destruction of these cells accompanied by abundant haemorrhages, especially in the periportal region. Prolonged ingestion of the toxin gives rise to lesions in the perilobular cells with the appearance of monolobular cirrhotic nodules. A tendency to form adenomas and hepatomas is associated with the toxicosis arising from luteoskyrin. Electron microscopic studies of the liver reveal hypertrophy of the mitochondria with destruction of the mitochondrial ridges and partial pycnosis, dilatation of the endoplasmic reticulum close to the injured mitochondria with an irregular distribution and reduction in the number of ribosomes, as well as a partial destruction of the cytoplasmic membranes. The bile ducts show a clear dilatation (739, 1161, 1323, 1697, 2678, 2684, 2690).

Erythroskyrin is also toxic having an LD_{50} of 60 mg/kg by intraperitoneal injection into mice. Its consumption provokes paralysis followed by coma and then death. The disorders associated with this toxin appear to be centrolobular degeneration of liver cells, modifications of the kidneys, and a change in the lymphatic system (2668).

After prolonged application, rugulosin also causes liver lesions like luteoskyrin (2692).

There is evidence that chronic poisoning with *Penicillium islandicum* induces the formation of liver cancer, mainly through the activity of islanditoxin (1700, 1754, 2692).

Mode of action of luteoskyrin – The work of Tatsuno (2553) has demonstrated that luteoskyrin acts at the level of the respiratory chain in the liver cell. It does not inhibit glycolysis but does inhibit the Krebs' tricarboxylic acid cycle by blocking the oxidation of succinic acid. This block in the pathway by which $NADH_2$ is oxidized occurs in the mitochondria and some remarkable lesions in the mitochondria of liver cells have been revealed by electron microscopy. Research using tritium labelled luteoskyrin shows that its affinity for mitochondria is greater than that for other cell particles (2911). According to Tatsuno (2553), luteoskyrin inhibits the enzymes of the mitochondrial respiratory system at concentrations greater than 10^{-4}–10^{-5} M and the process of oxidative phosphorylation at lower concentrations in the region of 10^{-6} M.

Recent research (851, 2673) has shown that luteoskyrin will not bind to native DNA but does firmly bind the two chains of denatured DNA, at least in the presence of bivalent metals (Mg, Mn) (2113, 2675). Luteoskyrin does bind to the surface of RNA polymerase (2676) and seems to be specific for chains beginning with guanine rather than those starting with adenine.

CHAPTER 7

DIVERSE PENICILLIUM TOXICOSES

INTRODUCTION

The penicillia are, with the aspergilli, the moulds most frequently implicated in mycotoxicoses. In 1894, Zipfel (2935), suspecting the toxicity of moulds, fed various animals with large quantities of *'Penicillium glaucum'* and concluded that it was quite inoffensive. Today, it is considered that the term *P. glaucum* covers a large number of species and it is not known with any certainty which one had been tested by Zipfel.

Similarly, the nature of an unidentified *Penicillium,* growing on forage, and reported in 1893 to be associated with intoxication of cattle in the United States (670), is not known.

In the preceding chapter, the most serious of the *Penicillium* toxicoses, that associated with *Penicillium islandicum,* has been described, but there are numerous other species of *Penicillium* thought to be toxic. Some have received only little attention, either because they are of low toxicity, or because they occur so rarely that they have caused no serious outbreaks. It is perhaps worth mentioning them, however, because new research may, in some cases, show that their role is not, in fact, negligible.

The penicillia are well known, particularly from the work of Thom (2589) and Raper and Thom (2106). A review of Nicot (1852) outlines the features which distinguish this genus from several genera which it closely resembles such as *Gliocladium, Paecilomyces, Scopulariopsis,* etc. A beautiful study of their spores using the electron microscope has been reported (1068).

This chapter follows the systematic analysis of the genus outlined by Raper and Thom (2106) and an illustrated table (Table 29) gives the main structure of this classification, drawing attention to those species known to be toxic.

TOXICOSES ASSOCIATED WITH THE *MONOVERTICILLATA*

P. frequentans and P. roseo-purpureum – *P. frequentans* Westl. is very widespread in the soil and occurs on a variety of vegetable substrates. It has been found on mouldy rice and is common on wheat grain. The species is best known for the production of frequentic acid, $C_{14}H_{10}O_7.2H_2$), also known as citromycetin (Figure 105), which has also been isolated from *P. roseo-purpureum* Dierckx (1083).

Table 29. Key to the sections and subsections of *Penicillium* according to Raper and Thom (2106)

Section	Toxic Species	
MONOVERTICILLATA	Penicillus unbranched, phialides in a single verticil.	
	P. frequentans Westl.	
	P. spinulosum Thom	
	P. lividum Westl.	
	P. aurantio-violaceum Biourge	
	P. Chermesinum Biourge	
	P. decumbens Thom	
	P. citreo-viride Biourge (= *P. toxicarium* Miyake)	
	P. roseo-purpureum Dierckx	
	P. terlikowskii Zal.	
ASYMMETRICA	A. Penicillus divergent: *Divaricata*	
	P. ochrosalmoneum Udagawa	
	P. melinii Thom	
	B. Penicillus non-divergent	
	(1) Thallus velvety: *velutina*	
	P. corylophilum Dierckx	*P. notatum* Westl.
	P. citrinum Thom	*P. oxalicum* Currie and Thom
	P. sartoryi Thom	*P. brevicompactum* Dierckx
	P. steckii Zal.	*P. stoloniferum* Thom
	(2) Thallus lanose or floccose: *lanata*	
	P. commune Thom	
	(3) Thallus with aerial hyphae often aggregated: *funiculosa*	
	P. terrestre Jensen	
	(4) Thallus mealy, aerial hyphae fasciculate, in tufts or coremia: *fasciculata*	
	P. viridicatum Westl.	
	P. palitans Westl.	
	P. cyclopium Westl.	*P. urticae* Bain.
	P. puberulum Bain.	*P. griseo-fulvum* Dierckx
	P. martensii Biourge	*P. claviforme* Bain.
	P. expansum (Link) Thom	*P. divergens* Bain and Sart.
BIVERTICILLATA SYMMETRICA	Penicillus branched, biverticillate and symmetrical	
	P. islandicum Sopp	
	P. brunneum Udagawa	
	P. concavo-rugulosum Abe	
	P. piceum Raper and Fennel	
	P. purpurogenum Stoll	
	P. rubrum Stoll	
	P. variabile Sopp.	
	P. rugulosum Thom	
	P. tardum Thom	
	P. novae zelandiae v. Beyma	
	P. atrovenetum G. Smith	
POLYVERTICILLATA	Penicillus branched several times in verticils	
RELATED GENERA	*Byssochlamys fulva* Olliv. and Smith	
	Paecilomyces varioti Bain. (=*Penicillium divaricatum* Thom)	
	Gliocladium virens Miller, Giddens and Foster	

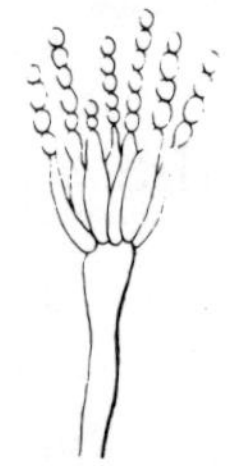

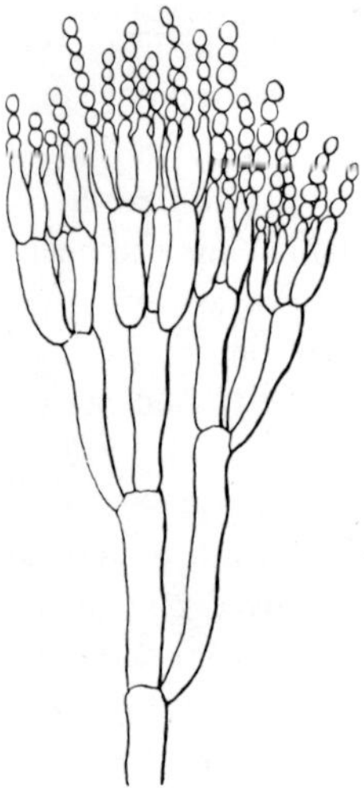

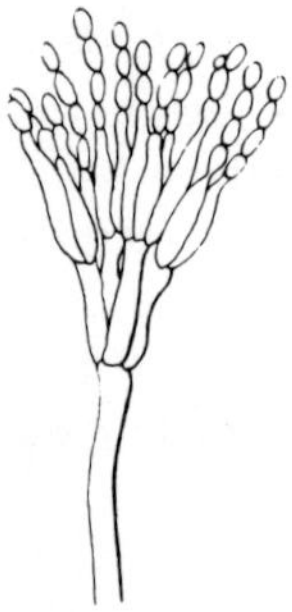

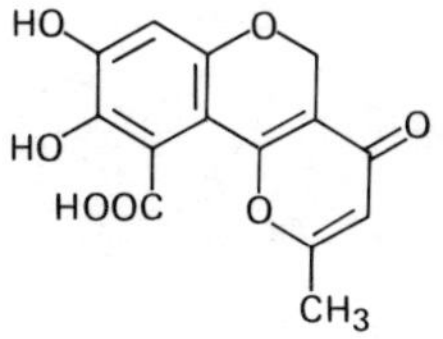

Frequentic acid (= citromycetin)

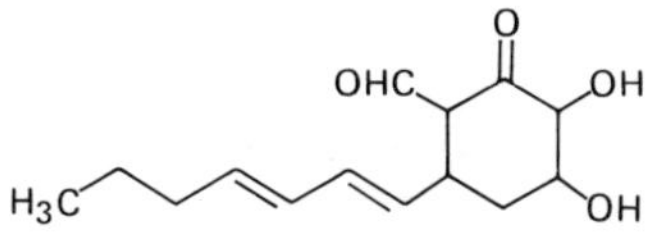

Frequentin

Figure 105. Metabolites of *Penicillium frequentans*

Frequentic acid has an LD_{50} of 1,500 mg/kg when injected intravenously into mice (2450) and so cannot be considered as a toxin. This leaves open the question of what causes the kidney lesions recorded in animals following ingestion of feed contaminated with *P. frequentans* (789). Perhaps other metabolites are responsible. Frequentin is another antibiotic produced by *P. frequentans*. It is an aldehyde with an empirical formula $C_{14}H_{20}O_4$ (Figure 105) but does not seem to be toxic to animals (573).

Other metabolites isolated from cultures of *P. frequentans* (2498), and which may contribute to its toxicity, include:

– methylemodin or questin, $C_{16}H_{12}O_5$, m.p. 301–303 °C, and the corresponding alcohol, questinol, $C_{16}H_{12}O_6$, m.p. 280–282 °C (Figure 106). These anthraquinone derivatives are not far removed from clavorubin, produced by *Claviceps purpurea,* and averantin, produced by *Aspergillus versicolor.*

– (+) dechlorogeodin, $C_{17}H_{16}O_7$, m.p. 170–173 °C, $[\alpha]^{20}_{5461}$ +230 ° (Figure 107). The dichloro compound, geodin, is produced by *Aspergillus terreus.*

Penicillium frequentans also produces an amino derivative called hadacidin, $C_3H_5O_4N$, m.p. 119–120 °C (Figure 108). This compound is considered to be an

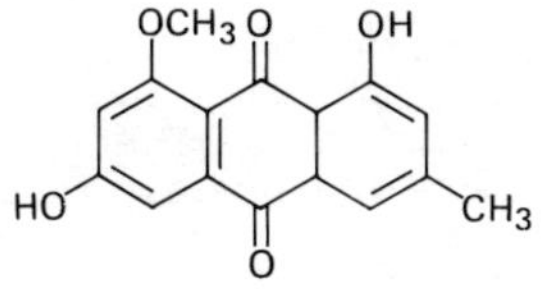

Questin

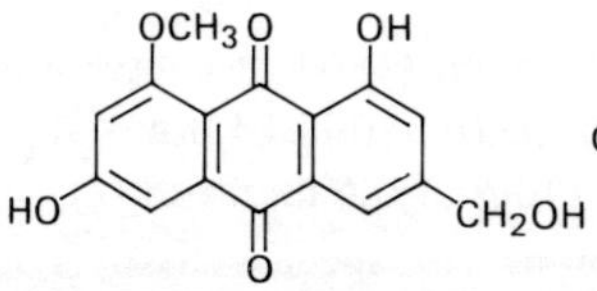

Questinol

Figure 106. Anthraquinone metabolites of *Penicillium frequentans*

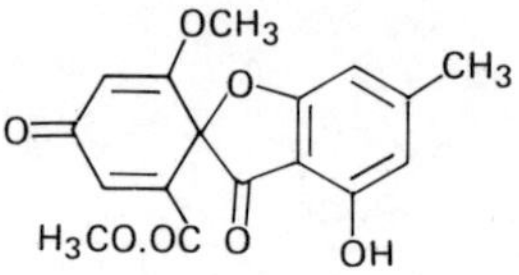

Figure 107. (+) Dechlorogeodin

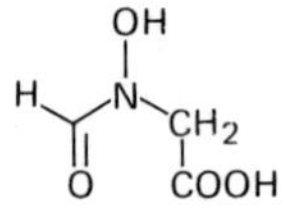

Figure 108. Hadacidin

inhibitor of adenocarcinomas in man (1268), acting as an analogue of aspartic acid and inhibiting the synthesis of adenylic acid and deoxyadenylic acid (2360, 2361).

P. spinulosum Thom. This species is also common in soil and sometimes abundant on seeds, was the source of spinulosin which was isolated in 1931 before being isolated from *A. fumigatus.* The yield is improved by the presence of ferrous sulphate in the medium (207). Some strains also produce fumigatin as well as spinulosin (2355).

P. lividum Westl. This is considered to be toxic because of its ability to produce citrinin, a metabolite more commonly associated with *P. citrinum.*

P. decumbens Thom. *P. decumbens* is a frequent contaminant of maize, produces the toxic metabolite decumbin, $C_{16}H_{24}O_4$ (2400). This metabolite has also been called brefeldin A and cyanein (Figure 109), having been isolated from filtrates of

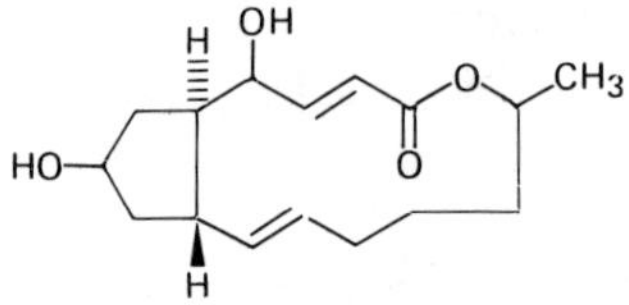

Figure 109. Decumbin (= Brefeldin A, = Cyanein)

P. brefeldianum (1032) and *P. cyaneum* (196). It causes a loss of appetite in rats, diarrhoea, lethargy, cyanosis, and death within 24 hours. The fish *Carassius auratus* is killed by concentrations of the order of 0.6 mg/l (2832).

Cultures of *P. chermesinum* Biourge, a species closely related to *P. decumbens,* yield costaclavin, one of the alkaloids isolated from the ergot of rye (19) (Figure 110).

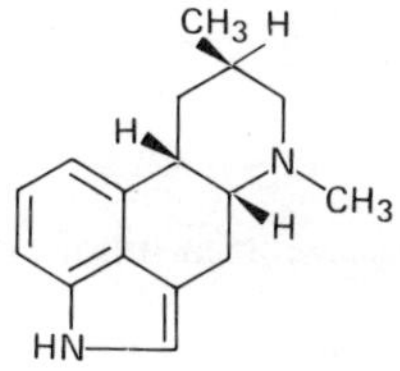

Figure 110. Costaclavin

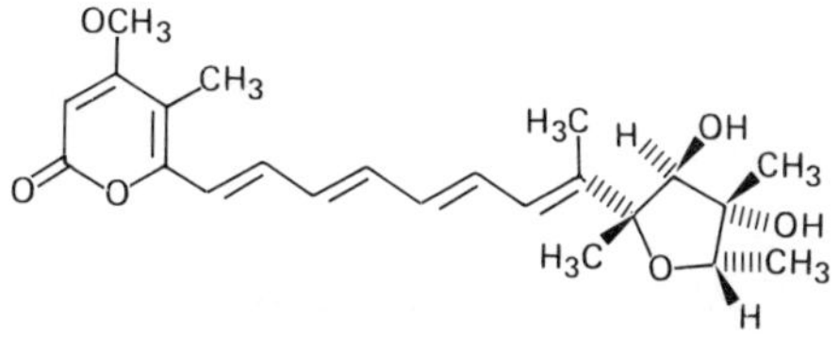

Figure 111. Citreoviridin

P. citreo-viride Biourge. This species, isolated from leather in 1923, has since been isolated from rice by Miyake (1694–1696), who called it *P. toxicarium.* Both Hirata (1095) and Sakabe, Goto, and Hirata (2220) have shown that the toxic metabolite, citreoviridin, may be produced by this species. Numerous toxicological studies have confirmed its toxicity (1694, 1696, 1801, 2223, 2645, 2648, 2688). Citreoviridin, $C_{23}H_{30}O_6$, m.p. 107–111 °C, contains a pyrone nucleus and a tetrahydrofuran ring joined by an unsaturated chain (Figure 111). Citreoviridin is most abundantly produced on rice at a low temperature, 12–18 °C (2666) and has an LD_{50} of 20 mg/kg by subcutaneous injection into mice (1801).

At low doses citreoviridin causes congestion of the veins, haemorrhage of the liver and spleen, and an abnormal swelling of the kidneys. At higher concentrations it causes paralysis of the ascending type by affecting the motor nerves of the spinal cord, causing difficulties in breathing and eventually death (2222). Citreoviridin is a neurotoxin (2666).

Some of the symptoms recall those of beri-beri and there has perhaps been some confusion between them. The compound has a strong yellow fluorescence under u.v. light and post-mortem examination under u.v. shows the citreoviridin to be located in the central nervous system, the adrenal cortex, and the kidneys (2622).

P. ochrosalmoneum Udagawa and *P. pulvillorum* Turfitt which are members of the Asymmetrica Divaricata, have also been shown to produce citreoviridin among their metabolites, and *P. citreo-viride* itself can produce citrinin (2647).

P. terlikowski and P. waksmani. These both produce gliotoxin, a weakly toxic antibiotic (292, 1782). Isolates of *P. waksmani* Zal. have been shown to be toxic in laboratory tests (1235) but the nature of the toxic agent is not known.

Penicillia of subsection Divaricata

Among the penicillia of this subsection implicated in mycotoxicoses are *P. ochrosalmoneum* Udagawa, isolated from rice, and *P. chrzaszczi* Zaleski, which is perhaps a synonym for *P. jenseni* Zaleski. Both these species produce citreoviridin, whereas *P. melinii* Thom has been reported to produce patulin.

The culture filtrates and extracts of the mycelium of *P. lilacinum* Thom are toxic to mice (2216) and, in South Africa, citreoviridin has been found among the metabolites of an isolate of *Penicillium* from maize, erroneously reported to be *P. lapidosum* Raper and Fennel, subsequently determined as *P. pulvillorum* Turfitt of the *P. raistrickii* series (1801).

Penicillia producing citrinin

Citrinin is an important toxin production of which is mainly associated with *P. citrinum* and related species. In 1953, Tsunoda (2642) observed *P. citrinum* Thom (not *P. citrinum* Sopp. which is an unidentifiable species perhaps belonging to the *raistrickii* series!) on rice imported from Thailand. It was later also recovered from rice from Burma, Italy, Egypt, and the United States (1314). It is thus a widely distributed mould occurring in the soil and on a variety of substrates such as decomposing vegetable matter. It has often been recorded as growing on leather and has been found on optical instruments and glass fibre, which may be spoiled, especially in tropical climates. The species is frequently isolated from mouldy groundnuts (656). It has been isolated from various respiratory infections, but its pathogenicity needs to be confirmed. There has been an instance when it was isolated from the urine of a patient suffering from a kidney disease.

Penicillium citrinum (Figure 112) forms grey-green colonies which become grey, or olive green, as the spores mature. The underside of the colony in culture is a beautiful lemon yellow becoming paler with age. The conidiophores are erect, 50–200 x 2–3 μm, not very branched and bear a verticil of 3–4 metulae each with 6–10 phialides. The chains of conidia are produced in distinctive columns which may be 100–150 μm long, individual spores being subglobose with a diameter of 2.5–3 μm. Although typically asymmetric, the conidial heads may often appear to be symmetrical from which it is possible to misidentify isolates.

P. citrinum produces the antibiotic citrinin and some strains have been reported as producing penicillin, gluconic acid, citric acid, and even aflatoxin (1365). Citrinin is also produced by *P. sartoryi* Thom, often considered to be simply a variety of *P. citrinum, P. lividum* Westl., *P. aurantio-violaceum* Biourge, *P. implicatum* Biourge, *P. chrzaszszi* Zal., *P. viridicatum* Westl., *P. steckii* Zal., and *P. corylophilum* Dierckx (1167), although Bilai (207) considers that the last two species do not produce citrinin. *P. corylophilum* has been reported to produce corylophilin, shown to be identical with notatin, an antibiotic isolated from *P. notatum* to which it is related, and gliotoxin (292, 1782). Citrinin is also found among the metabolites of *Aspergillus terreus* Thom, *A. niveus* Blochwitz, and

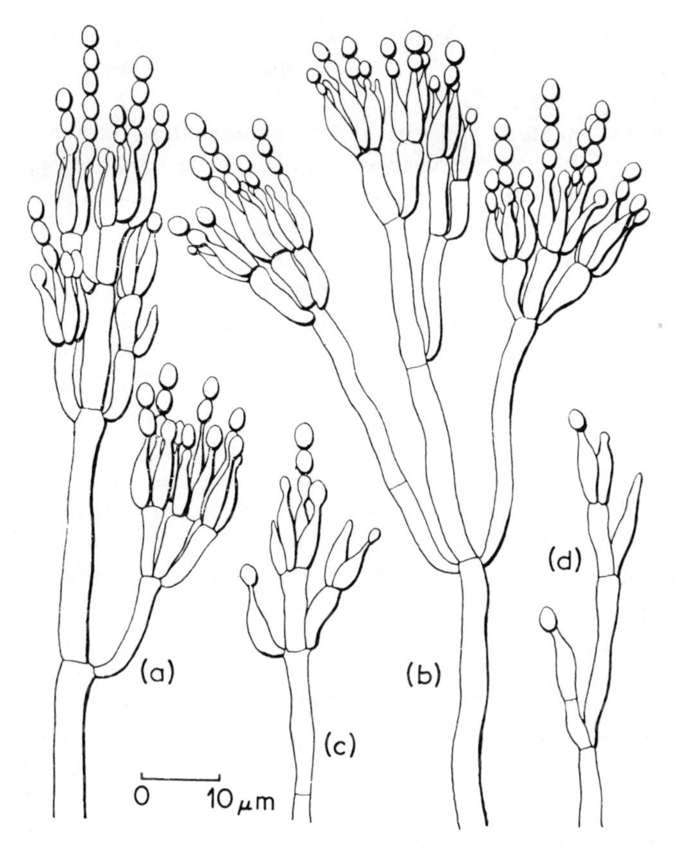

Figure 112. *Pencillium citrinum* Conidial head. (a), (b) Mature conidiophores. (c), (d) Young conidiophores

perhaps several aspergilli of the candidus group (2608). It may also be present in the phanerogams such as *Crotalaria crispata* (758).

Citrinin (Figure 113) has been known since 1931 (553, 1083) but its chemical structure was not defined until much later (324, 1091, 1252, 2467). A rapid method for the production of citrinin has been described by Wyllie (2886), Pollock (2012), and Bastin (160). It can be isolated by treating an alcoholic extract of moulded rice with benzene and ether. Citrinin remains in the alcoholic fraction. It is a monobasic benzopyran derivative, $C_{13}H_{14}O_5$, the antibiotic properties of which were first demonstrated by Raistrick and Smith (2090) in 1941. Its

Figure 113. Citrinin

antibacterial and antifungal properties were confirmed subsequently by several workers (944, 1905, 2606, 2608). The antibiotic properties are destroyed if the heterocyclic ring is opened (2355).

Although the antibiotic properties seemed quite exciting, it is too toxic to be of any pharmacological use. The subcutaneous LD_{50} in mice is 35 mg/kg (2450), and Ambrose and de Eds (50) have shown that mice, rats, and guinea-pigs are killed by the parental administration of 50 mg/kg. The LD_{50} for the rabbit, by intravenous injection, has been reported as 19 mg/kg (207).

The toxicity of citrinin has been studied by Ohmori and colleagues (1880) and it has been shown to be inactivated by u.v. light. It is essentially a toxin of the liver and the kidneys (1799) and is particularly responsible for kidney lesions in which there is considerable enlargement of the collecting tubules (2222, 2790). It is thought to act by destroying the epithelium of the tubules and a subsequent diffuse formation of the connective tissues (interstitial fibrosis) has been reported (1359). Excretion of urine is up to two and a half times greater than normal (2647).

Chu (501) has demonstrated that when the sodium salt of citrinin is applied to the skin or mucous membranes of man and animals, a slight irritation is produced. Citrinin also inhibits the growth of some plants, the pea for example, and influences the nitrogen concentration of most plants (1677).

Other penicillia of the subsection Velutina

Penicillium notatum Westl., from which penicillin was first isolated before the use of the related mould *P. chrysogenum* Thom, produces notatin which may upset oxygen transport in the blood and provoke methaemoglobinaemia. The LD_{50} of this toxin is 3 mg/kg when given intraperitoneally to the mouse (2450). Penicillin itself is sometimes responsible for epileptic convulsions in patients to whom it has been administered (108, 2766). This species also produces xanthocillin X which has been studied both as an antibiotic and as a toxin.

P. oxalicum Currie and Thom occurs on a variety of cereal products and some strains are toxic to the rat, mouse, and duckling (2315). The major toxic metabolite produced by this species is secalonic acid D (2495). In the mouse it causes severe congestion of the cortex and glomerules of the kidney with serious haemorrhage. The liver is also very congested and shows considerable dilation of the portal veins and the sinusoidal spaces (2717).

Mycophenolic acid, $C_{17}H_{20}O_6$ (Figure 114), was isolated in 1913 as a product of the metabolism of *P. stoloniferum* Thom (229) and has since been found from

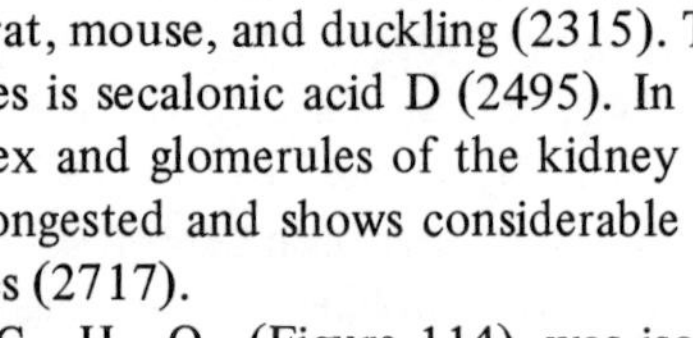

CH3 OH O
HOOC O
H3CO
CH3

Figure 114. Mycophenolic acid

numerous species of the *brevicompactum* series. Species of this group are widespread on many substrates such as cereals, milk products, soil, etc.

Bilai (207) reported the LD_{50} of mycophenolic acid to be 500 mg/kg for the mouse when given intravenously, and 2,000 mg/kg *per os.* The LD_{50} by oral administration is 2,500 mg/kg for the mouse, and 450 mg/kg for the rat. These figures seem to indicate a weak toxicity but mycophenolic acid is particularly toxic to leucocytes, which explains why it was thought to be incriminated in alimentary toxic aleukia, a disease usually associated with species of *Fusarium.* It is also the reason why mycophenolic acid has been considered in the treatment of leukaemia (2830). Chronic exposure results in an anaemia which may lead to death.

P. brevicompactum also produces a number of other metabolites related to mycophenolic acid (399) and these may have a more serious role in toxicosis associated with this species.

It has been reported that culture filtrates of *P. stoloniferum* have antiviral activity and two factors, M5–8450 and statolon, have been isolated (1099).

The brutal death of cattle at Kobe has been attributed to silage infected with *Penicillium roqueforti* Thom (2640). It is, of course, this species which has been used since the fifteenth century in the manufacture of the celebrated cheese of Roquefort. Some strains, notably MR 212-2 isolated from rice flour, do however seem to be toxic and three toxins have been demonstrated (1274):

– the first, which is a lactone, has a relatively low toxicity and a single intraperitoneal injection of 20 mg in mice causes a reduction in growth but not death;
– the second toxin gives rise to serious changes in the liver with haemorrhages in the digestive tract;
– the third toxin is also very mild in its effects.

Penicillium commune

In the subsection lanata, *P. commune* Thom, has been shown to be responsible for gastrointestinal upsets. This species is frequently found on 'mochi' or cakes prepared from rice and is sometimes used in the manufacture of certain cheeses. Culture filtrates of *P. lanosum* Westl., also isolated from rice, have been reported to cause liver lesions in mice (2216).

Penicillium terrestre

Patulin has been shown to be among the metabolic products of *P. terrestre* Jensen, which is a member of the subsection *funiculosa.* This species also produces terrestric acid which is considered to have cardiotoxic properties (227) like carolic acid from *P. charlessii*, penicillic acid from *P. cyclopium*, and viridicatic acid from *P. viridicatum* (2832) (Figure 115).

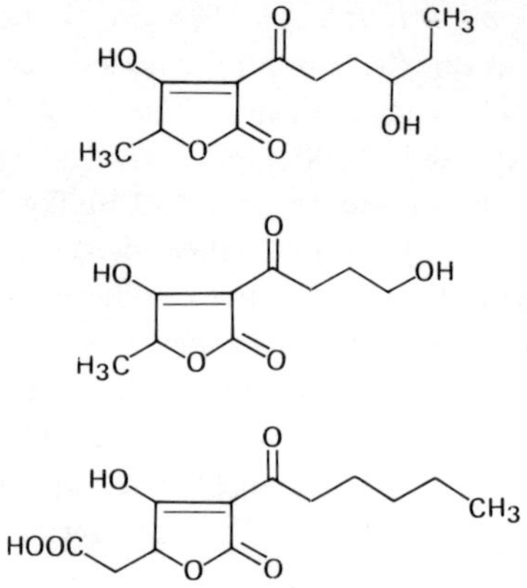

Terrestric acid

Carolic acid

Viridicatic acid

Figure 115. Tetronic acids produced by penicillia

Penicillia of the subsection Fasciculata

Penicillium viridicatum Westl. is a common isolate from the soil and actively decomposing organic matter. It is also found on grain, especially maize, barley, and wheat which have been stored at high humidity.

Degeneration of the kidney in pigs has been recognized in Denmark since 1928 as being associated with the consumption of mouldy feed. *Penicillium viridicatum* is considered to be the principal agent in porcine nephrotoxicosis, but *Aspergillus candidus, A. fumigatus,* and *A. flavus* may also produce similar symptoms (1358). The principal agent is citrinin, with oxalic acid as a secondary toxin (849, 1359) and an hepatotoxin which has been misidentified as brevianamide A (220, 2832). The antibiotic viridicatin, $C_{15}H_{11}O_2N$ (Figure 116) has been isolated from *P. viridicatum, P. cyclopium,* and *P. palitans* (505) and it is possible that it may also play a role in these toxicoses.

Using culture extracts of *P. viridicatum*, Carlton and Tuite (416–419) have produced symptoms of toxicity in mice, rats, guinea-pigs, and pigs. On autopsy they observed hepatic necrosis with bile duct cell hyperplasia and kidney lesions with changes in the epithelium of the tubules.

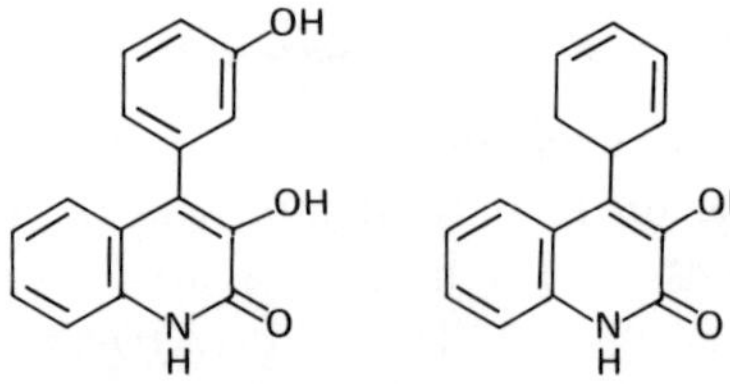

Viridicatol

Viridicatin

Figure 116. Metabolites of *Penicillium viridicatum*

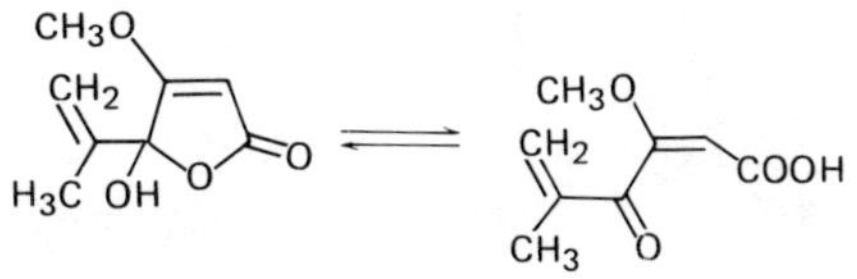

Figure 117. Tautomeric forms of penicillic acid

Van Walbeek *et al.* (2726) have isolated the highly toxic metabolite, ochratoxin A, from both mycelium and culture filtrates of an isolate of *P. viridicatum.*

P. palitans Westl., which produces both palitantin, $C_{14}H_{22}O_4$, and viridicatin, has been implicated in a toxicosis of cattle consuming mouldy grain (28). Following the death of some milking cows, Ciegler (502, 503) demonstrated the production of two tremorgenic mycotoxins, the tremortins A and B, from a strain of *P. palitans* (1128). The LD_{50} for mice is 1.05 mg/kg for tremortin A and 5 mg/kg for tremortin B. A third metabolite, tremortin C, has subsequently been isolated and purified.

Penicillium cyclopium Westl., and the related species *P. puberulum* Bainier, are both very common on grain and flour and both produce penicillic acid, $C_8H_{10}O_4$ (Figure 117) (49, 190, 265, 1319), especially at low temperatures of the order of 1–10 °C (507). This acid, which is produced by *P. cyclopium* when grown on Raulin-Thom medium but not Czepak-Dox, is present in two tautomeric forms in aqueous solution (228, 798, 973, 2716).

The LD_{50} of this acid in mice is 110 mg/kg subcutaneously, 250 mg/kg intravenously, and 600 mg/kg intraperitoneally. It has an antidiuretic activity and acts on the heart in a manner similar to digitalin. It has also been suggested that it is carcinogenic (643) and has, in any case, a strong mitotic activity at concentrations greater than 10 μg/ml (1811). Penicillic acid is also produced by *Aspergillus ochraceus, Penicillium martensii, P. fenelliae, P. olivino-viride* (1321), as well as *P. palitans* (507, 1380).

Several other compounds have been isolated from *P. cyclopium*:

– a tremorgenic and diuretic compound (2832, 2840) with an empirical formula $C_{37}H_{44}O_6NCl$, molecular weight 633, and a steroid-like nucleus;

– the alkaloids, 2,3-dioxy-4-phenylchinoline and cyclopiamine, $C_{25}H_{33}O_5N_3$, the toxicological significance of which is unknown (990, 1116);

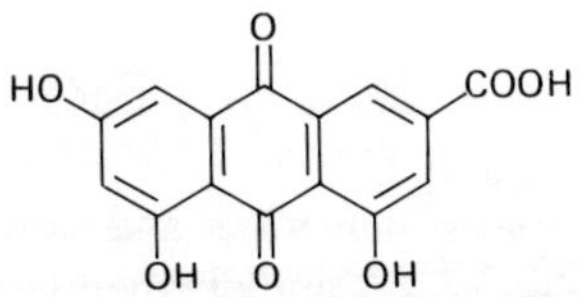

Figure 118. Emodic acid

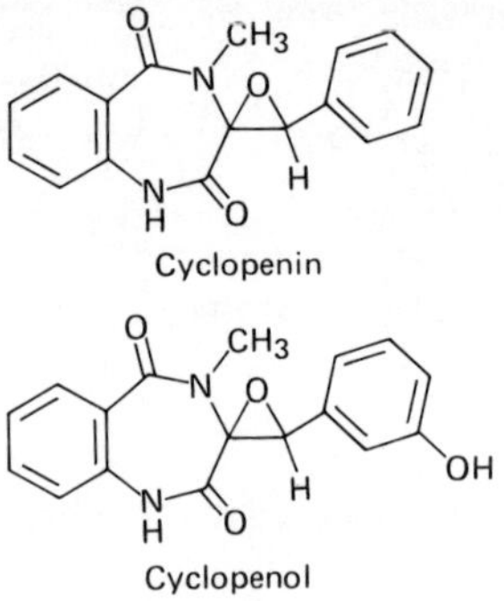

Cyclopenin

Cyclopenol

Figure 119. Cyclopenin, cyclopenol

– frequentin, also produced by *P. frequentans,* and palitantin, a metabolite of *P. palitans* (573);

– ergosterol; i-erythritol;

– emodic acid, $C_{15}H_8O_7$ (Figure 118), an anthraquinone pigment with an orange colour;

– cyclopenin (207, 282, 1706), $C_{17}H_{14}O_3N_2$, derived from anthranilic acid and L-phenylalanine (Figure 119);

– cyclopine, a nitrogenous antiviral metabolite (1099);

– cyclopolic acid, $C_{11}H_{12}O_6$ (Figure 120), which is readily oxidized into cyclopaldic acid;

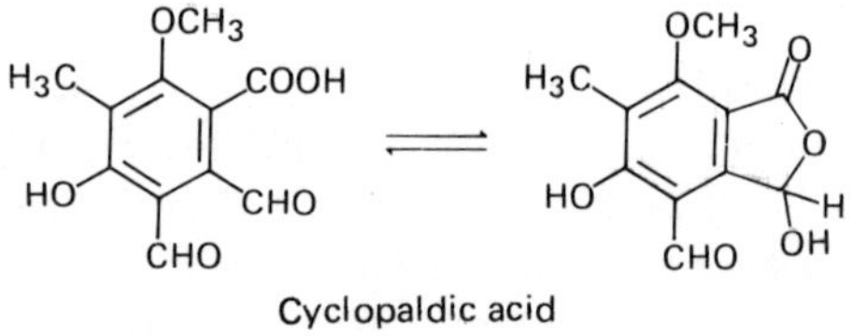

Cyclopolic acid

Cyclopaldic acid

Figure 120. Tautomeric forms of cyclopolic and cyclopaldic acids

– α- and β-cyclopiazonic acids, which are considered by some to be the main toxins elaborated by *P. cyclopium* (1115, 1116, 2709). They are neurotoxins, ingestion of which also produces diarrhoea and convulsions in the duck, the chicken, and the rat. The LD_{50} is 2.3 mg/kg when given intraperitoneally to the male rate (1116) (Figure 121).

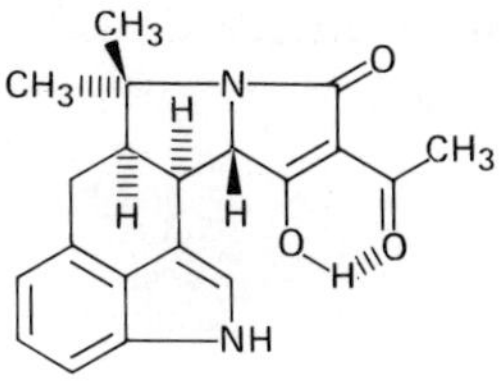

α-Cyclopiazonic acid

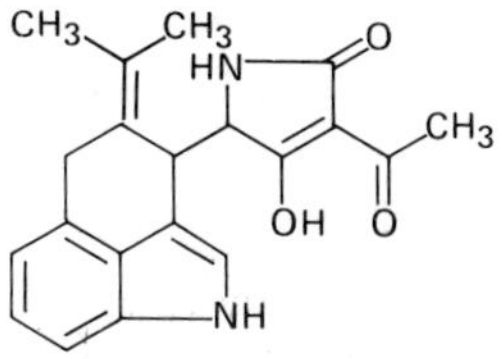

β-Cyclopiazonic acid

Figure 121. Toxic metabolites of *Penicillium cyclopium*

P. puberulum produces, among its metabolites, puberulic acid $C_8H_6O_6$ (Figure 122), which is also produced by *P. martensii* Biourge, and puberulonic acid, $C_9H_4O_7$. It has also been suggested that it may produce aflatoxins (1101, 2833), penicillic acid (2832), and viridicatin.

The toxicities of and the interactions between the components of this great variety of metabolites when they are ingested by animals requires a deeper study. It has been considered, in consequence, that *P. cyclopium* is a very important agent of mycotoxicoses and may be responsible for the death of sheep and horses. It causes congestion of the renal glomerules and cell nuclei tend to be pycnotic. Large variations in the dimensions of the nuclei of hepatic cells are observed and the cytoplasm shows signs of degeneration (2717).

The mouse, sheep, cow, and horse are very sensitive to the tremorgenic/diuretic toxin and the effects in the rat are very severe and prolonged. This toxicosis is

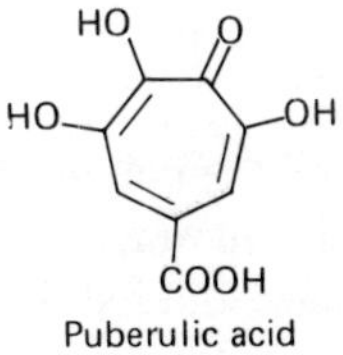

Puberulic acid

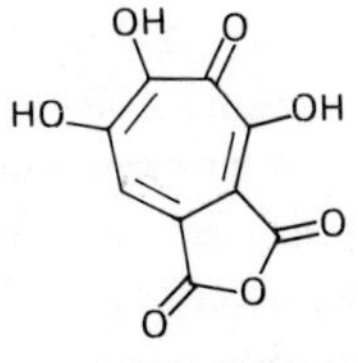

Puberulonic acid

Figure 122. Metabolites of *Penicillium puberulum*

reminiscent of the nervous disorders called 'Ilesha shakes' observed in humans in Nigeria and the outbreak of polyuria affecting 150 people in India after eating mouldy millet (2830).

Penicillia of the subsection Fasciculata producing patulin

Many members of the Fasciculata are able to produce patulin. *P. expansum* Link probably corresponds to the blue-green *Penicillium* referred to by early authors as 'P. glaucum' and it has also sometimes been referred to as *P. leucopus.* This species has a rapid growth rate often producing somewhat indefinite colonies. The conidiophores appear either individually or in groups in very thick coremia, from whence the name *P. crustaceum* is sometimes given to strains. The thallus is greenish blue with a sterile white margin. The conidiophores, which may have a smooth or roughened wall, are 150–400 x 3–3.5 μm. The penicillus, which is asymmetric and usually branched once or twice, ends in a chain of spores of 150–200 μm. The branches of the conidiophore, 15–25 x 2.5–3.5 μm, are typically divaricate and the metulae, 10–15 x 2–3 μm grouped in verticils of 3–6, are generally all at the same level. The phialides, in groups of 5–9, are 8–12 x 2–2.5 μm and bear elliptical, smooth-walled conidia, 3–3.5 μm, which appear yellow-green in mass (Figure 123).

It is an extremely common fungus and can be isolated from numerous substrates but its growth on previously damaged apples is spectacular, and probably not without danger. In fact, Brian, Elson, and Lowe (297) have definitely demonstrated the presence of patulin in apples attacked by *P. expansum* and this may pose a danger to the consumer. It can grow on other fruits, bananas for example (1258), and its spores are readily transported by air. Culture filtrates of some strains, when fed to miniature Japanese quail, *Coturnix coturnix japonica,* produce characteristic liver lesions which are peribiliary and perivascular and resemble a lymphocytic hepatitis (1675).

Mycelial extracts and culture filtrates of *P. italicum* Wehmer, the organism responsible for blue mould of citrus and closely related to *P. expansum,* are toxic to mice (2216).

P. urticae Bainier (= *P. patulum* Bainier), also appears in the subsection Asymmetrica Fasciculata of Thom. The thallus is granular, frequently producing coremia, at least around the edge of the colony. It is grey in colour, although the mass of spores is green. Some strains have a characteristic smell and the reverse of the colony is dirty yellow, then orange and reddish.

The conidial heads are loose, divergent, often large (40–50 μm), and carry chains of spores which may be 50–100 μm long. Conidiophores may be isolated, or in fascicles, smooth walled, and measure 400–500 x 3–4 μm. Branching is fairly divergent and there may be both primary and secondary branches (12–20 x 3–3.5 μm), the metulae (7–9 x 3–3.5 μm) and phialides (4.5–6.5 x 2.0–2.5 μm) are relatively short. Each branch bears a verticil of 2–4 metulae, each of which gives rise to 8–10 phialides which, in turn, produce elliptical to subglobose, smooth conidia, 2.5–3 μm in diameter.

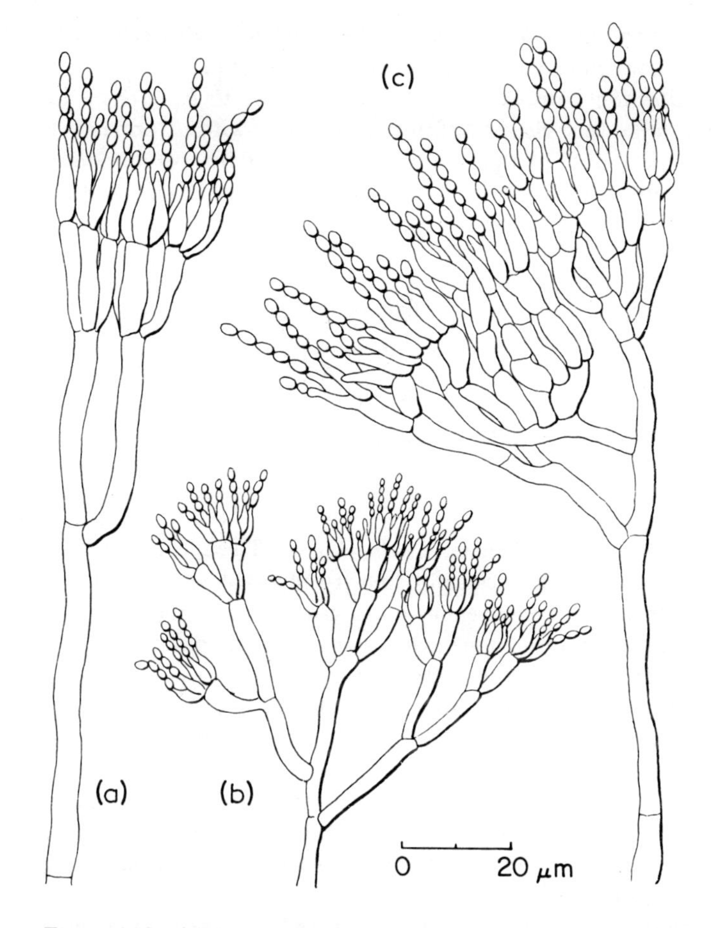

Figure 123. (a) *Penicillium expansum*. (b) *P. urticae*. (c) *P. claviforme* conidial heads

Although fairly widely distributed, *P. urticae* is not a particularly abundant mould. It is soil borne and may be isolated from decomposing vegetable matter, sheep's faeces and a number of other substrates (1868). This species also produces patulin (1293, 2549) and, in 1955, more than 100 cows were poisoned in the province of Kobe in Japan, after consumption of malt moulded by this fungus (1126, 1127, 1314, 2679).

OH

CH_2OH

Figure 124. *m*-Hydroxybenzylic alcohol

As well as patulin *P. urticae* also produces *m*-hydroxybenzylic alcohol, $C_7H_8O_2$ (Figure 124), which may be a precursor of patulin (2112).

P. griseofulvum Dierckx, at first classified close to *P. terrestre* Jensen, is now considered to be very close to, if not synonymous with, *P. urticae* (2589). Among its many metabolic products are found, not only patulin and its possible precursors, but other antibiotics such as griseofulvin, $C_{17}H_{17}O_6Cl$ (Figure 125),

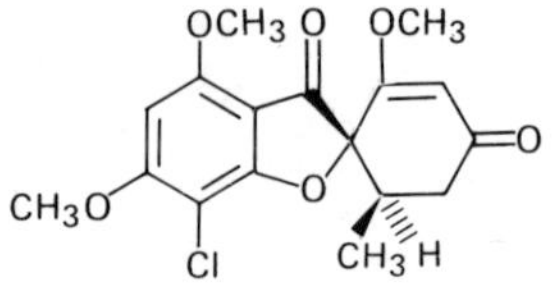

Figure 125. Griseofulvin

isolated by Oxford, Raistrick, and Simonart (1907), dechlorogriseofulvin, $C_{17}H_{18}O_6$, and mycelianamide, $C_{22}H_{28}O_5N_2$. These compounds are not very toxic to man and the rare cases of toxicity reported seem to be associated with large intraperitoneal or intravenous doses given to rats. Such toxicity arises from nuclear disturbances in the bone marrow and seminal epithelium (294, 1910). The intravenous LD_{50} for the rat is 400 mg/kg.

Griseofulvin is also produced by *Penicillium albidum, P. raciborskii, P. melinii, P. patulum, P. raistrickii, P. brefeldianum, P. viridi-cyclopium, P. brunneostoloniferum* (2146), *P. nigricans,* and perhaps also from *Trichoderma viride.* It has a characteristic activity against fungi having chitin in their cell walls, causing the tips of the hyphae to curl. These interesting antifungal properties, which have been useful in the treatment of dermatitic mycoses, gave rise to the name 'curling factor' when the antibiotic was first studied.

P. claviforme Bainier, which is closely related to *P. expansum* and *P. urticae,* is characterized by the presence of abundant coremia, which are aggregates of conidiophores, at the top of which the penicillus heads bearing the spores are tightly interwoven. The coremia frequently occur in concentric circles. This species also produces patulin and extracts of cultures are also known to provoke leaf necrosis of the vine (1279).

TOXICOSES ASSOCIATED WITH THE *BIVERTICILLATA SYMMETRICA*

Penicillium from mouldy rice related to *P. islandicum*

The serious toxicosis associated with *P. islandicum* has been dealt with in a previous chapter. There are at least five other species which are taxonomically related to *P. islandicum*, isolated from rice and considered to be toxigenic: *P. brunneum* Udagawa, *P. rugulosum* Thom, *P. concavo-rugulosum* Abe, *P. tardum* Thom, and *P. wortmanni* Klocker. *P. brunneum,* isolated first in 1959 from imported rice (2644, 2646) and then from indigenous rice, is particularly dangerous. Mice fed with moulded rice show atrophy and changes in the fatty tissue of the liver, followed by nodular hyperplasias and cirrhosis, often complicated by icterus. The kidneys show nephrosis which is sufficiently severe to be often the cause of death.

P. brunneum, like the other four species, produces rugulosin. It also produces emodin and skyrin (1096), but these may perhaps only be degradation products of rugulosin (2553), or they may also be the precursors of the more toxic metabolites.

P. tardum produces an antibiotic referred to as tardin, which is tolerated by mice at doses of 40 mg/kg when given intravenously, and 125 mg/kg when given subcutaneously. *In vitro,* human leucocytes are destroyed within 30 minutes by concentrations of tardin of 0.5 mg/ml (207).

Penicillia of the *Purpurogenum* series

Penicillium rubrum and rubratoxicosis

P. rubrum Stoll is very widespread in nature and is frequently isolated from soil and from organic material. It colours the substrate on which it is growing a deep red. It has been isolated from grain and is suspected, both alone and in association with *Aspergillus flavus,* of being responsible for a serious toxicosis of pigs as well as other animals, such as horses, geese, rats, and mice (357, 718, 1567, 2408, 2409). The toxin produced is stable at 70 °C (2409) and was considered to possibly be the red pigment phoenicin, $C_{14}H_{10}O_6$ (Figure 126) (2031), a compound which contains two toluquinone residues and was first isolated from *P. phoeniceum* v. Beyma (847). Phoenicin is very similar to oosporein, a metabolite of *Oospora colorans* v. Beyma, *Verticillium psalliotae* Treschow, a species of *Chaetomium,* as well as two isolates of sterile fungi which have not yet been identified (1325). Oosporein is related to phoenicin in the same manner as spinulosin is to fumigatin.

Growing *P. rubrum* on a Raulin-Thom medium enriched with 2.5% malt extract, Townsend, Moss, and Peck (2623) demonstrated the presence of two toxic compounds which they called rubratoxins A and B. An improved method of production and purification has since been described (1047). The culture filtrate, first concentrated on a rotary evaporator under reduced pressure, was continuously extracted with diethyl ether for 48 hours. The resulting semi-crystalline precipitate was filtered, dried, and analysed by thin layer chromatography and infrared spectroscopy.

Using n.m.r. and mass spectroscopy it has been shown that the rubratoxins are

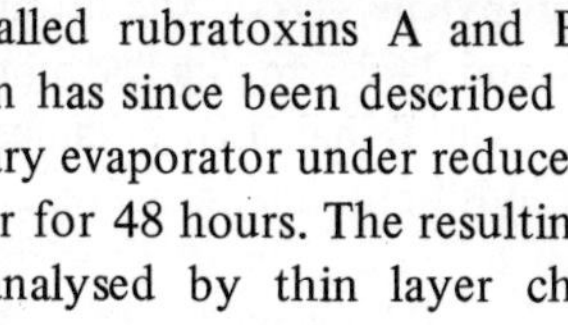

O OH O
H_3C CH_3
O HO O

Figure 126. Phoenicin

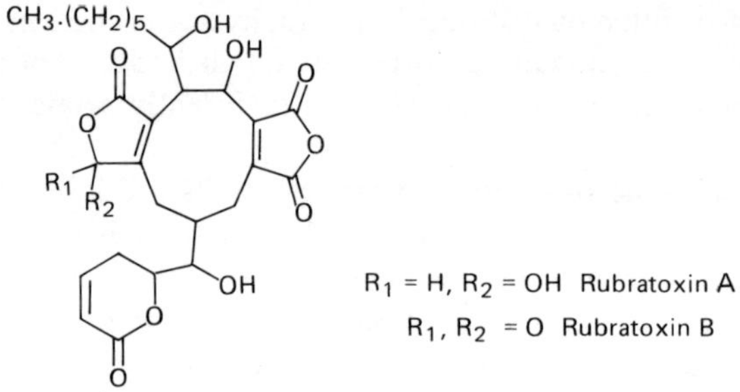

$R_1 = H, R_2 = OH$ Rubratoxin A
$R_1, R_2 = O$ Rubratoxin B

Figure 127. The rubratoxins

chemically related to glauconic and glaucanic acids, metabolites of *Penicillium purpurogenum,* on the one hand, and to byssochlamic acid, a metabolite of *Byssochlamys fulva,* on the other. They are C_{26} metabolites formed by the condensation of two C_{13} units, each of which is thought to be produced by the condensation of a decanoic acid derivative and oxaloacetic acid followed by decarboxylation and dehydration (1772).

Rubratoxin A, $C_{26}H_{32}O_{11}$ (Figure 127), has an R_f of 0.56 on TLC plates developed with acetic acid/methanol/chloroform (2:20:80), whereas rubratoxin B, $C_{26}H_{30}O_{11}$, has an R_f of 0.7 (1769). Such chromatograms also show artefacts believed to be hydrolysis products from the opening of the anhydride rings present in these molecules. They give rise to spots at R_f values of 0.12 and 0.2 respectively (1047).

The rubratoxin molecule contains two anhydride rings fused to a nine membered carbocyclic ring and the rubratoxins thus belong to a family of metabolites known as the nonadrides (1770–1773). The structure of rubratoxin B was confirmed by work in Japan (1812) where the compound was isolated from a strain of *P. purpurogenum.*

The LD_{50} of the rubratoxins appears to depend on the nature of the solvent in which it is given. Thus, in propylene glycol, the intraperitoneal LD_{50} for mice is 6–7 mg/kg for rubratoxin A and 3–3.5 mg/kg for rubratoxin B (1767) and for the dog and the chicken, rubratoxin B has LD_{50} values of 5 mg/kg and 4 mg/kg respectively (2858). In dimethylsulphoxide, however, the intraperitoneal LD_{50} of rubratoxin B has been shown to be 0.27 mg/kg for mice, 0.35 mg/kg for the rat, 0.48 mg/kg for the guinea-pig, and 0.2 mg/kg for the cat (2858).

The pig, after consuming grain infected with *P. rubrum,* shows enhanced salivation, periods of depression alternating with hyperirritation, incoordination of movement, and development of superficial reddish purple erythemas around the abdomen and hindquarters (800).

Histopathological examination (800) reveals that the most pronounced changes occur in the stomach, liver, kidneys, and central nervous system. The mucosa of the stomach is desquamated and the submucosa is congested and oedematous.

Histological changes can be observed in the guinea-pig after administration of a dose of 800 μg rubratoxin B, so the toxicity is comparable to that of aflatoxin G_2

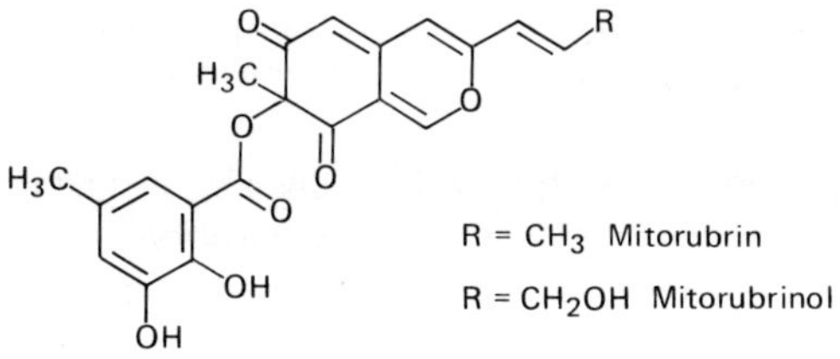

R = CH3 Mitorubrin
R = CH2OH Mitorubrinol

Figure 128. Orange pigments of *Penicillium rubrum*

(889). Detailed description of post-mortem and histological observations of animals poisoned by the toxins of *P. rubrum* are given by Wilson and Wilson (2838).

Penicillium rubrum is also known to produce allergies in people sensitive to asthma and hay fever (1020, 1956). A correlation with certain types of otitis has been demonstrated (1560).

Among other metabolites produced by *P. rubrum* are the orange pigments mitorubrin, $C_{21}H_{18}O_7$ (Figure 128), m.p. 218 °C, $[\alpha]_D^{25}$ −405°, and mitorubrinol, $C_{21}H_{18}O_8$, m.p. 219–221 °C, $[\alpha]_D^{25}$ −375°.

An uncharacterized antibiotic, fungistatic to a number of species of plant pathogens, including *Endothia parasitica* and *Ceratocystis minor,* has been described from a strain of *P. rubrum* isolated from a chestnut tree in Yugoslavia (1360).

Penicillium purpurogenum

P. purpurogenum Stoll is frequently isolated from rice, wheat, and a variety of cereal products. It has also been isolated from miso obtained by the fermentation of soya (2219). Ingestion of contaminated seed causes haemorrhagic syndromes in

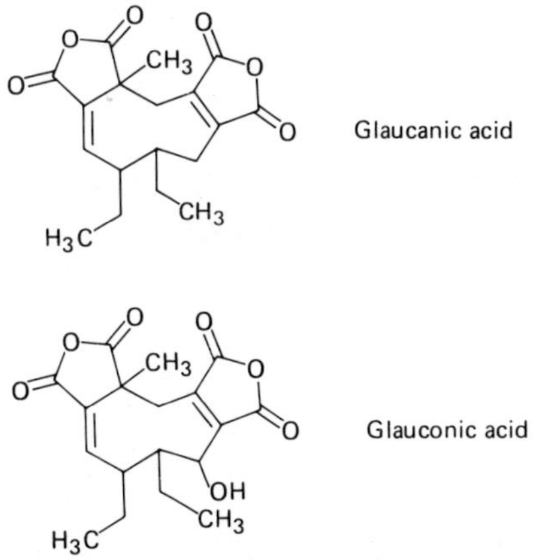

Glaucanic acid

Glauconic acid

Figure 129. Nonadrides produced by *Penicillium purpurogenum* chickens (808) and, in laboratory experiments, hepatic lesions in the mouse (418). Natori *et al.* (1812) have isolated rubratoxin B from a strain obtained from wheat in Japan.

P. purpurogenum produces two other nonadride metabolites similar to byssochlamic acid, the latter thought to be responsible for the toxicity of *Byssochlamys fulva,* which have been called glauconic acid, $C_{18}H_{20}O_7$ (Figure 129), and glaucanic acid, $C_{18}H_{20}O_6$ (125, 148). The first has a melting point of 202 °C and the second 186 °C. These nonadrides (239) are considered to result from the condensation of two C_9 units, themselves arising from the condensation of a derivative of hexanoic acid and oxaloacetic acid, followed by decarboxylation, a biosynthetic pathway similar to that proposed for the C_{26} rubratoxins.

It is possible to reduce the haemorrhages in chickens consuming feed contaminated with penicillia by adding vitamin K to the feed (806), and the use of antifungal compounds, such as 8-hydroxyquinoline, as a feed additive has been suggested (803).

Other penicillia of the Biverticillata Symmetrica

Scott (2315) has demonstrated that extracts of *P. piceum* Raper and Fennel are toxic to ducklings, perhaps due to the production of helenin (207), an antibiotic also produced by the closely related species, *P. funiculosum* Thom (1099). *P. variabile* Sopp., which has been reported to produce aflatoxins (1365), is also toxic to ducklings (2315), and may inhibit the growth of rabbits, producing mild kidney lesions (2717).

The unusual metabolite, β-nitropropionic acid (BNPA), the presence of which in plants such as *Indigofera endecaphylla* Jacq. and *Viola odorata* L. is thought to make them toxic, is produced by *P. atrovenetum* G. Smith growing on the plants (2091).

P. novae zelandiae v. Beyma is able to produce patulin.

CHAPTER 8

FUSARIOTOXICOSES

Poisoning due to moulds of the genus *Fusarium,* members of which are very widespread especially in the soil and on the seed of cereals, may be designated as fusariotoxicoses (1338, 2555). The specific identification of these fungi is often difficult because two nomenclatures are frequently used. On the one hand Wollenweber (205, 2081, 2865) recognized about 150 species, varieties, and forms distributed into 16 groups, and on the other hand, Snyder and Hansen (2437–2440) suggested that Wollenweber's species reflected a large degree of genetic variability of the species and reduced the number to 10. The systematics of *Fusarium* have been clarified by the recent studies of Booth (263) and Messiaen and Cassini (1652).

Many species of *Fusarium* may be associated with perfect states among the ascomycetes and belong to such genera as *Hypomyces* and *Nectria*, which have two-celled ascospores, or *Gibberella, Calonectria,* and *Griphosphaeria,* which have triseptate ascospores (see Table 30).

TOXICOSES ASSOCIATED WITH *FUSARIUM NIVALE*

Toxicosis of giant fescue and of rice

Fusarium nivale (Fr.) Ces. (*Griphosphaeria nivalis* (Schaffnit) Müll. and v. Arx) grows on grasses. Sporadic poisoning of cattle consuming hay containing giant fescue, *Festuca arundinacae* Schreb., has been described (2914) as a weakening of the hindquarters and gangrene of the extremities, such as the tail and hooves. These symptoms occur particularly when it is cold whereas fever usually occurs during warmer periods (1186).

Because this toxicosis is seasonal in its occurrence, and the concentration of alkaloids in fescue also varies with the season (2915), it was first considered that the plant alkaloids played a role in the appearance of these symptoms. The results of these early studies were negative and even the main alkaloid isolated, festucin, is only weakly toxic to the mouse (2917, 2921). The next phase of the study was to systematically isolate all the moulds present on the toxic hay (1296). Among the isolates, *Cladosporium cladosporioides, Epicoccum nigrum, Mucor fragilis,* and especially *Fusarium nivale* all gave a positive skin reaction on the rabbit, causing oedema and necrosis.

Table 30. The perfect states of *Fusarium* and the corresponding groups of Snyder and Hansen compared with those of Wollenweber, indicating those species, based on the latter classification, which are reputed to be toxic

Perfect State	*Species of Snyder and Hansen*	*Groups of Wollenweber*	*Toxic Species*	*Microconidia*	*Chlamydospores*
Nectria	F. episphaeria	Eupionnotes Macroconia			
Calonectria	F. rigidiusculum F. ciliatum	Spicarioides Submicrocera Pseudomicrocera			
Griphosphaeria	F. nivale	Arachnites	*F. nivale* (Fr.) Ces.	+	+
	F. tricinctum	Sporotrichiella	*F. poae* (Peck.) Wr. *F. sporotrichioides* Sherb.	+	+
	F. roseum	Roseum	*F. graminum* Corda *F. avenaceum* (Fr.) Sacc.		+
		Arthrosporiella	*F. semitectum* Berk. and Rav. *F. diversisporum* Sherb.		+
		Gibbosum	*F. equiseti* (Cda) Sacc. *F. scirpi* Lamb. and Fautr.		+
Gibberella		Discolor	*F. sambucinum* Fuck. *F. culmorum* (W.G. Sm.) Sacc. *F. graminearum* Schwabe		+
	F. lateritium	Lateritium	*F. lateritium* Nees		
	F. moniliforme	Liseola	*F. moniliforme* Sheld.	+	–
Hypomyces	F. oxysporum	Elegans	*F. oxysporum* Schl. *F. redolens* Wr.	+	+
	F. solani	Martiella Ventricosum		+ +	+ +

The poisoning of horses consuming oats infected with *Fusarium nivale* has been reported (2645) and this species, which is very common on wheat (2557), may cause nausea and vomiting in man, but its ingestion is not usually fatal. In Japan the considerable toxicity for animals of rice infected with *F. nivale* has been recorded (2649) and several cases of poisoning of cattle have been observed.

Toxic metabolites of *Fusarium nivale*

Once the relationship between toxicoses and the presence of *Fusarium nivale* has been established (2919), the next stage in the investigation is the characterization of the toxic metabolites of this species. The strain NRRL 3249 of *Fusarium* isolated by Keyl *et al.* (1296), and referred to as *F. nivale,* is considered by Snyder to be *Fusarium tricinctum* (2916). Because of this confusion some of the characteristics ascribed to *F. nivale* may, in fact, belong to *F. tricinctum.*

The most abundant toxic metabolite is a butenolide lactone, the γ-lactone of 4-acetamido-4-hydroxy-2-butenoic acid, $C_6H_7NO_3$ (Figure 130) (723, 2918, 2919, 2920).

The butenolide has an LD_{50} for the mouse of 43.6 ± 1.24 mg/kg.

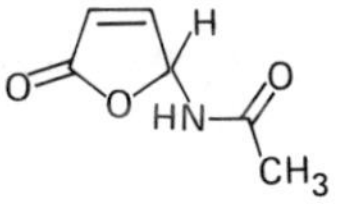

Figure 130. Butenolide of *Fusarium nivale*

Several very toxic sesquiterpenes are produced by *F. nivale* growing on rice including nivalenol, $C_{15}H_{20}O_7$ (3α,4β,7α,15-tetrahydroxyscirp-9-en-8-one), its monoacetate fusarenone (3α,7α,15-trihydroxy 4β-acetoxyscirp-9-en-8-one), a diacetate, $C_{19}H_{24}O_9$ (3α,7α–dihydroxy 4β,15-diacetoxyscirp-9-en-8-one), and a compound referred to as T_2 toxin, which is more common among the metabolites of *F. tricinctum* (974, 2556, 2920) (see Figure 131).

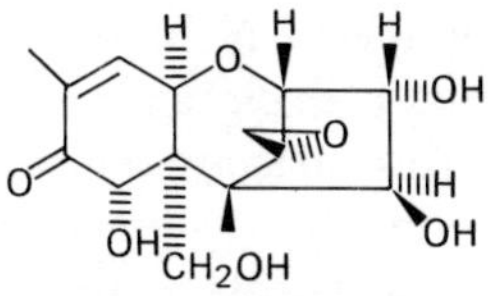

Nivalenol

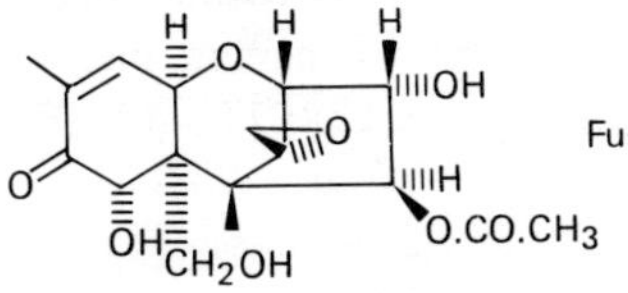

Fusarenone

Figure 131. Epoxytrichothecenes of *Fusarium nivale*

Another toxin, fusarenone X, $C_{17}H_{22}O_8$, has also been reported, having almost the same R_f as fusarenone itself, but some reports give different melting points and elemental analyses (2672, 2674), whereas others report that they are identical (1753). The general consensus is that fusarenone and fusarenone X are, in fact, identical.

The properties of these toxins have been intensively studied, especially in Japan (1755, 1881, 1883, 2218, 2558, 2649, 2665, 2667, 2669, 2670, 2674). The following properties have been reported for them:

nivalenol, m.p. 222–223 °C, $[\alpha]_D^{24}$ (ethanol) + 21.5°;
fusarenone, m.p. 78–80 °C, $[\alpha]_D^{24}$ (ethanol) + 29.6°;
fusarenone X, m.p. 91–92 °C (87–90 °C according to Morooka *et al.*(1753)).

Their intraperitoneal acute toxicities for the mouse are all very similar, the LD_{50} being 4.0 mg/kg for nivalenol, 3.5 mg/kg for fusarenone and 3.56 mg/kg for fusarenone X. Subcutaneously the LD_{50} values are 4.5 and 4.0 mg/kg and by the oral route 4.5 mg/kg for fusarenone X (2667).

Both nivalenol and fusarenone inhibit protein biosynthesis in reticulocytes of rabbits, but the former is ten times more active than the latter. The site of this inhibition appears to be the ribosomes.

Toxicity in the whole animal is due to changes in the cells during the development of haematopoietic tissue in the bone marrow, the spleen, the lymphatic ganglia, as well as cells of the intestinal epithelium. Spermatogenesis is also upset (2217).

With fusarenone X, the symptoms reported in mice include a dilation of the intestines with haemorrhage, atrophy, and ecchymosis of the thymus, hyperaemia of the peripheral lobes of the liver, all accompanied by diarrhoea (2667).

TOXICOSES ASSOCIATED WITH *FUSARIUM TRICINCTUM*

Alimentary toxic aleukia

Alimentary toxic aleukia (commonly abbreviated to ATA) is a disease of human beings who have ingested moulded grain, or its products, after storage in the open field over winter. It has also been described in the horse (115).

The disease is neither contagious nor infectious and the symptoms are very varied including, leucopoenia, agranulocytosis, necrotic angina, haemorrhagic diathesis, and exhaustion of the bone marrow. This toxicosis, which is frequently fatal, has received various names such as; aplastic anaemia, haemorrhagic aleukia, agranulocytosis, septic angina, alimentary mycotoxicosis, panmyelotoxicosis, etc.

Mayer (1634) has reviewed 239 original papers, mainly in Russian, relating to this disease and there are reviews by Gajdusek (862) and Joffe (1247).

Historic

Reported for the first time in the extreme east of Siberia in 1913, ATA seems then to have reappeared again during 1932 in several regions of eastern Siberia.

Whole families died and the populations of some villages were decimated. Since that period, the disease sprang up regularly in spring in the agricultural regions of the Urals, in Ukraine, central Asia and eastern Siberia. During the 1941–45 war years, ATA reoccurred, affecting 10% of the population (1237).

Before recognition of the true cause, this disease was attributed to bacterial infections, avitaminosis, especially of vitamins B, C, and riboflavin, to the auto-oxidation of cereal proteins with the formation of toxic amines, to alkaloids or glucosides, to oxidation of unsaturated fatty acids, etc. It was Murashkinskii (1793) who, in 1934, showed that a species of *Fusarium* infecting cereals, especially millet, but also wheat and barley, was responsible for ATA (1232, 1233, 1237, 1686). During the later years *Fusarium sporotrichioides* was isolated frequently (2239). Although this species occurs throughout the world, it would seem that ATA has only been recognized in the USSR between the latitudes 50–60° north and longitudes 40–140° east, in an area of badly drained clay soil where the winter temperature is of the order of –10 to –15 °C, and the summer temperature 15 to 25 °C.

Causes of ATA

A mild winter, with abundant snow, followed by a spring in which there are alternating periods of freezing and thawing, seem to be the most favourable conditions for the development of ATA. On the contrary, a hard winter and an absence of alternating freezing and thawing followed by a dry period seem to be the most unfavourable.

The grain eaten during the winter is not toxic. It is during the spring that it becomes toxic, corresponding to the period of active growth of the responsible fungi.

Although *F. sporotrichioides* is generally considered to be the principal agent of ATA (2241), several other species of fungi, frequently isolated at the same time, may play a significant role (1238, 1241). Thus *F. poae* (Peck) Wr., *F. lateritium* Nees, *F. graminum* Cda., *F. oxysporum* Schl., *F. redolens* Wr., *F. scirpi* Lamb. and Fautr., *F. sambucinum* Fuck, *F. semitectum* Berk. and Rav., and *F. equiseti* (Cda.) Sacc. may all cause the deaths of guinea-pigs 5–21 days after experimental feeding with infected grain, and that of rabbits after 8–24 days.

Equally, species of genera other than *Fusarium* may be present on overwintered grain and some are reputed to be toxigenic (286, 1238, 1337): *Cladosporium epiphyllum* (Pers.) Mart., *C. gracile* Cda., *C. fagi* Oud., *C. fuligineum* Bon., *C. penicillioides* Preuss, *Alternaria humicola* Oud., *A. tenuis* Nees, *Penicillium steckii* Zal., *P. brevicompactum* Dier., *P. notatum* Westl., *P. jenseni* Zal., *P. crustosum* Thom, *Trichothecium roseum* Link, *Verticillium lateritium* Rabenh., as well as several members of the Mucorales such as *Piptocephalis freseniana* de Bary, *Rhizopus nigricans* Ehrh., *Mucor albo-ater* Naoumov, *M. corticolus* Hag., *M. racemosus* Fres., and *Thamnidium elegans* Link.

In all cases culture extracts of these fungi, when administered orally to either guinea-pigs, rabbits or cats, have given rise to serious disorders which sometimes ended in death of the animal.

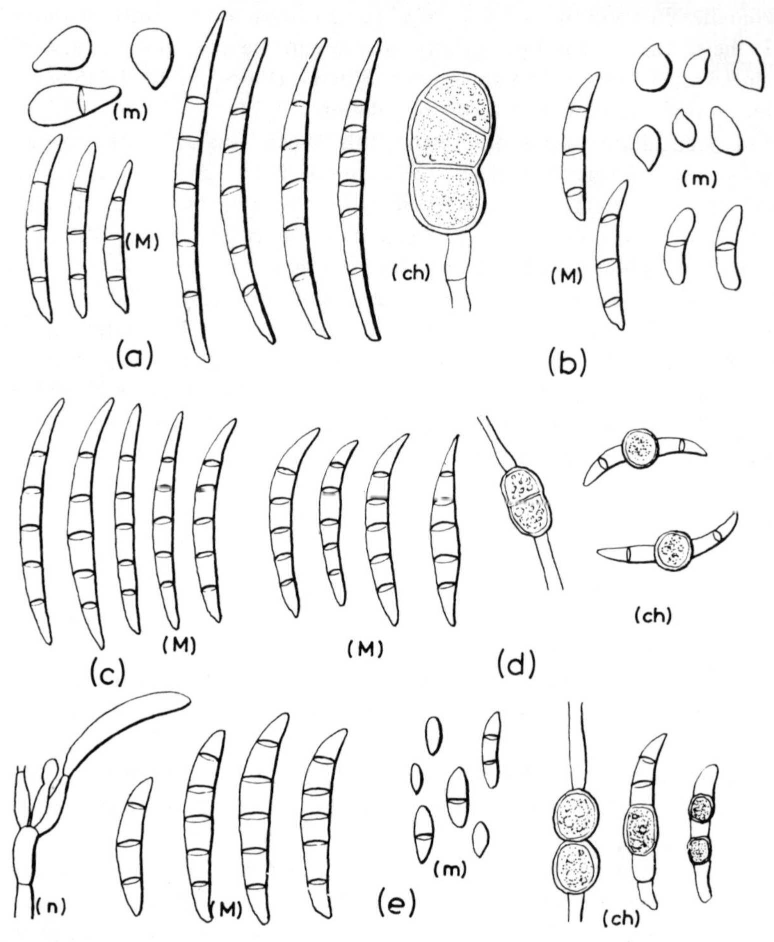

Figure 132. (a) *Fusarium tricinctum* (*sporotrichioides*). (b) *Fusarium tricinctum* (*poae*). (c) *F. roseum* (*graminearum*). (d) *F. roseum* (*sambucinum*), (e) *F. roseum* (*culmorum*). M = macroconidia, m = microconidia, ch = chlamydospores, n = production of macroconidia

'Fusarium sporotrichioides' – This is the name which is generally given to the principal agent of ATA corresponding to the species *F. sporotrichioides* Sherb. It has sometimes been referred to as *F. sporotrichiella,* an erroneous designation because the name *sporotrichiella* is not a specific epithet but a name given by Wollenweber to one of his groups. Another name used has been *F. sporotrichioides* Sherb. var. *tricinctum* (Cda.) Raillo, and according to Snyder and Hansen, they should all be *F. tricinctum* (Cda.) Sn. H. Seemüller (2331) has recently revised the systematics of this group.

The distinctive characteristic of *Fusarium tricinctum* (Figure 132) is the presence of pyriform microconidia, which are sometimes limoniform, produced by lageniform conidiophores.

The macroconidia are produced on sporodochia (in this the species differs from *F. roseum*) and in pionnotes, which are slimy or gelatinous, effuse sporodochia. Chlamydospores are plentiful and the culture is carmine red, purple or yellow ochre. An ascospore stage is unknown.

Fusarium sporotrichioides Sherb. *sensu stricto* is, according to Wollenweber and Reinking (2865), widespread on cereals and their products, stone fruits (in which it causes a rot) in Europe, Asia (Japan and Siberia), and North America. It may survive in the soil and on decaying vegetable matter. It has been isolated by the author from diseased stems of *Quercus borealis* in the south-west of France.

The species is very resistant to cold and grows, albeit slowly, at temperatures as low as −10 °C, although its optimum temperature for growth is about 24 °C. A period of four months at −20 °C does not kill it, although it is destroyed by heating to 100 °C for 10 minutes (979, 2243).

Toxins

It is usually during spring, after the cold of the winter, that the grain harbouring the fungus is most toxic. The optimum temperature for toxin production appears to be between 1.5 and 4 °C.

The intensity of the symptoms of poisoning depends on the proportion of infected grain in the food and it seems that more toxin develops on millet than on wheat or other grains. The oil-rich embryo is the part of the grain which becomes most toxic.

The toxicity of infected grain is not destroyed by normal cooking and, in fact, the toxin is stable at 125 °C for a period of 30 minutes and at 110 °C for 18 hours or longer. Mouldy grain stored for six years retained its toxicity.

The precise chemical nature of the myelotoxin of ATA was unknown for a long time. Okunev (1634) considered that it was a lipoprotein which, on hydrolysis, gave a protein and a substance similar to an oxidized coumarin which was responsible for ATA. Gajdusek (862), however, thought that the toxin was simply dissolved in the lipids of the grain. Gularev and Gulareva considered that the toxin consisted of two fractions when extracted into ether; one was considered to be present among the rancid fatty acids and to cause a localized dermatitic response in rabbits, the other was considered to be present in the non-saponifiable fraction and to be responsible for a dermatitic necrosis as well as being toxic orally.

It has been suggested that the myelotoxin may only be a phytotoxin containing nitrogen which interacts with immunoproteins and antibodies.

It was Olifson (1890, 1891) who first isolated and characterized a toxin from *F. sporotrichioides*, suggesting that it was a saponin, which was called sporofusarin and to which the formula $C_{65}H_{96}O_{25}$ was ascribed. On acid hydrolysis it was said to give a sapogenin, sporofusariogen, to which the empirical formula $C_{24}H_{30}O_4$ was given. From *Fusarium poae*, he isolated a glycoside, poaefusarin which was shown to contain xylose and a steroid aglycone referred to as poaefusariogen, $C_{24}H_{28}O_5$

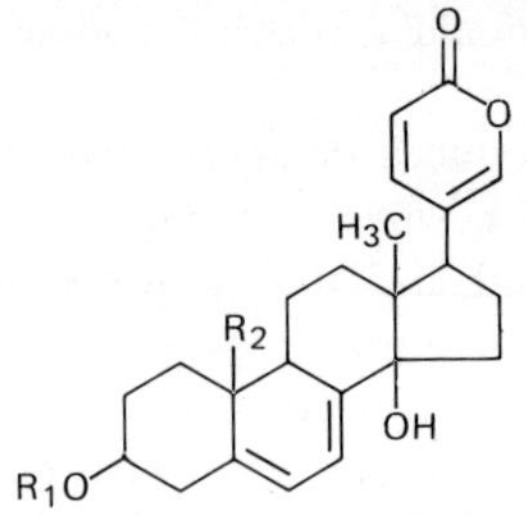

R$_1$ = oligosaccharide, R$_2$ = CH_3 sporofusarin
R$_1$ = oligosaccharide, R$_2$ = CHO poaefusarin

Figure 133. The structures suggested for sporofusarin and poaefusarin. (There has always been considerable confusion concerning the validity of these structures and it should be noted that a sample of poaefusarin has been analysed by American scientists, who could not confirm the presence of steroid-like compounds but did find both T-2 tetraol and zearalenone in the sample as well as T-2 toxin itself. See C. J. Mirocha and S. Pathre, 1973, *Applied Microbiology*, **26**, p. 719.)

which differed from sporofusariogen by the replacement of a methyl group with an aldehyde function. The structures which best fit all the information available are shown in Figure 133, although they are not consistent with the empirical formulae suggested!

A method for detecting toxic grain consists of a flotation test using 10–25% sodium chloride solution, which allows the recognition, not only of grain infected with *F. sporotrichioides,* but all moulded grain. Another test is based on the observation that the toxic compounds react with stannic chloride, in the presence of hydrochloric acid, to form a red colour, the intensity of which is related to the quantity of toxin present.

A biological test has also been proposed based on the observation that the growth of *Saccharomyces cerevisiae* is inhibited by the toxin. It has also been noted that a drop of an aqueous or alcoholic extract of the toxic grain causes chrysanthemums to wilt.

A general test used to detect the fusarium toxins consists of preparing two extracts of 50 g of suspect grain using anhydrous diethyl ether. The extractions are carried out continuously in a Soxhlet, one for six hours, the other for three days. Samples from the two extracts are then applied to a 9 mm^2 zone of the shaved skin of several rabbits. If toxin is present it causes a dermatitic inflammatory reaction, characterized by a pronounced hyperaemia, within two hours, followed by necroses several days later. The chick embryo test has also been used (1163).

Symptoms of ATA

In those cases localized in the USSR alimentary toxic aleukia mainly affected poor rural communities, probably because it is they who consume the most millet. The disease usually occurred between the end of April and the end of June.

Clinical studies – Forgacs and Carll (803) recognize and have described several stages in the development of the disease:

1st Stage – The first symptoms noticed were changes in the buccal cavity and the gastrointestinal tract. Soon after consuming food prepared from toxic grain, the patient complained of burning sensations in the mouth, on the tongue, on the palate, in the throat, the oesophagus and the stomach. This sensation resulted from the action of the toxin on the mucous membranes. The tongue sometimes became hard and swollen because a hyperaemia formed in the mucosa of the buccal cavity. After several days gastroenteritis developed, characterized by diarrhoea, nausea, vomiting, and accompanied by heavy breathing but without any fever. Fever only occurred in serious cases, the temperature reaching 39 °C, and was accompanied by excessive salivation, acute oesophagitis, gastritis or gastroenteritis, abdominal suffering, tachycardia, mild cyanosis, vertigo, headaches, and a sensation of coldness at the extremities. On rare occasions convulsions occurred.

If the toxic food was removed from the diet at this stage, these symptoms only persisted for two to three days. If, on the other hand, ingestion of toxic food was prolonged, the symptoms continued for five to nine days and then spontaneously disappeared, even though consumption of the dangerous food was continued. This clinically quiescent period was then followed by the symptoms of the second stage.

2nd Stage – Also known as the leucopoenic stage. At this stage the patient appeared clinically normal but the toxin begins to exert its destructive action on the haematogenic systems of the bone marrow. Haematological examination of the peripheral blood revealed serious changes showing a progressive leucopoenia, as well as granulopoenia and a relative lymphocytosis. These changes were associated with a reduction of erythrocytes and haemoglobin. The number of leucocytes decreased gradually and the neutrophils showed an abnormal granulation.

This stage normally lasted for three to four weeks but it sometimes extended over the period of two to eight weeks. Although there were usually no external symptoms, this phase of the disease was sometimes accompanied by changes in the nervous system. A weakening of the patient was noticed who showed fatigue, vertigo, headaches, palpitations, and mild asthma. The skin and the mucosa sometimes showed signs of icterus, the pupils were sometimes dilated, the pulse feeble, and the arterial pressure decreasing.

Even at this stage, if the toxic food was removed, a complete recovery could be hoped for. But if consumption of the toxic food continued then the third stage was inevitable.

3rd Stage, the angino-haemorrhagic stage – The change from the second to the third stage was sudden, often influenced by an accentuation of secondary symptoms such as fatigue. The first visible sign of the third stage was the appearance of red haemorrhagic petechia on the skin of the chest and abdomen, in the groin, and the armpits, as well as on the inner surfaces of the arms and thighs. Similar reddening also occurred on the buccal mucosa, the tongue, and the tonsils. The capillaries having become very fragile, the slightest trauma caused haemorrhages which spread and changed from a rose colour to red and then to violet.

Catarrhal changes intervened and necroses of 5–7 mm diameter appeared in the mouth. Gangrene of the pharynx, having something of the appearance of diphtheria, formed in the throat. The necrotic lesions sometimes spread to the palate, the uvula, the mucosa of the cheeks, the larynx, and the vocal chords. They often became infected with bacteria which, although non-virulent, caused decay producing a putrid odour from the mouth of the sick person.

The necrotic ulcers were round, flat, grey, and surrounded by a zone of hyperaemia and there was haemorrhage in the membranes close to the necrotic regions, but not with any well defined demarcation. The necrotic regions often coalesced and reached a considerable size, often 20–30 mm, and they often involved, not only cutaneous tissue, but also subcutaneous tissue and even the underlying muscle tissue. They could appear on the lips and, occasionally, on the skin of the nose, the jaws, and the fingers.

The necrotic lesions developed further in the buccal cavity; the petechia and ecchymoses, observed at the beginning of the third stage, became increasingly abundant, large, and confluent. Bleeding from the nose, mouth, stomach, and intestines could also occur. There was considerable vascular fragility and the mucosa of the gastrointestinal tract bled with the slightest shock.

Tissues close to the lymph nodes became oedematous and frequently the submaxillary and cervical glands were so swollen and the surrounding tissue so oedematous that the patient had difficulty opening the mouth. The lesions of the oesophagus and changes in the epiglottis led to oedema of the larynx and loss of the voice. In some cases death followed strangulation and 30% of the deaths of people reaching this third stage were due to stenosis of the glottis.

The abnormalities of the blood, observed during the second stage, were intensified during the third. The number of leucocytes was reduced to 100, or even less, per cubic millimetre, and the remaining neutrophils were stuffed with granules. Lymphocytosis could reach 90% and the number of erythrocytes decreased to less than one million per cubic millimetre. The concentration of haemoglobin was often less than 10% of normal and the number of thrombocytes decreased to about 25,000 per cubic millimetre, and there were cases where this figure was as low as 3,000. The blood sedimentation rate increased, reaching 70–90 mm at the end of an hour, most sedimentation occurring during the first 30 minutes. A deficiency in plasma fibrinogen was also noticed.

The blood bilirubin and uric nitrogen were normal but urinary excretion of vitamins C and B_1 was strongly reduced. The secretion of gastric hydrochloric acid varied from complete achylia to hyperacidity. No abnormalities were to be found in the spinal fluid.

In some cases patients suffered from acute parenchymatous hepatitis accompanied by jaundice. Mineral metabolism, as well as that of glucose and proteins, were severely disturbed in the liver and an increase in blood calcium was recorded.

During this stage the temperature could fluctuate up to 39–40 °C and then return to normal.

In the fully developed aleukia, hypotony became apparent and the average arterial pressure in adults was 8/5. The pulse reached 180 per minute but was feeble and a functional systolic murmur could be detected although there was no increase in the dimensions of the heart.

Pulmonary complications, usually bronchopneumonia, haemorrhages and abscesses were frequently observed. They could be aggravated by the appearance of secondary infections by micro-organisms of low virulence.

Surprisingly, the affected person usually retained all his reason, despite these distressing conditions. This third stage could last from 5–20 days.

When recovery occurred, the duration of convalescence depended on the severity of the toxicosis. About two weeks treatment was necessary to cure the necrotic lesions, and for the haemorrhagic diathesis, the bacterial infections, and the accompanying fever to disappear. It could take two months or more before the bone marrow started producing normal blood.

Mode of action of the toxin – The toxin acts essentially by suppressing the proliferation of the myeloid and lymphoid cells of the haematopoietic system. When it was badly affected the bone marrow appeared to be devoid of myeloid tissue. The toxin thus clearly has an antileukaemic activity. Moreover, attempts have been made to treat leukaemia with this fungal toxin (725–727, 1527, 1528, 1833).

Several researchers have observed a variety of disturbances of the nervous system during the development of the disease; changes in reflexes, meningitis, general hyperaesthesia, lack of orientation, dermographism, psychasthenia, and depressive states. Others have noted encephalitis, cerebral haemorrhages, and a disequilibrium of the parasympathetic nervous system. Histopathological studies have demonstrated haemorrhages and destructive changes of the neurones of the third ventricle and in the sympathetic ganglia.

Experimental toxicology – The first tests on animals were a failure. Cattle, sheep, poultry, horses, pigs, rabbits, and mice were all fed on infected grain, or the solvent extracts of them had been given by either the oral or parenteral routes. In no case could the symptoms observed in man be entirely reproduced.

Cattle, sheep, and chickens are able to eat a large quantity of toxic grain without showing any morbid effects. Horses consuming the same grain develop local reactions on the lips and in the mouth, resembling stachybotryotoxicosis, but without any changes in the blood, even in dead animals. A leucopoenia may be established in the pig and the rabbit after subcutaneous injection of toxic extracts.

When toxic extracts are given orally to the mouse or the rabbit, convulsions and paralysis, followed by death were observed, and 0.5 g of toxic millet ingested during five days is sufficient to kill a mouse. Geese which have ingested toxic grain showed disturbances of the central nervous system characterized by paralysis, vertigo or a staggering walk; mortality was high in poisoned geese.

Symptoms similar to those observed in man have been produced by feeding cats, guinea-pigs, monkeys, and dogs with millet infected with *F. sporotrichioides.* In the cat, depending on the individual, it required 29–103 g of infected millet, given over a period of 8–17 days to cause death (862, 1634).

Prophylaxis and treatment – Immunization is not possible. Preventive measures consist of informing the rural community of the dangers which may occur.

The only measure required during the first stage of the disease is the removal of the toxic grain from the diet of the patient. During the second stage, it is desirable to give blood transfusions and to administer preparations based on nucleic acids, sulphonamides, antibiotics, vitamins (especially C and K), and calcium. During the third stage, intensive therapy with sulphonamides and antibiotics is required and all these measures must be accompanied by a really good diet.

Other toxicoses of *Fusarium tricinctum*

In 1956, a toxin was isolated in the United States from grain infected with *F. tricinctum* which caused the rapid death of cattle (906). But it was not until ten years or more later that Bamburg *et al.* (127, 128) recognized the role of mould metabolites of the scirpene group:

– diacetoxyscirpenol, also one of the principal toxic metabolites of *Fusarium scirpi*;

– HT_2 toxin isolated from a strain of *F. tricinctum* and shown to be, 3,4,-dihydroxy-15-acetoxy-8-(3-methylbutyryloxy)-12,13 epoxy-Δ9-trichothecene (129);

– T_2 toxin, $C_{24}H_{34}O_9$, the most important of them, which is 8α-(3-methylbutyryloxy)-4β,15-diacetoxyscirp-9-en-3-αol (figure 134).

The strain NRRL 3299 is reported to produce at least 9,000 mg of T_2 toxin when grown for three weeks at 15 °C on 1,200 g of white maize grits (349). T_2 toxin is characterized by m.p. 150–151 °C, $[\alpha]_D^{24}$ –50 ° (2920), and has an oral LD_{50} for the rat of 4 mg/kg (2416), and for the trout 6.1 mg/kg. When given intraperitoneally to the mouse the LD_{50} is 3.04 mg/kg (2920).

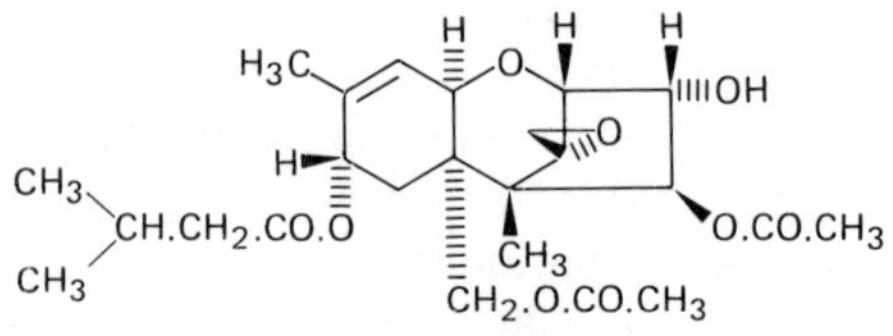

Figure 134. T_2-toxin

It has been shown that T_2 toxin will inhibit the growth of *Rhodotorula rubra* (350), and at a concentration of 2.5 ppm it induces a blighting of the pea, a phenomenon which may be a means of assaying the toxin (130, 1596).

A lethal or sublethal dose given to the rat causes changes in the concentration of prothrombin in the blood; the cardiac rhythm and the arterial pressure diminish and then increase again passing their normal values before gradually diminishing once more; respiration follows the same pattern; gastrointestinal problems and haemorrhages are frequent (2416).

In experimental chronic toxicity tests, lasting over a period of 12 months, T_2 toxin induces hepatomas in the trout at concentrations of 200–400 μg/kg. Doses of 5–15 mg/kg cause a severe inflammation of the intestinal mucosa and a fatty degeneration of the liver in albino rats (1595).

A diet containing 1% of maize contaminated with strain 2061 of *Fusarium tricinctum,* given to turkeys, caused the death of 13% of the animals in 35 days. If the diet contained 2%, 60–83% of the turkeys died and if it contained 5, 10 or 20% of the toxic maize, all the turkeys died between 5 and 15 days (485). They showed necrotic lesions at the corner of the beak.

The relationship between T_2 toxin and alimentary toxic aleukia has not yet been clearly established and it is still an open question as to whether these cases represent aspects of the same toxicosis or are distinct types.

Fusarium poae (Peck) Wr., sometimes referred to as *F. sporotrichioides* var. *poae* or even *F. sporotrichiella* var. *poae*, is also considered by Snyder and Hansen to be a form of *Fusarium tricinctum* (Cda.) Sn. and H. It is thus very closely related to *F. sporotrichioides* Sherb. and it may have been recorded under this name, giving rise to some confusion. Elpidina (725, 727) has suggested that the antibiotic poïn should be considered as one of the toxic metabolites of *Fusarium.* It is toxic to a number of microfungi, certain protozoa, and also inhibits malignant tumours. It is cytotoxic, affecting the synthesis of DNA.

TOXICOSES ASSOCIATED WITH *FUSARIUM ROSEUM*

Among the species of *Fusarium* regrouped by Snyder and Hansen under the name *Fusarium roseum,* several may be implicated as agents of mycotoxicoses including those which cause gastrointestinal upsets accompanied by considerable haemorrhaging, and others associated with a remarkable oestrogenic effect.

'Mouldy corn toxicosis' – diacetoxyscirpenol

Fungi incriminated – Under the single name 'mouldy corn toxicosis' there are a number of toxicoses associated with several fungi:

– in some cases *Aspergillus flavus* is the dominant organism and the disease is a typical aflatoxicosis;

– in other cases *Penicillium rubrum* plays the dominant role through the production of rubratoxins, a disease which may be referred to as rubratoxicosis;

– sometimes, however, a large number of species of fungi are isolated, such as

Fusarium roseum (which may include *F. avenaceum, F. diversisporum, F. equiseti, F. scirpi, F. sambucinum, F. culmorum,* and *F. graminearum*), *F. tricinctum, F. moniliforme, Epicoccum nigrum, Trichoderma viride, Trichothecium roseum, Cladosporium herbarum, Alternaria tenuis, Nigrospora* spp, *Paecilomyces varioti, Acremoniella atra, Papulospora* spp, (*Melanospora* spp), many *Aspergillus* spp, especially *A. versicolor, A. fumigatus* and *A. candidus,* several *Penicillium* spp such as *P. cyclopium* and *P. oxalicum* as well as members of the Mucorales such as *Mucor fragilis, M. alternans, Rhizopus nigricans,* and *Absidia corymbifera* (2417), but among all these fungi *Fusarium roseum* appears to occupy the principal role.

Toxicity in the pig – (97, 253, 795, 807, 1385, 1579, 1882, 2241, 2252, 2408, 2409, 2742). The mortality of sick pigs in the first outbreaks observed in the United States (2409) varied from 5–55% with a mean of 22%, young individuals being more sensitive than older animals.

In its acute form the disease is characterized by anorexia, depression, vertigo, and pallor of the mucous membranes. Death follows within two to five days. The body temperature, haemoglobin concentration, and that of prothrombin remain normal but the plasma concentration of vitamin A diminishes.

The chronic form associated with pigs consuming low quantities of infected grain over extended periods, gives rise to depression, anorexia, and a general debility. The animals have a stooping head, the posterior region is arched, and the flanks swollen, and they move with a stiff gait. The concentration of red blood corpuscles is low and icterus frequently appears.

The carcass of an animal killed by this type of mouldy corn toxicosis decomposes very slowly and perhaps this is because of the presence of antibacterial compounds which inhibit decomposition (2409).

The sick pig has abundant haemorrhages in several tissues and it is these which often bring about death. The abdominal cavity and the small intestine may become full of blood. Numerous petechia are visible in the muscle and on the surface of the liver; the spleen may be normal or it may also be haemorrhaged. In chronic cases the liver is yellow in colour and a straw-coloured fluid fills the abdomen and thoracic cavity, the kidneys being pale and oedematous.

Histological studies of the liver show a fatty degeneration with necrosis and subacute hepatitis. If intoxication is slow, an intralobular cirrhosis, proliferation of the bile ducts, and necrosis of the central region of the lobes, which sometimes extends to the periphery, are all recorded. In the kidney, there is atrophy of the glomerules, tubular dilations, oedema, and fatty degeneration of the epithelial cells (803).

Fusarium graminearum – In 1891, Woronin reported symptoms of vertigo, headache, and fatigue accompanied by fever in people who had consumed bread prepared from grain infected with *Fusarium graminearum* Schwabe (= *Gibberella zeae* (Schwein.) Petch). A vomiting agent has been isolated from this species, which is frequent on rye, oats, and barley, but rare in rice (493, 691).

According to Zeleny (2932), such grain and its products are highly toxic and should never be used in human foods, or even in that of animals. It is known, for example, that numerous pigs died in Germany, Belgium, and Holland after consuming barley infected with this fungus, and which had been imported by ship from the New World (187, 493, 1642), where this species is fairly frequent (1786, 1787). In parts of America it sometimes infects as much as 80% of the harvest (2174), causing a toxicosis in cattle, especially in the states of Indiana, Ohio, Illinois, Wisconsin, and Iowa (1575).

The infection of barley may be detected by a discoloration of the spikelet which becomes yellowish and then brownish. Sometimes a red efflorescence of mycelium is visible on the surface, followed by tiny black perithecia of the ascomycete stage. The infected spike is sterile and may be subsequently invaded by a wide range of other moulds.

Barley in which more than 5% of the seed is infected with *F. graminearum* is refused by pigs unless they are very hungry. Should they accept and consume it, they soon begin to vomit and, if the quantity ingested is sufficient, they become somnolent and lie down. Several minutes later they get up but recommence vomiting followed by sleeping, a process which may be repeated several times. Depending on the intensity of poisoning, the symptoms may continue to increase in severity, or they may gradually diminish. The toxin, which is soluble in water, may be readily extracted from infected grain and it is possible to reproduce the symptoms by injecting such a solution by stomach tube. An extract from 15 g of infected grain is sufficient to induce vomiting in a pig of 45 kg body weight when administered in this way (493).

Horses and mules show a repugnance for barley and oats moulded with *Fusarium,* as do the rat and the guinea-pig in laboratory trials. However an aqueous extract of infected barley seems to have no effect on rabbits, rats or mice, although it will cause the death of suckling mice within 24 hours (383).

Ruminants do not seem to be sensitive to the toxic effects of this species, even if their diet contains 56% of infected grain (2174). In poultry, the results of experimental intoxication are equivocal, although there is usually a loss of appetite and reduction in weight.

Fusarium graminearum, which is common in Japan on barley and wheat during humid years (1151), may occasionally attack rice (1370, 2645) and several cases of poisoning of sheep have been reported by Hokkaido during the period 1951–53 (2543). The infected grain was toxic and its consumption led to nausea, vomiting, a febrile state, and sometimes even death. In 1948, Hirayama and Yamamoto (1097) recorded a case of poisoning by *F. graminearum* growing on pasta. Not all strains are equally toxigenic (1376).

A search has been made for the toxic metabolites of *F. graminearum* (2650, 2691), and several toxins have been associated with this species (572). According to Hoyman (1134), it is a polysaccharide which is responsible for the problems of the nervous system, the violent nausea and the irritation of the stomach muscles, and intestines accompanied by diarrhoea. This same emetic principle has been recovered from *F. culmorum* (W. G. Smith) Sacc. and *F. avenaceum* (Fr.) Sacc. (2033).

More recent work (603) has demonstrated the formation of a toxic metabolite

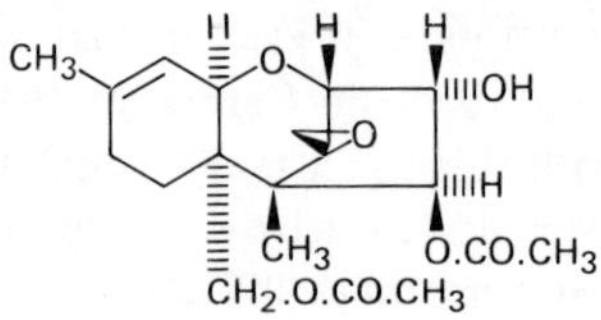

Figure 135. Diacetoxyscirpenol

variously called 'toxin B-24' and diacetoxyscirpenol, first by *F. scirpi* and *F. equiseti,* then from other species of *Fusarium.* It is a sesquiterpene, the structure (Figure 135) of which was elucidated by Flury *et al.* (792), Dawkins *et al.* (603), and Sigg *et al.* (2384) as 4,15-diacetoxyscirp-9-en-3-ol or, using more recent nomenclature, 3-hydroxy-4,15-diacetoxy-12,13-epoxy-Δ9-trichothecene.

The LD_{50} of diacetoxyscirpenol for the mouse is 7.3 mg/kg by the oral route and 10 mg/kg given intravenously.

Exposure of the larvae of the brine shrimp (*Artemia salina*) to 0.47 μg/ml for 16 hours causes 50% mortality (1038).

Opening the epoxide ring reduces, or even destroys, the toxic activity. Related compounds are nivalenol and fusarenone from *F. nivale*, T_2 toxin from *F. tricinctum* and 4β,8α,15-triacetoxy-12,13-epoxytrichothec-9-ene-3α,7α-diol, $C_{21}H_{28}O_{10}$ (974). It is this group of fungi which is frequently isolated from the feed implicated in mouldy corn toxicosis. It would seem likely that there is a relationship between the chemical structures of the toxins and the symptoms observed.

The symptoms of toxicity in the cow involve, first, mainly disorders of the digestive tract accompanied by diarrhoea, often with bloody faeces. There is a reduction in milk yield, loss of weight, and a general loss of appetite. Nervous symptoms with encephalitis or leuco-encephalomalacia and death are less common. On the other hand, frequent haemorrhagic lesions are noted in the stomach, the heart, the intestines, the lungs, the bladder, and the kidneys (28, 803, 2409).

In the guinea-pig a dose of 25 μg of diacetoxyscirpenol is sufficient to cause histological changes, a level of toxicity which is comparable to that of aflatoxin G_1 (889).

The anthraquinone pigment, rubrofusarin, $C_{15}H_{12}O_5$ (Figure 136), has also been isolated from strains of *F. culmorum* (W. G. Smith) Sacc. (2510, 2545) and its toxicity merits some study.

This species has been isolated from maize associated with a reduction in the milk

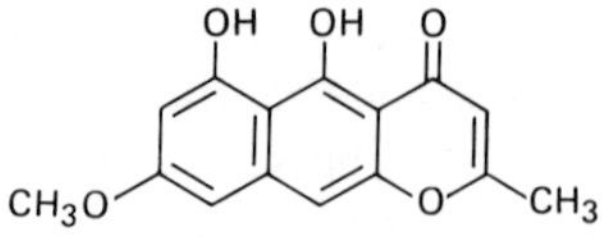

Figure 136. Rubrofusarin

yield of cows (785). This symptom has also been observed when dairy cows have been consuming forage in the prairie which is infected with *F. culmorum* (1728).

It appears to be difficult to control *Fusarium* although recent laboratory trials have demonstrated that the silicate of methoxyethyl mercury may give some measure of control due to the gaseous compounds which it produces (411)

Oestrogenic effects: zearalenone

In 1928, McNutt *et al.* (1561) described cases of vaginal prolapse, which they called vulvo-vaginitis, in sows in Wisconsin. The same phenomenon was described several years later in other parts of the United States of America, in Australia, Ireland, and, more recently, in France, Rumania, Hungary, and Canada (303, 378, 1330, 1454, 1551, 1913, 2043, 2285). Changes in the genital organs of young male pigs in Russia after consumption of mouldy maize or barley have also been recorded (1385).

Mirocha (1678) has obtained evidence of the occurrence of this type of toxicosis in central Europe and the author reported many cases in the south-west of France.

The visible symptoms appear to be considerable hypertrophy of the vulva with prolapse of the vagina and, secondarily, of the rectum. Young females have oedematous vulvas and the mammary glands are swollen. Atrophy of the ovaries and abortions are also a frequent response to the disease. Histological sections show that the vaginal cells are up to 15 times larger than normal (1379).

In the male, either a more or less pronounced swelling of the prepuce occurs which may sometimes be sufficiently severe as to constrict the flow of urine (1917), or mammary disorders have been recorded (1330).

Pigs are usually affected because their feed is often rich in maize, but such disorders have been reported in other farm animals often leading to abortion and sterility. The poisoning of turkeys and geese in this manner has been described in Hungary, although chickens do not seem to be as sensitive (1649).

Several fungi have been isolated from the mouldy feeds causing such oestrogenic disorders. The most common is *Fusarium graminearum* (= *Gibberella zeae*), but also *Penicillium, Trichothecium roseum,* and *Rhizopus* (2476).

In 1955, Christensen, Nelson, and Mirocha (487) isolated an oestrogenic compound which they called F-2 toxin from an unidentified isolate of *Fusarium.* Since then, several very similar compounds, F-2, F-3, F-4, F-5, have been isolated from cultures of *F. graminearum.* Toxin F-2, which has also been called FES (fermentation oestrogenic substance), RAL, and is now usually referred to as zearalenone, has the highest oestrogenic activity (1680, 1681, 1682, 1685).

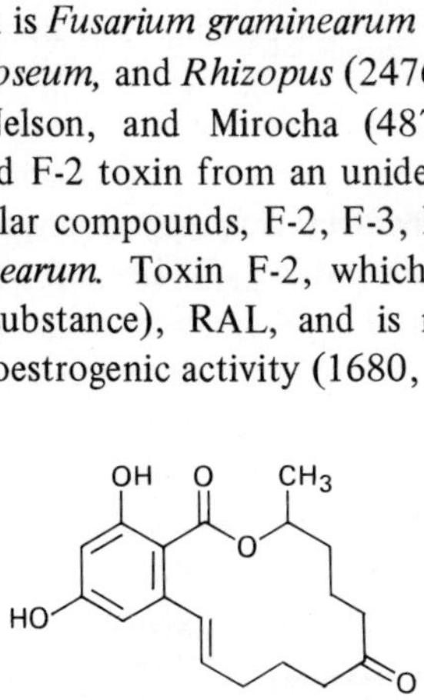

Figure 137. Zearalenone

Zearalenone, $C_{18}H_{22}O_5$ (Figure 137) is a lactone derived from resorcylic acid, or more specifically, 6-(10-hydroxy-6-keto-*trans* 1-undecenyl)β-resorcylic acid (1808, 2696). It has a molecular weight 318, m.p. 163–165 °C, and the u.v. spectrum shows λ^{EtOH}_{max} 314, 274, 236 nm, the compound producing an intense blue fluorescence when irradiated with u.v. light.

Two other toxic metabolites, called F-5-3 and F-5-4, have also been isolated. The first has m.p. 198–199 °C, the second m.p. 168–169 °C. They both have the same empirical formula, $C_{18}H_{22}O_6$, and molecular weight 334, but differ in their stereochemical configuration (1679).

Metabolites with related structures have been isolated from other fungi (1681): for example, curvularin from a *Curvularia* sp, *Penicillium steckii*, and *P. expansum*; radicicol from *Nectria radicicola*; and monorden from a *Monosporium* sp.

When injected into female white virgin rats, zearalenone gives an oestrogenic response at a dose of 20 μg (1680, 1682). When a dose of 650 μg is given over a period of one week the fresh weight of the uterus triples. These results are similar to those obtained in the pig (486) and the hen (2451).

Zearalenone appears to be produced by *Fusarium graminearum* growing on stored grain but not by the same species growing on maize in the open field (386). In the laboratory, not all culture media give the same yield (2026), but a good yield is obtained on either maize containing at least 45% water, or on rice at 12 °C (753, 1684). At 25 °C toxin production is nil and, moreover, a preliminary passage at a low temperature seems to be necessary to obtain any yield at temperatures greater than 12 °C.

When ingested by *Tribolium confusum* and *Alphitobius diaperinus,* two insects which live in flour, zearalenone persists during their metamorphosis (754), but does not affect subsequent egg production (1028).

It would seem that zearalenone has an influence on the production of the perfect stage of *Fusarium* (386, 752, 1823). A high concentration in the culture medium limits the production of perithecia by a number of ascomycetes, whereas low concentrations act as a stimulant. Monoascosporic cultures of *F. graminearum* itself which are forming perithecia produce less zearalenone than those cultures which do not produce perithecia.

Zearalenone, of related compounds, may have a useful role in pharmacy for regulating problems during menopause and tests have been carried out in Mexico (1678). Some of these derivatives may act as contraceptives, others, added to animal feed, may increase the growth of farm animals. Because of this interest a number of patents exist on the preparation of zearalenone and its derivatives, especially by the Commercial Solvents Corporation. The synthesis of zearalenone has been achieved (see, for example, the patent of Merck & Co. US 667,361).

Detoxifying infected grain in animal feeds is a problem and the use of sodium carbonate to hydrolyse the lactone ring does reduce the oestrogenic activity. Similar reactions may be carried out biologically by increasing the humidity of the grain and allowing the growth of micro-organisms which break down zearalenone.

It has been demonstrated that the addition of 20 mg of dichlorvos (dimethyl 2,2-dichlorovinyl phosphate) per kilogram in 10 μl of dimethylsulphoxide to a culture of *F. roseum* (*graminearum*), although it does not affect growth, inhibits the formation of zearalenone (1481).

TOXICOSES ASSOCIATED WITH *FUSARIUM MONILIFORME*

Fusarium moniliforme Sheld. (*Gibberella fujikuroi* (Saw.) Wr.), well known for the production of gibberellins and other tetracyclic hydroaromatic compounds, is considered to cause problems when it infects grain used in animal feeds (2315, 2465, 2491). It has sometimes been isolated in association with *Aspergillus flavus* and *Penicillium rubrum* (357). In some instances it may have oestrogenic activity (1678).

Associated with *F. roseum* and *F. solani* it has been incriminated in a typical interstitial pneumonia (695).

In the mouse, it may cause hyperaemia of the renal glomerules accompanied by a vacuolar degeneration in the proximal tubules (2717).

When 3–4.5 kg of infected grain is given orally to horses, a congestion of the nasal membranes, depression, anorexia, incoordination of movement, repeated spasms, vomiting, difficulty in breathing, and prostration have all been reported, death following in 34 hours to five days (357). A horse receiving 500 g of toxic grain during a period of ten days, developed anorexia at the fourth day, icterus and general weakness at the seventh day, diarrhoea at the ninth, depression and abdominal pain at the tenth, and death followed on the eleventh day.

After autopsy, haemorrhagic lesions in various tissues were noted; gastro-intestinal ulceration and other visceral lesions were also visible. Although the brain and spinal cord were oedematous and congested, with several necroses, there was no liquefaction. The leucoencephalomyelitis has been observed in other circumstances by Biester *et al.* (199). This disease has recently also been reported in donkeys in Egypt (114, 2830).

OTHER FUSARIOTOXICOSES

A number of cases of poisoning involving species of *Fusarium* have been described, thus, for example, culture extracts of *Fusarium solani* (Mart.) Sacc.

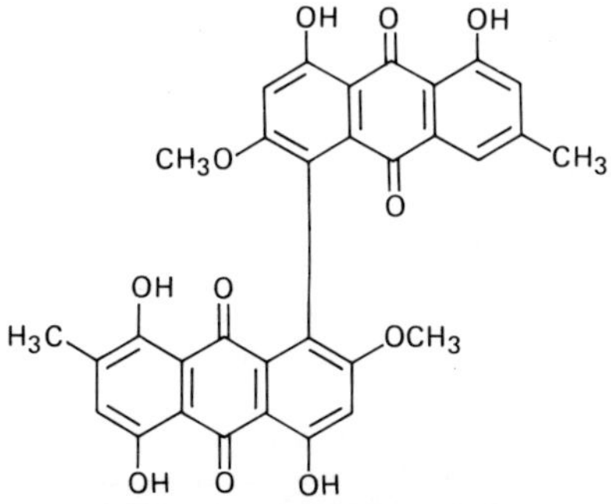

Figure 138. Fusaroskyrin

caused the death of a cow after ingestion (695) and a number of outbreaks of disease of cattle in the USSR (627, 1378) and Bulgaria (2477) have been associated with *Fusarium.* The metabolite fusaroskyrin, $C_{32}H_{22}O_{12}$ (Figure 138), which resembles luteoskyrin isolated from *Penicillium islandicum,* has been isolated from a strain of *Fusarium* (852). It has a m.p. greater than 300 °C.

A review of the toxic metabolites of the genus *Fusarium* was presented by Chilton (449) during a conference at California.

CHAPTER 9

SPORIDESMIOTOXICOSIS

INTRODUCTION

The name 'sporidesmiotoxicosis' refers to a mycotoxicosis associated with a photosensitization caused by the toxins produced in mouldy forage by *Pithomyces chartarum.* The mechanism of inducing such dermatoses, together with secondary reactions, has been described in Chapter 2.

In 1882 according to some (514), or 1897 according to others (571), a disease of sheep, to which the name facial eczema or facial dermatosis (1124) was given, was described in New Zealand. It was, however, only in 1958–59 that the relationship with the presence of a mould on the forage on which the animals were feeding was established (1958, 2393, 2596, 2598, 2599).

The incidence of facial eczema varies from year to year and the disease is seasonal, developing particularly in the autumn when this follows a particularly dry summer (312, 1692).

In New Zealand, the most toxic forages are those of the North Island, consisting mainly of *Lolium perenne* and the fungus usually sporulates on short grass, that is to say, on the most highly grazed pasture land (2420, 1474). The disease is neither contagious nor infectious but the toxicosis appears to be associated with certain prairies.

Although this photosensitization reaction mainly affects sheep in New Zealand (514), nevertheless, the disease, associated with the same fungus, has been recorded in sheep in Australia (793, 1214) and horses and colts in the USSR (2226). Thus, despite the fact the *Pithomyces chartarum* is widespread in its distribution, the conditions for toxin production seem to occur only in rather limited regions.

On the other hand, diseases very similar to sporidesmiotoxicosis have been observed in cattle, as far apart as the south-east of the United States, in animals consuming Bermuda grass (*Cynodon dactylon*) moulded by *Periconia minutissima* (900, 901, 902, 1299, 1300, 1301, 1302), and Switzerland in the presence of an unknown mould (918).

PITHOMYCES CHARTARUM

First described in error as *Stemphylium botryosum,* the fungus responsible for this disease had been determined to be a species of *Sporidesmium* (1958) (Saccardo used the spelling *Sporodesmium*). The name *Pithomyces chartarum* (Berk. and Curt.) M. B. Ellis was finally given to this dematiaceous hyphomycete by Ellis (724). The organism has been described under the following names:

Sporidesmium chartarum Berkeley and Curtis;
Piricauda chartarum (Berk. and Curt.) R. T. Moore;
Sporidesmium echinulatum Speg.;
Scheleobrachea echinulata (Speg.) Hughes;
Sporidesmium bakeri Sydow.

Morphology – According to the description given by Ellis (724), the species forms punctiform, black colonies with a diameter of up to 0.5 mm, but sometimes becoming confluent. The mycelium is superficial, made up of a network of branched and anastomosing septate hyphae which are subhyaline to pale olive, smooth or sometimes verruculose and 2–5 μm thick. The conidiophores are borne laterally on the hyphae and are straight or slightly curved, cylindrical, hyaline or subhyaline, and measure 2.5–10 x 2–3.5 μm.

The conida are formed singly at the apex of each conidiophore; they are ovoid with 3–4 (usually 3) transverse septa, the middle cells usually being further divided by longitudinal septa; at maturity they are dark brown, verruculose to echinulate and measure 18–29 x 10–17 μm. When they become detached, the conidia usually

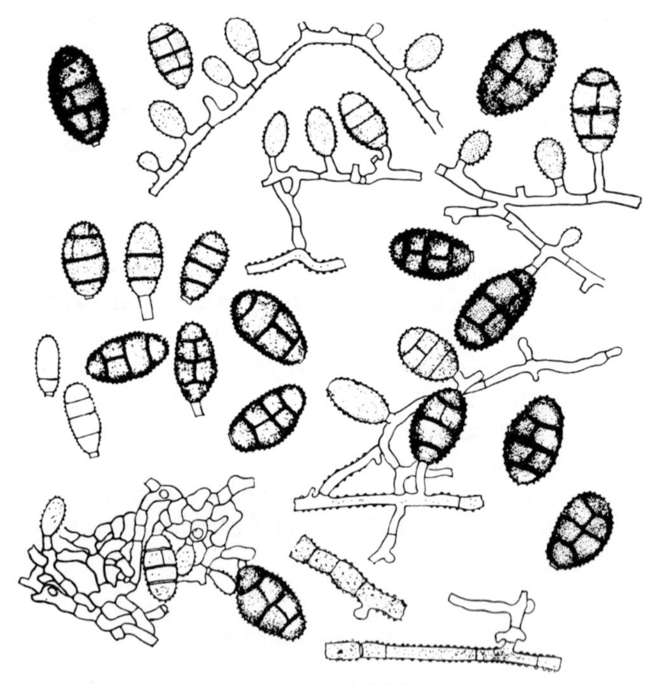

Figure 139. *Pithomyces chartarum*. Mycelium and spores at various stages of maturity (after Ellis (724))

retain the upper portion of the conidiophore as a small colourless collarette (Figure 139).

Ecology – *Pithomyces chartarum* is essentially saprophytic being found on dead vegetable matter, and even on paper (as indicated by its specific name). It seems, however, that, at least in New Zealand, it prefers to grow on the rye grasses of pasture land (1959). Even there it can also be found on dead Gramineae and Papilionaceae (311, 2598, 2599).

According to the specimens deposited in the herbarium of the CMI at Kew, *P. chartarum* has been found on representatives of the following genera: *Arachis, Bridelia, Cajanus, Calopogonium, Centrosema, Foeniculum, Holcus, Ipomoea, Jatropha, Musa, Newbouldia, Nicotiana, Oryza, Pueraria, Sorghum, Trifolium, Triticum,* and *Zea.* It has, moreover, been possible to isolate it from the atmosphere over a banana plantation in Queensland (724) and in England above a field of *Holcus lanatus* on which it grew (965).

Described in Europe in 1879 by Spegazzini, it has been isolated in Britain in 1964 (965). Although it is particularly common in New Zealand and Australia, Ellis (724) and Dingley (669) have recorded it on various plants in Ghana, Sierra Leone, Sudan, Nyasaland, Rhodesia, Mauritius, Malaya, Philippines, Jamaica, and the United States (2538). It has been recorded on hay in the USSR (2226) and has recently been found on the caryopses of *Dactylis glomerata* and on cherries in Oregon (1429).

Conditions of growth – *P. chartarum* grows fairly rapidly and the conidia germinate readily if conditions are favourable. It requires a relative humidity approaching 100% and the minimum temperature at which growth will occur is 13 °C (678). The temperature for optimum growth is 24 °C (311). It is understandable then why contamination in the field occurs mainly in summer and autumn, when the minimum temperature is of the order of at least 18 °C. Rain favours the growth of the fungus (311) and three days is sufficient for *Pithomyces* to undergo its full life cycle if conditions are favourable (312).

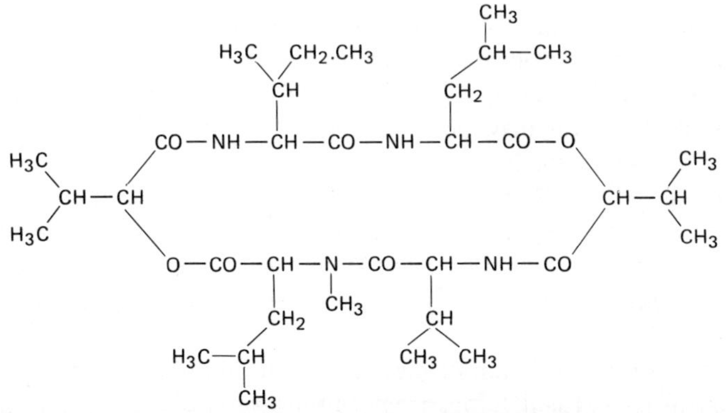

```
                                                      CH3
                                                       |
                        H3C  CH2.CH3                  CH—CH3
                           \ /                        /
                           CH                      CH2
                            |                       |
                  CO—NH—CH—CO—NH—CH—CO—O
     H3C         /                                      \              CH3
        \       /                                        \            /
         CH—CH                                            CH—CH
        /       \                                        /            \
     H3C         \                                      /              CH3
                  O—CO—CH—N—CO—CH—NH—CO
                        |    |        |
                        |   CH3       CH
                       CH2           /  \
                      /           CH3   CH3
               H3C—CH
                    |
                   CH3
```

Figure 140. Sporidesmolide II

Compounds produced – The first compound isolated was called sporidesmin followed by the isolation of two very closely related metabolites called respectively sporidesmin B and C. These three sporidesmins are toxic.

P. chartarum also produces a number of depsipeptides which have insignificant toxicity. They include sporidesmolide II, $C_{33}H_{60}O_8N_4$ (Figure 140) (195), the sporidesmolides III and IV (1902), pithomycolide (300) and angolide (1646).

THE SPORIDESMINS

Evidence for the toxicity of plants in the field – The best demonstration of toxicity is to determine if the sheep grazing in the field show any signs or not of liver damage.

The guinea-pig and some breeds of rabbit, being as sensitive as the sheep, may serve as test animals in laboratory trials (515, 678, 755, 1963, 2204). Unfortunately, difficulty arises in the length of time required to produce these lesions. For example, the first method due to Perrin (1963), using guinea-pigs, requires 35–40 days. Inspired by the techniques used by White (2810) for the chemical extraction of the toxin from grass, a second method known as the beaker test was proposed (1964). It consists of an empirical chemical procedure using a series of solvents from which the final fraction is evaporated in a small beaker. This procedure leaves a white deposit if the sample is toxic.

Sandos *et al.* (2233) have used this procedure with some success, but it has been criticized (2811). The method of Russell (2203, 2204) is preferred using a titration procedure involving iodine and sodium arsenite. A bioassay based on the inhibition of germination of clover seed by the sporidesmins has also been proposed (2881). The presence of sporidesmins can also be demonstrated by their activity against the cells of tissue cultures (1761).

Production and isolation of the toxins – The toxic metabolites of *P. chartarum* are present in both the mycelium and the spores of the fungus (1958), but in fact the occurrence of liver damage appears to be proportional to the quantity of spores ingested (313, 516).

The production of toxins by *P. chartarum* depends on the strain, the culture medium, and the time of incubation. In the laboratory, optimum production has been obtained using a medium based on potato and carrot and, apparently, production of the toxin may occur in the light or the dark (2596). Optimum temperature for production is 20 °C (602).

When the spores of *Pithomyces chartarum* are moistened, their concentration of sporidesmins decreases rapidly (516, 1598), which explains the low toxicity of fields contaminated at the end of the wet season, despite the large number of spores (2421). Sporidesmin concentration seems on the other hand to increase with temperature (602).

Experiments on the isolation of the toxin from infusions of rye-grass have long been unfruitful (2810) and it was from cultures of *P. chartarum* that Synge and White (2534, 2535) first isolated a metabolite to which they gave the name sporidesmin. The sporidesmin obtained was crystalline (688, 1102, 2188, 2203) and, when given experimentally to sheep, it produced the symptoms of facial eczema (686).

Chemical structure and properties of the sporidesmins – The sporidesmins are peptide derivatives belonging to a class of compounds referred to as epipolythio-diketopiperazines for they contain atoms of sulphur within a nucleus of diketopiperazine as part of their structure (Figure 141).

There are as many as eight compounds now recognized (845, 1065, 2188, 2211, 2562):

– Sporidesmin A, $C_{18}H_{20}ClN_3O_6S_2$, was the first to be isolated and is not very stable, m.p. 179 °C;

– sporidesmin B, $C_{18}H_{20}ClN_3O_5S_2$, m.p. 183 °C; treatment of its acetate with boron trifluoride gives anhydrosporidesmin B;

Sporidesmin A

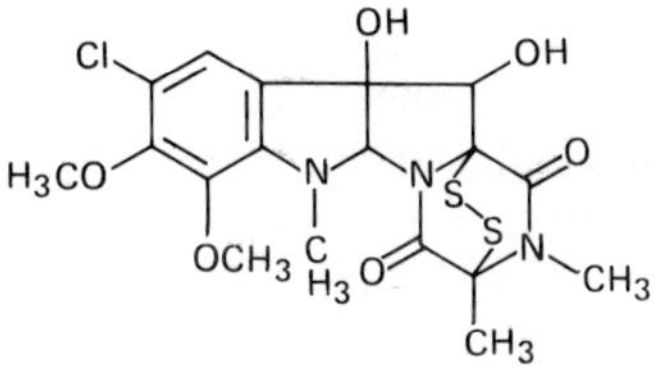

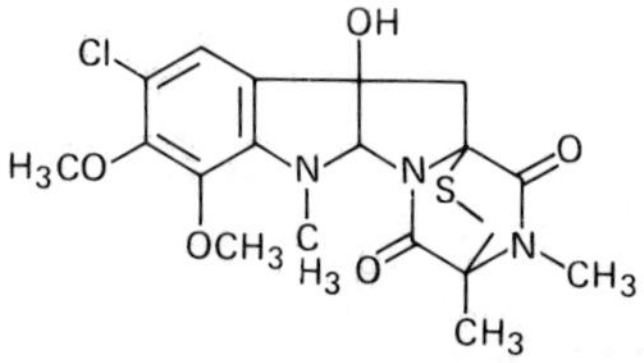

Sporidesmin B

Sporidesmin C

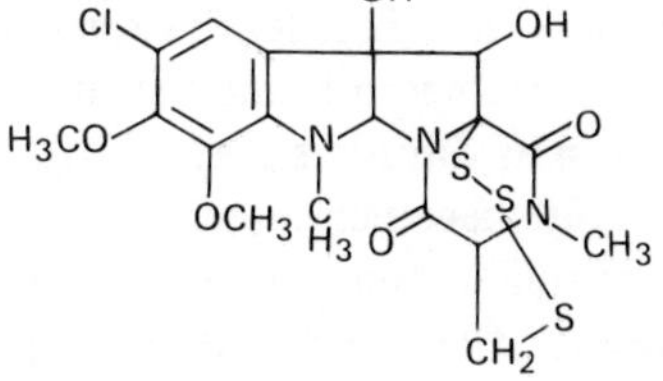

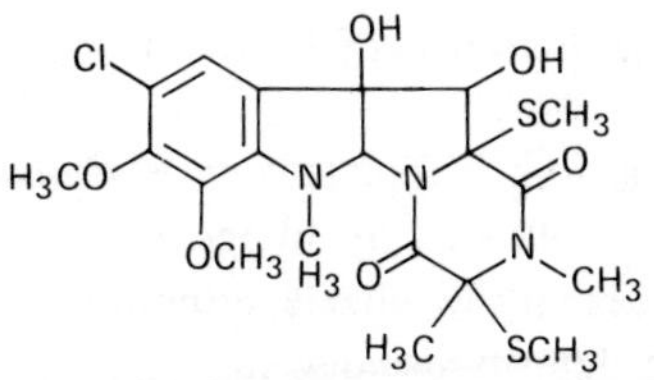

Sporidesmin D

Figure 141. The sporidesmins

– sporidesmin C, $C_{18}H_{20}ClN_3O_6S_3$, only known as its diacetate;
– sporidesmin D, $C_{20}H_{26}ClN_3O_6S_2$;
– sporidesmin F, $C_{18}H_{20}ClN_3O_6S_3$, known only as its etherate, the m.p. of which is 180–185 °C;
– sporidesmin F, $C_{19}H_{22}ClN_3O_6S$;
– sporidesmin G, $C_{18}H_{20}ClN_3O_6S_4$;
– sporidesmin H, $C_{18}H_{20}ClN_3O_4S_2$.

In the same family of compounds are found chaetocin and chetomin, produced by *Chaetomium,* and gliotoxin, produced by *Gliocladium virens.*

MYCOTOXICOSIS OF *PITHOMYCES CHARTARUM*

Sensitivity of various animals – The sheep is undoubtedly the animal which suffers tho worst effects of sporidesmiotoxicosis and the LD_{50} of sporidesmin *per os* is 1.5–2 mg/kg. Other animals such as cattle are not, however, unaffected (571) and it has been found that horses can also suffer (2226).

Experimentally, the guinea-pig (755, 1963, 2974), the mouse (2874), the rat (2413), as well as the rabbit (515, 677, 1969) all show some sensitivity to sporidesmins, however, it is the rabbit which is sufficiently sensitive to be used as a test animal. In the young rat, the toxicity of sporidesmin is inversely related to the concentration of protein in the food, thus rats receiving diets with either 9.22% or 37% protein gave values for the oral LD_{50} of 4.8 and 16 mg/kg respectively (1130).

Symptoms of poisoning – In the sheep, poisoning is at first characterized by a hypersensitivity of the animal with inflammation in the eye and nasal discharge. The sheep shakes its head violently, shows signs of itching and urinates frequently. It also shuns the sun and seeks out shade.

Then there rapidly follows an oedema of the ears, the eyelids, the face, and the lips, frequently also of the vulva. Oedema of the ears may be such as to cause them to hang. This appearance of sunburn develops on those parts of the body not protected by the wool and these lesions weep and form scabs. After several days, the regions of scabs become necrotic and scarified.

This period of skin disorders may last for several weeks then seems to subside when the temperature becomes cooler.

In the terminal stages a more or less pronounced icterus develops depending on the individual animal. General debility is frequently observed, followed by death of the animal within several weeks.

Examination of the blood reveals the presence of bilirubin and leucocytosis (1759).

When a milking cow consumes grass contaminated with *P. chartarum,* a reduction in milk yield is first observed, followed by hyperaemia, sensitivity to touch, and mucous discharges. Burns appear on the muzzle, around the eyes, on the tongue, and in the region of the perineum. Icterus also develops.

Histopathology – This disease essentially involves changes in the liver and oedema of the skin (803). On autopsy, the liver appears to be granular with a greenish colour, which may be diffuse or in spots. The walls of the bile ducts are white, thick, and fibrous; their lumen is partially obstructed with thick bile and cell debris. Sporidesmin is secreted, in fact, in the bile within several hours after the start of the toxicosis (1430, 1762).

In some cases, fibrosis of the tissues is particularly important. The left lobe of the liver is almost completely atrophied and the whole organ becomes cirrhotic and distorted. These lesions precede the photosensitization or may be independent of it. A generalized icterus occurs in the worst cases and bilirubinaemia is always present in photosensitized animals.

The kidneys are swollen, with haemosiderosis. The skin shows zones of congestion during the first stages of poisoning and an increase in the permeability of the capillaries occurs (2413), the dermal tissues surrounding them being necrosed. The lymphatic spaces are dilated and contain a bile-coloured fluid.

In the cow the histopathological symptoms are very similar to those observed in the sheep. Gibbons (902) particularly described deposits of haemosiderin in the liver and the kidneys, attributing this to the destruction of the erythrocytes by the fungal toxin. Deposits of lipids occur in the liver cells and are usually first reported in the portal region.

During toxicity tests in the guinea-pig what is observed is essentially a cholangitis with hyperplasia of the bile ducts (755), fibrosis, and atrophy of the liver parenchyma.

According to Work (2874) the toxin induces an inflammatory process in the bile ducts of the mouse.

Recent observations of Mortimer (1760) have added to these histological results and in all cases a fatty infiltration into the liver was noted.

Mode of action of sporidesmins – Once the toxin has been absorbed from the intestinal tract it penetrates the liver via the portal vein (2875). Indeed, intraportal injections of toxic extracts cause the characteristic liver lesions associated with facial eczema, whereas injections into the jugular vein do not. As soon as the sporidesmins are injected into the sheep, they provoke an inhibition in the secretion of triglycerides in the liver (1965) associated with disturbances in the synthesis of proteins in the mitochondria of hepatic cells (1969). In fact, sporidesmin is recognized as a powerful inhibitor *in vitro* of the enzyme systems of mitochondria and it seems to act in the same way *in vivo* on the metabolism of the liver. A consequence of the action of sporidesmin which is most clearly seen is the accumulation of bile acids in the liver (1969). Sporidesmin will inhibit the respiration of homogenates of the liver from guinea-pig or sheep (2879).

PROPHYLAXY AND TREATMENT

The prevention of facial eczema is difficult because it is not easy to detect the presence of the toxin in the field and *Pithomyces chartarum* is not always obvious.

In order to destroy the mould, it has been advocated that thc fields be spread with copper sulphate powder. Results have been satisfactory but, in New Zealand, preference has been given to spreading a mixture of C_9–C_{11} fatty acids and a non-ionic surface active agent such as lissopol N (2600). Trials of spraying with thiabendazole have also been carried out (2394).

The irradiation of solutions of sporidesmins using a mercury vapour arc lamp considerably reduces their toxicity (517). Concerning the sick animal, it appears to be beneficial to give intravenous injections of cortisone acetate (900, 902) or a solution of sodium thiosulphate (1302).

A sophisticated forecasting service functions in New Zealand and, by providing regular predictions of the possibility of outbreaks of *Pithomyces chartarum* on pastures, allows the possibility of selective control on this species in the field (314).

CHAPTER 10

STACHYBOTRYOTOXICOSIS

INTRODUCTION

Stachybotryotoxicosis is a disease associated with the ingestion of feed on which *Stachybotrys atra* Cda. has grown and produced a toxin. Although it mainly affects horses, other animals and man himself may also be affected.

History – The first case of toxicoses in the horse were recorded in the Ukraine in 1931 (803). It appeared to be caused by *Stachybotrys atra,* although the symptoms resembled those of a disease observed in Hungary and which veterinarians attributed to *Bacillus viscosus* (2758). In fact, according to Vertinskii (2741), this bacterium is frequently isolated from the blood and a variety of organs of horses which have died of stachybotryotoxicosis, its growth being possible because of the exhaustion of the myelopoietic centres of the animal.

Until the causative agent had been identified, this disease was referred to in the USSR, where it usually occurs, either by the initials 'NZ' (nevis volenie zabolivanie = disease of unknown aetiology) or, after 1938, 'MZ' (massovie zabolivanie = great disease) because of the high mortality rate involved.

Early studies demonstrated that the disease was neither contagious nor infectious and healthy animals in contact with sick animals remained in good health so long as their feed was different from that of the affected horses. The blood, or the macerated tissues from a dead horse, were not capable of provoking the appearance of the disease when injected into healthy horses.

It required the efforts of many researchers throughout the USSR (1766, 2226, 2741) and, much later, some in Germany (1104, 2803) to demonstrate that this disease was due to the presence of the mould *Stachybotrys atra* growing on the hay on which the horses were fed.

In some crops in Uzbekistan, 88% of the cotton grown is infected with *Stachybotrys atra,* giving rise to considerable danger if the cottonseed meal or oil is incorporated into animal feeds (917).

Recently this mycotoxicosis has been reported in the pig, in Finland, after consumption of barley and wheat contaminated with *S. atra* (1339).

Development of the disease – Stachybotryotoxicosis is seasonal in its occurrence, the first cases appearing in autumn if the animals are stabled. The number of sick animals increases during the winter months, reaching a maximum number in February and April, and then the toxicosis disappears as the animals are returned to pasture and no longer consume hay and straw.

The toxicosis remains endemic and, on those farms where a bad outbreak occurs, it reoccurs several years running. The rate of development of the illness varies considerably from one farm to another. There have been cases where, in one week, 60% of a herd were affected and, a week later, all the animals were sick. Most frequently, it is observed that a few sporadic cases of isolated sick animals occur during the first three to four months followed by a massive and severe increase as the disease affects the whole herd.

STACHYBOTRYS ATRA

This mould, which would be placed in the Adélomycétes in the system of classification used by Langeron, is a member of the dematiaceous Hyphomycetes. It was first described by Corda in 1837 and has subsequently received several names, especially *S. alternans* Bonorden. Although Oudemans (1901), and then Salikov (2226) recognized some morphological differences between *S. alternans* and *S. atra,* Bisby (230) regrouped these species under the name *S. atra* Cda.

Morphology – (Figure 142) The mycelium, which readily fragments, bears erect conidiophores which are well differentiated, simple, very rarely branched, and sympodial, often swollen at the base and 25–100 μm long. Each sporophore carries at its head 6–15 ampulliform phialides which produce initially smooth spores which become echinulate, elliptical-oblong, and measure 8–11 x 4–8 μm. When the spores become detached from the phialides they remain agglomerated in a mucilaginous mass. The upper part of the sporophore, the phialides, and the spores are sooty to opaque black in colour (1733).

Ecology – *S. atra* is a saprophytic mould and is considered to be a major cellulose decomposer. It has a worldwide distribution and is usually found in the soil (1230, 1651, 1733) but has been isolated from many substrates such as paper (1733, 2207) and vegetable debris. It is particularly common on straw and is not rare on grain (1585). The spores are readily disseminated by air currents through the atmosphere.

Conditions of growth and sporulation – Growth of this species is optimum between 20–25 °C and can occur between 2–40 °C but not at 0 °C. A period of one hour at 80 °C at low humidity is sufficient to destroy *S. atra,* whereas at a high humidity one hour at 60–65 °C or five minutes at 100 °C is sufficient. However the spores can survive long periods at a temperature of about –40 °C.

Thus the spores may survive passage through the gastrointestinal tract but may be subsequently destroyed during the biological heating of the faeces (803).

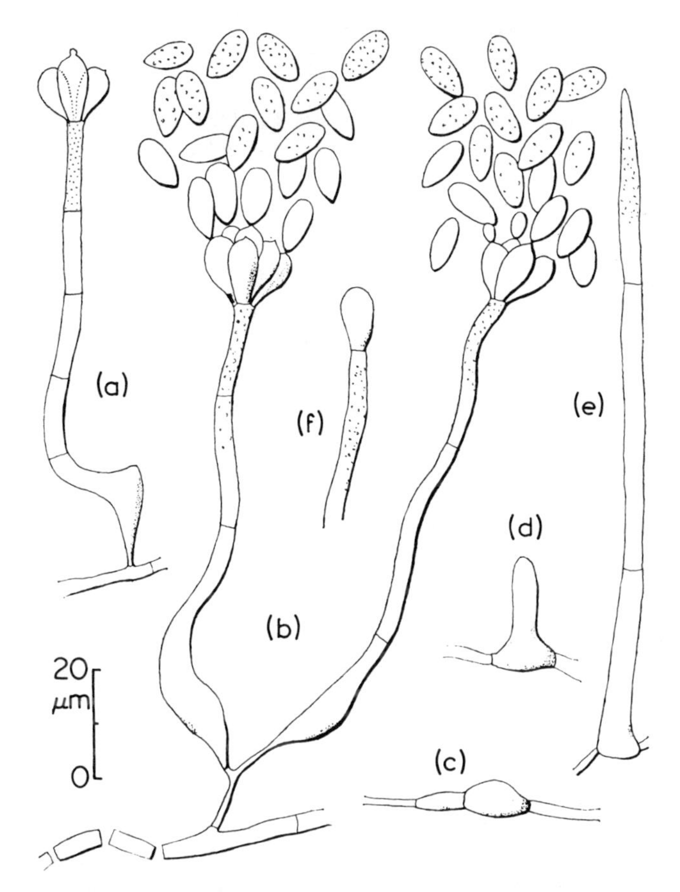

Figure 142. *Stachybotrys atra.* (a) Young sporophore with 3 phialides. (b) Older sporophores and spores at various stages of maturation. (c)–(e) Successive stages in the formation of a sporophore. (f) Terminal of a young sporophore with a single phialide

THE TOXIN OF *STACHYBOTRYS ATRA*

The present facts – There are strains of *S. atra* which are toxic and others which are harmless. Ether extracts from the substrate on which the toxic strain has been growing cause an inflammatory skin reaction when applied to the skin of a rabbit or calf (806), but the extracts from the non-toxic strain do not cause such a reaction.

S. atra will grow on damp straw and it is possible to demonstrate a skin reaction after the tenth day of culture. The reaction is most clear between days 19–74 but it is still possible to demonstrate it after 414 days of culture.

The toxin is very powerful and as little as 0.00175 μg of the residue from an ether extract of toxic hay, when suspended in 0.125 ml of olive oil, will produce hyperaemia when applied to the skin of a rabbit (806).

Characteristics – The toxin of *S. atra* is resistant to u.v. light, sunlight, and X-rays as well as being thermostable, thus it is not destroyed by autoclaving for an hour at 120 °C. It is not inactivated by organic or inorganic acids at concentrations of 2%, but it is destroyed by alkali (674, 2742, 2758). It is insoluble in water but very soluble in ethanol, dichloromethane, ether, chloroform, and fats (1912).

The LD_{50} by intraperitoneal administration is 44.5 mg/kg for the albino rat, 51.6 mg/kg for the mouse, 62.4 mg/kg for the guinea-pig and 92 mg/kg for the one-day-old chick. The toxin will also kill cells of *Paramecium caudatum* (1911).

According to Drobotko (699, 700), Fialkov, of the Academy of Sciences of Moscow, had isolated the toxin in a crystalline state and determined its empirical formula. The name stachybotryotoxin has been given to the compound (1693).

It has been considered that two different toxins exist in the straw after infection with *S. atra*, one giving the skin reaction, the other acting after oral ingestion, but this assertion has not been verified.

MYCOTOXICOSIS OF *STACHYBOTRYS ATRA*

Sensitivity of various animals – It would appear that the horse is most susceptible to poisoning by *Stachybotrys atra* (1104, 1766, 2226, 2741, 2803). Nevertheless, stachybotryotoxicosis has been reported in the USSR among all farm animals consuming straw contaminated with *Stachybotrys*, especially on those farms where horses had previously been ill (810, 1162, 1340, 1377, 1468, 1631, 1802). It is, moreover, possible to produce experimental toxicosis in horses, sheep, pigs, and calves (806). In the laboratory, it has been possible to reproduce the clinical symptoms of the disease in white mice, guinea-pigs, rabbits, dogs (1335, 2244), and chickens (2305).

The toxic factor does not appear to be secreted from the mammary glands, for a colt drinking the milk from a poisoned mare does not show any signs of toxicosis itself.

The possibility cannot be excluded that, in some cases, human beings may also be poisoned (699, 700).

Clinical study of chronic poisoning in the horse – According to Forgacs and Carll (803), three stages may be recognized:

1st stage – This develops progressively during the prolonged ingestion of sublethal doses of toxic feed and is first characterized by stomatitis accompanied by cracks in the corners of the mouth. An inflammatory process of the mucosa and buccal oedema rapidly appear. The mouth of the horse begins to take on the appearance of that of a hippopotamus. Salivation becomes accentuated, the submaxillary lymph nodes become swollen and painful to touch. Rhinitis and conjunctivitis also develop.

The duration of this first stage varies from 8–12 days and sometimes lasts for a month. If the toxic feed is removed at this stage these localized lesions remain for several weeks.

2nd stage – This stage is characterized by internal symptoms in the sick animal, especially thrombocytopoenia. Instead of 40–60 thrombocytes per 3,000 erythrocytes, it is difficult to detect more than 8–12, and sometimes as few as 2 or 3. The blood coagulates only with difficulty. At the same time leucopoenia and agranulocytosis develop, the number of leucocytes falling to 2,000 or less per cubic millimetre of blood. The respiratory and cardiac processes remain normal, as does the body temperature, although it is possible that very slight disturbances of the intestine may appear. This stage lasts for 15–20 days.

3rd stage – The onset of this stage is marked by the appearance of fever, the temperature rising to 41.5 °C. The numbers of thrombocytes and leucocytes rapidly decrease even further, and there may sometimes be only 100 leucocytes per cubic millimetre of blood (700). The blood becomes quite unable to form clots and the animal becomes very depressed, the pulse feeble, and irregular diarrhoea is often frequent. The necrotic areas around the mouth continue to grow, reaching the lips, cheeks, palate, and tongue, and they are often infected with bacteria.

The third stage, which usually lasts for one to six days, is generally fatal, death usually being brief and quiet, although occasionally it may be prolonged and violent. Pregnant mares usually abort shortly before their death.

In those cases reported from Romania in 1969 (698), a symptom particularly noted was a pronounced stripping of the epithelia around the nose and the mouth with ulcerative lesions forming on the buccal mucosa accompanied by oedema. These lesions turned into nasty cracks and the animals became subfebrile.

Acute poisoning in the horse – Acute intoxication, following the ingestion of a large quantity of very toxic feed, is very much less frequent than chronic poisoning. The symptoms of buccal inflammation and necroses of the lips associated with chronic cases are not seen in acute cases. The visible signs of poisoning, which appear 72 hours after ingestion of the toxic feed, are essentially nervous disorders. The sick horses lose their reflex actions, become hypersensitive and show hyperaesthesia, following a loss of vision. Indeed they become completely blind and have difficulty in coordinating their movements. There is also anorexia, although the horses drink avidly, but have difficulty in swallowing.

In some cases, 10–12 hours after ingestion of a large quantity of toxin, a high fever with temperature of 41 °C may occur associated with a weak pulse, panting respiration, cyanosis, and haemorrhage. All the animals die rapidly.

Poisoning of other animals – Forgacs *et al.* (806) inoculated sterilized, humidified wheat straw with *Stachybotrys atra* and allowed it to grow for 60 days at ambient temperature. The straw was then pulverized, suspended in water, and administered to animals with the help of a stomach tube.

– A cow of 110 kg received 3 kg of toxic straw over a period of 14 days. Four days after the start of the experiment, the animal became slightly depressed and after six days became anorexic. At the ninth day agranulocytosis became evident, the number of leucocytes, which was initially 4,500 per cubic millimetre fell to 150 by the fifteenth day. Coagulation of the blood, which took 9 minutes during the first day, required 19 minutes by the fourteenth day. The body temperature, which remained normal for the first 14 days, rose rapidly to 41.8 °C on the fifteenth day. Respiratory problems and dyspepsia appeared and on the seventeenth day the cow died.

– A ewe of 60 kg received a single dose of 170 g of toxic straw. Twelve hours later the animal was prostrate, respiring with pain, continued to weaken, and died after 25 hours.

– A pig of 56 kg, having been given 100 g of toxic straw, died within 12 hours. The pigs in the outbreak of poisoning reported from Finland showed haemorrhages of the stomach, duodenum, and the meninges on post-mortem examination (1339).

The observations of Fortuskny and her collaborators (810) on a variety of farm animals in the Ukraine confirmed previous findings. In one case concerning a cow, they reported a reduction of milk yield, anorexia, general debility, hyperaemia of the mucous membranes in the buccal cavity, the nose, and the vagina, an increase in temperature, excessive salivation, and an increase in nasal discharge.

Korneev (1335) produced poisoning in the white mouse using 2 g of oats on which the mould had grown. Depression, general debility, muscular incoordination, and a panting respiration were all noted during the first two days and death followed on the third day. The grey mouse is equally sensitive.

Guinea-pigs fed *ad lib.* with toxic oats showed necrotic areas on the lips, a diffuse necrosis of the buccal mucosa, and leucocytosis. Death followed within three to five days.

Similar symptoms have been recorded in the rabbit and the dog, whereas in chickens there is a reduction of growth and necrotic lesions in the oral mucosa and the crop (2305). The LD_{50} for six-day-old chickens has been recorded as 40.4 mg (1911).

Toxicity to man – Stachybotryotoxicosis may affect man as well as other animals (699, 700). It is those people who handle the mouldy straw who are most often affected, showing irritation, particularly of those parts of the body which perspire the most, such as the armpits. These irritations spread and frequently become a wet dermatitis with abundant exudation. At the same time, catarrhal angina and serious pharyngitis may occur and the patient complains of a sensation of burning in the nose accompanied by bleeding and a moderate to severe cough. Leucopoenia may also develop at the same time in individuals who are particularly sensitive.

In man, it would appear that the toxin is ingested mainly by the respiratory route.

Histopathology – The major histopathological findings with stachybotryotoxicosis are an abundant haemorrhage and necrosis in various tissues. The haemorrhages resemble those caused by strong contusions and ecchymoses, in spots or in lines, are observed in the pleura, the diaphragm, the mesenteries, and the spleen. Numerous haemorrhages appear in the lymph nodules, in the parenchyma of the lungs and the liver, the suprarenal, the brain, and the meninges.

Necroses of the mucous membranes appear along practically the whole length of the digestive tract and are very characteristic of this toxicosis. Only the oesophagus is rarely affected. It is usually on the large intestine that these lesions are seen, either as large greyish lumps on the surface of the mucosa, or as cavities spreading into the underlying tissues.

The changes observed in the liver, the kidneys, and the cardiac muscle are degenerative processes of the albuminoid or adipose-dystrophic types. Fatty infiltration is particularly evident in the liver cells.

When looked at microscopically, the haemorrhagic regions are seen to contain corpuscles of various forms which appear to be aggregated in the epithelium and the walls of the capillaries of various organs. It has sometimes been suggested that these corpuscles are fungal in origin (1473), however, the symptoms of stachybotryotoxicosis may be observed without it being possible to isolate *Stachybotrys* from any of the tissues of the poisoned animal (806).

Histopathological examination of a number of different, experimentally poisoned, animals has confirmed the earlier observations made on the horse.

Mode of action of the toxin – The observations of Forgacs *et al.* (806) on the death of horses with acute stachybotryotoxicosis have led to the idea that the toxin present in the infected hay reacts with the gastric juices of the stomach and is then assimilated in a water-soluble form. Confirmation of this hypothesis comes from the fact that, at low pH, pepsin can react with toxic hay to give a water-soluble fraction which gives a positive skin reaction, whereas the insoluble hay remaining has lost its toxicity.

Examination of the blood of poisoned horses shows a clear eosinophilia and the toxin may well have very specific sites of action, as well as a more general, non-specific activity, by exciting the neurohypophyseal complex with a response of the suprarenal glands (698).

PROPHYLAXIS AND TREATMENT

Prevention of this toxicosis consists essentially of storing straw in a dry location in order to prevent the germination of the spores of *Stachybotrys atra* and its subsequent growth.

Detoxification of straw to be used as feed has been attempted by Dobrynin (674) making use of the instability of the toxin to alkali.

The treatment of affected animals depends on the stage of toxicosis reached. When the first symptoms are recognized it is obviously essential to remove the suspect feed and to change the litter. A number of medications for both external and internal application have been recommended but, according to Moseliani (1766), none has any beneficial effect.

Human beings who are likely to be exposed to the fungus should wear a face mask when handling infected straw. Local washing with warm soapy water and dusting with talcum powder are recommended for those parts of the skin which have become affected (1104).

CHAPTER 11

OTHER MYCOTOXICOSES

A number of fungi have been implicated as the causative agents in several cases of toxicosis, although in a number of other cases the role of fungi has not yet been definitely demonstrated. Some species, which do not produce toxic metabolites, nevertheless deserve mention because they are able to transform harmless compounds into toxic derivatives.

Also described are those cases of typical mycotoxicoses for which it has not been possible to define the role of each of the numerous species of fungi associated with them.

Finally there are a number of toxicoses which may be confused with mycotoxicoses and the agents of which are now well known and are not, in fact, fungi.

TOXICOSES ASSOCIATED WITH THE MUCORALES

Although the Mucorales have rarely been considered as causes of mycotoxicoses, it is possible that they play an important and unsuspected role. There is no doubt that mucormycoses, sometimes referred to as phycomycoses, which were not widely recognized in animals until recently, are actually widespread. They may be due to:

– *Mucor* (*M. pusillus* Lindt, *M. spinosus* v. Tiegh., *M. javanicus* Wehm., *M. circinelloides* v. Tiegh.);

– *Absidia* (*A. lichtheimii* (Lucet and Cost.) Lend. (= *A. corymbifera* (Cohn) Sacc. and Trott.), *A. ramosa* (Lindt) Lend.);

– *Rhizopus* (*R. cohnii* Berl. and de Toni, *R. microsporus* v. Tiegh., *R. arrhizus* Fischer, *R. oryzae* Went and Prinsen Geerlings) (1416, 1756, 1757).

Lesions are most frequently found in the digestive tract, but they sometimes occur in the nasal cavity and the brain (164). If the central nervous system becomes infected by these fungi, the resulting mycosis almost always has a fatal outcome. The diabetic is more susceptible to infection (122, 1142, 1480).

Extracts of the cultures of several of the Mucorales (*Rhizopus nigricans* Ehrb., *Mucor albo-ater* Naumov, *M. corticolus* Hagem, *M. racemosus* Fres., *Thamnidium elegans* Link, and *Piptocephalis freseniana* de Bary), when administered orally to guinea-pigs, rabbits or cats, have given rise to poisoning and often death according to Joffe (1238).

Growth of *Thamnidium elegans* on lucerne, which is often very abundant, appears to have caused the death of rabbits on farms in northern France, death being preceded by nervous symptoms.

Mucor

Paralysis in the hindquarters of the hamster may be induced by injecting a suspension of spores of *Mucor circinelloides* into the heart. However, it is improbable that this is a result of the diffusion of a toxic substance but rather to a simple localized embolism associated with the passage of the spores along the femoral artery (1757). However, *M. circinelloides* v. Tiegh has frequently been isolated from foods which have caused toxicoses. For example, as many as 4.5×10^3 spores per gram have been isolated from some feeds and in a case where mouldy barley seemed to be responsible for cardiac ulcers in pigs, 45% of the seed was infected. Feed infected with this organism has been considered to cause digestive upsets in cows and sheep, haemorrhagic syndrome in poultry, and a reduction of milk yield in cows (1944).

Diarrhoea associated with congestive lesions of the duodenal mucosa in poultry may perhaps be associated with feed contaminated with this species.

Mucor pusillus Lindt, which is common on flour and bread, has been considered to be responsible for abortions in cows (2000). Pidoplichka (1473) has also isolated toxic strains of *Mucor albo-ater* Naum. and *M. hiemalis* Wehm.

Absidia

Organic acids, some of which have found practical application, especially in the Far East, are among the metabolic products of members of the Mucorales. *Absidia* spp produce mainly oxalic acid which may confer some toxicity (995).

Absidia lichtheimii has often been isolated from feeds causing digestive upsets in cattle and, more rarely, laying hens (1730, 1944). This species is well known as an agent of mycotic infections under the name *Absidia corymbifera* or *Mucor corymbifera* Cohn. It grows optimally at 37 °C, but can also tolerate temperatures below 20 °C and above 50 °C and is thus similar in its temperature requirements to such thermotolerant fungi as *Aspergillus fumigatus* (1662).

Absidia ramosa (Lindt) Lend., which is common on grain, has been found in association with cases of abortion in cattle (101, 2000).

Rhizopus

Blakeslee and Gortner (235, 236) isolated a water-soluble compound from *Rhizopus nigricans* Ehrb. which, even at very low concentrations, could cause the death of rabbits following intravenous injection (500, 946). It would seem, however, that no cases of toxicity have subsequently been recognized.

Perhaps it is worth reporting some mishaps occurring after the consumption of tempeh, in the manufacture of which *Rhizopus* may play an important part.

Tempeh, which is a popular food in South East Asia and is known also as bongkrek, semaji or tempeh hedelee, is prepared by inoculating pressed coconut cake (or sometimes groundnut meal), grated fresh coconut, and soya with a strain of *Rhizopus.* In some areas crab flesh is also added to the mixture. It may be served cooked in a soup, fried in vegetable oil or grilled. *Rhizopus oryzae* is the most commonly used fungus (1650, 2485), an organism which Hesseltine (1070) considers to be more appropriately named *Rhizopus oligosporus* Saito. It is, moreover, possible to prepare tempeh with any of 39 strains of *Rhizopus* including *R. nigricans* (1080) and sometimes *Neurospora sitophila* (2724) or *Aspergillus oryzae* (348) may be used. In Java, several outbreaks of food poisoning, often fatal, have been noted following the consumption of tempeh (1650). After several hours the usual symptoms were headache, giddiness, vertigo, a staggering walk, perspiration and then drowsiness, convulsions, coma, and cyanosis. Death usually occurred one to two days after the meal and on autopsy pulmonary oedema was found. Such material could also poison cattle and pigeons. It was always tempeh made from pressed coconut meal, with or without the addition of crab meat, which caused these outbreaks, whereas soya tempeh, or groundnut tempeh were rarely toxic.

Extracts from toxic tempeh, injected intraperitoneally, killed monkeys and rats and it has even been wondered whether this material might have any connection with certain types of cancer. Van Veen and Mertens (2722, 2723) have researched into the aetiology of this type of poisoning and it is now considered to be due to micro-organisms contaminating the tempeh. In fact, some strains of *Rhizopus oryzae* themselves produce toxic compounds on the medium on which they are grown and, when culture filtrates are injected into mice, a fatty degeneration and focal necrosis of the liver results (1314).

Two toxins, a yellow metabolite called toxoflavin, and a colourless one referred to as bongkrekic acid have been isolated by Van Veen and Mertens (2723) from a bacterium present in the toxic tempeh.

It is perhaps worth recording that *Rhizopus nigricans* is one of the most active agents in the transformation of progesterone, and this may explain certain phenomena reported after ingestion of food contaminated by this species. A haemolysin has also been extracted from the mycelium of *Rhizopus nigricans* and *R. arrhizus* (853, 854). *Rhizopus arrhizus* appears to be particularly toxic to mice, causing a periportal necrosis with haemorrhages, discoloration of the epithelium of the renal tubules with vacuolar degeneration (2717).

Cunninghamella

Some strains of *C. elegans* may cause hepatic and renal lesions with congestion and a moderate pycnosis of cellular constituents in mice (2717).

Byssochlamys and *Paecilomyces*

Byssochlamys fulva Olliver and Smith is an ascomycete, the presence of which in fruit juice not only causes spoilage of the juice itself, but may also be harmful to the consumer. Its spores may survive at a temperature of 88 °C for 30 minutes, or even much higher temperatures for a shorter period of time, thus explaining the survival of the fungus in heat treated juice (1892, 1893, 2468).

This species produces the metabolite byssochlamic acid, $C_{18}H_{20}O_6$ (Figure 143) (1016, 1656, 2087, 2315), which is known to be toxic to mice. It has a m.p. of 163.5 °C and is structurally related to glauconic and glaucanic acids.

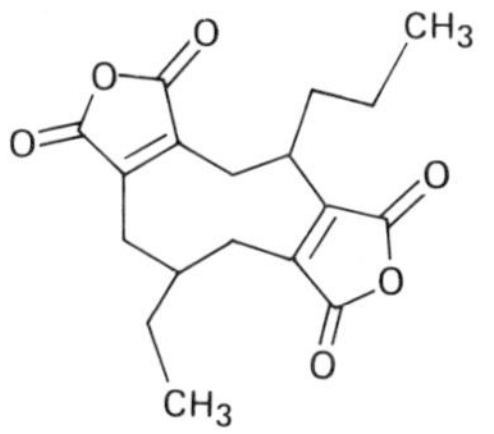

Figure 143. Byssochlamic acid

At a concentration of 10^{-2} M byssochlamic acid completely inhibits adenosine deaminase, alcohol dehydrogenase, and isocitrate dehydrogenase (2272).

The possible importance of *Byssochlamys fulva* in human foods has led to research into practical methods for detecting byssochlamic acid using thin layer chromatography (2271, 1656). This toxic metabolite is only formed in foods containing fatty acids if there is free glycerol present. Thus it is not formed in margarine, olive oil or ham, whereas a metabolite very similar to byssochlamic acid may be formed in butter (2271).

The closely related species *B. nivea* Westl. also infects foods and produces spores which tolerate both low and high temperatures (75 °C for five minutes). This species is even able to grow in an atmosphere containing 90% carbon dioxide (2913) and has been isolated from silage. It has been considered that *Paecilomyces varioti* Bain. (= *Penicillium divaricatum* Thom) is the imperfect stage of *Byssochlamys nivea* (321), but recent observations prove that this is not so (2503). This conidial form is common on food materials such as eggs, margarine, soya products, sauerkraut, as well as a variety of other substrates. It would appear that it may be the agent of a toxicosis of poultry to the same extent as *Aspergillus* and *Penicillium* (803, 976). Whether there is any relationship between the toxic activity of this mould and its ability to produce the antibiotic variotin, $C_{17}H_{25}O_3N$ (Figure 144) (3), assymetrin (2930), and pyrenophorol (1316) remains to be established. Some strains may produce patulin (1316). Chu has observed the production of a fluorescent metabolite which he considers may be the toxin (495) and Meyer (1656) has studied the production of byssochlamic acid by *Paecilomyces varioti.*

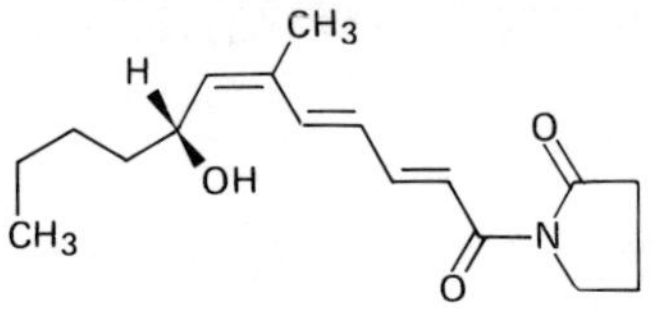

Figure 144. Variotin

Chaetomium

When grain infected with monoascosporic cultures of *Chaetomium* was fed to rats, Christensen *et al.* (489) found that more than half of such cultures of *Chaetomium globosum* Kunze caused the death of the animals within four to six days. The maximum yield of toxic metabolites was obtained after six to eight weeks incubation of the grain.

The symptoms preceding death consisted of changes in the central nervous system although autopsy showed lesions such as haemoglobinuria and haemorrhagic enteritis. The grain which was toxic for the rat was not so when fed to pigs even over a period of six weeks.

A toxic compound can be isolated, after defatting with petroleum ether, by extraction with acetone followed by purification by eluting from columns of silica gel with acetone and 5% acetone in chloroform. It is a piperazine derivative related to gliotoxin and the sporidesmins and has been referred to as chetomin (2651). A suggested molecular formula is $C_{31}H_{30}N_6O_6S_4$ and the possible structure is shown in Figure 145. The compound has $[\alpha]_D^{22}$ 360° and λ_{max}^{EtOH} 278, 287, 297 nm. A

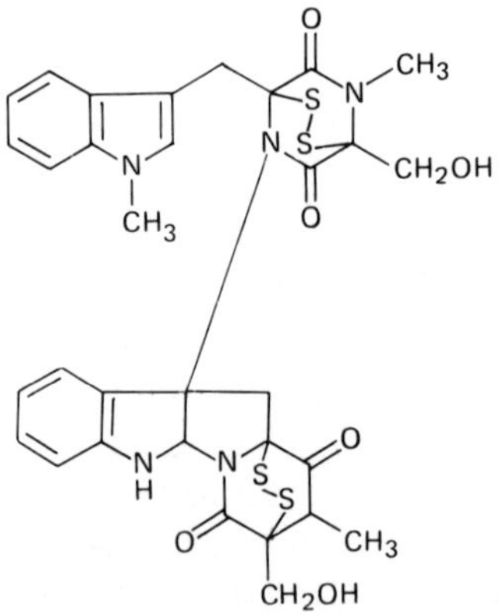

Figure 145. Chetomin. A structure proposed by S. Safe, A. Taylor in Sporidesmins. Part XIII. Ovine Ill-thrift in Nova Scotia. Part III. The characterization of Chetomin a toxic metabolite of *Chaetomium cochlioides* and *C. globosum*. *J. Chem. Soc. Perkin Transactions*, **I**, pp. 472–479 (1972)

related compound is the equally toxic metabolite chaetocin, $C_{30}H_{28}N_6O_6S_4$ (2652).

The production of oosporein by a species of *Chaetomium* has already been discussed. It has also been shown that *Chaetomium cochlioides* Palliser is toxic to the mouse in laboratory experiments (2216).

Neurospora sitophila

Neurospora sitophila Shear and Dodge, which belongs to the family Sordariaceae, is a member of a specialized flora found on burnt wood and vegetable debris. It is also found on a variety of other substrates (1745), especially bread and, indeed, the 'red mould' of bread has been known since 1831 when it was first described at Chartres, and since then it has been found many times on a variety of bakery products.

According to Decaisne (609), it has been shown that an outbreak of food poisoning in Italy was associated with the consumption of bread infected with *Neurospora.* This is in contradication with the conclusions of a report presented by the Ministry of War on the consumption of such bread by the army. Tests carried out on young cats indicated that the effects produced depended on the disposition of the individual consumer.

It would however seem that *Neurospora* may be associated with allergies, asthma or hay fever and sometimes even with otomycosis.

Gloeotinia temulenta

This fungus has been known under a variety of names, the imperfect stage as *Endoconidium temulentum* Prillieux and Delacroix (2035), the perfect stage as *Phialea temulenta* Prillieux and Delacroix, *Stromatinia temulenta* Prill. and Del., *Sclerotinia secalincola* Rehm, *Sclerotinia temulenta* (Prill. and Del.) Rehm and *Gloeotinia temulenta* (Prill. and Del.) Wilson, Noble and Gray. It parasitizes the ovaries of rye and numerous other grasses (*Agrostis, Festuca, Lolium, Poa,* etc.) growing subsequently as a saprophyte on the caryopsis at the expense of the starch. The seeds thus infected are small, wrinkled, and entirely altered into a sclerotium, the disease being known as blind seed disease in English speaking countries. Gray (956) reported that most of the damage to *Lolium* was caused by a related species *Phialea mucosa.*

Gloeotinia temulenta produces conidiophores which are borne laterally, at regular intervals, on the hyphae of the mycelium spreading over the surface of the seed. When a conidium has formed, it separates from the conidiophore and a new one develops pushing away the first. As the conidia remain attached, they form a significant mass at the tip of each conidiophore. The conidia are smooth, colourless, allantoid, biguttulate and measure 9–18 x 2.5–4 μm (1583).

The apothecia of the perfect stage are produced on the seed and ejection of the ascospores occurs during the day between 9.30 and 15.30, maximum release occurring between 11.00 and 13.30. Some of the flowers of *Lolium* are open from 8.30 and others again at 15.00, so this represents a remarkable adaptation allowing infection (1257). *Lolium multiflorum* escapes infection because of its late anthesis (1026).

This fungus, described in France in 1891 (2035), appears to be widespread throughout the whole of Europe, the United States, and even Australia and New Zealand (cf. Map No. 347 of the Commonwealth Mycological Institute Distribution Maps of Plant Diseases).

It has been established that bread made with infected rye gave rise to a toxicosis resembling food poisoning in man and domesticated animals. The symptoms consisted of apathy, giddiness, and vertigo accompanied by nausea, vomiting, stomach pains, and diarrhoea. Sometimes convulsions and numbness occur and, in some cases, end in death (548, 1057).

It has been known for a long time that consumption of *Lolium temulentum* L. by grazing animals could give rise to similar symptoms, hence the name 'intoxicating tares' sometimes given to this grass. The Bible, Ovid, Virgil, Plato, Dioscorides and Shakespeare all refer to such intoxications (1446). It has been suggested that the toxicity of the grass is due to the production of an alkaloid called temulin.

In fact, it is only the seeds of *Lolium* which are poisonous and it is very frequently parasitized by *Gloeotinia temulenta,* which moreover considerably reduces the level of germination of the seed (1426). It is now considered that the real cause of these toxicoses could be this parasitic fungus (837, 1057). Some recent experiments confirm this opinion whereas others seem to cast some doubt on it (2492). In New Zealand, for example, *Gloeotinia temulenta* has been repeatedly observed as an endophyte of *Lolium perenne* L. and yet no cases of poisoning have been reported (1818, 1819). Similarly, although common in the United States (nearly 90% of *Lolium perenne* may be infected in western Oregon (1027), no typical cases of poisoning have been reported (1916). It is mainly in Europe that these cases have been described and even there they are, it seems, very rare (809). So, some doubt continues to exist concerning the relationship between the poisoning resulting from the consumption of infected rye-grass and the presence of *Gloeotinia temulenta.* It appears that immersion of the seed in water at 50 °C is sufficient to destroy the fungus (1585).

Research on varieties of rye-grasses which are not susceptible to *Gloeotinia* has begun in N. Ireland (2878).

TOXICOSES ASSOCIATED WITH VARIOUS FUNGI IMPERFECTI

Cladosporium

Cladosporium herbarum (Pers.) Link (referred to as *C. epiphyllum* (Pers.) Nees) as well as *C. fagi* have been incriminated in some forms of alimentary toxic aleukia (1238). They are frequently isolated from seed which has been exposed to the cold during winter and are considered to be toxic due to the production of epicladosporic and fagicladosporic acids (Figure 146) produced by each respectively (1891).

$$CH_3.(CH_2)_9.CH{=}CH.(CH_2)_{13}.C\begin{smallmatrix}{=}O\\ \diagdown SH\end{smallmatrix}$$ Epicladosporic acid

$$CH_3.(CH_2)_9.CH{=}CH.(CH_2)_9.C\begin{smallmatrix}{=}O\\ \diagdown SH\end{smallmatrix}$$ Fagicladosporic acid

Figure 146. Metabolites of *Cladosporium*

The toxicity of *C. exoasci* Link, *C. fuligineum* Bon., *C. gracile* Cda., *C. molle* Cke., *C. penicillioides* Preuss, and *C. sphaerospermum* Penz. has also been demonstrated but the possible role of an anthraquinone pigment, cladofulvin, associated with the genus, has not yet been determined (442).

Alternaria and *Stemphylium*

A species of *Alternaria* may be associated with a particular toxicosis of poultry but there is, as yet, little information on the subject (803). Alternaric acid, which is one of the metabolites of members of the genus, is toxic to plants and may also have some toxicity for animals. Joffe (1235) has shown in the laboratory that *A. humicola* Oud. and *A. tenuis* Nees can cause toxicoses.

Of 44 monosporic strains of *A. longipes* (Ell. and Ev.) Tisdale and Wadkins, pathogenic to tobacco plants, only 14 could be considered as non-toxic to chickens, whereas, of 52 strains which were not pathogenic to tobacco, 39 were also not toxic to chickens (697, 2441). Other tests have shown that *Alternaria* can be toxic to mice (1013).

Meronuck and Christensen (1648) have estimated that 68% of isolates of *Alternaria* from seed, flour, and products derived from them, are toxic to the rat when tested in the laboratory. They cause anorexia, loss of weight, haemoglobinuria, intestinal haemorrhages, and even death.

A yellow quinone pigment, stemphone, has been isolated from *Stemphylium sarcinaeforme* (Cav.) Wiltshire and has been shown to be toxic in the chick embryo test (2321).

Epicoccum

Lesions of the liver and the kidney have sometimes been observed in the autumn in chickens fed with wheat recently infected with *Epicoccum nigrum* Link (101). This may be due to the production of flavipin, 3,4,5-trihydroxy-6-methylphthalaldehyde, a toxic metabolite which has been isolated from *Epicoccum* (131), as well as *Aspergillus flavipes* and *A. terreus.*

Cephalosporium

Cephalosporium acremonium Cda. is a widespread Hyphomycete isolated from a variety of substrates, and is known to produce a number of antibiotics. The

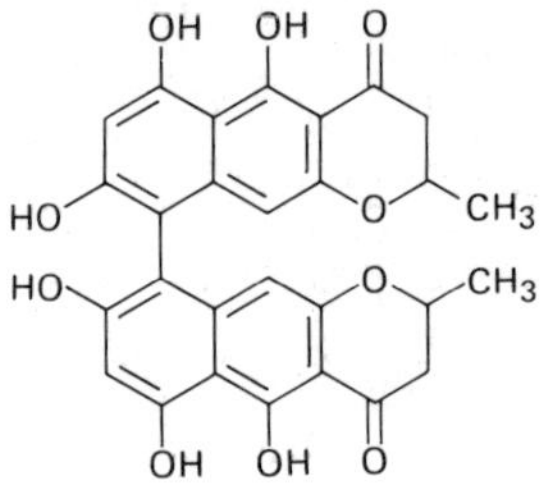

Figure 147. Cephalochromin

cephalosporins P_1–P_5 and C are soluble in butyl acetate, whereas cephalosporin N is insoluble in organic solvents. Cephalosporin N is, in fact, a penicillin and cephalosporin C is a member of a related family of β-lactam antibiotics for which the name cephalosporins has been retained. The cephalosporins P are steroid-like compounds having no structural relationship with the other antibiotics.

The LD_{50} of cephalosoporin P_1 for the mouse is 500 mg/kg when given intravenously but a dose of 5 mg administered orally at 12 hourly intervals for five days has no toxic effect (207). Thus it is practically non-toxic as are the chemically related metabolites, fusidic acid, isolated from *ACREMONIUM fusidioides* (Nicot) W. Gams reported in error to be *Fusidium coccineum* Tubaki, and ramycin, isolated from *Mortierella ramaniana* (Möller) Linnemann (920, 921, 1853, 2700, 2713, 2714).

The rare cases of poisoning which have been reported as possibly involving species of *Cephalosporium* may perhaps, in fact, be due to the microconidial stage of *Fusarium*, unless the pigment referred to as cephalochromin, $C_{28}H_{22}O_{10}$ (Figure 147), turns out to be toxic (2574).

Dendrodochium

Dendrodochium toxicum Pidoplichko and Bilai is a frequent saprophyte on straw and has been implicated in the poisoning of horses in the USSR (862). Ether or acetone extracts of cultures are just as toxic to plants, in which they produce leaf necrosis, and animals, causing skin hyperaemia in rabbits (203). Such extracts are particularly toxic to the rat (1580) giving lesions similar to those produced by the toxins of *Stachybotrys atra*.

The active agents have been isolated and are referred to as dendrodochins 1, 2, 3, and 4 (211, 1298). A number of fungi and bacteria are able to inactivate dendrodochin to a non-toxic product if they are inoculated into the culture medium in which the toxin has been formed. The toxin may even be degraded by the mycelium of *Dendrodochium toxicum* itself with a concomitant increase in the absorption of oxygen (213).

Trichothecium

Trichothecium roseum Link is a mould which produces a rose pink coloration and grows on a variety of substrates, especially food products (945, 2153). In

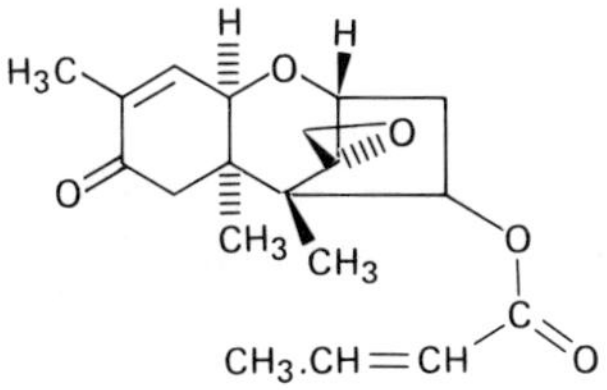

Trichothecin

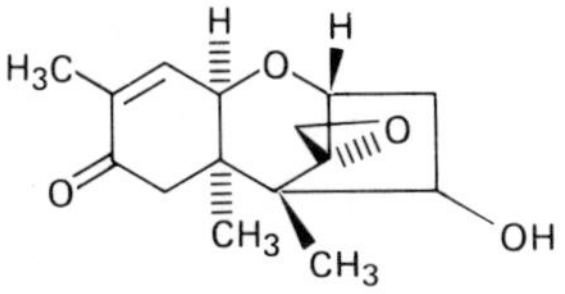

Trichothecolone

Figure 148. Metabolites of *Trichothecium*

particular it causes a bitter rot of apples and pears. This species produces the antibiotic trichothecin, $C_{19}H_{24}O_5$ (Figure 148), m.p. 243–244 °C, which is a member of the scirpene family of metabolites among which are toxic metabolites produced by *Fusarium, Trichoderma, Myrothecium,* etc. (127, 840, 841, 2151, 2152). Production, isolation, and biosynthesis of trichothecin have been studied by Silaev *et al.* (2385).

Mice can tolerate the intravenous injection of 5 mg trichothecin/kg, but are affected by a dose of 12.5 mg/kg and killed by 500 mg/kg. The LD_{50} for the rabbit is 250 mg/kg when given subcutaneously (207, 838).

Trichothecin is the isocrotonic ester of the ketonic alcohol, trichothecolone, $C_{15}H_{20}O_4$, which has also been isolated from *Trichothecium roseum,* and has m.p. 183–184 °C, $[\alpha]_D^{19.5}$ 22.5° (Figure 148) (839, 922).

Trichothecium roseum also produces other metabolites, some of which are irritants and may cause stomach ulcers. Such compounds are isorosenolic acid, $C_{20}H_{30}O_3$, m.p. 193 °C (Figure 149) (2312), and rosenonolactone, $C_{20}H_{30}O_3$, m.p. 186°C, $[\alpha]_D^{18}$ +6.3° (Figure 149) (2312).

The presence of *T. roseum* on wheat may be associated with the loss of appetite and reduction of egg yield recorded in laying hens (1944). Other phenomena which may be associated with this organism are a paralysis observed in ducks (2315) and a tremorgenic activity, with incoordination of movement and convulsions, followed by death within 10–20 minutes in mice (2153).

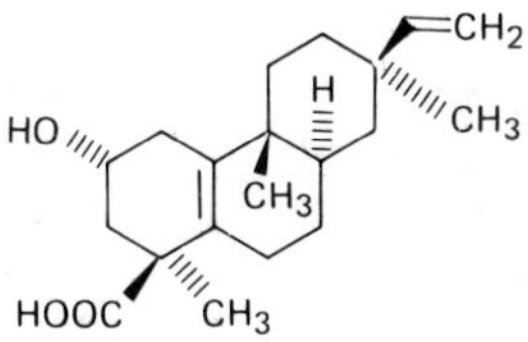

Isorosenolic acid

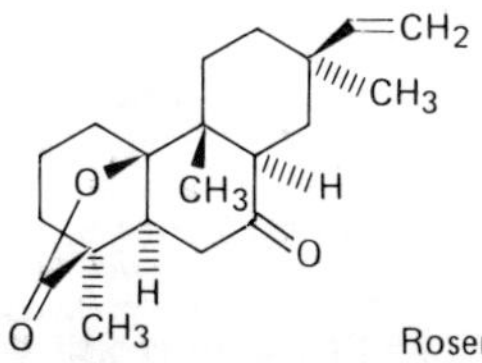

Rosenonolactone

Figure 149. Metabolites of *Trichothecium roseum*

Wallemia

Wallemia ichtyophaga Johan-Olsen (= *Hemispora stellata* Vuill.) is a chocolate brown mould which grows vegetatively very slowly, it is common on wheat, flour, and their products such as manufactured bread, and on dry forage. It is an osmophilic species and, although it has not been possible to attribute any specific toxicosis to this species with certainty, the frequency with which it is associated with various digestive upsets of cattle makes it suspect (1944). In laboratory trials it can provoke cell lesions (2216).

It may equally be considered as pathogenic and the mycoses which it causes resemble those of sporotrichoses (286). It is said to produce sporidesmolides like those of *Pithomyces chartarum* (2215).

Scopulariopsis

Some outbreaks of food poisoning of cows and pigs appear to correlate with the presence of *Scopulariopsis brevicaulis* (Sacc.) and *S. candida* (Gueguen) Vuill., these two species being the dominant components of the flora isolated from incriminated feed (1944). *S. brevicaulis,* moreover, is pathogenic to man (286). It is worth recording that strains of this mould can transform the green pigment, copper arsenate, into highly toxic and volatile derivatives of arsine.

Myrothecium

Myrothecium verrucaria (Alb. and Schw.) Ditmar produces verrucarol and verrucarin, two further examples of the scirpene group of metabolites (127) as well as muconomycins A and B, two antibiotics with inflammatory properties (977, 978).

When the culture filtrates of this fungus are ingested by sheep they cause haemorrhages (668). Vertinskii *et al.* (2743) have reported cases of death among sheep consuming plants in a pasture infected with *Myrothecium.*

Gliocladium

Weindling and Emerson (2796) isolated from a soil fungus a metabolite showing antagonism towards *Rhizoctonia solani* (2792, 2793, 2794) to which they gave the name gliotoxin (2795). It seems that there has been some confusion over the identity of fungi producing gliotoxin (2788). At first it was referred to as *Trichoderma lignorum* (Tode) Harz (2792), which itself is considered to be a synonym of *Trichoderma viride* (Pers.) ex Fr., and was then determined as being *Gliocladium fimbriatum* G. and A., now called *Myrothecium verrucaria* (Alb. and Schw.) Ditmar.

Now, Brian (291), examining a strain originating from Weindling, and considering it to be *Trichoderma viride* (298), found it to produce, not only gliotoxin, but another antibiotic which he called viridin (295, 299) (see Figure 150).

The perfect and conidial forms of the genus *Hypocrea* were studied and revised

Figure 150. Viridin

by Webster (2787), who confirmed that *Hypocrea rufa* has as its imperfect stage *Trichoderma viride,* but noticed that only some strains of this organism produce an antibiotic compound, and in no case was he able to find gliotoxin or viridin, an observation which was confirmed by Bilai (207). These metabolites are produced by another *Hypocrea,* the imperfect stage of which is the fungus of Weindling and of Brian, and which is *Gliocladium virens* Miller, Giddens, and Foster, a species described from the isolates made from forest soils of Georgia (1667).

Gliocladium virens differs from *Trichoderma viride* by its elliptical and smooth-walled conidia, instead of the globose, echinulate conidia of the latter species. In *Trichoderma*, the conidiophores are often straggly and frequently branched, the phialides being borne singly or in clusters at a significant angle to the conidiophore. In *Gliocladium virens,* on the other hand, the phialides are bunched

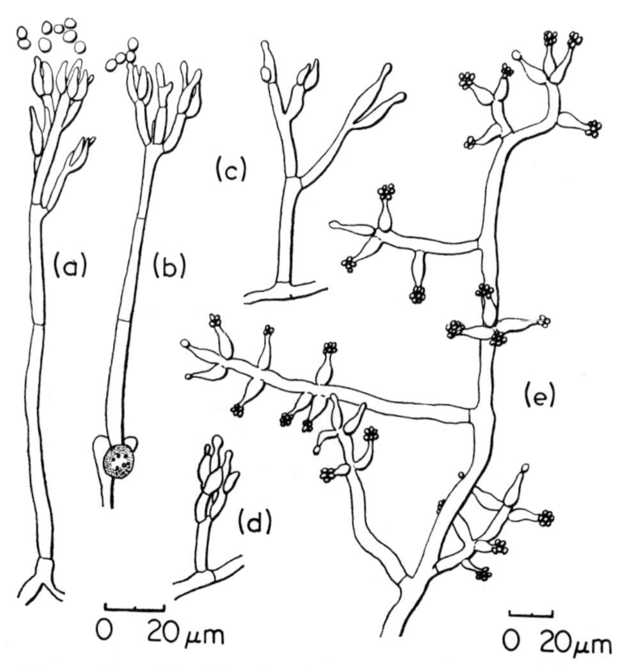

Figure 151. (a)–(d) *Gliocladium virens* (after Webster and Lomas (2788)). (e) *Trichoderma viride* (after the original drawing of Tuslasne (1865))

together much more in the manner of a *Penicillium* and the conidiophores are not very branched. In both genera the spores remain agglomerated in mucilage in balls at the ends of the phialides, a feature which distinguishes them from *Penicillium* (see Figure 151).

The structure and absolute configuration of gliotoxin (Figure 12) have been determined (179) and its toxicity is farily low. The LD_{50} for mice varies from 45–65 mg/kg depending on the route of administration. This toxin is destroyed by heating at 100 °C for 10 minutes (1314).

Gliotoxin is produced by a number of fungi other than *Gliocladium virens,* for example, *Aspergillus fumigatus, A. chevalieri* (2826), *Penicillium terlikowskii, P. corylophilum* (1782).

Viridin, $C_{19}H_{14}O_6$ (Figure 150), is a yellow pigment produced in good yield when the fungus is growing on a medium containing ammonium tartrate and glucose. Two isomers, α-viridin and β-viridin, are known (292). Unlike gliotoxin, viridin is strongly fungistatic and has very little antibacterial activity, whereas the former is antagonistic to both fungi and bacteria. Viridin is also more toxic than gliotoxin, a dose of 5 mg/kg being toxic to mice when given intraperitoneally (207).

Gliocladium roseum frequently produces derivatives of paraquinone, $C_8H_8O_4$ and $C_9H_{10}O_4$ (Figure 152), similar to those produced by *Aspergillus fumigatus.*

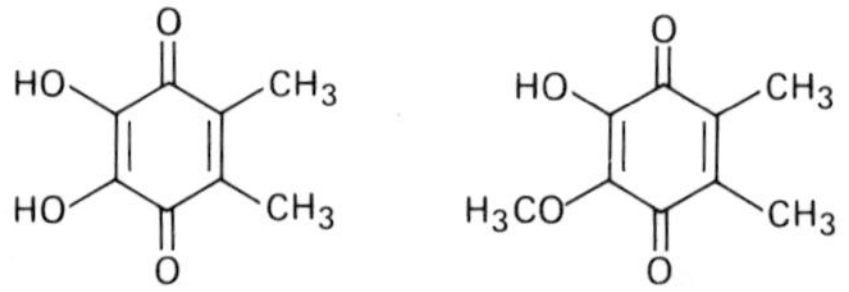

Figure 152. Quinone metabolites of *Gliocladium roseum*

Trichoderma

As has been seen it was in error that *Trichoderma viride* (Pers.) ex. Fr. was reported as producing gliotoxin and viridin, both metabolites of *Gliocladium virens* (2787).

Trichoderma viride does, however, produce trichodermin, $C_{17}H_{24}O_4$ (Figure 153), m.p. 45–46 °C, $[\alpha]_D^{20}$ −11°. This metabolite is very similar to trichothecolone and trichothecin, both produced by *Trichothecium roseum* Link.

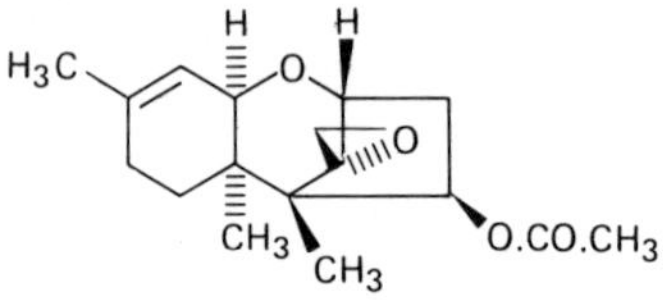

Figure 153. Trichodermin

A cyclic peptide, trichotoxin A, has been found to be produced by *T. viride* growing on maize (1129), and a number of antibiotic compounds, especially polypeptides, have been isolated from this species which is known for its antagonistic properties against other micro-organisms (629).

A leucoencephaly of horses has been attributed to *T. viride* (2307), but it may be due to another fungus (26). A toxic strain of *T. lignorum* (Tode) Harz, growing on the stems of melon, has been shown to cause abortions in guinea-pigs (2478).

Diplodia

It has been reported that *Diplodia zeae* (Schw.) Lev. (= *Phaeostagonosporopsis zeae* (Schw.) Woronichin), growing on maize, causes a toxicosis of cattle in South Africa. The disorder is characterized by excessive salivation, incoordination of movement, paralysis, and sometimes even death (1690, 2784). Although this member of the Sphaeropsidales has been encountered in other countries, it does not seem to have been implicated in toxicoses except in South Africa. Maize which has been artificially infected with *Diplodia zeae* has been found to be toxic to sheep and cows but not to pigs or horses (2582).

Rhizoctonia

Hay, containing red clover infected by *Rhizoctonia leguminicola* Gough and Elliot, has been responsible for a disease in cattle, the most apparent characteristics of which were excessive salivation and a reduction in milk yield (564, 565).

In the laboratory such enhanced salivation has been induced in sheep, pigs, chickens, and rats. Guinea-pigs react vigorously to the toxin for, as well as excessive salivation, they may weep and suffer from diuresis and frequently defaecate. Just before death they have difficulty with breathing. On autopsy, pulmonary emphysema and a centrolobular necrosis of the liver are revealed (566).

This toxic activity is associated with a parasympathomimetic metabolite, slaframine, $C_{10}H_{18}O_2N_2$ (Figure 154) (99, 2082). It is possible to treat poisoned animals with atropine or with methantheline bromide and hexamethylammonium bromide (566).

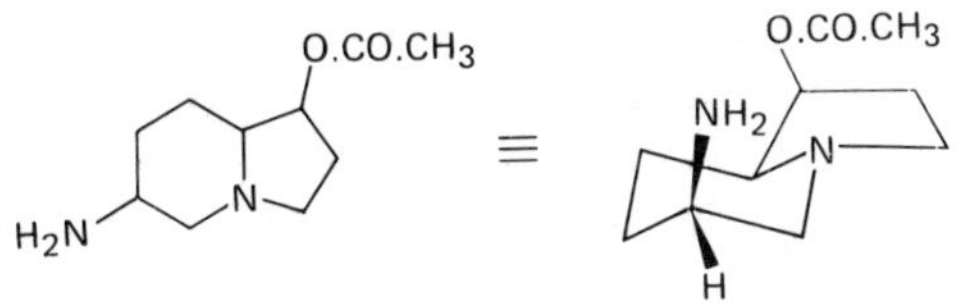

Figure 154. Slaframine

Others

From *Dactylium dendroides*, a parasite of mushroom beds, Staron and Allard (583) have isolated a compound which caused insular hepatic necroses and renal hypertrophy in mice.

Among those fungi parasitic to insects, *Beauveria bassiana* may owe its toxicity to the production of oxalic acid as well as the beauvericin which it produces and *Metarrhizium anisopliae* to the production of destruxins A and B (2164).

MYCOTOXICOSES INVOLVING SEVERAL FUNGI OR OF ILL-DEFINED AETIOLOGY

It is probable, as Brook and White note (315), that the number of confirmed and characterized mycotoxicoses, in which the fungus responsible has been identified and the toxin characterized, is a relatively low percentage of all those that occur.

The literature frequently contains references to, on the one hand fungi whose toxicity has not been proven but which have been isolated from feeds reputed to be toxic, and on the other hand fungi present on foods which have not been incriminated but are known to produce toxic metabolites in the laboratory. Finally, there are many cases in which several species have been isolated without it being possible to attribute the toxicity of the food to one or the other of them.

'Haemorrhagic syndrome' of poultry

The name of this toxicosis is due to Baker and Jacquette (121) based on the predominant character of a disease of poultry known since 1931 (2191) in which the consumption of mouldy feed, especially when it had been previously treated with sulphonamides, played an important part. A number of authors have reported such toxicoses, especially among chickens (264, 552, 802, 808, 943, 957, 1982, 2782).

Several fungi could be involved: among the aspergilli, *A. clavatus, A. flavus, A. fumigatus, A. glaucus, A. sydowii*; among the penicillia, *P. citrinum, P. purpurogenum, P. rubrum;* as well as *Paecilomyces varioti,* a *Scopulariopsis,* and an *Alternaria* (802, 943, 2191). *Mucor circinelloides, Absidia lichtheimii,* and *Gliocladium roseum* may also play a part in this disorder (1944). All these fungi may be isolated both from the chicken feed as well as the litter of their runs. It has been suggested that members of the *glaucus* group of the aspergilli play a particularly dominant role (2306).

Although the disease mainly affects four- to seven-week-old chickens, it may occur in birds of all ages, and is neither contagious nor infectious, but a typical toxicosis. The mortality of affected birds varies from 1–30% depending on a large number of factors and, in particular, it varies from one year to another. At the end of 1956 more than 80% of the chickens affected by this disorder in east Texas died (392).

The first clinical symptoms are depression, diarrhoea, in which the faeces are often mixed with blood, loss of appetite, anorexia, and dull plumage. Haematological examination reveals a reduction in the number of red cells (about 340,000/mm^3), resulting from corresponding changes in the bone marrow. The coagulation time of the blood is increased (957) and a thrombocytopoenia and leucopoenia become established.

The presence of petechia and enormous haemorrhages in a variety of tissues is noted, first occurring in subcutaneous tissues of the legs, the thighs, and the chest accompanied by inflammation of the mucosa of the gizzard and small intestine, as well as of the liver, spleen, heart, and the kidneys. Sometimes haemorrhages occur in the anterior chamber of the eye. The bone marrow becomes a rose to yellow chamois colour.

Histological examination of the tissues of affected birds reveals swelling and vacuolation, a precocious fibroblastic proliferation, and areas of necrosis in the liver, as well as the frequent presence of thrombosis in the blood vessels. Changes observed in the kidneys consist of a congestion of the tubular epithelium, which becomes an acute glomerularnephritis, tubular necrosis and, occasionally, a widespread infiltration of lymphocytes. In the spleen, congestion, necrosis, haemosiderosis, and an increase in fibroblastic activity are all recorded. There is sometimes a necrosis and hyalinization of the surface of the adenoids and the lymph follicles. Haemorrhages and necroses are abundant in the intestine and are associated with the villi and glandular epithelium. A vascular infiltration of lymphocytes is sometimes observed in all parts of the brain. The haematopoietic function of the bone marrow gradually diminishes whereas fatty cells increase (803). There seems to be a relationship between this disorder and a vitamin K deficiency (56, 463, 968, 2032).

In many ways, the haemorrhagic syndrome resembles stachybotryotoxicosis and alimentary toxic aleukia (802). It is possible for the disease to be confused with a haemorrhagic syndrome associated with excess sulphonamides, which is not uncommon (641).

Paralysis of the pharynx

Several cases of paralysis of the pharynx have been observed following the ingestion of mouldy plants and seed (902) but the nature of the moulds has not been reported. Occasionally the paralysis becomes general and is followed by the death of the animal.

Grassland tetany

The terms grassland tetany, plant tetany, plant paralysis, green wheat poisoning, hypomagnesium tetany, have been applied to a syndrome which has occurred in herbivores in several regions.

The disease affects mainly cows, sometimes sheep and goats, and only rarely horses (2638). Grassland tetany has been reported from many countries and has, it seems, recently spread to France (579, 1423, 2751) where departments in the north are particularly affected (Figure 155). Indeed it seems to be those departments north of the April 10° isotherm and with an April rainfall greater than 50 mm which are mainly affected. In 1956, it was estimated that 1–2% of all milking cows in Holland were affected (2579) and Great Britain (975, 1088, 2851) was equally badly affected as was Germany where it was reported that, in some regions, 10% of

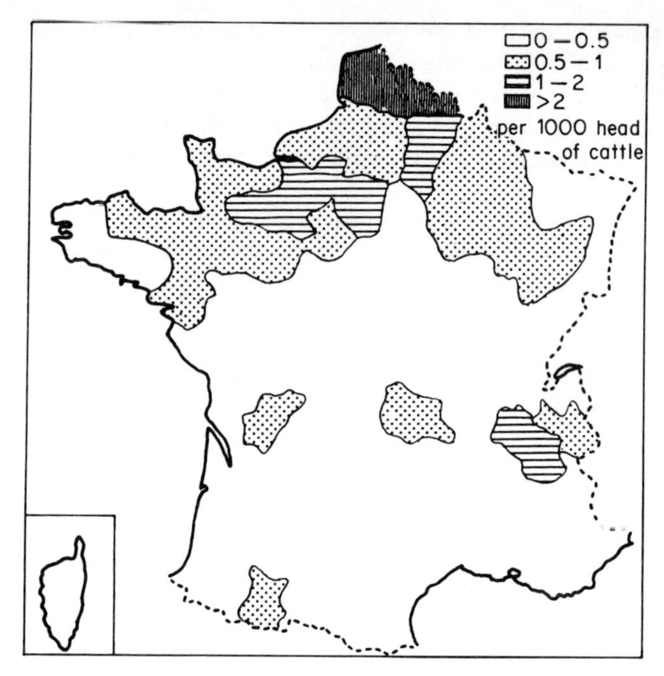

Figure 155. Frequency of grassland tetany in France by departments (1435)

the cows were ill (1813, 1966). Similar reports came from Scandinavia (1143, 2388).

In the south-western United States, tetany appears mainly in bullocks or cows grazing on young spring pasturage (355, 1654, 1866, 2395), and in sheep consuming oats and rye in winter (145, 355). In Texas and Oklahoma, tetany is associated with the consumption of green wheat in winter (561, 2118, 2390, 2391, 2392), and in the north-east, where the prairies are luxuriant in the spring, it is grass tetany in the spring which is the worst (1138, 1794).

Similar diseases occur in New Zealand where heavy losses of as high as 20% on one farm, have been reported (1125, 2364, 2531).

The botanical nature of the plants making up the pasture seem to be of little importance and the disease may occur on a natural grassland as well as on a cultivated pasture (32, 147).

Grassland tetany appears in animals as convulsions or paralysis, and such disorders may have a complex aetiology (2751). The basic syndrome characteristic of grassland tetany is a disorder in the metabolism of magnesium. The first visible symptoms of disease to appear after one to two weeks on the pasture are a stiff gait: the animal falls over and its limbs become completely stiff for a few seconds. Such periods of prostration alternate with fits during which the animal becomes belligerent, salivating abundantly, rolling its eyes and grinding its teeth. It is generally incapable of getting back onto its feet on its own. The temperature and

Table 31. A comparison of the mean concentrations of magnesium, calcium, and phosphorus in the blood serum of normal cows with those suffering from grass tetany (2412)

State of the cows	*Concentration in blood serum* (mg/100 ml)		
	Ca	Mg	P
Normal	9.35	1.66	4.57
With tetany	6.65	0.45	4.33

the heart beat are normal at first but tend to increase. In the acute form, death, preceded by a coma, occurs six to ten hours after the first appearance of these symptoms (1088, 1125). In the chronic form the symptoms are spaced out over several months (32, 2664) and only the demonstration of hypomagnesia allows a clinical diagnosis.

It is lactating females which are particularly affected and they seem to be especially sensitive between the first and twelfth week following calving, especially during the sixth week (2411). Grassland tetany is always accompanied by a reduction in the concentration of magnesium in the blood serum and is often associated with a reduction in the concentration of calcium, although the concentration of mineral phosphate does not show any detectable variation (cf. Table 31) (1422, 1866, 2118).

However it is essential to recognize that the magnesium concentration in blood serum of cows does vary during the year in animals on pasture, normally reaching 2.25 mg/100 ml in mid-summer and falling to 1.3 mg/100 ml at the end of winter (46), which may explain the reappearance of tetany during this period of the year.

On the other hand, an old animal has a lower magnesium concentration in its blood serum than a young animal, thus in young ewes, for example, the average concentration of magnesium in serum is 2.26 mg/100 ml compared with 1.62 mg/100 ml in old ewes (1060). This would seem to indicate why very old individuals are more sensitive to tetany than very young (1602).

From the research carried out on the basic cause of hypomagnesia a number of suggestions have emerged:

– Some consider that, because of an excessive production of ammonium ions, which may occur when plants rich in proteins are degraded by the micro-organisms of the gut flora, precipitation or chelation of magnesium ions may occur in the rumen before the food reaches the intestine (1048). Certainly the urinary excretion of magnesium decreases considerably when the production of ammonia by the animal increases (1049) but, simply giving the animal large quantities of urea does not induce hypomagnesiaemia (494, 2846).

– Others consider that grassland tetany may originate from a putrid fermentation of the food in the gastrointestinal tract and, in fact, a nauseating form of diarrhoea is often noted in sick animals. The disturbances in the composition of the blood plasma may thus only be a secondary symptom associated with the reaction of the animal to the influence of toxins produced in the digestive tract (2330).

– Others consider that changes in the mineral equilibrium of the grass brought about by the misuse of fertilizers, especially potassium fertilizers (2751), may be responsible. Potassium acts as an antagonist to magnesium in the plant (2770). Nevertheless, although tetanigenic grass is characterized by a very low concentration of magnesium in some cases, this does not appear to be always the case (1424). It has even been considered by some authors that the mineral composition of the plant may only have a remote connection with the appearance of tetany (1874, 1966, 2192, 2802). For all that there do exist definite variations in the concentration of magnesium in plants (2595). On the other hand, the evolution of magnesiaema in sheep on pasture follows a course parallel to that of the concentration of cellulose in the plant and inversely to that of the total concentration of nitrogenous material (1156), observations which are not accepted by all (1814).

Be that as it may, Bartlett *et al.* (147) consider that there exists in the very young plant a factor which triggers off tetany, an observation which seems to be confirmed by experiments by Pulss (2044). When the juice of a plant causing tetany is added to the food of female rats, he noticed a reduction of 20% in the basal metabolism, compared with a reduction of only 6% when the juice of a non-tetanigenic plant was used. This particular factor contained in the toxic grass may be an inhibitor of specific enzymes (2045).

Hypomagnesiaemia may result from (1425):

– either an increase in the removal of magnesium from the exchangeable pool by such processes as urinary excretion and endogenous accretion;

– changes in the distribution of the different compartments of the exchangeable pool;

– or reduction in the liberation of magnesium from a slowly exchangeable fraction.

In fact, it now seems certain that hypomagnesiaemia is not the consequence of a reduction of magnesium entering the exchangeable pool and thus probably originates from:

– either a reduction in the intestinal absorption of magnesium, which represents the classical hypothesis;

– or a reduction in the rate of liberation of magnesium from those tissues with a slow exchange, a mechanism which, according to Larvor and Violette (1425), appears to play an important role in the sheep.

According to some recent, but not yet verified, information resulting from the observations of Breton and Larvor (290), there may exist a connection between grass tetany and the presence of fungi of the Mucorales in the region of the rhizosphere of the grasses, toxic metabolites from which are secreted into the young plants. It is, in fact, known that members of the Mucorales may be particularly frequent in the region of the root hairs of young plants (444). It is suggested that these toxins, which are absorbed by the animal at the same time as the grass, disturb the metabolism of magnesium, perhaps by chelation. It is thus possible that grass tetany may be a mycotoxicosis and the resemblance of the external symptoms to those reported following the action of patulin give some credence to this hypothesis.

The treatment of grass tetany, which consists of the injection of calcium gluconate, often in association with magnesium, is often effective if it is started within the first two to three hours of the appearance of symptoms (145, 1602, 2392, 2664). There are however cases where this treatment is insufficient (2395). It is sometimes possible to rectify the problems of grass tetany by the addition of food supplements which complement the disequilibria (2751).

It is worth noting that in Oklahoma (2813) there has been reported a poisoning of cattle associated with the consumption of *Cynodon dactylon* (Bermuda grass) where the external symptoms are similar to those of grass tetany but in which no disturbance in magnesium metabolism was noted. It has been demonstrated that this intoxication was caused by moulds, which have not yet been identified, growing in the apical region of the stem. They have been isolated and it has been possible to induce nervous reactions with them experimentally. In contrast, those fungi isolated from other regions of the stem do not have the same activity.

TOXICOSES READILY CONFUSED WITH MYCOTOXICOSES

It is often difficult, in the case of the poisoning of cattle for example, to distinguish a mycotoxicosis from poisoning involving a quite unrelated agent. Among the phanerogams present in the fields which may be consumed by cattle, there are several which produce known toxic compounds (cf. the recent review by Kingsbury (1312, 1052). It is equally possible for those plants which are normally non-toxic to become so by the accumulation in their tissues of substances such as nitrates. The presence of more than 1.5% of nitrate in a plant, in the form of KNO_3 for example, is known to be lethal to domestic animals (233, 283, 2521). Ruminants in particular become ill after consuming a quantity of nitrate equivalent to 0.05% of their own weight (1144) and several hundred cows and sheep have died in some years in the United States (591, 1350). The toxicity of nitrate is the reason why feeding the tops of beetroots to cattle is avoided for they are very rich in nitrates (2249). The same toxicosis appears to be associated with the steps in the reduction of nitrates to ammonium via nitrites, the latter being ten times more toxic than nitrates (1475). Nitrites are responsible for the transformation of haemoglobin into methaemoglobin and thus inhibit the transport of oxygen by the blood. Antibiotics, which inhibit the conversion of nitrate into nitrite in the rumen, protect the animal from the effects of the accumulation of nitrite in their blood (730). In this instance the presence of micro-organisms producing such antibiotics in their feed may be desirable.

Selenium and molybdenum figure equally among the principal inorganic elements which may be responsible for poisoning of cattle. Those plants such as grasses growing on seleniferous or molybdeniferous soils accumulate these elements and, when the concentration passes 5 mg/kg, there is a danger of serious poisoning (146, 174, 304, 570, 2194, 2818).

It is also well recognized that herbicides may be responsible for toxicoses by cumulative effects. Thus the addition of 1 g/l of 2,4-D or 2,4,5-T to drinking water causes nephrotoxic effects if consumed over a period of seven months (747)

TRANSLATOR'S NOTE: Two specific instances in which mycotoxins were suspected and subsequently found to be absent have been brought to my attention (3000). In one case a large number of pigs were suffering from gastroenteritis in which no infectious agent seemed to be involved. Mycotoxins were suspected but analysis using the multimycotoxin method (3045) was negative. An examination of the formulation of the diet showed that it contained 55% wheat instead of the recommended 30%. An excess of wheat gives a high gluten content in the diet which was almost certainly the factor responsible and a change in the diet led to an eradication of the condition.

In the second case history a large poultry unit was suffering a drop in egg production and feed samples were analysed for mycotoxins with negative results. Amino acid analysis of the feed demonstrated statistically significant low levels of methionine which was considered to be the cause of the condition.

CHAPTER 12

PERSPECTIVES OF THE STRUGGLE

The therapeutics of mycotoxicoses is as yet poorly understood and may be seen from several points of view:

– on the one hand the treatment of poisoned animals or individuals in order to secure their recovery;
– on the other hand the detoxification of mouldy feeds;
– finally, research on practical measures for destroying moulds in feeds (disinfection) or, better still, in preventing their growth (conservation).

The problems are considered here briefly in general terms, having been touched upon earlier for each toxicosis, although the relative problems of disinfection and conservation merit further consideration.

TREATMENT OF POISONED ANIMALS

There is an appropriate therapy corresponding to each type of toxicosis and only a doctor or a vet may judge which is best to apply. It will be necessary to diagnose the type of poisoning, to recognize the organs affected, and act accordingly. Measures which may be used in the treatment of poisoning are essentially (761):

1. to evacuate or eliminate the poison from the organism;
2. to destroy or neutralize the poison by administering appropriate antidotes;
3. to combat the effects produced by the poison.

There are unfortunately not many specific remedies recognized and it is often necessary to be content with enhancing the natural defence mechanisms of the body.

It has been possible, for example, to bring about a regression in the rat of hepatomas associated with aflatoxin by using cortisone acetate and hydrocortisone, provided that treatment was started before the disappearance of parenchymatous tissues (280). Vitamin therapy may bring relief in some cases (1202) and blood transfusions, and even plasma tranfusions, are sometimes recommended.

The organism is sometimes able to defend itself naturally against toxic substances. Thus it has been shown that the rumen of bovines may be able to detoxify, either by modifying or destroying toxic substances such as certain insecticides, gossypol from cottonseed, and digitalin. A search for antibiotics formed in the rumen has been made but few results have been obtained as yet.

PREVENTIVE DETECTION OF TOXINS AND DETOXIFICATION OF MOULDED FOODS

Before taking efficient action against a toxin present in a food it is advisable to first characterize the toxin. The principal methods, using modern techniques of chemical analysis and competent identification of suspect moulds, have already been indicated in earlier chapters. Control should be made in a systematic manner. It has already been found, in the case of milk for example, that the fixing of a purchase price which takes into account, not only the concentration of proteins, fats, etc., but also of microbiological qualities has a controlling influence. It then becomes more or less necessary for there to be near each food industry a laboratory equipped, not only for biochemical and bacteriological examination, but also for mycological work.

International regulations are beginning to appear for maximum tolerated levels of aflatoxins in animal cake. There need to be good analytical methods which allow determinations to be carried out easily, accurately, and rapidly and work continues in several countries to this end (251).

Experiments on detoxification have often been disappointing. For the removal of aflatoxin either treatment of contaminated food with ammonia, or extraction with solvents has been recommended, but these processes result in some denaturation of the proteins. Future solutions may well depend on the use of micro-organisms carrying out specific bioconversions of mycotoxins such as aflatoxin to less toxic compounds.

PRESERVATION OF FOODS

As stressed by both Spensley (2453) and Gilman (913), the best protection against the formation of toxic metabolites by moulds consists of the storage and transportation of food products under conditions that reduce the risks of contamination and do not allow the growth of any contaminant. It is thus necessary to know with some precision the parameters which control the growth of each species incriminated. Such studies will include the function of the physico-chemical factors of the substrate which control spore germination, growth, and sporulation. Even if completely effective treatments do not arise from such studies, they should at least suggest improvements.

Sometimes relatively simple changes are sufficient, thus for example, the conservation of grain in silos is improved when aeration is more evenly distributed and, instead of letting warm air rise to the surface which may cause condensation in the upper parts of the silos, it is advisable to force it to leave at the base (1111).

The different means of control which may be used in the food industry either directly or indirectly are considered briefly.

Heat sterilization

Of the methods for sterilization, heat, and in particular steam, under pressure, is probably the oldest, simplest, and the most efficient process where it is possible to use it (325, 521).

Heat acts essentially by inactivating cell proteins, especially the enzymes, and this in itself is sufficient to cause the death of micro-organisms (Table 32). Wet heat is much more efficient in this process than dry heat, thus egg albumin containing 50% water coagulates at 56 °C, whereas a temperature of 170 °C is required to coagulate albumin without the addition of water.

The processes for sterilization most widely used in the laboratory involve autoclaving, in the presence of water, at 115–120 °C for 20–30 minutes, or a period in a dry oven at 170 °C of at least one hour. Such sterilization, which destroys or eliminates all viable micro-organisms, is realized in practice in the ovens of industrial bakeries where the temperature may reach 150 °C, although the centre of a loaf may only reach 100 °C for about 15–30 minutes. Having left the oven, the bread should remain sterile so long as it is not recontaminated during packing (1741).

Pasteurization, which is often used for milk products and a variety of beverages, traditionally consists of maintaining the product at a temperature of between 60–80 °C for anything up to an hour but higher temperatures for very much shorter times are also used. Such treatment is sufficient to kill many bacteria and a large number of fungi giving the treated product a certain degree of stability. However it may be ineffective for species such as *Aspergillus repens, A. flavus,* and *A. fumigatus* (2590).

Tyndallization consists of three or more successive heat treatments at 60–100 °C for a period of 30–60 minutes, incubating for up to 24 hours between

Table 32. Activity of wet heat against moulds (2525)

Species	*Stage*	*Time required to kill at:*			
		50 °C	55 °C	60 °C	100 °C
Neurospora tetrasperma	ascospores	–	–	4 h	–
	conidia	30 min	–	–	5 min
Byssochlamys fulva	ascospores	–	–	–	10 min
Botrytis cinerea	spores and mycelium	several min	–	–	–
Penicillium chrysogenum	conidia	–	3 days	–	5 min
Verticillium albo-atrum	conidia and mycelium	10 min	–	–	–

each treatment. The heat shock of the first treatment, although not sufficient to kill microbial spores, may induce them to germinate during the period of incubation making them susceptible to subsequent heat treatment.

The Therlak method, currently used in Bulgaria (2478), consists of systematically treating all suspect maize to 400–450 °C before it is fed to animals.

Use of low temperatures

The widespread use of refrigerated warehouses has opened up new possibilities for the food industry (1157, 2682), and the inhibitory effect of low temperatures on the growth of moulds has already been stressed. Maintenance of food products at a temperature close to 0 °C, deep freezing at –18 °C, or even –30 °C when the product to be stored is compatible with such treatment, usually prevents the growth of moulds (1816, 2631, 2768).

Some products do not survive such low temperatures, thus bananas cannot be transported at temperatures below 11 °C. In many cases there exists an optimum temperature for the conservation of a product. This is known for a number of varieties of apples, for example 8 °C for Reinette Clochard and 1 °C for Golden Delicious (2683).

Refrigeration processes require a careful study so that they can be adapted to each particular problem (860). Thus, in the case of meat, very rapid refrigeration immediately after slaughter is recommended. This provokes the formation of a thin skin which reduces evaporation and impedes colonization by moulds (523). In some cases it may be advisable to make use of a prerefrigeration stage.

It is essential not to forget that, although low temperatures may impede the growth of fungi, they do not usually kill them, for the spores remain viable and germinate as soon as they are placed in a more favourable environment, as when a chain of refrigeration is broken.

Although of considerable interest, refrigeration is not the answer to all food preservation problems involving moulds, and indeed the breakdown of refrigeration equipment at certain stages in the food chain may be very prejudicial.

Dehydration, desiccation, and lyophilization

Reducing the concentration of water in a product below the critical level required for germination of spores is often sufficient to limit the growth of contaminating moulds (2015). Drying agents such as calcium chloride or silica gel may be useful, especially under reduced pressure and with the application of a gradual increase in temperature. The seed of dry cereals is very much less contaminated than damp seed.

The production of dried fruit presents a very easy method of conservation and, in many cases, a product proportionally very much richer in sugars is obtained with the result that only a few osmophilic species can grow on it.

The addition of sugar or salt to increase the osmotic pressure of the substrate will reduce the number of species capable of growing on it. Thus the transformation of fruits to jams gives a long storage life and pickling must be one of the oldest processes for protecting food products from microbial spoilage (771, 1555).

Lyophilization is a process which is not yet widely used because of the cost but will certainly become much more widespread (346, 1816, 2145). Lyophilization is essentially the rapid freezing of the material followed by direct sublimation of the ice to leave a dry product of high organoleptic quality. The drying of a frozen product, protected from oxygen and at a low temperature, gives a dry product which is not denatured, can be kept for a long time so long as it is protected from changes in the ambient temperature, from air, and from water vapour (547).

Lyophilization was first used for the preservation of blood plasma, then for pharmaceutical products, and is gradually taking its place in the food industry (1867). In 1964, the United States produced 45,000 tonnes of freeze-dried food products and in Europe several factories for freeze-drying fruits and vegetables have been established. The process is being increasingly used for the conservation of shrimps, scampi, eggs, certain fruits such as strawberries, fruit juices, and soups based on chicken, vegetables, and mushrooms. So long as the product is packed in a material which is impervious to oxygen, water vapour, and light, freeze-dried products have a remarkably long shelf life.

Only economic considerations may prevent the widest use of freeze-drying for, according to Bird (223), it costs a significant amount of money to remove a kilogram of water from a fresh food product.

Changes of pH

The efficiency of acids as inhibitors of mould growth may be due as much to the concentration of hydrogen ions as to the nature of the acid radical. This will usually be so when mineral acids are used, but organic acids owe much of their activity to the undissociated molecule.

Acidification of a medium to about 3.9 will inhibit the growth of many moulds although there are some which can grow even in strong sulphuric acid! It is because of the increase in acidity that sauerkraut is a better way of preserving cabbage than the fresh form.

Because weak organic acids are frequently more active in the undissociated form pH often plays an indirect role in the use of such compounds as preservatives. Thus benzoic acid and the esters of *p*-hydroxybenzoic acid are two to three times more active at pH 3.0 than at pH 6.0 (740).

Sterilization by radiation

Ionizing radiation

The term ionizing radiation is applied strictly to X-rays, α-rays, cathodic rays, β-rays, and relatively heavy particles such as neutrons and protons.

During the last few years, considerable progress has been made in the practical use of such radiations (931), however there is public disaffection with such

Table 33. A comparison of the sensitivity of mould spores and bacteria to radiation (2525)

A. Ultraviolet of 253.7 nm. Dose in ergs/cm^2 x 10^2 required to inhibit the colony formation from 90% of the organisms treated.
B. γ-rays. Dose in rads x 10^{-6} required to inhibit the growth of all colonies treated.
C. X-rays. Dose in rads x 10^{-3} required to kill 90% of cells treated.

	Organism	*Stage*	*Dose*
	Bacillus anthracis	spores	452
	Bacillus subtilis	vegetative	520
	Bacillus subtilis	spores	900
A	*Neurospora crassa*	microconidia	1,360
	Neurospora crassa	macroconidia	2,800
	Aspergillus niger	conidia	23,000
	Bacillus pumilus	spores	2.1
B	*Streptococcus pneumoniae*	vegetative	0.5
	Aspergillus niger	conidia	0.32
	Bacillus brevis	dormant spores	47.0
C	*Neurospora crassa*	microconidia	48.0
	Neurospora tetrasperma	ascospores	40.0
	Aspergillus terreus	conidia	30.0

processes because of the suspicion that the treated products will be tainted and rigorous legislation does not permit as wide a use of such methods as they may perhaps merit.

The United States have made a considerable investment in these processes and, in one centre alone, a cobalt-60 source of more than one million curies has been installed (1469, 2387). Work in Russia has been summarized in the reviews of Sosedov and Vakar (2447) on wheat, and Chmyr' (451) on maize.

In Western Europe, research has been carried out by several organizations, notably the Atomic Energy Commission. Research mainly concerns the preservation of potatoes, cereals, and fruit (2006, 2212, 2213).

There are reviews on the present state of food preservation using ionizing radiation (2301, 2746, 2888, 2889), the toxicology of the products (1349), and the sensitivity of micro-organisms to such radiation (Table 33). Ionizing radiation affects cell metabolism, modifying the structure of essential metabolites such as amino acids, nucleic acids, and various proteins (915).

Doses of the order of 1 x 10^6 rad are sufficient to sterilize most food products, although the complete disinfection of wheat is only obtained at 3 x 10^6 rad. *Alternaria* and *Cladosporium* seem to be particularly difficult to destroy (2001).

Recent studies (2444) have confirmed earlier observations on the advantages of heat treatment following irradiation by γ-rays for the protection of fruit against a variety of rots (see Figure 156).

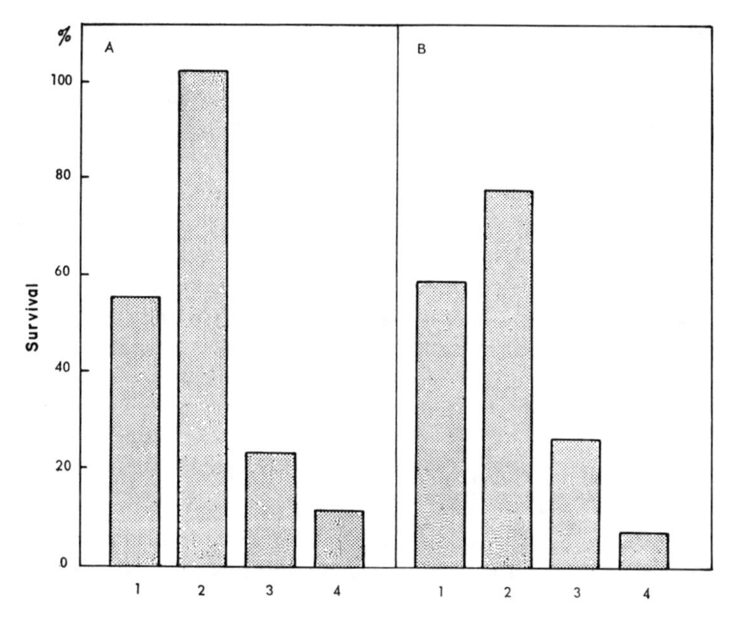

Figure 156. Survival of conidia after treatment by heat and irradiation. 1 Irradiation alone. 2. Heat treatment alone. 3. Irradiation treatment after heat treatment. 4. Heat after irradiation treatment. (A) *Botrytis cinerea* at 44 °C for 4 minutes or a dose of 75 krad or both. (B) *Penicillium expansum* at 56 °C for 4 minutes or a dose of 20 krad (2444)

The fairly high cost of ionizing radiations, the precautions required, the changes in organoleptic qualities which they may cause (2931), and the fact that the harmlessness of the product is still often doubted, all contribute to the resistance in the use of this means of preserving foods, although use of ionizing radiations in the pharmaceutical industry is widespread (77, 260).

Other radiations

1. Infrared radiation does not have any recognized antifungal activity if the heating effect is discounted.

2. Ultraviolet radiation, between 200–400 nm, can kill micro-organisms exposed to it, the activity varying according to the wavelength. A wavelength of 253.7 nm is often used because it appears to be relatively effective and is easy to produce commercially.

Ultraviolet light is more active against bacteria than fungi and, whereas a dose of 3,000–7,000 $\mu W/s/cm^2$ may be enough to kill bacteria, at least 15,000 may be required to have any effect on most moulds and 2,000,000 may be required to destroy spores. It is possible to either increase the intensity of the lamp, or to prolong the period of exposure to obtain the desired effect (896, 1152). Good results have been obtained in small rooms where germicidal lamps can be left on for a long time in the absence of personnel (611), a fact that is taken advantage of in cheese making (633). The selective activity of u.v. radiation has been commented on (2190) and there may be inconvenience in the rapid loss of energy flux in some lamps.

It should be remembered that u.v. light does have mutagenic activity and also precautions must be taken to protect the eyes of those who use it.

3. Electrostatic precipitation and ultrasonics are relatively recent innovations and their range of application is still very limited. Again, bacteria are more sensitive than fungi. Thus treatment at a frequency of 26 kc/s and 500 W for 20 minutes is ineffective against the mycelium and ascospores of *Aspergillus nidulans* and only the conidiospores are partially destroyed after ten minutes (594).

Chemical preservatives

The addition of antiseptics to foodstuffs is controlled in most countries. In France the 'Service de la Répression des Fraudes' is based on the law of 1 August 1905 which replaced Article 423, which in 1810 prohibited deception only in the nature of the merchandise (612, 1720). Commenting on this law, Souverain (2448) pointed out that 'in principle, any chemical additive is a compound foreign to the food. Recourse should be made to the use of chemical additives only for serious reasons and in the minimum quantities compatible with needs, not in basic foodstuffs if possible and based on three reservations for the protection of people:

(a) Banning the use of compounds which could be harmful to health;
(b) Banning the use of substances which reduce the nutritional value;
(c) Banning substances employed simply as agents of deception.'

These are very reasonable measures, the aim of which is to protect the consumer. In practice some antimicrobial agents such as those used in pharmaceutical products are tolerated (in France these are covered by ammendment No. 16 to the Codex of 1949, which came into effect on 29 April 1960):

– propionic acid, sorbic acid, and glycerol;
– in limited concentrations, ethanol (20% v/v), sulphur dioxide, sulphites, bisulphites, and metabisulphites (1‰ as SO_2); benzoic acid, sodium benzoate (2‰ as the acid); parahydroxybenzoic acid, its ethyl, methyl, and propyl esters and their sodium salts (1.5‰ for a single compound or their mixture); phenol (5‰); cresol (3‰); *p*-chorometacresol (3‰);
– subject to restriction; chlorobutanol; hydroxyquinoline; the acetate, borate, and nitrate of phenyl mercury; orthochloromercuriphenol; sodium ethyl mercurithiosalicylate.

The United States Pharmacopoeia and the British Pharmaceutical Code contain similar recommendations.

Those which are bacteriostatic may not be very active against fungi. Propionic acid, at 0.5%, has provided a convenient means of preserving damp maize for up to two months, as well as a variety of cereals and damp fodder (1139, 1955, 2399, 2461), although its smell makes it less appetizing to animals and it may be corrosive to machinery (2887).

Sorbic acid, CH_3.CH=CH.CH=CH.COOH, and its sodium, potassium, and calcium salts are probably among the more fungistatic compounds permitted in foods. Used at concentrations of between 0.015–0.2% they give good results in the preservation of many different foods (1103), especially those containing fats. A derivative, sorbohydroxamic acid, is more active than the parent compound (2635), and sorboyl palmitate at 0.13% has also been recommended in the preservation of bread (970).

Among the esters of *para*hydroxybenzoic acid, the methyl ester is reputed to be the most fungistatic, but is less active against yeasts than the propyl ester.

To this very limited list may be added a number of other substances which have been adopted by various international agencies (Joint FAO-WHO Committee in 1958, European Codex Alimentaire from 1959–63, European Economic Commission, 27 January 1964) often as a temporary measure (formic acid and its salts, organoborate compounds, boric acid and its salts, lactic acid, hexamethylene tetramine, isopropyl citrate, ethylene and propylene oxides, sodium biacetate), or for very specific purposes:

– chlorine and its derivatives, silver and its derivatives, and ozone for water;
– sodium and potassium nitrite and nitrate for curing;
– diphenyl, orthophenylphenol, sodium orthophenylphenate for the surface treatment of plums;
– water glass (sodium and potassium silicates) for the shells of eggs.

Several of Eckert's papers (714, 715, 716) describe products which are recommended for the treatment of fruits after harvest.

With Jumel (1266) we may conclude that, whatever confusion seems to arise from the study of such diverse legislation, allowance must be made for the different eating habits throughout the world and use must be made of the competence of local expertise.

As well as the few widely used compounds, work has been in progress for several years in the search for new synthetic fungicides with low mammalian toxicity, but such studies are still too fragmentary for any general conclusions to be made from them.

There have been several studies on amines, butylamine in particular, dodecyl-di (aminoethyl) glycine, the derivatives of benzimidazole, 3-pyridine thiol, and a variety of antibiotics. The latter include nisin from *Streptococcus lactis,* a natural product in some cheeses and recommended for addition to others; subtilin produced by *Bacillus subtilis,* which may be useful in tinned fruits and vegetables; chlortetracycline and oxytetracycline, which have been used for poultry and fish; nystatin (= mycostatin), produced by *Streptomyces noursei*, and pimaricin (= tennecetin) produced by *S. natalensis* (1219, 2797).

Indirect means of tackling the problem

It is possible in many cases, with a minimum of elementary precautions, to restrict the sources of contamination. The maintenance of rigorous hygiene in areas where foods are stored and handled is the basis of any fight against the growth of moulds in foodstuffs.

Walls, ceilings, and floors ought to be regularly disinfected and hypochlorite, cresol, and the sodium salt of cresol have all been recommended for this. The use of whitewash or fungicidal paints is highly desirable (71).

Sometimes it is the wrappings which are the source of contamination of food products and it may be desirable to treat paper or plastic wrappings with a fungicide to overcome these problems.

The atmosphere carries mould spores and is one of the most important sources of contamination (1724, 1737, 2457). If such contamination comes from outside it is possible to filter the incoming air to remove spores and, moreover, by maintaining a slightly higher pressure in an important food handling area, it is possible to prevent air entering from the outside through opened doors.

Air may be purified either by treatment with mists or smokes of disinfecting substances, by gaseous disinfectants, or with ultraviolet light. Simply spraying water or bleach into the atmosphere cleanses by a mechanical precipitation of micro-organisms, and is sometimes used in the laboratory.

The following have been used in medicine for some time (1861):

– a 10% solution of resorcinol at the rate of 0.2 ml/m^3 of air;

– gylcols such as diethylene glycol and triethylene glycol by evaporation or in aerosols;

– aromatic oils, either alone or associated with the preceding materials.

Although these substances have some bacteriostatic activity (424), they are not sufficiently effective against moulds for their use to be recommended.

Formaldehyde diffuses well and is an excellent fungicide, especially in the presence of an atmosphere saturated with water vapour. It may be used (2067), either as a spray of formalin at the rate of 10 g of a 3% solution per cubic metre of air, or by the release of gaseous formaldehyde vapour from formalin (for this latter procedure several processes have been recommended. For example, per litre of air, the following mixtures may be used:

40% formalin	20 ml
water	20 ml
potassium permanganate	8 g

or

40% formalin	30 ml
boiling water	180 ml
quicklime	30 g

Alternatively, formaldehyde may be released into the atmosphere by depolymerizing trioxymethylene using heat. Special containers are manufactured for this purpose and about 75 g of trioxymethylene is adequate for fumigating 20 m^3 of air. The very disagreeable odour, which can be removed to some extent by spraying with ammonia after the formaldehyde has been allowed to act, and the irritant action on mucous membranes, limit its use despite its efficiency and low cost. Certain recently published formulations may, however, overcome these snags.

Lactic aerosols have been recommended (2354) at dose rates of 300–500 mg/m^3, but, although active against bacteria, they have little activity against mould spores.

Among the cationic antiseptics of the quaternary ammonium group, lauryl-dimethyl-carbetoxy-methylammonium bromide has an unquestionable fungicidal activity (1723) but, the irritation which it provokes on the mucous membranes makes its use as an aerosol difficult.

On the other hand, mists of organoborate complexes, which are less toxic than the borates themselves, suspended in the form of extremely fine particles of a few microns diameter, have produced good results.

Recently, a new technique for the dispersal of insecticides by the use of azeotropic mixtures (549, 550) has been published. Using the same method, it is possible to disperse various fungicides in the form of smokes and this may present a very useful means of readily disinfecting rooms used for storing foodstuffs.

Gaseous compounds, which are usually oxidizing or alkylating agents, have been recommended for purifying the air (331, 1105):

– ozone seems to give satisfactory results as a bacteriocide (981), but, such high concentrations are required for fungicidal activity that its use is very laborious;

– propylene oxide, which is active at concentrations of 800–2,000 mg/l, has been recommended for superficial disinfection, especially of powdered products (332);

– methyl bromide has a greater penetrating ability but its efficiency is low. Minimal activity is shown at 3,500 mg/l and it has a significant toxicity to man (1577, 1795);

– ethylene oxide, which is active at concentrations of 400–1,000 mg/l, is the most frequently used gaseous compound both for hospital equipment (842, 1059) and for food commodities (848, 1433, 1987). It has an optimum activity at a particular relative humidity (30%). It has been criticized as having an antivitamin activity and its use does require considerable care, for it is inflammable, which is why it is often diluted with inert gases (especially as a mixture of 10% ethylene oxide and 90% carbon dioxide);

– β-propiolactone (BPL) in the form of its vapour is more efficient than ethylene oxide (1106, 2463) and may be used at concentrations of 1–1.5 mg/l air. Its vesicant and lachrymatory properties, associated with a possible carcinogenicity, considerably reduce its practical use (2329, 2849).

Very recently, the use of diethylpyrocarbonate (DEPC) (2028) has been recommended. It breaks down to alcohol and carbon dioxide on hydrolysis and has been used successfully to sterilize wine (2601).

Sulphur dioxide, which is a good insecticide, has also been used as a bacteriostatic and fungistatic agent in, for example, the preservation of table grapes (1937).

The impregnation of wrapping paper with diphenyl or cardboard containers which slowly release ammonia have been used to increase the storage life of plums and analogous processes could be envisaged for other foods.

TRANSLATOR'S NOTE: Legislation covering the chemical preservation of food has changed rapidly during the last few years and has become more complex in Europe as individual countries within the European Economic Community attempt to harmonize their regulations. Compounds such as diethylpyrocarbonate have enjoyed a short spell in use but have now been prohibited as food preservatives as a result of more searching toxicological studies. In 1962 Great Britain completely revised her regulations to widen the classes of foods to which preservatives could be added and the antibiotics tetracycline, nisin, and nystatin could be used for specified foods under strictly limited conditions. However, in 1974, the Statutory Instrument 1974/1119 revoked permission to use tetracyclines in fish and nystatin is no longer used as a fungicide on the skin of bananas, although nisin may still be used in processed cheese.

An excellent review of the changes occurring in the legislation covering the chemical preservation of food in Great Britain, and its relationship with legislation in other countries, has been compiled by Jarvis and Burke (3013).

CONCLUSION

As the British mycologist, Dr. Austwick, recently wrote 'Our knowledge of the mycotoxicoses, associated with microfungi, is still very limited and new research is required to define the role of the fungi in many cases as well as the conditions of production of toxic compounds.'

Only cooperation, at an international level, between mycologists, doctors and veterinarians, biochemists and toxicologists will allow the accumulation of effective results in this field of study.

One purpose of this book is to draw attention to the importance of this problem, which concerns us directly.

Based in part on the programme suggested by Miyaki (1701) in Japan, the following programme of further research is recommended:

1. To study the ecology of mycotoxin-producing fungi which affect crops or contaminate food commodities.
2. Research on precise and rapid chemical and biological assays for mycotoxins in foods.
3. Research on mycotoxins as possible agents of human diseases as yet poorly understood.
4. An extension of the means of preventing the growth and dispersal of moulds.
5. Development of practical means of detoxifying foods suspected as containing mycotoxins.

As stressed by Butler (366); In countries said to be developed 'the presence of minute traces of pesticide residues in foods often causes considerable emotion, although the toxicity of some of them is even doubtful. It is advisable not to neglect those compounds such as mycotoxins which, as trace amounts in foods, may be harmful to the health of man.' A choice can be made.

CHAPTER 13

RECENT DEVELOPMENTS IN THE STUDY OF MYCOTOXINS

The last decade has seen a considerable increase in the international interest in mycotoxins, both as environmental factors in the health of man, and as the agents of disease in farm animals. Limits to the levels of mycotoxins acceptable in foods and animal feeds are beginning to appear in our legislation; thus Instrument No. 840 (1976), the Fertilizers and Feeding Stuffs (amendment) Regulations of the British Ministry of Agriculture, Food, and Fisheries, sets out limits for aflatoxin B_1. In ordinary feeding stuffs for cattle, sheep, and goats this level is 0.05 mg/kg; for pigs, poultry, young calves, and lambs 0.02 mg/kg; and for piglets and chickens, 0.01 mg/kg.

Although it may be hoped that such undoubted mycotoxicoses of man as alimentary toxic aleukia and ergotism are now mainly of historic interest, there is continued and justifiable concern about the role of aflatoxin in the aetiology of liver cancer. However, the description of aflatoxin B_1 as one of the most potent carcinogens in certain animal species must not be allowed to mask the fact that the aflatoxins are potent acute toxins, as the suffering of a large number of people in India currently bears witness to. Some of the case histories of acute aflatoxin poisoning in man, such as those recorded in Thailand (3064) and Uganda (2343), have been reviewed in the context of diseases of unknown aetiology such as Reye's syndrome (3034). The symptoms described in the case of a three-year-old boy from north-east Thailand, just before he died, corresponded very closely to those described by Reye, Morgan, and Boral (3049), namely fever, convulsions, vomiting, and disturbed respiratory rhythm. Observations from autopsy were also similar including cerebral swellings and damage of the liver. Although there is no doubt about the implication of aflatoxin and *Aspergillus flavus* in the cases studied in Thailand, a number of other toxigenic moulds have since been isolated from the foods involved in this case of poisoning, and the possibility cannot be ruled out that the symptoms observed arose from the interaction of a multiplicity of toxins.

Balkan (endemic) nephropathy

This strange disease, apparently confined to areas of Yugoslavia, Romania, and Bulgaria, was first described over twenty years ago. This illness, of uncertain aetiology, has caused considerable distress and, in the rural areas close to the Danube and its tributaries, it may claim over 20,000 victims. There has never been any evidence to implicate a bacterial infection, although there have been suggestions that an arborvirus or coronavirus may be involved. Several theories concerning the aetiology of Balkan nephropathy have been proposed and the theory that fungal toxins may be involved was first suggested by Dimitrov (2978) in 1960, taken up again by Barnes (2950) in 1967 and critically reexamined by Austwick (2949) in 1975. A careful study of the epidemiology of the disease has led to the suggestion that it may be associated with the mould *Penicillium verrucosum* var. *cyclopium* (2951). Cultures of a strain of this organism, isolated from maize stored in an area in Bulgaria where the disease occurs, have consistently induced kidney damage when force fed to rats over a period of 20 days. The detailed changes described in the kidney of the rat are said to closely resemble those observed in people suffering from Balkan nephropathy who, in fact, suffer from a slowly progressive renal failure.

The mould implicated, which is one of the commonest species found on the food commodities of the endemic areas, is known to be capable of producing the neurotoxic penitrems, the hepato- and nephrotoxic ochratoxins and cyclopiazonic acid, as well as penicillic acid, which has been suspected as being carcinogenic. Elling and Krogh (2980) describe Balkan nephropathy as a chronic interstitial nephritis characterized by hyalinization of glomeruli, tubular loss, and interstitial fibrosis. They suggest that ochratoxin A is a likely culprit as the toxin most involved, for it is known to give rise to a similar pattern in porcine nephropathy in Denmark (3019).

Using ^{14}C labelled ochratoxin A, Chang and Chu (2963) have demonstrated that a considerable proportion of the toxin is bound to serum proteins in the rat. As well as being an acutely poisonous compound, ochratoxin A is a relatively stable molecule and may remain in animal tissues for some time. Indeed, there has been some concern that ochratoxin A may find its way into the human food chain via animal products. It has been demonstrated that as much as 67 ppb of ochratoxin A may be present in the meat and organs of pigs which have suffered mycotoxic nephropathy (3020), although contamination of meat with this particular mycotoxin may be avoided by feeding the pigs ochratoxin-free diets during the last four weeks before slaughter, that being the time that the tissues take to clear the toxin (3021).

Although there is still a great deal to learn about the detailed toxicology of the now large number of toxic fungal metabolites, studies have entered a second stage of complexity as the possibilities for interactions of mycotoxins with each other, and with nutritional factors in general, are appreciated. It has certainly been demonstrated that rubratoxin B will potentiate the acute toxicity of the aflatoxins (718) and Hamilton (3002) has reviewed some of the interactions known to occur between dietary factors and the effects of mycotoxins such as aflatoxins. Thus a high lipid content may have a sparing effect on mortality from aflatoxicosis, although deficiencies in vitamins A, D, or riboflavin usually increase the sensitivity of animals to the effect of aflatoxin.

Recent approaches to the analysis of mycotoxins

A certain indication of the seriousness with which mycotoxins are now considered is the considerable attention given to their analysis. The requirements for sensitive and specific analyses arise, not only from the need to assess the presence of these compounds in our environment, but also because the legislative machinery of several countries has set upper limits to their presence in human foods and animal feeds. Modern techniques of chemical analysis have made it possible to separate, detect, and quantitatively estimate the majority of known mycotoxins. Thin layer chromatography, combined with processes for removing many of the impurities extracted with the toxic metabolites, and a preliminary separation of the mycotoxins from each other, makes it possible to analyse complex mixtures of mycotoxins in a wide range of commodities.

The multimycotoxin screening method described by Roberts and Patterson (3052), combined with confirmatory tests, allows the detection of any of 13 mycotoxins which may be present in animal feeds. A less complex screening method has been described for the detection of aflatoxin, ochratoxin, patulin, sterigmatocystin, and zearalenone in cereals (3014), whereas a combination of TLC and ultraviolet densitometric estimation may be used for the analysis of citreoviridin in foods (2982).

High pressure liquid chromatography (HPLC) has been widely applied, with considerable success, to the analysis of mycotoxins. A complete quantitative analysis of aflatoxins B_1, B_2, G_1, and G_2 in cottonseed products may be achieved in less than one and a half hours by an initial extraction with aqueous acetone, clean up with lead acetate, partition into methylene chloride, purification on a small silica gel column, and final analysis on a microparticulate silica gel column using water-saturated chloroform/cyclohexane/acetonitrile as the eluting agent on an HPLC coupled to a detector using u.v. radiation of 365 nm (3046). Recovery of aflatoxins at the parts per billion level has been satisfactorily achieved from wines by a reversed phase HPLC and the method used was sensitive down to 0.02 μg/l (3068). Using HPLC, Engstrom *et al.* (2983) have shown that it is possible to detect as little as 5 ng rubratoxin B, 1 ng zearalenone, 0.04 ng ochratoxin A and 1 ng each of aflatoxins B_1 and G_1 in a mixture containing these mycotoxins as well as patulin and penicillic acid, and by coupling this technique to laser-induced fluorescence detection it is possible to detect incredibly minute quantities of fluorescent compounds (2977). As little as 7.5×10^{-12} g of aflatoxin could be quantitatively estimated by this method. Aflatoxin B_1 has a poor fluorescent quantum yield in solution but, by converting it into the B_{2a} derivative with trifluoroacetic acid or hydrochloric acid, the fluorescence is considerably enhanced in hydrogen bonded solvents. After eluting the derivative from a reverse phase column, the 325 nm output of a helium–cadmium laser is focused into a suspended droplet of eluant and the resulting fluorescence measured using a phase sensitive detector. It is interesting to note that the analysis of rubratoxin B by HPLC may give rise to a second peak presumably corresponding to the artefact which frequently arises during the TLC analysis of this mycotoxin (1769).

It is probably even more important that care be taken over sample preparation when using HPLC and clean up procedures have been described for a number of commodities (3023).

A particularly sophisticated approach to the analysis of mycotoxins has been in the use of field desorption mass spectrometry (3069), a technique which allows the formation and detection of the parent molecular ion of even relatively non-volatile compounds with very little, if any, fragmentation of the molecule. The sample to be analysed is deposited from a solution onto a tungsten wire on which has been grown a large number of highly branched fine carbonaceous microneedles. The sample is introduced into the ion source of the mass spectrometer through a vacuum lock and is maintained at a considerable voltage difference from the negatively charged extraction plate. The very high electric field which then occurs at the tips of the microneedles is considered to cause a quantum mechanical tunnelling of an electron from the previously neutral molecules of the sample leaving them as positively charged ions which are emitted and can be focused by the traditional methods of a mass spectrometer. The power of this technique is demonstrated by the ability to obtain the parent molecular ion, $C_{26}H_{30}O_{11}^{+}$, at m/e 518 for rubratoxin B, the presence of which is very difficult to demonstrate by traditional mass spectrometry because of the instability of the molecule at the temperatures normally required to produce a spectrum (3035). The normal mass spectrum of most compounds usually contains a complex array of peaks derived from fragmentation of the molecule (indeed, this information in invaluable in the deduction of the structure of a pure compound) and the spectrum derived from a mixture would usually be very difficult to interpret. If, however, it is known that the peaks observed are due to the parent molecular ions of constituents of a mixture, then the presence or absence of particular mycotoxins may be established.

The wide range of analytical methods now available have been used to screen a large number of samples in surveys covering large areas, especially for the presence of zearalenone, ochratoxin, and aflatoxin (3007, 3066, 3075). Such surveys can lead to a reassessment of animal husbandry techniques and feed storage practices. Indeed, the survey of wheat and soyabeans harvested in a number of American states during 1975 led Shotwell *et al.* (3066) to ask whether it would not be sensible to replace crops such as wheat and maize with soyabeans in those areas where mycotoxin producing fungi were a particular problem. They were unable to detect zearalenone, aflatoxin, or ochratoxin in any sample of soyabean although zearalenone was detected in, for example, 19 out of 42 samples of wheat collected from Virginia.

When analytical methods are required to back up legislative standards, then reproducibility, sensitivity, precision, and specificity are all essential and a number of detailed procedures are described in the twelfth edition of the *Official Methods for Analysis* issued in 1975. It is frequently necessary to have different extraction and clean up methods for different commodities and, although it would be desirable to have a method applicable to a wide range of foods, food commodities, and animal feeds, it has been demonstrated that the method giving good results for one commodity may give poor results when applied to another (3065).

Advances in analytical techniques, coupled with an increasing sophistication in the use of labelling with ^{3}H, ^{13}C, ^{14}C, and the development of ^{13}C nuclear magnetic resonance spectroscopy, have led to some interesting observations on mycotoxins. Nowhere have these advances been so spectacular as in the study of aflatoxin and its biosynthesis, on the one hand, and mode of action on the other.

Biosynthesis of aflatoxins and related compounds

The elegant studies of the ^{13}C nuclear magnetic resonance spectrum of aflatoxin B_1 produced by using either 1-^{13}C or 2-^{13}C labelled acetate as a precursor, have confirmed the polyketide origin of this metabolite (3010) and similar studies have confirmed the polyketide origin of averufin and versicolorin B (2953).

For some time now there have been speculations about the interrelationships between the two groups of anthraquinone metabolites, i.e. those containing an unbranched C_6 side chain and those containing the fused difuran ring system derived from a branched C_4 side chain, the xanthone metabolites, such as sterigmatocystin and the austocystins, and the aflatoxins (3033).

By feeding ^{14}C labelled aflatoxins to cultures of *Aspergillus flavus* it has been possible to confirm that the G series of aflatoxins are derived from the B series (3005). The use of mutants of *Aspergillus parasiticus,* blocked at various stages in

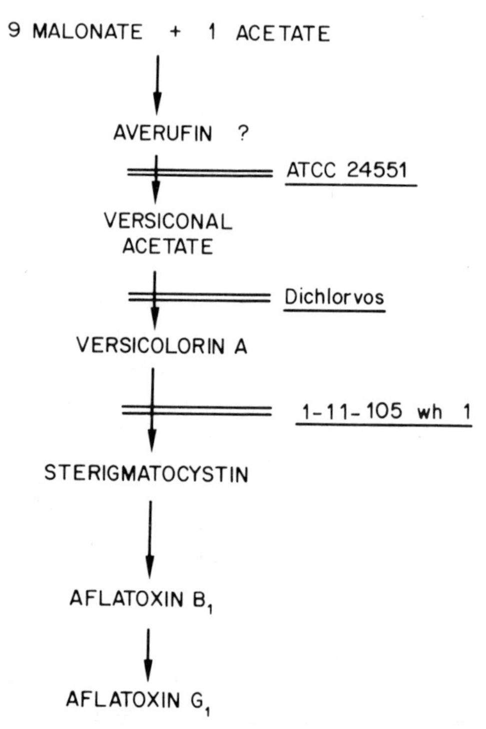

Figure 157. Some steps in the biosynthesis of aflatoxins

the biosynthesis of aflatoxins, coupled with the use of dichlorvos (dimethyl-2,2-dichlorovinyl phosphate), which causes the accumulation of versiconal acetate, has provided confirmation of the overall scheme indicated in Figure 157 (3067).

The mutant, ATCC 24551, which accumulates averufin, is able to efficiently incorporate ^{14}C labelled versiconal acetate, versicolorin A, and sterigmatocystin into aflatoxins B_1 and G_1. Dichlorvos inhibits the incorporation of versiconal acetate, but not versicolorin A or sterigmatocystin.

The suggestion that the unbranched C_6 side chain could be the precursor of the branched C_4 fused difuran ring system, by a rearrangement combined with the loss of a C_2 fragment was rationalized by Thomas who suggested a possible, but hypothetical, mechanism (3033). Evidence that a C_{20} anthraquinone metabolite, and hence a decaketide rather than a nonaketide chain, is the precursor of all the C_{18} metabolites containing the fused difuran ring was provided by the failure to incorporate a number of labelled model C_{18} compounds, considered as possible intermediates between a nonaketide chain and the aflatoxins, whereas the C_{20} compound, norsolorinic acid (Figure 158), was incorporated into aflatoxin (3009).

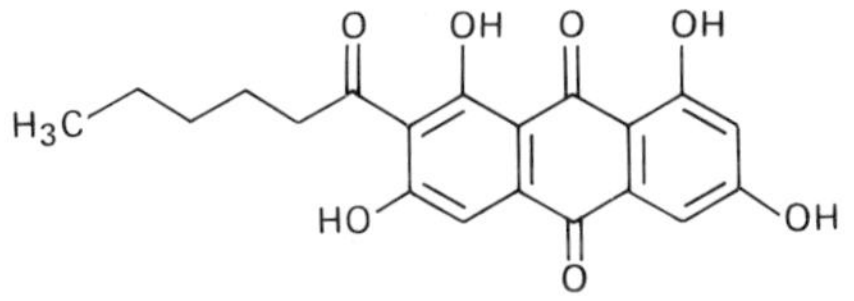

Figure 158. Norsolorinic acid

Versiconal acetate (Figure 159) is a key intermediate between the C_{20} group of anthraquinone derivatives, such as averufin, and the C_{18} group such as the versicolorins. It is an orange pigment found to accumulate when aflatoxin production by strains of *Aspergillus flavus* and *A. parasiticus* is inhibited by the addition of dichlorvos.

Studies using 1-^{13}C, 2-^{13}C and 1,2^{13}C labelled acetate have provided ^{13}C labelled versiconal acetate which has been used for a detailed examination by nuclear magnetic resonance spectroscopy (2972, 2989). These studies have not only confirmed the structure of this important intermediate, but have added considerable evidence to the evidence for the polyketide route of biosynthesis, and the uniform intensities of the peaks observed in the ^{13}C n.m.r. indicate that all the

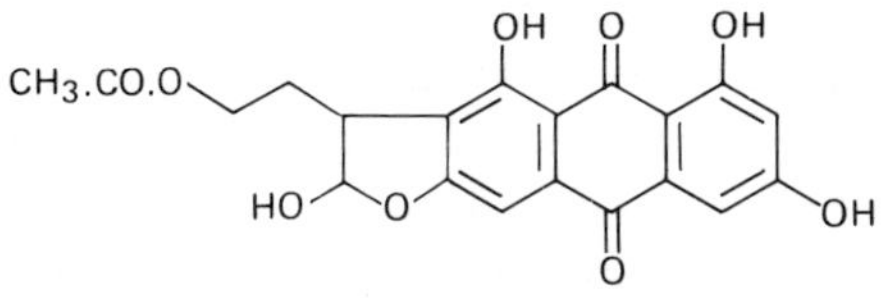

Figure 159. Versiconal hemiacetal acetate

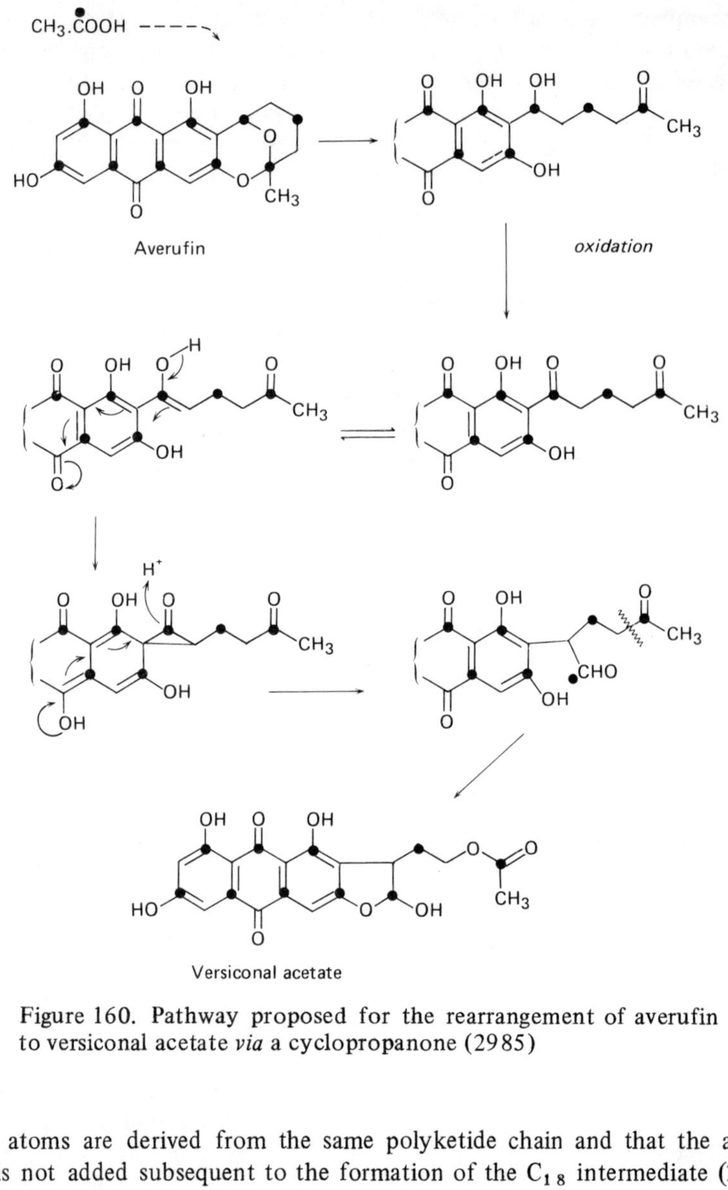

Figure 160. Pathway proposed for the rearrangement of averufin to versiconal acetate *via* a cyclopropanone (2985)

carbon atoms are derived from the same polyketide chain and that the acetate group is not added subsequent to the formation of the C_{18} intermediate (2972). These studies have thus provided substance for hypotheses concerning the generation of the branched C_4 chain of the bisfuran ring system from the straight C_6 chain. One such hypothesis (2989) involves a rearrangement to a cyclopropanone with subsequent cleavage of the three membered ring (Figure 160). A similar mechanism had previously been suggested by Tanabe *et al.* (3078) based on observations on the polyketide biosynthesis of the branched chain metabolite, aspyrone, by *Aspergillus melleus* (Figure 161). However, the mechanism which the latter workers proposed for the biosynthesis of versiconal acetate gives rise to a

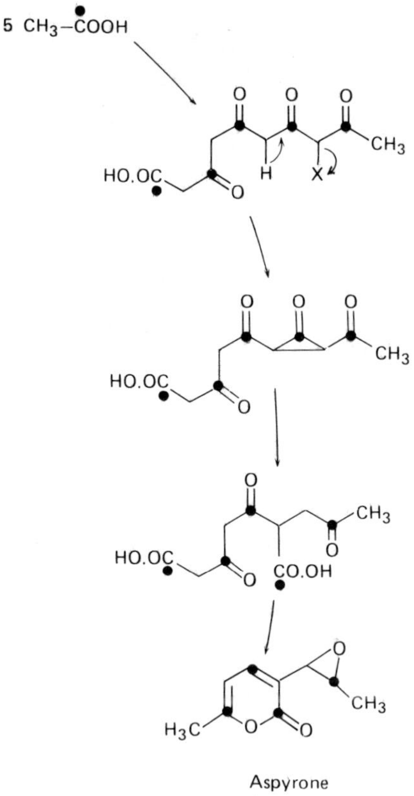
5 CH_3–COOH
HO.OC
CH_3
H
X
CO.OH
H_3C
Aspyrone

Figure 161. Aspyrone and its biosynthesis

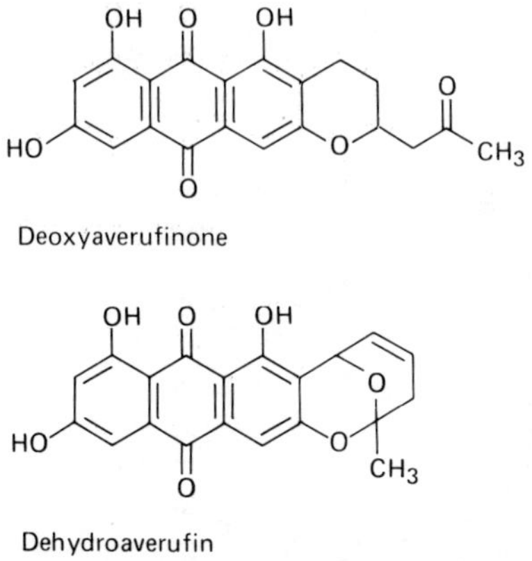
OH
HO
CH_3
Deoxyaverufinone
Dehydroaverufin

Figure 162. Recently isolated metabolites from *Aspergillus versicolor*

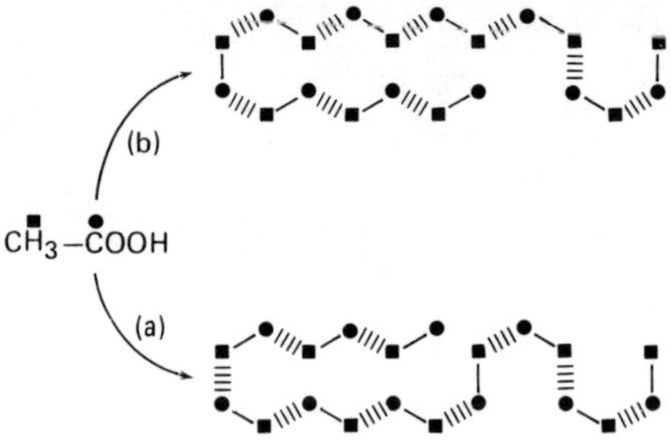

CH3–COOH

(b)

(a)

Figure 163. Possible folding patterns for for decaketide chain

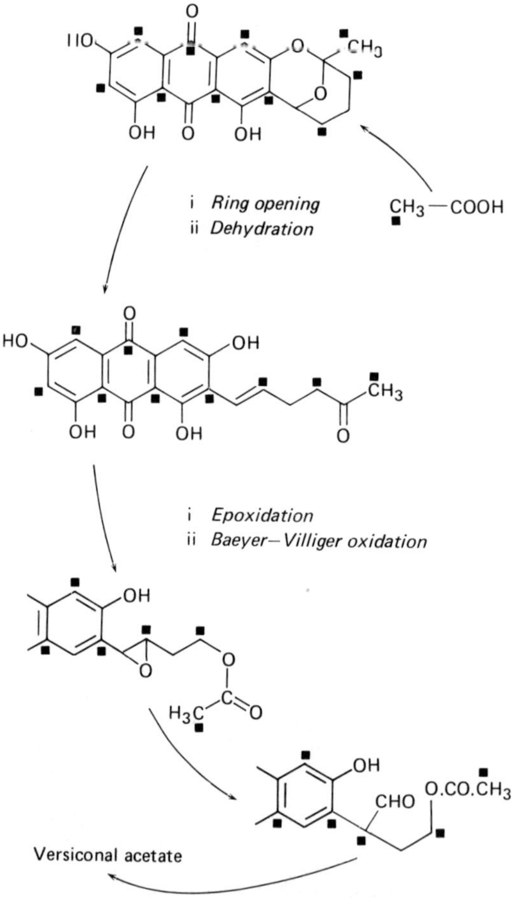

HO

OH O OH

O

CH3

i *Ring opening*
ii *Dehydration*

CH3—COOH

HO

OH

OH O OH

CH3

O

i *Epoxidation*
ii *Baeyer–Villiger oxidation*

OH

O

O

H3C

C=O

OH

CHO O.CO.CH3

Versiconal acetate

Figure 164. Rearrangement of averufin to versiconal acetate via an epoxide intermediate (2993).

carboxylic acid which then has to be reduced to provide the aldehyde moiety in versiconal acetate hemiacetal, whereas the similar hypothesis of Fitzell *et al.* (2989) leads directly to the generation of an aldehyde function. Both hypotheses involve the cleavage of the acetate group as a separate stage independent of the rearrangement to a branched chain. The absence of any known metabolites with a branched C_6 chain indicates that the cleavage of the acetate group may be concerted with the rearrangement itself, and may even be the driving force for it. The continuing search for metabolites which may be intermediates has yielded deoxyaverufinone and dehydroaverufin, Figure 162 (2952).

The debate concerning the polyketide origin of the aflatoxins has thus moved on to the details concerning the rearrangement of a decapeptide chain and one of those details is the initial pattern of folding which can yield an averufin-like anthraquinone. Of the two possible folding patterns (Figure 163), Seto, Cary, and Tanabe (3062) concluded that route (a) was the most likely from their studies of the ^{13}C–^{13}C coupling patterns in the n.m.r. spectrum of sterigmatocystin produced from $^{13}CH_3.^{13}COOH$. It was, however, subsequently demonstrated that these conclusions were probably based on false assignments of signals in the n.m.r. spectrum (3039) and the consensus of opinion seems to favour folding pattern (b) (2998).

An alternative mechanism for generating the branched C_4 system of the difuran rings, involving the intermediacy of an epoxide (Figure 164) has been proposed (2997).

The ^{13}C nuclear magnetic resonance spectra of twelve difuran metabolites related to either aflatoxin or sterigmatocystin, including aflatoxin D_1 (Figure 165), have now been recorded (2973) and undoubtedly this valuable technique will continue to provide new insight into the details of the biosynthesis of these important fungal metabolites.

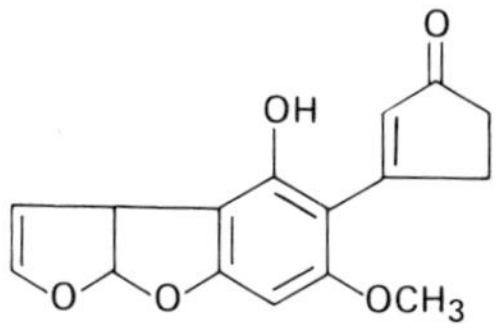

Aflatoxin D_1

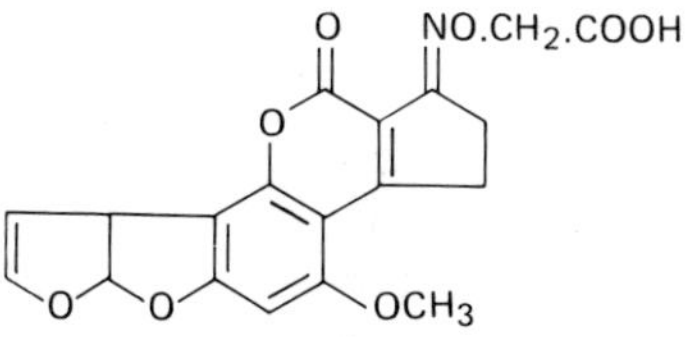

Aflatoxin B_1-1-(*O*-carboxymethyl)oxime

Figure 165. Newly described derivatives of aflatoxin B_1

Genetics of aflatoxin biosynthesis

The genetics of aflatoxin biosynthesis has been difficult to study because the *Aspergillus flavus* group does not have any known sexual stages. However, the parasexual cycle has been described in *A. flavus* itself (3040) and linkage groups have been established (3041). Mutants producing different levels of aflatoxins were first obtained by u.v. irradiation of a single parent culture (3025) and, although none produced aflatoxin G, production of B_1 was always greater than B_2, both being produced over a wide range of concentrations (B_1 1.3–967 μg/ml and B_2 0–30 μg/ml). Treatment with nitroguanidine did eventually yield a stable mutant consistently producing more B_2 than B_1 (3043) and a genetic analysis of this mutant revealed that a single gene was probably responsible for the difference between strains in which $B_2 < B_1$ and those in which $B_2 > B_1$, the former being dominant to the latter (3042). Although a wild type isolate producing only aflatoxin B_2 had previously been described (3059), during subculturing it rapidly lost its ability to produce any aflatoxin at all.

Immunological assays of aflatoxins

Although most studies involving the analysis of aflatoxins and related metabolites necessitate an extraction and purification of the compounds before quantitative analysis, some progress has been made in detecting the presence and distribution of aflatoxins in tissues by the preparation of antibody against aflatoxin B_1 after binding it to a serum protein (2967) and using such specific antibodies for radioimmunoassay (3022) or immunocytochemical techniques (3024). Aflatoxin may be found to serum protein by using the *O*-carboxymethyl oxime (Figure 165), a method for the preparation of which has been described (2966). Antibody obtained from rabbits after injecting them with bovine serum albumin conjugate of aflatoxin B_1 showed the greatest binding to aflatoxin B_1 itself, less against B_2 and G_1, least against G_2, M_1, and aflatoxicol, and did not bind to sterigmatocystin, coumarin or 4-hydroxycoumarin.

Studies on the mode of action of aflatoxins

The toxicology of aflatoxin has been intensively investigated, particularly the relationship between the acute and chronic toxicity, with the implications for carcinogenicity of the latter. Further evidence that the aflatoxins may be carcinogenic to man arises indirectly from epidemiological studies (3063) and from reports of cases of primary liver cancer in primates such as Rhesus monkeys (2943, 2944, 2996, 3080).

As long ago as 1938, Haddow suggested that an early stage in chemical carcinogenesis may involve the induction of cells resistant to the acute toxicity of a carcinogen and such a phenomenon has been demonstrated in the case of aflatoxin and the rat. Judah, Legg, and Neal (3015) showed that rats fed on a diet containing 4–5 ppm of aflatoxin B_1 for six weeks, followed by a return to a control diet, developed hepatocarcinoma in 100% of cases. However, this process occurred in two stages for, during the first three weeks histological examination showed that the liver suffered an acute toxicosis and, if the animals were removed from the toxic diet during this period, insignificant neoplasia occurred although there was significant damage to the liver. During the second three weeks of exposure to the toxic diet there was a proliferation of both parenchymal and non-parenchymal cells to replace the dead cells. These new cells were found to be resistant to the acute toxicity of aflatoxin B_1, but changes occurred leading to the induction of carcinoma. It was also shown that animals fed on subacute levels of aflatoxin were much more resistant to doses of aflatoxin which killed control animals.

Some animals, such as the mouse, are relatively insensitive to aflatoxin and it has been shown that in the mouse, although treatment with a sublethal dose of aflatoxin B_1 results in an immediate inhibition of RNA synthesis in the liver, recovery of synthesis is rapid and is even followed by a mild stimulation of protein synthesis in the liver (3036).

It has also been known for some time that some physiological changes in animals may protect them from liver cancer induced by aflatoxin. Thus hypophysectomized rats are more resistant (938) and treatment with adrenocorticotropin has the same effect (2964). Exposure of rats to a temperature of 4–5 °C made them considerably more resistant to the acute toxicity of aflatoxin compared with rats maintained at 20–21 °C (3081). Alterations of hormone balance affect, not only the metabolism of compounds such as aflatoxins, but also phenomena such as the binding of ribosomes to the endoplasmic reticulum and the formation of polysomes involved in balanced protein biosynthesis. Data have been provided which suggest an interference in the maturation of certain forms of ribosomal RNA by aflatoxin (3068).

Protein deficiency in the diet has been shown to influence the *in vivo* binding of aflatoxin B_1 to the macromolecules of rat liver (3047). Animals fed on protein-deficient diets containing aflatoxin B_1 consistently showed a 70% decrease in the binding of aflatoxin to such nuclear macromolecules as chromatin protein and DNA, this being associated with an increased toxicity but decreased carcinogenicity of aflatoxin B_1.

Many of these observations can be explained if the aflatoxin molecule itself is not the carcinogen, or even the acute toxin, but has to be metabolized in some way to an activated derivative. Thus the well documented sex difference in the sensitivity of the rat to the carcinogenic activity of aflatoxin may be due to the ability of microsomes from male animals to metabolize and activate aflatoxins more readily than those isolated from females (2999). It was demonstrated in 1973 that

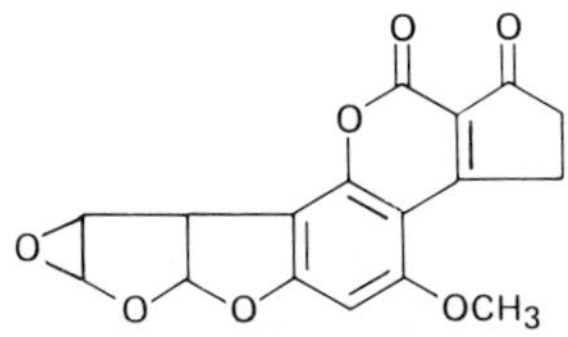

O O O O O OCH3

Figure 166. 2,3-Epoxide of aflatoxin B_1

liver microsomes brought about a change in the molecule of aflatoxin B_1 which produced a derivative with the property of inducing mutations in DNA repair deficient bacteria (2994). It had already been shown by Garner *et al.* (2993) that such a metabolite was actually lethal to *Salmonella typhimurium* TA 1530. Having shown that neither aflatoxin B_1, nor any of its then known metabolites had the same effect, these authors suggested that the activity was due to the formation of a new derivative which they suggested could be the 2,3-expoxide (Figure 166), a suggestion made a few years earlier by Schoental (2280). A combination of activation by a liver homogenate and genetic changes in a sensitive bacterium has been proposed as a useful test system for mutagens and possibly carcinogens (2945). Metabolites of aflatoxin B_1 have also been shown to promote the reactivation of u.v. damaged phage λ in the cells of *Escherichia coli* K 12, another phenomenon which lends itself as a possible test system for the presence of potential mutagens and carcinogens (3057).

Patterson (3044) provided some interesting data as the basis for the hypothesis that differences in metabolism were a factor determining the observed differences among animal species in their response to aflatoxins (1933, 1934, 3045). Roebuck and Wogan (3054) have shown that the supernatant fraction of the livers from a wide range of animal species, including man, differed very much from species to species in the *in vitro* metabolism of aflatoxin B_1, but they were unable to demonstrate any consistent pattern for correlating these differences with the known toxic response *in vivo.*

It has now been shown that one of the metabolites of aflatoxin B_1, the B_{2a} compound, readily reacts with free amino groups of proteins (2948). It is

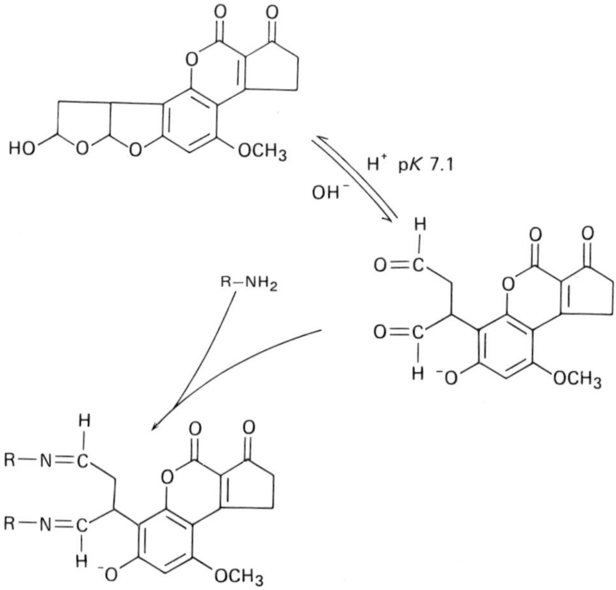

Figure 167. Schiff base formation by aflatoxin B_{2a}

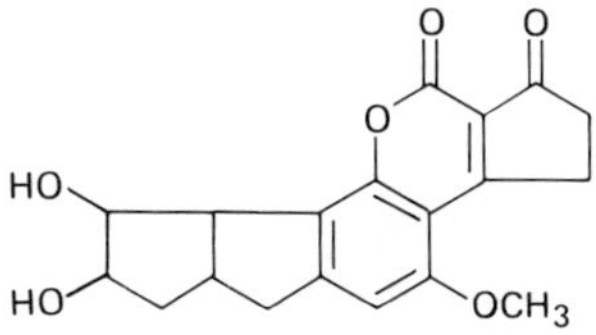

Figure 168. 2,3-Dihydro-2,3-dihydroxyaflatoxin B_1

suggested that this derivative is in equilibrium with the dialdehyde form which may react to give Schiff bases (Figure 167). Another metabolite, which is almost certainly the epoxide, interacts with rat liver DNA and ribosomal RNA to form a compound which, on mild acid hydrolysis, gave 2,3-dihydro-2,3 dihydroxyaflatoxin B_1 (Figure 168) (3076). A study of DNA synthesis and repair in human diploid fibroblasts showed that an interaction with aflatoxin B_1 would only occur in the presence of an activating system containing rat liver microsomal enzymes (3058).

Further evidence for the involvement of the epoxide in the carcinogenic properties of aflatoxin B_1 has come from the demonstration that peracid oxidation of aflatoxin B_1 could replace the microsomal activation system, now considered to involve a mixed function oxidase, in the production of nucleic acid adducts (3029). Although the epoxide itself has not been identified and characterized, an elegant study has led to the isolation and structural determination of the major DNA adduct as a covalent derivative between aflatoxin and guanine (Figure 169) (2986).

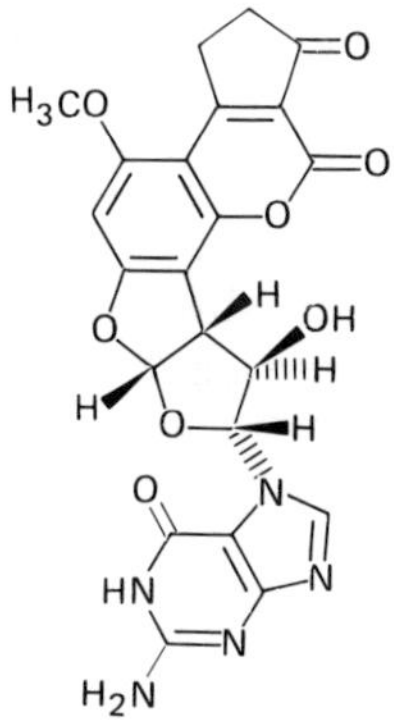

Figure 169. Guanine-aflatoxin derivative

One of the products of the metabolism of aflatoxin B_1 by animal tissues is aflatoxicol (Figure 170) and the toxicology and further metabolism of this compound have been studied. These studies have been made easier by the description of an efficient method for the preparation of aflatoxicol by reduction of aflatoxin B_1 with zinc borohydride (3008). It has been shown that some animals, including the monkey and man, are able to reconvert aflatoxicol to aflatoxin B_1 by a microsomal enzyme system linked to NADP (3056), whereas a

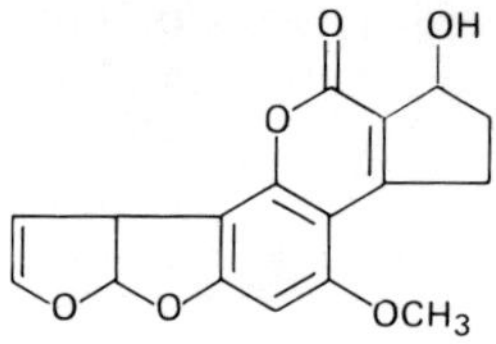

Figure 170. Aflatoxicol

dog liver microsomal preparation will convert aflatoxicol into aflatoxicol M_1 (Figure 171), a derivative that could equally be produced by a reductase in the cytosol of rabbit liver (3055).

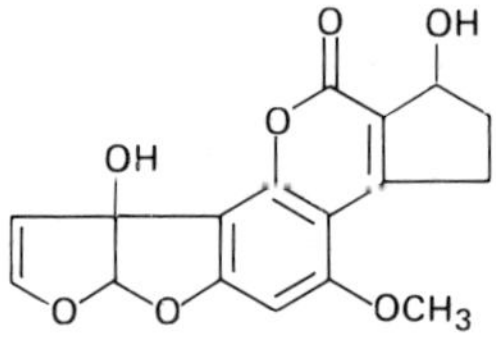

Figure 171. Aflatoxicol M_1

Important though it is to have a thorough understanding of the mode of action of aflatoxin, it is also necessary to know whether it is possible to treat contaminated animal feeds so as to render them innocuous without destroying their nutritional value. The ammonification process would still seem to be an acceptable method of treating commodities such as maize. Using the rainbow trout, which is extremely sensitive to the carcinogenic activity of aflatoxin, it has been demonstrated that maize, contaminated with 180 μg/kg, could be rendered harmless without any loss of nutritional value by treating with sufficient ammonia to give a concentration of 1.5% based on dry weight (2956). The maize was then adjusted to a water content of 12–17%, held overnight at 25 °C and then for 12 days in a forced convection dryer at 49 °C. Aqueous ammonia reacts with coumarin

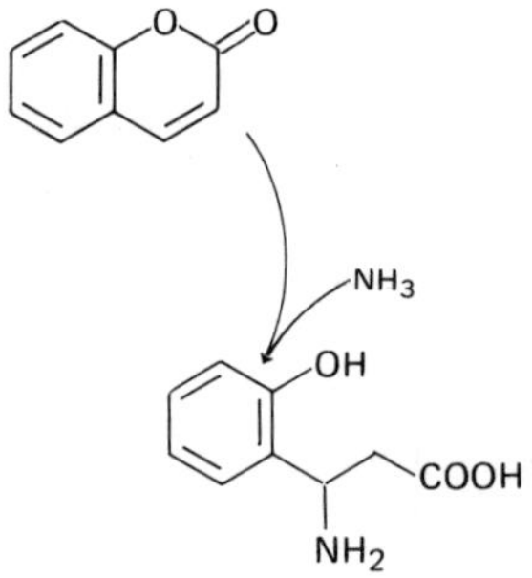

Figure 172. The formation of β-amino hydrocoumaric acid

to form β-aminohydrocoumaric acid (Figure 172), a reaction which has been suggested as a model for the reaction of ammonia with aflatoxin (2954).

Tremorgenic mycotoxins

One of the growth areas in the study of mycotoxins has been the discovery and characterization of tremorgenic mould metabolites (Table 34). Wilson and Wilson (2839) first described the initiation of body tremors in mice receiving crude extracts from a strain of *Aspergillus flavus* and subsequently reported the isolation of a compound with similar neurotoxic properties from *Penicillium cyclopium* (2840). Wilson (2832) reviewed the studies concerning the production of the tremorgen produced by a group of fungi, including *Penicillium cyclopium, P. crustosum,* and *P. palitans.* The molecular formula which he and his colleagues established for this tremorgen, which has been called penitrem A, is $C_{37}H_{44}ClNO_6$ and this has since been confirmed by Malaiyandi *et al.* (3027) who

Table 34. Fungi known to produce tremorgenic mycotoxins

Organism	*Tremorgen*	*Formula*
Aspergillus caespitosus	verruculogen	$C_{27}H_{33}N_3O_7$
	fumitremorgin B	$C_{27}H_{33}N_3O_5$
Aspergillus clavatus	tryptoquivaline	$C_{29}H_{30}N_4O_7$
	nortryptoquivaline	$C_{28}H_{28}N_4O_7$
	deoxytryptoquivaline	$C_{29}H_{30}N_4O_6$
	deoxynortryptoquivaline	$C_{28}H_{28}N_4O_6$
	nortryptoquivalone	$C_{26}H_{24}N_4O_6$
	deoxynortryptoquivalone	$C_{26}H_{24}N_4O_5$
Aspergillus flavus	aflatrem	$C_{32}H_{39}NO_4$
	aflavinine	$C_{28}H_{39}NO$
Aspergillus fumigatus	fumitremorgin A	$C_{32}H_{41}N_3O_7$
	fumitremorgin B	$C_{27}H_{33}N_3O_5$
	verruculogen	$C_{27}H_{33}N_3O_7$
	TR-2	$C_{22}H_{27}N_3O_6$
Claviceps paspali	paspalinine	$C_{27}H_{31}NO_4$
Penicillium crustosum	penitrem A	$C_{37}H_{44}ClNO_6$
Penicillium cyclopium	penitrem A	$C_{37}H_{44}ClNO_6$
Penicillium palitans	penitrem A	$C_{37}H_{44}ClNO_6$
Penicillium paraherquei	verruculogen	$C_{27}H_{33}N_3O_7$
Penicillium paxilli	paxilline	$C_{27}H_{33}NO_4$
Penicillium piscarium	verruculogen	$C_{27}H_{33}N_3O_7$
	fumitremorgin B	$C_{27}H_{33}N_3O_5$
Penicillium verruculosum	verruculogen	$C_{27}H_{33}N_3O_7$

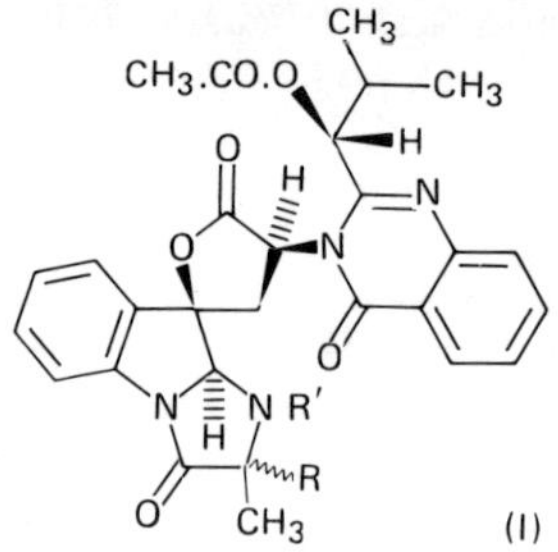

(I)

I	R	R′	
	CH_3	OH	Tryptoquivaline
	H	OH	Nortryptoquivaline
	CH_3	H	Deoxytryptoquivaline
	H	H	Deoxynortryptoquivaline

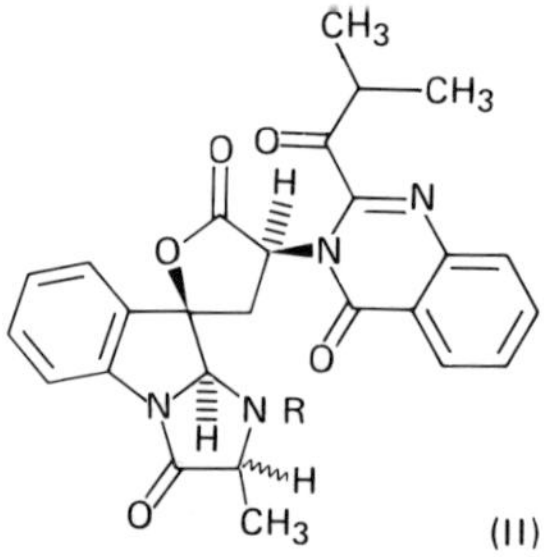

(II)

II	R	
	OH	Nortryptoquivalone
	H	Deoxynortryptoquivalone

Figure 173. Tryptoquivaline and related metabolites

investigated the large-scale production of penitrem A and demonstrated the production of a second toxin, $C_{37}H_{44}ClNO_5$. Penitrem A was first isolated from three distinct strains, two of which were associated with outbreaks of disease in farm animals, sheep on the one hand and horses on the other.

Since those early reports a number of tremorgenic metabolites have been characterized including tryptoquivaline and related compounds (Figure 173) from *Aspergillus clavatus* (2958, 2968), the fumitremorgins (Figure 174) of *Aspergillus fumigatus* (2979, 3087), and verruculogen (Figure 175) from *Penicillium verruculosum* (2988) and *P. paraherquei,* a species which appears to be particularly widespread in soils and from agricultural commodities of Papua, New Guinea, and other countries of the tropical and subtropical areas of the pan-Pacific (3088). Verruculogen has recently been identified as one of the toxic metabolites isolated from strains of *Aspergillus fumigatus* growing in moulded maize silage (2970). Both *A. caespitosus* and *Penicillium piscarium* are able to produce the two tremorgens,

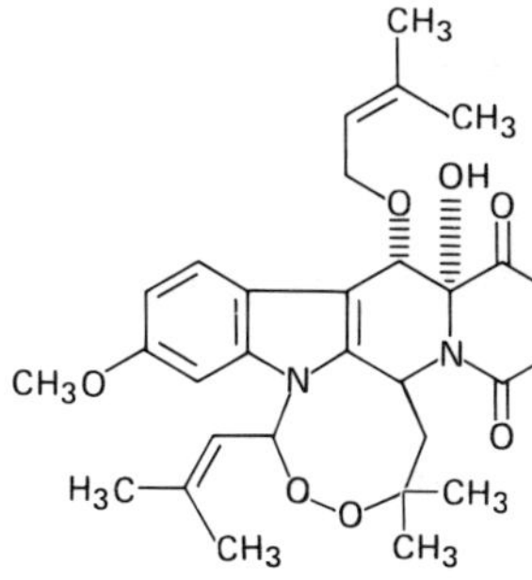

Fumitremorgin A

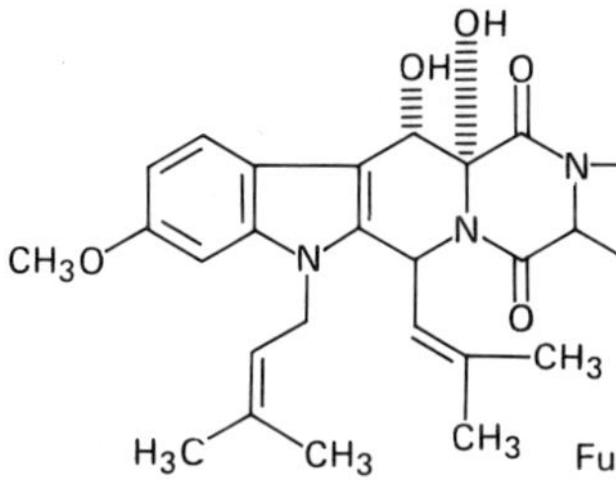

Fumitremorgin B

Figure 174. The fumitremorgins

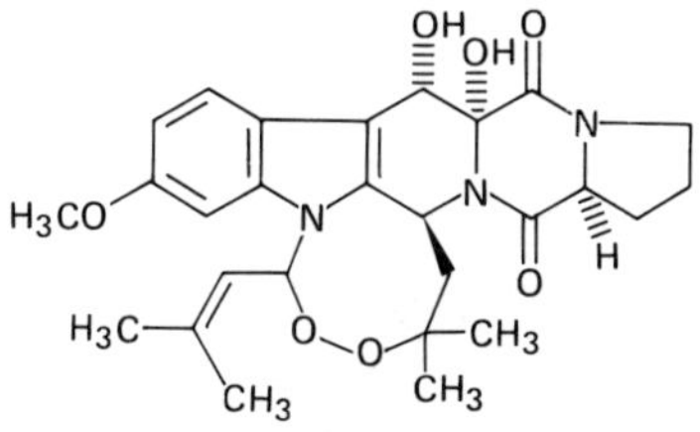

Figure 175. Verruculogen

verruculogen and fumitremorgin B, at the same time and it has been suggested that the coproduction of the two metabolites by relatively unrelated species indicates a common biosynthetic origin (2991). These metabolites are all complex derivatives of indole and presumably involve tryptophan in their biosynthesis.

The tremorgenic toxin from *Aspergillus flavus,* known as aflatrem, $C_{32}H_{39}NO_4$, has now also been characterized (Figure 176) and shows a remarkable similarity to paspalinine, $C_{27}H_{31}NO_4$, (Figure 177) a tremorgenic metabolite of the ergot fungus, *Claviceps paspali* (2990). *Penicillium paxilli,* isolated from insect damaged pecans, also produces a related tremorgenic metabolite, paxilline, $C_{27}H_{33}NO_4$ (Figure 178), reported to be active at a dose of 25 mg/kg (2971). It is interesting to note that the tremorgenic activity of aflatrem, paspalinine, and paxilline is

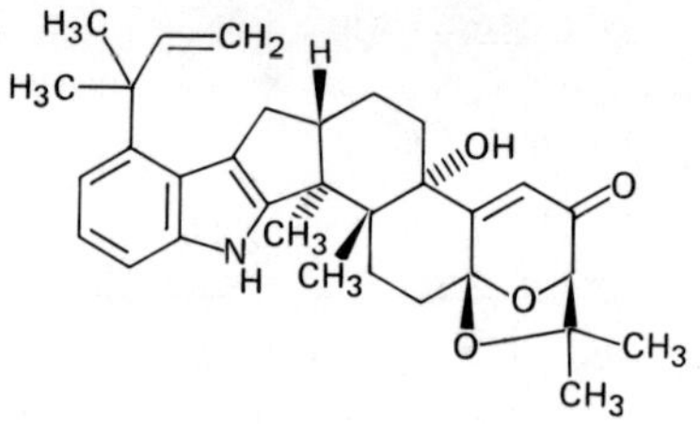

H3C
CH2
H3C
H
OH
O
N
H
CH3
CH3
O
O
CH3
CH3

Figure 176. Aflatrem

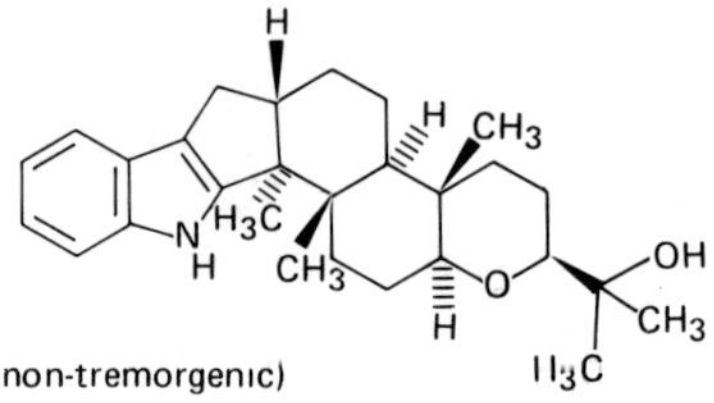

Paspaline (non-tremorgenic)

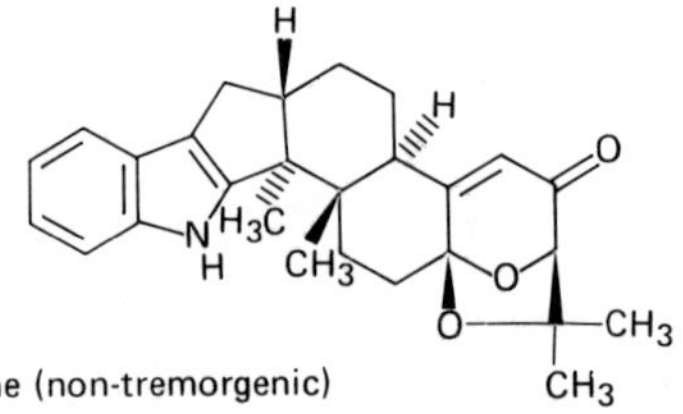

Paspalicine (non-tremorgenic)

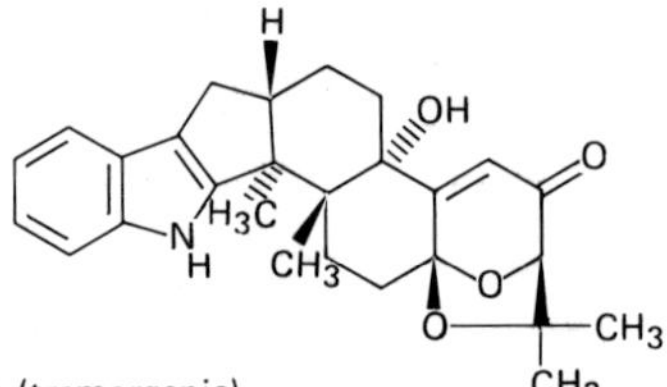

Paspalinine (tremorgenic)

Figure 177. Metabolites of *Claviceps paspali*

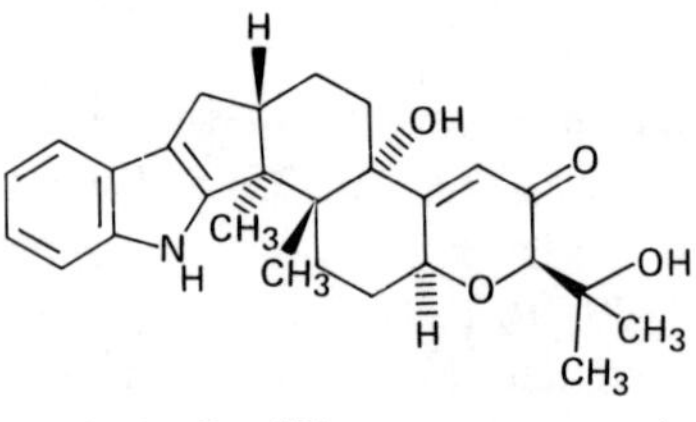

Figure 178. Paxilline, a tremorgenic metabolite of *Penicillium paxilli*

associated with the presence of a tertiary hydroxyl group which is absent from the non-tremorgenic, but structurally related compounds paspaline and paspalicine.

Claviceps paspali is considered to be responsible for paspalum staggers, also known as dallisgrass poisoning, which occurs in cattle grazing on infected *Paspalum dilatatum.* The disease was particularly extensive in Louisiana during 1976, compared with previous years, and it is considered that the increased incidence may be associated with a lower than normal rainfall in most parts of the state (2969). *Paspalum dilatatum* is also a common constituent of New Zealand pastures and paspalum staggers has been reported in that country. Mantle, Mortimer, and White (3028) carried out a comparative study of the tremorgenic agents in ergots of *C. paspalum,* responsible for staggers, and penitrem A isolated from *Penicillium cyclopium.* They found that the ergot alkaloids, mainly D-lysergic acid and α-hydroxyethylamide and its isolysergic acid isomer, were not involved in the toxicosis but that another group of indole derivatives, presumably those identified by Cole *et al.* (2969), did give symptoms identical to those described for paspalum staggers.

Indole derivatives involving mevalonate in their biosynthesis seem now to be a widespread group of metabolites among moulds, especially among aspergilli and penicillia. As well as aflatrem, strains of *Aspergillus flavus* produce an intriguing indole mevalonate derivative referred to as aflavinine, $C_{28}H_{39}NO$ (Figure 179), the structure of which has been confirmed by X-ray crystallography (2992) and gives a clearer indication of the biosynthetic origin of these compounds.

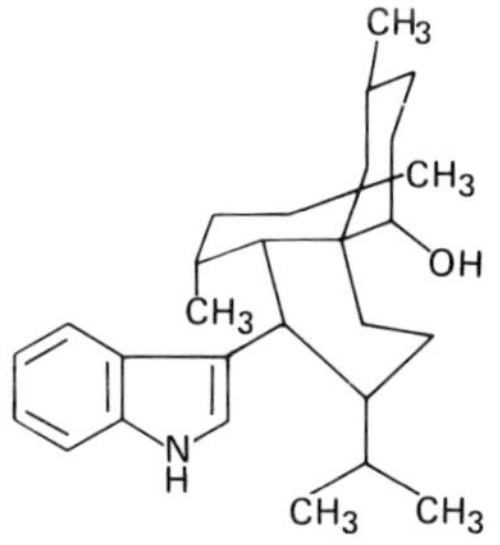

CH_3 CH_3 OH CH_3 N H CH_3 CH_3

Figure 179. Aflavinine, a metabolite from *Aspergillus flavus*

An outbreak of acute toxicosis in beef cattle led to the isolation of a number of strains of *Aspergillus fumigatus* as the predominant mould in the maize silage on which they were fed. A detailed study by Cole *et al.* (2970) revealed the presence of the alkaloids, fumigaclavines A and C (Figure 180), as well as a number of tremorgens, including verruculogen, TR-2 (Figure 181) and three tentatively referred to as SM-S, SM-R and SM-Q.

Individual species of moulds show a remarkable ability to produce a diverse range of metabolites, thus *Aspergillus clavatus* produces, not only patulin, ascladiol, and the tryptoquivalines, but also cytochalasin E (Figure 182) (2957, 2995), the dimethyl ether of xanthocillin X (Figure 183), and kotanin (Figure 184) (2958).

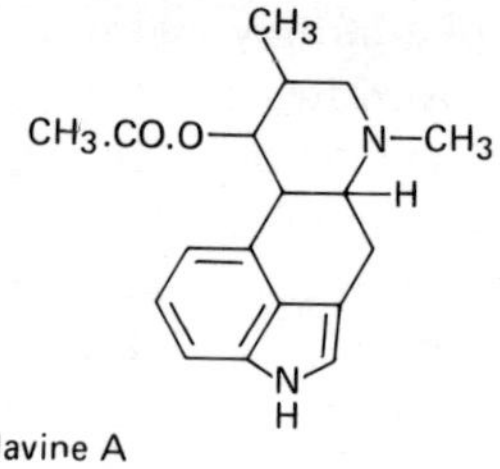

Fumigaclavine A

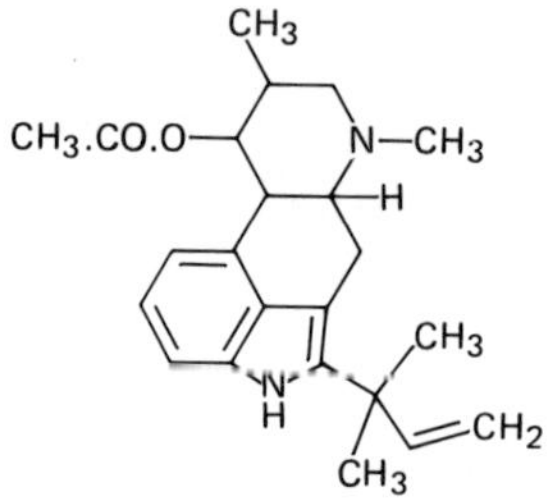

Fumigaclavine C

Figure 180. The fumigaclavines, metabolites from *Aspergillus fumigatus*

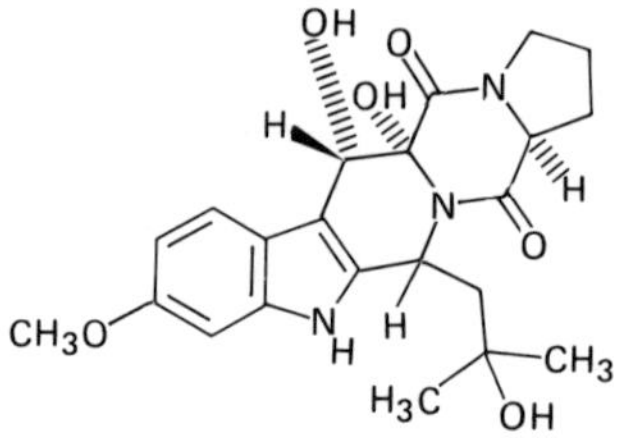

Figure 181. Compound Tr-2, isolated from the tremorgenic fraction of *Aspergillus fumigatus*

Cytochalasin E belongs to a group of metabolites which may affect the process of cell division producing binucleate, or even multinucleate cells (2962). Even such an extensive knowledge of the toxic metabolites of any particular species may not make it possible to account for its role in a specific mycotoxicosis. Thus a tremorgenic neurotoxicosis affecting a herd of cattle in the north Transvaal during February of 1975 was shown to be due to *Aspergillus clavatus.* A complex range of symptoms were noticed, including hypersensitivity, incoordination, a strange stiff-legged gait of the hind limbs, severe generalized tremors of the skeletal muscles, progressive paresis, paralysis, and constipation, all of which could not be

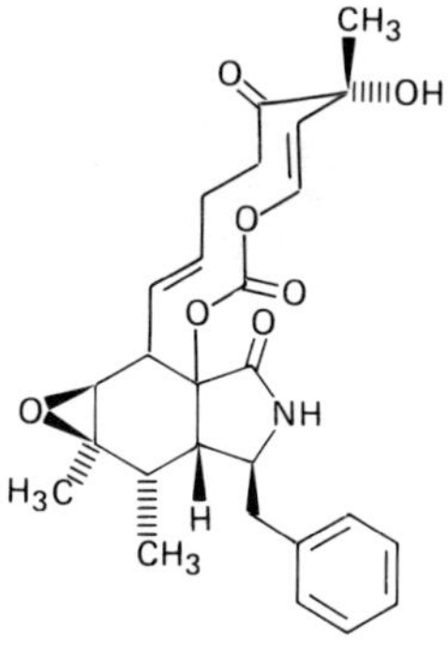

Figure 182. Cytochalasin E, a metabolite of *Aspergillus clavatus*

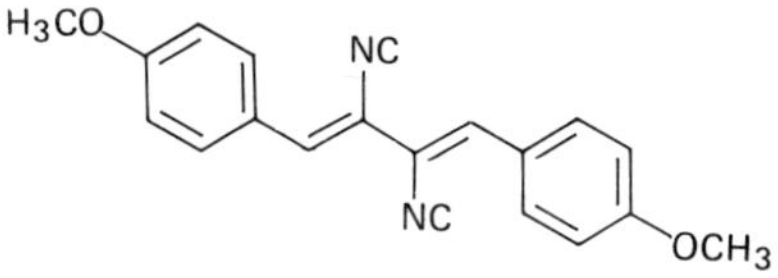

Figure 183. Xanthocillin X dimethyl ether

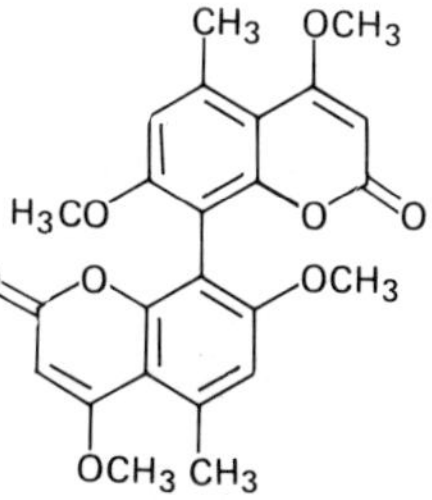

Figure 184. Kotanin

accounted for by the presence of the known toxic metabolites of this mould (3016). However, it must be accepted that our knowledge of the manner in which several mycotoxins interact in producing illness is little understood and is only just beginning to be studied. Such studies require large quantities of pure toxins and improvements in the yields of complex mycotoxins have to be deliberately sought (2975).

Toxic metabolites of *Aspergillus ustus*

This species also has the ability to produce a diverse range of metabolites, many of which are toxic. Austin, $C_{27}H_{32}O_9$ (Figure 185), an interesting polyisoprenoid

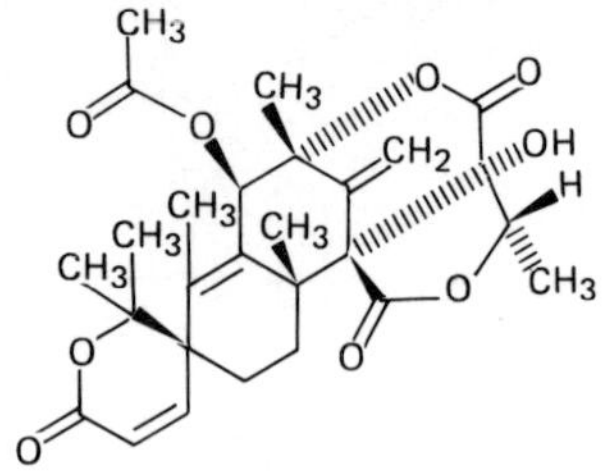

Figure 185. Austin, a polyisoprenoid toxin from *Aspergillus ustus*

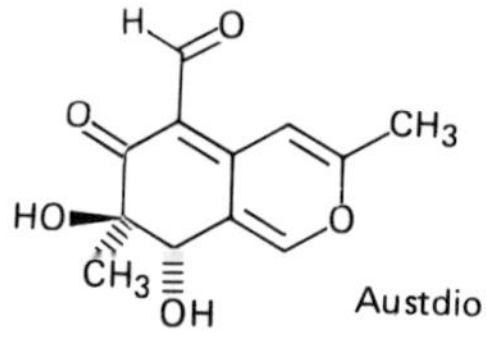

Austdiol

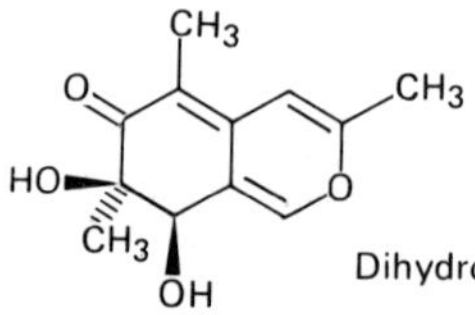

Dihydrodeoxy-8-epi-ausdiol

Figure 186. Azaphilone metabolites from *Aspergillus ustus*

compound with a relatively low toxicity, was isolated from a strain found on stored black-eyed peas, *Viga sinensis*, (2965). Austdiol, $C_{12}H_{12}O_5$, and dihydrodeoxy-8-epi-austdiol, $C_{12}H_{14}O_5$ (Figure 186), which are members of the group of metabolites referred to as azaphilones, were isolated from maize meal causing acute toxicosis in day-old chickens (3082, 3073). Austdiol itself was the major toxin and is considered to be particularly associated with disturbances of the gastrointestinal system. Austamide, $C_{21}H_{21}N_3O_3$ (Figure 187), is a diketopiperazine metabolite of

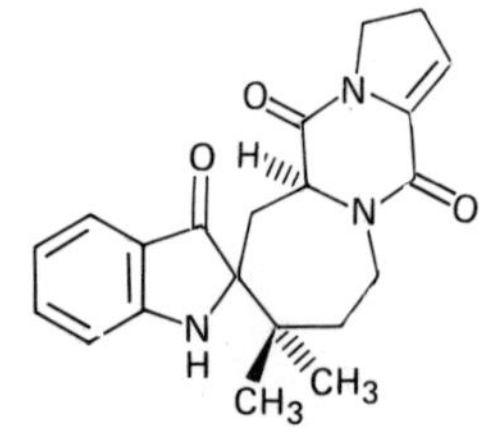

Figure 187. Austamide, a toxic indole derivative from *Aspergillus ustus*

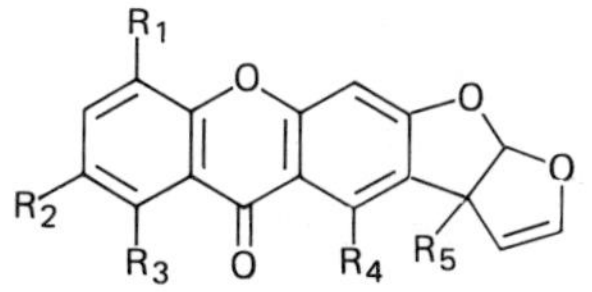

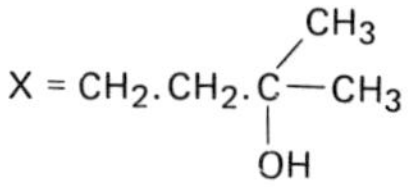

$X = CH_2.CH_2.C(CH_3)_2$—OH

	R_1	R_2	R_3	R_4	R_5
A	H	Cl	OCH_3	OCH_3	H
B	X	H	OH	OH	H
C	X	H	OH	OCH_3	H
D	X	H	OH	OH	OH
E	X	H	OH	OCH_3	OH
F	H	H	OH	OH	OH

Figure 188. The austocystins

A. ustus (3071) whereas the austocystins (Figure 188) form a group of closely related xanthones containing the fused difuran ring system found in the aflatoxins, sterigmatocystins, and versicolorins (3072).

PR-toxin and *Penicillium roqueforti*

Penicillium roqueforti, a species better known for its role in the maturation of certain cheeses, is a saprophyte which may also be isolated from decomposing

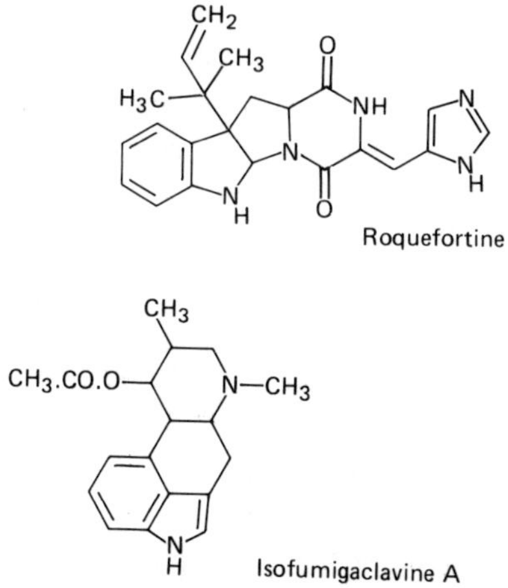

Roquefortine

Isofumigaclavine A

Figure 189. Roquefortine, a neurotoxin from *Pencillium roqueforti*, and isofumigaclavine A from the same organism

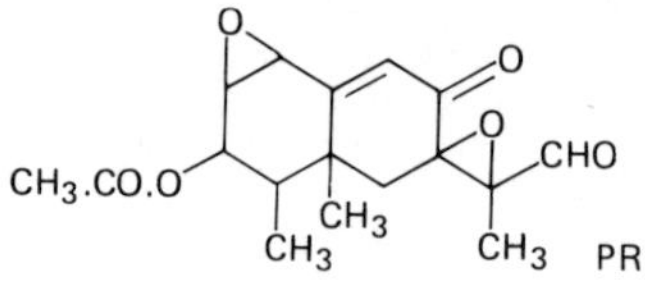

PR toxin

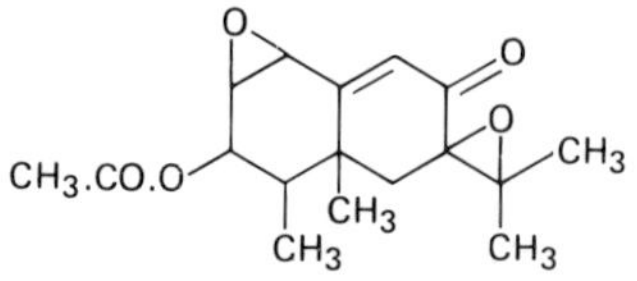

Eremofortin A

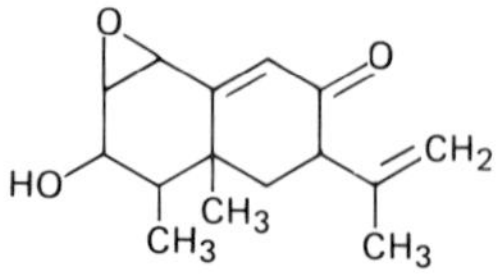

Eremofortin B

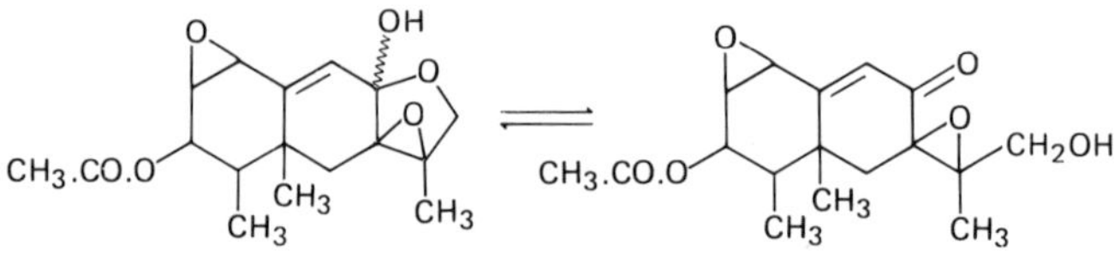

Eremofortin C

Figure 190. PR toxin and the eremofortins

organic material such as silage and compost heaps. Thus toxigenic strains have been isolated from silage and milled rice from Japan (1274) and the organism formed the major isolate from ground mouldy mixed grain and maize silage associated with bovine abortion and placental retention in Wisconsin, USA (3086). Some strains of this species are now known to produce several toxic metabolites. These include the indole derivatives roquefortine, $C_{22}H_{23}N_5O_2$, and isofumigaclavine A, $C_{18}H_{22}N_2O_2$, shown in Figure 189 (3060) as well as a number of sesquiterpenoid eremophilane epoxide derivatives such as the eremofortins (3031, 3032). The most important member of the latter group of compounds is the toxic metabolite known as PR toxin, $C_{17}H_{20}O_6$ (Figure 190), which has an intraperitoneal LD_{50} of 11 mg/kg (3085, 3086). PR toxin is said to impair the transcriptional process in rat liver cells, without the need for any activation process, by inhibiting both initiation and elongation of polynucleotide chains (3037).

This group of closely related compounds, along with the mycotoxin phomenone (Figure 191) produced by *Phoma exigua* var. *inoxydabilis* (3051), have provided a further opportunity to study the relationship between chemical structure and toxicity (3038). It seems that the aldehyde function, present in PR toxin itself, is essential for the expression of maximum toxicity.

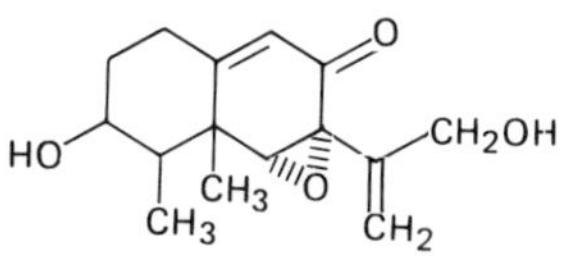

Figure 191. Phomenone, an eremophilane metabolite from *Phoma exigua*

Studies on rubratoxin B

A number of studies have continued to make a contribution to an understanding of the mode of action of rubratoxin B. Although rubratoxin inhibits oxygen consumption by mitochondria (3003), it does not appear to inhibit oxidative phosphorylation (2955). By using tetramethyl *p*-phenylenediamine, a reagent used in the detection of the presence of cytochrome oxidase in bacteria, it has been possible to locate a possible site of activity in the electron transport chain (3003). It has also been suggested that rubratoxin B may inhibit ATPase activity in the mouse and that this biochemical lesion may also be involved in its action as a toxin (2976).

Rubratoxin B has been shown to be both mutagenic and teratogenic in the mouse (2987), and when males were treated with 0.75–1.5 mg rubratoxin B/kg body weight, by intraperitoneal injection for five days, and then mated with untreated virgin females, a dose related increase in early foetal deaths was observed. Dihydrorubratoxin B, a derivative in which the α-β unsaturated δ-lactone ring is hydrogenated, had no effect on prenatal growth and development, neither was it mutagenic at doses of up to ten times the mutagenic dose of rubratoxin B itself.

Other effects of rubratoxin B at the subcellular level include the disaggregation of liver polyribosomes, thus affecting protein biosynthesis (3084) and a suppression of complement activity, associated with increased prothrombin time, in guinea-pig serum (3050). The description of such a diverse range of activities would indicate that we are still some way from understanding the nature of the most important biochemical lesion caused by rubratoxin.

Much of the work on this interesting metabolite is made more difficult by the absence of a sensitive routine analytical assay. Hayes and McCain (3004) have shown that the careful treatment of TLC plates at 200 °C for ten minutes, after chromatography in a solvent system consisting of $MeOH/CHCl_3$/glacial acetic acid/water (20:80:1:1) gave a fluorescent zone in the position of rubratoxin B allowing the detection of as little as 0.5 μg. Emeh and Marth (2981) have described many of the nutritional parameters controlling the biosynthesis of rubratoxin in laboratory cultures. Thus, for example, the presence of malonate or acetyl coenzyme A in the medium enhanced the formation of toxin. Although 10^{-3} M citrate also enhanced the formation of rubratoxins, higher concentrations had an inhibitory effect. It is tempting to speculate that low concentrations of citrate may induce an increased level of oxaloacetate, one of the postulated precursors of the rubratoxins, whereas higher concentrations of citrate may inhibit the citric acid synthetase-like activity which is required in the first stage of its biosynthesis (Figure 192) (1765).

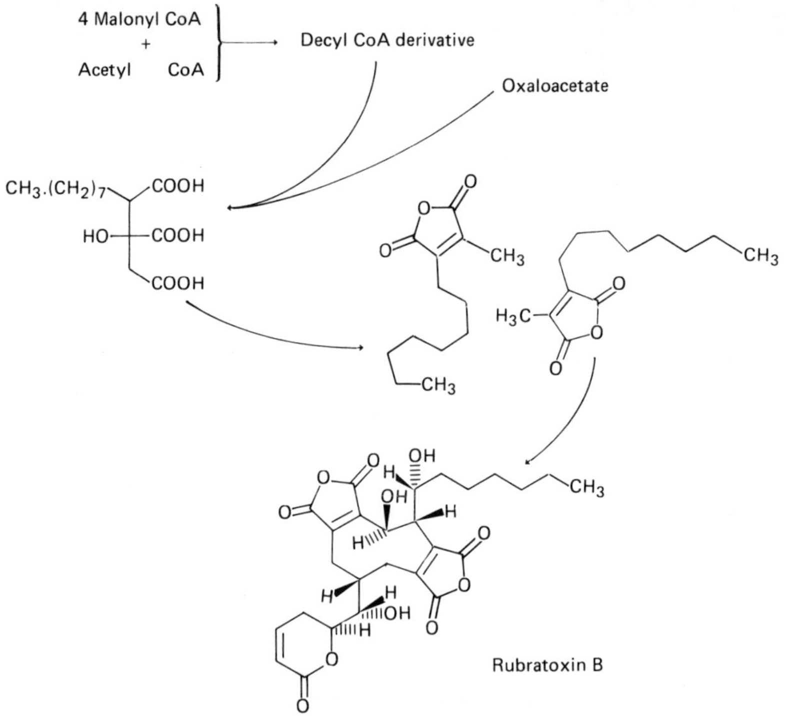

4 Malonyl CoA + Acetyl CoA → Decyl CoA derivative

Oxaloacetate

Rubratoxin B

Figure 192. Postulated stages in the biosynthesis of rubratoxin B

The enhanced formation of rubratoxin in the presence of malonate has not been confirmed by Rajkovic (3048), using different growth conditions for *Penicillium rubrum,* although Senn (3061) has demonstrated that 2-^{14}C labelled malonate was more effectively incorporated into rubratoxin B than was 2-^{14}C labelled acetate. The compound [3-$^{14}CH_3$]-2-(1′octenyl)-3-methylmaleic anhydride (Figure 193) has been synthesized and tested as a model C_{13} precursor in the biosynthesis of rubratoxin B. It was incorporated into the toxin by cultures of *P. rubrum* (3061). By the oxidation of rubratoxin B with chromium trioxide in sulphuric acid the 20 oxo-derivative is formed and this in turn yields n-hexyl methyl ketone on treatment with sodium hydroxide (Figure 194). The proposed biosynthetic route requires that the n-hexyl methyl ketone thus obtained should be free of ^{14}C label although it was, in fact, found to be slightly radioactive. It was reasonably suggested that this label could have arisen from the degradation of the C_{13}

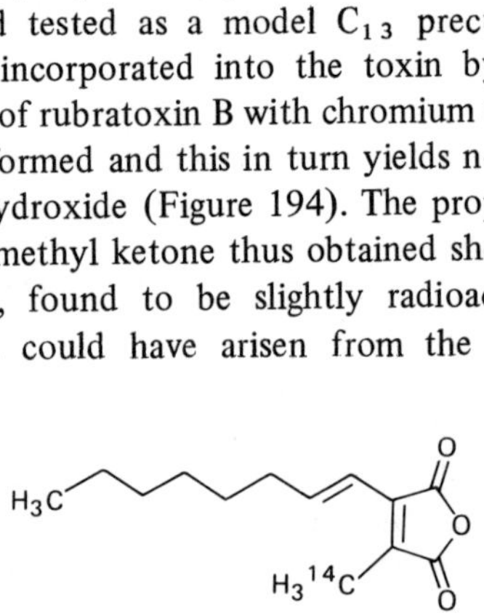

Figure 193. [3^{14}C]-2-(1′octenyl)-3-methylmaleic anhydride

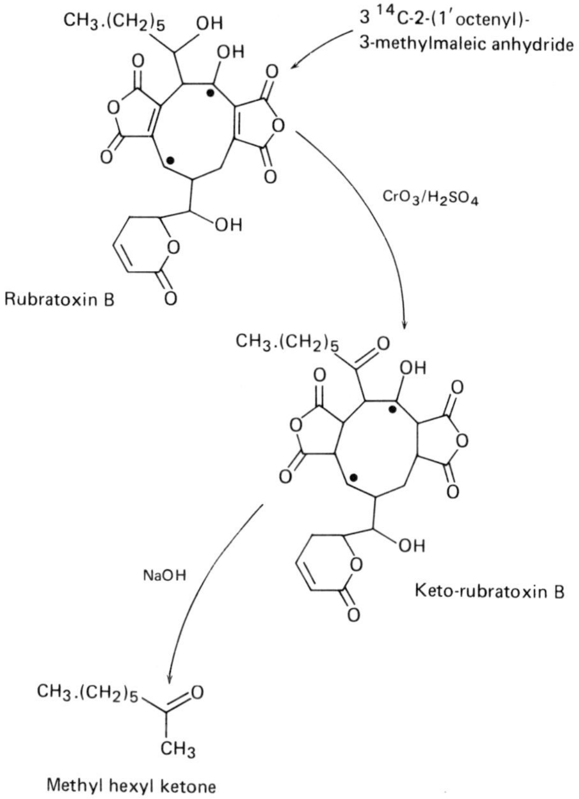

3 ^{14}C-2-(1′octenyl)-3-methylmaleic anhydride

Rubratoxin B

CrO_3/H_2SO_4

Keto-rubratoxin B

NaOH

Methyl hexyl ketone

Figure 194. Degradation of ^{14}C labelled rubratoxin.

precursor to smaller fragments which were then reincorporated into the toxin thus introducing some spurious label into the side chain. Comparison of the amount of radioactive label incorporated into the side chain when the C_{13} compound was the precursor, with that incorporated when 2-^{14}C malonate was the precursor indicated that at least 80% of the C_{13} compound which had been incorporated resulted from the intact precursor thus giving some confirmation of the original hypothesis for the biosynthesis of the rubratoxins.

Epoxy trichothecenes

Since Bamburg and Strong (130) made out a good case for considering that alimentary toxic aleukia was caused by epoxy trichothecenes, this important group of mould metabolites has also been implicated in the diseases caused by *Stachybotrys atra* and *Dendrodochium toxicum* (= *Myrothecium roridum*). The original toxic fractions from *Stachybotrys atra*, referred to as stachybotryotoxins, have been shown to be a complex mixture (3053) containing such compounds as ergosterol and mellein, which have no toxicity in the brine shrimp test, as well as

Figure 195. Satratoxin H, a toxic metabolite of *Stachybotrys atra*

five very toxic compounds labelled satratoxins C, D, F, G, and H. Satratoxin D has since been shown to be the known metabolite, roridin E, satratoxin C to be the same as verrucarin J, and satratoxin H to be a new roridin-like compound (Figure 195) (2985).

In a study of the manner in which the 12, 13-epoxytrichothecenes inhibit protein biosynthesis, Liao *et al.* (3026) showed that some members of this group, such as diacetoxyscirpenol (= anguidine), verrucarin A (= muconomycin A), and T-2 toxin, are all irreversible inhibitors, whereas crotocin and trichodermin are reversible. These authors suggested that the irreversible inhibitors probably affect the initiation of protein synthesis, causing a disruption of polyribosomes to monosomes, whereas the reversible inhibitors are thought to affect chain elongation of the growing protein molecule.

Studies of the metabolism of the trichothecenes in animal tissues have been facilitated by the synthesis of radioactively labelled toxins (3083), but studies on this group of toxins are held back by the lack of rapid, sensitive, and specific analytical techniques comparable to those available for the aflatoxins.

Diversity of toxin structures

The incredible diversity of structures which fungi may produce from a relatively small range of primary metabolites is indicated by the increasing number of new mycotoxins described in the current scientific literature.

Secalonic acid F (Figure 196) is a new member of a group of metabolites also known as ergochromes which, with secalonic acid D, are considered to account for the toxicity of a strain of *Aspergillus aculeatus* Iizuka (2947). The latter compound had previously been isolated from *Penicillium oxalicum,* secalonic acid A from *Aspergillus ochraceus,* A, B, and C from *Claviceps purpurea,* and A and E from *Phoma terrestris.*

The very complexity of secondary metabolites gave some hope that their study may provide some useful guidance on the taxonomic relationships between moulds,

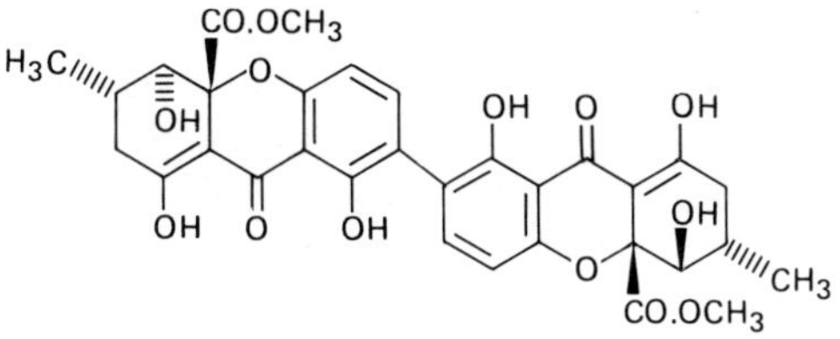

Figure 196. Secalonic acid F, a toxic metabolite of *Aspergillus aculeatus*

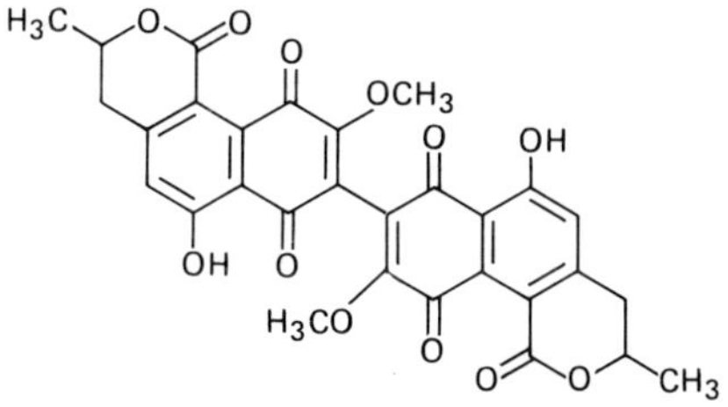

Xanthomegnin

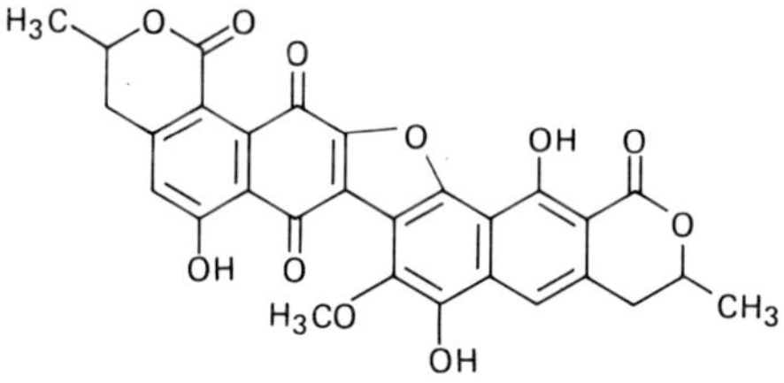

Viopurpurin

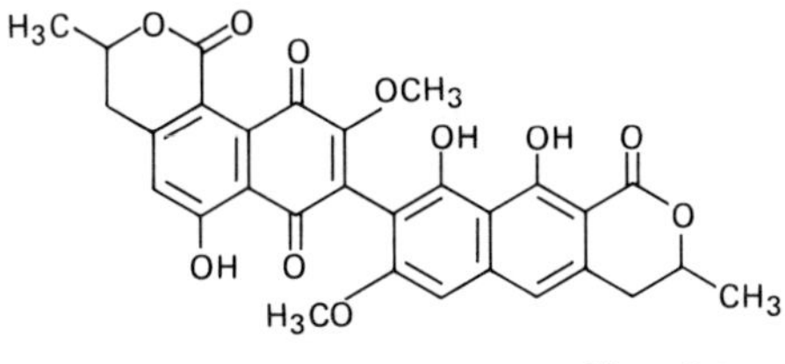

Viomellein

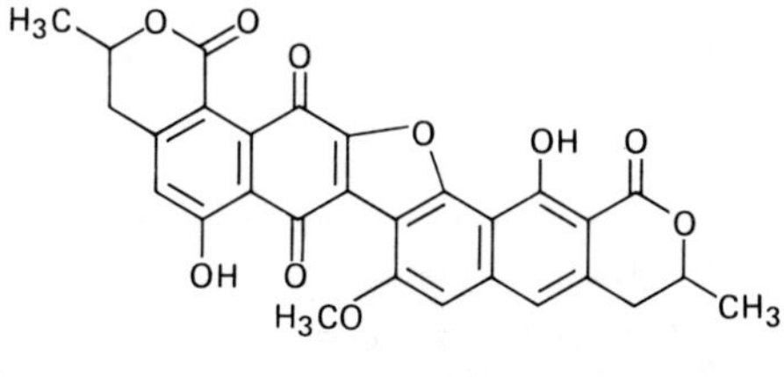

Rubrosulphin

Figure 197. Pigments isolated from *Penicillium viridicatum* but the four genera known to produce secalonic acids are not particularly closely related with the exception perhaps of *Aspergillus* and *Penicillium.*

Byssotoxin A is a new metabolite, with a possible empirical formula $C_{25}H_{26}N_2O_3$, isolated from *Byssochlamys fulva* (3018), a species already known to produce byssochlamic acid and patulin. The metabolite is said to be toxic to chickens.

Penicillium viridicatum is already known to produce a complex array of toxic metabolites including ochratoxin A, citrinin, viridicatin, viridicatol, cyclopenin, cyclopenol, viridicatic acid, mycophenolic acid, and brevianamide A, but apparently none of these toxins considered alone accounts for the observation that prolonged exposure of mice to this organism causes adenocarcinomas (3089). A number of pigments, identified as xanthomegnin, viopurpurin, viomellein, and rubrosulphin (Figure 197), have also been isolated from this organism (3070) and it has been demonstrated that xanthomegnin and viomellein may produce changes in the liver when fed to mice (2961), although these changes did not seem to include carcinomas.

Among the cyclic peptidolactones, the destruxins are best known as insecticidal metabolites of such insect pathogens as *Metarrhizium anisopliae* and *Ooospora destructor,* but destruxin B (Figure 198) is also known as a metabolite of a strain of *Aspergillus ochraceus* (3077); it is not understood whether or not this compound contributes to the mammalian toxicity of this strain.

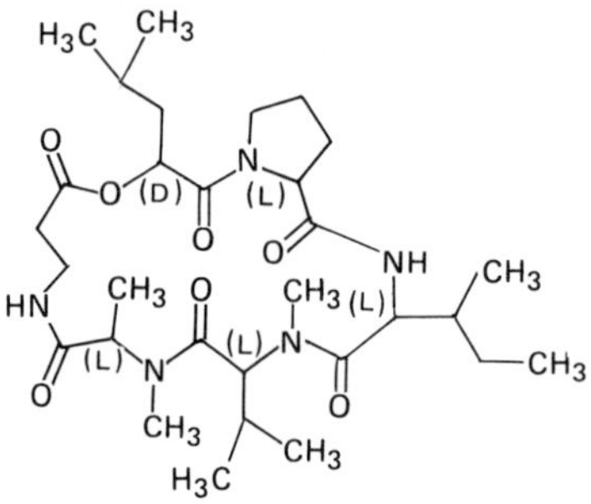

Figure 198. Destruxin B

A group of cyclic peptides, known as malformins because of their effects on the growth and morphology of plants (2974) are produced by some strains of *Aspergillus niger* and at least one of these metabolites, malformin C (Figure 199), has been shown to have a significant mammalian toxicity (2946). By growing toxigenic strains of *A. niger* on a solid medium made up of white wheat, as much as 1,800 mg crude toxin per kilogram of grain was obtained. This crude preparation contained about 15% malformin C, the mean lethal dose of which is 0.9 mg/kg. Like malformin A, this compound has a significant antibacterial activity against a variety of Gram positive and Gram negative bacteria (3017).

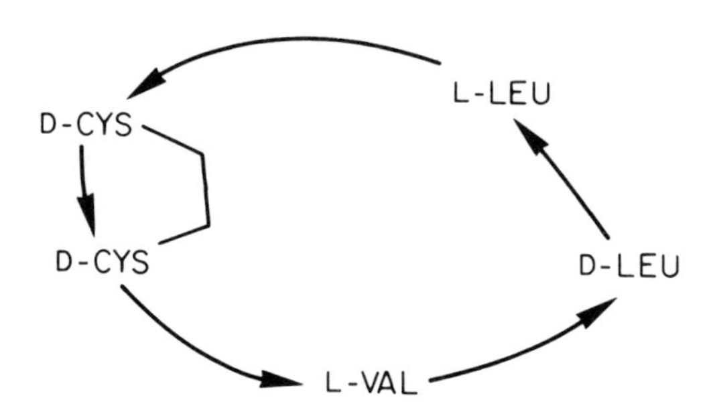

Figure 199. Malformin C, a toxic metabolite produced by *Aspergillus niger*

Patulin, one of the smallest of mycotoxin molecules, is of particular interest in the fruit juice and beverage industry because of the ability of *Penicillium expansum,* one of many species of fungi which produces it, to synthesize sufficient quantity in rotten apples to render them toxic. Patulin is stable in acid conditions and, although it will react with sulphur dioxide at high enough concentrations, such reactions are apparently insignificant at the levels of sulphur dioxide (< 200 ppm) used in processing apple juice and cider. Patulin is, however, destroyed during an active yeast fermentation (2960). One of the most sensitive methods of detecting the presence of patulin on TLC plates is to spray with 3-methyl-2-benzthiazolinone but unfortunately this is susceptible to interference by such compounds as 5-(hydroxymethyl) furfural, which may also be present in apple juice. A method involving the conversion of patulin to its 2,4-dinitrophenylhydrazone by passing a solvent extract through a microcolumn packed with celite containing 2,4-dinitrophenylhydrazine and phosphoric acid, separating the derivatives by TLC, spraying with alkali, and measuring the absorption at 375 nm, has been described (3074). It is said to give good recovery, in the range of 50–340 ppb, in apple juice.

The range of toxic metabolites produced by any one species of mould, and the range of different species producing a single toxic metabolite, are only two aspects of the complexity of interactions leading to toxic food or animal feed. Sweet potato toxicity is a situation in which a plant, *Ipomoea batatas,* produces several

Ipomeamarone

Ipomeamaronol

Figure 200. Hepatotoxins produced by *Ipomoea batata* as a result of fungal infection

Ipomoea batatas

Damage by fungi
e.g. *Ceratocystis fimbriata*

Fusarium solani
Fusarium oxysporum

Reduction by *F. solani*
F. oxysporum and *C. fimbriata*

Figure 201. The generation of lung toxins from 4-hydroxymyoporone

toxic metabolites in response to a number of stresses, including fungal infection. The production of ipomeamarone and ipomeamaronol (Figure 200) as a result of infection of *Ipomoea* by *Ceratocystis fimbriata*, and other fungi, has been shown to account for toxicity to liver, but not the production of oedema of the lung, by infected sweet potatoes (2830). The production of the lung-oedema toxin has now been found to involve a further microbial process, the breakdown of 4-hydroxymyoporone by species of *Fusarium,* especially *F. solani* and, to a lesser extent, *F. oxysporum* (2959). This retroaldol reaction appears to be specifically carried out by these species of *Fusarium* and not by *Ceratocystis fimbriata,* although the latter is able to participate in the subsequent reduction of the diketone to the lung toxins (Figure 201) which are responsible for the observed severe pulmonary oedema. The LD_{50} of the diol is 20–70 mg/kg.

The role of mycotoxins in disease of man

There can be little doubt about the role of mycotoxins in specific syndromes such as those of alimentary toxic aleukia, ergotism, and yellow rice disease, and there is increasing evidence implicating mycotoxins in disorders such as Balkan nephropathy, liver damage in Thailand and East Africa as well as India, and even liver cancer in some parts of the world, but it is still pertinent to ask whether mycotoxins present a health hazard in countries such as the United Kingdom.

In considering this question Jarvis (3011) pointed out the difficulties in arriving at a straightforward answer but suggests that an approach could be made by looking at both the incidence and the levels of contamination in foods. In this, and a subsequent review (3012), Jarvis concludes that the available evidence favours the view that, at least in the UK and at the present time, mycotoxins do not constitute a major health hazard to the majority of consumers. However, continuous awareness of their potential hazard is essential for, as Hesseltine (3006) has pointed out, problems in the natural occurrence of mycotoxins in some agricultural commodities may arise from changes in agricultural technology, and such changes take place with increasing speed and on an increasing scale. In reviewing the role of mycotoxins in dairy products, Moreau (3030) also concludes that the danger of these compounds should be carefully considered and continuously monitored, but not exaggerated.

A detailed knowledge of the biosynthetic potential of the fungi should help to sharpen our awareness of the toxicological possibilities arising from the growth of moulds on foods. A detailed knowledge of the chemical structures of these metabolites provides the necessary background for the development of specific and sensitive analytical methods for monitoring the presence of mycotoxins in our food environment. A detailed knowledge of the toxicology of these compounds should make it easier to diagnose illnesses associated with the presence of mycotoxins, although we have relatively little information on the possible interactions of complex mixtures of toxic metabolites.

The most sensible point at which the control of mycotoxin production should be tackled is in the handling, harvesting, and storage of foods and animal feeds, and here a knowledge of the ecology, physiology, and taxonomy of the moulds themselves should provide some assistance.

REFERENCES

(References (1)-(2938) have been reproduced direct from the French edition)

(1) ABADJIEFF W., BÄR H. J., BEYER J., EHRENTRAUT W., ELZE K., FRITZSCH W., LEIPNITZ W., LINDNER K., NEUNDORF R., PANNDORF H., PRIBOTH W., SCHÜPPEL K. F., STOLZENBURG P., ULBRICH M. et VOIGT O. — Intoxikationen bei Rindern durch Verfütterung Pilzhaltiger Malzkeime. *Monatsheft f. Veterinärmedizine*, t. **XXI**, p. 452-458, 1966.

(2) ABBOT E. V. — The occurrence and action of fungi in soils. *Soil Sci.*, t. **XVI**, p. 207-216, 1923.

(3) ABE S., TAKEUCHI S. et YONEHARA H. — Studies on variotin. *Journ. Antibiotics*, t. **XII**, p. 201-202, 1959.

(4) ABDEL KADER M. M., EL-AASER A. B. A., EL-MERZABANI M. M. et KING L. J. — The metabolism of aflatoxins in the rat. *Acta Physiol. Acad. Sci. Hung.*, t. **XXXIX**, p. 375-381, 1971.

(5) ABEDI Z. H. et MC KINLEY W. P. — Zebra fish eggs and larvae as aflatoxin biossay test organisms. *J. Ass. Offic. Anal. Chem.*, t. **LI**, p. 902-905, 1968.

(6) ABEDI Z. H. et SCOTT P. M. — Detection of toxicity of aflatoxins, sterigmatocystin and other fungal toxins by lethal action on zebra fish larvae. *J. Ass. Offic. Anal. Chem.*, t. **LII**, p. 963-969, 1969.

(7) ABERCROMBIE M. et HARKNESS R. D. — The growth of cell populations and the properties in tissue culture of regenerating liver of the rat. *Proc. Roy. Soc.* (London), B, t. **CXXXVIII**, p. 544-561, 1951.

(8) ABRAHAM E. P. et NEWTON G. G. F. — Structure of cephalosporin C. *Biochem. J.*, t. **LXXXIX**, p. 377-393, 1961.

(9) ABRAHAMSSON S. et NILSSON B. — Direct determination of the molecular structure of trichodermin. *Proc. Chem. Soc.*, p. 188, 1964.

(10) ABRAMS L. — Mycotoxicoses. *J. S. Afr. veter. med. Ass.*, t. **XXXVI**, p. 5-13, 1965.

(11) ADLER M. et WINTERSTEINER O. — A reinvestigation of flavacidin, the penicillin produced by *Aspergillus flavus*. *J. Biol. Chem.*, t. **CLXXVI**, p. 873-891, 1948.

(12) ADRIAN J. — Aflatoxins in the peanut. *Rev. Port. Cienc. Vet.*, t. **LXII**, p. 215-232, 1967.

(13) ADRIAN J. — Les aflatoxines. III. Les moyens de prévention et de détoxification. *Oléagineux*, t. **XXIV**, p. 155-161, 1969.

(14) ADRIAN J. et LUNVEN P. — Les aflatoxines. I. Les agents responsables. *Oléagineux*, t. **XXIV**, p. 31-35, 1969.

(15) ADRIAN J. et LUNVEN P. — Les aflatoxines. II. Les manifestations de la toxicité. *Oléagineux*, t. **XXIV**, p. 83-86, 1969.

(16) ADYE J. et MATELES R. I. — Incorporation of labeled compounds into aflatoxins. *Biochim. Biophys. Acta*, t. **LXXXVI**, p. 418-420, 1964.

(17) AFRIDI M. M. R. K. — Effect of some antibiotics on the induced formation of nitrate reductase in higher plants. *Hindustan Antibiot. Bull.*, t. **V**, p. 51-54, 1962.

(18) AGTHE C., LIJINSKY W. et OREMUS D. — Determination of aflatoxins in food by absorption spectometry. *Food. Cosmet. Toxicol.*, t. **VI**, p. 627-631, 1968.

([19]) Agurell S. L. — Costaclavine from *Penicillium chermesinum*. *Experientia*, t. **XX**, p. 25-26, 1964.

([20]) Aibara K. et Miyaki K. — Aflatoxin. *Shokuhin Esisei Kenkyu*, t. **XV**, p. 19-25, 1965.

([21]) Aibara K. et Yamagishi S. — Effects of ultraviolet irradiation on the destruction of aflatoxin B_1. *in* Herzberg M., *Toxic Micro-organisms*, p. 211-221, 1970.

([22]) Aibara K. et Yoshikuya Y. — Synchronized cultures as an assay system for aflatoxin. *Biochim. Biophys. Acta*, 1968.

([23]) Ainsworth G. C. et Austwick P. K. C. — A survey of animal mycoses in Britain. Mycological aspects. *Trans. Brit. Mycol. Soc.*, t. **XXXVIII**, p. 369-386, 1955.

([24]) Ainsworth G. C. et Austwick P. K. C. — Fungal diseases of animals. *Commonwealth Bureau of Animal Health Review Series* n° 6, 148 p., 1959.

([25]) Ainsworth G. C. et Rewell R. E. — The incidence of aspergillosis in captive wild birds. *J. Comp. Pathol. Therap.*, t. **LIX**, p. 213-224, 1949.

([26]) Aitken W. A. — Cornstalk poisoning. *J. Am. Vet. Med. Ass.*, t. **CXXI**, p. 402-403, 1952.

([27]) Akao M., Kuroda K. et Wogan G. N. — Aflatoxin B_1 : the kidney as a site of action in the mouse. *Life Sci.*, II, t. X, p. 495-501, 1971.

([28]) Albright J. L., Aust S. D., Byers J. H., Fritz T. E., Brodie B. O., Olsen R. E., Link R. P., Simon J., Rhoades H. E. et Brewer R. L. — Moldy corn toxicosis in cattle. *J. Am. Vet. Med. Ass.*, t. **CXLIV**, p. 1013-1019, 1964.

([29]) Aleksandrowicz J. et Smyk B. — Mycotoxins and their role in oncogenesis, with special reference to blood disease. *Pol. Med. Sci. Hist. Bull.*, t. **XIV**, p. 25-30, 1971.

([30]) Alfin-Slater R. B., Aftergood L., Hernandez H. J., Sterne E. et Melnick D. — Studies of long term administration of aflatoxin to rats as a natural food contaminant. *J. Amer. Oil Chem. Soc.*, t. **XLVI**, p. 493-497, 1969.

([31]) Allard C. — Modalités de la production des aflatoxines. *Phytiatrie-Phytopharmacie*, t. **XIV**, p. 81-87, 1965.

([32]) Allcroft R. — Hypomagnesaemia in cattle. *Vet. Res.*, t. **LXVI**, p. 517, 1954.

([33]) Allcroft R. — Aspects of aflatoxicosis in farm animals. *Mycotoxins in Foodstuffs*, p. 153-162, 1965.

([34]) Allcroft R. — Aflatoxicosis in farm animals. *in* Goldblatt L. A., *Aflatoxin*. Academic Press, p. 237-264, 1969.

([35]) Allcroft R. et Carnaghan R. B. A. — Groundnut toxicity — *Aspergillus flavus* toxin (aflatoxin) in animal products: preliminary communication. *Vet. Rec.*, t. **LXXIV**, p. 863-864, 1962.

([36]) Allcroft R. et Carnaghan R. B. A. — Groundnut toxicity: an examination for toxin in human food products from animals fed toxic groundnut meal. *Vet. Rec.*, t. **LXXV**, p. 259-263, 1963.

([37]) Allcroft R. et Carnaghan R. B. A. — Toxic products in groundnuts: biological effects. *Chemistry and Industry*, t. **II**, p. 50-53, 1963.

([38]) Allcroft R., Carnaghan R. B. A., Sargeant K. et O'Kelly J. — A toxic factor in Brazilian groundnut meal. *Vet. Rec.*, t. **LXXIII**, p. 428-429, 1961.

([39]) Allcroft R. et Lancaster M. C. — Proc. *Third Int. Meeting Forensic Immunology. Medicine. Pathology and Toxicology*, London, 1963.

([40]) Allcroft R. et Lewis G. — Groundnut (meal) toxicity in cattle: experimental poisoning of calves and a report on clinical effects in older cattle. *Vet. Rec.*, t. **LXXV**, p. 487-494, 1963.

([41]) Allcroft R. et Lewis G. — Aflatoxicosis in animal caused by a mycotoxin present in some batches of peanuts (*Arachis hypogaea*). *Biochem. J., G. B.*, t. **LXXXVIII**, p. 58, 1963.

([42]) Allcroft R. et Roberts B. A. — Aflatoxin in milk. *Food Cosmet. Toxicol.*, t. **V**, p. 597-598, 1967.

([43]) ALLCROFT R. et ROBERTS B. A. — Toxic groundnut meal: the relationships between B_1 intake by cows and excretion of aflatoxin M_1 in milk. *Vet. Rec.*, t. **LXXXII**, p. 116-118, 1968.

([44]) ALLCROFT R., ROBERTS B. A. et LLOYD M. K. — Excretion of aflatoxin in a lactating cow. *Food Cosmet. Toxicol.*, t. **VI**, p. 619-625, 1968.

([45]) ALLCROFT R., ROGERS H., LEWIS G., NABNEY J. et BEST P. E. — Metabolism of aflatoxin in sheep: excretion of the « milk toxin ». *Nature, G. B.*, t. **CCIX**, p. 154-155, 1966.

([46]) ALLCROFT W. M. — Seasonal hypomagnesaemia of the bovine without clinical symptoms. *Vet. J.*, t. **CIII**, p. 75, 1947.

([47]) ALLINGER N. L. et COKE J. L. — Structure of helvolic acid. III. *J. Organic Chem.*, t. **XXVI**, p. 4522-4529, 1961.

([48]) ALPERT E., SERCK-HANSSEN A. et RAJAGOPOLAN B. — Aflatoxin-induced hepatic injury in the African monkey. *Arch. Environ. Health*, t. **XX**, p. 723-728, 1970.

([49]) ALSBERG C. L. et BLACK O. F. — Contributions to the study of maize deterioration; biochemical and toxicological investigations of *Penicillium puberulum* and *Penicillium stoloniferum. U. S. Dept Agric. Bureau Plant Ind.*, Bull. n° 270, 47 p., 1913.

([50]) AMBROSE A. M. et DE EDS F. — Acute and sub-acute toxicity of pure citrinin. *Proc. Soc. Exp. Biol. med.*, t. **LIX**, p. 289-291, 1945.

([51]) AMERICAN ASSOCIATION OF CEREAL CHEMISTS. — *Cereal Laboratory methods.* 7e éd., St Paul, Minnesota, 1962.

([52]) AMERICAN PUBLIC HEALTH ASSOCIATION. — Recommended methods for the microbiological examination of foods. New York, 1958.

([53]) AMLA I., KAMALA C. S., GOPALAKRISHNA G. S., JAYARAJ A. P., SREENIVASAMURTHY V. et PARPIA H. A. B. — Cirrhosis in children from peanut meal contaminated by aflatoxin. *Am. J. Clin. Nutr.*, t. **XXIV**, p. 609-614, 1971.

([54]) AMONKAR S. V. et NAIR K. K. — Pathogenicity of *Aspergillus flavus* Link to *Musca domestica nebulo* Fabricius. *J. Invert. Path.*, t. **VII**, p. 513-514, 1965.

([55]) ANANTHARAMAN K. et CARPENTER K. J. — The effect of heat treatment of the limiting amino acids of groundnut flour for the chick. *Proc. Nutr. Soc. (Engl. Scot.)*, t. **XXIV**, p. 32, 1965.

([56]) ANDERSON G. C., HARE J. H., BLETNER J. K., WEAKLEY C. E. et MASON J. A. — A hemorrhagic condition in chicks fed simplified rations. *Poult. Sci.*, t. **XXXIII**, p. 120-126, 1954.

([57]) ANDRAUD G., AUBLET-CUVELIER A. M., COUQUELET J., CUVELIER R. et TRONCHE P. — Activité comparée sur la respiration cellulaire de la patuline naturelle et d'un isomère de synthèse. *C. R. Soc. Biol.*, t. **CLVII**, p. 1444-1446, 1963.

([58]) ANDRELLOS P. J. et REID G. R. — Confirmatory tests for aflatoxin B_1. *J. Ass. Offic. Anal. Chem.*, t. **XLVII**, p. 801-803, 1964.

([59]) ANDRES V. de, ANDRES J. de et CALET C. — Action de l'aflatoxine sur les facultés reproductrices de *Gallus gallus. Cah. Nutr. Diét.*, t. **V**, p. 55-57, 1970.

([60]) ANSLOW W. K. et RAISTRICK H. — Studies in the biochemistry of microorganisms. LVII. Fumigatin (3-hydroxy-4-methoxy-2:5-toluquinone) and spinulosin (3:6-dihydroxy-4-methoxy-2:5-toluquinone), metabolic products respectively of *A. fumigatus* Fres. and *P. spinulosum* Thom. *Biochem. J.*, 696, t. **XXXII**, p. 687-696, 1938.

([61]) ANSLOW W. K. et RAISTRICK H. — Studies in the biochemistry of micro-organisms. LVIII. Synthesis of spinulosin (3:6-dihydroxy-4-methoxy-2:5-toluquinone) a metabolic product of *Penicillium spinulosum* Thom. *Biochem. J.*, t. **XXXII**, p. 803-806, 1938.

([62]) ANSLOW W. K. et RAISTRICK H. — Studies in the biochemistry of micro-organisms. LIX. Spinulosin (3:6-dihydroxy-4-methoxy-2:5-toluquinone) a metabolic product of a strain of *A. fumigatus* Fres. *Biochem. J.*, t **XXXII**, p. 2288-2289, 1938.

([63]) ANSLOW W. K., RAISTRICK H. et SMITH G. — Antifungal substances from moulds. Part I. Patulin (anhydro-3-hydroxymethylen tetrahydro-1:4-pyrone-2-carboxylic acid) a metabolic product of *Penicillium patulum* Bainier, and *P. expansum* (Link) Thom. *J. Soc. Chem. Ind.*, t. **LXII**, p. 236-238, 1943.

(64) ANTJUKOV M. A. — (Effets de substances toxiques de champignons du genre *Aspergillus* sur les porcelets). *Veterinarija*, t. **XLII**, p. 33-36, 1966.

(65) APINIS A. E. — Thermophilous fungi of coastal grasslands. *Soil Organisms. Proc. Colloquium on soil fauna, soil microflora and their relationships*, p. 427-438, North Holland Publishing Co, 1963.

(66) ARGEN R. J., LESLIE E. V. et LESLIE M. B. — Intracavitary fungus ball-pulmonary aspergillosis. Report of a case. *J.A.M.A.*, t. **CLXXIX**, p. 944-947, 1962.

(67) ARMBRECHT B. H., GELETA J. N., SHALKOP W. T. et DURBIN C. G. — A sub acute exposure of beagle dogs to aflatoxin. *Toxicol. Appl. Pharmacol.*, t. **XVIII**, p. 579-585, 1971.

(68) ARMBRECHT B. H., HODGES F. A., SMITH H. R. et NELSON A. A. — Mycotoxins. I. Studies on aflatoxin derived from contaminated peanut meal and certain strains of *Aspergillus flavus*. *Ass. Off. Agr. Chem. J.*, t. **XLVI**, p. 805-817, 1963.

(69) ARMBRECHT B. H., SHALKOP W. T., WISEMAN H. G., JACKSON J. E. et ZIMMERMAN H. E. — The effects produced in brood sows by feeding aflatoxin. *Toxicol. Appl. Pharmacol.*, t. **XIV**, p. 649-650, 1969.

(70) ARMOLIK N. et DICKSON J. G. — Minimum humidity requirement for germination of conidia of fungi associated with storage of grain. *Phytopathology*, t. **XLVI**, p. 462-465, 1956.

(71) ARNOLD M. H. M. et CLARKE H. J. — The evaluation of some fungistats for paint. *J. Oil and Colour Chemists' Ass.*, p. 900-908, 1956.

(72) ARNOLD W. R. et PETTIT R. E. — Interrelationship between peanut kernel moisture and storage gases with growth of *Aspergillus flavus* and aflatoxin production. *Phytopathology*, t. **LIX**, p. 111, 1969.

(73) ARNSTEIN H. R. V. et BENTLEY R. — Biosynthesis of kojic acid. *Biochem. J.*, t. **LIV**, p. 493-522, 1953.

(74) ARRHENIUS E. et HULTIN T. — Effects of carcionogenic amines on amino acid incorporation by liver systems. I. Secondary increase in microsomal activity after aminofluorene treatment. *Cancer Res.*, t. **XXII**, p. 823-834, 1962.

(75) ARSECULERATNE S. N., DE SILVA L. M., BANDUNATHA C. H. S. R., TENNEKOON G. E., WIJESUNDERA S. et BALASUBRAMANIAM K. — The use of tadpoles of *Bufo melanostictus* (Schneider), *Rhacophorus leucomystax maculatus* (Grey) and *Uperodon sp.* in the bioassay of aflatoxins. *Brit. J. Exp. Pathol.*, t. L, p. 285-294, 1969.

(76) ARSECULERATNE S. N., DE SILVA L. M., WIJESUNDERA S. et BANDUNATHA C. H. S. R. — Coconut as a medium for the experimental production of aflatoxin. *Appl. Microbiol.*, t. **XVIII**, p. 88-94, 1969.

(77) ARTANDI C. — Orientations importantes en matière de radiostérilisation. *J. Biol. Méd. Nucl. A.T.E.N.*, n° 78, p. 21-27, 1969.

(78) ASAO T., BUCHI G., ABDEL KADER M. M., CHANG S. B., WICK E. L. et WOGAN G. N. — Aflatoxins B and G. *J. Am. Chem. Soc.*, t. **LXXXV**, p. 1706-1707, 1963.

(79) ASAO T., BUCHI G., ABDEL KADER M. M., CHANG S. B., WICK E. L. et WOGAN G. N. — The structures of aflatoxin B_1 and G_1. *Mycotoxins in Foodstuffs*, p. 265-273, M.I.T. Press, 1965.

(80) ASHLEY J. N., RAISTRICK H. et RICHARDS T. — Studies in the biochemistry of microorganisms. 62. The crystalline colouring matters of species in the *Aspergillus glaucus* series. *Bioch. J.*, t. **XXXIII**, p. 1291, 1939.

(81) ASHLEY L. M. — Histopathology of rainbow trout aflatoxicosis. *U.S. Fish Wildlife Serv. Res. Rep.*, t. **LXX**, p. 48-55, 1967.

(82) ASHLEY L. M. et HALVER J. E. — Hepatomagenesis in Rainbow Trout hepatoma. *Fed. Proc.*, t. **XX**, p. 290, 1961.

(83) ASHLEY L. M. et HALVER J. E. — Multiple metastasis of rainbow trout hepatoma. *Trans. Am. Fish. Sa.*, t. **XCII**, p. 365-371, 1963.

([84]) ASHLEY L. M., HALVER J. E., GARDNER W. K. et WOGAN G. N. — Crystalline aflatoxins cause trout hepatoma. *Feder. Proc. U. S. A.*, t. **XXIV**, p. 627, 1965.

([85]) ASHWORTH L. J. et LANGLEY B. C. — The relationship of pod damage to kernel damage by molds in Spanish peanut. *Plant Disease Reptr.*, t. **XLVIII**, p. 875-878, 1964.

([86]) ASHWORTH L. J. et MCMEANS J. L. — Association of *Aspergillus flavus* and aflatoxins with a greenish yellow fluorescence of Cotton seed. *Phytopathology*, t. **LVI**, p. 1104-105, 1966.

([87]) ASHWORTH L. J., MCMEANS J. L. et BROWN C. M. — Infection of cotton by *Aspergillus flavus* : the influence of temperature and aeration. *Phytopathology*, t. **LIX**, p. 669-673, 1969.

([88]) ASHWORTH L. J., MCMEANS J. L. et BROWN C. M. — Infection of cotton by *Aspergillus flavus*: time of infection and the influence of fiber moisture. *Phytopathology*, t. **LIX**, p. 383-385, 1969.

([89]) ASHWORTH L. J., MCMEANS J. L., PYLE J. L., BROWN C. M., OSGOOD J. W. et PONTON R. E. — Aflatoxins in Cotton seeds: influence of weathering on toxin content of seeds and on a method for mechanically sorting seed lots. *Phytopathology*, t. **LVIII**, p. 102-107, 1968.

([90]) ASHWORTH L. J., RICE R. E., MCMEANS J. L. et BROWN C. M. — The relationship of insects to infection of cotton bolls by *Aspergillus flavus*. *Phytopathology*. t. **LXI**, p. 488-493, 1971.

([91]) ASHWORTH L. J., SCHROEDER H. W., LANGLEY B. C. — Aflatoxins. Environmental factors governing occurrence in Spanish peanuts. *Science*, t. **CXLVIII**, fasc. 3674, p. 1228-1229, 1965.

([92]) ASHWORTH L. J. et THAMES W. H. — Comparative pathogenicity of *Sclerotium rolfsii* and *Rhizoctonia solani* to Spanish peanut. *Phytopathology*, t. **LI**, p. 600-605, 1961.

([93]) ASPLIN F. D. et CARNAGHAN R. B. A. — The toxicity of certain groundnut meals for poultry, with special reference to their effect on ducklings and chickens. *Vet. Rec.*, t. **LXXIII**, p. 1215-1219, 1961.

([94]) ATKINSON N. — Antibacterial substances produced by moulds. 1. Penicidin, a product of the growth of a *Penicillium*. *Aust. J. Biol. Med. Sci.*, t. **XX**, p. 287-288, 1942.

([95]) ATKINSON N. et STANLEY N. F. — Antibacterial substances produced by molds. 4. The detection and occurrence of suppressors of penicidin activity. *Aust. J. Biol. Med. Sci.*, t. **XXI**, p. 249-253, 1943 — 5. The mechanism of the action of some penicidin suppressors. *Id.*, t. **XXI**, p. 255-257, 1943.

([96]) AUCAMP J. L. — The role of mites in the development of aflatoxin in groundnuts. *1st Int. Congr. Pl. Pathol.*, Londres, juil. 1968.

([97]) AUST S. D., ALBRIGHT J. L., OLSEN R. E., BYERS J. H. et BROQUIST H. P. — Observations on moldy corn toxicosis. *J. Anim. Sci.*, t. **XXII**, p. 831-832, 1963.

([98]) AUST S. D. et BROQUIST H. P. — Isolation of a parasympathomimetic alkaloid of fungal origin. *Nature*, t. **CCV**, p. 204, 1965.

([99]) AUST S. D., BROQUIST H. P. et RINEHART K. L. — Slaframine, a parasympathomimetic from *Rhizoctonia leguminicola*. *Biotechnol. Bioeng.*, t. **X**, p. 403-412, 1968.

([100]) AUSTIN M. L., WIDMAYER D. et WALKER L. M. — Antigenic transformation as an adaptive response of *Paramecium aurelia* to patulin; relation to cell division. *Physiol Zool*, t. **XXIX**. p. 261-287, 1956.

([101]) AUSTWICK P. K. C. — Deterioration of moist grain in silos. Memorandum on veterinary aspects, 2 p. ronéot., *Central Vet. Lab.*, 1964.

([102]) AUSTWICK P. K. C. — *Pathogenicity in* Raper et Fennel. *The genus Aspergillus*, p. 82-126, 1965.

([103]) AUSTWICK P. K. C. — The significance of fungi in animal feeding stuffs. *Proc. Symp. Agric. Coll. Norway*, 1967, 9 p. ronéot., 1967.

([104]) AUSTWICK P. K. C. — Mycotoxins-Introductory survey. *1st Int. Congr. Pl. Pathol-Londres, juil.* 1968, *Abstr.*, p. 7, 1968.

([105]) Austwick P. K. C. et Ayerst G. — Toxic products in groundnuts : Groundnut microflora and toxicity. *Chemistry and Industry*, t. **II**, p. 55-61, 1963.

([106]) Austwick P. K. C. et Elphick J. J. — The occurrence of toxin-producing isolates in the *Aspergillus flavus-oryzae* series. *10th Int. Botan. Congr., Abstr.*, p. 69, Edinburgh, 1964.

([107]) Austwick P. K. C. et Venn J. A. J. — Mycotic abortion in England and Wales 1954-1960. *Proc. 4th. Intern. Congr. Animal Reproduct.*, t. **III**, p. 562-568, 1962.

([108]) Ayala G. F., Lins S. et Vasconetto C. — Penicillin as epileptogenic agent: its effect on an isolated neuron. *Science*, t. **CLXVII**, p. 1257-1259, 1970.

([109]) Ayres J. L., Lee D. J. et Sinnhuber R. O. — Preparation of ^{14}C— and ^{3}H— labeled aflatoxins. *J. Ass. Offic. Anal. Chem.*, t. **LIV**, p. 1027-1031, 1971.

([110]) Ayres J. L., Lee D. J., Wales J. H. et Sinnhuber R. O. — Aflatoxin structure and hepatocarcinogenicity in rainbow trout (*Salmo gairdneri*). *J. Nat. Cancer Inst.*, t. **XLVI**, p. 561-564, 1971.

([111]) Ayres J. L. et Sinnhuber R. O. — Fluorodensitometry of aflatoxin on thin-layer plates. *J. Amer. Oil Chem. Soc.*, t. **XLIII**, p. 423-424, 1966.

([112]) Bababunmi E. A. et Bassir O. — The effect of aflatoxin on blood clotting in the rat. *Brit. J. Pharmacol.*, t. **XXXVII**, p. 497-500, 1969.

([113]) Babudieri B. — L'azione antibatterica di alcuni antibiotici, studiata col microscopio elettronico. *Rend. Ist. Super. Sanita* (Rome), t. **XI**, p. 577-598, 1948.

([114]) Badiali L., Abou-Youssef M. H., Radway A. I., Hamby F. M. et Hildebrandt P. K. — Moldy corn poisoning as the major cause of encephalomalacia syndrome in Egyptian equidae. *Amer. Vet. Res.*, t. **XXIX**, p. 2029-2035, 1968.

([115]) Baies A., Contiu I., Petrica L., Dancea I. et Diaconescu I. — Observatii asupra unui focar de micotoxicoza la coi cu « Fusarium sporotrichioides » in regiunea Banat. *Stud. Cerc. Biol. Sti. Agric., Timisoara*, t. **VIII**, p. 73-88, 1961.

([116]) Bailey J. C. et Fulmer J. M. — Aspergillosis of orbit. Report of a case treated by the newer antifungal antibiotic agents. *Am. J. Ophthalmol.*, t. **LI**, p. 670-675, 1961.

([117]) Bailey W. S. et Groth A. H. — The relationship of hepatitis X of dogs and moldy corn poisoning of swine. *J. Am. Vet. Med. Assoc.*, t. **CXXXIV**, p. 514-516, 1959.

([118]) Baird B. M., Halsall T. G., Jones E. E. H. et Lowe G. — Cephalosporin P_1. *Proc. Chem. Soc.*, p. 257-258, 1961.

([119]) Bakalivanov D. — Vurkhu stimulatsionnite i toksichnite svoistva na nyajoi pochveni mikroscopichni gubi. *Izv. Inst. Fiziol. Rast. Sof.*, t. **XV**, p. 133-152, 1966.

([120]) Baker G. J., Greer E. N., Hinton J. J. C., Jones C. R. et Stevens D. J. — The effect on flour color of *Cladosporium* growth on wheat. *Cereal Chem.*, t. **XXXV**, p. 260-275, 1958.

([121]) Baker H. R. et Jacquette D. S. — Observations concerning the « hemorrhagic syndrome » of poultry. *Proc. 25th Ann. Conf. Lab. Workers in Pullorum disease Control, Amherst, Mass.*, 1953.

([122]) Baker R. O. — Pulmonary mucormycosis. *Am. J. Path.*, t. **XXXII**, p. 287-313, 1956.

([123]) Balabanov V. A. — Toksicheskie griby kukuruzy v Moldavii. *Tr. Vses. Nauchissled Inst. Vet. Sanit.*, t. **XXVIII**, p. 48-54, 1967.

([124]) Balasubramaniam K., Wijesundera S., Arseculeratne S. N. et Tennekoon G. E. — Effect of aflatoxins on rat liver lysosomes. *Toxicon*, t. **VII**, p. 159-161, 1969.

([125]) Baldwin J. E., Barton D. H. R., Bloomer J. R., Jackman L. M., Rodriguez-Hahn L. et Sutherland J. K. — The constitution of glauconic, glaucanic and byssochlamic acids. *Experientia*, t. **XVIII**, p. 345-352, 1962.

([126]) Balis M. E. — Antagonists and nucleic acid. (*Frontiers of Biology*, t. X), 293 p., North Holland Publ. Co, Amsterdam, 1968.

([127]) Bamburg J. R., Marasas W. F., Riggs N. V., Smalley E. B. et Strong F. M. — Toxic spiroepoxy compounds from Fusaria and other Hyphomycetes. *Biotechnol. Bioengin.*, t. **X**, p. 445-455, 1968·

([128]) Bamburg J. R., Riggs N. V. et Strong F. M. — The structures of toxins from two strains of *Fusarium tricinctum. Tetrahedron*, t. **XXIV**, p. 3329-3336, 1968.

([129]) Bamburg J. R. et Strong F. M. — Mycotoxins of the trichothecane family produced by *Fusarium tricinctum* and *Trichoderma lignorum. Phytochemistry*, t. **VIII**, p. 2405-2410, 1969.

([130]) Bamburg J. R. et Strong F. M. — 12, 13-Epoxytrichothecenes. *in* Kadis S. et al. Microbial Toxins, t. **VII**, p. 207-292, 1971.

([131]) Bamford P. C., Norris G. L. F. et Ward G. — Flavipin production by *Epicoccum* spp. *Trans. Brit. Mycol. Soc.*, t. **XLIV**, fasc. 3, p. 354-356, 1961.

([132]) Bampton S. S. — Growth of *Aspergillus flavus* and production of aflatoxin in groundnuts. *Trop. Sci.*, t. V, p. 74-81, 1963.

([133]) Banks J. C. et Lynd J. Q. — Infrared detection of *Aspergillus parasiticus* growth on moist sorghum grain. *Agron. J.*, t. **LXIII**, p. 340-342, 1971.

([134]) Băr H. — Untersuchungen über die Wirkungsweise der Fusarinsäure. *Phytopath. Z.*, t. **XLVIII**, p. 149-177, 1963.

([135]) Barber R. S., Braude R., Mitchell K. G., Harding J. D. J., Lewis G. et Loosmore R. M. — The effects of feeding toxic groundnut meal to growing pigs and its interaction with high-copper diets. *Brit. J. Nutr.*, t. **XXII**, p. 535-554, 1968.

([136]) Barbosa M. A. de F. — Sobre a influencia dos fungos do amendoim armazenado na qualidade da semente e de oleo. *Garcia de Orta*, t. **X**, p. 655-660, 1962.

([137]) Barnes G. L. — Mycoflora of developing peanut pods in Oklahoma. *Mycopathol. Mycol. Appl.*, t. **XLV**, p. 85-92, 1971.

([138]) Barnes G. L., Nelson G. L. et Manbeck H. B. — Effects of drying, storage gases, and temperature on development of mycoflora and aflatoxins in stored high-moisture peanuts. *Phytopathology*, t. **LX**, p. 581, 1970.

([139]) Barnes G. L. et Young H. C. — Relationship of harvesting methods and laboratory drying procedures to fungal populations and aflatoxins in peanuts in Oklahoma. *Phytopathology*, t. **LXI**, p. 1180-1184, 1971.

([140]) Barnes J. M. — Toxic fungi with special reference to aflatoxin. *Trop. Sci.*, t. **IX**, p. 64-74, 1967.

([141]) Barnes J. M. et Butler W. H. — Carcinogenic activity of aflatoxin to rats. *Nature*, t. **CCII**, p. 1016, 1964.

([142]) Barnes J. M. et Magee P. N. — Some toxic properties of dimethylnitrosamine. *Brit. J. Industr. Med.*, t. **XI**, p. 167, 1954.

([143]) Barnum C. C. — The production of substances toxic to plants by *Penicillium expansum* Link. *Phytopathology*, t. **XIV**, p. 238-243, 1924.

([144]) Baron, Buzas, Clement et Dufour. — Etude d'une substance antibiotique : la mycoïne. *Bull. Soc. Chim. Fr.*, t. **XVIII**, p. 526, 1951.

([145]) Barrentine B. F. et Morrison E. G. — Grass tetany in sheep grazing winter forages. *Proc. Assoc. Southern Agr. Workers*, t. **L**, p. 64, 1953.

([146]) Barshad I. — Molybdenum content of pasture plants in relation to toxicity to cattle. *Soil. Sci.*, t. **LXVI**, p. 187, 1948.

([147]) Bartlett S., Brown B. B., Foot A. S., Rowland S. J., Allcroft R. et Parr W. H. — The influence of fertilizer treatment of grasslands on the incidence of hypomagnesaemia in milking cows. *Brit. Vet. J.*, t. **CX**, p. 3, 1954.

([148]) Barton D. H. R., De Mayo P., Morrison G. A., Shaeppi W. H. et Raistrick H. — Some observations on the constitution of herqueinone and related compounds. *Chem. and Ind.*, p. 552-553, 1956.

([149]) Barton-Wright E. C. — Studies on the storage of wheaten flour. III. Changes in the flora and the fats and the influence of these changes on gluten character. *Cereal Chem.*, t. **XV**, p. 521-541, 1938.

([150]) BARUAH H. K. — The air spora of a cowshed. *J. Gen. Microbiol.*, t. XXV, p. 483-491, 1961.

([151]) BASAPPA S. C., JAYARMAN A., SREENIVASAMURTHY V. et PARPIA H. A. B. — Effect of B-group vitamins and ethyl alcohol on aflatoxin production by *Aspergillus oryzae. Indian J. Exp. Biol.*, t. V, p. 262-263, 1967.

([152]) BASEDEN S. et ALDRICK S. — Toxigenic *Aspergillus flavus* in northern Australia. *J. Aust. Inst. Agr. Sci.*, t. **XXXVI**, p. 237-240, 1970.

([153]) BASSETT E. et TANNENBAUM S. — The metabolic products of *Penicillium patulum* and their probable interrelationship. *Experientia*, t. **XIV**, p. 38-40, 1958.

([154]) BASSIR O. — The use of Nigerian foods containing cultures of toxic strains of *Aspergillus flavus.* (*in.* Proc. Symp. Chem. and Nigerian Food Probl., avr. 1964) *West Afr. J. Biol. Appl. Chem.*, t. **VIII**, p. 3-15, 1964.

([155]) BASSIR O. et ADEKUNLE A. — Two new-metabolites of *Aspergillus flavus. Febs. lett.*, t. **II**, p. 23-25, 1968.

([156]) BASSIR O. et ADEKUNLE A. — Teratogenic action of aflatoxin B_1, palmotoxin Bo and palmotoxin Go on the chick embryo. *J. Pathol.*, t. **CII**, p. 49-51, 1970.

([157]) BASSIR O. et ADEKUNLE A. — Production of aflatoxin B_1 from defined natural cultures of *Aspergillus flavus* Link. *Mycopathol. Mycol. Appl.*, t. **XLVI**, p. 241-246, 1972.

([158]) BASSIR O. et EMAFO P. O. — Oxidative metabolism of aflatoxin B_1 by mammalian liver slices and microsomes. *Biochem. Pharmacol.* t. **XIX**, p. 1681-1687, 1970.

([159]) BASSIR O. et OSIYEMI F. — Biliary excretion of aflatoxin in the rat after a single dose (*Aspergillus flavus*). *Nature, G. B.*, t. **CCXV**, p. 882, 1967.

([160]) BASTIN R. — Antibiotic citrinin. *Rev. Ferm. Indust. Alim.*, t. **VII**, p. 1, 1952.

([161]) BAUCH R., SEIDLEIN H. J., VALENTIN J. — Metabolic products of higher fungi in connection with ergot and corn smut investigations. **I.** *Pharmazie*, t. **XIV**, p. 582-596, 1959. II. *Pharmazie*, t. **XV**, p. 719-721, 1960.

([162]) BAUDYS E. — Die Sporen der Getreidebrandpilze sind nicht giftig. *Zeitschr. f. Pflanzenkrankh.*, t. **XXXI**, p. 24-27, 1921.

([163]) BAUER D., LEE D. J. et SINNHUBER R. O. — Acute toxicity of aflatoxins B_1 and G_1 in the rainbow trout (*Salmo gairdneri*). *Toxicol. Appl. Pharmacol.*, t. XV, p. 415-419, 1969.

([164]) BAUER H., AJELLO J., ADAMS E. L. et HERNANDEZ D. H. — Cerebral mucormycosis pathogenesis. *Amer. J. med.*, t. **XVIII**, p. 822-831, 1955.

([165]) BAUER L. et LEISTNER L. — Die Wirkung von Aflatoxin B_1 auf Amöben. *Arch. Hyg.*, t. **CLIII**, p. 397-402, 1969.

([166]) BAUER L. et MÜLLER E. — Aflatoxin-induzierte nukleolare Kappen in Molluskenzellen. *Naturwissenschaften*, t. **LVII**, p. 457-458, 1970.

([167]) BAWDEN F. C. et FREEMAN G. G. — The nature and behaviour of inhibitors of plant viruses produced by *Trichothecium roseum* Link. *J. Gen. Microbiol.*, t. **VII**, p. 154-168, 1952.

([168]) BEAN G. A., KLARMAN W. L., RAMBO G. W. et SANFORD J. B. — Dimethyl sulfoxide inhibition of aflatoxin synthesis by *Aspergillus flavus. Phytopathology*, t. **LXI**, p. 380-382, 1971.

([169]) BEAN G. A., KLARMAN W. L. et SANFORD J. B. — Inhibition of aflatoxin synthesis by *Aspergillus flavus* using dimethyl sulfoxide. *Phytopathology*, t. **LIX**, p. 1017-1018, 1969.

([170]) BEAN G. A., RAMBO G. W. et KLARMAN W. L. — Influence of dimethyl sulfoxide on conidial pigmentation, and consequent ultraviolet light sensitivity of aflatoxin-producing strains of *Aspergillus flavus. Phytopathology*, t. **LIX**, p. 1555, 1969.

([171]) BEARD R. — Mycotoxin : a cause of death in milkweed bugs. *J. Invert. Pathol.* t. **X**, p. 438-439, 1968.

([172]) BEARD R. et WALTON G. — An *Aspergillus* toxin lethal to larvae of house fly. *J. Invert. Pathol.*, t. **VII**, p. 522-523, 1965.

([173]) Beard R. L. et Walton G. S. — Kojic acid as an insecticidal mycotoxin. *J. Invert. Pathol.*, t. **XIV**, p. 53-59, 1969.

([174]) Beath O. A. — Economic potential and botanic limitation of some Selenium-bearing plants. *Wyoming Agr. Expt. Stn. Bull.*, n° 360, 1959.

([175]) Becker Z. E., Suprun T. P., Dmitrieva S. W. et Nesterenko E. J. — Morphogenesis and metabolism of fungi with special attention to the nucleic acids and their antimetabolites. *Abstr. 10th Int. Bot. Congr.*, p. 38-39, 1964.

([176]) Beckwith A. C. et Stoloff L. — Fluorodensitometric measurement of aflatoxin thin layer chromatograms. *J. Ass. Offic. Anal. Chem.*, t. **LI**, p. 602-608, 1968.

([177]) Becquerel L. — La vie latente des spores des bactéries et des moisissures. *Trav. Crypt. dèdiés a L. Mangin, Muséum Hist. Nat.*, Paris, p. 303-307, 1931.

([178]) Bedi K. S. — Factors affecting viability of sclerotia of *Sclerotinia sclerotiorum* (Lib.) de Bary. *Indian J. Agr. Sci.*, t. **XXXI**, p. 236-245, 1964.

([179]) Beecham A. F., Fridrichsons J. et McL. Matieson A. — The structure and absolute configuration of gliotoxin and the absolute configuration of sporidesmin. *Tetrahedron Lett.*, t. **XXVII**, p. 3131-3138, 1966.

([180]) Bekker Z. E. et Silaev A. B. — Fumagillin from a producer isolated in the U. S. S. R. *Antibiotiki*, t. **VI**, p. 4, 1958.

([181]) Beljaars P. R. — Some methods of analysis for the determination of aflatoxin B_1 in peanuts and peanut products. *Vitatron, Application Rept. TLD 100*, n° 2, p. 13-24, 1970.

([182]) Bell D. K. — Fungi from hypocotyls and senescent cotyledons of peanuts from fungicide treated seed planted in two soils. *Plant Dis. Reptr.*, t. **L**, p. 162-166, 1966.

([183]) Bell D. K. — Pathogenicity of fungi to peanut seedlings in known fungal culture at four temperatures. *Oléagineux*, t. **XXII**, p. 373-375, 1967.

([184]) Bell D. K. et Crawford J. L. — A botran-amended medium for isolating *Aspergillus flavus* from peanuts and soil. *Phytopathology*, t. **LVII**, p. 939-941, 1967.

([185]) Bell H. G. — Fungus and bacterial growth on stored flour. *Amer. Miller. Processor*, t. **XXXVII**, p. 280-281, 1909.

([186]) Bell M. R., Johnson J. R., Wildi B. S. et Woodward R. B. — The structure of gliotoxin. *J. Am. Chem. Soc.*, t. **LXXX**, p. 1001, 1958.

([187]) Beller K. et Wedemann W. — Untersuchung über die Schadwirkung amerikanischer Futtergerste (sog. Barley Federal Nr II). *Z. Infektionskrkh., parasit. Krankh. u. Hyg. d. Haustiere*, t. **XXXVI**, p. 103-129, 1929.

([188]) Bendixen C. H. et Plum N. — Schimmelpilze (*Aspergillus fumigatus* und *Absidia ramosa*) als Abortursache beim Rinde. *Acta path. microbiol. Scan.*, t. **VI**, p. 252-322, 1929.

([189]) Benjamin C. R. — The changing face of mycology. *Mycologia*, t. **XL**, p. 1-8, 1968.

([190]) Bentley R. et Keil J. G. — Tetronic acid biosynthesis in molds II. Formation of penicillic acid in *Penicillium cyclopium*. *J. Biol. Chem.*, t. **CCXXXVII**, p. 867-873, 1962.

([191]) Bergel F., Morrison A. L., Moss A. R., Klein R., Rinderknecht H. et Ward J. L. — An antibacterial substance from *Aspergillus clavatus* and *Penicillium claviforme* and its probable identity with patulin. *Nature, G. B.*, t. **CLIII**, p. 750, 1943.

([192]) Bergel F., Morrison A. L., Moss A. R. et Rinderknecht H. — An antibacterial substance from *Aspergillus clavatus*. *J. Chem. Soc.*, p. 415-421, 1944.

([193]) Bernhard W. — *Intervention au Colloque aflatoxines.* Villejuif, 7 mars 1966.

([194]) Bernhard W., Frayssinet C., Lafarge C. et Le Breton E. — Lésions nucléolaires précoces provoquées par l'aflatoxine dans les cellules hépatiques du rat *C. R. Acad. Sci.*, t. **CCLXI**, p. 1785-1788, 1965.

([195]) Bertaud W. S., Probine M. C., Shannon J. S., et Taylor A. — Isolation of a new depsipeptide from *Pithomyces chartarum*. *Tetrahedron*, t. **XXI**, p. 677-680, 1965.

([196]) Betina V., Nemec P., Kovak S., Kjaer A. et Shapiro R. H. — *Acta Chem. Scand.*, t. **XIX**, p. 519, 1965.

([197]) Beutner E. — Immunofluorescent staining: the fluorescent antibody method. *Bacteriol. Rev.*, t. **XXV**, p. 49-76, 1961.

([198]) Biester H. E. et Schwarte L. H. — *Diseases of poultry.* 5e éd., 1382 p., Iowa State University Press, 1965.

([199]) Biester H. E., Schwarte L. H. et Reddy C. H. — Further studies on moldy corn poisoning (leucoencephalomalacia) in horses. *Vet. Med.*, t. **XXXV**, p. 636-639, 1940.

([200]) Bieth J. — Les aflatoxines. propriétés physico-chimiques et effets biologiques. *Prod. et Prob. pharm.*, t. **XXII**, p. 243-252, 1967.

([201]) Bigot S. K. et Schramm L. C. — Investigations of *Gibberella zeae* — toxins. *Lloydia*, t. **XXIX**, p. 380, 1966.

([202]) Bihaly A., Kostyak J. et Orosz D. — Aflatoxinnal fertözött takarmannyal vegzett etetési kiserletek serteseken. *Allattenyésztés, Magyar*, t. **XIV**, p. 243-249, 1965.

([203]) Bilai V. I. — (Effet d'extraits de champignons toxiques sur des tissus animaux et végétaux.) *Mikrobiologiya*, t. **XVII**, p. 142-147, 1948.

([204]) Bilai V. I. — Yadovitye griby na zerne khlebnykh zlakov (Champignons toxiques sur céréales). *Izdatel'stvo Akad. Nauk Ukrain S.S.R.*, 93 p., Kiev, 1953.

([205]) Bilai V. I. — Fusaria. 320 p., Kiev, 1955.

([206]) Bilai V. I. — Mikotoxikozy celoveka i selskochozjajstvennych zivotnych. *Izdatel'stvo Akademii Nauk Ukrainskoi S.S.R.*, 168 p., Kiev, 1960.

([207]) Bilai V. I. — *Antibiotic-producing microscopic fungi.* 215 p., Elsevier Publ. Co, 1963.

([208]) Bilai V. I. — Toxin producing fungi on cereal seeds and fodder, and some problems of health protection. *1st Int. Congr. Pl. Pathol.*, Londres, juil. 1968.

([209]) Bilai V. I. — Aflatoksyny. *Mikrobiol. Zh.*, t. **XXI**, p. 123-128, 1969.

([210]) Bilai V. I. — Phytopathological and hygienic significance of representatives of the section *Sporotrichiella* in the genus *Fusarium* Lk. *Ann. Acad. Sci. Fenn.*, A. t. **CLXVIII**, p. 19-24, 1970.

([211]) Bilai V. I., Mikhailovnina A. A. et Stepanov F. N. — The active principale dendrodochine. *Dokl. Akad. Nauk SSSR*, Chem. Sect., t. **CXLIV**, p. 105-107, 1962.

([212]) Bilai V. I. et Pidoplichko N. M. — *Toksinobrazuvushchie mikroskopicheskie griby i vyzyvaemye imi zabolevaniya cheloveka i zhivotnykh.* 291 p., Kiev, 1970.

([213]) Bilai V. I. et Proskuriakova N. S. — Microbial detoxification of dendrodochine *Mikrobiologiya*, t. **XXXIX**, p. 293-299, 1970.

([214]) Billet D. — Progrès récent dans le chimie des xanthones naturelles: *Actualités de phytochimie fondamentale*, 2° série, p. 35-43, 1966.

([215]) Biollaz M., Büchi G. et Milne G. — Biosynthesis of aflatoxin. *J. Am. Chem. Soc.*, t. XC, p. 5017-5019, 1968.

([216]) Biollaz M., Büchi G. et Milne G. — The biogenesis of bisfuranoids in the genus *Aspergillus. J. Am. Chem. Soc.*, t. **XC**, p. 5019-5020. 1968.

([217]) Biollaz M., Büchi G. et Milne G. — Biosynthesis of the aflatoxins. *J. Am. Chem.* Soc., t. **XCII**, p. 1033-1055, 1970.

([218]) Birch A. J. — *Ciba foundation Symposium on amino-acids and peptids with antimetabolic activity*, 247 p. Londres, 1958.

([219]) Birch A. J., Cameron D. W., Holloway P. W. et Rickards R. W. — Further examples of biological C-methylation. Novobiocin and actinomycin. *Tetrahedron Letters*, t. **XXV**, p. 26-31, 1960.

([220]) Birch A. J. et Wright J. J. — *Chem. commun.*, p. 644. 1969.

([221]) Birch A. J. et al. — Ipomeamarone and ngaione. *Chem. and Ind.*, t. **XXIX**, p. 902, 1954.

([222]) Birchir I. G. — Possible Warfarin poisoning. *Vet. Med.*, t. **XLVI**, p. 416, 1951.

([223]) BIRD K. — Selected writings on freezedrying of foods. *U.S. Dept. Agr.*, 1964.

([224]) BIRKINSHAW J. H., BRACKEN A., GREENWOOD M., GYE W. E., HOPKINS W. A., MICHAEL S. E. et RAISTRICK H. — Patulin in the common cold. *Lancet*, t. **II**, p. 625-635, 1943.

([225]) BIRKINSHAW J. H., BRACKEN A. et RAISTRICK H. — Studies in biochemistry of micro-organisms. 72. Gentisyl alcohol (2:5-dihydroxybenzen alcohol) a metabolic product of *Penicillium patulum* Bainier. *Biochem. J.*, t. **XXXVII**, p. 726-728, 1943.

([226]) BIRKINSHAW J. H. et HAMMADY I. M. H. — Studies in the biochemistry of micro-organisms. 99. Metabolic products of *Aspergillus versicolor* (Vuillemin) Tiraboschi. *Biochem. J.*, t. **LXV**, p. 162-165, 1957.

([227]) BIRKINSHAW J. H., LUCKNER M., MOHAMMED Y. S. et STICKINGS C. E. — Biochemistry of micro-organisms CXIV. Viridicatol and cyclopenol metabolites of *Penicillium viridicatum* and *Penicillium cyclopium*. *Biochem. J.*, t. **LXXXIX**, p. 196, 1963.

([228]) BIRKINSHAW J. H., OXFORD A. E. et RAISTRICK H. — Studies in biochemistry of micro-organisms, 48. Penicillic acid, a metabolic product of *Penicillium puberulum* Bain and *P. cyclopium* West. *Bioch. J.*, t. **XXX**, p. 394, 1936.

([229]) BIRKINSHAW J. H., RAISTRICK H., ROSS D. J. et STICKINGS C. E. — Cyclopolic and cyclopaldic acids, metabolic products of *Penicillium cyclopium* West. *Bioch. J.* t. **L**, p. 610, 1952.

([230]) BISBY G. R. — *Stachybotrys. Trans. Brit. Mycol. Soc.*, t. **XXVI**, p. 133-143, 1943.

([231]) BITANCOURT A. A. — Podridnoes da castanha do para. *Biologico*, t. **VII**, p. 303-312, 1941.

([232]) BIXLER E. et LOPEZ L. C. — Estudios preliminares en aves sobre la toxicidad de los granos atacados por *Aspergillus flavus. Tecnica Pecuaria en Mexico*, n° 2, p. 27-29, 1963.

([233]) BJORNSON C. B., MCILWAIN P., EVELETH D. F. et BOLIN F. M. — Sources of nitrate intoxication. *Vet. Med.*, t. **LVI**, p. 198, 1961.

([234]) BLACK H. S. et ALTSCHUL A. M. — Gibberellic acid-induced lipase and α-amylase formation and their inhibition by aflatoxin. *Biochem. Biophys. Res. Commun.* t. **XIX**, p. 661-665, 1965.

([235]) BLAKESLEE A. F. et GORTNER R. A. — On the occurrence of a toxin in juice expressed from the bread mould, *Rhizopus nigricans* (*Mucor stolonifer*). *Biochem. Bull.*, t. **II**, p. 542-544, 1913.

([236]) BLAKESLEE A. F. et GORTNER R. A. — Reaction of rabbits to intravenous injections of mould spores. *Biochem. Bull.*, t. **IV**, p. 45-51, 1915.

([237]) BLINC M. et JOHANIDES V. — Antibiotics from aspergilli with special regard to species isolated in Yugoslavia. *Bull. Sci. Conseil acad. R. P. F. Yougoslavie*, t. **II**, p. 99, 1956.

([238]) BLOCHWITZ A. — Luftmyzelbildungen bei Schimmelpilzen. *Botan. Centr. Beih.*, Abt. A., t. **XLVIII**, p. 176-182, 1931.

([239]) BLOOMER J. L., MOPPET C. E. et SUTHERLAND J. K. — The nonadrides V. Biosynthesis of glauconic acid. *J. Chem. Soc.* (C), p. 588-591, 1968.

([240]) BLOUNT W. P. — « Disease » of turkey poults. *Vet. Rec.*, t. **LXXII**, p. 786, 1960.

([241]) BLOUNT W. P. — Turkey « X » disease and the labelling of poultry foods. *Vet. Rec.*, t. **LXXIII**, p. 227, 1961.

([242]) BLOUNT W. P. — Turkey « X » disease. *Proc. 10th Ann. Congr. Brit. Turkey Fed.*, p. 52-61, 1961.

([243]) BLOUNT W. P., FRASEE D. C. K., KNIGET D. et DOWLING W. M. — The use of ducklings for the detection of aflatoxin. *Vet. Rec.*, t. **LXXV**, p. 35, 1963.

([244]) BLUM H. F. — Photodynamic action and diseases caused by light. *Am. Chem. Soc., Monograph Ser.*, 310 p., Rheinhold, Publ. Corp. New York, 1941.

([245]) BODIN E. et GAUTIER L. — Note sur une toxine produite par l'*Aspergillus fumigatus. Ann. Inst. Pasteur.*, t. **XX**, p. 209, 1906.

([246]) BOESENBERG H. — Das Auftreten fluoreszierender Abkommlinge von Aflatoxin. *Arch. Hyg. Bakteriol.*, t. **CLIII**, p. 37-40, 1969.

([247]) BOESENBERG H. — Zur Bedeutung der Aflatoxine. *Naturwissenschaften*, t. **LVI**, p. 350-352, 1969.

([248]) BOESENBERG H. — Effect and importance of mycotoxins. *Zentralbl. Bakteriol.* Abt. I, t. **CCXII**, p. 222-224, 1970.

([249]) BOESENBERG H. et EBERHARDT E. — Untersuchungen ueber den Verderb von Lebensmittel durch Schimmelpilze in Supermaerkten. *Med. Ernähr.*, t. **X**, p. 12-13, 1969.

([250]) BOESENBERG H. et GERKE G. — Möglichkeiten der Gewinnung von Aflatoxin. *Arch. Hyg. Bakteriol.*, t. **CLIII**, p. 33-36, 1969.

([251]) BOGORODITSKAYA V. P. — Biologicheskie metody opredeleniya toksichnosti zerna v svyazi s porazheniem ego toksigennymi mikroskopicheskimi gribami. *Vop. Pitan*, t. **XXVII**, p. 73-76, 1968.

([252]) BOHOSIEWICZ M. et BUBIEN Z. — Toksyczność aflatoksyny dla zwierząt. *Med. Wet.*, t. **XXIII**, p. 705-708, 1967.

([253]) BOIARSKII I. V. — Otravlenie swinei kormame porozhennymi plesnin. *Veterinariya*, t. **XXVII**, p. 38-39, 1950.

([254]) BOITEAU P., PASICH B. et RAKOTO RATSIMAMANGA A. — *Les triterpénoïdes en physiologie végétale et animale.* 1370 p., Gauthier-Villars édit., 1964.

([255]) BOLLAG W. — Die Wirkung von Patulin auf das myelo-und lymphopoetische System der Maus. *Experientia*, t. **V**, p. 447-448, 1949.

([256]) BOLLER R. A. et SCHROEDER H. W. — Aflatoxin-producing potential of *Aspergillus flavus-oryzae* isolates from rice. *Cereal Science Today*, t. **XI**, p. 342-344, p. 433-435, 1966.

([257]) BOLLER R. A. et SCHROEDER H. W. — Accumulation of aflatoxins in stored rice in relation to competition between *Aspergillus parasiticus* and *A. chevalieri. Phytopathology*, t. **LVII**, p. 725-726, 1968.

([258]) BOMPEIX G. — Contribution à l'étude de la maladie des taches lenticellaires des pommes « Golden Delicious » en France. *Mém. Dipl. Et. Sup. Fac. Sciences, Rennes*, 120 p., 5 mars 1966.

([259]) BONDOUX P. — Les principales maladies cryptogamiques des poires et des pommes en conservation. *Bull. Techn. Inf. des S. A.*, p. 235-249, 1963.

([260]) BONET-MAURY P. — La radiostérilisation dans l'industrie pharmaceutique. *J. Biol. Med. Nucl. A. T. E. N.*, n° 78, p. 7-16, 1969.

([261]) BONILLA-SOTO O., ROSE N. R. et ARBESMAN C. E. — Allergenic molds; antigenic and allergenic properties of *Alternaria tenuis. J. Allerg.*, t. **XXXII**, p. 246-270, 1961.

([262]) BONNER R. D. et FERGUS C. L. — The fungus flora of cattle feeds. *Mycologia*, t. **LI**, p. 855-863, 1959.

([263]) BOOTH C. — The genus *Fusarium.* 237 p., C. M. I., Kew, 1971.

([264]) BORCHERS R. et PELTIER G. L. — Molded foodstuffs. I. Supplementary value in chick starting rations. *Poultry Sci.*, t. **XXVI**, p. 194-197, 1947.

([265]) BORKER E., INSALATA N. F., LEVI C. P. et WITZEMAN J. S. — Mycotoxins in feeds and foods. *Adv. Appl. Microbiol.*, t. **VIII**, p. 315-351, 1966.

([266]) BÖRNER H. — Untersuchungen über die Bildung antiphytotischer und antimikrobieller Substanzen durch Mikroorganismen im Boden und ihre mögliche Bedeutung für die Bodenmüdigkeit beim Apfel (*Pirus malus* L.). I. Bildung von Patulin und einer phenolischen Verbindung durch *Penicillium expansum* auf Wurzel und Blattrückständen des Apfels. II. Der Einfluss verschiedener Faktoren auf die Bildung von Patulin und einer phenolischen Verbindung durch *Penicillium expansum* auf Blatt und Wurzelrückständen des Apfels. *Phytopathologische Zeitschrift*, t. **XLVIII**, fasc. 4, p. 370-396, 1963; t. **XLIX**, p. 1-28, 1963-1964.

([267]) BORUT S. Y. et JOFFE A. Z. — *Aspergillus flavus* Link and other fungi associated with stored groundnuts kernels in Israel. *Israel J. Bot.*, t. **XV**, p. 112-120, 1966.

([268]) Boudreaux H. B. et Frampton V. L. — *Arch. Biochem. Biophys.*, t. **LXXXIX**, p. 276, 1960.

([269]) Boudreaux H. B. et Frampton V. L. — A peanut factor for haemostasis in haemophilis. *Nature*, t. **CLXXXV**, p. 469-470, 1960.

([270]) Bouhot D. — Contribution à l'étude des maladies des gousses et des graines d'arachides dues au *Macrophomina phaseoli. Agron. Trop.*, t. **IX**, p. 864-887, 1967.

([271]) Bouriquet G. et Jaubert P. — Une affection des graines de l'arachide du Sénégal causée par *Sclerotium bataticola. Agron. Trop.*, t. **IX**, p. 197-200, 1954.

([272]) Boutibonnes P. et Jacquet J. — Recherches sur la production de toxine par *Aspergillus flavus. Bull. Acad. Vet. Fr.*, t. **XXXIV**, p. 393-403, 1967.

([273]) Boutibonnes P. et Jacquet J. — Une moisissure de grande actualité biologique : *Aspergillus flavus* Link. *Rev. Immunol.*, t. **XXXI**, p. 293-315, 1967.

([274]) Boutibonnes P. et Jacquet J. — Sur la recherche des produits de sécrétion des microbes. Application aux *Aspergillus. C. R. Soc. Biol.*, t. **CLXII**, p. 118-121, 1969.

([275]) Boutibonnes P. et Jacquet J. — Sur la fréquence des spores et des *Aspergillus* dans les aliments. *C. R. Soc. Biol.*, t. **CLXII**, p. 1119-1124, 1969.

([276]) Boutibonnes P. et Jacquet J. — Effets biologiques des flavacoumarines d'*Aspergillus parasiticus* ATCC 15517. I. Animaux. *Rev. Immunol.*, t. **XXXV**, p. 103-120, 1971.

([277]) Boutibonnes P. et Jacquet J. — Recherches complémentaires sur les effets biologiques des flavacoumarines. *Rev. Immunol.*, t. **XXXVI**, p. 85-101, 1972.

([278]) Boutibonnes P., Jacquet J. et Teherani A. — Chromatographie rapide en couche mince des flavatoxines contenues dans les aliments. *Bull. Acad. Vét.*, t. **XLII**, p. 825-833, 1969.

([279]) Bouwmeester P. C. et Koopmans H. J. — Fluorescence measurements with TLD 100. Quantitative estimation of aflatoxin B_1. *Vitatron, Application Rep. TLD 100*, n° 2, p. 3-10, 1970.

([280]) Boy J. et Chany E. — Influence de la cortisone sur les caractères morphologiques de l'hépatome expérimental du rat. *C. R. Acad. Sci.*, t. **CCL**, p. 3912-3914, 1960.

([281]) Brack A. — Isolierung von Gentisilalcohol neben Patulin aus den Kulturfiltrat eines *Penicillium*-Stamm über einige Derivate des Gentisilalcohols. *Helvet. Chim. Acta*, t. **XXX**, p. 1, 1947.

([282]) Bracken A., Pocker A. et Raistrick H. — Studies in the biochemistry of microorganisms. 93. Cyclopenin, a nitrogen-containing metabolic product of *Penicillium cyclopium. Biochem. Jour.*, t. **LVII**, p. 587-595, 1954.

([283]) Bradley W. B., Eppson H. F. et Beath O. A. — Livestock poisoning by oat hay and other plants containing nitrate. *Wyoming Agr. Expt. Stn. Bull.*, n° 241, 1940.

([284]) Brancato F. P. et Golding N. S. — The diameter of the mold colony as a reliable measure of growth. *Mycologia*, t. **XLV**, p. 848-864, 1953.

([285]) Braun A. C. et Pringle R. B. — Pathogen factors in the physiology of disease. Toxins and other metabolities. *Plant Pathology. Problems and Progress 1098-1958*, p. 88-99, 1959.

([286]) Braun W. — Zur Menschenpathogenität der Schimmelpize. *Mykosen*, t. **X**, p. 141-150, 1967.

([287]) Brechbuhler S., Büchi G. et Milne G. — The absolute configuration of the aflatoxins. *J. organic Chem.*, t. **XXXII**, p. 2641-2642, 1967.

([288]) Brenet M., Centeleghe J. L., Milliere J. B., Ramet J. P. et Weber F. — Etude d'un accident en fromagerie de type « Camembert » causé par des Mucorales. *Lait*, t. **LII**, p. 141-148, 1972.

([289]) Bressani R. — The use of cottonseed protein in human food. *Food Technol.*, t. **XIX**, p. 51-58, 1965.

([290]) Breton A. et Larvor P. — *In litt.*, 1966.

([291]) BRIAN P. W. — Production of gliotoxin by *Trichoderma viride*. *Nature*, t. **CLIV**, p. 667-668, 1944.

([292]) BRIAN P. W. — Production of gliotoxin by *Penicillium terlikowskii* Zal. *Trans. Brit. Mycol. Soc.*, t. **XXIX**, p. 211-219, 1946.

([293]) BRIAN P. W. — Effects of antibiotics on plants. *Ann, Rev. Plant Physiol.*, t. **VIII**, p. 413-426, 1957.

([294]) BRIAN P. W. — Griseofulvin. *Trans. Brit. Mycol. Soc.*, t. **XLIII**, p. 1-13, 1960.

([295]) BRIAN P. W., CURTIS P. J., HEMMING H. G. et MCGOWAN J. C. — The production of viridin by pigment-forming strains of *Trichoderma viride*. *Ann. Appl. Biol.*, t. **XXXIII**, p. 190-200, 1946.

([296]) BRIAN P. W., DAWKINS A. W., GROVE J. F., HEMMING H. G., LOWE D. et NORRIS G. L. F. — Phytotoxic compounds produced by *Fusarium equiseti*. *J. Exptl. Botany*, t. **XII**, p. 1-12, 1961.

([297]) BRIAN P. W., ELSON G. W. et LOWE D. — Production of patulin in apple fruits by *Penicillium expansum*. *Nature*, t. **CLXXVIII**, p. 263-264, 1956.

([298]) BRIAN P. W. et HEMMING H. G. — Gliotoxin, a fungistatic product of *Trichoderma viride*. *Ann. Appl. Biol.*, t. **XXXII**, p. 214, 1945.

([299]) BRIAN P. W. et MCGOWAN J. C. — Viridin : a highly fungistatic substance produced by *Trichoderma viride*. *Nature*, t. **CLVI**, p. 144-145, 1945.

([300]) BRIGGS L. H., COLEBROOK L. D., DAVIS B. R. et LE QUESNE P. W. — Chemistry of fungi. Part I. Pithomycolide, a novel depsipeptide from *Pithomyces chartarum*. *J. Chem. Soc.*, p. 5626-5633, 1964.

([301]) BRINK R. A. et ROBERTS W. L. — The coumarin content of *Melitotus dentata*. *Science*, t. **LXXXVI**, p. 41, 1937.

([302]) BRISARD A. — Etude des moisissures de la farine et de l'atmosphère des boulangeries. *Thèse Fac. Médecine Marseille*, 184 p., 1971.

([303]) BRISTOL F. M. et DJURICKOVIC S. — Hyperestrogenism in female swine as the result of feeding mouldy corn. *Can. Vet. J.*, t. **XII**, p. 132-135, 1971.

([304]) BRITTON I. W. et GOSS H. — Chronic Molybdenum poisoning in cattle. *J. Am. Vet. Med. Assoc.*, t. **CVIII**, p. 176, 1946.

([305]) BROADBENT D. — Antibiotics produced by fungi. *The Bot. Rev.*, t. **XXXII**, p. 219-242, 1966.

([306]) BROADBENT J. A. — Microbiological deterioration of maize used as poultry and live stock feed at farms near Ibadan during the wet season. *Nigerian Stored Prod. Res. Inst. Ann. Rept.*, p. 115-118, 1966.

([307]) BROADBENT J. H., CORNELIUS J. A. et SHONE G. — The detection and estimation of aflatoxin in groundnuts and groundnut materials. II. Thin-layer Chromatographic method. *Analyst*, t. **LXXXVIII**, p. 214, 1963.

([308]) BROCE D., GRODNER R. M., KILLEBREW R. L. et BONNER F. L. — Ochratoxins A and B confirmation by microbiological assay using *Bacillus cereus mycoides*. *J. Ass. Offic. Anal. Chem.*, t. **LIII**, p. 616-619, 1970.

([309]) BROCQ-ROUSSEU D. et FABRE R. — *Les toxines végétales*. 242 p., Paris 1947.

([310]) BROERMAN A. — Poisoning of cattle by sweet clover hay. *J. Am. Vet. Med. Assoc.*, t. **LXVII**, p. 367, 1925.

([311]) BROOK P. J. — Ecology of the fungus *Pithomyces chartarum* (Berk. et Curt.) M. B. Ellis in pasture in relation to facial eczema disease of sheep. *New Zealand J. Agr. Res.*, t. **VI**, p. 147-228, 1963.

([312]) BROOK P. J. — Growth cycle of the fungus *Pithomyces chartarum* (Berk. et Curt.) M. B. Ellis. *New Zealand J. Agr. Res.*, t. **VII**, p. 87-89, 1964.

([313]) BROOK P. J. — *Pithomyces chartarum* and facial eczema of sheep and cattle in New Zealand. *C. R. 10e Congr. Int. Bot.*, p. 68-69, Edimbourg, 1964.

([314]) BROOK P. J. — *Pithomyces chartarum* in pasture, and measures for prevention of facial eczema. *1st Congr. Int. Pl. Pathol.*, Londres, Juil. 1968.

([315]) BROOK P. J. et WHITE E. P. — Fungus toxins affecting mammals. *Ann. Review of Phytopathology*, t. **IV**, p. 171-194, 1965.

([316]) BROOKES D., TIDD B. K. et TURNER W. B. — Avenaciolide, an antifungal lactone from *Aspergillus avenaceus*. *J. Chem. Soc.*, p. 5385-5391, 1963.

([317]) BROOKS F. T. et HANSFORD C. G. — Mould growth upon cold store meat. *Trans. Brit. Mycol. Soc.*, t. **VIII**, p. 113-114, 1923.

([318]) BROOM W. A., BULBRING E., CHAPMAN C. J., HAMPTON J. W. F., THOMSON A. M., UNGAR J., WIEN R. et WOOLFE G. — The pharmacology of patulin. *Brit. J. Exp. Pathol.*, t. **XXV**, p. 195-207, 1944.

([319]) BROQUIST H. P. et SNYDER J. J. — *Rhizoctonia* Toxin. *in* Kadis S. et al., Microbial Toxins, t. **VII**, p. 319-333, 1971.

([320]) BROUNST G. et ALLAME F. — Un cas primaire d'aspergillome urinaire observé au Liban. *Bull. Acad. Nat. Méd.*, t. **CXLVII**, p. 50-55, 1963.

([321]) BROWN A. H. S. et SMITH G. — The genus *Paecilomyces* Bainier and its perfect stage *Byssochlamys* Westling. *Trans. Brit. Mycol. Soc.*, t. **XL**, p. 17, 1957.

([322]) BROWN J. M. et ABRAMS L. — Biochemical studies on aflatoxicosis. *Onderstepoort J. Vet. Res.*, t. **XXXII**, p. 119-146, 1965.

([323]) BROWN J. M., SAVAGE A. et ROBINSON A. D. — A report on some investigations into the cause of sweet clover (*Melilotus*) disease. *Sci. Agr.*, t. **XIII**, p. 561, 1933.

([324]) BROWN J. P., ROBERTSON A., WHALLEY W. B. et CARTWRIGHT N. J. — The constitution of citrinin. *J. Chem. Soc.*, p. 867-879, 1949.

([325]) BROWN M. R. W. et MELLING J. — Inhibition and destruction of microorganisms by heat. *in* HUGO W. B., *Inhibition and destruction of the microbial cell*, p. 1-37, Academic Press, 1971.

([326]) BROWN R. F. — The effect of some mycotoxins on the brine shrimp *Artemia salina*. *J. Am. Oil Chem. Soc.*, t. **XLVI**, p. 119, 1969.

([327]) BROWN R. F. — Some bioassay methods for mycotoxins. *in* Herzberg M. , *Toxic Micro-organisms*, p. 12-18, 1970.

([328]) BROWN R. F., WILDMAN J. D. et EPPLEY R. M. — Temperature — dose relationships with aflatoxin on the brine shrimp. *Artemia salina*. *J. Ass. Offic. Anal. Chem.*, t. **LI**, p. 905-906, 1968.

([329]) BRUCE H. M. et PARKES A. S. — Feeding and breeding of laboratory animals. III. Observations on the feeding of guinea pig. *J. Hyg., Camb.*, t. **XLV**, p. 70-87, 1947.

([330]) BRUCE W. F., DUTCHER J. D., JOHNSON J. R. et MILLER L. L. — Gilotoxin, the antibiotic principle of *Gliocladium fimbriatum*. II. General chemical behaviour and crystalline derivatives. *Amer. Chem. Soc. J.*, t. **LXVI**, p. 614-616, 1944.

([331]) BRUCH C. W. — Gaseous sterilization. *Ann. Rev. Microbiology*, t. **XV**, p. 245-262, 1961.

([332]) BRUCH C. W. et KOESTERER M. G. — Microbicidal activity of gaseous propylene oxide and its application to powdered or flaked foods. *J. Food Sci.*, t. **XXVI**, p. 428-435, 1961.

([333]) BRUNE W., DELWICHE C. C., DAL POZZO ARZOLLO J. et MALAVOLTA E. — The occurrence of kojic acid in cultures of *Aspergillus wentii*. *Naturwissenschaften*, t. **XLV**, p. 113, 1958.

([334]) BÜCHI G., FOULKES D. M., KURONO M., MITCHELL G. F. et SCHNEIDER R. S. — The total synthesis of racemic aflatoxin B_1. *J. Am. Chem. Soc.*, t. **LXXXIX**, p. 6745-6753, 1967.

([335]) BÜCHI G. et RAE I. D. — The structure and chemistry of the aflatoxins. *in* GOLDBLATT L. A., *Aflatoxin*, *Academic* Press, p. 55-75, 1969

([336]) BUCHI G. et WEINREB S. M. — The total synthesis of racemic aflatoxin M_1 (milk toxin). *J. Am. Chem. Soc.*, t. **XCI**, p. 5408-5409, 1969.

([337]) Büchi G., White J. D. et Wogan G. N. — The structure of mitorubrin and mitorubrinol. *J. Am. Chem. Soc.*, t. **LXXXVII**, p. 3484-3489, 1965.

([338]) Buckley S. S. et Maccallan W. G. — Acute hemorragic encephalitis prevalent among horses in Maryland. *Am. Vet. Rev.*, t. **XXV**, p. 99, 1901.

([339]) Bullerman L. B. et Ayres J. C. — Aflatoxin producing potential of fungi isolated from cured and aged meats. *Appl. Microbiol.*, t. **XVI**, p. 1945-1946, 1968.

([340]) Bullerman L. B., Hartman P. A. et Ayres J. C. — Extraction and analysis of aflatoxins from cured and aged meats. *J. Ass. Offic. Anal. Chem.*, t. **LII**, p. 638-641, 1969.

([341]) Bullock E. Kirkaldy D., Roberts J. C. et Underwood J. G. — Studies in mycological chemistry. **XII.** Two new metabolites from a variant strain of *Aspergillus versicolor* (Vuillemin) Tiraboschi. *J. Chem. Soc.*, fasc. 155, p. 829-835, 1963.

([342]) Bullock E., Roberts J. et Underwood J. G. — Studies in mycological chemistry XI. The structure of isosterigmatocystin and an amended structure for sterigmatocystin. *J. Chem. Soc.*, fasc. 808, p. 4179-4183, 1962.

([343]) Bullough W. S. et Johnson M. — Energy relation of mitotic activity in adult mouse epidermis. *Proc. Roy. Soc.* (Londres), B, t. **CXXXVIII**, p. 562-575, 1951.

([334]) Bunting R. H. — Fungi occurring in Cacao beans. *Gold Coast Dept. Agric., Yearbook 1928* (Bull. 16), p. 44-57, 1929.

([345]) Burdon K. L. et Williams R. P. — *Microbiology*. 5e éd., Mac Millan Co., N.Y., 1964.

([346]) Burke R. et Decareau R. — Recent advance in the freeze-drying of food products. *Advances in Food Research*, t. **XIII**, 1964.

([347]) Burkhardt H. J. et Forgacs J. — O-methylsterigmatocystin, a new metabolite from *Aspergillus flavus* Link ex Fries. *Tetrahedron*, t. **XXIV**, p. 717-720, 1968.

([348]) Burkill I. H. — *A dictionary of the economic products of the Malay Peninsula*, 1080 p., Crown Agents, Londres, 1935.

([349]) Burmeister H. R. — T-2 toxin production by *Fusarium tricinctum* on solid substrate. *Appl. Microbiol.*, t. **XXI**, p. 739-742, 1971.

([350]) Burmeister H. R., Ellis J. J. et Yates S. G. — Correlation of biological to chromatographic data for two mycotoxins elaborated by *Fusarium. Appl. Microbiol.*, t. **XXI**, p. 673-675, 1971.

([351]) Burmeister H. R. et Hesseltine C. W. — Survey of the sensitivity of microorganisms to aflatoxin. *Appl. Microbiol.*, t. **XIV**, p. 403-404, 1966.

([352]) Burmeister H. R. et Leistner L. — Aflatoxin in Fleischwaren. *Jahresbericht Bundesanstalt für Fleischforschung* 1968, p. G 48-G 49, 1968.

([353]) Burmeister H. R. et Leistner L. — Aflatoxin-bildung in Fleischwaren. *Fleischwirtschaft.* t. **L**, p. 685, 1970.

([354]) Burns A. C. — Investigations on raw cotton. Deterioration of cotton during damp storage. *Min. Agr. Egypt. Tech. Sci. Bull.* n° 71, 92 p., 1927.

([355]) Burns M. J. — Grass tetany (Hypomagnesaemia). *Auburn Vet.*, t. **XIII**, p. 118, 1957.

([356]) Burnside C. E. — Fungous Diseases of the Honey Bee. *U.S. Dept. Agr.*, Tech. Bull. n° 149, 43 p., 1930.

([357]) Burnside J. E., Sippel W. L., Forgacs J., Carll W. T., Atwood B. et Doll E. R. — A disease of swine and cattle caused by eating moldy corn. II. Experimental production with pure cultures of molds. *Am. J. Vet. Research*, t. **XVIII**, p. 817-824, 1957.

([358]) Burrell N. J., Grundey J. K. et Harkness C. — Growth of *Aspergillus flavus* and production of aflatoxin in groundnuts. Part V. *Trop. Sci.*, t. **VI**, p. 74-90, 1964.

([359]) Bush M. T. et Goth A. — Flavicin : an antibacterial substrace produced by an *Aspergillus flavus. J. Pharmacol. Exptl. Therap.*, t. **LXXVIII**, p. 164-169, 1943.

([360]) Bustinza L. F. et Lopez A. C. — Preliminary tests in the study of influence of antibiotics on the germination of seeds. *Rep. Proc. 4th Int. Congr. Microbiol.* (1947), p. 160-161, 1949.

([361]) Buszewicz G. — Reports on forest research for the years ending March 1952, 1953, 1954. *H. M. Stationary Office* (London) ,Pub. n° 6, 429 p., 1955.

([362]) Butler F. C. — Ear, cob and grain rots of maize. *Agr. Gaz. N. S. W.*, t. **LVIII**, p. 144-151, 1947.

([363]) Butler T. — Notes on a feeding experiment to produce leucoencephalitis in a horse with positive results. *Am. Vet. Rev.*, t. **XXVI**, p. 748, 1902.

([364]) Butler W. H. — Acute toxicity of aflatoxin B_1 in rats. *Brit. J. Cancer*, t. **XVIII**, p. 756-762, 1964.

([365]) Butler W. H. — Acute liver injury to ducklings as a result of aflatoxin poisoning. *J. Pathol. Bacteriol.*, t. **LXXXVIII**, p. 189-196, 1964.

([366]) Butler W. H. — Liver injury and aflatoxin. *Mycotoxins in Foodstuffs*, p. 175-186, 1965.

([367]) Butler W. H. — Aflatoxicosis in laboratory animals. *in* Goldblatt L. A., *Aflatoxin*, Academic Press, p. 223-236, 1969.

([368]) Butler W. H. — The toxicology of aflatoxin. *in* Purchase I. F. H., *Mycotoxins in human health*, p. 141-151, 1971.

([369]) Butler W. H. — Further ultrastructural observations on injury of rat hepatic parenchymal cells induced by aflatoxin B_1. *Chem.-Biol. Interactions*, t. **IV**, p. 49-65, 1971.

([370]) Butler W. H. et Barnes J. M. — Toxic effects of groundnut meal containing aflatoxin to rats and guinea pigs. *British J. Cancer*, t. **XVII**, p. 699-710, 1964.

([371]) Butler W. H. et Barnes J. M. — Carcinoma of the glandular stomach in rats given diets containing aflatoxin. *Nature, G. B.*, t. **CCIX**, p. 90, 1966.

([372]) Butler W. H. et Barnes J. M. — Carcinogenic action of groundnut meal containing aflatoxin in rats. *Foods Cosmet. Toxicol.*, t. **VI**, p. 135-141, 1968.

([373]) Butler W. H. et Clifford J. L. — Extraction of aflatoxin from rat liver. *Nature*, t. **CCVI**, p. 1045-1046, 1965.

([374]) Butler W. H., Greenblatt M. et Lijinsky W. — Carcinogenesis in rats by aflatoxins B_1, G_1 and B_2. *Cancer Res.*, t. **XXIX**, p. 2206-2211, 1969.

([375]) Butler W. H. et Lijinsky W. — Acute toxicity of aflatoxin G_1 to the rat. *J. Pathol.*, t. **CII**, p. 209-212, 1970.

([376]) Butler W. H. et Wigglesworth J. S. — The effects of aflatoxin B_1 on the pregnant rat. *Brit. J. exp. Pathol.*, t. **XLVII**, p. 242-247, 1966.

([377]) Buu-Hoi NG. PH. et Zajdela F. — Is luteoskyrin the hepatotoxic factor in yellow rice ? *Med. Exptl.*, t. **VI**, p. 29-32, 1962.

([378]) Buxton E. A. — Mycotic vaginitis in gilts. *Vet. Med.*, t. **XXII**, p. 428, 451-452, 1927.

([379]) Byckroft B. W., Dobson T. A. et Roberts J. C. — Mycological chemistry. VIII. Structure of flavasperone (« asperxanthone ») a metabolite of *Aspergillus niger*. *J. Chem. Soc.*, p. 40-44, 1962.

([380]) Bycroft B. W., Hatton J. R. et Roberts J. C. — Studies in mycological Chemistry. XXV. Experiments directed towards a synthesis of aflatoxin G_2: synthesis of the coumarinolactone system. *J. Chem. Soc., Sect. C, Org. Chem.*, t. **II**, p. 281-284, 1970.

([381]) Bye A. et King H. K. — The biosynthesis of 4-hydroxycoumarin and dicoumarol by *Aspergillus fumigatus* Fresenius. *Biochem. J.*, t. **LXVII**, p. 237-245, 1970.

([382]) Byrns R. V. — Dermatitis of white pigs associated with lucerne and trefoil grazing. *Agr. Gaz. New S. Wales*, t. **XLVIII**, p. 214, 1937.

([383]) Byung-Ryul C. H. O. — Toxicity of water extracts of scalby barley to suckling mice˙ *Amer. J. Veter. Res.*, t. **XXV**, fasc. 107, p. 1267-1270, 1964.

([384]) Calderwood D. L. et Schroeder H. W. — Aflatoxin development and grade of undried rough rice following prolonged storage in aerated bins. *U. S. Dept. Agr., Agr. Res. Serv. Rep.*, 32 p. 1968.

[385] CALDWELL R. L. — Aflatoxin. Some problems and progress. *Prog. Agric. Arizona*, t. **XXIII**, p. 3-16, 1971.

[386] CALDWELL R. W. et TUITE J. — Zearalenone production in field corn in Indiana. *Phytopathology*, t. **LX**, p. 1696-1697, 1970.

[387] CALDWELL R. W., TUITE J., STOB M. et BALDWIN R. — Zearalenone production by *Fusarium* species. *Appl. Microbiol.*, t. **XX**, p. 31-34, 1970.

[388] CALET C. — Action de l'aflatoxine sur le développement embryonnaire du poulet et sur les premiers jours de la vie du canard. *Colloque Aflatoxines, Villejuif*, 7 mars 1966.

[389] CALVERT E. L. et MUSKETT A. E. — Blindseed disease of rye-grass. *Nature* (Lond.) 1944.

[390] CALVERT E. L. et MUSKETT A. E. — Blindseed of rye-grass (*Phialea temulenta* Prill. et Delacr.). *Ann. Appl. Biol.*, t. **XXXII**, p. 329-343, 1945.

[391] CAMERON G. R. et KARUNARATNE W. A. E. — Carbon tetrachloride cirrhosis in relation to liver regeneration. *J. Path. Bacteriol.* t. **XLII**, p. 1, 1936.

[392] CAMP A. A. — The field hemorrhagic-anemic syndrome. New treatment. *Feedstuffs*, t. **XXIX**, p. 20, 23, 24, 1957.

[393] CAMPBELL A. D. — Mycotoxins. *Cereal Sci. Today*, t. **XII**, p. 119, 1967.

[394] CAMPBELL A. D. — Chemical methods for mycotoxins. in HERZBERG M., *Toxic Microorganisms*, p. 36-42, 1970.

[395] CAMPBELL A. D., DORCEY E. et EPPLEY R. M. — Rapid procedure for extraction of aflatoxin from peanuts, peanut meal, and peanut butter for bioassay. *J. Ass. Off. anal. Chemists*, t. **XLVII**, p. 1002-1003, 1964.

[396] CAMPBELL A. D. et FUNKHOUSER J. T. — Collaborative study on the analysis of aflatoxins in peanut butter. *J. Ass. Off. Anal. Chem.*, t. **XLIX**, p. 730-739, 1966.

[397] CAMPBELL H. A. et LINK K. P. — Studies on the hemorrhagic sweet clover disease. IV. *J. Biol. Chem.*, t. **CXXXVIII**, p. 21, 1941.

[398] CAMPBELL H. A., ROBERTS W. L., SMITH W. K. et LINK K. P. — Studies on the hemorrhagic sweet clover disease. I. *J. Biol. Chem.*, t. **CXXXVI**, p. 47, 1949; II. *J. Biol. Chem.*, t. **CXXXVIII**, p. 1, 1941.

[399] CAMPBELL I. M., CALZADILLA C. H. et MCCORKINDALE N. J. — Some new metabolites related to mycophenolic acid (from *Penicillium brevicompactum* antifungal activity). *Tetrahedron Lett.*, t. **XLII**, p. 5107-5111, 1966.

[400] CAMPBELL J. G. — *Proc. Roy. Soc. Edinburgh*, t. **LXI**, p. 11-129, 1955-1957.

[401] CAMPBELL T. C. — Mycotoxins in the food chain. *Va. J. Sci.*, t. **XVIII**, p. 67-74, 1967.

[402] CAMPBELL T. C., CAEDO J. P., BULATAO-JAYME J., SALAMAT L. et ENGEL R. W. — Aflatoxin M_1 in human urine. *Nature. G. B.*, t. **CCXXVII**, p. 403-404, 1970.

[403] CAMPBELL T. C. et SALAMAT L. — Aflatoxin ingestion and excretion by humans. *in* PURCHASE I. F. K., *Mycotoxins in human health*, p. 271-280, 1971.

[404] CANTINI G. — Le micotossicosi degli animali domestici. *Veter. ital.*, t. **XV**, p. 964-980, 1964.

[405] CANTINI G. et SCURTI J. C. — Prime indagini sulla tossicita di isolati da mangimi. *Allionia*, t. **XI**, p. 29-40, 1965.

[406] CANTINI G., SCURTI J. C., MARCHISIO F. V. et JORIO M. — Relazione tra tossicita de antibiose di miceti isolati da mangimi e da insilati. *Allionia*, t. **XIV**, p. 97-111, 1968.

[407] CAO G. — Oidien und Oidiomykose. *Zeitschr. f. Hygiene und Infektionskrankheiten*, t. **XXXIV**, p. 282-340, 1900.

[408] CAPEK A., HANC O. et TADRA M. — Microbial transformations of steroids. *Biologia et Industria*, 250, p., Junk édit., La Haye, 1966.

[409] CAPITAINE R. — Recherches sur la production de métabolites toxiques par des moisissures. Premières expérimentations sur les filtrats de culture de l'*Aspergillus clavatus* Desm. *Rapp. D. E. A. Chimie structurale, Fac. Sci. Brest*, p. 11-24, 1970.

([410]) CAPPUCI D. T. — Aflatoxin and chromosomal studies (*Aspergillus flavus*). *Bull. Environ. Contam. Toxicol.*, t. **I**, p. 205-207, 1966.

([411]) CARDIN L. et PONCHET J. — Essais de traitement du *Fusarium graminearum* Schw. *Phytiat. — Phytopharm.*, t. **XIV**, p. 127-131, 1965.

([412]) CARLL W. T. et FORGACS J. — The significance of fungi in hyperkeratosis. *Military Surgeon*, t. **CXV**, p. 187-193, 1954.

([413]) CARLL W. T., FORGACS J. et HERRING A. S. — Toxic Aspergilli isolated from a food concentrate. *Vet. Med.*, t. **XLVIII**, p. 324, 1953.

([414]) CARLL W. T., FORGACS J. et HERRING A. S. — Toxicity of fungi isolated from a food concentrate. *Am. J. Hyg.*, t. **LX**, p. 8-14, 1954.

([415]) CARLL W. T., FORGACS J., HERRING A. S. et MAHLANDT B. G. — Toxicity of *Aspergillus fumigatus* substances to animals. *Vet. Med.*, t. **L**, p. 210, 1955.

([416]) CARLTON W. W. et TUITE J. — Toxicosis in miniature swine induced by corn cultures of *Penicillium viridicatum. Toxicol. Appl. Pharm.*, t. **XIV**, p. 636, 1969.

([417]) CARLTON W. W. et TUITE J. — Mycotoxicosis induced in guinea pigs and rats by corn cultures of *Penicillium viridicatum. Toxicol. Appl. Pharmacol.*, t. **XVI**, p. 345-361, 1970.

([418]) CARLTON W. W., TUITE J. et MISLIVEC P. — Investigations of the toxic effects in mice of certain species of *Penicillium. Toxicol. Appl. Pharmacol.*, t. **XIII**, p. 372-387, 1968.

([419]) CARLTON W. W., TUITE J. et MISLIVEC P. — Pathology of the toxicosis produced in mice by corn cultures of *Penicillium viridicatum. in* HERZBERG M., *Toxic Micro-organisms*, p. 94-106, 1970.

([420]) CARNAGHAN R. B. A. — (Observations non publiées) 1960-1962.

([421]) CARNAGHAN R. B. A. — Hepatic tumours and other chronic liver changes in rats following a single oral administration of aflatoxin. *Brit. J. Cancer*, t. **XXI**, p. 811-814, 1967.

([422]) CARNAGHAN R. B. A., HARTLEY R. O. et O'KELLY J. — Toxicity and fluorescence properties of the aflatoxins. *Nature*, t. **CC**, p. 1101, 1963.

([423]) CARNAGHAN R. B. A., HERBERT C. N., PATTERSON D. S. P. et SWEASY D. — Comparative biological and biochemical studies in hybrid chicks. 2. Susceptibility to aflatoxin and effects on serum protein constituents. *Brit. Poult. Sci.*, t. **VIII**, p. 279-284, 1967.

([424]) CARRAZ M. — Activité antibactérienne des huiles essentielles. *Colloque Antiseptiques et Désinfectants*, Inst. Pasteur, mai 1964.

([425]) CARTER E. P. et YOUNG G. Y. — Role of fungi in the heating of moist wheat. *U. S. Dept. Agr. Circ.* n° 838, 26 p., 1950.

([426]) CAVALLITO C. J. et BAILEY J. H. — Preliminary note on the inactivation of antibiotics. *Science*, t. **C**, p. 390, 1944.

([427]) CAVALLITO C. J., BAILEY J. H., HASKELL T. H., MCCORMICK J. R. et WARNER W. F. — The inactivation of antibacterial agents and their mechanism of action. *J. Bacteriol.*, t. **L**, p. 61, 1945.

([428]) CAVALLITO C. J. et HASKELL T. H. — The mechanism of action of antibiotics. *J. Am. Chem. Soc.*, t. **LXVII**, p. 1991, 1945.

([429]) CAWLEY E. P. — Aspergillosis and the Aspergilli : report of a unique case of the disease. *Arch. internat. Med.*, t. **LXXX**, p. 423-434, 1947.

([430]) CENI C. — Di alcune nuove muffe velenose in rapporto coll' etiologia della pellagra. *Centralbl. f. Bakt.*, Abt. I, t. **XLI**, p. 407-408, 1908.

([431]) CENI C. et BESTA C. — Ueber die Toxine von *Aspergillus fumigatus* und *Aspergillus flavescens* und deren Beziehungen zur Pellagra. *Centralbl. f. allgem. Pathologie u. pathol. Anatomie*, p. 930, 1902.

([432]) CENI C. et BESTA C. — Di alcuni caratteri biologici dei penicilli in rapporto colle stagioni e col ciclo annuale della pellagra. *Resoconto del. 3. Congresso della societa italiana di Patologia*, fasc. 5, 1905.

([433]) CENI C. et BESTA C. — Ulteriori ricerche sul ciclo biologico dei penicilli vredi in rapporto colle stagioni dell'anno e colla pellagra. *Atti della Societa italiana di patologia Quarta riunione tenuta in Pavia dal 1 al 4 ottobre 1906. Supplemento all'annuata 20. Boll. Soc. Med. Chir. Pavia.* 1906.

([434]) CENI C. et BESTA C. — Di un aspergillo bruno gigante e delle sue proprieta tossiche in rapporto colla pellagra. *Rivista sperimentale di freniatria*, t. **XXXIII**, fasc. 1, 1908.

([435]) CENI C. et BESTA C. — Sulle modificazioni dei caratteri fisiologici dei Pencicilli verdi in rapporto colla laro proprieta tossica. *Centralbl. f. Bakt.*, Abt. I., t. **XLI**, p. 407, 1908.

([436]) CHAFFEE V. W., EDDS G. T., HIMES J. A. et NEAL F. C. — Aflatoxicosis in dogs. *Amer. J. Vet. Res.*, t. **XXX**, p. 1737-1749, 1969.

([437]) CHAIN E., FLOREY H. W. et JENNINGS M. A. — An antibacterial substance produced by *Penicillium claviforme. Brit. J. exp. Path.*, t. **XXIII**, p. 202, 1942.

([438]) CHAIN E., FLOREY H. W. et JENNINGS M. A. — Identity of patulin and claviformin. *Lancet*, t. **CCXLVI**, p. 112-114, 1944.

([439]) CHAIN E., FLOREY H. W., JENNINGS M. A. et WILLIAMS T. I. — Helvolic acid, an antibiotic produced by *Aspergillus fumigatus* mut. *helvola* Yuill. *Brit. J. Exp. Pathol.*, t. **XXIV**, p. 108-118, 1943.

([440]) CHANG C. M. et LYND J. Q. — UV aflatoxin quantitation with polaroid recordings. *Agron. J.*, t. **LX**, p. 582-584, 1968.

([441]) CHANG S. B., ABDEL KADER M. M., WICK E. L. et WOGAN G. N. — Aflatoxin B_2: chemical identity and biochemical activity. *Science*, t. **CXLII**, p. 1191-1192, 1963.

([442]) CHARI V. M., NEELAKANTAN S. et SESHADRI T. R. — Structure of cladofulvin. *Tetrahedron Lett.*, t. **XI**, p. 999-1002, 1967.

([443]) CHARPIN J. — Les syndromes respiratoires allergiques. *Les monographies médicales et scientifiques*, t. **XIV**, n° 97, 1962.

([444]) CHESTERS C. G. C. in PARKINSON D. et WAID J. S. — *The ecology of soil fungi.* Liverpool Univ. Press, 1960.

([445]) CHEUNG K. K. et SIM G. A. — Aflatoxin G_1 : direct determination of the structure by the method of isomorphous replacement. *Nature*, t. **CCI**, p. 1185-1188, 1964.

([446]) CHEYMOL J., BOURILLET F. et ROCH-ARVEILLER M. — Le terme toxine a-t-il une signification précise? *Bull. Acad. Nat. Méd.*, t. **CLIII**, p. 412-416, 1969.

([447]) CHILDS E. A., AYRES J. C. et KOEHLER P. C. — Fluorometric measurement of aflatoxins. *J. Am. Oil Chem. Soc.*, t. **XLVII**, p. 461-462, 1970.

([448]) CHILDS V. A. et LEGATOR M. S. — Induction of thymidine kinase by aflatoxin. *Life Sci., G. B.*, t. **V**, p. 1053-1056, 1966.

([449]) CHILTON W. S. — Toxic metabolites of Fusaria. *Abstr. Papers of Western Exp. Sta. Collaborators Conf. on importance of mold metabolites in Agricultural Products, 1-3 March, Berkeley, Calif.*, p. 12, 1965.

([450]) CHINDEMI A. et CULTRERA A. — Considerazioni sui metodi di determinazione dell' aflatossina B_1 nei semi di arachide e nei prodotti derivati. *Bol. Lab. Chimico Provinciale* t. **XVII**, p. 59-75, 1966.

([451]) CHMYR' A. D. — Effect of the γ-rays of the radioisotope Co^{60} upon certain biochemical and physiological properties of maize grain during storage. *Acad. Sci. U. S. S. R.*, Israel Program for Scientific Translation, p. 154-160, 1965.

([452]) CHOLLET M.-M. — Les pigments de l'*Aspergillus mangini* (Mangin) Thom et Raper. *Bull. Soc. Myc. Fr.*, t. **LXXIX**, p. 429-455, 1963.

([453]) CHOLLET M.-M. et MOREAU C. — Pigments et métabolisme glucidique chez quelques *Aspergillus* du *gr. glaucus*. Comparaison avec l'*A. wentii* et le *Penicillium wortmanni. C. R. Acad. Sci.*, t. **CCLXII**, p. 451-453, 1966.

([454]) CHOLLET M.-M. et VIEILLARD J. — Variations du métabolisme glucidique de l'*Aspergillus mangini*, en relation avec la nature du sucre utilisé comme source de carbone. *C. R. Acad. Sci.*, t. **CCLII**, p. 936-938, 1961.

([455]) CHONG Y. H. — Aflatoxins in groundnuts and groundnut products. *Far East Med. J.*, t. **II**, p. 228-230, 1966.

([456]) CHONG Y. H. et BENG C. G. — Aflatoxins in unrefined groundnut oils. *Med. J. Malaya*, t. **XX**, p. 49-50, 1965.

([457]) CHONG Y. H. et PONNAMPALAM J. T. — The effect on duckling of a diet containing moulded soybeans. *Med. J. Malaya*, t. **XXII**, p. 104-109, 1968.

([458]) CHOUDHARY P. G. et MANJREKAR S. L. — Preliminary observations on biological activity of pure aflatoxin B_1 in chick embryo. *Indian Vet. J.*, t. **XLIV**, p. 543-548, 1967.

([459]) CHOUDHARY P. G. et MANJREKAR S. L. — Toxicity of Aspergilli isolated from groundnuts and groundnut cakes. *Indian J. Vet.*, t. **XLIV**, p. 359-365, 1967.

([460]) CHOUDHURY H., CARLSON C. W. et SEMENIUK G. — A study of ochratoxin toxicity in hens. *Poult. Sci.*, t. **L**, p. 1855-1859, 1971.

([461]) CHOU MING-WU et TUNG TA-CHANG — Aflatoxin B_1 in the excretion of aflatoxin poisoned rats. *J. Formosan Med. Ass.*, t. **LXVIII**, p. 389-391, 1969.

([462]) CHRANOWSKA H. — Microflore du seigle et de la farine de seigle. *Industr. aliment. agric.*, t. **LXXXIV**, p. 24-27, 1967.

([463]) CHRISMAN J. — Dehydrated alfalfa in hemorrhagic control. *Feed Age*, t. **V**, p. 42, 1955.

([464]) CHRISTENSEN C. M. — The quantitative determination of molds in flour. *Cereal Chem.*, t. **XXIII**, p. 322-329, 1946.

([465]) CHRISTENSEN C. M. — Molds and bacteria in flour and their significance to the baking industry. *Baker's Digest*, t. **XXI**, p. 21-23, 1947.

([466]) CHRISTENSEN C. M. — Bread molds. *Baker's Dig.*, t. **XXIII**, p. 25-26 et 30, 1949.

([467]) CHRISTENSEN C. M. — *The Molds and Man. Univ. Minn.* Press, Minneapolis, Minn. 1951.

([468]) CHRISTENSEN C. M. — Grain storage studies. XXI. Viability and moldiness of commercial wheat in relation to the incidence of germ damage. *Cereal Chem.*, t. **XXXII**, p. 507-518, 1955.

([469]) CHRISTENSEN C. M. — Deterioration of stored grains by fungi. *Botanical Review*, t. **XXIII**, p. 108-134, 1957.

([470]) CHRISTENSEN C. M. — Invasion of stored wheat by *Aspergillus ochraceus*. *Cereal Chem.*, t. **XXXIX**, p. 100-106, 1962.

([471]) CHRISTENSEN C. M. — Influence of small differences in moisture content upon the invasion of hard red winter wheat by *Aspergillus restrictus* and *A. repens*. *Cereal Chemistry*, t. **XL**, p. 385-390, 1963.

([472]) CHRISTENSEN C. M. — Fungi in cereal grains and their products. *Mycotoxins in Foodstuffs*, p. 9-14, 1965.

([473]) CHRISTENSEN C. M. — Influence of moisture content, temperature, and time of storage upon invasion of rough rice by storage fungi. *Phytopathology*, t. **LIX**, p. 145-148, 1969.

([474]) CHRISTENSEN C. M. — Moisture content, moisture transfer, and invasion of stored sorghum seeds by fungi. *Phytopathology*, t. **XL**, p. 280-283, 1970.

([475]) CHRISTENSEN C. M. — Invasion of sorghum seeds by storage fungi at moisture contents of 13,5-15% and condition of samples from commercial bins. *Mycopathol. Mycol. Appl.*, t. **XLIV**, p. 277-282, 1971.

([476]) CHRISTENSEN C. M. — Evaluating condition and storability of sunflower seeds. *J. stored Prod. Res.*, t. **VII**, p. 163-169, 1971.

([477]) CHRISTENSEN C. M. et COHEN M. — Numbers, kinds and source of molds in flour. *Cereal Chem.*, t. **XXVII**, p. 178-185, 1950.

([478]) CHRISTENSEN C. M. et DORWORTH C. E. — Influence of moisture content, temperature, and time on invasion of soybeans by storage fungi. *Phytopathology*, t. **LVI**, p. 412-418, 1966.

[479] CHRISTENSEN C. M., FANSE H. A., NELSON G. H., BATES F. et MIROCHA C. J. — Microflora of black and red Pepper. *Appl. Microbiol.*, t. **XV**, p. 622-626, 1967.

[480] CHRISTENSEN C. M. et GORDON D. R. — The mold flora of stored wheat and corn and its relation to heating of moist grain. *Cereal Chem.*, t. **XXV**, p. 40-51, 1948.

[481] CHRISTENSEN C. M. et KAUFMANN H. H. — Deterioration of stored grains by Fungi. *Ann. Review of Phytopathology*, t. **III**, p. 69-84, 1965.

[482] CHRISTENSEN C. M. et KAUFMANN H. H. — *Grain storage. The role of fungi in quality loss.* 153 p., Univ. Minnesota Press, Minneapolis, 1969.

[483] CHRISTENSEN C. M. et KENNEDY B. W. — Filamentous fungi and bacteria in macaroni and spaghetti products. *Appl. Microbiol.*, t. **XXI**, p. 144-146, 1971.

[484] CHRISTENSEN C. M. et LOPEZ L. C. — Damage to fungi to stored grain in Mexico. *Foll. Tec. Inst. Nac. Agr.* (Mexico), n° 44, 29 p., 1962.

[485] CHRISTENSEN C. M., MERONUCK R. A., NELSON G. H. et BEHRENS J. C. — Effects on turkey poults of rations containing corn invaded by *Fusarium tricinctum* (Cda) Sny. et Hans. *Appl. Microbiol.*, t. **XXIII**, p. 177-179, 1972.

[486] CHRISTENSEN C. M., MIROCHA C. J., NELSON G. H. et QUAST J. F. — Effect on young swine of consumption of rations containing corn invaded by *Fusarium roseum. Appl. Microbiol.*, t. **XXIII**, p. 202, 1972.

[487] CHRISTENSEN C. M., NELSON G. H. et MIROCHA C. J. — Toxicity to animals of feeds invaded by fungi. I. Increase in weight of uteri of white rats due to a toxin produced by *Fusarium* and isolation of the toxin. *Appl. Microbiol.*, t. **XIII**, p. 653-659, 1965.

[488] CHRISTENSEN C. M., NELSON G. H., MIROCHA C. J. — Work on mycotoxins at the University of Minnesota. *1st Congr. Pl. pathol.*, Londres, juil. 1968, abstr. p. 31, 1968.

[489] CHRISTENSEN C. M., NELSON G. H., MIROCHA C. J., FARN BATES et DORWORTH C. E. — Toxicity to rats of corn invaded by *Chaetomium globosum. Appl. Microbiol.*, t. **XIV**, p. 774-777, 1966.

[490] CHRISTENSEN, C. M., OLAFSON J. H. et GEDDES W. F. — Grain storage studies. VIII. Relation of molds in moist stored cotton-seed to increased production of carbon dioxide, fatty acids and heat. *Cereal Chem.*, t. **XXVI**, p. 109-128, 1949.

[491] CHRISTENSEN C. M., PAPAVIZAS G. C. et BENJAMIN C. R. — A new halophilic species of *Eurotium. Mycologia*, t. **LI**, p. 636-640, 1959.

[492] CHRISTENSEN J. J. — Longevity of fungi in barley kernels. *Plant Disease Reptr.*, t. **XLVII**, p. 639-642, 1963.

[493] CHRISTENSEN J. J. et KERNKAMP H. C. H. — Studies on the toxicity of blighted barley to swine. *Minn. Agr. Exp. Stn. Bull.*, n° 113, 1936.

[494] CHRISTIAN K. R. et WILLIAMS V. J. — Attempts to produce hypomagnesaemia in dry nonpregnant sheep. *N. Z. J. Agr. Res.*, t. **III**, p. 389, 1960.

[495] CHU F. S. — Studies on the fungus *Byssochlamys fulva, in Byssochlamys Seminar Abstracts, Dept. Food Sci. and Technol.*, Cornell Univ., Circ. n° 20, p. 3-4, 1969.

[496] CHU F. S. et BUTZ M. E. — Spectrofluorodensitometric measurement of ochratoxin A in cereal products. *J. Ass. Off. Anal. Chem.*, t. **LIII**, p. 1253-1257, 1970.

[497] CHU F. S. et CHANG C. C. — Sensitivity of chicks to ochratoxins. *J. Ass. Off. Anal. Chem.*, t. **LIV**, p. 1032-1034, 1971.

[498] CHU F. S., NOH I, et CHANG C. C. — Structural requirements for ochratoxin intoxication. *Life Sci.*, t. **XI**, p. 503-508, 1972.

[499] CHUNG-MIN CHANG et LYND J. Q. — Soil microfloral interactions with aflatoxin synthesis and degradation. *Mycologia*, t. **LXII**, p. 878-896, 1970.

[500] CHURCH M. B. et BUCKLEY J. S. — Laboratory feeding of molds to animals. *N. Am. Vet.*, t. **IV**, p. 7, 1923.

[501] CHU WEI CHANG. — Miscellaneous pharmacologic actions of citrinin. *J. Lab. Clin. Med.*, t. **XXXI**, p. 72-78, 1946.

([502]) CIEGLER A. — Tremorgenic toxin from *Penicillium palitans. Bacteriol. Proc.* t. **LXIX**, p. 12, 1969.

([503]) CIEGLER A. — Tremorgenic toxin from *Penicillium palitans. Appl. Microbiol.*, t. **XVIII**, p. 128-129, 1969.

([504]) CIEGLER A., DETROY R. W. et LILLEHOJ E. B. — Patulin, penicillic acid and other carcinogenetic lactones. *in* CIEGLER A. et *al.*, *Microbial Toxins*, t. **VI**, p. 409-434, 1971.

([505]) CIEGLER A. et HOU C. T. — Isolation of viridicatin from *Penicillium palitans. Arch. Microbiol.*, t. **LXXIII**, p. 261-267, 1970.

([506]) CIEGLER A., KADIS S. et AJL J. — *Fungal toxins* (Microbial Toxins ; *a comprehensive treatise*, vol. **VI**). 586 p., Academic Press, Londres, 1971.

([507]) CIEGLER A. et KURTZMANN C. P. — Penicillic acid production by blue-eye fungi on various agricultural commodities. *Appl. Microbiol.*, t. **XX**, p. 761-764, 1970.

([508]) CIEGLER A., LILLEHOJ E. B., PETERSON R. E. et HALL H. H. — Microbial detoxification of aflatoxin. *Appl. Microbiol.*, t. **XIV**, p. 934-939, 1966.

([509]) CIEGLER A. et PETERSON R. E. — Aflatoxin detoxification : hydroxydihydroaflatoxin B_1. *Appl. Microbiol.*, t. **XVI**, p. 665-666, 1968.

([510]) CIEGLER A., PETERSON R. E., LAGODA A. A. et HALL H. H. — Aflatoxin production and degradation by *Aspergillus flavus* in 20-liter fermentors. *Appl. Microbiol.*, t. **XIV**, p. 826-833, 1966.

([511]) CIFFERRI R. — Studies on cacao. *J. Dept. Agr. Puerto Rico*, t. **XV**, p. 223-286, 1931.

([512]) CIFFERRI R. et CIFFERRI F. — Defectos y alteractiones del Cacao en granos. *Mat. Veg., La Haye*, t. **I**, p. 148-166, 1953.

([513]) CLARE N. T. — *Photosensitization in diseases of domestic animals. Commonwealth Agric. Bureaux*, p. 24-28, Farnham Royal, Bucks, 1952.

([514]) CLARE N. T. — Photosensitization in animals. *Advances in Veterinary Science*, t. II, p. 182-211, Academic Press Publ., New York et Londres, 1955.

([515]) CLARE N. T. — Photosensitivity diseases in New Zealand. XIX. The susceptibility of New Zealand white rabbits to facial eczema liver damage. *New Zealand J. Agr. Research*, t. II, p. 1249-1256, 1959.

([516]) CLARE N. T. et GUMBLEY J. M. — Some factors which may affect the toxicity of spores of *Pithomyces chartarum* (Berk, et Curt.) M. B. Ellis collected from pasture. *New Zealand J. Agric. Res.*, t. V, p. 36-42, 1962.

([517]) CLARE N. T. et MORTIMER P. H. — The effect of mercury arc radiation and sunlight on the toxicity of water solutions of sporidesmin to rabbits. *New Zeal. J. Agr. Res.*, t. **VII**, p. 258-263, 1964.

([518]) CLARK F. E. — The concept of competition in microbial ecology. In BAKER K. F. et SNYDER W. C. *Ecology of soil-borne plant pathogens*, p. 339-345, Univ. California Press, 1963.

([519]) CLARKE J. H. — Fungi in stored products. *Trop. stored Prod. Inf.*, t. **XV**, p. 1-14, 1968.

([520]) CLAUGHTON W. P. et CLAUGHTON H. D. — Vetch seed Poisoning. *Auburn Vet.*, t. **X**, p. 125, 1954.

([521]) CLAVEAU J. — L'opération unitaire de destruction des micro-organismes par la chaleur. *Chim. Industr., Génie chimique*, t. **XCVI**, p. 1359-1370, 1966.

([522]) CLEGG F. G. et BRYSON H. — An outbreak of poisoning in store cattle attributed to Brazilian groundnut meal. *Vet. Rec.*, t. **LXXIV**, p. 992-994, 1962.

([523]) CLÉMENT P. — La réfrigération rapide des viandes. *Colloque Inst. int. Froid*, Commissions 4-5, 7 p., Bologne, 6-9 juin 1966.

([524]) CLEMENTS N. L. — Note on a microbiological assay for aflatoxin B_1 : a rapid confirmatory test by effect on growth of *Bacillus megaterium. J. Ass. Off. Anal. Chem.*, t. **LI**, p. 611-612, 1968.

([525]) CLEMENTS N. L. — Rapid confirmatory test for aflatoxin B_1 using *Bacillus megaterium*. *J. Ass. Off. Anal. Chem.*, t. **LI**, p. 1192-1194, 1968.

([526]) CLIFFORD J. I. et REES K. R. — Aflatoxin : a site of action in the rat liver cell. *Nature, G. B.*, t. **CCIX**, p. 312-313, 1966.

([527]) CLIFFORD J. I. et REES K. R. — The action of aflatoxin B_1 on the rat liver. *Biochem. J.*, t. **CII**, p. 65-75, 1967.

([528]) CLIFFORD J. I. et REES K. R. — The interaction of aflatoxins with purines and purine nucleosides. *Biochem. J.* ,t. **CIII**, p. 467-471, 1967.

([529]) CLIFFORD J. I., REES K. R. et STEVEN M. E. M. — The effect of the aflatoxin B_1, G_1, and G_2 on protein and nucleic acid synthesis in rat liver. *Biochem. J.*, t. **CIII**, p. 258-261, 1967.

([530]) CLINTON P. K. S. — Seed-bed pathogens of groundnuts in the Sudan and an attempt at control with an artificial testa. *Empire J. Exp. Agr.*, t. **XXVIII**, p. 211-22, 1960.

([531]) COADY A. — Aflatoxin. *Brit. med. J.*, t. **I**, p. 1510, 1964.

([532]) COCHRANE V. W. — *Physiology of fungi.* 524 p., John Willy and Sons, New York, 1958.

([533]) CODEGONE M. K., PROVANA A. et GHITTINO P. — Evoluzione dell'epatoma precoce della trota iridea. *Tumori*, t. **LIV**, p. 419-425, 1968.

([534]) CODNER R. C., SARGEANT K. et YEO R. — Production of aflatoxin by the culture of strains of *Aspergillus flavus-oryzae* on sterilized peanuts. *Biotechnol. and Bioeng.*, t. **V**, p. 185-192, 1963.

([535]) COHEN M. et CHRISTENSEN C. M. — Incidence of molds in wheat flour. *Am. J. Bot.*, t. **XXXIV**, p. 593-594, 1947.

([536]) COIC Y., FAUCONNEAU G. et TERROINE E. F. — Influence des conditions de production, de récolte et de stockage sur la composition biochimique des céréales. Répercussion des variations sur la valeur nutritionnelle. *Ann. Nutrit. Aliment.*, t. **XX**, p. 147-156, 1966.

([537]) COLE R. J. et KIRKSEY J. W. — Aflatoxin G_1 metabolism by *Rhizopus* species. *J. Agr. Food Chem.*, t. **XIX**, p. 222-223, 1971.

([538]) COLE R. J., MOORE J. H., DAVIS N. D., KIRKSEY J. W. et DIENER U. L. — 4-Hydroxymellein : a new metabolite of *Aspergillus ochraceus. J. Agric. Food Chem.*, t. **XIX**, p. 909-911, 1971.

([539]) COMIANT J. P. — Contribution à l'étude de l'étiologie de la tétanie d'herbage. Rôle de l'alcalinité volatile, applications thérapeutiques. *Thèse Doct. Vét. Lyon*, 86 p., 1966.

([540]) COMMISSION DES COMMUNAUTES EUROPEENNES. — *Projet d'un règlement du conseil concernant les substances et produits indésirables dont la présence est tolérée dans les aliments des animaux*. 7 p., Bruxelles, 20 déc. 1968.

([541]) COOKE W. B. et BRAZIS A. R. — Occurrence of molds and yeasts in dairy products. *Mycopathol. Mycol. Appl.*, t. **XXXV**, p. 281-289, 1968.

([542]) COOMES T. J., CROWTHER P. C., FEUELL A. J. et FRANCIS B. J. — Experimental detoxification of groundnut meals containing aflatoxin. *Nature, G. B.*, t. **CCIX**, p. 406-407, 1966.

([543]) COOMES T. J., CROWTHER P. C., FRANCIS B. J. et STEVENS L. — The detection and estimation of aflatoxin in groundnuts and groundnut materials. IV. Routine assessment of toxicity due to aflatoxin, B_1. *Analyst*. t. **XC**, p. 492-496, 1965.

([544]) COOMES T. J. et SANDERS J. C. — The detection and estimation of aflatoxin in groundnuts and groundnut materials. I. Paper chromatographic procedure. *Analyst*, t. **LXXXVIII**, p. 209-213, 1963.

([545]) COONEY D. G. et EMERSON R. — *Thermophilic fungi.* 188 p., Freeman and Co, San Francisco, 1964.

([546]) CORDIER F. S. — *Les Champignons. Histoire, description, culture, usages des espèces comestibles, vénéneuses, suspectes employées dans les arts, l'industrie, l'économie domestique, la médecine.* Rotschild, Paris, 1876.

([547]) CORNAZ G. — La lyophilisation en général et la lyophilisation alimentaire en particulier. *Rev. Gén. Froid*, t. **LVII**, p. 1343-1348, 1966.

([548]) CORNEVIN C. — *Des plantes vénéneuses*, Firmin-Didot et Cie, Paris, 1893.

([549]) COURTIER A. J. — Une nouvelle technique de dispersion d'insecticides. *Cong. Protect. Cultures Tropicales, Marseille*, p. 55-58, mars 1965.

([550]) COURTIER A. J. — Résultats d'une nouvelle méthode de dispersion. *C. R. Acad. Agric. Fr.*, p. 1142—1152, 1967.

([551]) COVENEY R. D., PECK H. M. et TOWNSEND R. J. — S.C.I. Monograph n° 23. Recent advances in mycotoxicoses, p. 31, 1966.

([552]) COVER M. S., MELLEN W. J. et GILL E. — Studies of hemorrhagic syndromes in chickens. *Cornell Vet.*, t. **XLV**, p. 366-386, 1955.

([553]) COYNE F. P., RAISTRICK H. et ROBINSON R. — Studies in the biochemistry of microorganisms. XV. The molecular structure of citrinin. *Roy. Soc. London, Phil. Trans.*, sér. B, t. **CCXX**, p. 297-301, 1931.

([554]) CRADDOCK V. M. — Methylation of transfer RNA and of ribosomal RNA in rat liver in the intact animal and the effect of carcinogens. *Biochem. Biophys. Acta*, t. **CXCV**, p. 351-369, 1969.

([555]) CRAIB W. H. — Address delivered at formal opening. *S. Afr. med. J.*, t. **XXIX**, p. 761-762, 1968.

([556]) CRAWFORD M. A. — Epidemiological interactions. *in* PURCHASE I. F. H., *Mycotoxins in human health*, p. 231-244, 1971.

([557]) CRAWLEY W. E., MORTIMER P. H. et SMITH J. D. — Characteristic lesions of facial eczema produced in sheep by dosing material containing *Pithomyces chartarum* collected from pasture. *N. Z. J. Agric. Res.*, t. **IV**, p. 552-559, 1961.

([558]) CRISAN E. V. — 2,4-dinitrophenylhydrazine spray for the identification of aflatoxin B_1 on thin-layer chromatoplates. *Contr. Boyce Thompson Inst.*, **XXIV**, p. 37-38, 1968.

([559]) CRISAN E. V. et GREFIG A. T. — The formation of aflatoxin derivatives. *Contr. Boyce Thompson Inst.*, t. **XXIV**, p. 3-8, 1967.

([560]) CRISAN E. V. et MAZZUCCA E. — Separation of aflatoxin on selectively deactivated silicic acid. *Contr. Boyce Thompson Inst.*, t. **XXIII**, p. 361-365, 1967.

([561]) CROOKSHANK H. H. et SIMS F. H. — Serum values in wheat pasture poisoning cases. *J. Animal Sci.*, t. **XIV**, p. 964, 1955.

([562]) CRUICKSHANK I. A. M. — Phytoalexins. *Ann. Rev. Phytopathology*, t. **I**, p. 351-374, 1963.

([563]) CRUICKSHANK I. A. M. — Pisatin studies: The relation of phytoalexins to disease reaction in Plants, in *Ecology of soilborne plant pathogens*, p. 325-334, U. Calif. Press, 1965.

([564]) CRUMP M. H. — The study of an animal mycotoxicosis. *Dissert. Abstr.*, t. **XXV**, p. 6129, 1965.

([565]) CRUMP M. H., SMALLEY E. B., HENNING J. N. et NICHOLS R. E. — Mycotoxicosis in animals fed legume hay infested with *Rhizoctonia leguminicola*. *J. amer. veter. med. Ass.*, t. **CXLIII**, fasc. 9, p. 996-997, 1963.

([566]) CRUMP M. H., SMALLEY E. B., NICHOLS R. E. et RAINEY D. P. — Pharmacologic properties of a slobber-inducing mycotoxin from *Rhizoctonia leguminicola*. *Am. J. Vet. Res.*, t. **XXVIII**, p. 865-874, 1967.

([567]) CUCULLU A. F., LEE L. S., MAYNE R. Y. et GOLDBLATT L. A. — Determination of aflatoxins in individual peanuts and peanut sections. *J. Am. Oil Chem. Soc.*, t. **XLIII**, p. 89-92, 1966.

([568]) CUCULLU A. F., LEE L. S., PONS W. A. et GOLDBLATT L. A. — Determination of aflatoxins in peanut and cotton seed soapstock. *J. Amer. Oil Chem. Soc.*, t. **XLVII**, p. 226-228, 1970.

([569]) CUDKOWICZ G. et SCOLARI C. — Un tumore primitivo epatico a diffusione epizootica nella iridea di allevamento (*Salmo irideus*). *Tumori*, t. **XLI**, p. 524-537, 1955.

([570]) CUNNINGHAM H. M., BROWN J. M. et EDIE A. E. — Molybdenum poisoning of cattle in the Swan River valley of Manitoba. *Canad. J. Sci. Agr.*, t. **XXXIII**, p. 254, 1953.

([571]) CUNNINGHAM I. J., HOPKIRK C. S. M. et FILMER J. F. — Photosensitivity diseases in New Zealand. I. Facial eczema : its clinical, pathological and biochemical characteristics. *New Zealand J. Sci. Technol.*, sér. A, t. **XXIV**, p. 185-198, 1942.

([572]) CURTIN T. M. et TUITE J. — Emesis and refusal of feed in swine associated with *Gibberella zeae* infected corn. *Life Sci.*, t. V, p. 1937-1944, 1966.

([573]) CURTIS P. J., HEMMING H. G. et SMITH W. K. — Frequentin ; an antibiotic produced by some strains of *Penicillium frequentans* Westling. *Nature*, t. **CLXVII**, p. 557-558, 1951.

([574]) CUTHBERTSON W. F. J., LAURSEN A. C. et PRATT D. A. H. — Effect of groundnut meal containing aflatoxin on Cynomolgus monkeys. *Brit. J. Nutr.*, t. **XXI**, p. 893-908, 1967.

([575]) CYSEWSKI S. J., PIER A. C., ENGSTROM G. W., RICHARD J. L., DOUGHERTY R. W. et THURSTON J. R. — Clinical pathologic feature of acute aflatoxicosis of swine. *Amer. J. Vet. Res.*, t. **XXIX**, p. 1577-1590, 1968.

([576]) DACK G. M. — *Food poisoning*. 3e éd., The University of Chicago Press, Chicago, 1956.

([577]). DADE H. A. — Internal moulding of prepared Cacao. *Gold Coast Dept. Agric., Yearbook 1928* (Bull. 16), p. 74-100, 1929.

([578]) DALEZIOS J., WOGAN G. N. et WEINREB S. M. — Aflatoxin P_1 : a new aflatoxin metabolite in monkeys. *Science*, t. **CLXXI**, p. 584-585, 1971.

([579]) DALLYN H. et EVERTON J. R. — The xerophilic mould, *Xeromyces bisporus*, as a spoilage organism. *J. Food Technol.*, t. **IV**, p. 399-403, 1969.

([580]) DANIEL P., LASFARGUES E. et DELAUNAY A. — Sur la dissociation possible du pouvoir antibiotique et du pouvoir antimitotique de la patuline. *C. R. Soc. Biol.*, t. **CXLIX**, p. 18-19, 1955.

([581]) DANIELS M. R. — *In vitro* assay systems for aflatoxin. *Brit. J. Exper. Pathol.*, t. **XLVI**, fasc. 2, p. 183-188, 1965.

([582]) DARKEN M. et SJOLANDER N. — Production of patulin by submerged fermentation. *Antibiot. and Chemoth.*, t. **I**, p. 573-578, 1951.

([583]) DARPOUX H. — Les grandes voies de recherches. *Journées fr. ét. inf. A.C.T.A.* p. 611-642, 1969.

([584]) DATTA P. R. et GAJAN R. J. — Plasma protein index of aflatoxin-fed ducklings. *Feder. Proc.*, U.S.A., t. **XXIV**, fasc. 2, p. 392, 1965.

([585]) DATTA S. — The role of Aspergilli in carcinogenesis. *Ind. J. Cancer*, t. **IV**, p. 412-426, 1967.

([586]) DAUBEN H. J. et WEISENBORN F. L. — The structure of patulin. *J. Am. Chem. Soc.*, t. **LXXI**, p. 3853, 1949.

([587]) DAVID H. — Occurrence of two entomophagous fungi on sugarcane pests in Tansorea area of Madras St. *Curr. Sci.*, t. **XXXIII**, p. 349, 1964.

([588]) DAVIDSON C. S. — Some contributions of geographic study to understanding pathogenesis of cirrhosis. *Progr. in Liver Dis.*, t. **I**, p. 1-13, 1961.

([589]) DAVIDSON C. S. — Plants and fungi as etiologic agents of cirrhosis. *New Engl., J. Med.*, t. **CCLXVIII**, fasc. 19, p. 1072-1073, 1963.

([590]) DAVIDSON J. — Action of retrorsine in rat's liver. *J. Path. Bacteriol.*, t. **XL**, p. 285, 1935.

([591]) DAVIDSON W. B., DOUGHTY J. H. et BOLTON L. — Nitrate poisoning of livestock. *Canad. J. Comp. Med.*, t. V, p. 303, 1941.

([592]) DAVIES J. E., KIRKALDY D. et ROBERTS J. C. — Mycological chemistry. VII. Sterigmatocystin, a metabolic of *Aspergillus versicolor* (Vuillemin) Tiraboschi. *J. Chem. Soc.*, p. 2169-2178, 1960.

([593]) DAVIES J. E., ROBERTS J. C. et WALLWORK L. C. — Sterigmatocystin, a metabolic product of *Aspergillus versicolor*. *Chem. et Ind.*, p. 178, 1956.

([594]) DAVIES R. — Use of ultrasound waves for the disruption of microorganisms. *Biochem. Biophys. Acta*, t. **XXXIII**, p. 481-493, 1959.

([595]) DAVIS N. D. et DIENER U. L. — Inhibition of aflatoxin (carcinogen) synthesis by p-amino-benzoic acid, potassium sulfite and potassium fluoride (*Aspergillus parasiticus* Speare var. *globosum* Murikami). *Appl. Microbiol.*, t. **XV**, p. 1517-1518, 1967.

([596]) DAVIS N. D. et DIENER U. L. — Growth (in vitro) and aflatoxin production by *Aspergillus parasiticus* from various carbon sources. *Appl. Microbiol.*, t. **XVI**, p. 158-159, 1968.

([597]) DAVIS N. D. et DIENER U. L. — Environmental factors affecting the production of aflatoxin. *in* HERZBERG M., *Toxic micro-organisms*, p. 43-47, 1970.

([598]) DAVIS N. D., DIENER U. L. et AGNIHOTRI V. P. — Production of aflatoxins B_1 and G_1 in chemically defined medium (by *Aspergillus flavus*). *Mycopathol. Mycol. Appl.*, t. **XXXI**, p. 251-256, 1967.

([599]) DAVIS N. D., DIENER U. L. et ELDRIDGE D. W. — Production of aflatoxins B_1 and G_1 by *Aspergillus flavus* in a semisynthetic medium. *Appl. Microbiol.*, t. **XIV**, p. 378-380, 1966.

([600]) DAVIS N. D., SANSING G. A., ELLENBURG T. V. et DIENER V. L. — Medium-scale production and purification of ochratoxin A, a metabolite of *Aspergillus ochraceus*, *Appl. Microbiol.*, t. **XXIII**, p. 433-453, 1972.

([601]) DAVIS N. D., SEARCY J. W. et DIENER U. L. — Production of ochratoxin A by *Aspergillus ochraceus* in a semisynthetic medium. *Appl. Microbiol.*, t. **XVII**, p. 742-744, 1969.

([602]) DAVISON S. et MARBROOK J. — The effect of temperature on the toxicity of spores of *Pithomyces chartarum* (Berk. et Curt.) M. B. Ellis. *New Zealand J. Agric. Res.*, t. **VIII**, p. 126-130, 1965.

([603]) DAWKINS A. W., GROVE J. F. et TIDD B. K. — Diacetoxyscirpenol and some related compounds. *Chem. Comm.*, p. 27-28, 1965.

([604]) DEAN F. M., ERNI A. D. T. et ROBERTSON A. — Chemistry of fungi. **XXVI**. Dechlorornornidulin. *J. Chem. Soc.*, p. 3545-3548, 1956.

([605]) DEAN F. M., ROBERTS J. C. et ROBERTSON A. — Chemistry of fungi. XXII. Nidulin and nornidulin (ustin): chlorine-containing metabolic products of *Aspergillus nidulans*. *J. Chem. Soc.*, p. 1432-1439, 1954.

([606]) DEAN F. M., ROBERTSON A., ROBERTS J. C. et RAPER K. B. — Nidulin and ustin, two chlorine-containing metabolic products of *Aspergillus nidulans*. *Nature*, t. **CLXXII**, p. 344, 1953.

([607]) DE ANDRES V. et CALET C. — Influence de l'aflatoxine sur les facultés reproductrices du coq et de la poule *Gallus gallus* L. *Symposium INSERM Mycotoxines et Alimentation, Le Vésinet*, 24 oct. 1969.

([608]) DEBRÉ R. et NÉVOT A. — Sur le problème de l'étiologie de l'acrodynie infantile. *Annales pediatrici*, 1939.

([609]) DECAISNE E. — Sur l'*Oidium aurantiacum* du pain. *C.R. Acad. Sci.*, t. **LXXIII**, p. 507-508, 1871.

([610]) DEFOORT R., KAIVERS R. et VANBREUSEGHEM R. — Aspergillose vésicale. *Acta Urol. Belg.*, t. **XXIII**, p. 100-102, 1955.

([611]) DEGOIX P. — Emploi des lampes germicides dans les chambres froides. *Rev. Prat. Froid*, t. **XXI**, p. 47-49, 1968.

([612]) DEHOVE R. A. — *La réglementation des produits alimentaires et non alimentaires. Répression des fraudes et contrôle de la qualité.* 6e éd., 936 p., Commerce-éditions, Paris, 1967.

([613]) DE IONGH H., BEERTHUIS R. K., VLES R. O., BARRETT C. B. et ORD W. O. — Investigation of the factor in groundnut meal responsible for « turkey X disease », *Biochim. Biophys. Acta*, t. **LXV**, p. 548-551, 1962.

(614) De Iong H., Van Pelt J. G., Ord W. O. et Barrett C. B. — A semi quantitative determination of aflatoxin B_1 in groundnut meal, groundnuts and peanut butter. *Vet. Rec.*, t. **LXXVI**, p. 901-903, 1964.

(615) De Iongh H., Vles R. Q., Barrett C. et Van Pelt J. G. — Milk of mammals fed an aflatoxin-containing diet. *Nature*, t. **CCII**, p. 466-467, 1964.

(616) De Iongh H., Vles R. O. et De Vogel P. — The occurrence and detection of aflatoxin in food. *Mycotoxins in Foodstuffs*, p. 235-245, 1965.

(617) Delage J. — Contribution à l'étude des répercussions de la présence d'aflatoxine dans des régimes d'animaux domestiques. *Colloque Aflatoxines, Villejuif*, 7 mars 1966.

(618) Delage J. et Bernage L. — Les problèmes posés par l'emploi de tourteaux d'arachide pollués par l'*Aspergillus flavus* dans l'alimentation des bovins. *Ind. Anim.*, t. **XVII**, p. 37-41, 1966.

(619) Delage S. et Fehr P. — Contribution à l'étude des répercussions de la présence d'aflatoxine dans les régimes d'animaux domestiques. *Ind. Alim. Anim.*, n° 183, p. 44-47, 1967.

(620) Delassus M. — Intervention au *Colloque aflatoxines, Villejuif*, 7 mars 1966.

(621) Delassus M. — Etudes phytosanitaires sur l'arachide et les céréales alimentaires au Sénégal. *Agron. Trop.*, t. **XXII**, p. 1227-1234, 1967.

(622) Delassus M. et Goarin P. — Préservation des arachides des infections fongiques : cas de l'*Aspergillus flavus. Comm. Congr. Prot. Cult. trop.*, Marseille, mars 1965.

(623) Delassus M., Goarin P. et Goarin S. — Revue d'études mycologiques et agronomiques faites sur l'aflatoxine. *Agron. Trop.*, t. **XXI**, p. 1398-1406, 1966.

(624) Delaunay A., Daniel P., De Roquefeuil C. et Henon M. — Effets exercés par la patuline et des mélanges patuline-cystéine sur les propriétés physiologiques des cellules phagocytaires. *Ann. Inst. Pasteur*, t. **LXXXVIII**, p. 699-712, 1955.

(625) Del Prado F. A. et Christensen C. M. — Grain storage studies XII. The fungus flora of stored rice seed. *Cereal Chem.*, t. **XXIX**, p. 456-462, 1952.

(626) De Luca H. F. — Diet and aflatoxin toxicity. *Nutr. Rev.*, p. 181-182, 1971.

(627) Demakov G. P. — (Sur la fusariotoxicose de bovins). *Veterinarja*, t. **XLI**, p. 59-60, 1964.

(628) Demayer E. M. et al. — Current research in aflatoxin. WHO/FAO/UNICEF. *Protein advisory Group News Bull.*, n° 5, p. 76-82, 1965.

(629) Dennis C. et Webster J. — Antagonistic properties of species-groups of *Trichoderma*. I. Production of non-volatile antibiotics. *Trans. Brit. Mycol. Soc.*, t. **LVII**, p. 25-39, 1971.

(630) Denoix. — Le cancer ; quelques-uns de ses aspects. *Bull. M.G.E.N.*, t. **XIX**, fasc. 6, p. 10-11, 1966.

(631) Dentice di Accadia F. — Tavola rotonda sulle Micotossine. *Ann. Ist. Super. Sanita*, t. **III**, p. 327-357, 1967.

(632) Derzsy D., Meszaros J., Prokopovitsch L. et Toth-Baranyi I. — (Virus hepatitis and toxic hepatitis caused by groundnut meal in duck in Hungary). *Mag. allator. Lapja*, t. **XVII**, p. 49-53, 1962.

(633) Desfleurs M. — Contribution à la connaissance du genre *Penicillium*. Application à la fabrication de fromages à pâte molle, et notamment du camembert. 78 p. *Thèse Fac. Sci. Caen*, juin 1966.

(634) Desikachar H. S. R., Majunder S. K., Pignale S. V. et Subrahmanyan V. — Discoloration in rice : some studies on its nature and effect on nutritive value. *Cereal Chem.*, t. **XXXVI**, p. 78-83, 1959.

(635) Detroy R. W. et Hesseltine C. W. — Isolation and biological activity of a microbial conversion product of aflatoxin B_1. *Nature*, t. **CCIX**, p. 967, 1968.

(636) Detroy R. W. et Hesseltine C. W. — Transformation of aflatoxin B_1 by steroid-hydroxylating fungi. *Canad. J. Microbiol.*, t. **XV**, p. 495-500, 1969.

([637]) DETROY R. W. et HESSELTINE C. W. — Secondary biosynthesis of aflatoxin B_1 in *Aspergillus parasiticus*. *Can. J. Microbiol.*, t. **XIV**, p. 959-963, 1970.

([638]) DETROY R. W., LILLEHOJ E. B. et CIEGLER A. — Aflatoxin and related compounds. *in* CIEGLER A. et al., *Microbial Toxins*, t. **VI**, p. 4-178, 1971.

([639]) DEVERALL J. — Substances produced by pathogenic organisms that induce symptoms of disease in higher plants *in* Smith H. et Taylor J. Microbial behaviour «in vivo » and « in vitro », p. 165-186, Cambridge University Press, 1963.

([640]) DE VOGEL P., VAN RHEE R. et KOELENSMID W. A. A. — A rapid screening test for aflatoxin synthesizing Aspergilli of the *flavus-oryzae* group. *J. Appl. Bact.*, t. **XXVIII**, p. 213-220, 1965.

([641]) DEVOS A., STAELENS M. et VIAENE N. — Het hemorrhagisch anemisch syndroom by kuikens. *Vlaams diergeneeskd, T.*, t. **XXXIV**, p. 90-99, 1965.

([642]) DHAR A. et BOSE S. — Studies on versicolin, a new antifungal antibiotic from *Aspergillus versicolor*. 3. Relationship between antibiotic synthesis and basic cellular metabolism. *Acta microbiologica polonica*, t. **XVII**, p. 327-330, 1968.

([643]) DICKENS F. — *Carcinogenesis. A broad critique.* p. 447-470, The Williams and Wilkins Co, Baltimore, 1967.

([644]) DICKENS F. et JONES H. E. H. — Carcinogenic activity of a series of reactive lactones and related substances. *Brit. J. Cancer*, t. **LI**, p. 85, 1961.

([645]) DICKENS F. et JONES H. E. H. — The carcinogenic action of aflatoxin after its subcutaneous injection in the rat. *Brit. J. Cancer*, t. **XVII**, p. 691-698, 1963.

([646]) DICKENS F. et JONES H. E. H. — Carcinogenic action of pure aflatoxin. *Rep. Br. Exemp. Canc. Compt.*, t. **XLI**, p. 8, 1965.

([647]) DICKENS F., JONES H. E. H. et WAYNFORTH H. B. — Oral, subcutaneous and intratracheal administration of carcinogenic lactones and related substances: the intratracheal administration of cigarette tar in the rat. *Brit. J. Cancer*, t. **XX**, p. 134, 1966.

([648]) DICKENS J. W. et PATTEE H. E. — Time-Temperature-Moisture effects on aflatoxin production in peanuts inoculated with a toxic strain of *Aspergillus flavus*. *Rept. to Peanut Improvement Working Group Meeting, Washington, 27-28 avr.*, 13 p., 1956.

([649]) DICKENS J. W. et PATTEE H. E. — The effects of time, temperature, and moisture on aflatoxin production in peanuts inoculated with a toxic strain of *Aspergillus flavus*. *Trop. Sci.*, t. **VIII**, p. 11-22, 1966.

([650]) DICKENS J. W. et SATTERWHITE J. B. — Diversion program for farmers' stock peanuts with high concentrations of aflatoxin. *Oléagineux*, t. **XXVI**, p. 321-328, 1971.

([651]) DICKENS J. W. et WELTY R. E. — Detecting farmers' stock peanuts containing aflatoxin by examination for visible growth of *Aspergillus flavus*. *Mycopathol. Mycol. Appl.*, t. **XXXVII**, p. 65-69, 1969.

([652]) DICKINSON L. — Bacteriophages of *Pseudomonas pyocyanea*. I. The effect of various substances upon their development. *J. Gen. Microbiol.*, t. **II**, p. 154-161, 1948.

([653]) DIEHL P. — Substituent effects in proton magnetic resonance spectra of substituted benzenes. *Helv. Chim. Acta*, t. **XLIV**, p. 829, 1961.

([654]) DIENER U. L. — Qualitative and quantitative determinations in stored peanut seed. *J. Alabama Acad. Sci.*, t. **XXIX**, p. 77-78, 1957.

([655]) DIENER U. L. — The mycoflora of stored peanuts. *J. Alabama Acad. Sci.*, t. **XXX**, p. 5-6, 1958.

([656]) DIENER U. L. — The mycoflora of peanuts in storage. *Phytopathology*, t. **L**, p. 220-223, 1960.

([657]) DIENER U. L. — Relation of *Aspergillus flavus* invasion to maturity of peanuts at harvest. *J. Alabama Acad. Sci.*, t. **XXXV**, 1965.

([658]) DIENER U. L. et DAVIS N. D. — Aflatoxin production by isolates of *Aspergillus flavus*. *Phytopathology*, t. **LVI**, p. 1390-1393, 1966.

([659]) DIENER U. L. et DAVIS N. D. — Effect of environment on aflatoxin production by *Aspergillus flavus* in sterile peanuts. *Proc. Nat. Peanut Res. Conf., Tifton, Ga*, p. 93-94, 1966.

([660]) DIENER U. L. et DAVIS N. D. — Effect of temperature on aflatoxin production by isolates of *Aspergillus flavus* and *A. parasiticus. Proc. Nat. Peanut Res. Conf. Tifton, Ga*, p. 95, 1966.

([661]) DIENER U. L. et DAVIS N. D. — Limiting temperature and relative humidity for growth and production of aflatoxin and free fatty acids by *Aspergillus flavus* in sterile peanuts. *J. Amer. Chem. Soc.*, t. **XLIV**, p. 259-263, 1967.

([662]) DIENER U. L. et DAVIS N. D. — Production of aflatoxin on peanuts under controlled environments. *1st Int. Congr. Pl. Pathol.*, Londres, juil. 1968.

([663]) DIENER U. L. et DAVIS N. D. — Effect of environment on aflatoxin production *Trop. Sci.*, t. X, p. 22-28, 1968.

([664]) DIENER U. L. et DAVIS N. D. — Aflatoxin formation by *Aspergillus flavus. in* Goldblatt L. A., Aflatoxin, Academic Press, p. 13-54, 1969.

([665]) DIENER U. L., DAVIS N. D., SALMON W. D. et PRICKETT C. O. — Toxin producing *Aspergillus* isolated from domestic peanuts. *Science*, t. **CXLII**, p. 1491-1492, 1963.

([666]) DIENER U. L., JACKSON C. R., COOPER W. E., STIPES R. J. et DAVIS N. D. — Invasion of peanut pods in the soil by *Aspergillus flavus. Pl. Dis. Reptr.*, t. **XLIX**, p. 931-935, 1965.

([667]) DIERCHEN W. — Mikroflora des Getreides im Reinigungs. Vermahlungsdiagram. *Die Mühle*, t. **XXVII**, p. 408-410, 1953.

([668]) DI MENNA M. E. — The mycoflora of leaves of pasture plants in New Zealand. *in* PREECE T. F. et DICKINSON C. H., *Ecology of leaf surface microorganisms*, p. 159-174, Academic Press, 1971.

([669]) DINGLEY J. M. — *Pithomyces chartarum*, its occurrence, morphology and taxonomy. *New Zealand J. Agr. Res.*, t. V, p. 49-61, 1962.

([670]) DINWIDDIE R. R. — Unsound corn and forage as a cause of disease of livestock. *Arkansas Agr. Exp. Stn., Bull* 25, 1893.

([671]) DI PALMA J. R. — *Drill's Pharmacology in Medicine.* Mc Graw-Hill Book Co, New York, 1965.

([672]) DIPAOLO J. A., ELIS J. et ERWIN H. — Teratogenic response by hamsters, rats and mice to aflatoxin B_1. *Nature, G. B.*, t. **CCXV**, p. 638-639, 1967.

([673]) DOBROSKY T. M. — *Grain storage newsletter*, n° 1, FAO, 1959.

([674]) DOBRYNIN V. P. — Obesvrezhevanie kormovoi slomi pre porazhenii eyo pathogenimi gribami. *Konevadstvo*, t. **XVI**, p. 27-30, 1946.

([675]) DOBSON N. — The toxicity of the spores of *Tilletia tritici* to animals. *Trans. Brit. Mycol. Soc.*, t. **XI**, p. 82-91, 1926.

([676]) DOCTOR V. et KERUR L. — *Penicillium* mycelium waste as protein supplement in animals. *Appl. Microbiol.*, t. **XVI**, p. 1723-1726, 1968.

([677]) DODD D. C. — Photosensitivity diseases in New Zealand. **XXI.** The susceptibility of the rabbit to facial eczema. *New Zealand J. Agr. Research*, t. **III**, p. 491-497, 1960.

([678]) DODD D. C. — Facial eczema in ruminants. *Mycotoxins in Foodstuffs*, p. 105-110, 1965.

([679]) DOERING W. E., DUBOS R. J., NOYCE D. S. et DREYFUS R. — Metabolic products of *Aspergillus ustus. J. Am. Chem. Soc.*, t. **LXVIII**, p. 725-726, 1946.

([680]) DOLIMPIO D., GREEN S., LEGATOR M. et JACOBSON C. B. — Aflatoxin induced chromosomal aberrations from various *in vitro* systems. *Int. Congr. Human Gen. Abstr. Contrib. Pap.*, n° 3, p. 26, 1966.

([681]) DOLIMPIO D., JACOBSON C. et LEGATOR M. — Effect of aflatoxin on human leucocytes. *Proc. Soc. Exp. Biol. Med.*, t. **CXXVII**, p. 559-562, 1968.

([682]) DOLLAR A. M. et KATZ M. — *Res. Fisheries*, t. **CXXXIX**, p. 23-25, 1962.

([683]) Dollear F. G. — Detoxification of aflatoxins in foods and feeds. *in* Goldblatt L. A., *Aflatoxin*, Academic Press, p. 359-391, 1969.

([684]) Dollear F. G. et Gardner H. K. — Inactivation and removal of aflatoxin. *Proc. IVth Nation. Peanut Res. Conf. Tifton, Ga*, p. 72-81, 1966.

([685]) Dollear F. G., Mann G. E., Codifer L. P., Gardner K. H., Koltun S. P. et Vix H. L. E. — Elimination of aflatoxins from peanut meal. *J. Amer. Oil Chem. Soc.*, t. **XLV**, p. 862-865, 1968.

([686]) Done J., Mortimer P. H. et Taylor A. — Some observations on field cases of facial eczema : liver pathology and determinations of serum bilirubin, cholesterol, transaminase and alkaline phosphatase. *Res. Vet. Sci.*, t. **I**, p. 76-83, 1960.

([687]) Done J., Mortimer P. H. et Taylor A. — The experimental intoxication of sheep with sporidesmin, a metabolic product of *Pithomyces chartarum*. II. *Res. Vet. Sci.*, t. **III**, p. 161, 1962.

([688]) Done J., Mortimer P. H., Taylor A. et Russell D. W. — The production of sporidesmin and sporidesmolides by *Pithomyces chartarum. J. Gen. Microbiol.*, t. **XXVI**, p. 207-222, 1961.

([689]) Donkersloot J. A., Hsieh D. P. H. et Mateles R. I. — Biosynthesis of aflatoxins. *J. Am. Chem. Soc.*, t. **XC**, p. 5017, 1968.

([690]) Dos Santos J. A. — Aflatoxina e cancer hepatico. *Pesquis Agropecuar Brasil*, t. **I**, p. 75-85, 1966.

([691]) Dounin M. — The fusariosis of cereal crops in European Russia in 1923. *Phytopathology*, t. **XVI**, p. 305, 1926.

([692]) Doupnik B. — Factors associated with aflatoxins in farmers' stock peanuts in Georgia in 1967. *Phytopathology*, t. **LIX**, p. 112-113, 1969.

([693]) Doupnik B. — Aflatoxins produced on peanut varieties previously reported to inhibit production. *Phytopathology*, t. **LIX**, p. 1554, 1969.

([694]) Doupnik B. et Bell D. K. — Toxicity to chicks of *Aspergillus* and *Penicillium* species isolated from moldy pecans. *Appl. Microbiol.*, t. **XXI**, p. 1104-1106, 1971.

([695]) Doupnik B., Jones O. H. et Peckham J. C. — Toxic Fusaria isolated from moldy sweet potatoes involved in an epizootic of atypical interstitial pneumonia in cattle. *Phytopathology*, t. **LXI**, p. 890, 1971.

([696]) Doupnik B. et Peckman J. C. — Mycotoxicity of *Aspergillus ochraceus* to chicks. *Appl. Microbiol.*, t. **XIX**, p. 594-597, 1970.

([697]) Doupnik B. et Sobers E. K. — Mycotoxicosis : toxicity to chicks of *Alternaria longipes* isolated from tobacco. *Appl. Microbiol.*, t. **XVI**, p. 1596-1597, 1900.

([698]) Draghici C., Moldoveanu C. et Comsa E. — Considérations cliniques, pharmacologiques et physiopathologiques concernant quelques cas de stachybotryotoxicose. *C. R. 5e Congr. ISHAM*, p. 218-219, Paris, 6 juil. 1971.

([699]) Drobotko V. G. — Stachybotryotoxicosis, a new disease of horse and humans. *Am. Rev. Soviet Med.*, t. **II**, p. 238-242, 1945.

([700]) Drobotko V. G., Marushenko P. E., Aizeman B. E., Kolesnik N. G., Kudlai D. B., Iatel P. D. et Melnichenko V. D. — Stakhibotriotoksikoz novoe zabolevanie loshadei i liudei. *Vrachebnoe Dela*, t. **XXVI**, p. 125-128, 1946.

([701]) Drouhet E. — L'évaluation des antifongiques pour les essais thérapeutiques des mycoses humaines et animales. *C. R. 6e Congr. int. Standardisation microbiol.*, p. 331-369, Wiesbaden, 1960.

([702]) Dulong De Rosnay C., Martin-Dupont C. et Jensen R. — Etude d'une substance antibiotique, la « mycoïne C ». *J. Med. Bordeaux*, t. **CXXIX**, p. 189-199, 1952.

([703]) Dumontell M. — Technologie de la fabrication des aliments du bétail. 131 p., Vigot édit., Paris, 1966.

([704]) Dupont de Dinechin B. — Les moisissures toxiques de l'arachide au Sénégal. *Rapport stage Centre de Rech. agron. Bambey*, 1963.

([705]) Durbin C. G., Geleta J. N., Shalkop W. T. et Armbrecht B. H. — The feeding of aflatoxin to beagle dogs. *Toxicol. Appl. Pharmacol.*, t. **XIV**, p. 649, 1969.

([706]) Durbin R. D. — The possible relationship between *Aspergillus flavus* and albinism in citrus. *Pl. Dis. Reptr.*, t. **XLIII**, p. 922-923, 1959.

([707]) Dutcher J. D. — The chemical nature of gliotoxin : a microbicidal compound produced by the fungus *Gliocladium fimbriatum. Jour. Bact.*, t. **XLII**, p. 815-816, 1941.

([708]) Dutcher J. D., Johnson J. R. et Bruce W. F. — Gliotoxin, the antibiotic principle of *Gliocladium fimbriatum.* III. The structure of gliotoxin-degradation by hydroiodic acid. IV. The action of selenium. *Amer. Chem. Soc. J.*, t. **LXVI**, p. 617-621, 1944.

([709]) Dutton M. F. et Heathcote J. G. — Two new hydroxyaflatoxins. *Biochem. J.*, t. **CI**, p. 21-22, 1966.

([710]) Dutton M. F. et Heathcote J. G. — The structure, biochemical properties and origin of the aflatoxins B_{2a} and G_{2a}. *Chem. Ind.*, p. 418-421, 1968.

([711]) Dutton M. F. et Heathcote J. G. — O-Alkyl derivatives of aflatoxins B_{2a} and G_{2a}. *Chem. and Ind.*, p. 983-986, 1969.

([712]) Duyvene de Wit J. J., Jaarsveld A., Jansen B. C. P., Van Luyk A., Luyken R., Oosterhuis H. K. et Wibrans J. R. — The isolation of a bactericidal and fungicidal substance from a penicillate mould. *Nederland. Tijdschr. voor Geneeskunde*, t. **LXXXVIII**, n° 31-32, 1944.

([713]) Dwarakanath C. T., Rayner E. T., Mann G. E. et Dollear F. G. — Reduction of aflatoxin levels in cottonseed and peanut meals by ozonization. *J. Amer. Oil Chem. Soc.*, t. **XLV**, p. 93-95, 1968.

([714]) Eckert J. W. — Chemical treatments for control of postharvest diseases. *World Review Pest Control*, t. **VIII**, p. 116-137, 1969.

([715]) Eckert J. W. et Kolbezen M. J. — Fumigation of fruits with 2-aminobutane to control certain postharvest diseases. *Phytopathology*, t. **LX**, p. 545-550, 1970.

([716]) Eckert J. W. et Sommer N. F. — Control of diseases of fruits and vegetables by post harvest treatment. *Ann. Rev. Phytopathol.*, t. **V**, p. 391-432, 1967.

([717]) Eckles C. H., Fitch C. P. et Seal J. L. — Molds in silage and their significance in the production of diseases among livestock. *J. Am. Vet. Med. Assoc.*, t. **LXIV**. p. 716-722, 1924.

([718]) Edwards G. S. et Wogan G. N. — Acute and chronic toxicity of rubratoxin in rats. *Fed. Proc.*, t. **XXVII**, p. 552, 1968.

([719]) Eggins M. O. W. et Coursey D. G. — The industrial significance of the biodeterioration of oilseeds. *Int. Biodet. Bull.*, t. **IV**, p. 29-38, 1968.

([720]) Eldridge D. W. — Nutritional factors influencing the synthesis of aflatoxin B_1 by *Aspergillus flavus.* M. S. Thesis, Auburn Univ., Alabama.

([721]) Elis J. et Dipaolo J. Å. — Aflatoxin B_1 : Induction of malformations. *Arch. Pathol.*, t. **LXXXIII**, p. 53-57, 1967.

([722]) Ellis J. J. — An orange-yellow mutant of *Aspergillus parasiticus* produces aflatoxin. *Mycologia*, t. **LXI**, p. 651-653, 1969.

([723]) Ellis J. J. et Yates S. G. — Mycotoxins of fungi from fescue. *Econ. Bot.*, t. **XXV**, p. 1-5, 1971.

([724]) Ellis M. B. — Dematiaceous Hyphomycetes. I. *Mycological Papers*, n° 76, 36 p., C. M. I., Kew, 1960.

([725]) Elpidina O. K. — Antiblastic properties of poin on the basis of experimental data. *Kazansk, med. zhurn.*, t. **IV**, 1958.

([726]) Elpidina O. K. — Antibiotic and antiblastic properties of poin. *Antibiotiki*, t. **IV**, p. 46, 1959.

([727]) Elpidina O. K. — Toxic and antibiotic properties of poin. *Mycotoxicoses of Man and Farm Animals.* Izd-vo AN UKSSR, Kiev, 1960.

[728] ELRIDGE D. W., DAVIS N. D. et DIENER U. L. — Aflatoxin content and fatty acid composition of peanuts inoculated with *Aspergillus flavus. Phytopathology*, t. **LV**, p. 1057, 1965.

[729] ELSWORTHY G. C., HOLKER J. S. E., MCKEOWN J. M., ROBINSON J. B. et MULHEIRN L. J. — The biosynthesis of the aflatoxins. *J. Chem. Soc.* (D), t. **XVII**, p. 1069-1070, 1970.

[730] EMERICK R. J. et EMBRY L. B. — Effect of chlortetracycline on methemoglobinemia resulting from the ingestion of sodium nitrate by ruminants. *J. Anim. Sci.*, t. **XX**, p. 844, 1961.

[731] EMMONS C. W. — The Jekyll-Hydes of Mycology. *Mycologia*, t. **LII**, p. 669-680, 1960.

[732] EMMONS C. W. — Mycology and Medicine. *Mycologia*, t. **LIII**, fasc. 1, p. 1-10, 1961.

[733] ENGEBRECHT R. H., AYRES J. L. et SINNHUBER R. O. — Isolation and determination of aflatoxin B_1 in cottonseed meals. *J. Ass. off. agric. Chemists*, t. **XLVIII**, p. 815-818, 1965.

[734] ENGEL B. G. et BRZESKI W. — Über die Isolierung eines Chinhydrons von Gentisinalcohol und Oxymethyl-p-bensochinon (Gentisinchinon) aus dem Kulturfiltrat von *Penicillium urticae* Bainier. *Helv. Chim. Acta*, t. **XXX**, p. 1472-1478, 1947.

[735] ENGELBRECHT J. C. — The effects of sterigmatocystin on a primary cell culture. *S. Afr. Med. J.*, t. **XLIV**, p. 153-159, 1970.

[736] ENGELBRECHT J. C. — The effects of aflatoxin B_1 and sterigmatocystin on two different types of cell cultures. *in* PURCHASE I. F. H., *Mycotoxins in human health*, p. 215-222, 1971.

[737] ENGLISH W. H. — Taxonomic and pathogenicity studies of the fungi which cause decay of pears in Washington. *Res. Stud. St. Coll. Wash.*, t. **VIII**, p. 127-128, 1940.

[738] ENGSTROM G. W. — New solvent systems for thin-layer chromatography of aflatoxins. *J. Chromatogr.*, t. **XLIV**, p. 128-132, 1969.

[739] ENOMOTO M. — Histopathological studies on adenomatous nodules of liver induced in mice by *Penicillium islandicum* Sopp. *Acta Pathol. Jap.*, t. **IX**, p. 189-215, 1959.

[740] ENTREKIN D. N. — Relation of pH to preservative effectiveness. I. Acid media, *J. Pharm. Sci.*, t. **L**, p. 743-746, 1961.

[741] EPPLEY R. M. — Note on a developer for the thin layer chromatography of aflatoxins. *J. Ass. Off. Anal. Chemists*, t. **XLIX**, p. 473-474, 1966.

[742] EPPLEY R. M. — Aflatoxins : a versatile procedure for assay and preparatory separation of aflatoxins from peanut products. *J. Ass. Off. Anal. Chem.*, t. **XLIX**, p. 1218-1223, 1966.

[743] EPPLEY R. M. — Screening methods for zearalenone, aflatoxin and ochratoxin. *J, Ass. Off. Anal. Chem.*, t. **LI**, p. 74-78, 1968.

[744] EPPLEY R. M., STOLOFF L. et CAMPBELL A. D. — Collaborative study of « a versatile procedure for assay of aflatoxins in peanuts products » including preparatory separation and confirmation of identity. *J. Ass. Off. Anal. Chem.*, t. **LI**, p. 67-73, 1968.

[745] EPSTEIN S. M., BARTUS B. et FARBER E. — Renal epithelial neoplasms induced in male Wistar rats by oral aflatoxin B_1. *Cancer Res.*, t. **XXIX**, p. 1045-1050, 1969.

[746] EPSTEIN E., STEINBERG M. P., NELSON A. I. et WEI L. S. — Aflatoxin production as affected by environmental conditions. *J. Food Sci.*, t. **XXXV**, p. 389-391, 1970.

[747] ERNE K. et BJÖRKLUND N. E. — Nephrotic effects of phenoxy acetic herbicides. *VIIe Congr. Int. Prot. Pl., Rés. Communic.*, p. 768-769, 1970.

[748] ESCOULA L. — Mycoflore des ensilages et élaboration de métabolites toxiques par les espèces fongiques. *Mém. D. E. A., Fac. Sci. Toulouse*, 41 p. ronéot., juin 1971.

[749] ESCOULA L. et LE BARS J. — La mycoflore des ensilages. Croissance d'espèces fongiques en anaérobiose. *Ann. Rech. Vét.*, t. **IV**. p. 253-264, 1973.

[750] ESCOULA L., LE BARS J. et LARRIEU G. — Etudes sur la mycoflore des ensilages. Mycoflore des fronts de coupe d'ensilages de Graminées fourragères. *Ann. Rech. Vét.*, t. **III**, p. 469-481, 1972.

[751] ESSELEN W. B. et LEVINE A. S. — Bacterial food poisoning and its control. *Univ. of Mass., Bull.* n° 493, 1957.

([752]) EUGENIO C. P. — Effect of the fungal estrogen F-2 on the production of perithecia by *Fusarium roseum « graminearum »*. *Phytopathology*, t. **LVIII**, p. 1050-1051, 1968.

([753]) EUGENIO C. P., CHRISTENSEN C. M. et MIROCHA C. J. — Factors affecting production of the mycotoxin F-2 by *Fusarium roseum*. *Phytopathology*, t. **LX**, p. 1055-1057, 1970.

([754]) EUGENIO C., DE LAS CASAS E., HAREIN P. R. et MIROCHA C. J. — Detection of the mycotoxin F-2 in the confused flour beetle and the lesser mealworm. *J. Econ. Entomol.*, t. **LXIII**, p. 412-415, 1970.

([755]) EVANS J. V., MC FARLANE D., REID C. S. W. et PERRIN D. D. — Photosensitivity diseases in New Zealand. IX. The susceptibility of the guinea pig to facial eczema. *New Zealand J. Sci. Technol.*, sér. A., t. **XXXVIII**, p. 491-496, 1957.

([756]) EVANS M. M. et POOLE R. F. — Some parasitic fungi harbored by peanut seed stock. *J. Elisha Mitchell Sci. Soc.*, t. **XLIII**, p. 190, 1939.

([757]) EVERETT G. M., BLOCKUS L. E. et SHEPPERD I. M. — Tremor induced by tremorine and its antagonism by anti-Parkinson drugs. *Science*, t. **CXXIV**, p. 79, 1956.

([758]) EWART A. J. — The presence of citrinin in *Crotalaria crispata*. *Ann. Bot.*, t. **XLVII**, p. 913-915, 1933.

([759]) FABIAN F. W. — The preservative action of salt. *Food and Food Products. Interscience Publishers*, New York, t. **III**, p. 1897-1900, 1951.

([760]) FABRE R. et TRUHAUT R. — *Toxicologie des produits phytopharmaceutiques*. 272 p., Sedes édit., Paris, 1954.

([761]) FABRE R. et TRUHAUT R. — Précis de Toxicologie, 2 t., 721 p., Sedes édit., 1965.

([762]) FAGARD P. et DEKEYSER J. — A propos de deux cas d'aspergillose généralisée chez des Poulets. *Bull. Agr. Congo Belge*, t. **L**. p. 1321-1328, 1959.

([763]) FALK H. L., THOMPSON S. J. et KOTIN P. — Metabolism of aflatoxin B_1 in the rat. *Proc. Amer. Ass. Cancer Res.*, t. **VI**. p. 18, 1965.

([764]) FALKOVA T. D. et MISHUSTIN E. H. — The fungus of the genus *Monilia* responsible for the bluing of macaroni during drying. *Microbiologia*, t. IX, p. 54-57, 1940.

([765]) FARBER E. — Carcinoma of the liver in rats fed ethionine. *Arch. Path.*, t. **LXII**, p. 445-453, 1956.

([766]) FARLEY T. M. — The chemistry of mold products. Part II. Chrysophanic acid and pachybasin from *Aspergillus crystallinus*. Ph. D. Thesis, Univ. Wisconsin, 1965.

([767]) FEHR P. M. et DELAGE J. — Effet de l'aflatoxine sur les fermentations du rumen. *Symposium INSERM Mycotoxines et Alimentation, Le Vésinet* 24 oct. 1969. *Cah. Nutr. Diét.*, t. V, p. 59-61, 1970.

([768]) FEHR P. M., DELAGE J. et RICHIR C. — Répercussions de l'ingestion d'aflatoxine sur le lapin en croissance. *Cah. Nutr. Diét.*, t. **V**. p. 62-64, 1970.

([769]) FERGUS C. L. — Thermophilic and thermotolerant molds actinomycetes of mushroom compost during peak heating. *Mycologia*, t. **LVI**, p. 267-284, 1964.

([770]) FERRANDO R. — Conférence au *Colloque sur l'industrialisation des élevages*. RENNES, 3 févr. 1967.

([771]) FERRANDO R. et HENRY N. — Les conséquences pathologiques des aliments infestés par *Aspergillus flavus*. *Bull. Acad. Nat. Méd.*, t. **CXLIX**, p. 94-103, 1965 ; *Presse Méd.*, t. **LXXIII**, p. 1111-1112, 1965.

([772]) FERRANDO R. et HENRY N. — *Détermination microscopique des composants des aliments du bétail*. 112 p., Vigot édit., Paris. 1966.

([773]) FERRANDO R., LALLOUETTE P., BOURDERON G. et FROGET J. — Influence d'un facteur de croissance du colibacille extrait d'*Aspergillus flavus* sur le gain de poids et la flore intestinale. *C. R. Acad. Sci., t.* **CCXLIII**, p. 537-538, 1956.

([774]) FERREIRA N. P. — Recent advances in research on ochratoxin. Part 2. Microbiological aspects. *Biochemistry of some Food-borne Microbial Toxins, Symp. Amer. Chem. Soc. New York, sept.* 1966., p. 157-168, M. I. T. Press, 1967.

([775]) FERREIRA N. P. — The effect of amino acids on the production of ochratoxin A in chemically defined media. *Antonie van Leeuwenhoek, J. Microb. Ser.*, t. **XXXIV**, p. 433-440, 1968.

([776]) FEUELL A. J. — Aflatoxin in groundnuts (*Aspergillus flavus*). IX. Problems of detoxification. *Trop. Sci.*, t. **VIII**, p. 61-70, 1966.

([777]) FEUELL A. J. — Symposium on toxic factors in foods. Toxic factors of mould origin. *Canad. Medic. Ass. J.*, t. **XCIV**, p. 574-581, 1966.

([778]) FEUELL A. J. — Types of mycotoxins in foods and feeds. *in* GOLDBLATT L. A., *Aflatoxin*, Academic Press, p. 187-221, 1969.

([779]) FIELDS R. W. et KING R. H. — Influence of storage fungi on deterioration of stored pea seed. *Phytopathology*, t. **LII**, p. 336-339, 1962.

([780]) FINE B. S. — Intraocular mycotic infections. *Lab. Invest.*, t. **XI**, p. 1161-1171, 1962.

([781]) FISCHBACH H. — Report on the joint AOAC-AOCS aflatoxin Committee. *J. Ass. Off. Anal. Chem.*, t. **LII**, p. 970-975, 1969.

([782]) FISCHBACH H. et CAMPBELL A. L. Note on detoxification of the aflatoxins. *J. Assoc. Off. Agr. Chem.*, t. **XLVIII**, p. 1-28, 1965.

([783]) FISCHER G. W. — The blind-seed disease of rye grass (*Lolium* spp.) in Oregon. *Phytopathology*, t. **XXXIV**, p. 934-935, 1944.

([784]) FISHBEIN L. et FALK H. L. — Chromatography of mold metabolites. I. Aflatoxins, ochratoxins and related compounds. *Chromatogr. Rev.*, t. **XII**, p. 42-87, 1970.

([785]) FISHER E. E., KELLOCK A. W. et WELLINGTON N. A. M. — Toxic strain of *Fusarium culmorum* (W. G. Sm.) Sacc. from *Zea mays* L. associated with sickness in dairy cattle. *Nature, G. B.*, t. **CCXV**, p. 322-323, 1967.

([786]) FITCH C. P. — Disease in cattle caused by feeding sweetclover hay. *Rec. Proc. Am. Soc. Anim. Production 1923*, p. 37, 1924.

([787]) FLANNIGAN B. — Microflora of dried barley grain. *Trans. Brit. Mycol. Soc.*, t. **LIII**, p. 371-379, 1969.

([788]) FLATLA J. L. — Feeding experiments with aflatoxin-contaminated groundnut meal to bulls. *Proc. Symp. Agric. Coll. Norway*, 4p. ronéot., 1967.

([789]) FLOREY H. W., CHAIN E., HEATLEY N. G., JENNINGS M. A., SANDERS A. G., ABRAHAM E. P. K., et FLOREY M. E. — *Antibiotics*. 2 vol., 1774 p., Oxford Univ. Press, 1949.

([790]) FLOREY H. W., JENNINGS M. A. et PHILPOT F. J. — Claviformin from *Aspergillus giganteus* Wehm. *Nature*, t. **CLIII**, p. 139, 1944.

([791]) FLOYD L. R., UNUMA T. et BUSCH H. — Effects of aflatoxin B_1 and other carcinogens upon nucleolar RNA of various tissues in the rat. *Exp. Cell. Res.*, t. **LVII**, p. 423-438, 1968.

([792]) FLURY E., MAULI R. et SIGG H. P. — The constitution of diacetoxyscirpenol. *Chem. Comm.*, p. 26-27, 1965.

([793]) FLYNN D. M., HORE D., LEAVER D. D. et FISHER E. E. — Facial eczema. *J. Agr., Victoria*, t. **LX**, p. 49-58, 1962.

([794]) FOLDI Z., FODOR G. et DEMJEN I. — Investigations relating to the synthesis of patulin. *J. Chem. Soc.*, p. 1295-1299, 1948.

([795]) FOMENCHEV B. F. — Sluchai massovoho otravleniia sveneoi gribam *Fusarium sporotrichioides. Veterinariya*, t. **XXXVI**, p. 73, 1959.

([796]) FONSECA H. — Aflatoxinas G_1 e G_2 degradas na presenca do inseto *Trogium pulsatorium* (L.) *Rev. Brasil Biol.*, t. **XXIX**, p. 241, 1969.

([797]) FORBISHER M. — *Fundamentals of microbiology*. 7e éd., W. B. Saunders Co édit., Philadelphia, London, 1962.

([798]) FORD J. H., JOHNSON A. R. et HINMAN J. W. — The structure of penicillic acid. *J. Am. Chem. Soc.*, t. **LXXII**, p. 4529-4531, 1950.

([799]) FORGACS J. — Mycotoxicoses, neglected diseases. *Feedstuffs*, t. **XXXIV**, fasc. 18, p. 124-134, 1962.

([800]) FORGACS J. — Stachybotryotoxicosis and moldy corn toxicosis. *Mycotoxins in Foodstuffs*, p. 87-104, 1965.

([801]) FORGACS J. — Types of mycotoxicity occurring in feeds and foods. *Food Technol.*, t. **XX**, p. 45-50, 1966.

([802]) FORGACS J. et CARLL W. T. — Preliminary mycotoxic studies on hemorrhagic disease in poultry. *Vet. Med.*, t. **L**, p. 172, 1955.

([803]) FORGACS J. et CARLL W. T. — Mycotoxicosis. *Advances in Veterinary Science*, t. VII, p. 273-382, New York Academic Press, 1962.

([804]) FORGACS J. et CARLL W. T. — Mycotoxicoses : Toxic fungi in tobaccos. *Science*, t. **CLII**, p. 1634-1635, 1966.

([805]) FORGACS J., CARLL W. T. et HERRING A. S. — Toxic fungus isolated from feed pellets. *Vet. Med.*, t. **XLVIII**, p. 410, 1953.

([806]) FORGACS J., CARLL W. T., HERRING A. S. et HINSHAW W. R. — Toxicity of *Stachybotrys atra* for animals. *Trans. N. Y. Acad. Sci.*, t. **XX**, p. 787-808, 1958.

([807]) FORGACS J., CARLL W. T., HERRING A. S. et MAHLANDT B. G. — A toxic *Aspergillus clavatus* isolated from feed pellets. *Am. J. Hygiene*, t. **LX**, p. 15, 1954.

([808]) FORGACS J., KOCH H., CARLL N. T. et WHITESTEVENS R. M. — Additional studies on a relationship of mycotoxicose to the poultry hemorragic syndrom. *Amer. J. Vet. Res.*, t. **XIX**, p. 744-753, 1958.

([809]) FORSYTH A. A. — British Poisonous Plants. *Ministry of Agr., Fisheries, and Food (London), Bull.* n° 161, 1954.

([810]) FORTUSKNY V. A., GOVROV A. M., TEBYBENKO I. Z., BIOCHENKO A. S. et KALITENKO E. T. — Stakhibotriotoksikoz krupnogo rogatogo skota i ego iechenie. *Veterinariya*, t. **XXXVI**, p. 67-70, 1959.

([811]) FOSTER J. W. — *Chemical activities of fungi.* 548 p. Acad. Press, New York, 1949.

([812]) FOXALL C. D. et MORGAN J. W. W. — Extractives of *Afrormosia elata. J. Chem. Soc.* p. 5573-5575, 1963.

([813]) FOY H., GILMAN T., KONDI A. et PRESTON J. K. — Hepatic injuries in riboflavin and pyridoxin deficient baboons — Possible relations to aflatoxin induced hepatic cirrhosis and carcinoma in Africans. *Nature, G. B.*, t. **CCXII**, p. 150-153, 1966.

([814]) FRANK H. K., Aflatoxin in Lebensmitteln. *Arch. Lebensmittel-Hyg. Dtsch.*, t. **XVII**, p. 237-242, 1966.

([815]) FRANK H. K. — Mykotoxine und ihre Produzenten. *Med. Klin.*, t. **LXII**, p. 1933-1941, 1967.

([816]) FRANK H. K. — Diffusion of aflatoxins in foodstuffs. *J. Food. Sci.*, t. **XXXIII**, p. 98-100, 1968.

([817]) FRANK H. K. — Occurrence and demonstration of mycotoxins. *Zentralbl. Bakteriol., Abt. I*, t. **CCXII**, p. 216-221, 1970.

([818]) FRANK H. K. et EYRICH W. — Ueber den Nachweis von Aflatoxinen und das Vorkommen Aflatoxinvortäuschender Substanzen in Lebensmitteln. *Z. Lebensmittel-Unters. Forsch.*, t. **CXXXVIII**, p. 1-10, 1968.

([819]) FRANK H. K. et GRUNEWALD T. — Radiation resistance of aflatoxin. *Irradiat. Aliments*, t. **XI**, p. 15-20, 1970.

([820]) FRANK H. K., MUENTZER R. et DIEHL J. F. — Response of toxigenic and nontoxigenic strains of *Aspergillus flavus* to irradiation. *Sabouraudia, J. Int. Soc. Hum. Anim. Mycol.*, t. **IX**, p. 21-26, 1971.

([821]) FRASER C. M. et NELSON J. — Sweet Clover poisoning in newborn calves. *J. Am. Vet. Med. Assoc.*, t. **CXXXV**, p. 283, 1959.

([822]) FRAYSSINET C. — Recherches concernant le pouvoir carcinogène des toxines de l'*Aspergillus flavus* Link. *Rapport d'activité du C. N. R. S., oct. 1962-oct. 1963*, p. 460, 1963.

([823]) FRAYSSINET C. — Le contrôle de la teneur en aflatoxine : techniques recommandées par l'Union de chimie pure et appliquée. *Colloque aflatoxines, Villejuif,* 7 mars 1966.

([824]) FRAYSSINET C. — Inhibition par l'aflatoxine des synthèses d'acide ribonucléique chez le rat. *Colloque Aflatoxines, Villejuif,* 7 mars 1966.

([825]) FRAYSSINET C., BERNHARD W. et LAFARGE C. — Action de l'aflatoxine sur la cellule hépatique du rat. Aspect morphologique. *Symposium INSERM Mycotoxines et Alimentation, Le Vésinet,* 24 oct. 1969.

([826]) FRAYSSINET C. et LAFARGE C. — Action de l'aflatoxine sur la cellule hépatique du rat. *Cah. Nutr. Diét.,* t. **V**, p. 67-69, 1970.

([827]) FRAYSSINET C., LAFARGE C., de RECONDO A. M. et LE BRETON E. — Inhibition de l'hypertrophie compensatrice du foie chez le rat par les toxines d'*Aspergillus flavus. C. R. Acad. Sci.,* t. **CCLIX**, p. 2143-2146, 1964.

([828]) FRAYSSINET C., LAFARGE C. et coll. — Etude du pouvoir carcinogène de l'aflatoxine. *Rapport d'activité du C. N. R. S., oct. 1963-oct. 1964,* p. 544-545, 1964.

([829]) FRAYSSINET C., LAFARGE C. et coll. — Mécanisme d'action de l'aflatoxine. *Rapport activité C. N. R. S., oct. 1964-oct. 1965,* p. 351, 1966.

([830]) FRAYSSINET C. et LAFONT P. — Les mycotoxines : un nouvel aspect de l'hygiène des aliments. *Cahiers de nutrition et de diététique,* t. **I**, p. 21-29, 1966.

([831]) FRAYSSINET C. et LAFONT P. — La toxine B_3d'*Aspergillus flavus. Symposium INSERM Mycotoxines et Alimentation, Le Vésinet,* 24 oct. 1969.

([832]) FRAYSSINET C. et LAFONT P. — Production par des *Aspergillus* de mycotoxines différentes des aflatoxines. *Ann. Inst. Pasteur,* t. **CXVI**, p. 331-340, 1969.

([833]) 412. FRAYSSINET C., LE BRETON E. et BOY J. — *Rapport d'activité du C. N. R. S., oct. 1961-oct. 1962,* p. 391, 1962.

([834]) FRAZIER W. C. — An outbreak of moldy bread. *Baking Technol.,* t. **II**, p. 184-187, 1923.

([835]) FRAZIER W. C. — Moldy bread outbreak due to infected flour. *Wis. Exp. Stn. Bull.,* n° 352, p. 74-75, 1923.

([836]) FRAZIER W. C. — *Food Microbiology.* Mc Graw Hill Book Co, New York, 1958.

([837]) FREEMAN E. M. — The seed fungus of *Lolium temulentum,* the Darnel. *Proc. Roy. Soc. (London),* t. **LXXI**, p. 27, 1902.

([838]) FREEMAN G. G. — Further biological properties of trichothecin, an antifungal substance from *Trichothecium roseum,* and its derivatives. *J. Gen. Microbiol.,* t. **XII**, p. 213-221, 1955.

([839]) FREEMAN G. G., GILL J. E. et WARING W. S. — The structure of trichothecin and its hydrolysis products. *J. Chem. Soc.,* p. 1105-1132, 1959.

([840]) FREEMAN G. G. et MORRISON R. I. — Trichothecin : an antifungal metabolic product of *Trichothecium roseum. Nature, G. B.,* t. **CLXII**, p. 30, 1948.

([841]) FREEMAN G. G. et MORRISON R. I. — The isolation and chemical properties of trichothecin, an antifungal substance from *Trichothecium roseum.* Link. *Biochem. J.,* t. **XLIV**, p. 165, 1949.

([842]) FREEMAN M. A. R. et BARWELL C. F. — Ethylene oxide sterilization in hospital practice. *J. Hyg. Camb.,* t. **LVIII**, p. 337-344, 1960.

([843]) FREERKSEN E. et BONICKE R. — Die Inaktivierung des Patulins *in vivo. Z. Hyg. Infektionskrkh.,* t. **CXXXII**, p. 274-291, 1951.

([844]) FRIAS L. C. L. — Factors affecting invasion of grains and seeds by *Aspergillus flavus* and effects of the fungus on germination of the seeds. *Diss. Abstr.,* t. **XXVII**, B, p. 4208, 1967.

([845]) FRIDRICHSONS J. et MATHIESON A. McL. — The structure of sporidesmin, causative agent of facial eczema in sheep. *Tetrah. Lett.,* n° 26, p. 1265-1268, 1962.

([846]) FRIED J., THOMAS R. W. et KLINGSBERG A. — Oxidation of steroids by microorganisms. III. Side chain degradation, ring D cleavage, and dehydrogenation in ring A. *J. Am. Chem. Soc.,* t. **LXXV**, p. 5764-5765, 1953.

([847]) FRIEDHEIM E. A. M. — Sur la fonction respiratoire du pigment rouge de *Penicillium phoeniceum. C. R. Soc. Biol., Paris,* t. **CXII**, p. 1030-1032, 1933.

([848]) FRIEDL J. L., ORTENZIO L. F. et STUART L. S. — The sporicidal activity of ethylene oxide as measured by the A. O. A. C. sporicide test. *J. Assoc. Offic. Agr. Chemists,* t. **XXXIX**, p. 480-483, 1956.

([849]) FRIIS P., HASSELAGER E. et KROGH P. — Isolation of citrinin and oxalic acid from *Penicillium viridicatum* Westling and their nephrotoxicity in rats and pigs. *Acta Path. Microbiol. Scand.,* t. **LXVII**, p. 559-560, 1969.

([850]) FRISCHBIER et RICHTESTEIGER. — Bildung von Oxalsäure durch *Aspergillus niger* in Brot und in der Streu. *Z. Veterinärk.,* t. **LIII**, p. 391, 1941.

([851]) FROMAGEOT P. — Influence d'un pigment carcinogène, la lutéoskyrine, sur la transcription *in vitro* du DNA. *Conference Fac. des Sciences, Brest,* 10 mars 1967.

([852]) FUJISE S., HISHIDA S., SHIBATA M. et MATSUEDA S. — Structure of fusaroskyrin, a pigment of *Fusarium* species : a pathogen of the soybean purple speck disease. *Chem. and Ind.,* p. 1754-1755, 1961.

([853]) FUJIWARA A., LANDAU J. W. et NEWCOMER V. D. — Hemolytic activity of *Rhizopus nigricans* and *Rhizopus arrhizus. Mycopathol. Mycol. Appl.,* t. **XL**, p. 131-138, 1970.

([854]) FUJIWARA A., LANDAU J. W. et NEWCOMER V. D. — Preliminary characterization of the hemolysin of *Rhizopus nigricans. Mycopathol. Mycol. Appl.,* t. **XL**, p. 139-144, 1970.

([855]) FULLER G., WALKER H. G., MOTTOLA A. C., ZUZMICKY D. D., KOHLER G. O. et VOHRA P. — Potential for detoxified castor meal. *J. Amer. Oil Chem. Soc.,* t. **XLVIII**, p. 616-618, 1971.

([856]) FULTON N. D. et BOLLENBACHER R. — Pathogenicity of fungi isolated from diseased cotton seedlings in Arkansas. *Phytopathology,* t. **XLVIII**, p. 343, 1958.

([857]) FUSEY P. et NICOT J. — Quelques exemples de dégradation par les Champignons. *La Mycothèque. Catalogue Collections Muséum Paris,* 1er Suppl., p. 53-59, 1951.

([858]) GABLICKS J., SCHAEFFER W., FRIEDMAN L. et WOGAN G. N. — Effect of aflatoxin B_1 on cell cultures. *J. Bact.,* t. **XC**, p. 720-723, 1965.

([859]) GABRIEL M. — Recherches sur les pigments des Agaricales. V. *Bull. Soc. Mycol. Fr.,* t. **LXXVII**, p. 262-272, 1961.

([860]) GAC A. — La congélation des denrées. Principes et possibilités. *Rev. Gén. Froid,* t. **LVII**, p. 655-668, 1966.

([861]) GAGNE W. E., DUNGWORTH D. L. et MOULTON J. E. — Pathologic effects of aflatoxin in pigs. *Pathol. Vet.,* t. **V**, p. 370-384, 1968.

([862]) GAJDUSEK D. C. — Acute infections hemorrhagic fevers and mycotoxicoses in the Union of Soviet Socialist Republics. *Med. Sci. Publ. n° 2, Army Medical Service Graduate School, Walter Reed Army Med. Center, Washington,* p. 107-111, 1953.

([863]) GALIKEEV K. L. — Izuchenie biologickeskikh svoistv aflatoksinov gribov *Aspergillus flavus,* vydelennykh iz otrubei i shelukhi zemlyanogo orekha. *Vop. Pitan,* t. **XXVII**, p. 83-84, 1968.

([864]) GALIKEEV K. L. — Khimicheskaya priroda i biologicheskaya aktivnost' aflatoksinov. *Gig. Sanit.,* t. **XXXIII**, p. 80-83, 1968.

([865]) GALIKEEV K. L., RAIPOV O. R. et MANYASHEVA R. A. — Vliyanie aflatoksina na dinamiku obrazovaniya antitel. *Byul. Eksp. Biol. Med.,* t. **LXV**, p. 88-90, 1968.

([866]) GALLAGHER C. H. — *Nutritional factors and enzymological disturbances in animals.* Crosby Lockwood and Sons London. 1964.

([867]) GALLAGHER C. H. — The effect of sporidesmin on liver enzyme systems. *Biochem. Pharmacol.,* t. **XIII**, p. 1017-1026, 1964.

([868]) GARBOWSKI L. — Przyczynek do znajomości mikroflory grzybnej nasion drzew leśnych. *Prace Wydz. Chor. Rośl. Państw. Inst. Nauk. Gosp. Wiejsk. Bydogoszczy,* t. **XV**, p. 5-30, 1936.

([869]) GARDINER M. R. — Liver cell abnormalities induced in mice by the fungal toxin causing lupinosis. *J. Pathol. Bact.,* t. **XCIV**, p. 452-455, 1967.

([870]) GARDINER M. R. et OLDROYD B. — Avian Aflatoxicosis. *Austral. Veter. J.*, t. **XLI**, p. 272-276, 1965.

([871]) GARDNER H. K., KOLTUN S. P., DOLLEAR F. G. et RAYNER E. T. — Inactivation of aflatoxin in peanut and cottonseed meals by ammoniation. *J. Amer. Oil Chem. Soc.*, t. **XLVIII**, p. 70-73, 1971.

([872]) GARDNER H. K., KOLTUN S. P. et VIX H. K. E. — Solvent extraction of aflatoxins from oilseed meals. *J. Agr. Food Chem.*, t. **XVI**, p. 990-993, 1968.

([873]) GARGUS J. L., PAYNTER O. E. et REESE W. H. — Utilization of newborne mice in the bioassay of chemical carcinogens. *Toxicol. Appl. Pharmacol.*, t. **XV**, p. 552-559, 1969.

([874]) GARREN K. H. — Peanut (groundnut) microfloras and pathogenesis in peanut pod rot. *Phytopath. Z.*, t. **LV**, p. 359-367, 1966.

([875]) GARREN K. H. — The mycotoxin potential-peanuts (groundnuts), the USA view-point. *1st Int. Congr. Pl. Pathol.*, Londres, juil. 1968.

([876]) GARREN K. H. et HIGGINS B. B. — Fungi associated with runner peanut seeds and their relation to concealed damage. *Phytopathology*, t. **XXXVII**, p. 512-522, 1947.

([877]) GARREN K. H., HIGGINS B. B. et FUTRAL J. G. — Blue black discoloration of Spanish peanuts. *Phytopathology*, t. **XXXVII**, p. 669-679, 1947.

([878]) GARREN K. H. et WILSON C. — The Peanut, the unpredictable legume. 262 p., *U. S. Nat. Fert. Ass.*, Washington, 1951.

([879]) GARRETT W. N., HEITMAN H. et BOOTH A. N. — Aflatoxin toxicity in beef cattle. *Proc. Soc. Exp. Biol. Med.*, t. **CXXVII**, p. 188-190, 1968.

([880]) GASTROCK E. A., D'AQUIN E. L., KEATING E. J., KRISHNAMOORTHI V. et VIX H. L. E. — A mixed solvent-extraction process for cottonseed. *Cereal Sci. Today*, t. **X**, p. 572-574, 1965.

([881]) GATTANI M. L. — The quantitative determination of moulds in sweet potato and wheat flours. *Indian Phytopathol.*, t. **III**, p. 148-152, 1951.

([882]) GATTANI M. L. — Control of damping off of safflower with antibiotics. *Pl. Dis. Reptr.*, t. **XLI**, p. 160-164, 1957.

([883]) GAUCHER E. et SERGENT E. — Un cas de pseudotuberculose aspergillaire simple chez un gaveur de pigeons. *Bull. mém. soc. méd. hop. Paris*, sér. III, t. **XI**, p. 512-521, 1894.

([884]) GÄUMANN E. — Toxines et maladies des plantes. *Endeavour*, t. **XIII**, p. 198-204, 1954.

([885]) GÄUMANN E. et VON ARX J. A. — Antibiotika als pflanzliche Plasmagifte. II. *Ber. Schweiz. Bot. Gesellsch.*, t. **LVII**, p. 174-183, 1947.

([886]) GÄUMANN E. et JAAG O. — Die physiologischen Grundlagen des parasitogenen Welkens. I-III. *Ber. Schweiz. Bot. Gesellsch.*, t. **LVII**, p. 132-227, 1947.

([887]) GÄUMANN E., JAAG O. et BRAUN R. — Antibiotika als pflanzliche Plasmagifte. *Experientia*, t. **III**, p. 80-71, 1947.

([888]) GEDEK B. — Zur Bedeutung von Mykotoxinen für den Menschen. *Dtsch. Med. Wschr.*, t. **XCIII**, p. 2397-2399, 1968.

([889]) GEDEK B. et HOFMANN H. — Tierexperimentelle Untersuchungen zur hepatotoxischen Wirkung von Schimmelpilzgiften. 1. Mitteilung : Akute Veränderungen beim Meeresschweinchen. *Zentralbl. Veterinärmed.*, B. t. **XVII**, p. 658-666, 1970.

([890]) GEIGER W. B. et CONN J. E. — The mechanism of the antibiotic action of clavacin and penicillic acid. *J. Amer. Chem. Soc.*, t. **LXVII**, p. 112-116, 1945.

([891]) GEILINGER H. — Experimentelle Beiträge zur Mikrobiologie der Getreidemehle. *Mitt. Gebiete Lebensmitt. Hyg.*, t. **XII**, p. 49-81, 105-119, 231-262, 1921.

([892]) GELBOIN H. V., WORTHAM J. S., WILSON R. G., FRIEDMAN M. et WOGAN G. N. — Rapid and marked inhibition of rat-liver RNA polymerase by aflatoxin B_1. *Science*, t. **CLIV**, p. 1025-1026, 1966.

([893]) GENEST C. et SMITH D. M. — A note on the detection of aflatoxins in peanut butter. *J. Ass. Off. Agr. Chem.*, t. **XLVI**, p. 817-818, 1963.

([894]) Gerlach A. C. — *Handbuch der gerichtlichen Thierheilkunde*. Berlin, 1862.

([895]) Gerstner W. — Möglichkeiten der Bekämpfung des Weizensteinbrands *Tilletia tritici* Wint. und der Streifenkrankheit der Gerste (*Helminthosporium* Rabh.) mit Antibiotica von Penicillien. *Arch. f. Microb.*, t. **XXIII**, p. 1-4, 1956.

([896]) Geslin R. C. — Les lampes germicides. 40 p., Cie des lampes, 1964.

([897]) Ghittino P. — Eziologia, patogenesi e tentativi di trasmissione della « Degenerazione Lipoidea Spatica » nella trota iridea (*Salmo gairdnerii. Veterin. Ital.*, t. **XII**, p. 3-16, 1961.

([898]) Ghittino P. — Caso di epatoma nel salmerino di allevamento (*Salvelinus fontinalis*). *Atti Soc. Ital. Sci. Vet.*, t. **XVII**, p. 574-579, 1963.

([899]) Ghittino P. et Ceretto F. — Studio sulla eziopatogenesi dell'epatoma della trota iridea di allevaments. *Tumori*, t. **XLVIII**, p. 393-409, 1962.

([900]) Gibbons W. J. — Photosensitization in cattle. *Auburn Vet.*, t. **XI**, p. 177-179, 1953.

([901]) Gibbons W. J. — Green oats intoxication and icterogenic photosensitization in cattle. *Vet. Med.*, t. **LIII**, p. 297-300, 1958.

([902]) Gibbons W. J. — Forage poisoning. *Mod. Vet. Pract.*, t. **XL**, fasc. 15, p. 41 ; fasc 16, p. 43, 1959.

([903]) Gibson E. A. et Harris A. H. — Disease of turkey poults. *Vet. Rec.*, t. **LXXIII**, p. 150, 1961.

([904]) Gibson I. A. S. et Clinton P. K. S. — Pre-emergence seed-bed losses in groundnuts at Urambo, Tanganyika Territory. *Empire J. Exp. Agr.*, t. **XXI**, p. 226-235, 1953.

([905]) Gibson W. W. C. — Turkey « X » disease and the labelling of poultry foods. *Vet. Rec.*, t. **LXXIII**, p. 150-151, 1961.

([906]) Gilgan M. W., Smalley E. B. et Strong F. M. — Isolation and partial characterization of a toxin from *Fusarium tricinctum* on moldy corn. *Arch. Biochem. Biophys.*, t. **CXIV**, p. 1-3, 1966.

([907]) Gill-Carey D. E. — Antibiotics from Aspergilli. *Brit. J. Exp. Pathol.*, t. **XXX**, p. 114-118, 1949.

([908]) Gill-Carey D. E. — The nature of some antibiotics from Aspergilli. *Brit. J. Exp. Pathol.*, t. **XXX**, p. 119-122, 1949.

([909]) Gillier P. — Recherches de l'I. R. H. O. sur l'aflatoxine dans l'arachide de bouche. *Oléagineux*, t. **XXV**, p. 467, 1970.

([910]) Gilliver K. — Inhibitory action of antibiotics on plant pathogenic bacteria and fungi. *Ann. Bot.*, t. **X**, p. 271-282, 1946.

([911]) Gilman G. A. — Black and mouldy groundnuts in the Gambia. *Comm. Phytopath. News*, fasc. 4, p. 1-2, 1965.

([912]) Gilman G. A. — Fungus spoilage of stored food in the tropics and its possible effect on man. *Commonwealth Phytopathol. News*, p. 1-3, 1970.

([913]) Gilman G. A. — Storage surveys and how they may be used both to detect and estimate fungal contamination in the diet. *in* Purchase I. F. H., *Mycotoxins in human health*, p. 133-140, 1971.

([914]) Gilman H. L. et Birch R. R. — A mold associated with abortion in cattle. *Cornell Vet.*, t. **XV**, p. 81, 1925.

([915]) Ginoza W. — Effects of ionizing radiation on nucleic acids. *Ann. Review of Nuclear Science*, t. **XVII**, 1967.

([916]) Gip L. et Palsson G. — On the hemolytic activity of geophilic Dermatophytes. *Mycopathol. Mycol. Appl.*, t. **XL**, p. 221-223, 1970.

([917]) Gitman L. S. — Toksicheskie varianty griba *Stachybotrys alternans* Bon. na khlopke. *Mikol. Fitopatol.*, t. **III**, p. 361-364, 1969.

([918]) Glenn B. L., Panciera R. J. et Monlux A. W. — A hepatogenous photosensivity disease of cattle. II. Histopathology and pathogenesis of the hepatic lesions. *Pathol. Veter., Suisse*, t. **II**, fasc. 1, p. 49-67, 1965.

([919]) GLINOS A. D., BUCHER N. L. R. et AUB J. C. — The effect of liver regeneration on tumor formation in rats fed 4-dimethylaminoazobenzene. *J. Exp. Med.*, t. **XCIII**, p. 313, 1951.

([920]) GODTFREDSEN W. O., JAHNSEN S., LORCK H., ROHOLT K. et TYBRING L. — Fusidic acid ; a new antibiotic. *Nature*, t. **CXCIII**, p. 987, 1962.

([921]) GODTFREDSEN W. O. et VANGEDAL S. — The structure of fusidic acid. *Tetrahedron Letters*, t. **XVIII**, p. 1029-1048, 1962.

([922]) GODTFREDSEN W. O. et VANGEDAL S. — Trichodermin, a new antibiotic related to trichothecin. *Proc. Chem. Soc.*, p. 188-189, 1964.

([923]) GOEFFRION R. — Les plantes malades sont-elles dangereuses? *Phytoma*, n° 195, p. 35-40, 1968.

([924]) GOLDBLATT L. A. — Removal of aflatoxin from peanut products with acetone-hexane water solvent. *Mycotoxins in Foodstuffs*, p. 261-263, M. I. T. Press, 1965.

([925]) GOLDBLATT L. A. — Some approaches to the elimination of aflatoxin from protein concentrates. *Adv. Chem. Ser.*, n° 57, p. 216-227, 1966.

([926]) GOLDBLATT L. A. — Aflatoxin and its control. *Economic botany*, t. **XXII**, p. 51-62, 1968.

([927]) GOLDBLATT L. A. — *Aflatoxin. Scientific background, control, and implications*, 472 p., Academic Press, 1969.

([928]) GOLDBLATT L. A. — Objective determination of aflatoxins B_1, B_2, G_1 and G_2 by fluorodensitometry. *in* Herzberg M., Toxic micro-organisms, p. 30-35, 1970.

([929]) GOLDBLATT L. A. — Control and removal of aflatoxin. *J. Am. Oil Chem. Soc.*, t. **XLVIII**, p. 605-610, 1971.

([930]) GOLDBLATT L. A. et ROBERTSON J. A. — Extraction of *Aspergillus flavus* aflatoxin from groundnut meal with acetone-hexane-water azeotrope. *Int. Biodet. Bull.*, t. **I**, p. 41-42, 1965.

([931]) GOLDBLITH S. A. — The inhibition and destruction of the microbial cell by radiations. *in* HUGO W. B., *Inhibition and destruction of the microbial cell*, p. 285-305, Academic Press, 1971.

([932]) GOLDING N. S. — The gas requirements of molds. IV. A preliminary interpretation of the growth rates of four common mold cultures on the basis of absorbed gases. *J. Dairy Sci.*, t. **XXVIII**, p. 737-750, 1945.

([933]) GOLUMBIC C. — Fungal spoilage in stored food crops. *Mycotoxins in Foodstuffs*, p. 49-67, 1965.

([934]) GOLUMBIC C. et KULIK M. M. — Fungal spoilage in stored crops and its control. *in* Goldblatt L. A., *Aflatoxin*, Academic Press, p. 307-332, 1969.

([935]) GOMEZ DE MORAES G. W. — Estudos sobre aflatoxina. *Arqu. Esc. Veter., Univ. Minas Gerais*, t. **XVI**, p. 281-330, 1964.

([936]) GONYE E. R. — Mycotoxicoses : the new frontier. *Amer. Med. Technol. J.*, t. **XXVIII**, p. 555-560, 1966.

([937]) GONZALES CANCHO F. — Storage of olives before pressing. IV. Microbe population of the piles. *Grasas y aceites*, t. **VIII**, p. 258-266, 1957.

([938]) GOODALL C. M. et BUTLER W. H. — Aflatoxin carcinogenesis : inhibition of liver cancer induction in hypophysectomized rats. *Int. J. Cancer*, t. **IV**, p. 422-429, 1969.

([939]) GOODMAN J. J. et CHRISTENSEN C. M. — Grain storage studies XI. Lipolytic activity of fungi isolated from stored corn. *Cereal Chem.*, t. **XXIX**, p. 299-308, 1952.

([940]) GOPAL T., SYED ZAKI, NARAYANASWAMY M. et PREMLATA S. — Aflatoxicosis in dairy cattle. *Indian Vet. J.*, t. **XLV**, p. 707-712, 1968.

([941]) GOPLEN B. P. et BELL J. M. — Dicoumarol studies. IV. Antidotal and antagonistic properties of vitamins K_1 and K_3 in cattle. *Canad. J. animal. Sci.*, t. **XLVII**, p. 91-100, 1967.

([942]) GORBACH G., VICARI M. et DEDIC G. — Zur Kenntnis der Hemmstoffbildung und des Hemmstoffes von *Aspergillus clavatus* und *Penicillium expansum*. *Monatsh.*, t. **LXXXIII**, p. 377-385, 1952.

([943]) GORCICA H. J., PETERSON W. H. et STEENBOCK H. — The nutritive value of fungi. II. The vitamin B_1, G, and B_4 content of the mycelium of *Aspergillus sydowi*. *J. Nutrition*, t. **IX**, p. 691-714, 1935.

([944]) GORE T. S., PANSE T. B. et VENKATARAMAN K. — Citrinin. *Nature*, t. **CLVII**, p. 333, 1946.

([945]) GORLENKO M. V. — The toxins of moulds. *Am. Rev. Sov. Med.*, t. **V**, p. 168-169, 1948.

([946]) GORTNER R. A. et BLAKESLEE A. F. — Observations on the toxin of *Rhizopus nigricans*. *Am. J. Physiol.*, t. **XXXIV**, p. 353, 1914.

([947]) GOTH A. — The antitubercular activity of aspergillic acid and its probable mode of action. *J. Lab. Clin. Med.*, t. **XXX**, p. 899-902, 1945.

([948]) GOTTLIEB D. et SHAW P. D. — *Antibiotics*. I. Mechanism of action. 785 p. II. Biosynthesis. 466 p., Springer edit., 1967.

([949]) GOTTLIEB D. et SINGH J. — The mechanism of patulin inhibition of fungi. *Riv. Patol. Veg., Ital.*, t. **IV**, fasc. 4, p. 455-479, 1964.

([950]) GOULD B. S. et RAISTRICK H. — Studies in the biochemistry of microorganisms. 40. The crystalline pigments of species in the *Aspergillus glaucus* series. *Bioch. J.*, t. **XXVIII**, p. 1640, 1934.

([951]) GRAHAM R. — Forage poisoning in horses, cattle and mules, so-called cerebrospinal meningitis and commonly termed « staggers ». *Kentucky Agr. Exp. Stn., Bull.* 167, p. 369, 1913.

([952]) GRAHAM R. — Cornstalk disease investigations. *Vet. Med.*, t. **XXXI**, p. 46-50, 1938.

([953]) GRAVES R. R. et HESSELTINE C. W. — Fungi in flour and refrigerated dough products. *Mycopathologia et Mycol. Appl.*, t. **XXIX**, p. 277-290, 1966.

([954]) GRAVES R. R., HESSELTINE C. W. et ROGERS R. F. — Microbiology of wheat flour and wheat. *Bact. Proc.*, A 27, p. 5, 1965.

([955]) GRAVES R. R., ROGERS R. F., LYONS A. J. et HESSELTINE C. W. — Bacterial and Actinomycete Flora of Kansas-Nebraska and Pacific Northwest wheat and wheat flour. *Cereal Chem.*, t. **XLIV**, p. 288-299, 1967.

([956]) GRAY E. G. — *Phialea mucosa* sp. nov., the blind-seed fungus. *Trans. Brit. Mycol. Soc.*, 1942.

([957]) GRAY J. E., SNOEYENBOS G. H. et REYNOLDS I. M. — The hemorrhagic syndrome of chickens. *J. Am. Vet. Med. Assoc.*, t. **CXXV**, p. 144-151, 1954.

([958]) GRAY W. D. et BUSHNELL W. R. — Biosynthetic potentialities of higher fungi. *Mycologia*, t. **XLVII**, p. 646-663, 1955.

([959]) GRAY W. D. et KARVE M. D. — Fungal protein for food and feeds. V. Rice as a source of carbohydrate for the production of fungal protein. *Econ. Bot. USA*, t. **XXI**, p. 110-114, 1967.

([960]) GRCEVIC N. et MATTHEWS W. F. — Pathologic changes in acute disseminated aspergillosis, particularly involvement of the central nervous system. *Am. J. Clin. Pathol.*, t. **XXXII**, p. 536-551, 1959.

([961]) GREGORY P. H., GUTHRIE E. J. et BUNCE M. E. — Experiments on splash dispersal of fungus spores. *J. Gen. Microbiol.*, t. **XX**, p. 328-354, 1959.

([962]) GREGORY P. H. et LACEY M. E. — Isolation of thermophilic Actinomycetes. *Nature, Lond.*, t. **CXCV**, p. 95, 1962.

([963]) GREGORY P. H. et LACEY M. E. — Mycological examination of dust from mouldy hay associated with farmer's lung disease. *J. Gen. Microbiol*, t. **XXX**, p. 75, 1963.

(964) GREGORY P. H. et LACEY M. E. — Liberation of spores from mouldy hay. *Trans. Brit. Mycol. Soc.*, t. **XLVI**, p. 73-80 1963.

(965) GREGORY P. H. et LACEY M. E. — The discovery of *Pithomyces chartarum* in Britain. *Trans. Brit. Mycol. Soc.*, *t.* **XLVII**, fasc. 1, p. 25-30, 1964.

(966) GREGORY P. H., LACEY J. et LACEY M. E. — Moulding of hay in relation to farmer's lung disease. 7e *Congr. Int. Bot.*, p. 69, Edinburgh 1964.

(967) GREGSON A. E. W. et LA TOUCHE C. J. — Otomycosis : a neglected disease. *J. Laryngol. et Otol.*, t. **LXXV**, p. 45-69, 1961.

(968) GRIMINGER P., FISHER H., MORRISON W. D., SNYDER J. M. et SCOTT H. M. — Factors influencing blood clotting time in the chick. *Science*, t. **CXVIII**, p. 379-380, 1963.

(969) GRISHAN J. W. et HARTROFT W. S. — Morphologic identification by electron microscopy of « oval » cells in experimental hepatic degeneration. *Lab. Invest.*, t. **X**, p. 317, 1961.

(970) GROLL D. et LÜCK E. — Wirkung von Sorbinsäure und Sorboylpalmitat auf die Aflatoxinbildung im Brot. *Z. Lebensm. — Unters. u. — Forsch.*, t. **CXLIV**, p. 297-300, 1970.

(971) GROSSBARD E. — Antibiotic production by fungi on organic manures and in soil. *J. Gen. Microbiol.*, t. **VI**, p. 295-310, 1952.

(972) GROSSER A. et FRIEDRICH W. — Über die Bildung von 6. Methylsalicylsäure durch *Penicillium claviforme* Bain. *Z. Naturforschg*, t. **III** b, p. 380-381, 1948.

(973) GROVE J. F. — Infra-red spectroscopy and structural chemistry. II. Isoclavacin and patulin. *J. Chem. Soc.*, p. 883-885, 1951. III. Gladiolic acid. *Id.*, p. 3345-3354, 1952.

(974) GROVE J. F. — Phytotoxic compounds produced by *Fusarium equiseti.* Part V. Transformation products of 4β, 15-diacetoxy— 3α, 7α-dihydroxy — 12, 13-epoxytrichothec-9-en-8-one and the structure of nivalenol and fusarenone. *J. Chem. Soc.* (C), p. 375-378, 1970. Part **VI**. *Id.*, p. 378-379, 1970.

(975) GRUNSEL. — A statistical survey of veterinary practice. *Vet. Rec.*, t. **LXVII**, p. 974-979, 1955.

(976) GUALLINI L. — Sulla caratterizzazione biologica dell'attività micotossica di un ceppo di *Paecilomyces varioti* Bainier 1907. *Clinica veterinaria*, t. **XCI**, p. 7-15, 1968.

(977) GUARINO A. M. et DEFEO J. J. — Inhibition of creatinine phosphokinase as a possible mechanism for creatinuria produced by two toxic antibiotics, muconomycin A and B. *Biochem. Pharmacol.*, t. **XVII**, p. 1683-1687, 1968.

(978) GUARINO A. M., MENDILLO A. B. et DE FEO J. J. — Toxic and inflammatory properties of two antibiotics : muconomycin A and B. *Biotechnol. Bioeng.*, t. **X**, p. 457-467, 1968.

(979) GUBIN I. E. — Razvitié toksichekikh shtammov *Fusarium sporotrichiella* Bilai var. *sporotrichioides* (Sherb.) Bilai pri razlichnykh usloviyakh kul'tivirovaniya. *Sb. Nauch Tr. Ryazan Sel'skokhoz Inst.*, t. **XVII**, p. 172-175, 1967.

(980) GUDAUSKAS R. T., DAVIS N. D. et DIENER U. L. — Sensitivity of *Heliothis virescens* larvae to aflatoxin in *ad libitum* feeding (*Aspergillus flavus*). *J. Invertebr. Pathol.*, t. **IX**, p. 132-133, 1967.

(981) GUÉRIN B. — Possibilités offertes par les solutions concentrées d'ozone dans la stérilisation des objets de pansement et du matériel chirurgical. *Thèse Fac. Pharmacie Paris*, 119 p., avr. 1963.

(982) GUILBOT A. — *Hygrométrie et teneur en eau. Conférence Centre de formation technique et de perfectionnement de l'Union inter-syndicale, des Industries francaises de biscuiterie, biscotterle.* Paris, 1960.

(983) GUILBOT A. — Production de toxines d'*Aspergillus flavus* sur milieux céréaliers, influence de différentes conditions. *Colloque aflatoxines, Villejuif*, 7 mars 1966.

(984) GUILBOT A. et JEMMALI M. — Conditions de production de l'aflatoxine. Influence des composants du milieu de culture sur la production de la toxine. *Symposium INSERM Mycotoxines et Alimentation, Le Vésinet*, 24 oct. 1969. *Cah. Nutr. Diet.*, t. **V**, p. 51-52, 1970.

([985]) GUILLON J. C. et RENAULT L. — Apparition d'hépatomes chez des canes. *Bull. Acad. Vét. Fr.*, t. **XXXIV**, p. 93-97, 1961.

([986]) GULLION G. W. — Some diseases and parasites of American coots. *Calif. Fish Game*, t. **XXXVIII**, p. 421-423, 1952.

([987]) GUMBMANN M. R. et WILLIAMS S. N. — Biochemical effects of aflatoxin in pigs. *Toxicol. Appl. Pharmacol.*, t. **XV**, p. 393-404, 1969.

([988]) GUMBMANN M. R. et WILLIAMS S. N. — Effect of phenobarbital pretreatment on the ability of aflatoxin B_1 to inhibit ribonucleic acid synthesis in rat liver. *Biochem. Pharmacol.*, t. **XIX**, p. 2861-2866, 1970.

([989]) GUPTA S. R., VISWANATHAN L. et VENKITASUBRAMANIAN T. A. — Incorporation of ^{32}P-orthophosphate into phospholipids by a toxigenic and a nontoxigenic strain of *Aspergillus flavus. Mycopathol. Mycol. Appl.*, t. **XLII**, p. 137-144, 1970.

([990]) GUSEVA I. I., KOZLOVSKY A. G. et BEZBORODOV A. M. — Synthesis of 2, 3-dioxy — 4-phenyl chinoline by *Penicillium cyclopium. Prikl. Biokhim. Mikrobiol.*, t. **VIII**, p. 259-261, 1972.

([991]) HADDON W. F., WILEY M. et WAISS A. C. — Aflatoxin detection by thin layer chromatography — mass spectrometry. *Annal. Chem.*, t. **XLIII**, p. 268-270, 1971.

([992]) HADDOW A. et BLAKE I. — Neoplasms in fish : a report of six cases with a summary of the literature. *J. Path. Bact.*, t. **XXXVI**, p. 41-47, 1933.

([993]) HADLOK R. — Schimmelpilze bei Fleischerzeugnissen. *Fleischwirtschaft*, t. **XLIX**, p. 455-460, 1969.

([994]) HADLOK R. — Aflatoxine bei Fleischprodukten und Untersuchungen uber die Häufigkeit der Aflatoxinbildung durch *Aspergillus flavus* — Stämme. *Fleischwirtschaft*, t. **L**, p. 1499-1502, 1970.

([995]) HAGEM O. — Untersuchungen über Norwegische Mucorineen. II. *Christiana Vidensk. — Selsk. Skrift.* I. *Math.-naturw. Kl.*, t. **IV**, p. 1-152, 1910.

([996]) HAGEN P. O. — The effect of low temperatures on microorganisms : conditions under which cold becomes lethal. *in* HUGO W. B., *Inhibition and destruction of the microbial cell*, p. 39-76, Academic Press, 1971.

([997]) HAIG D. A. — (Communication personnelle à Carnaghan et Sargeant), 1961.

([998]) HALD B. et KROGH P. — Forekomst of aflatoksin i importerede bomuldsfroprodukter. *Nord. Vet. — Med.*, t. **XXII**, p. 39-47, 1970.

([999]) HALL E. A., KAVANAGH F. et ASHESHOV I. N. — Action of forty-five antibacterial substances on bacterial viruses. *Antibiotics and Chemotherapy*, t. **I**, p. 369-378, 1951.

([1000]) HALLIDAY D. — Relationship between discoloured kernels, free fatty acid and aflatoxin content of groundnuts. *Nigerian stored Prod. Res. Inst. Ann. Rep.*, p. 67-69, 1966.

([1001]) HALSALL T. G., JONES E. R. H. et LOWE G. — The molecular formula of cephalosporin P_1. *Proc. Chem. Soc.*, p. 16, 1963.

([1002]) HALVER J. E. — Nutrition of salmonoid fishes. III. Water soluble requirements of chinook salmon. *J. Nutr.*, t. **LXII**, p. 225-243, 1957.

([1003]) HALVER J. E. — *Report Nat. Inst. Health*, avril 1962.

([1004]) HALVER J. E. — Hepatomas in fish. In Primary hepatoma, Univ. Utah Press, Salt Lake City, p. 103-112, 1965.

([1005]) HALVER J. E. — Aflatoxicosis and Rainbow Trout Hepatoma. *Mycotoxins in Foodstuffs*, p. 209-234, 1965.

([1006]) HALVER J. E. — Aflatoxicosis and trout hepatoma. *in* GOLDBLATT L. A., *Aflatoxin*, Academic Press, p. 264-306, 1969.

([1007]) HALVER J. E. — Crystalline aflatoxin carcino and other vectors for trout hepatoma. *U. S. Fish Wildlife Serv. Res. Rep.*, t. **LXX**, p. 78-102, 1967.

([1008]) HALVER J. E., JOHNSON C. L. et ASHLEY L. M. — Dietary carcinogens induce fish hepatoma, *Fed. Proc.*, t. **XXI**, p. 390, 1962.

([1009]) HALVER J. E., LA ROCHE G. et ASHLEY L. M. — Experimental hepatocellular carcinoma. *Abstr. VIth. Int. Congr. Nutrition*, p. 85, Edinburgh, 1963.

([1010]) HAMASAKI T., HATSUDA Y., TERASHIMA N. et RENBUTSU M. — Studies on the metabolites of *Aspergillus versicolor* (Vuillemin) Tiraboschi. V. Isolation and structures of three new metabolites, versicolorins A, B and C. *Agr. Biol. Chem.*, t. **XXXI**, p. 11-17, 1967.

([1011]) HAMILTON E. D. — Airborne spores as allergens with particular reference to *Cladosporium* and *Alternaria*. *7e Congr. Int. Bot.*, p. 69-70, Edinburgh, 1964.

([1012]) HAMILTON P. B. et GARLICH J. D. — Failure of vitamin supplementation to alter the fatty liver syndrome caused by aflatoxin. *Poult. Sci.*, t. **LI**, p. 688-692, 1972.

([1013]) HAMILTON P. B., LUCAS G. B. et WELTY R. E. — Mouse toxicity of fungi of tobacco. *Appl. Microbiol.*, t. **XVIII**, p. 570-574, 1969.

([1014]) HAMMADY I. M. — Isolemento della patulina e della griseofulvina da una specie egiziana non identificata di *Penicillium*. *G. Bioch.*, t. **XV**, p. 127-132, 1966.

([1015]) HAMMERL H. — Die Bakterien der menschlichen Faeces nach Aufnahme von vegetabilischer und gemischter Nahrung. *Zeitschr. f. Biologie*, t. **XXXV**, p. 355-376, 1897.

([1016]) HAMOR T. A., PAUL I. C., ROBERTSON J. M. et SIM G. A. — The structure of byssochlamic acid. *Experientia*, t. **XVIII**, p. 352-354, 1962.

([1017]) HANLIN R. T. — Fungi in developing peanut fruits. *Mycopathol. Mycol. Appl.*, t. **XXXVIII**, p. 93-100, 1969.

([1018]) HANLIN R. T. — Invasion of peanut fruits by *Aspergillus flavus* and other fungi *Mycopathol. Mycol. Appl.*, t. **XL**, p. 341-348, 1970.

([1019]) HANNA K. L. et CAMPBELL T. C. — Note on a rapid thin layer chromatographic method for the isolation of aflatoxin B_1. *J. Ass. Off. Anal. Chem.*, t. **LI**, p. 1197-1199, 1968.

([1020]) HANSEL F. K. — Hay fever. *Mo. State Med. Assoc. Journ.*, t. **XXXVII**, p. 241-246, 1940.

([1021]) HANSEN A. A. — The latest developments in the stock-poisoning plants situation in Indiana. *J. Am. Vet. Med. Assoc.*, t. **LXXIII**, p. 471, 1928.

([1022]) HANSON et LOPER. — Leafspot and plant hormone. *U. S. Dept. Agr., Res.*, t. **XIII**, n° 2, 1964.

([1023]) HANSSEN E. — Ergebnisse der Untersuchung einiger Lebensmittel auf Aflatoxin B_1. *Mitt. GDCh-Fachgr. Lebensmittelchem. Gerichtl. Chem.*, t. **XXII**, p. 83-88, 1968.

([1024]) HANSSEN E. et HAGEDORN G. — Untersuchungen über Vorkommen und Wanderung von Aflatoxin B_1 und seine Veränderungen bei einigen lebensmitteltechnologischen Prozessen. *Z. Lebensmittel-Unters. u.-Forsch.*, t. **CXLI**, p. 129-145, 1969.

([1025]) HARDING J. D. J., DONE J. T., LEWIS G. et ALLCROFT R. — Experimental groundnut poisoning in pigs. *Res. Vet. Sci.*, t. **IV**, p. 217-229, 1963.

([1026]) HARDISON J. R. — Susceptibility of Gramineae to *Gloeotinia temulenta*. *Mycologia*, t. **LIV**, fasc. 2, p. 201-216, 1962.

([1027]) HARDISON J. R. — Control of *Gloeotinia temulenta* in seed fields of *Lolium perenne* by cultural methods. *Phytopathology*, t. **LIII**, fasc. 4, p. 460-464, 1963.

([1028]) HAREIN P. K., DE LAS CASES E., EUGENIO C. P. et MIROCHA C. J. — Reproduction and survival of confused flour beetles exposed to metabolites produced by *Fusarium roseum* var. *graminearum*. *J. Econ. Entomol.*, t. **LXIV**, p. 975-976, 1971.

([1029]) HARKNESS C., MCDONALD D., STONEBRIDGE W. C., A'BROOK J. et DARLING H. S. — The problem of mycotoxins in ground nuts and other food crops in tropical Africa. *Food technol.*, t. **XX**, p. 72-78, 1966.

([1030]) HARKNESS W. D., LOVING W. L. et HODGES F. A. — Pyrexia in rabbits following the injection of typical mold cultures. *J. Amer. Pharm. Ass. Sci.*, t. **XXXIX-XL**, p. 502-504, 1950-1951.

([1031]) HARLEY E. H., REES K. R. et COHEN A. — A comparative study of the effect of aflatoxin B_1 and actinomycin D on HeLa cells. *Biochem. J.*, t. **CXIV**, p. 289-298, 1969.

([1032]) Harri E., Loeffler W., Sigg H. P., Stahelin H. et Tamm C. — *Helv. Chim. Acta,* t. **XLVI**, p. 1235, 1963.

([1033]) Harrison J. — Mycotoxicoses. *Naas. Quart. Rev.*, t. **LXXVIII**, p. 78-85, 1968.

([1034]) Harrison J. W. — Thermogenesis in hay-inhabiting fungi. *Iowa State Coll. J. Sci.* t. **IX**, p. 37-60, 1934.

([1035]) Hart J. R. et Neustadt M. H. — Application of the Karl Fischer method to grain moisture determination. *Cereal Chem.*, t. **XXXIV**, p. 26-37, 1957.

([1036]) Hartill W. F. T. — The influence of temperature and humidity on the development of yellow mould of tobacco. *Rhodesia Zamb. malawi J. Agric. res.*, t. **V**, p. 61, 1967.

([1037]) Hartley R. O., Nesbitt B. F. et O'Kelly J. — Toxic metabolites of *Aspergillus flavus. Nature*, t. **CXCVIII**, p. 1056-1058, 1963.

([1038]) Harwig J. et Scott P. M. — Brine shrimp (*Artemia salina* L.) larvae as a screening system for fungal toxins. *Appl. Microbiol.*, t. **XXI**, p. 1011-1016, 1971.

([1039]) Hassall C. H. et Jones D. W. — The biosynthesis of phenols. IV. A new metabolic product of *Aspergillus terreus. J. Chem. Soc.*, p. 4189-4191, 1962.

([1040]) Hassall C. H. et Mc Morris T. C. — Constitution of geodoxin, a metabolic product of *Aspergillus terreus* Thom. *J. Chem. Soc.*, p. 2831-2834, 1959.

([1041]) Hasselager E. — Nephrotoxicity of mouldy barley. Experimental studies in swine and rats. *Proc. Symp. Agric. Coll. Norway*, 1967, 4p. ronéot.

([1042]) Hatsuda Y., Hamasaki T., Ishida M., Matsui K. et Hara S. — Dihydrosterigmatocystin and dihydrodemethylsterigmatocystin, new metabolites from *Aspergillus versicolor. Agric. Biol. Chem.*, t. **XXXVI**, p. 521-522, 1972.

([1043]) Hauson F. et Edle T. — Patent U. S. A. n° 2652356. *Chem. Abstr.*, t. **XLVIII**, p. 14 1954.

([1044]) Havens W. P. — Cholestatic jaundice in patients treated with erythromycin estolate. *J. A. M. A.*, t. 180, p. 30-32, 1962.

([1045]) Hawker L. E. — Physiology of fungi. 360 p. Univ. London Press, 1950.

([1046]) Hayes A. W., Davis N. D. et Diener U. L. — Effect of aeration on growth and aflatoxin production by *Aspergillus flavus* in submerged culture. *Appl. Microbiol.*, t. **XIV**, p. 1019-1021, 1966.

([1047]) Hayes A. W. et Wilson B. J. — Bioproduction and purification of rubratoxin B. *Appl. Microbiol.*, t. **XVI**, p. 1163-1167, 1968.

([1048]) Head M. J. et Rook J. A. F. — Hypomagnesaemia in dairy cattle and its possible relationship to ruminal Ammonia production. *Nature*, t. **CLXXVI**, p. 262, 1955.

([1049]) Head M. J. et Rook J. A. F. — Some effects of spring grass on rumen digestion and metabolism of the dairy cow. *Proc. Nutr. Soc.*, t. **XVI** p. 25, 1957.

([1050]) Heathcote J. G., Child J. J. et Dutton M. F. — Possible role of kojic acid in the production of aflatoxin by *Aspergillus flavus. Biochem. J.*, t. **XCV**, p. 23, 1965.

([1051]) Heathcote J. G. et Dutton M. F. — New metabolites of *Aspergillus flavus. Tetrahedron*, t. **XXV**, p. 1497-1500, 1969.

([1052]) Hedayat H., Khayetian H., Ghavifeker M. et Donoso G. — Substances toxiques naturelles des aliments. Cas de la fève et du favisme. *Cah. Nutr. Diét.*, t. **V**, p. 23-29, 1970.

([1053]) Hedge U. C., Chandra T. et Shanmugasundarum E. R. B. — Toxicity of different diets contamined with various fungi to rice moth larvae (*Corcyra cephalonica* St.). *Can. J. Comp. Med. Vet. Sci.*, t. **XXXI**, p. 106-163, 1967.

([1054]) Hedge U. C. et Shanmugasundarum E. R. B. — A study of free amino acids in rice moth larvae during mycotoxicosis. *Can. J. Comp. Med. Vet. Sci.*, t. **XXXII**, p. 392-394, 1968.

([1055]) Heiberg B. C. et Ramsey G. B. — Fungi associated with diseases of Peanuts on the market. *Phytopathology*, t. **XLIII**, p. 474, 1953.

([1056]) HEIM R. — Sur une aspergillose du grain de café. *C. R. Acad. Agric. Fr.*, p. 1-5, 5 juin 1946.

([1057]) HEIM R. — *Les Champignons toxiques et hallucinogènes.* 327 p., Boubée édit., Paris 1963.

([1058]) HEINTEZLER I. — Das Wachstum der Schimmelpilze in Abhängigkeit von den Hydraturverhältnissen unter verschiedenen Aussenbedingungen. *Arch. Mikrobiol.*, t. X, p. 92-132, 1939.

([1059]) HEISS W. et SCHMIDT-MENDE M. — Erfahrungen mit einem neuen Kaltsterilisationsgerät. *Münch. Med. Wschr.*, t. **XII**, p. 560-562, 1962.

([1060]) HEMINGWAY R. G., INGLIS J. S. S. et RITCHIE N. S. — Factors involved in hypomagnesaemia in sheep. *Conf. on Hypomagnesaemia*, p. 58-74, Londres, 1960.

([1061]) HENRICI A. T. — An endotoxin from *Aspergillus fumigatus. J. Immunol.*, t. **XXXVI**, p. 319-338, 1939.

([1062]) HERMAN C. M. et SLADEN W. J. L. — Aspergillosis in waterfowl. *Trans. North Am. Wildlife Conf.*, t. **XXIII**, p. 187-191, 1958.

([1063]) HERNANDEZ E. —Intoxicaciones alimentarias. IV. Aflatoxina. *Rev. Agroquim. Technol. Aliment., Esp.*, t. **V**, p. 425-428, 1965.

([1064]) HERRICK J. A. — Antifungal properties of clavacin. *Proc. Soc. Exp. Biol. Med.*, t. **LIX**, p. 41-42, 1945.

([1065]) HERRMANN H., HODGES R. et TAYLOR A. — Sporidesmins. Part V. The stereochemistry of the bridged dioxopiperazine ring in sporidesmins and gliotoxin. *J. Chem. Soc.*, p. 4315-4319, 1964.

([1066]) HERROLD K. M. — Aflatoxin induced lesions in Syrian hamsters. *Brit. J. Cancer*, t. **XXIII**, p. 665-660, 1969.

([1067]) HERZBERG M. — *Toxic Micro-organisms. Mycotoxins. Botulism.* 490 p. (U. S. — Japan Cooperative Program in the natural resources, Honolulu, 7-10 oct. 1968). Washington, 1970).

([1068]) HESS W. M., SASSEN M. M. A. et REMSEN C. C. — Surface characteristics of *Penicillium* conidia. *Mycologia*, t. **LX**, p. 290-303, 1968.

([1069]) HESS W. M. et STOCKS D. L. — Surface characteristics of *Aspergillus* conidia. *Mycologia*, t. **LXI**, p. 560-571, 1969.

([1070]) HESSELTINE C. W. — Research at Northern Regional Research Laboratory on fermented foods. *Proc. Conf. on Soybean Products for Protein in Human Foods*, p. 67, 1961.

([1071]) HESSELTINE C. W. — Aflatoxins and other mycotoxins. *Health Laboratory Science*, t. **IV**, p. 222-228, 1967.

([1072]) HESSELTINE C. W. — Flour and wheat : research on their microbiological flora. *Bakers Digest*, t. **XLII**, p. 40-42, 1968.

([1073]) HESSELTINE C. W. — Microbiological research on wheat and flour. *9th Ann. Symposium Amer. Ass. Cereal Chem., St. Louis, Miss.*, 25 p., 16 fev. 1968.

([1074]) HESSELTINE C. W. — Mycotoxins. *Mycopathol. Mycol. Appl.*, t. **XXXIX**, p. 371-383, 1969.

([1075]) HESSELTINE C. W. et GRAVES R. R. — Microbiological research on wheat and flour. *Rep. 2nd Nat. Conf. Wheat Util. Res., Peoria, Ill.*, p. 170-200, 1964.

([1076]) HESSELTINE C. W. et GRAVES R. R. — Microbiology of flours. *Economic Botany*, t. **XX**, p. 156-168, 1966.

([1077]) HESSELTINE C. W., SHOTWELL O. L., ELLIS J. J. et STUBBLEFIELD R. D. — Aflatoxin formation by *Aspergillus flavus. Bact. Rev.*, t. **XXX**, p. 795-805, 1966.

([1078]) HESSELTINE C. W., SHOTWELL O. L., SMITH M., ELLIS J. J., VANDEGRAFT E. et SHANNON G. — Production of various aflatoxins by strains of the *Aspergillus flavus* series. *in* HERZBERG M., *Toxic Micro-organisms*, p. 202-210, 1970.

([1079]) HESSELTINE C. W., SHOTWELL O. L., SMITH M. L., SHANNON G. M., VANDEGRAFT E. E. et GOULDEN M. L. — Laboratory studies on the formation of aflatoxin in forages. *Mycologia*, t. **LX**, p. 304-312, 1968.

([1080]) HESSELTINE C. W., SMITH M., BRADIE B. et KO SWAN DJIEN. — Investigation on tempeh, an Indonesian food. *Developments Indust. Microbiol.*, t. **IV**, p. 275, 1963.

([1081]) HESSELTINE C. W., SORENSON W. G. et SMITH M. — Taxonomic studies of the aflatoxin-producing strains in the *Aspergillus flavus* group. *Mycologia*, t. **LXII**, p. 123-132, 1970.

([1082]) HESSELTINE C. W., VANDEGRAFT E. E., FENNELL D. I., SMITH M. L. et SHOTWELL O. L. — Aspergilli as ochratoxin producers. *Mycologia*, t. **LXIV**, p. 539-550, 1962.

([1083]) HETHERINGTON A. C. et RAISTRICK H. — Studies in the biochemistry of microorganisms. XIV. On the production and chemical constitution of a new yellow colouring matter, citrinin produced from glucose by *Penicillium citrinum* Thom. *Phil. Trans. Royal Soc.*, t. **CCXX** b, p. 269-297, 1931.

([1084]) HEUSINKWELD M. R., SHERA C. C. et BAUR F. J. — Note on aflatoxin analysis in peanuts, peanut meals and peanut products. *J. Ass. Off. Agric. Chemists*, t. **XLVIII**, p. 448-449, 1965.

([1085]) HIGGINS B. B. — Les maladies de l'Arachide aux Etats-Unis. *Oléagineux*, t. **XI**, p. 213-220, 1956.

([1086]) HIGGINSON J. — Primary carcinoma of the liver in Africa. *Brit. J. Cancer*, t. X, p. 609-622, 1956.

([1087]) HIGGINSON J. — The geographical pathology of primary liver cancer. *Cancer Res.*, t. **XXIII**, p. 1624, 1963.

([1088]) HIGNETT S. L. — Hypomagnesaemia and some other metabolic disorders of the grazing animal. *J. Farmers'Club*, t. **II**, 1961.

([1089]) HILL K. R. — The world wide distribution of Seneciosis in Man and Animals. *Proc. Roy. Soc. Med.*, t. **LIII**, p. 281, 1960.

([1090]) HILL K. R. — Comment on the histological appearance in serial liver biopsies and *post mortem* specimens. *Vet. Rec.*, t. **LXXV**, p. 493, 1963.

([1091]) HILL K. R. et GARDELLA L. A. — The absolute configuration of citrinin. *J. Org. Chem.*, t. **XXIX**, p. 766-767, 1964.

([1092]) HILTNER L. — Ueber die Beziehungen verschiedener Bakterien- und Schimmelpilz-Arten zu Futtermitteln und Samen. *Landw. Vers. Sta.*, t. **XXXIX**, p. 471-476, 1891.

([1093]) HINTZ H. F., BOOTH A. N., CUCULLU A. F., GARDNER H. R. et HEITMAN H. Jr. — Aflatoxin toxicity in swine. *Proc. Soc. Exp. Biol. Med.*, t. **CXXIV**, p. 266-268, 1967.

([1094]) HINTZ H. F., HEITMAN M., BOOTH A. N. et GAGNE N. — Effects of aflatoxin on reproduction in swine. *Proc. Soc. Exper. Biol. Med.*, t. **CXXVI**, p. 146-148, 1967.

([1095]) HIRATA Y. — On the products of mould. Substances from mould rice (Part 3). Structure and properties I. *J. Chem. Soc. Japan*, t. **LXVIII**, p. 104-105, 1957. (Part 4). Structure and properties **II**. *Id.*, t. **LXVIII**, p. 105-106, 1957.

([1096]) HIRAYAMA S., KURATA H., SAKABE F., INAGAKI N., UDAGAWA S. et IKETANI S. — Cultivation of *Penicillium concavorugulosum*, isolation and purification of metabolic products. *Eisei Shikenjo Hokoku*, n° 74, p. 305-312, 1956.

([1097]) HIRAYAMA S. et YAMAMOTO M. — Biological studies on the poisonous wheat flour. II. *Eisei Shikenjo Hokoku*, t. **LXVII**, p. 117-121, 1950.

([1098]) HISCOCKS E. S. — The importance of molds in the deterioration of tropical foods and feedstuffs. *Mycotoxins in Foodstuffs*, p. 15-26, 1965.

([1099]) HO M., FANTES K. H., BURKE D. C. et FINTER N. B. — Interferons or interferon-like inhibitors induced by non-viral substance. *in* FINTER N. B., *Interferons*, p. 181-201, North Holland Publ. Co édit., 1966.

([1100]) HOBBS B. — *Food poisoning and food hygiene*. 2e éd. Edward Arnold Ltd, London, 1968.

([1101]) HODGES F. A., ZUST J. R., SMITH H. R., NELSON A. A., ARMBRECHT B. H. et CAMPBELL A. D. — Mycotoxins : aflatoxin isolated from *Penicillium puberulum*. *Science*, t. **CXLV**, p. 1439, 1964.

([1102]) HODGES R., RONALDSON J. W., TAYLOR A. et WHITE E. P. — Sporidesmin and sporidesmin B. *Chem. Ind.*, p. 42-43, 1963.

([1103]) HOECHST. — L'acide sorbique comme agent de conservation. Revue de littérature. 30 p., 1959.

([1104]) HOFFERBER O. — Ueber die Stachybotryotoxikose der Pferde. *Z. Veterinärkunde*, t. **VIII**, p. 230-234, 1942.

([1105]) HOFFMAN R. K. — Toxic gases. *in* HUGO W. B., *Inhibition and destruction of the microbial cell*, p. 226-258, Academic Press, 1971.

([1106]) HOFFMAN R. K. et WARSHOWSKY B. — Beta-propiolactone vapor as a disinfectant. *Appl. Microb.*, t. **VI**, p. 358-362, 1958.

([1107]) HOFMANN K., MINTZLAFF H. J., ALPERDEN I. et LEISTNER L. — Untersuchung über die Inaktivierung des Mykotoxins Patulin durch Sulfhydrylgruppen. *Fleischwirtschaft*, t. **LVII**, p. 1534-1536, 1539, 1971.

([1108]) HOGEBOOM G. H. et CRAIG L. C. — Identification by distribution studies. **VI**. Isolation of antibiotic principles from *Aspergillus ustus*. *J. Biol. Chem.*, t. **CLII**, p. 363-368, 1946.

([1109]) HOLADAY C. E. — Rapid method for detecting aflatoxins in peanuts. *J. Amer. Oil Chem. Soc.*, t. **XLV**, p. 680-682, 1968.

([1110]) HOLKER J. S. E. et UNDERWOOD J. G. — Synthesis of a cyclopentenocoumarin structurally related to aflatoxin B. *Chem. Ind.*, p. 1865-1866, 1964.

([1111]) HOLMAN L. E. et SNITZLER J. R. — Transporting, handling and storing seeds. *Yearbook of Agriculture*, U. S. Dept. Agr., p. 338-347, 1961.

([1112]) HOLSCHER H. A. — Ueber den Nachweis von Dehydrasen der Tumorzelle mittels Tetrazoliumsalzen. *Z. Krebsforsch.*, t. **LVI**, p. 587-595, 1950.

([1113]) HOLSCHER H. A. — Einige Beobachtungen an Tumorzellen und den darin enthaltenen Granula. *Z. Krebsforsch.*, t. **LVII**, p. 634-636, 1951.

([1114]) HOLTMAN D. F. — The microbiological content of soft wheat flour. *J. Bacteriol.*, t. **XXX**, p. 359-361, 1935.

([1115]) HOLZAPFEL C. W. — The isolation and structure of cyclopiazonic acid, a toxic metabolite of *Penicillium cyclopium* Westling. *Tetrahedron*, t. **XXIV**, p. 2101-2199, 1968.

([1116]) HOLZAPFEL C. W. — Cyclopiazonic acid and related toxins. *in* Ciegler A. et al., Microbial Toxins, t. **VI**, p. 435-457, 1971.

([1117]) HOLZAPFEL C. W. et STEYN P. S. — The isolation and structure of a new diterpene lactone from *Trichothecium roseum* Link. *Tetrahedron*, t. **XXIV**, p. 3321-3328, 1965.

([1118]) HOLZAPFEL C. W., STEYN P. S., et PURCHASE I. F. H. — Isolation and structure of aflatoxins M_1 and M_2. *Tetrahedron Letters*, t. **XXV**, p. 2799-2803, 1966.

([1119]) HONNA Y. et SHIRAI K. — Cystoma found in the liver Rainbow Trout (*Salmo gairdnerii irideus*). *Gibbons Bull. Jap. Soc. Sci..Fish*, t. **XXIV**, p. 966-970, 1959.

([1120]) HOOPER I. R., ANDERSON H. W., SKELL P. et CARTER M. E. — The identity of clavacin with patulin. *Science*, t. **XCIX**, p. 16, 1944.

([1121]) HOPKINS J. C. F. — Annual report of the branch of plant pathology for the year ending 31 December 1934. *Rhodesia Agr.* J., t. **XXXII**, p. 397-405, 1935.

([1122]) HOPKINS J. C. F. — Diseases of tobacco in southern Rhodesia (Suppl. I, 1932 to 1938). *Rhodesia Agr. J.*, t. **XXXVI**, p. 97-119, 1939.

([1123]) HOPKINS W. A. —Patulin in the common cold. IV Biological properties : extended trial in the common cold. *Lancet*, t. **XI**, p. 631-634, 1943.

([1124]) HOPKIRK C. S. M. — Facial dermatitis in sheep in New Zealand. *New Zealand J. Agr.* t. **LII**, p. 98-103, 1936.

([1125]) HOPKIRK C. S. M., MARSHALL D. et BLAKE T. A. — Grass tetany of dairy cows. *Vet. Rec.*, t. **XIII**, p. 355-361, 1933.

([1126]) HORI M. et YAMAMOTO T. — Studies on the chemical constituants of poisonous substances from *Penicillium* (Hori-Yamamoto strain). I. Morphology and characters of the *Penicillium* separated. *J. Pharmacol. Soc. Jap.* t. **LXXIII**, p. 1097-1101, 1953.

([1127]) HORI M., YAMAMOTO T., OZAWA A, MATSUKI Y., HAMAGUCHI A. et SORAOKA H. — Studies on a fungus species isolated from the malt-feed which caused massdeath of cows. III. Classification of the isolated fungus, mechanism of its toxin production and chemical nature of the toxin. *Jap. J. Bact.*, t. **IX**, p. 1105-1111, 1954.

([1128]) HOU C. T., CIEGLER A. et HESSELTINE C. W. — Tremorgenic toxins from Penicillia. I. Colorimetric determination of tremortins A and B. *Anal. Biochem.*, t. **XXXVII**, p. 422-428, 1970. II. A new tremorgenic toxin tremortin B from *Penicillium palitans. Can. J. Microbiol.*, t. **XVII**, p. 599-603, 1971. III. Tremortin production by *Penicillium* species on various agricultural commodities. *Appl. Microbiol.*, t. **XXI**, p. 1101-1103, 1971.

([1129]) HOU C. T., CIEGLER A. et HESSELTINE C. W. — New mycotoxin, trichotoxin A, from *Trichoderma viride* isolated from southern leaf blight-infected corn. *Appl. Microbiol.*, t. **XXIII**, p. 183-185, 1972.

([1130]) HOVE E. L. et WRIGHT D. E. — Casein, phosphopeptones and phosphoserine protect rats against the mycotoxin, sporidesmin. *Life Sci.*, t. **VIII**, p. 545-550, 1960.

([1131]) HOWARD B. H. et RAISTRICK H. — Studies in the biochemistry of microorganisms. **LXXX**. The coloring matters of *Penicillium islandicum*. 1. *Biochem. Jour*,, t. **XLIV**, p. 227-233, 1949.

([1132]) HOWARD B. H. et RAISTRICK H. — Studies in the biochemistry of microorganisms. **XCIV**, *Biochem. J.*, t. **LIX**, p. 475-486, 1955.

([1133]) HOWELL R. W. — Mycotoxin research in oilseeds. *in* HERZBERG M., *Toxic Micro-organisms*, p. 61-66, 1970.

([1134]) HOYMAN W. G. — Concentration and characterization of the emetic principle present in barley infected with *Gibberella saubinetii. Phytopathology*, t. **XXXI**, p. 871-885, 1941.

([1135]) HSIEH D. P. H. et MATELES R. I. — The relative contribution of acetate and glucose to aflatoxin biosynthesis. *Biochim. Biophys. Acta*, t. **CCVIII**, p. 482-486, 1970.

([1136]) HUEBNER C. F. et LINK K. P. — Studies on the hemorrhagic sweet clover disease. VI. *J. Biol. Chem.*, t. **CXXXVIII**, p. 529, 1941.

([1137]) HUEPER W. C. et PAYNE W. W. — Observations on the occurence of hepatomas in Rainbow Trout. *J. Nat. Cancer Inst.*, t. **XXVII**, p. 1123-1143, 1961.

([1138]) HUGHES J. P. et CORNELIUS C. E. — An outbreak of grass tetany in lactating beef cattle. *Cornell Vet.*, t. **L**, p. 26, 1960.

([1139]) HUITSON J. J. — Cereals preservation with propionic acid. *Process Biochemistry*, t. **III**, p. 31-32, 1968.

([1140]) HUMMEL B. C. W. — Isolation and partial characterization of flavicidic acid, phytotoxic metabolite of *Aspergillus flavus. Dissertation Abstr.*, t. **XVI**. p. 1332-1333, 1956.

([1141]) HUNTER J. M. et TUITE J. F. — The growth of storage fungi and aflatoxin production in corn as related to moisture and temperature. *Phytopathology*, t. **VII**, p. 816, 1967.

([1142]) HUTTER R. V. P. — Phycomycetous infection (Mucormycosis) in cancer patients: a complication of therapy. *Cancer*, t. **XII**, p. 330-350, 1959.

([1143]) HVIDSTEN H., OEDELIEN M., BAERUG R. et TOLLERSRUD S. — The influence of fertilizer treatment of pastures on the mineral composition of the herbage and the incidence of hypomagnesaemia in dairy cows. *Act. Agric. Scand.*, t. **IX**, p. 261-290, 1959.

([1144]) HYAMS T. A. et MESLER R. J. — Effets of a synthetic nitrate concentrate administered orally to cattle. *J. Am. Vet. Med. Assoc.*, t. **CXXXVII**, p. 477, 1960.

([1145]) HYDE M. B. et GALLEYMORE H. B. — The subepidermal fungi of cereal grains. II. The nature, identity and origin of the mycelium in wheat. *Ann. Appl. Biol.* t. **XXXVIII**, p. 348-356, 1951.

(1146) HYUN J. S. — Development of storage fungi in polished rice infested with weevil, *Sitophilus oryzae* L. *Seoul Univ. J., ser. B*, t. **XIII**, p. 77-86, 1963.

(1147) ICHINOE M., UDAGAWA S. I., TAZAWA M. et KURATA H. — Some considerations on a biological method for the detection of mycotoxins in Japanese foods. *in* HERZBERG M., *Toxic Micro-organisms*, p. 191-197, 1970.

(1148) IIZUKA H. — Studies on the micro-organisms found in Thai and Burma rice. Part II. On the microflora of Burma rice. *J. Gen. Appl. microbiol.*, t. **IV**, p. 108-119, 1958.

(1149) IIZUKA H. et IIDA M. — Maltoryzine, a new toxic metabolite produced by a strain of *A. oryzae* var. *microsporus* isolated from the poisonous malt sprout. *Nature*, t. **CXCVI**, p. 681-682, 1962.

(1150) IKAWA M., MA D. S., MEEKER G. B. et DAVIS R. P. — Use of *Chlorella* in mycotoxin and phycotoxin research. *J. Agr. Food. Chem.*, t. **XVII**, p. 425-429, 1969.

(1151) IKEDA Y., OMORI Y., FURUYA T. et ICHINOE M. — Experimental studies on some causal *Fusarium* for the wheat and barley scab. IV. Feeding test in mice. *Eisei Shikenjo Hokoku*, t. **LXXXII**, p. 130-132. 1964.

(1152) ILNICKA-OLEJNICZAK O. et SOZYNSKI J. — Ueber den Einfluss von UV — Strahlen auf die Ueberlebensrate von Sporen verschiedener *Aspergillus* Arten. *Mycopathol. Mycol. Appl.*, t. **XXX**, 129-136, 1966.

(1153) IMAN-FAZLI S. F. et AHMED Z. A. — Fungus organisms associated with jute seeds and their effect on germinating seeds and seedlings. *Agr. Pakist.*, t. **III**, 14 p. 1960.

(1154) INAGAKI N. — Study on fungi in foodstuff from standpoint of food sanitation. *Japan J. Pub. Health.*, t. **XII**, p. 1123-1136, 1960.

(1155) INAGAKI N. et IKEDA M. — Studies on the fungi isolated from foods. II. Identification of Penicillia and Aspergilli isolated from flours. *Bull. Nat. Hyg. Lab.* (Tokyo), t. **LXXVII**, p. 347-366, 1959.

(1156) INGLIS J. S. S. — Studies on hypomagnesaemia in ruminants. *8th Internat. Grassland Congr., Reading*, Sect. 58, p. 17-23, 1960.

(1157) INSTITUT INTERNATIONAL DU FROID. — Conditions recommandées pour l'entreposage frigorifique des produits périssables. 2e Edit., *Inst. int. Froid*, 104 p. 1967.

(1158) IRVING G. W. — Perspectives on the mycotoxin problem in the United States. *in* HERZBERG M., *Toxic Micro-organisms*, p. 3-7, 1970.

(1159) ISACHENKO B. L. et MALCHEVSKAJA N. N. — Biogenic spontaneous heating of peat. *Dokl. Akad. Nauk. USSR*, t. **XIII**, p. 377, 1936.

(1160) ISHIKO T. — Histological studies in the injuries of various organs of mice and albino rats except liver, fed with « yellowed rice » artificially polluted by *Penicillium islandicum* Sopp. *Acta Pathol. Jap.*, t. **VII**, p. 368, 1957.

(1161) ISHIKO T. — Histopathological studies on the injuries of various organs of mice and rats fed with « yellowed rice » by *Penicillium islandicum* Sopp. especially on the pathological findings of the other organs associated with the liver injuries. *Trans. Soc. Path. Japan*, t. **XLVIII**, p. 867-892, 1959.

(1162) ISMAILOV I. A. et MOROSHKIN B. F. — (The etiology and pathogenesis of stachybotryotoxicosis in cattle.) *Veterinariya*, t. **IV**, p. 27-28, 1962.

(1163) IVANOV A. T. — Izuchenie vozmozhnosti ispol'zovaniya kurinykh embrionov dlya opredekniya toksichnosti griba *Fusarium sporotrichiella* var. *poae*. *Tr. Vses. Nauch.-Issled Inst. Vet. Sanit.*, t. **XXVIII**, p. 41-47, 1967.

(1164) IWATA K. et YOSIOKA I. — Terrecin a new antibiotic substance produced by *Aspergillus terreus*. *J. Antibiotics* (*Japan*), t. **III**, p. 192-197, 1950.

(1165) IYENGAR M. R. S. et HORA T. S. — Nitrite oxidation by soil fungi. *Naturwissenschaften*, t. **XLVI**, p. 211, 1959.

(1166) IYENGAR M. R. S. et STARKEY R. L. — Synergism and antagonism of auxin by antibiotics. *Science*, t. **CXVIII**, p. 357-358, 1953.

(1167) JABBAR A. et RAHIM A. — Citrinin from *Penicillium steckii*. *Pharm. Sci.*, t. **LI**, p. 595-596, 1962.

([1168]) JACKMAN L. M. — *Applications of nuclear magnetic resonance Spectroscopy*. Pergamon Press, N. Y., 88 p., 1959.

([1169]) JACKSON C. R. — Peanut kernel infection and growth *in vitro* by four fungi at various temperatures. *Phytopathology*, t. **LV**, p. 46-48, 1963.

([1170]) JACKSON C. R. — Seed-borne fungi from peanut seed stocks. *Georgia Coastal Plain Exp. Stn.* Mimeogr. Ser., N. S. 166, 16 p., 1963.

([1171]) JACKSON C. R. — Peanut infection by soil-borne fungi. *Proc. third Nat. Peanut Res. Conf., Auburn, Alabama*, p. 111-113, 1964.

([1172]) JACKSON C. R. — Location of fungal contamination or infection in peanut kernels from intact pods. *Plant Dis. Reptr.*, t. **XLVIII**, p. 980-983, 1964.

([1173]) JACKSON C. R. — A review of information on peanut invasion by toxin-producing fungi during production and storage. *Comm. pers.*, 13 p. mimeogr., juin 1965.

([1174]) JACKSON C. R. — A list of fungi reported on peanut pods and kernels. *U. Georgia Coastal Plain Exp. Sta., N. S.*, n° 234, 6 p., 1965.

([1175]) JACKSON C. R. — Peanut kernel infection and growth *in vitro* by four fungi at various temperatures. *Phytopathology*, t. **LV**, p. 46-48, 1965.

([1176]) JACKSON C. R. — Reduction of *Sclerotium bataticola* infection of peanut kernels by *Aspergillus flavus*. *Phytopathology*, t. **LV**, p. 934, 1965.

([1177]) JACKSON C. R. — Growth of *Aspergillus flavus* and other fungi in windrowed peanuts in Georgia. Part VII. *Trop Sci.*, t. **VI**, p. 27-34, 1965.

([1178]) JACKSON C. R. — Laboratory evaluation of fungicides for control of some fungi found on peanuts. *Plant Dis. Reptr.*, t. **XLIX**, p. 928-931, 1965.

([1179]) JACKSON C. R. — Peanut pod mycoflora and kernel infection. *Plant and soil*, t. **XXIII**, p. 203-212, 1965.

([1180]) JACKSON C. R. — Some effects of harvesting methods and drying conditions on development of aflatoxins in peanut. *Phytopathology*. t. **LVII**, p. 1270-1271, 1967.

([1181]) JACKSON C. R. — Factors affecting aflatoxin development during growth and harvest of peanut by mechanized culture. *1st Int. Congr. Pl. Pathol.*, Londres, juil. 1968.

([1182]) JACKSON C. R. et BELL D. L. — *Diseases of peanut (groundnut) caused by fungi*. 137 p., Univ. Georgia, Coll. Afric. Exper. Stns, Bull. 56, 1969.

([1183]) JACKSON C. R. et PRESS A. F. — Changes in mycoflora of peanuts stored at two temperatures in air or in high concentrations of nitrogen or carbon dioxide. *Oléagineux*, t. **XXII**, p. 165-168, 1967.

([1184]) JACKSON E. W., WOLF H. et SINNHUBER R. O. — The relationship of hepatoma in rainbow trout to aflatoxin contamination and cottonseed meal. *Cancer Res.*, t. **XXVIII**, p. 987-991, 1968.

([1185]) JACOB F., BRENNER S. et CUZIN F. — On the regulation of DNA replication in bacteria. *Cold Spring Harb. Symp. quant. Biol.*, t. **XXVIII**, p. 329, 1964.

([1186]) JACOBSON D. R., MILLER W. M., SEATH D. M., YATES S. G., TOOKEY H. L. et WOLFF I. A. — Nature of fescue toxicity and progress toward identification of the toxic entity. *J. Dairy Sci.*, t. **XLVI**, p. 416-422, 1963.

([1187]) JACOBSON W. C., HARMEYER W. C. et WISEMAN H. G. — Determination of aflatoxin B_1 and M_1 in milk. *J. Dairy Sci.*, t. **LIV**, p. 21-24, 1971.

([1188]) JACQUET J. — Les maladies des fromages. *Rev. Path. Gén. et Physiol. Clin.*, n° 742, p. 1007-1044, 1962.

([1189]) JACQUET J. et BOUTIBONNES P. — Sur les propriétés antibiotiques et toxiques d'*Aspergillus clavatus* Desmazières. *C. R. Acad. Agric. Fr.*, t. **XLIX**, p. 368-373, 1963.

([1190]) JACQUET J. et BOUTIBONNES P. — Recherches sur les *Aspergillus* pathogènes. Les espèces majeures. *A. fumigatus, A. flavus, A. clavatus. Bull. Acad. Vet.*, t. **XXX**, p. 169-180, 1967.

([1191]) JACQUET J. et BOUTIBONNES P. — Recherches sur les mycotoxines, spécialement la flavatoxine : intérêt pour la microbiologie alimentaire. *Bull. Acad. Nat. Med.*, t. **CLI**, p. 561-566, 1967.

([1192]) JACQUET J. et BOUTIBONNES P. — Recherches sur la production de toxines par *Aspergillus flavus. Bull. Acad. Vet.*, t. **XXX**, p. 387-397, 1967.

([1193]) JACQUET J. et BOUTIBONNES P. — Procédé de détection rapide des flavatoxines par chromatographie en couche mince. Application à la microbiologie alimentaire. *C. R. Acad. Agric. Fr.*, t. **LIII**, p. 1244-1251, 1967.

([1194]) JACQUET J. et BOUTIBONNES P. — Fluorescence des aflatoxines et des *Aspergillus. C. R. Soc. Biol.*, t. **CLXIX**, p. 2289-2293, 1969.

([1195]) JACQUET J. et BOUTIBONNES P. — Actions de l'aflatoxine sur les cellules microbiennes. *C. R. Soc. Biol.*, t. **CLXIII**, p. 2574-2577, 1969.

([1196]) JACQUET J. et BOUTIBONNES P. — Sur la fréquence et la répartition des spores d'*Aspergillus.* Conséquences hygiéniques. *Bull. Acad. Nat. Méd.*, t. **CLIII**, p. 647-653, 1969.

([1197]) JACQUET J. et BOUTIBONNES P. — Les conidiospores des *Aspergillus.* Intérêt pour la classification. *Rev. Immunol.*, t. **XXXIII**, p. 47-72, 1969.

([1198]) JACQUET J. et BOUTIBONNES P. — Effets nocifs de l'aflatoxine sur les cellules microbiennes. *Cah. Nutr. Diét.*, t. V, p. 65-66, 1970.

([1199]) JACQUET J. et BOUTIBONNES P. — Action des flavacoumarines (aflatoxines) sur les petits crustacés d'eau douce. Possibilité de création d'un test biologique. *Bull. Acad. Vet. Fr.*, t. **XLIII**, p. 299-308, 1970.

([1200]) JACQUET J. et BOUTIBONNES P. — Recherches sur les flavatoxines ou mieux flavacoumarines. *Rev.. Immunol.*, t. **XXXIV**, p. 245-274, 1970.

([1201]) JACQUET J. et BOUTIBONNES P. — Nouvelles recherches sur les effets des flavacoumarines (aflatoxines) sur les animaux. *Bull. Acad. Vet. Fr.*, t. **XLIV**, p. 65-72, 1971.

([1202]) JACQUET J., BOUTIBONNES P. et CICILE J. P. — Observations sur la toxicité d'*Aspergillus clavatus* pour les animaux. *Bull. Acad. Vét.*, t. **XXXVI**, p. 199-208, 1963.

([1203]) JACQUET J., BOUTIBONNES P. et SAINT S. — Effets biologiques des flavacoumarines d'*Aspergillus parasiticus* ATCC 15517. II. Végétaux. *Rev. Immunol.*, t. **XXXV**, p. 159-186, 1971.

([1204]) JACQUET J., BOUTIBONNES P. et TEHERANI A. — Sur la présence des flavatoxines dans les aliments des animaux et dans les aliments d'origine animale destinés à l'homme. *Bull. Acad. Vét.*, t. **XLIII**, p. 35, 1970.

([1205]) JACQUET J., BOUTIBONNES P. et TEHERANI A. — Fréquence actuelle des flavatoxines dans les aliments du bétail. *C. R. Acad. Agric. Fr.*, t. **LVI**, p. 187, 1970.

([1206]) JACQUET J., BOUTIBONNES P. et TEHERANI A. — Methode simple de recherche et appréciation quantitative rapides des flavacoumarines (aflatoxines). *Ind. Alim. Agric.*, t. **LXXXVIII**, p. 5-14, 1971.

([1207]) JACQUET J., BOUTIBONNES P., TEHERANI A. et TANTAOUI A. — Sur la fréquence des moisissures du genre *Aspergillus.* Apercu sur la présence des flavacoumarines (aflatoxines) dans les aliments. *Bull. Acad. Nat. Médicine*, t. **CLV**, p. 268-273, 1971.

([1208]) JACQUET J. et DESFLEURS M. — *Cladosporium herbarum* Link, agent d'accidents tardifs du « bleu » sur les fromages à pâtes molles, spécialement le camembert. *Le lait*, t. **XLVI**, p. 485-497, 1966.

([1209]) JACQUET J. et DESFLEURS M. — *Scopulariopsis brevicaulis* Bainier, ou mieux *Scopulariopsis fusca* Zach., agent des taches superficielles brun-violet des fromages à pâtes molles. *Le lait*, p. 241-253. 1966.

([1210]) JALALUDDIN M. et SINHA S. — Microorganisms associated with stored food grains and milled products from Dacca city godowns. *Mycopathol. Mycol. Appl.*, t. **XLIII**, p. 61-64, 1971.

([1211]) JAMES N. et LEJEUNE A. R. — Microflora and the heating of damp stored wheat, *Can. J. Bot.*, t. **XXX**, p. 1-8, 1952.

([1212]) JAMES N. et SMITH K. N. — Studies on the microflora of flour. *Canad. J. Research*, t. **XXVI** C, p. 479-485, 1948.

([1213]) JAMES N., WILSON J. et STARK E. — The microflora of stored wheat. *Can. J. Res., Sect. C*, t. **XXIV**, p. 224-233, 1946.

([1214]) JANES B. S. — *Sporidesmium bakeri* recorded from Victoria, Australia. *Nature*, t. **CLXXXIV**, p. 1327, 1959.

([1215]) JANICKI J. — Aflatoxins in grain. *Deutsch. Lebensmittel-Rundsch.*, t. **LXIII**, p. 276-277, 1967.

([1216]) JANICKI J., SZEBIOTKO K., CHELKOWSKI J., KOKORNIAK M. et WIEWIOROWSKA M. — Aflatoxine in Lebensmitteln. *Nahr. Chem. Biochem., Mikrobiol. Technol.*, t. **XVI**, p. 85-98, 1972.

([1217]) JARVIS B. — Factors affecting the production of mycotoxins. *J. Appl. Bacteriol.*, t. **XXXIV**, p. 199-213, 1971.

([1218]) JARVIS F. G. et JOHNSON M. J. — The role of the constituents of synthetic media for penicillin production. *J. Am. Chem. Soc.*, t. **LXIX**, p. 3010-3017, 1947.

([1219]) JARVIS B. et MORISETTI M. D. — The use of antibiotics in food preservation. *Int. Biodetn. Bull.*, t. **V**, p. 39-61, 1969.

([1220]) JAY J. M. — *Modern food microbiology*. 328 p., Van Nostrand Reinhold Co édit., New York, 1970.

([1221]) JAYRAJ A. P., SHANKAR MURTI A., SREENIVASAMURTHY V. et PARPIA H. A. B. — Toxic effects of aflatoxin on the testis of albino rats. *J. Anat. Soc. India*, t. **XVII**, p. 101-104, 1968.

([1222]) JAYARAMAN A., HERBST E. J. et IKAWA M. — The bioassay of aflatoxins and related substances with *Bacillus megaterium* spores and chick embryos. *J. Amer. Oil Chem. Soc.*, t. **XLV**, p. 700-102, 1968.

([1223]) JEMMALI M. — Inhibition d'un bactériophage virulent par l'aflatoxine B_1. *Ann. Inst. Pasteur*, t. **CXVI**, p. 288-291, 1969.

([1224]) JEMMALI M. et GUILBOT A. — Influence de l'irradiation γ des spores d' *A. flavus* sur la production d'aflatoxine B_1. *C. R. Acad. Sc.*, D, t. **CCLXIX**, p. 2271-2273, 1969.

([1225]) JEMMALI M. et GUILBOT A. — Influence de divers traitements physiques des spores d'*Aspergillus flavus* sur l'aptitude des cultures à produire des toxines. *Symposium INSERM Mycotoxines et alimentation, Le Vésinet*, 24 Oct. 1969.

([1226]) JEMMALI M. et GUILBOT A. — Conditions de production de l'aflatoxine : influence de divers traitements physiques des spores d'*A. flavus* sur l'aptitude des cultures à produire des toxines. *Cah. Nutr. Diét.*, t. **V**, p. 47-49, 1970.

([1227]) JEMMALI M. et LAFONT P. — Evolution de l'aflatoxine B_1 au cours de la panification. *Comm. Coll. Detmold*, juin 1971.

([1228]) JEMMALI M., POISSON J. et GUILBOT A. — Production d'aflatoxines dans les produits céréaliers. Influence de différentes conditions. *Ann. Nutr. Aliment.*, t. **XXIII**, p. 151-166, 1969.

([1229]) JENNINGS M. A. et WILLIAMS T. I. — Production of kojic acid by *Aspergillus effusus* Tiraboschi. *Nature*, t. **CLV**, p. 302, 1945.

([1230]) JENSEN H. L. — The microbiology of farmyard manure decomposition in soil. II. Decomposition of cellulose. *J. Agr. Sci.*, t. **XXI**, p. 81-100, 1931.

([1231]) JENSEN L. B. — *Microbiology of meats*. Garrard Press, Champaign, Ill., 1954.

([1232]) JOFFE A. Z. — The mycoflora of normal and overwintered cereals 1944-1945 *in* Alimentary-toxic aleukia, *Orenburg Inst. Epidemiol. Microbiol.*, t. **II**, p. 35-41, 1947.

([1233]) JOFFE A. Z. — The dynamics of toxin accumulation in overwintered cereals and their mycoflora in 1945-1946. *in* Alimentary-toxic aleukia, *Orenburg Inst. Epidemiol. Microbiol.*, t. **II**, p. 192, 1947.

([1234]) JOFFE A. Z. — Toxicity and antibiotic properties of some Fusaria. *Bull. Res. Counc. of Israel*, t. 8 **D**, p. 81-95, 1960.

([1235]) Joffe A. Z. — The mycoflora of overwintered cereals and its toxicity. *Bull. Res. Counc. of Israel*, t. 9 **D**, p. 101-126, 1960.

([1236]) Joffe A. Z. — Biological properties of some toxic fungi isolated from overwintered cereals. *Mycopathologia Mycologia Appl.*, t. **XVI**, p. 201-221, 1962.

([1237]) Joffe A. Z. — Toxicity of overwintered cereals. *Plant and Soil*, t. **XVIII**, p. 31-44, 1963.

([1238]) Joffe A. Z. — Toxin production by cereal fungi causing toxic alimentary aleukia in man. *Mycotoxins in foodstuffs*, p. 77-85, 1965.

([1239]) Joffe A. Z. — The effect of soil inoculation with *Aspergillus flavus* on the mycoflora of groundnut soil, rhizosphere and geocarposphere. *Mycologia*, t. **LX**, p. 908-914, 1968.

([1240]) Joffe A. Z. — Mycoflora of surface sterilized groundnut kernels. *Pl. Dis. Reptr.*, t. **LII**, p. 608-611, 1968.

([1241]) Joffe A. Z. — Toxic properties and effects of two *Fusaria* and of *Aspergillus flavus*. *1st Int. Congr. Pl. Pathol.* Londres, juil. 1968.

([1242]) Joffe A. Z. — Effects of *Aspergillus flavus* on groundnuts and on some other plants. *Phytopathol. Z.*, p. 321-326, 1969.

([1243]) Joffe A. Z. — Relationships between *Aspergillus flavus*, *A. niger* and some other fungi in the mycoflora of groundnut kernels. *Pl. Soil*, t. **XXXI**, p. 57-64, 1969.

([1244]) Joffe A. Z. — The mycoflora of groundnut rhizosphere, soil and geocarposphere on light medium and heavy soils and its relations to *Aspergillus flavus*. *Mycopathol. Mycol. Appl.*, t. **XXXVII**, p. 150-160, 1969.

([1245]) Joffe A. Z. — The mycoflora of fresh and stored groundnut kernels in Israel. *Mycopathol. Mycol. Appl.*, t. **XXXIX**, p. 255-264, 1969.

([1246]) Joffe A. Z. — Feeding tests with ducklings, turkey, chicks and rabbits and the effects of aflatoxin on these animals. *Mycopathol. Mycol. Appl.*, t. **XL**, p. 49-61, 1970.

([1247]) Joffe A. Z. — Alimentary toxic aleukia. *in* Kadis S. et al., Microbial Toxins, t. **VII**, p. 139-189, 1971.

([1248]) Joffe A. Z. et Borut S. Y. — Soil and kernel mycoflora of groundnut fields in Israel. *Mycologia*, t. **LVIII**, p. 629-640, 1966.

([1249]) Joffe A. Z. et Lisker N. — Effect of soil fungicides on development of fungi in soil and on kernels of groundnut. *Pl. Dis. Reptr.*, t. **LII**, p. 718-721, 1968.

([1250]) Joffe A. Z. et Ungar H. — Cutaneous lesions produced by topical application of aflatoxin to rabbit skin. *J. Invest. Dermatol.*, t. **LII**, p. 504-507, 1969.

([1251]) John D. W. et Miller L. L. — Effect of aflatoxin B_1 on net synthesis of albumin, fibrinogen, and α 1-acid glycoprotein by the isolated perfused rat liver. *Biochem. Pharmacol*, t. **XVIII**, p. 1135-1146, 1969.

([1252]) Johnson D. H., Robertson A. et Whalley W. B. — Citrinin. *J. Chem. Soc.*, p. 2971-2975, 1950.

([1253]) Johnson H. C., Walker A. E., Case T. J. et Kollros J. J. — Effects of antibiotic substances on the central nervous system. *Arch. Neurol. Psychiat., Chicago*, t. **LVI**, p. 184-197, 1946.

([1254]) Johnson J. R., Bruce W. F. et Dutcher J. D. — Gliotoxin, the antibiotic principle of *Gliocladium fimbriatum*. I. Production, physical and biological properties. *Amer. Chem. Soc. J.*, t. **LXV**, p. 2005-2009, 1943.

([1255]) Johnson J. R., Mc Crone W. C. et Bruce W. F. — Gliotoxin, the antibiotic principle of *Gliocladium fimbriatum*. *Amer. Chem. Soc. J.*, t. **LXVI**, p. 501, 1944.

([1256]) Johnson R. M., Greenaway W. T. et Dolan W. P. — Estimation of aflatoxin in corn by « dockage » assay. *J. Ass. Off. Anal. Chem.*, t. **LII**, p. 1304-1306, 1969.

([1257]) Johnston M. E. H., Matthews D. et Harrison S. C. — The diurnal periodicity in the ejecton of ascospores of *Gloeotinia* (*Phialea*) *temulenta* (Prill. et Delacr.) Wilson, Noble et Gray. *N. Z. J. Agric. Res.*, t. VII, p. 639-643, 1964.

[1258] JOLY P. — Les flores de dégradation des bananes. *Rev. de Mycol.*, t. **XXVI**, p. 101-117, 1961.

[1259] JONES D. — The effect of antibiotic substances upon bacteriophages. *J. Bacter.*, t. **L**, p. 341-348, 1945.

[1260] JONES F. T. et LEE K. S. — Note on aflatoxin M_1 : optical and X-ray crystallographic data. *J. Ass. Off. Anal. Chem.*, t. **LI**, p. 610, 1968.

[1261] JONES H., RAKE G. et HAMRE D. M. — Studies on *Aspergillus flavus*. I. Biological properties of crude and purified aspergillic acid. *J. Bacteriol.*, t. **XLV**, p. 461-469, 1943.

[1262] JOOSTE W. J. — The effect of different crop sequences on the rhizosphere fungi of wheat. *S. Afr. J. Agr. Sci.*, t. **IX**, p. 127-136, 1966.

[1263] JORGENSEN A. — Det mugne korn. *Medlemsbl. danske dyrlaegef.*, t. **XLVII**, p. 1-5, 1964.

[1264] JOUIN C., MONTREUIL J., MONSIGNY, MERCIER C., POISSON J., GUILBOT A. et CALET C. — Influence de l'échauffement, du stockage en atmosphère confinée et du mode de séchage sur la composition du maïs-grain, son amylolyse *in vitro* et sur sa valeur nutritionelle globale. *C. R. travaux actions concertées 1963-1964, CERDIA*, 1964.

[1265] JUHASZ S. et GRECZI E. — Extracts of mould-infected groundnut samples in tissue culture. *Nature, G. B.*, t. **CCIII**, p. 861-862, 1964.

[1266] JUMEL G. — Les agents conservateurs utilisés dans les industries alimentaires. *Fruits*, t. **XX**, p. 153-156, 1965.

[1267] JUST NIELSEN A. — Svampebefaengt (muggen) byg sammenlignet met prima byg. *Bilag Landok. Forsogslav. efterarsmode (arbog)* 1965, p. 96-105, 1965.

[1268] KACZKA E. A., GITTERMAN C. O., DULANEY E. L. et FOLKERS K. — Hadacidin, a new growth-inhibitory substance in human tumor systems. *Biochemistry*, t. **I**, p. 340-343, 1962.

[1269] KADIS S., CIEGLER A. et AJL J. — Algal and fungal toxins (Microbial toxins ; a comprehensive treatise, vol. **VII**), 418 p., Academic Press, Londres, 1972.

[1270] KAHN J. B. — Effects of various lactones and related compounds on cation transfer in incubated cold stored human erythrocytes, *J. Pharmacol. Exp. Therap.*, t. **CXXI**, p. 234-251, 1957.

[1271] KALASHNIKOV E. Y. et NIKOLAEVA N. M. — A study of the proteolytic enzyme system of *Aspergillus flavus*. *Tekhnika*, p. 150-158, 1965.

[1272] KAMIBAYASHI A. et MATSUI M. — Metabolic products of *Ustilago Maydis*. I. Isolation of organisms producing ustilaginic acid. *Kogyo Gijutsuin Hagko Kenkyusho Kenkyu Hokoku*, t. **XIX**, p. 89-96, 1961.

[1273] KAMINSKI E., LAZANAS J., WOLFSON L., FLETCHER O. et CALANDRA J. — Fate of aflatoxins in cigarette tobacco. *Beitr. Tabakforsch.*, t. **V**, p. 189-192, 1970.

[1274] KANOTA K. — Studies on toxic metabolites of *Penicillium roqueforti. in* Herzberg M., *Toxic Micro-organisms*, p. 129-132, 1970.

[1275] KARDEVAN A. et PALYUSIK M. — Untersuchungen ueber die pathogene Wirkung von *Aspergillus flavus* an Gaensen. *Acta Vet. Acad. Sci. Hung.*, t. **XVII**, p. 301-310.

[1276] KAROW E. O. et FOSTER J. W. — An antibiotic substance from species of *Gymnoascus* and *Penicillium*. *Science*, t. **XCIX**, p. 265-266, 1944.

[1277] KAS V. — Plisne-skudci sucheho Tabaku. *Ochrana Rostlin*, t. **VI**, p. 55-58, 1926.

[1278] KASATKINA I. D. et ZHELTOVA E. T. — Selection of aconidial races of *Aspergillus flavus* forming proteolytic enzymes. *Izv. Akad. Nauk. SSSR. Ser. Biol.*, t. **I**, 128-133, 1967.

[1279] KATAR'YAN B. T. — Nekotorye priznaki proyavleniya mikotoksikoza vinogradnoi lozy. *Mikol. Fitopatol.*, t. **III**, p. 228-231, 1969.

[1280] KATZMAN P. A., HAYS E. E., CAIN C. K., van WYCK J. J., REITHEI F. J., THAYERS S. A. DOISY E. A. et al. — Clavacin, an antibiotic substance from *Aspergillus clavatus*. *J. Biol. Chem.*, t. **CLIV**, p. 475-486, 1944.

[1281] KAVANAGH F. — Activities of 22 antibacterial substances against 9 species of Bacteria. *J. Bacteriol.*, t. **LIV**, p. 761-766, 1947.

([1282]) KEEN P. — Mycological studies of foodstuffs in southern Africa in relation to the incidence of cancer. *1st Int. Mycol. Congr. Exeter, 7-16 sept.* 1971, *Abst.* p. 49-50, 1971.

([1283]) KEEN P. et MARTIN P. — The toxicity and fungal infestation of foodstuffs in Swaziland in relation to harvesting and storage. *Trop. Geogr. Med.*, t. **XXIII**, p. 35-43, 1971.

([1284]) KEEN P. et MARTIN P. — Is aflatoxin carcinogenic in man ? The evidence in Swaziland. *Trop. Geogr. Med.*, t. **XXIII**, p. 44-53, 1971.

([1285]) KEILHOLTZ F. J. — Spoiled feed shown to be the cause of pig trouble. *Ill. Agr. Exp. Stn.*; *Ann. Rept.*, n° 41, p. 144, 1928.

([1286]) KEILLING J., CASALIS J., SOUIGNAC G. et DUBREUIL G. — Sur une altération superficielle des fromages par *P. funiculosum. Le lait*, n° 355-356, p. 241-250, 1956.

([1287]) KEILOVA-RODOVA H. — Effect of patulin on tissue cultures. *Experientia*, t. V, p. 242, 1949.

([1288]) KEILOVA-RODOVA H. — Effect of antibiotics on tissue cultures. *Casopis lékaru ceskych*, t. **XC**, p. 38-42, 1951.

([1289]) KELLNER O. — *The scientific feeding of animals.* Duckworth et Co édit., Londres, 1915.

([1290]) KENNEDY B. W. — Moisture content, mold invasion, and seed viability of stored soybeans. *Phytopathology*, t. **LIV**, p. 771-774, 1964.

([1291]) KENNEL J. von, KIMMI G. et LEMBKE A. — Die Mycoïne, eine neue Grupp therapeutisch wirksamer Substanzen aus Pilzen. *Klin. Wochsch.*, n° 16-17, p. 321, 1943.

([1292]) KENSLER C. J. et NATOLI D. J. — Processing to ensure wholesome products. *in* GOLDBLATT L. A. , *Aflatoxin, Academic Press*, p. 333-358, 1969

([1293]) KENT J. et HEATLEY N. G. — Antibiotics from moulds. *Nature, G. B.*, t. **CLVI**, p. 295-296, 1946.

([1294]) KERN H. et NAEF-ROTH S. — Zur Bildung phytotoxischer Farbstoffe durch Fusarien der Gruppe *Martiella. Phytopath. Zeitsch.*, t. **LIII**, p. 45-64, 1965.

([1295]) KEYL A. C., BOOTH A. N., MASRI M. S., GUMBMANN M. R. et GAGNE W. E. — Chronic effects of aflatoxin in farm animal feeding studies. *in* HERZBERG M., *Toxic Microorganisms*, p. 72-75, 1970.

([1296]) KEYL A. C., LEWIS J. C., ELLIS J. J., YATES S. G. et TOOKEY H. L. — Toxic fungi isolated from tall fescue. *Mycopathol. Mycol. Appl.*, t. **XXXI**, p. 326-331, 1967.

([1297]) KEYL A. C. et MASRI M. S. — The effects of grade levels of aflatoxins on various physiological parameters in swine. *Abstr. 152nd Meeting Am. Chem. Soc., New York*, 11-16 sept. 1966.

([1298]) KHARCHENKO S. M. — The pharmacological and toxicological properties of dendrodochins. *Mikrobiolohiya dlya narodnho hospodarstva imedyssyny. Nauk. Dumka, Kiev*, p. 146-149, 1966.

([1299]) KIDDER R. W. et BEARDSLEY D. W. — Moldy grass may cause cattle sunburn. *Agr. Expt. Stn. Univ. Florida*, Research Dept. n° 6, p. 15-16, 1961.

([1300]) KIDDER R. W., BEARDSLEY D. W. et ERWIN T. C. — Photosensitis control is now possible, Everglades experiment station says. *Florida Cattleman*, 1950.

([1301]) KIDDER R. W., BEARDSLEY D. W. et ERWIN T. C. — Photosensitization in cattle grazing Bermuda grass. *Proc. 48th Ann. Conv. Assoc. Southern Agr. Workers, Memphis, Tenn.*, p. 80, 1951.

([1302]) KIDDER R. W., BEARDSLEY D. W. et ERWIN T. C. — Photosensitization in cattle grazing frosted common Bermuda grass. *Univ. Florida Agr. Exp. Sin. Bull.*, n° 630, p.1-21, 1961.

([1303]) KIERMEIER F. — Zum Nachweis von Aflatoxin in Käse. *Z. Lebensmittel-Untersuch.-Forsch.*, t. **CXLIV**, p. 293-297, 1970.

([1304]) KIERMEIER F. et BÖHM S. — Zur Aflatoxinbildung in Milch und Milchprodukten. V. Anwendung des Hühnerembryo-Testes zur Sicherung des dünnschichtchromatographischen Aflatoxin-Nachweises in Käsen. *Z. Lebensmittel-Untersuch.-Forsch.*, t. **CXLVII**, p. 61-64, 1971.

([1305]) KIERMEIER F. et GROLL D. — Zur Aflatoxin B_1-Bildung in Käsen. *Z. Lebensmittel-Untersuch.-Forsch.*, t. **CXLIII**, p. 81-88, 1970.

([1306]) KIGER J. L. et KIGER J. G. — *Techniques modernes de la biscuiterie pâtisserie-boulangerie industrielles et artisanales et des produits de régime*. Vol. 1, 700 p., Dunod, Paris, 1967.

([1307]) KILLIAN C. et FEHR D. — Recherches sur les phénomènes microbiologiques des sols sahariens. *Ann. Inst. Pasteur*, t. **LV**, p. 573-622, 1935.

([1308]) KIM CHAN-JO et SONG SUK-HOON. — Studies on harmful microorganisms in aged rice. Part 1. On the detection of microbes and its metabolic products. *Bull. Sc. Research Inst. Korea*, t. **IV**, p. 64-72, 1959.

([1309]) KING W. A., CAMPBELL H. A., RUPEL I. W., PHILLIPS P. H. et BOHSTEDT G. — The effect of alfalfa lipids on the progress of sweet clover poisoning in cattle. *J. Dairy Sci.*, t. **XXIV**, p. 1, 1941.

([1310]) KING W. H., KUCK J. C. et FRAMPTON V. L. — Properties of cottonseed meals prepared with acetone-petroleum ether-water azeotrope. *J. Am. Oil. Chem. Soc.*, t. **XXXVIII**, p. 19-21, 1961.

([1311]) KING A. M. Q. et NICHOLSON B. H. — The interaction of aflatoxin B_1 with polynucleotides and its effect on ribonucleic acid polymerase. *Biochem. J.*, t. **CXIV**, p. 679-687, 1969.

([1312]) KINGSBURY J. M. — *Poisonous plants of the United States and Canada*. 626 p., Prentice Hall édit., 1964.

([1313]) KINOSITA R., SEITO T. et SUMMER P. — Toxicity of koji (*Asp. oryzae*). Feder. Proc. *U. S. A.*, t. **XXIV**, p. 627, 1965.

([1314]) KINOSITA R. et SHIKATA T. — On toxic mo!dy rice, *in Mycotoxins in Foodstuffs*, p. 111-132, Mass. Inst. Technol., 1965.

([1315]) KIRSCHSTEIN R. L. et SIDRANSKY H. — Mycotic endocarditis of the tricuspid valve due to *Aspergillus flavus*. Report of a case. *A. M. A. Arch. Pathol.*, t. **LXII**, p. 103-106, 1956.

([1316]) KIS Z., FURGER P. et SIGG H. P. — Isolation of pyrenophorol. *Experientia*, t. **XXV**, p. 123-124, 1969.

([1317]) KLEMMER H. W., RIKER A. J. et ALLEN O. N. — Inhibition of crown gall by selected antibiotics. *Phytopathology*, t. **XLV**, p. 618-625, 1955.

([1318]) KLOPOTEK A. Von. — Ueber das Vorkommen und Verhalten von Schimmelpilzen bei der Kompostierung städtischer Abfallstoffe. *Antonie von Leeuwenhoek J. Microbiol. Serol.*, t. **XXVIII**, p. 141-160, 1962.

([1319]) KLOSA J. — Antibiotical Berlin, 1952.

([1320]) KNAPSTEIN H. — Bestimmung von Aflatoxin in Futtermitteln. *Landwirt. Forsch.*, t. **XXI**, p. 164-171, 1968.

([1321]) KOBAYASHI H., TSUNODA H. et TATSUNO T. — Recherches toxicologiques sur les mycotoxines qui polluent le fourrage artificiel du porc. *Chem. Pharm. Bull.*, t. **XIX**, p. 839-842, 1971.

([1322]) KOBAYASHI Y., URAGUCHI K., SAKAI F. et MIYAKE I. — Pyrogen producing molds and other microbes existing in glucose powder. *Jap. J. Pharmacol.*, t. **I**, p. 1, 1951.

([1323]) KOBAYASHI Y., URAGUCHI K., SAKAI F., TATSUNO T., TSUKIOKA M., SAKAI Y., SATO T., MIYAKE M., SAITO M., ENOMOTO M., SHIKATA T. et ISHIKO T. — Toxicological studies on the yellowed rice by *Penicillium islandicum* Sopp. I. Experimental approach to liver injuries by long term feedings with the noxious fungus on mice and rat. *Proc. Japan. Acad.*, t. **XXXIV**, p. 139-144, 1958. II. Isolation of the two toxic substances from the noxious fungus, and their chemical and biological properties. *Proc. Japan Acad.*, t. **XXXIV**, p. 736-741, 1959. III. Experimental verification of primary hepatic carcinoma of rats by long term feeding with the fungus-growing rice. *Proc. Japan Acad.*, t. **XXXIV**, p. 501-506, 1959.

([1324]) KOBERT R. — *Lehrbuch der Intoxicationen*. 2e éd., Ferdinand Enke édit., Stuttgart, 1906.

([1325]) KOEGL F. et VAN WESSEM G. C. — Fungus dyestuffs. XIV. Oosporein, the dyestuff of *Oospora colorans*. *Rec. Trav. Chim. Pays-Bas*, t. **LXIII**, fasc. 5-12, 1944.

([1326]) KOEHLER B. — Pathologic significance of seed coat injury in dent corn. *Phytopathology*, t. **XXV**, p. 24, 1935.

([1327]) KOEHLER B. — Fungus growth in shelled corn as affected by moisture. *J. Agr. Res.*, t. **LVI**, p. 291-307, 1938.

([1328]) KOEHLER B. — Pericarp injuries in seed corn. Prevalence in dent corn and relation to seedling blight. *Ill. Agr. Exp. Stn. Bull.*, n° 617. 72 p., 1957.

([1329]) KOEHLER B. et WOODWORTH C. M. — Corn seedling virescence caused by *Aspergillus flavus* and *Aspergillus tamarii*. *Phytopathology*, t. **XXVIII**, p. 811-823, 1938.

([1330]) KOEN J. S. et SMITH H. C. — An unusual cause of genital involvement in swine associated with eating moldy corn. *Vet. Med.*, t. **XL**, p. 131-133, 1945.

([1331]) KOHLBRUGGE J. H. F. — Der Darm und seine Bakterien. *Centralbl. f. Bakt.*, Abt. I, t. **XXX**, p. 10-26, 70-80, 1901.

([1332]) KÖHLER H. et SWOBODA R. — (Liver cirrhosis produced in ducks by groundnut meal). *Wien. Tierärtztl. Mschr.*, t. **XLIX**, p. 205-219, 1962.

([1333]) KOLK H. — The influence of storage temperature on the mycoflora of newly harvested moist feed grain. *Proc. Symp. Agric. Coll. Norway*, 1967, 7p. ronéot.

([1334]) KOOIMEN P. — Some properties of cellulase of *Myrothecium verrucaria* and some other fungi. II. *Enzymologia*, t. **XVIII**, p. 371-384, 1957.

([1335]) KORNEEV N. E. — Eksperimentalnyi stakhibotriotoksikoz u laboratornykh zhivotnykh. *Veterinariya*, t. **XXV**, p. 36-37, 1948.

([1336]) KOROBKIN M. et WILLIAMS E. H. — Hepatoma and groundnuts in the West Nile District of Uganda. *Yale J. Biol. Med.*, t. **XLI**, p. 69-78, 1968.

([1337]) KOROLEVA V. P. — Toksicheskie griby, porazhayushchie zerno v protsesse ego prorashchivaniya. *Mikol. Fitopatol.*, t. **I**, p. 82-84, 1967.

([1338]) KOROLEVA V. P. et YABLOCHNIK L. M. — Toksikomikologicheskoe issledovante fuzarioznoi pshenitsy. *Byull. Vses. Inst. Eksp. Vet.*, t. **II**, p. 77-79. 1967.

([1339]) KORPINEN E. L. et YLMÄKI A. — Discovery of toxigenic *Stachybotrys chartarum* strains in Finland. *Experientia*, t. **XXVIII**, p. 108-109, 1972.

([1340]) KOSHEVOI V. A. — (On stachybotryotoxicosis in the pig.) *Veterinarija*, Moscou, t. **XXXIX**, p. 32-33, 1962.

([1341]) KOTSKOVA-KRATOKHVALOVA A., GEBAUEROVA A. et GRDINOVA M. — (Production de composés arsenicaux volatils par les Champignons.) *Ceska Mykol.*, t. **X**, p. 77-78, 1956.

([1342]) KRAL F. — Skin diseases. *Advances in Veterinary Science*, t. **VII**, p. 183-224, Academic Press Publ., New York et Londres, 1962.

([1343]) KRANZ J. et PUCCI E. — Studies on soil borne rots of groundnuts (*Arachis hypogaea*). *Phytopathol. Z.*, t. **XLVII**, p. 101-112, 1963.

([1344]) KRATZER F. H., BANDY D., WILEY M. et BOOTH A. N. — Aflatoxin effects in poultry. *Proc. Soc. Exp. Biol. Med.*, t. **CXXXI**, p. 1281-1284, 1969.

([1345]) KRAYBILL H. F. — Rapport présente à la « *session on toxicological problems in Agriculture and Industry at International seminar on occupational health problems in developing countries* », Lagos, Nigeria, 1-6 avr. 1968.

([1346]) KRAYBILL H. F. — The toxicology and epidemiology of mycotoxins. *Trop. Geogr. Med.*, t. **XXI**, p. 1-18, 1969.

([1347]) KRAYBILL H. F. et SHAPIRO R. E. — Implications of fungal toxicity to human health. *in* Goldblatt L.A. , *Aflatoxin*, Academic Press, p. 401-441, 1969.

([1348]) KRAYBILL H. F. et SHIMKIN M. B. — Carcinogenesis related to foods contaminated by processing and fungal metabolits. *Advances in Cancer Research*, t. **VIII**, p. 191-248, 1964.

([1349]) KRAYBILL H. F. et WHITEHAIR L. A. — Toxicology and irradiated Foods. *Ann. Review of Pharmacology*, t. **VII**, 1967.

([1350]) KRETSCHMER A. E. — Nitrate accumulation in everglades forages. *Agron. J.*, t. **L**, p. 314, 1958.

([1351]) Krogh P. — Investigations on mycotoxicoses in Denmark. *Proc. Symp. Agric. Coll. Norway*, 1967, 6 p. ronéot.

([1352]) Krogh P. — The pathology of mycotoxins. *1st Congr. Pl. Pathol. Londres*, Abstr. p. 109, 1968.

([1353]) Krogh P. — The pathology of mycotoxicoses. *Nord. Vet.-Med.*, t. **XXI**, p. 342-346, 1969.

([1354]) Krogh P. — Foderstoffer kontamination med mykotoksiner. *Saertryk af ugeskrift for agronomer*, n° 9, p. 168-174, 1969.

([1355]) Krogh P. et Basse A. — The pathogenesis of bovine mycotic abortion : an experimental study. *C. R. 5e Congr. ISHAM*, p. 228-229, Paris, 6 juil. 1971.

([1356]) Krogh P. et Hald B. — Forekomst af aflatoksin i importerede jordnodprodukter. *Nord. Vet.-Med.*, t. **XXI**, p. 398-407, 1969.

([1357]) Krogh P., Hald B. et Korpinen E. L. — Occurrence of aflatoxin in groundnut and copra products imported to Finland. *Nord. Vet.-Med.*, t. **XXII**, p. 584-589, 1970.

([1358]) Krogh P. et Hasselager E. — Studies on fungal nephrotoxicity. *Kong. Vet. Landbohojsk. Arsskr.*, p. 198-214, 1968.

([1359]) Krogh P. et Hasselager E. — Svampebetingede sygdomme hos husdyr, med saerlig henblick pa mykotoksikoser hos danske svin. *Nord. Vet.-Med.*, t. **XXII**, p. 141-160, 1970.

([1360]) Krstic M. M. — On the fungistatic and fungicidal properties of an antibiotic produced by *Penicillium rubrum*. *Pl. Dis. Reptr.*, t. **LI**, p. 669-671, 1967.

([1361]) Krzywanski Z. et Borys M. — The influence of « toxins » produced by *Phytophthora infestans* de By. on the enzyme activity of leaves of late blight resistant and susceptible potato varieties. *Phytopath. Z.*, t. **LI**, fasc. 3, p. 262-266, 1964.

([1362]) Kuč J. — A biochemical study of the nature of disease resistance in plants. Ph. D. *Thesis, Purdue University, Lafayette, Indiana*, 1955.

([1363]) Kühl H. — Im Kampf gegen die durch das Wachstum von Schimmelpilzen und Bakterien bedingten Wertverminderungen von Brotgetreide und Mehl. *Mehl u. Brot*, t. **XL**, p. 505-506, 1940.

([1364]) Kühn J. — Die Schmarotzerpilze der Lupinenpflanze und die Bekämpfung der Lupinenkrankheit der Schafe. *Ber. Landw. Inst. Halle*, t. **II**, p. 115-128, 1880.

([1365]) Kulik M. M. et Holaday C. E. — Aflatoxin : a metabolic product of several fungi. *Mycopathol. Mycol. Appl.*, t. **XXX**, p. 137-140, 1966.

([1366]) Kulik Y. D. — A study of the lipids of peas infected by *Aspergillus flavus*. *Mikrobiol. Zhur. Akad. Nauk Ukr. R. S. R.*, t. **XIX**, p. 36-43, 1957.

([1367]) Kulikarni L. G. — « Asiriya mwitunde » groundnut gives good results at Hyderabad *Tropical Abstr.*, n° 22, p. 711, 1967.

([1368]) Kumar G. V. et Sampath S. R. — Aflatoxins : their nature and biological effects. *Indian J. Nutr. Diet.*, t. **IV**, p. 85-101, 1971.

([1369]) Kunkel H. O. — Quantitative studies on incidence and distribution of mycotoxins. *Rept. to Peanut Improvement Working Group Meeting, Washington, 27-28 avr.*, 5 p., 1965.

([1370]) Kurata H., Sakabe F., Udagawa S., Ichinoe M., Suzuki M., et Takahashi N. — A mycological examination for the presence of mycotoxin-producers on the 1954-1967's stored rice grains (en jap.). *Bull. Nat. Inst. Hyg. Sci.*, t. **LXXXVI**, p. 183-188, 1968.

([1371]) Kurata H., Tanabe H., Kanota K., Udagawa S. et Ichinoe M. — Studies on the population of toxigenic fungi in foodstuffs. IV. Aflatoxin producing fungi isolated from foodstuffs in Japan. *J. Food Hyg. Soc. Jap.*, t. **IX**, p. 29-34, 1968.

([1372]) Kurata H., Udagawa S., Ichinoe M., Kawasaki Y., Takada M., Tazawa M., Koizumi A. et Tanabe H. — Studies on the population of toxigenic fungi in foodstuffs. III. Mycoflora of milled rice harvested in 1965. *J. Food Hyg. Soc. Jap.*, t. **IX**, p. 23-28, 1968.

([1373]) Kurata H., Udagawa S., Ichinoe M., Kawasaki Y., Tazawa M., Tanabe H. et Okudaira M. — Studies on the population of toxigenic fungi in foodstuffs. VI. Histopathologic changes in mice caused by toxic metabolites of fungi isolated from domestic rice. *J. Food Hyg. Soc. Jap.*, t. **IX**, p. 385-394, 1968.

([1374]) Kurata H., Udagawa S., Ichione M., Kawasaki Y., Tazawa M., Tanaka J., Takada M. et Tanabe H. — Studies on the population of toxigenic fungi in foodstuffs V. Acute toxicity test for representative species of fungal isolates from milled rice harvested in 1965. *J. Food Hyg. Soc. Jap.*, t. **IX**, p. 379-384, 1968.

([1375]) Kurata H., Udagawa S., Ichinoe M., Natori S. et Sakaki S. — Field survey mycotoxin producing fungi contaminating foodstuffs in Japan, with epidemiological background. *in* Purchase I. F. H. , *Mycotoxins in human health*, p. 101-106, 1971.

([1376]) Kurata H., Udagawa S. et Sakabe F. — Experimental studies on some causal *Fusarium* of the wheat and barley scab. I-III. *Eisei Shikenjo Hokoku*, t. **LXXXII**, p. 123-130, 1964.

([1377]) Kurmanov I. A. — (Stachybotryotoxicosis in cattle.) *Veterinariya*, t. **X**, p. 41-44, 1961.

([1378]) Kurmanov I. A. — (Sur la fusariotoxicose des animaux de ferme.) *Veterinarija*, t. **XL**, fasc. 10, p. 55-56, 1963.

([1379]) Kurtz H. J., Nairn M. E., Nelson G. H., Christensen C. M. et Mirocha C. J. — Histologic changes in the genital tracts of swine fed estrogenic mycotoxin. *Amer. J. Vet. Res.*, t. **XXX**, p. 551-556, 1969.

([1380]) Kurtzmann C. P. et Ciegler A. — Mycotoxin from a blue-eye mold of corn. *Appl. Microbiol.* t. **XX**, p. 204-207, 1970.

([1381]) Kurung J. M. — *Aspergillus ustus. Science*, t. **CII**, p. 11, 1945.

([1382]) Küster E. et Locci R. — Studies on peat and peat microorganisms. II. Occurrence of thermophilic fungi in peat. *Arch. Mikrobiol.*, t. **XLVIII**, p. 319-324, 1964.

([1383]) Kwon T. W. et Ayres J. C. — The purity of aflatoxin G_1 and use of antioxydant and chelating agent on the purification of the toxin by thin-layer chromatography. *J. Chromatogr.*, t. **XXXI**, p. 420-426, 1967.

([1384]) Kysela V. — Onemocnini vepre plisnemi. *Zverolekarks obzor*, t. **XXVII**, p. 206-208, 1934.

([1385]) Kysela V. — Vergiftung von Schweinen durch Schimmelpilze. *Z. Schweinezucht*, t. **III**, p. 9-10, 1941.

([1386]) Lacassagne A. — Cancerogenic action of certain natural substances introduced into food. *Vitalst. Zivilisationskr.*, t. **XII**, p. 5-7, 1967.

([1387]) Lacassagne A., Rudali G., Fusey P., Sales L. — A propos d'une épidémie du syndrome entéro-hépatique par *Aspergillus flavus*, survenue dans un élevage de visons. *Bull. Ass. Fr. Et. Cancer*, t. **LI**, fasc. 4, p. 421-431, 1964.

([1388]) Lacey J. — Health hazards from mouldy fodder. *World Crops*, p. 211-215, 1969.

([1389]) Lacey J. — The microbiology of moist barley storage in unsealed silos. *Ann. Appl. Biol.*, t. **LXIX**, p. 187-212, 1971.

([1390]) Lacey J. et Lacey M. E. — Spore concentrations in the air of farm buildings. *Trans. Brit. Mycol. Soc.*, t. **XLVII**, p. 547-552, 1964.

([1391]) Lachiondo F. B. et Lopez A. C. — *Inst. Congr. Microbiol. Rep. Proc.*, t. **IV**, p. 160-161, 1947.

([1392]) Lafarge C. et Frayssinet C. — The reversibility of inhibition of RNA and DNA synthesis induced by aflatoxin in rat liver. A tentative explanation for carcinogenic mechanism. *Int. J. Cancer*, t. **VI**, p. 74-83, 1970.

([1393]) Lafarge C., Frayssinet C. et de Recondo A. M. — Inhibition par l'aflatoxine de la synthèse de RNA hépatique chez le rat. *Bull. Soc. Chim. Biol.*, t. **XLVII**, p. 1724-1725, 1965.

([1394]) Lafarge C., Frayssinet C. et Simard R. — Inhibition préférentielle des synthèses de RNA nucléolaire provoquée par l'aflatoxine dans les cellules hépatiques du rat. *C. R. Acad. Sci.*, D, t. **CCLXIII**, p. 1011-1014, 1966.

([1395]) Laffond Greletty J. — L'analyse mycologique des aliments du bétail. *Ind. Aliment. anim.*, fasc. 1, p. 9-22, 1972. *Econ. Méd. animales*, t. **XIII**, p. 223-235, 1972.

([1396]) LAFON. — Les intoxications fongiques chez le porc. *Bull. Mayenne-Sciences*, p. 20-23, 1963.

([1397]) LAFONT J. et FRAYSSINET C. — Mycotoxines élaborées par des *Aspergillus*. Leur activité sur l'embryon de poulet. *C.R. Soc. Biol.*, t. **CLXIII**, p. 1362-1364, 1969.

([1398]) LAFONT P. — Les mycotoxines produites par les *Aspergilli* différents d'*A. flavus*. *Colloque aflatoxines, Villejuif*, 7 mars 1966.

([1399]) LAFONT P. — Pollution par *Aspergillus flavus* de produits végétaux européens. *Colloque aflatoxines, Villejuif*, 7 mars 1966.

([1400]) LAFONT P. et FRAYSSINET C. — Des possibilités de contamination de produits alimentaires par les aflatoxines sous climat tempéré. *Proc. 7th Int. Congr. Nutr. Hamburg*, t. **I-V**, p. 3-8, 1966.

([1401]) LAFONT P. et FRAYSSINET C. — Isolement d'une nouvelle aflatoxine : l'aflatoxine 3 B. *Cah. Nutr. Diét.*, t. **V**, p. 78, 1970.

([1402]) LAFONT P. et LAFONT J. — Pollution par des *Aspergillus* de produits végétaux. *Ann. Inst. Pasteur*, t. **CXVI**, p. 237-245, 1969.

([1403]) LAFONT P. et LAFONT J. — Production d'aflatoxine par *Aspergillus flavus*. Link en culture statique. *Ann. Inst. Pasteur*, t. **CXVIII**, p. 340-348, 1970.

([1404]) LAFONT P. et LAFONT J. — Production d'aflatoxine en culture statique. *Cah. Nutr. Diét.*, t. **V**, p. 53-54, 1970.

([1405]) LAFONT P. et LAFONT J. — Production de nidulotoxine par des *Aspergillus* appartenant à diverses espèces. *Experientia*, t. **XXVI**, p. 807-808, 1970.

([1406]) LAFONT P. et LAFONT J. — Contamination de produits céréaliers et d'aliments du bétail par l'aflatoxine. *Food Cosmet. Toxicol.*, t. **VIII**, p. 403-408, 1970.

([1407]) LAFONT P. et LAFONT J. — Production d'aflatoxine par des souches d'*Aspergillus flavus* Link de différentes origines. *Mycopathol. Mycol. Appl.*, t. **XLIII**, p. 323-328, 1971.

([1408]) LAFONT P. et LAFONT J. — Action de l'aflatoxine sur les mycobactéries. *Cah. Nutr. Diét.*, t. **VI**, p. 75-77, 1971.

([1409]) LAFONT P., LAFONT J. et FRAYSSINET C. — La nidulotoxine. *Symposium INSERM Mycotoxines et Alimentation, Le Vésinet*, 24 oct. 1969. *Cah. Nutr. Diét.*, t. **V**, p. 79-80, 1970.

([1410]) LAGRANDEUR G. et POISSON J. — La microflore de maïs, son évolution en fonction des conditions hydriques et thermiques de stockage en atmosphère renouvelée. *Industr. alim. agr.*, t. **LXXXV**, p. 775-788, 1968.

([1411]) LAI M., SEMENIUK G. et HESSELTINE C. W. — Conditions for production of ochratoxin A by *Aspergillus* species in a synthetic medium. *Appl. Microbiol.* t. **XIX**, p. 542-544, 1970.

([1412]) LALAU-KERALY F., NIVIERE P. et TRONCHE P. — Sur la structure de la patuline naturelle. *C.R. Acad. Sci.*, t. **CCLXI**, p. 4028-4030, 1963.

([1413]) LALITHAKUMARI D. et GOVINDASWAMI C. V. — Role of aflatoxins in groundnut seed spoilage. *Curr. Sci.*, t. **XXXIX**, p. 308-309, 1970.

([1414]) LANCASTER M. C., JENKINS F. P. et MC L. PHILIP J. — Toxicity associated with certain samples of groundnuts. *Nature*, t. **CXCII**, p. 1095-1096, 1961.

([1415]) LANCILLOTTI F. et LUCISANO A. — Le aflatossine. *Ind. Conserve*, t. **XLVI**, p. 11-123, 1972.

([1416]) LANDAU J. W. et NEWCOMER V. D. — Acute cerebral phycomycosis (mucormycosis). *J. Pediat.* t. **LXI**, p. 363-385, 1962.

([1417]) LANDERS K. E. — The influence of carbon dioxide, nitrogen and oxygen on the growth sporulation and aflatoxin production in peanuts by *Aspergillus flavus*. *Ph. D. Thesis, Auburn Univ., Alabama*, 62 p., 1966 (*Diss, Abstr.*, t. **XXVII**, p. 2608, 1967.)

([1418]) LANDERS K. E., DAVIS N. D. et DIENER U. L. — Influence of atmospheric gases on aflatoxin production by *Aspergillus flavus* in peanuts. *Phytopathology*, t. **LVII**, p. 1086-1090, 1967.

([1419]) LAQUEUR G. L., MICKELSEN O., WHITING M. G. et KURLAND L. T. — Carcinogenic properties of nuts from *Cycas circinalis* L. indigenous to Guam. *J. Nat. Cancer Inst.*, t. **XXXI**, p. 919, 1963.

([1420]) LA ROCHE G., HALVER J. E., JOHNSON C. L. et ASHLEY L. M. — Hepatoma inducing agents in trout diets. *Fed. Proc.*, t. **XXI**, p. 300, 1962.

([1421]) LARSEN S. — Om kronisk degeneration of nyrerne (nefrose) hos svin forarsaget af skimlet rug. *Maanedsskr. Dyrl.*, t. **XL**, p. 259-284, 289-300, 1928.

([1422]) LARVOR P. — Relations entre la composition du plasma sanguin et les symptômes de tétanie d'herbage chez les bovins. *Ann. Zootech.*, t. **XI**, p. 134-149, 1962.

([1423]) LARVOR P., BROCHART M. et THÉRET M. — Enquête sur la fièvre vitulaire et la tétanie d'herbage des bovins en France. *Economie et médecine animales*, t. **II**, p. 5-38, 1961.

([1424]) LARVOR P. et GUÉGUEN L. — Composition chimique de l'herbe et tétanie d'herbage *Ann. Zootech.*, t. **XII**, p. 39-52, 1963.

([1425]) LARVOR P. et VIOLETTE C. — Influence de l'ingestion d'herbe tétanigène sur le métabolisme minéral (Mg, Ca, P, Na, K) et certains éléments du métabolisme énergétique (corps cétoniques, acides gras volatils) chez la brebis. Nouvelle hypothèse pathogénique sur la tétanie d'herbage. *Rech. Vét.*, fasc. 2, p. 27-44, 1969.

([1426]) LATCH G. C. M. — Fungous diseases of ryegrasses in New Zealand. II. Foliage, root, and seed diseases. *New Zeal. J. Agr. Res.*, t. **IX**, p. 808-819, 1966.

([1427]) LAXA O. — *Margarinomyces Bubaki*, ein Schädling der Margarine. *Centrabl. f. Bakt. u. Parasitenk.*, t. **LXXXI**, p. 392-396, 1930.

([1428]) LAYCOCK T. — Experiments on the fermentation and molding of cacao. *Nineth Ann. Bull. Agr. Dept., Nigeria*, p. 5-26, 1930.

([1429]) LEACH C. M. et TULLOCH M. — *Pithomyces chartarum*, a mycotoxin-producing fungus isolated from seed and fruit in Oregon. *Mycologia*, t. **LXIII**, p. 1086-1089, 1971.

([1430]) LEAVER D. D. — Sporidesmin poisoning in the sheep. I. Changes in bile secretion and the excretion of sporidesmin in bile. *Res. Vet. Sci.*, t. **IX**, p. 255-264, 1968. II. A comparison of some changes in clinical signs and serum constituents following sporidesmin poisoning and experimental obstruction of the common bile duct. *Id.*, t. **IX**, p. 265-273, 1968.

([1431]) LE BARS J. — Méthode d'obtention d'une série définie de dilutions de germes fongiques en vue de leur isolement et de leur numération. *Ann. Rech. Vét.*, t. **III**, p. 435-447, 1972.

([1432]) LE BARS J. et ESCOULA L. — Mycoflore des fourrages secs. Inventaire et fréquence des espèces. *Ann. Rech. Vét.*, t. **IV**, p. 273-282, 1973.

([1433]) LE BOURDELLES B. — Historique de la désinfection et de l'antisepsie. *Colloque Antiseptiques et Désinfectants, Inst. Pasteur*, mai 1964.

([1434]) LE BRETON E. — Quelques aspects de la recherche sur le cancer. *Bull. M.G.E.N.*, t. **XVIII**, p. 11-13, 1965.

([1435]) LE BRETON E. et BOY J. — Obtention de nombreux hépatomes chez le rat ingérant un régime « commercial » ne comportant pas d'adjonction de carcinogène chimique. *Abstr. Vth Cong. Nutrition*, p. 352, Washington, 1960.

([1436]) LE BRETON E., FRAYSSINET C. et BOY J. — Sur l'apparition d'hépatomes « spontanés » chez le rat Wistar. Rôle de la toxine de l'*Aspergillus flavus*. Intérêt en pathologie humaine et cancérologie expérimentale. *C. R. Acad. Sci.*, t. **CCLV**, p. 784-786, 1962.

([1437]) LE BRETON E., FRAYSSINET C., LAFARGE C., DE RECONDO A. M. — Aflatoxine. Mécanisme de l'action. *Food Cosmet. Toxicol.*, t. **II**, p. 675-677, 1964.

([1438]) LEE A. M. — Our newer knowledge of bovine hyperkeratosis (X-disease). *Proc. 56th Ann. Meeting U.S. Livestock Sanitary Assoc.*, p. 175-194, 1952.

([1439]) LEE E. G. H., TOWNSLEY P. M. et WALDEN C. C. — Effect of bivalent metals on the production of aflatoxins in submerged cultures. *Food Sci.*, t. **XXXI**, p. 432-436, 1966.

([1440]) LEE E. G. H. et TOWNSLEY P. M. — Chemical induction of mutation or variation in aflatoxin-producing cultures of *Aspergillus flavus*. *J. Food Sci.*, t. **XXXIII**, p. 420-423, 1968.

([1441]) LEE H. T. et LING K. H. — Application of inst. polyamide impregnated film thin layer chromatography for detection of aflatoxins. *J. Formosan Med. Ass.*, t. **LXVI**, p. 92-100, 1967.

([1442]) LEE L., YATSU L. Y. et GOLDBLATT L. A. — Aflatoxin Contamination. Electron microscopic evidence of mold penetration (*Aspergillus flavus*). *J. Amer. Oil Chem. Soc.*, t. **XVII**, p. 331-332, 1967. (*Oils and Oilseeds J.*, t. **XX**, p. 8, 1967).

([1443]) LEE L. S., CUCULLU A. F., FRANZ A. O. et PONS W. A. — Destruction of aflatoxins in peanuts during dry and oil roasting. *Agric. Food Chem.*, t. **XVII**, p. 451-453, 1969.

([1444]) LEE L. S., CUCULLU A. F., et GOLDBLATT L. A. — Appearance and aflatoxin contents of raw and dry roasted peanut kernels. *Food Technol.*, t. **XXII**, p. 1131-1134, 1968.

([1445]) LEE W. — Quantitative determination of aflatoxin in groundnut products. *Analyst*, t. **XC**, p. 305-307, 1965.

([1446]) LEEMAN A. C. — A short summary of our botanical knowledge of *Lolium temulentum* L. *Onderstepport J. Vet. Sci. Anim. Ind.* t. **I**, p. 213, 1933.

([1447]) LEGATOR M. — Biological effects of aflatoxin in cell culture. *Bacteriol. Rev.*, t. **XXX**, p. 471-477, 1966.

([1448]) LEGATOR M. S. — Biological assay for aflatoxins. *in* GOLDBLATT L. A., *Aflatoxin*, Academic Press, p. 107-149, 1969.

([1449]) LEGATOR M. S. — Mutagenic effects of aflatoxin. *J. Am. Vet. Med. Ass.*, t. **CLV**, p. 2080-2083, 1969.

([1450]) LEGATOR M. S. et WITHROW A. — Aflatoxin : effect on mitotic division in cultured embryonic lung cells. *J. Ass. Off. Agric. Chemists*, t. **XLVII**, p. 1007-1009, 1964.

([1451]) LEGATOR M. S., ZUFFANTE S. M. et HARP A. R. — Aflatoxin : effect on cultured heteroploid human embryonic lung cells. *Nature, G.B.*, t. **CCVIII**, p. 345-347, 1965.

([1452]) LEISTNER L. et AYRES J. C. — Schimmelpilze und Fleischwaren. *Fleischwirtschaft*, t. **XLVII**, p. 1320-1326, 1967.

([1453]) LEITCH J. M. — Food Science and Technology. *Proc. 1st Int. Congr. Food Science and Technology, Londres, sept. 1962*, 5 vol., Gordon et Breach Publ., 1965.

([1454]) LELIÈVRE J., BRÉMOND J. et REBOUR J. — Enzootie de vulvovaginite chez la truie. *Bull. Soc. Vet. Pratique*, t. **XLVI**, p. 18-19, 1962.

([1455]) LEMBKE A. et FRAHM F. — Untersuchungen über Mycoïne. II. *Zentralbl. f. Bakt. Parasit. Inf. u. Hyg. Sielbeck*, t. **I**, p. 152-231, 1947.

([1456]) LEMBKE A. et HAHN B. — The action of metabolic products of *Penicillium claviforme* on cells and tissues varying in their stage of development. *Kiel Michwirtsch. Forschungsber.*, t. **VI**, p. 41-58, 219-241, 1954.

([1457]) LEMBKE A., KORNELEIN M. et FRAHM H. — Beiträge zum Brucellose-Problem. I. Mitteilung. *Zentr. Bakteriol. Parasitenk.*, Abt. I, t. **CLV**, p. 16-31, 1950.

([1458]) LEPESME P. — Recherches sur une aspergillose des Acridiens, *Bull. Soc. Hist. Nat. Afrique Nord*, t. **XXIX**, p. 372-381, 1938.

([1459]) LE ROUX G. et AUDEBAUD G. — Toxicité de l'arachide due aux aflatoxines : son incidence au Cambodge. *Bull. Soc. Pathol. Exot.*, t. **LXIII**, p. 615-524, 1970.

([1460]) LEROUX M. — Influence des fractions lipidiques du germe de blé sur la production des aflatoxines. *Rapp. D. E. A. Fac. Sci. Paris*, 18 p., 1971.

([1461]) LETTRÉ H., SEIDLER E. et WRBA H. — Untersuchung der Hemmstoffe in Wirkung auf Verdopplungsgeschwindigkeit und Phosphatausnahme von Tumoren mit Hilfe von ^{32}P. *Z. Krebsforsch.*, t. **LX**, p. 86-90, 1954.

([1462]) LETTRÉ H., WRBA H. et SEIDLER E. — Charakterisierung von Tumorhemmstoffen mit Hilfe markierter Tumorzellen. *Naturwissenschaften*, t. **XLI**, p. 122-123, 1954.

([1463]) LEUKEL R. W. et MARTIN J. M. — Seed rot and seedling blight of sorghum. *Tech. Bull. U.S. Dept. Agr.*, n° 839, 26 p. 1943.

([1464]) Levaditi J., Besse P., Vibert R., Guillon J. C. et Nazimoff O. — Particularités actuelles de l'hépatome de la Truite arc-en-ciel d'élevage (*Salmo irideus*). *C.R. Acad. Sci.*, t. **CCLVII**, p. 1739-1741, 1963.

([1465]) Levaditi J., Besse P., Vibert R., Nazimoff O. — Sur les critères histopathologiques et biologiques de malignité propres aux tumeurs épithéliales hépatiques des salmonides. *C.R. Acad. Sci.*, t. **CCLI**, p. 608-610, 1960.

([1466]) Levenberg I. G. — Izuchenie toksicheskikh gribov i izyskanie mer preduprezhdeniya ikh razvitiya na myasnykh produktakh. *Vses Nauch. Issled. Inst. Vet. Sanit.*, t. **XXIX**, p. 153-157, 1967.

([1467]) Levenberg I. G. — Materialy k izucheniyu toksichnosti plesnevykh gribov na myase. *Vses Nauk-Issled Inst. Vet. Sanit.*, t. **XXIX**, p. 164-166, 1967.

([1468]) Levenberg I. G., Ivantsov L. I. et Prostakov M. P. — (Stachybotryotoxicosis in cattle). *Veterinariya*, t. **X**, p. 38-41, 1961.

([1469]) Lévêque P. — Irradiation des aliments ; état actuel des recherches dans le monde. *Bull. Inf. Scient. et Techn. C.E.A.*, n° 99, p. 35-38, 1965.

([1470]) Levi C. P. — Collaborative study on a method for detection of aflatoxin B_1 in green coffee beans. *J. Ass. Off. Anal. Chem.*, t. **LII**, p. 1300-1303, 1969.

([1471]) Levi C. P. et Borker E. — Survey of green coffee for potential aflatoxin contamination. *J. Ass. Off. Anal. Chem.*, t. **LI**, p. 600-602, 1968.

([1472]) Levine A. S. et Fellers C. R. — Action of acetic acid on food spoilage microorganisms. *J. Bact.*, t. **XXXIX**, p. 499-514 et t. **XL**, p. 255, 1940.

([1473]) Levitskii B. G. et Koniukhova V. B. — O toksichnosti kormov porazenni banalii plesenyami. *Veterinariya*, t. **XXIV**, p. 40-43, 1947.

([1474]) Levy E. B. et Smallfield P. W. — Photosensitivity diseases in New Zealand. II. Facial eczema : the influence of climate, pasture, composition and management. *New Zealand J. Sci. Technol.*, sér. A, t. **XXIV**, p. 198-214, 1942.

([1475]) Lewis D. — The metabolism of nitrate and nitrite in the sheep. II. *Biochem, J.*, t. **XLIX** p. 149, 1951.

([1476]) Lewis G., Markson L. M. et Allcroft R. — The effect of feeding toxic groundnut meal to sheep over a period of five years. *Vet. Rec.*, t. **LXXX**, p. 312-314, 1967.

([1477]) Lichtwardt R. W., Barron G. L. et Tiffany L. M. — Mold flora associated with shelled corn in Iowa. *Iowa State Coll. J. Sci.*, t. **XXXIII**, p. 1-11, 1958.

([1478]) Lie J. L. et Marth E. H. — Formation of aflatoxin in Cheddar cheese by *Aspergillus flavus* and *Aspergillus parasiticus*. *J. Dairy Sci.*, t. **L**, p. 1708-1710, 1967.

([1479]) Lie J. L. et Marth E. H. — Aflatoxin formation by *Aspergillus flavus* and *Aspergillus parasiticus* in a casein substrate at different pH values. *J. Dairy Sci.*, t. **LI**, p. 1743-1747, 1968.

([1480]) Lie-Kian-Joe, Njo-Injo, Sutoma Tjokrone-Goro Tjeoi Eng, SJ. Schaafma et Emmons C. W. — Phycomycosis of the central nervous system, associated with diabetes mellitus in Indonesia. *Am. J. Clin. Path.*, t. **XXXII**, p. 62-70, 1959.

([1481]) Lieberman J. R., Wolf J. C., Rao H. R. G. et Harein P. K. — Inhibition of F-2 (zearalenone) biosynthesis and perithecia production in *Fusarium roseum « graminearum »*. *Phytopathology*, t. **LXI**, p. 900, 1971.

([1482]) Liem D. H. et Beljaars P. R. — Note on a rapid determination of aflatoxins in peanuts and peanut products. *J. Ass. Off. Anal. Chem.*, t. **LIII**, p. 1064-1066, 1970.

([1483]) Liener I. E. — *Toxic constituents of plants foodstuffs*. 504 p., Academic Press édit., New York, 1969.

([1484]) Lijinsky W. — Aflatoxins and nitrosamines: cellular interactions and carcinogenesis. *N. Z. Med. J.*, t. **LXVII**, p. 100-109, 1968.

([1485]) Lijinsky W. — The preparation of aflatoxins labelled with tritium. *J. Labelled Compounds*, t. **VI**, p. 60-65, 1970.

([1486]) LIJINSKY W. et BUTLER W. H. — Purification and toxicity of aflatoxin G_1. *Proc. Soc. Exp. Biol. Med.*, t. **CXXIII**, p. 151-154, 1966.

([1487]) LILLARD J. S., HANLIN R. T. et LILLARD D. A. — Aflatoxigenic isolates of *Aspergillus flavus* from pecans. *Appl. Microbiol.*, t. **XIX**, p. 128-130, 1970.

([1488]) LILLEHOJ H. B. et CIEGLER A. — Inhibition of deoxyribonucleic acid synthesis in *Flavobacterium aurantiacum* by aflatoxin B_1. *J. Bacter.*, t. **XCIV**, p. 787-788, 1967.

([1489]) LILLEHOJ E. B. et CIEGLER A. — Biological activity of sterigmatocystin. *Mycopathol. Mycol. Appl.*, t. **XXXV**, p. 373-376, 1968.

([1490]) LILLEHOJ E. B. et CIEGLER A. — Aflatoxin B_1 binding and toxic effects on *Bacillus megaterium. J. Gen. Microbiol.*, t. **LIV**, p. 185-194, 1968.

([1491]) LILLEHOJ E. B. et CIEGLER A. — Aflatoxin B_1 effect on enzyme biosynthesis in *Bacillus cereus* and *Bacillus licheniformis. Canad. J. Microbiol.*, t. **XVI**, p. 1059-1108., 1970.

([1492]) LILLEHOJ E. B., CIEGLER A. et DETROY R. W. — Fungal toxins. *in* BLOOD F. R., *Essays in Toxicology*, t. **II**, p. 1-136, Academic Press, New York, 1970.

([1493]) LILLEHOJ E. B., CIEGLER A. et HALL H. H. — Fungistatic action of aflatoxin B_1. *Experientia*, t. **XXIII**, p. 187-190, 1967.

([1494]) LILLEHOJ E. B., CIEGLER A. et HALL H. H. — Aflatoxin B_1 uptake by *Flavobacterium aurantiacum* and resulting toxic effects. *J. Bacter.*, t. **XCIII**, p. 464-471, 1967.

([1495]) LILLEHOJ E. B., CIEGLER A. et HALL H. H. — Aflatoxin G_1 uptake by cells of *Flavobacterium aurantiacum. Canad. J. Microbiol.*, t. **XIII**, p. 629-633, 1967.

([1496]) LILLEHOJ E. B., STUBBLEFIELD R. D., SHANONN G. M. et SHOTWELL O. L. — Aflatoxin M_1 removal from aqueous solutions by *Flavobacterium aurantiacum. Mycopathol. Mycol. Appl.*, t. **XLV**, p. 259-266, 1971.

([1497]) LILLY L. L. — Induction of chromosome aberration by aflatoxin. *Nature, G.B.*, t. **CCVII**, p. 433-434, 1965.

([1498]) LILLY V. G. et BARNETT H. L. — *Physiology of the fungi.* 464 p., MC GRAW HILL, New York, 1951.

([1499]) LILY K. — Ecological studies on soil fungi I. Recolonization of steam sterilized soil by different microorganisms. *J. Indian Bot. Soc.*, t. **XLIV**, p. 276-289, 1965.

([1500]) LIM HAN KUO et YEAP GIM SAI. — The occurrence of aflatoxin in Malayan imported oil cakes and groundnut kernels. *Malay. Agric. J.*, t. **XLV**, p. 232-244, 1966.

([1501]) LINDEBERG G. — Toxin-producing fungi and their importance in animal feeding. *Symposium Agric. Coll. Norway*, 1967.

([1502]) LINDENFELSER L. A. et CIEGLER A. — Studies on aflatoxin detoxification in shelled corn by ensiling. *J. Agric. Food Chem.*, t. **XVIII**, p. 640-643, 1970.

([1503]) LINDSEY D. L. — Effect of *Aspergillus flavus* on peanuts grown under gnotobiotic conditions. *Phytopathology*, t. **LX**, p. 208-211, 1970.

([1504]) LING K. H., TUNG C. M., SHEH I. F., WANG J. J. et TUNG T. C. — An aflatoxin B_1 in unrefined peanut oil and peanut products in Taiwan. *J. Formosan Med. Ass.*, t. **LXVII**, p. 309-314, 1968.

([1505]) LINIÈRES E. de. — Editorial. *Cahiers de Nutrition et de Diététique*, t. **I**, fasc. 2, p. 3-4, 1966.

([1506]) LINK L. P. — The discovery of dicoumarol and its sequels. *Circulation*, t. **XIX**, p. 97, 1959.

([1507]) LINTON R. G. — *Animal nutrition and veterinary dietetics.* Green and Son Ltd édit., Edinburgh, 1927.

([1508]) LIPPINCOTT S. W., CALVO W. S., BAKER C. P., JESSEPH J. E. JANSEN C. R. et FARR L. F. — Effects of heavy high energy charged particles. *Arch. Path.*, t. **LXXVI**, p. 543, 1963.

([1509]) LISKER N., JOFFE A. Z. et FRANK Z. R. — Penetration of *Aspergillus flavus* and some other fungi into pods of various peanut varieties. *Oléagineux*, t. **XXV**, p. 347-348, 1970.

([1510]) LIST P. H. et WAGNER K. — Fungus constituents. XII. Ustilagin and other constituents of corn smut, *Ustilago maydis*. *Arzneimittel-Forsch.*, t. **XIII**, p. 36-41, 1963.

([1511]) LO J. T., LING K. M. et TUNG T. C. — Study on the isolation of aflatoxin. *J. Formosan Med. Ass.*, t. **LXVI**, p. 43-49, 1967.

([1512]) LOCHHEAD A. G., CHASE F. E. et LANDERKIN G. B. — Production of claviformin by soil Penicillia. *Canad. J. Res.*, t. **XXIV**, p. 1-9, 1946.

([1513]) LOIZELIER A. B. — Intoxicaciones en pollos producidas por el turto de cacahuete *Rev. patronato Biol. Anim.*, t. **VII**, p. 25-33, 1963.

([1514]) LONGREE K. et BLAKER G. G. — *Sanitary techniques in food service.* 225 p., John Wiley and Sons, N.Y., 1971.

([1515]) LOOSMORE R. M., ALLCROFT R., TUTTON E. A. et CARNAGHAN R. B. A. — The presence of aflatoxin in a sample of cottonseed cake. *Vet. Rec.*, t. **LXXVI**, p. 64-65, 1964.

([1516]) LOOSMORE R. M. et HARDING J. D. J. — A toxic factor in Brazilian groundnut causing liver damage in pigs. *Vet. Rec.*, t. **LXXIII**, p. 1362-1364, 1961.

([1517]) LOPEZ A. et CRAWFORD M. A. — Aflatoxin content of groundnuts sold for human consumption in Uganda (*Aspergillus flavus*). *Lancet*, n° 7530, p. 1351-1354, 1967.

([1518]) LOPEZ F. et CHRISTENSEN C. M. — Factors influencing invasion of sorghum seed by storage fungi. *Plant Dis. Reptr.*, t. **XLVII**, p. 597-601, 1963.

([1519]) LOPEZ L. C. et CHRISTENSEN C. M. — Effect of moisture content and temperature on invasion of stored corn by *Aspergillus flavus*. *Phytopathology*, t. **LVII**, p. 588-590, 1967.

([1520]) LOURIA D. B. — Deep-seated mycotic infections, allergy to fungi and mycotoxins (human). *N. Engl. J. Med.*, t. **CCLXXVII**, p. 1126-1134, 1967.

([1521]) LOURIA D. B. — Aflatoxin-like substances in tobacco. *C.R. 5e Congr. ISHAM*, p. 213, Paris, 6 juil. 1971.

([1522]) LOVE T. D. — Relation of temperature, time and moisture to the production of aflatoxin in fish meal. *Fish Ind. Res.*, t. **IV**, p. 139-142, 1968.

([1523]) LUCAS F. V., MONROE P., PHAM VAN NGA et TOWNSEND J. F. — Mycotoxin contamination of South Vietnamese rice. *J. Trop. Med. Hyg.*, t. **LXXIV**, p. 182-183, 1971.

([1524]) LUCISANO A., CAMPANINI M. et CASOLARI A. — Contributo allo studio dei fattori ambientali che condizionano la produzione di aflatossine negli alimenti. *Ind. Conserve*, t. **XLVII**, p. 27-31, 1972.

([1525]) LUCKNER M. et MOHAMMED Y. S. — Metabolic products of *Penicillium viridicatum* and *P. cyclopium*; synthesis of viridicatol, 3′-O-méthylviridicatol and N-méthyl-3′-O-viridicatol. *Tetrah. Lett.*, n° 29, p. 1987, 1964.

([1526]) LUTEY R. W. et CHRISTENSEN C. M. — Influence of moisture content, temperature and length of storage upon survival of fungi in barley kernels. *Phytopathology*, t. **LIII**, p. 713-717, 1963.

([1527]) LYASS L. S. — *Experimental studies on mycotherapy of leukemias.* Cand. diss., Moscou, 1950.

([1528]) LYASS L. S. — Experience in treating leukemia in mice with a preparation from the fungus *Fusarium*. *Voprosy onkologii*, t. **VI**, p.79, 1955.

([1529]) LYNCH B. T., GLASS R. L. et GEDDES W. F. — Grain storage studies. XXXII. Quantitative changes occurring in the sugars of wheat deteriorating in the presence and absence of molds. *Cereal Chem.*, t. **XXXIX**, p. 256-262, 1962.

([1530]) LYNCH G. P., TODD G. C., SHALKOP W. T. et MOORE L. A. — Responses of dairy calves to aflatoxin-contaminated feed. *J. Dairy Sci.*, t. **LIII**, p. 63-71, 1970.

([1531]) LYSO A. — Aflatoxicoses. A review. *Meld. Norges Landbrukshogsk*, t. **XLV**, n° 1, 17 p., 1966.

([1532]) MCCALLA T. M., GUENZI W. D. et NORSTADT F. A. — Microbial studies of phytotoxic substances in the stubble-mulch system. *Z. Allgem. Microbiol.*, t. **III**, p. 202-210, 1963.

([1533]) McCalla T. M., Guenzi W. D. et Norstadt F. A. — Phytotoxic substances in the stubble-mulch system. *Trans. 8th Int. Congr. Soil Sci., Bucarest*, t. **III**, p. 933-943, 1964.

([1534]) Mc Donald D. — Progress report on research into aflatoxin production in groundnuts in Northern Nigeria. *Inst. for Agr. Res., Ahmadu Bello Univ. Samaru Misc. Paper*, n° 2, Samaru, N. Nigeria, 13 p., 1964.

([1535]) Mc Donald D. — Research on the aflatoxin problem in groundnuts in Northern Nigeria, 1961-1965. *Samaru Miscellaneous Paper*, n° 14, 34 p., 1966.

([1536]) Mc Donald D. — *Aspergillus flavus* Link and its control in Nigeria. *1st Int. Congr. Pl. Pathol.*, Londres, juil. 1968.

([1537]) Mc Donald D. — Fungi associated with the fruit of *Arachis hypogaea* L. *Ph. D. Thesis, Ahmadu Bello Univ., Zaria*, 1968.

([1538]) Mc Donald D. — The influence of the developing groundnut fruit on soil mycoflora. *Trans. Brit. Mycol. Soc.*, t. **LIII**, p. 393-406, 1969.

([1539]) Mc Donald D. — Groundnut pod diseases. *Rev. Appl. Mycol.*, t. **XLVIII**, p. 465-474, 1969.

([1540]) Mc Donald D. — Fungal infection of groundnut fruit before harvest. *Trans. Brit. Mycol. Soc.*, t. **LIV**, p. 453-460, 1970.

([1541]) Mc Donald D. — Fungal infection of groundnut fruit after maturity and during drying. *Trans. Brit. Mycol. Soc.*, t. **LIV**, p. 461-472, 1970.

([1542]) Mc Donald D. et A'Brook J. — Growth of *Aspergillus flavus* and production of aflatoxin in groundnuts. Part. **III**. *Trop. Sci.*, t. V, p. 208-214, 1963.

([1543]) Mc Donald D. et Harkness C. — Growth of *Aspergillus flavus* and production of aflatoxin in groundnuts. Part **II**. *Trop. Sci.*, t. **V**, p. 143-154, 1963.

([1544]) Mc Donald D. et Harkness C. — Growth of *Aspergillus flavus* and production of aflatoxin in groundnuts. *Samaru Res. Bull.*, t. **XXXIII**, p. 143-154, 1964.

([1545]) Mc Donald D. Harkness C. — Growth of *Aspergillus flavus* and production of aflatoxin in groundnuts. Part IV. *Trop. Sci.*, t. **VI**, p. 12-27, 1964.

([1546]) Mc Donald D. et Harkness C. — Growth of *Aspergillus flavus* and production of aflatoxin in groundnuts. Part VIII. *Trop. Sci.*, t. **VII**, p. 122-137, 1965 et *Samaru Res. Bull.*, t. **XLIV**, p. 12-27, 1965.

([1547]) Mc Donald D. et Harkness C. — Aflatoxin in the groundnut crops at harvest in Northern Nigeria. *Trop. Sci.*, t. **IX**, p. 148-161, 1967.

([1548]) Mc Donald D., Harkness C. et Stonebridge W. C. — Growth of *Aspergillus flavus* and production of aflatoxin in groundnuts. Part VI. *Trop. Sci.*, t. **VI**, p. 131-154, 1964.

([1549]) Mc Donald J. C. — Biosynthesis of aspergillic acid. *J. Biol. Chem.*, t. **CCXXXVI**, p. 512-514, 1961.

([1550]) Mc Donald R. A. — Primary carcinoma of liver: clinicopathologic study of one hundred eight cases. *Arch. Int. Med.*, t. **XCIX**, p. 266-279, 1957.

([1551]) Mc Erlean B. A. — Vulvovaginitis of swine. *Vet. Record*, t. **LXIX**, p. 539-540, 1952.

([1552]) Mc Gee D. C. et Christensen C. M. — Storage fungi and fatty acids in seeds held thirty days at moisture contents of fourteen and sixteen per cent. *Phytopathology*, t. **LX**, p. 1775-1777, 1970.

([1553]) Mc Kee C. M. et Mc Phillamy H. B. — An antibiotic substance produced by submerged cultivation of *Aspergillus flavus*. *Proc. Soc. Exptl. Biol. Med.*, t. **LIII**, p. 247-248, 1943.

([1554]) Mc Kee C. M., Rake G. et Houck C. L. — Studies on *Aspergillus flavus*. II. The production and properties of a penicillin-like substance-flavicidin. *J. Bacteriol.*, t. **XLVII**, p. 187-197, 1944.

([1555]) Mc Lachlan T. — Salt as a preservative for food. *J. Food Technol., G.B.*, t. **II**, p. 149-151, 1967.

([1556]) Mc Laughlin J., Marliac J. P., Verret M. J., Mutchler M. K. et Fitzhugh O. G.— The injection of chemicals into the yolk sac of fertile eggs prior to incubation as a toxicity test. *Toxicol. Appl. Pharmacol.*, t. V, p. 760-771, 1963.

([1557]) Mc Lean A. E. M. et Marshall A. — Reduced carcinogenic effects of aflatoxin in rats given phenobarbitone. *Brit. J. Exp. Pathol.*, t. **LII**, p. 322-329, 1971.

([1558]) Mac Lean D. B. et Giese R. L. — Fungi associated with *Xyloterinus politus* (Say) (Coleoptera : Scolytidae). *J. Invertebr. Pathol.*, t. X, p. 185-189, 1968.

([1559]) Mc Means J. L., Ashworth L. J. et Pons W. A. — Aflatoxins in hull and meats of cottonseed. *J. Amer. Oil Chem. Soc.*, t. XLV, p. 575-576, 1968.

([1560]) Mc Murray J. — Some every day problems in otolaryngology. *Pa. Med. Journ.*, t. **XCIII**, p. 1690-1692, 1940.

([1561]) Mc Nutt S. H., Purwin P. et Murray C. — Vulvovaginitis in swine. Preliminary report. *J. Am. Vet. Med. Assoc.*, t. **LXXIII**, p. 484-492, 1928.

([1562]) Macy H., Coulter S. T. et Combs W. F. — Observations on the quantitative changes in the microflora during the manufacture and storage of butter. *Minn. Agr. Exp. Stn. Techn. Bull.*, n° 82, 1932.

([1563]) Madhavan T. V. et Gopalan C. — Effect of dietary protein on aflatoxin liver in jury in weanling rats. *Arch. Pathol., U.S.A.*, t. **LXXX**, p. 123-126, 1965.

([1564]) Madhavan T. V. et Gopalan C. — The effect of dietary protein on carcinogenesis of aflatoxin. *Arch. Pathol.*, t. **LXXXV**, p. 133-137, 1968.

([1565]) Madhavan T. V., Suryanarayana Rao K. — Hepatic infarction in ducklings in aflatoxin poisoning. *Arch. Pathol., U.S.A.*, t. **LXXXI**, p. 520-524, 1966.

([1566]) Madhavan T. V. et Suryanarayana Rao K. — Tubular epithelial reflux in the kidney in aflatoxin poisoning. *J. Pathol. Bacteriol.*, t. **XCIII**, p. 329-331, 1967.

([1567]) Madhavikutti K. et Shanmugasundaram E. R. B. — Toxicity of *Aspergillus flavus* Link ex Fries and *Penicillium rubrum* Stoll to Swiss albino mice. *Proc. Indian Acad. Sci.*, Sect. B, t. **LXVIII**, p. 261-267, 1968.

([1568]) Madsen A. — Some problems associated with the feeding of mouldy barley to pigs. *Proc. Sympos. Agric. Coll. Norway*, 1967, 8 p. ronéot.

([1569]) Madsen A., Lavrsen B. et Mortenson H. P. — Muggen bug. *Bilag Landok. Forsogslav. efterasmode (arbog) 1965*, p. 106-120, 1965.

([1570]) Madsen J. P. — Sygdomsstatistik for soroegnens Andelssvineslagteri 1960-1963. *Meddlemsbl. danske Dyrlaegef.*, t. **XLVII**, p. 900-913, 1964.

([1571]) Maficy K. et Carver D. H. — Cyclopin : a trypsin sensitive constituent of *Penicillium cyclopium* with antiviral properties. *Proc. Soc. Exp. Biol. and Med.*, t. **CXLIV**, p. 99, 1963.

([1572]) Magee P. N. — Toxic liver injury — Inhibition of protein synthesis in rat liver by dimethylnitrosamine *in vivo*. *Biochem. J.*, t. **LXX**, p. 606-611, 1958.

([1573]) Magee P. N. et Barnes J. M. — The production of malignant primary hepatic tumors in the rat by feeding dimethylnitrosamine. *Brit. J. Cancer*, t. **X**, p. 114, 1956.

([1574]) Magee P. N. et Hultin T. — Methylation of proteins of rat-liver slices by dimethylnitrosamine in vitro. *Biochem. J.*, t. **LXXXIII**, p. 106, 1962.

([1575]) Mains E. B., Vestal C. M. et Curtis P. B. — Scab of small grains and feeding trouble in Indiana in 1928. *Proc. Indiana Acad. Sci.*, t. **XXXIX**, p. 101, 1930.

([1576]) Majumder S. K. et al. — Studies on the storage of coffee beans. *Food Sci., India*, t. **X**, p. 321-338, 1961.

([1577]) Majumder S. K., Muthu M. et Narasimhan K. S. — Behaviour of ethylene dibromide, and their mixtures. *Food technol.*, t. **XVII**, p. 108, 1963.

([1578]) Majumder S. K., Narasimhan K. S. et Parpia H. A. B. — Microecological factors of microbial spoilage and the occurrence of mycotoxins on stored grains. Mycotoxins on stored grains. *Mycotoxins in Foodstuffs*, p. 27-47, M.I.T., Press, 1965.

([1579]) Maksimov Y. I. — On poisoning of cattle with uralite on pasture and on the use of the poisoned animals. *Trudy Baskhir. Selskokhoz. Inst.*, t. **VIII**, p. 363-371, 1957.

([1580]) MALASHENKO Y. R. — Study of some biological features of dendrodochine. Report I. The toxic effect of dendrodochine on the growth and weight of rats. *J. Microbiol., Kiev* t. **XXIII**, p. 25-30, 1961.

([1581]) MALAVOLTA E. et DE CAMARGO R. — Nitrification by soil fungi. *Bol. Inst Zimotec (Sao Paulo)*, t. **XIII**, p. 3-9, 1955.

([1582]) MALENÇON G. et RIEUF P. — Contribution à la flore mycologique du Maroc : les *Aspergillus. Bull. Soc. Sc. Nat. Maroc*, t. **XXIX**, p. 163-166, 1949.

([1583]) MALLIK K. C. B. — Mycotoxins as hepatotoxic agents. *J. Indian Med. Ass.*, t. **LII**, p. 201-204, 1969.

([1584]) MALLMANN W. L. et MICHAEL C. E. — The development of mold on cold storage eggs and methods of control. *Michigan State Coll. Techn. Bull.*, n° 174, 34 p., 1940.

([1585]) MALONE J. P. et MUSKETT A. E. — Seed-borne fungi. *C. R. Assoc. Int.essais de semences, Wageningen*, t. **XXIX**, p. 177-384, 1964.

([1586]) MALYAVIN I. — Poisoning of cattle by fodder infected by *Ustilago avenae. Veterinariya, Moscou*, t. **XL**, p. 45, 1963.

([1587]) MANABE M. et MATSUURA S. — Liquid chromatography of aflatoxins including aflatoxins B_2 and G_2. *Agric. Biol. Chem.*, t. **XXXV**, p. 417-423, 1971.

([1588]) MANABE M., MATSUURA S. et NAKANO M. — Isolation and quantitative analysis for aflatoxins B_1, B_2, G_1, G_2 by thin-layer and liquid chromatographies. *J. Agr. Chem. Soc. Jap.*, t. **XLI**, p. 592-598 ; 1967.

([1589]) MANABE M., MATSUURA S. et NAKANO M. — Isolation and quantitative analysis of four aflatoxins (B_1, B_2, G_1 and G_2) by thin-layer and liquid chromatography. *in* HERZBERG M., *Toxic Micro-organisms*, p. 23-29, 1970.

([1590]) MANN G. E., CODIFER L. P. et DOLLEAR F. G. — Effect of heat on aflatoxins in oilseed meals. *J. Agric. Food Chem.*, t. **XV**, p. 1090-1092, 1967.

([1591]) MANN G. E., CODIFER L. P., GARDNER H. K., KOLTUN S. P. et DOLLEAR F. G. — Chemical inactivation of aflatoxins in peanut and cottonseed meals. *J. Amer. Oil Chem. Soc.*, t. **XLVII**, p. 173-176, 1970.

([1592]) MANN T. B. — « Disease » of turkey poults. *Vet. Rec.*, t. **LXXII**, p. 715, 1960.

([1593]) MANN T. B. — Turkey « X » disease and the labelling of poultry foods. *Vet. Rec.*, t. **LXXIII**, p. 131-132 et p. 253, 1961.

([1594]) MANSO C. — O papel das aflatoxinas no desenvovimento do cancro do figado. *Rev. Estud. Gerais. Univ. Moçambique*, Ser. **III**, *Cienc. Med.*, t. **IV**, p. 309-323, 1967.

([1595]) MARASAS W. E. O., BAMBURG J. R., SMALLEY E. B., STRONG F. M., RAGLAND W. L. et DEGURSE P. E. — Toxic effect on trout, rats, and mice of T-2 toxin produced by the fungus *Fusarium tricinctum* (Cd.) Snyd. et Hans. *Toxicol. Appl. Pharmacol.*, t. **XV**, p. 471-482, 1969.

([1596]) MARASAS W. F. O., SMALLEY E. B., BAMBURG J. R. et STRONG F. M. — Phytotoxicity of T-2 toxin produced by *Fusarium tricinctum. Phytopathology*, t. **LXI**, p. 1488-1491, 1971.

([1597]) MARASAS W. F. O., SMALLEY E. B., DEGURSE P. E., BAMBURG J. R. et NICHOLS R. E. — Acute toxicity to rainbow trout (*Salmo gairdnerii*) of a metabolite produced by the fungus *Fusarium tricinctum. Nature*, t. **CCXIV**, p. 817-818, 1967.

[1598]) MARBROOK J. et MATTHEWS R. E. F. — Loss of sporidesmin from spores of *Pithomyces chartarum* (Berk. et Curt.) M. B. Ellis. *New Zealand J. Agr. Res.*, t. V, p. 232-236, 1962.

([1599]) MARSH P. B., BOLLENBACHER K., SAN ANTONIO J. P. et MEROLA G. V. — Observations on certain fluorescent spots in raw cotton associated with the growth of microorganisms. *Textile Res. J.*, t. **XXV**, p. 1007-1016, 1955.

([1600]) MARSH P. B., SIMPSON M. E., FERRETI R. J., CAMPBELL T. C. et DONOSO J. — Relation of aflatoxins in cotton seeds at harvest to fluorescence in the fiber. *J. Agr. Food Chem.*, t. **XVII**, p. 462-467, 1969.

([1601]) MARSH P. B., SIMPSON M. E., FERRETTI R. J., MEROLA G. V., DONOSO J., CRAIG G. O., TRUCKSESS M. W. et WORK P. S. — Mechanism of formation of a fluorescence in cotton fiber associated with aflatoxins in seeds at harvest. *J. Agr. Food Chem.*, t. **XVII**, p. 468-472, 1969.

([1602]) MARSHAK R. R. — Some metabolic derangements associated with magnesium metabolism in cattle. *J. Am. Vet. Assoc.*, t. **CXXXIII**, p. 539, 1958.

([1603]) MARSHALL K. C. — A study of the mechanism of nitrification by *Aspergillus flavus Dissertation Abstr.*, t. **XXII**, p. 975, 1961.

([1604]) MARSHALL K. C. et ALEXANDER M. — Nitrification by *Aspergillus flavus. J. Bacteriol.*, t. **LXXXIII**, p. 572-578, 1962.

([1605]) MARSTON R. Q. — Production of kojic acid from *Aspergillus lutescens. Nature*, t. **CLXIV**, p. 961, 1950.

([1606]) MARTIN P. — Einfluss der Kulturfiltrate von Mikroorganismen auf die Abgabe von Scopoletin aus den Keimwurzeln des Hafers (*Avena sativa*). *Arch. Mikrobiol.*, t. **XXIX**, p. 154-168, 1958.

([1607]) MARTIN P. — Untersuchungen über ein phytopathogenes Toxin von *Pythium irregulare* Buisman. *Phytopath. Z.*, t. **L**, fasc. 3, p. 235-249, 1964.

([1608]) MARTIN P. — Fungi of peanuts in relation to liver cancer. *S. Afr. J. Lab. Clin. Med.*, t. **XIV**, p. 23, 1968.

([1609]) MARTIN P. M. D., GILMAN G. A. et KEEN P. — The incidence of fungi in foodstuffs and their significance, based on a survey in the Eastern Transvaal and Swaziland. *in* PURCHASE I. F. H., *Mycotoxins in human health*, p. 281-290, 1971.

([1610]) MARTIN P. M. D., MEADLEY J., GILMAN G. A. et KEEN P. — Fungi in southern African foodstuffs and their significance. *1st Int. Mycol. Congr. Exeter, Abstr.* p. 62, 1971.

([1611]) MARUMO S. — Islanditoxin, a toxic metabolite produced by *Penicillium islandicum.* I. *Bull. Agr. Chem. Soc. Japan*, t. **XIX**, p. 258-261, 1955.

([1612]) MARUMO S. — Islanditoxin, a toxic metabolite produced by *Penicillium islandicum.* III. Structure of islanditoxin. *Bull. Agr. Chem. Soc. Japan*, t. **XXIII**, p. 428-437, 1959.

([1613]) MARUMO S., MIYAO K. et MATSUYAMA A. — Islanditoxin, a toxic metabolite produced by *Penicillium islandicum.* II. Acid hydrolysis of islanditoxin. *Bull. Agr. Chem. Soc. Japan*, t. **XIX**, p. 262-266, 1955.

([1614]) MARUMO S., MIYAO K., MATSUYAMA A. et SUMIKI Y. — Islanditoxin. *J. Agr. Chem. Soc. Japan*, t. **XXIX**, p. 913, 1955.

([1615]) MARUMO S. et SUMIKI Y. — Islanditoxin, a toxic metabolite produced by *Penicillium islandicum. J. Agr. Chem. Soc. Japan*, t. **XXIX**, p. 305, 1955.

([1616]) MASRI M. S. — Biochemical evaluation of aflatoxin. Western Experiment Stn. Collaborators Conference. Western Regional Research Laboratory. Albany, Calif, 3 mars 1965.

([1617]) MASRI M. S., GARCIA V. C. et PAGE J. R. — The aflatoxin M content of milk from cows fed known amounts of aflatoxin. *Vet. Rech.*, t. **LXXXIV**, p. 146-147, 1969.

([1618]) MASRI M. S., LUNDIN R. E., PAGE J. R. et GARCIA V. C. — Crystalline aflatoxin M_1 from urine and milk. *Nature, G. B.*, t. **CCXV**, p. 753-755, 1967.

([1619]) MASRI M. S., PAGE J. R. et GARCIA V. C. — Mycotoxins. Analysis for aflatoxin M in milk. *J. Ass. Off. Anal. Chem.*, t. **LI**, p. 594-600, 1968.

([1620]) MASRI M. S., PAGE J. R. et GARCIA V. C. — Modification and method for aflatoxins in milk. *J. Ass. Off. Anal. Chem.*, t. **LII**, p. 641-643, 1969.

([1621]) MATELES R. I. et ADYE J. — Production of aflatoxins in submerged culture. *Appl. Microbiol.*, t. **XIII**, p. 208-211, 1965.

([1622]) MATELES R. I. et FULD G. J. — Continuous hydroxylation of progesterone by *Aspergillus ochraceus. Antonie van Leeuwenhoek, J. Microbiol. Serol.*, t. **XXVII**, p. 33-50, 1961.

([1623]) MATELES R. I. et WOGAN G. N. — *Biochemistry of some foodborne microbial toxins.* 171, p., M. I. T. Press, Londres, 1967.

([1624]) MATELES R. I. et WOGAN G. N. — Aflatoxins. *Advances in Microbial Physiology*, t. **I**, p. 25-35, 1967.

([1625]) MATHEW C. P. — Chemistry of aflatoxins. *Chem. Ind.*, p. 913-917, 1970.

([1626]) MATHIEU C. M. et BARRÉ P. E. — L'utilisation digestive de quelques glucides par le veau préruminant. *C. R. travaux actions concertées 1963-1964, CERDIA*, 1964.

([1627]) MATHIS C. — Comparative Biochemistry of Hydroxyquinones, in SWAIN T., *Comparative Phytochemistry*, p. 245-270, Academic Press, 1966.

([1628]) MATSUMURA F. et KNIGHT S. G. — Toxicity and chemosterilizing activity of aflatoxin against insects. *J. Econ. Entomol.*, t. **LX**, p. 871-872, 1967.

([1629]) MATSUURA S., MASARU M. et SATO T. — Surveillance for aflatoxins of rice and fermented-rice products in Japan. *in* Herzberg M., *Toxic Microorganisms*, p. 48-55, 1970.

([1630]) MATTA A. — *Ovularia viciae* and *Botrytis sp.* on Vetches in Italy. *F.A.O. Plant Prot. Bull.*, t. **VII**, p. 69-70, 1959.

([1631]) MATUSEVICH V. G., FEKILSTOV M. H. et ROZHDESTVENSKII V. A. — (An outbreak of stachybotryotoxicosis in cattle). *Veterinariya*, t. **XXXVII**, p. 71, 1960.

([1632]) MAUREL G. — Problèmes actuels de la chaîne du froid. Comité parlementaire français du commerce. *Nutrition 67*, journée d'étude, 27 avr. 1967.

([1633]) MAYER A. M., POLIJAKOFF-MAYBER A., ROBINSON P. et SLOWATIZKY I. — A simple bioassay for detection of aflatoxin in milk. *Toxicon*, t. **VII**, p. 13-14, 1969,

([1634]) MAYER C. F. — Endemic panmyelotoxicosis in the Russian grain belt. I. The clinical aspects of alimentary toxic aleukia (ATA). A comprehensive review. II. The botany, phytopathology and toxicology of Russian cereal feed. *Military Surgeon*, t. **CXIII**, p. 173-189 et 295-315, 1953.

([1635]) MAYNE R. Y., HARPER G. A., FRANZ A. O., LEE L. S. et GOLDBLATT L. A. — Retardation of the elaboration of aflatoxin in cottonseed by impermeability of the seedcoats. *Crop. Sci.*, t. **IX**, p. 147-150, 1969.

([1636]) MAYNE R. Y., PONS W. A., FRANZ A. O. et GOLDBLATT L. A. — Elaboration of aflatoxin on cottonseed products by *Aspergillus flavus*. *J. Amer. Oil Chemists' Soc.*, t. **XLIII**, p. 251-253, 1966.

([1637]) MAYO N. S. — Enzootic cerebritis or « staggers » of horses. *Kansas Agr. Exp. Stn., Bull.*, n° 24, 1891.

([1638]) MAYO N. S. — Sweet clover hay poisoning. *J. Am. Vet. Med. Assoc.*, t. **LXV**, p. 229, 1924.

([1639]) MAYURA K. et SREENIVASAMURTHY V. — Quantitative method for the estimation of aflatoxins in peanut products. *J. Ass. Off. Anal. Chem.*, t. **LII**, p. 77-81, 1969.

([1640]) MAZUR P. — Studies on the effect of subzero temperatures on the viability of spores of *Aspergillus flavus*. I. The effect of rate warming. *J. Gen. Physiol.*, t. **XXXIX**, p. 869-888, 1956.

([1641]) MEDICAL RESEARCH COUNCIL. — Patulin Clinical Trial Committee. *Lancet*, t. **II**, p. 373, 1944.

([1642]) MEISSNER H. et SCHOOP G. — Ueber den Pilzbefall amerikanischer « Gifterste ». *Deutsche Tierärztl. Wochenschr.*, t. **XXXVII**, p. 67, 1929.

([1643]) MELERA A. — The constitution of helvolic acid and cephalosporin P_1. *Experientia* t. **XIX**, p. 565-566, 1963.

([1644]) MENTZER C. — *Actualités de Phytochimie fondamentale.* 2e Série. 320 p., Masson édit., Paris, 1966.

([1645]) MENTZER C. et FATIANOFF O. — *Actualités de Phytochimie fondamentale.* 266 p., Masson édit., 1964.

([1646]) MENTZER C., FATIANOFF O, et DESCHAMPS-VALLET C. — *Actualités de Phytochimie fondamentale.* 3e série. 336 p., Masson édit., 1968.

([1647]) Menzel A. E. O., Wintersteiner O. et Hoogerheide J. C. — The isolation of gliotoxin and fumigacin from culture filtrates of *Aspergillus fumigatus*. *J. Biol. Chem.*, t. **CLII**, p. 419-429, 1944.

([1648]) Meronuck R. A. et Christensen C. M. — Toxicity to rats of *Alternaria sp.* isolated from seeds, flour and feeds. *Phytopathology*, t. **LX**, p. 1303, 1970.

([1649]) Meronuck R. A., Garren K. H., Christensen C. M., Nelson G. H. et Bates F. — Effect on turkey poults and chicks of rations containing corn invaded by *Penicillium* and *Fusarium* species. *Am. J. Vet. Res.*, t. **XXXI**, p. 551-555, 1970.

([1650]) Mertens W. K. et Van Veen A. G. — De bongkrek vergiftigingen in Banjolmas. *Geneskundig Tijdschrift Nederland Indie*, t. **LXXIII**, p. 1223, 1933.

([1651]) Meshustin E. H. et Gromyko E. P. — Prochnost sosdavayenick mikroorganizmami makroagregatov pochvi. *Mikrobiologiya*, t. **XV**, p. 169-175, 1946.

([1652]) Messiaen C. M. et Cassini R. — Recherches sur les fusarioses. IV. La systématique des *Fusarium*. *Ann. Epiphyties*, t. **XIX**, p. 387-454, 1968.

([1653]) Messiaen C. M. et Lafon R. — *Les maladies des plantes maraichèrs*. 2 vol. 331 p., INRA, 1963-1965. 2e ed., 441 p., INRA, 1971.

([1654]) Metzger H. J. — A case of tetany with hypomagnesaemia in a dairy cow. *Cornell Vet.*, t. **XXVI**, p. 353, 1936.

([1655]) Meyer H. — Bibliographie der Aflatoxine. *Bundesanst. f. Fleischforsch. Inst. Bakter. Histol. Kulmbach*, 188 p. ronéot., 1969.

([1656]) Meyer H. — Zur Kenntnis des Stoffwechsels von Stämmen der Gattung *Byssochlamys* unter besonderer Berücksichtigung der Bildung des Mycotoxins Byssochlaminsäure. *Inaug. Dissert. Univ. Münster*, 133 p. 1971.

([1657]) Meyer H. et Leistner L. — Dokumentation der Mykotoxin-Literatur. I. Mitteilung : Aflatoxine. *Arch. f. Lebensmittelhyg.*, t. **XX**, p. 203-206, 1969.

([1658]) Meyer H. et Leistner L. — Tendenzen der Mykotoxin-Forschung in Vergangenheit und Gegenwart. *Arch. f. Lebensmittelhyg.*, t. **XXI**, p. 178-184, 1970.

([1659]) Meyer J., Sartory R., Malgras J. et Touillier J. — (Animal and Plant Control tests for antibiotics). *Bull. Ass. Dipl. Microbiol. Fac. Pharm. Nancy*, t. **XLVI**, p. 30-42, 1952.

([1660]) Michaels L. et Schubert M. P. — The theory of reversible two step oxidation involving free radicals. *Chem. Rev.*, t. **XXII**, p. 437-440, 1938.

([1661]) Michener J. C. — Cerebrospinal meningitis-fungosus toxicum paralyticus. *Amer. Vet. Rev.*, t. **VI**, p. 345-347, 1882.

([1662]) Miehe H. — *Die Selbsterhitzung des Heus. Eine Biologische Studie*. 127 p., G. Fischer edit., Iena, 1907.

([1663]) Miescher G. — Ueber die Wirkungsweise von Patulin auf höhere Pflanzen insbesondere auf *Solanum lycopersicum* L. *Phytopathol. Z.*, t. **XVI**, p. 369-397, 1950.

([1664]) Mietkiewski K., Janicki J., Malendowicz L., Urbanowicz M. et Filipiak B. — Histochemical studies on the action of aflatoxins. I. A cytological and cytochemical study of the liver of rats fed a diet contaminated with *Aspergillus flavus*. *Folio Histochem. Cytochem.*, t. **VII**, p. 379-405, 1969.

([1665]) Miller D. D. et Golding N. S. — The gas requirements of molds. V. The minimum oxygen requirements for normal growth and for germination of six mold cultures. *J. Dairy Sci.*, t. **XXXII**, p. 101-110, 1949.

([1666]) Miller J. A. — Tumorigenic and carcinogenic natural products. *in* Toxicants occurring naturally in foods, National Acad. Sci. Nat. Res. Counc. Washington. *Nat. Acad. Sci. Nat. Res. Counc. Publ.*, n° 1354, p. 24-39, 1967.

([1667]) Miller J. H., Giddens J. E. et Foster A. A. — A survey of the fungi of forest and cultivated soils of Georgia. *Mycologia*, t. **XLIX**, p. 779-808, 1957.

([1668]) Miller M. W. — *Pfizer Handbook of Microbial Metabolites*. 772 p., McGraw-Hill, New York, 1961.

(1669) MILLER P. A., TROWN P. W., FULMOR W., MORTON G. O. et KARLINER J. — An epidithiapiperazinedione antiviral agent from *Aspergillus terreus. Biochem. Biophys. Res. Commun.*, t. **XXXIII**, p. 219-221, 1968.

(1670) MILNE G., BIOLLAZ M. et BÜCHI G. — Biosynthesis of aflatoxins. *in* Herzberg M., *Toxic Micro-organisms*, p. 121-126, 1970.

(1671) MILNER M. et GEDDES W. F. — Grain storage studies III. The relation between moisture content, mold growth and respiration of soybeans. *Cereal Chem.*, t. **XXIII**, p. 225-247, 1946.

(1672) MILNER M. et GEDDES W. F. — Grain storage studies. IV. Biological and chemical factors involved in the spontaneous heating of soybeans. *Cereal Chem.*, t. **XXIII**, p. 449-470, 1946.

(1673) MINNE A. J., ADELAAR T. F., TERBLANCHE M. et SMITH J. D. — *Vet. Bull.*, t. **XXXIV**, p. 516, 1946.

(1674) MINTON N. A. et JACKSON C. R. — Invasion of peanut pods by *Aspergillus flavus* and other fungi in the presence of root knot nematodes. *Oléagineux*, t. **XXII**, p. 543-546, 1967.

(1675) MINTZLAFF H. J. — Die japanische Zwergwachtel (*Coturnix coturnix japonica*) als Versuchstier in der Mykotoxinforschung. *Fleischwirtschaft*, t. **LI**, p. 344-346, 1971.

(1676) MINTZLAFF H. J. et CHRIST W. — Biologischer Mykotoxin-Nachweis mit Hühnerküken unter besonderer Berücksichtigung von Leberschäden. *Fleischwirtschaft*, t. **LI**, p. 1802-1805, 1971.

(1677) MIRCHING T. G., BLAGOVESHCHENSKII V. S. et FEDOROV V. A. — Identifikatsiya toksina, obrazuemogo *Penicillium citrinum* n° 19 i izuchenie ego gerbitsidnykl svoistv. *Mikrobiologiya*, t. **XXXVI**, p. 1036-1040, 1967.

(1678) MIROCHA C. J. — *Recherches sur les mycotoxines effectuées a l'Université du Minnesota.* Conférence Paris, 1er juil. 1970.

(1679) MIROCHA C. J. — The estrogenic metabolites of *Fusarium. 1st Int. Mycol. Congr. Exeter. Abstr.* p. 66, 1971.

(1680) MIROCHA C. J., CHRISTENSEN C. M. et NELSON G. H. — Estrogenic metabolites produced by *Fusarium graminearum* in stored corn. *Appl. Microbiol.*, t. **XV**, p. 497-503, 1967.

(1681) MIROCHA C. J., CHRISTENSEN C. M. et NELSON G. H. — An estrogenic metabolite produced by *Fusarium graminearum* in stored corn. Biochemistry of some foodborne microbial toxins. *Symp. Amer. Chem. Soc. New York. sept.* 1966, p. 119-130, M.I.T. Press, 1967.

(1682) MIROCHA C. J., CHRISTENSEN C. M. et NELSON G. H. — Physiological activity of some fungal estrogens produced by *Fusarium. Cancer Res.*, t. **XXVIII**, p. 2319-2322, 1968.

(1683) MIROCHA C. J., CHRISTENSEN C. M. et NELSON G. M. — Toksyczne produkty przemiany materii grzybow wywolujacych mikotoksykozy. *Med. Wet.*, t. **XXIV**, p. 129-134, 1968.

(1684) MIROCHA C. J., CHRISTENSEN C. M. et NELSON G. H. — Toxic metabolites produced by fungi implicated in mycotoxicosis. *Biotechn. and Bioeng.*, t. **X**, p. 469-482, 1968.

(1685) MIROCHA C. J., CHRISTENSEN C. M. et NELSON G. H. — F-2 (Zearalenone) estrogenic mycotoxin from *Fusarium. in* Kadis S. et al. , *Microbial Toxins*, t. **VII**, p. 107-138, 1971.

(1686) MIRONOV S. G., JOFFE A. Z., BAKBARDINA M. K., FOK R. A. et DAVIDOWA V. L. — Phytopathological analysis of toxic samples of overwintered millet, in alimentary toxic aleukia. *Orenburg Inst. Epidemiol. Microbiol.*, t. **II**, p. 11-18, 1947.

(1687) MISLIVEC P. B., HUNTER J. H. et TUITE J. — Assay for aflataxin production by the genera *Aspergillus* and *Penicillium. Appl. Microbiol.*, t. **XVI**, p. 1053-1055, 1968.

(1688) MISLIVEC P. B. et TUITE J. — Species of *Penicillium* occurring in freshly-harvested and in stored dent corn kernels. *Mycologia*, t. **LXII**, p. 67-74, 1970.

(1689) MISLIVEC P. B. et TUITE J. — Temperature and relative humidity requirements of species of *Penicillium* isolated from yellow dent corn kernels. *Mycologia*, t. **LXII**, p. 75-88, 1970.

[1690] MITCHELL D. T. — A condition produced in cattle feeding on maize infected with *Diplodia zeae*. 7th and 8th Reports of the Director of Veterinary Research. U. of S. A., p. 425-437, 1918.

[1691] MITCHELL H. H. et BEADLES J. R. — The impairment in nutritive value of corn grain damaged by specific fungi. *J. Agr. Res.*, t. **LXI**, p. 135-141, 1950.

[1692] MITCHELL K. J., WALSHE T. O. et ROBERTSON N. G. — Weather conditions associated with outbreaks of facial eczema. *New Zealand J. Agr. Research*, t. **II**, p. 584-604, 1959.

[1693] MITROIU P., JIVOIN P. et ZAMFIR I. — Recherches expérimentales sur la toxicité des fourrages moisis. *C.R. 5e Congr. ISHAM*, p. 220-221, Paris, juil. 1971.

[1694] MIYAKE I. — *Penicillium toxicarium* growing on the yellow mouldy rice. *Niishin Igaku*, t. **XXXIV**, p. 161, 1947.

[1695] MIYAKE I. — Studien über die Pilze der Reispflanze in Japan. *J. Coll. Agr. Tokyo*, t. **XI**, p. 231-276, 1960.

[1696] MIYAKE I., NAITO H. et TSUNODA H. — Study on toxin production of stored rice by parasitic fungi. *Rice Utilization Res. Inst., Dept of Agric. and Forestry*, report n° 1, p. 1, 1940.

[1697] MIYAKE M. et SAITO M. — Liver injury and liver tumors induced by toxins of *Penicillium islandicum* Sopp growing on yellowed rice. *Mycotoxins in Foodstuffs*, p. 133-146, M.I.T. Press, 1965.

[1698] MIYAKE M., SAITO M., ENOMOTO M., ISHIKO T., URAGUCHI K., SAKAI F., TATSUNO T., TSUKIOKA M. et SAKAI Y. — Toxic liver injuries and liver cirrhosis induced in mice and rats through long-term feeding with *Penicillium islandicum* Sopp growing rice. *Acta Pathol. Jap.*, t. X, p. 75-123, 1960.

[1699] MIYAKE M., SAITO M., ENOMOTO M., SHIKATA T., ISHIKO T., URAGUCHI K., SAKAI F., TATSUNO T., TSUKIOKA M. et NOGUCHI Y. — Development of primary hepatic carcinoma in rats by long-term feeding with the yellowed rice by *Penicillium islandicum* Sopp-with study on influence of fungus growing rice on DAB carcinogenesis in rats. *Gann, Proc. Japan Cancer Assoc., 18th General Meeting, nov. 1959*, p. 117-118, 1960.

[1700] MIYAKE M., SAITO M., ENOMOTO M., URAGUCHI K., TSUKIOKA M., IKEDA Y. et OMORI Y. — Histological studies on the liver injury due to the toxic substances of *Penicillium islandicum* Sopp. *Acta Pathol. Jap.*, t. V, p. 208, 1955.

[1701] MIYAKI K. — Perspectives on the mycotoxin problems in Japan. *in* Herzberg M., *Toxic Micro-organisms*, p. 3-7, 1970.

[1702] MIYAKI K., AIBARA K. et MIURA T. — Resistance of aflatoxin to chemical and biological changes by gamma indication. *Microbiological problems in food preservation by irradiation, Vienne, I.A.E.A.*, p. 57-64, 1967.

[1703] MIYAKI K. et YAMAZAKI M. — Aflatoxins. *Ann. Rep. Inst. Food Microbiol. Chiba Univ.*, t. **XX**, p. 1-26, 1967.

[1704] MIYAKI K., YAMAZAKI M., HORIE Y. et UDAGAWA S. — Studies on the toxigenic fungi found in rice of Chiba prefecture. II. On the microflora of rice. *Ann. Rep. Inst. Food Microbiol. Chiba Univ.*, t. **XXII**, p. 41-45, 1969.

[1705] MIYAKI K., YAMAZAKI M., SUZUKI S., MAEBAYASHI Y. et SAKAKIBARA Y. — Studies on the toxigenic determination of *Aspergillus* metabolites. *Ann. Rep. Inst. Food Microbiol. Chiba Univ.*, t. **XXII**, p. 47-50, 1969.

[1706] MOHAMMED Y. S. et LUCKNER M. — Structure of cyclopenin and cyclopenol, metabolic products from *Penicillium cyclopium* and *Penicillium viridicatum*. *Tetrah. Lett.*, n° 28, p. 1953-1958, 1963.

[1707] MOHANTY G. P., CARLSON C. W., VOELKER M. M. et JORGENSEN N. A. — Extraction of possible toxic substances from molded alfalfa hay. *Proc. S.D. Acad. Sci.*, t. **XLVII**, p. 97-102, 1968.

[1708] MOHLER J. R. — Cerebrospinal meningitis (« forage poisoning »). *U.S.D.A. Dept. Bull.*, n° 65, 1914.

[1709] (MOHSEN EL-KHADEM. — Die Bedeutung von Aflatoxinen für die durch *Aspergillus flavus* verursachte Keimlingskrankheit der Erdnuss. *Phytopathol.* Z., **LI** t. p. 218-231, 1968.

([1710]) MOHSEN EL-KHADEM. — Einfluss von Fungiziden und Insektiziden auf den *Aspergillus*-Befall an Erdnusskeimlingen. I. Untersuchungen in vitro. *Z. f. Pflzkrkh. Pflsch.*, t. **LXXV**, p. 86-93, 1968. II. Beizversuche. *Id.*, t. **LXXV**, p. 151-160, 1968.

([1711]) MOHSEN EL-KHADEM, MENKE G. et GROSSMANN F. — Schädigung von Erdnusskeimlingen durch Aflatoxine. *Naturwissensch.*, t. **LIII**, p. 532-533, 1966.

([1712]) MOLINA C. — Les pneumopathies à précipitines en dehors de la maladie du poumon de fermier. *IXes journées nationales, Sté Fr. Allergologie, Clermont-Ferrand*, p. 43-53, 1970.

([1713]) MONJOUR L., GIORGI R., TOURY J. et QUENUM C. — Sur trois cas d'atrophie testiculaire dans un élevage de singes ingérant un contaminant toxique. *C. R. Soc. Biol.*, t. **CLXII**, p. 1454-1455, 1968.

([1714]) MONJOUR L. et OUDART J. L. — Aflatoxine et hormones. *Cah. Nutr. Diét.*, t. **V**, p. 71-72, 1970.

([1715]) MONNERON A., LAFARGE C. et FRAYSSINET C. — Induction par l'aflatoxine et la lasiocarpine d'amas de grains périchromatiniens dans le foie du rat. *C. R. Acad. Sci.*, D, t. **CCLXVII**, p. 2053-2056, 1968.

([1716]) MOODY D. P. — Biogenetic hypotheses of aflatoxin. *Nature*, t. **CCII**, p. 188, 1964.

([1717]) MOODY E. M. et MOODY D. P. — *Nature*, t. **CXCIII**, p. 294-295, 1963.

([1718]) MOORE J. H. et TRUELOVE B. — Ochratoxin A : inhibition of mitochondrial respiration. *Science*, t. **CLXVIII**, p. 1102-1103, 1970.

([1719]) MOORE V. A. — Cornstalk disease (Toxaemia maidis) in cattle. *U.S.D.A. Bur. Anim. Ind., Bull.*, n° 10, 1896.

([1720]) MOREAU C., BOURON H., KERN L., FABRE P. et PELISSIER P. — *La réglementation francaise des pesticides agricoles.* 251 p., A.C.T.A. édit., 1969.

([1721]) MOREAU C. — A propos d'*Ustilago esculenta* sur *Zizania* et de quelques autres parasites utiles pour l'alimentation. *Rev. Int. Bot. Appl. Agric. Trop.*, t. **XXX**, fasc. 327-328, p. 60-62, 1950.

([1722]) MOREAU C. — Nouvelles mycocécidies comestibles. *Rev. Int. Bot. Appl. Agric. Trop.*, t. **XXX**, fasc. 329-330, p. 222-223, 1950.

([1723]) MOREAU C. — Pollution de l'atmosphère d'entrepôts de fruits et désinfection par brouillard fongicide. *La Mycothèque*, t. **VI**, p. 36-40, 1953.

([1724]) MOREAU C. — Pollution fongique de l'atmosphère et altérations de denrées alimentaires. *Froid Informations*, fasc. 54, art. 268, p. 7-9, 1961 ; *Rev. embouteillage*, t. **XII**, p. 43-44, 1961 ; *Maroc-Fruits*, n° 71, p. 1, 6 ; n° 75, p. 4, 6, 1961.

([1725]) MOREAU C. — Causes, conditions de développement et traitement des pourritures des agrumes. *Rev. Gén. Froid*, t. **XXXIX**. fasc. 4, p. 425-434, 1962.

([1726]) MOREAU C. — Moisissures toxiques dans l'alimentation (1re éd.), 372 p., *Encyclopédie Mycologique*, t. **XXXV**, Lechevalier édit., Paris, 1968.

([1727]) MOREAU C. — Les moisissures des farines panifiables. *Ann. Nutr. Alim.*, t. **XXIV**, p. 117-127, 1970.

([1728]) MOREAU C. — Mycotoxicose chez des vaches laitières liée au développement du *Fusarium roseum* var. *culmorum* sur l'herbe d'une prairie. *C.R. Acad. Agric. Fr.*, t. **LVIII**, p. 383-387, 1972.

([1729]) MOREAU C. — Les moisissures des tourteaux de soja, leurs dangers. *Oléagineux*, t. **XXVII**, p. 321-324, 1972.

([1730]) MOREAU C. — L'*Absidia corymbifera* (Cohn) Sacc. et Trott., cause possible d'accidents chez les poules pondeuses. *Bull. Soc. Mycol. Fr.*, t. **LXXXIX**, p. 73-78, 1972.

([1731]) MOREAU C. — Contrôle mycologique d'une station de surgélation de plats cuisinés. *Rev. gén. Froid*, t. **LXIII**, p. p. 691-696, 1972.

([1732]) MOREAU C. — (sous presse).

([1733]) MOREAU C. et MOREAU M. — Sur quelques Hyphomycètes. *Bull. Soc. Linn. Norm.*, 9e sér., t. **VI**, p. 71-82, 1951.

([1734]) Moreau C. et Moreau M. — La mycoflore fimicole. *La Mycothèque*, 1er Suppl. : Micromycètes, p. 46-49, 1951.

([1735]) Moreau C. et Moreau M. — La « suie » des Sycomores à Paris. *Bull. Soc. Mycol. Fr.*, t. **LXVII**, p. 404-418, 1951.

([1736]) Moreau C. et Moreau M. — Alliances et antagonismes entre Champignons, leur intérêt pour la compréhension de certains problèmes phytopathologiques. *Bull. Soc. Mycol. Fr.*, t. **LXXII**, p. 250-253, 1956.

([1737]) Moreau C. et Moreau M. — Pollution fongique de l'atmosphère. Sa responsabilité dans les altérations de quelques denrées alimentaires. *Bull. Soc. Mycol. Fr.*, t. **LXXV**, fasc. 1, p. 72-79, 1959.

([1738]) Moreau C. et Moreau M. — Un danger pour le bétail nourri de plantules fourragères cultivées en germoirs : la pullulation d'une moisissure toxique, l'*Aspergillus clavatus*, cause des accidents mortels. *C.R. Acad. Agric. Fr.*, t. **XLVI**, fasc. 7, p. 441-445, 1960.

([1739]) Moreau C. et Moreau M. — Moisissures des plantules fourragères cultivées en germoirs et premières observations sur la pollution fongique d'une ferme d'élevage. *C.R. Acad. Agric. Fr.*, t. **XLVII**, fasc. 11, p. 554-557, 1961.

([1740]) Moreau C. et Moreau M. — Quelques moisissures toxiques des grains en stockage. *C.R. Acad. Agric. Fr.*, t. **XLVII**, p. 873-874, 1961.

([1741]) Moreau C. et Moreau M. — Observations sur l'origine des moisissures du pain de mie de fabrication industrielle. *C.R. Acad. Agric. Fr.*, t. **LXIX**, fasc. 11, p. 865-868, 1963.

([1742]) Moreau C., Moreau M. et Pelhate J. — Choix de milieux de culture sélectifs pour l'analyse des mycoflores osmophiles. *Bull. Soc. Mycol. Fr.*, t. **LXXX**, p. 234-246, 1964.

([1743]) Moreau C., Moreau M. et Pelhate J. — Comportement cultural de moisissures du blé en relation avec leur écologie sur grains. *C.R. Acad. Sci.*, t. **CCLV**, p. 1229-1231, 1965.

([1744]) Moreau F. et Moreau V. — Première contribution à l'étude de la microflore des dunes. *Rev. de Mycol.*, t. **VI**, p. 49-94, 1941.

([1745]) Moreau M. — Les Neurospora. *Bull. Soc. Bot. Fr.*, t. **CIII**, p. 678-738, 1956.

([1746]) Moreau M. — Etudes sur la moisissure verte des Prunes d'Ente. L'*Aspergillus Mangini* : ses exigences nutritives, ses conditions de développement. *Fruits*, t. **XIV**, p. 315-328, 1959.

([1747]) Moreau M. et Moreau C. — Recherches sur la sporulation de l'*Aspergillus clavatus* Desm. *C.R. Acad. Sci.*, t. **CCLI**, p. 1556-1557, 1960.

([1748]) Moreau M. et Pelhate J. — Influence de la température sur la croissance linéaire de quelques moisissures des grains. *Bull. Soc. Mycol. Fr.*, t. **LXXXII**, p. 467-472, 1966.

([1749]) Moreau M., Péresse M. et Pelhate J. — Etude sur la sporulation de quelques moisissures des grains. *Rev. de Mycol.*, t. **XXXI**, p, 225-232, 1966.

([1750]) Moreau M. et Trique B. — Recherche des exoenzymes dans les filtrats de culture de deux moisissures des grains : l'*Aspergillus versicolor* (Vuill.) Tiraboschi et le *Penicillium cyclopium* Westl. *C.R. Acad. Sci.*, D, t. **CCLXIII**, p. 239-241, 1966.

([1751]) Moreno-Martinez E. et Christensen C. M. — Differences among lines and varieties of maize in susceptibility to damage by storage fungi. *Phytopathology*, t. **XLI**, p. 1498-1500, 1971.

([1752]) Moro E. — Ueber den *Bacillus acidophilus* n. sp. Ein Beitrag zur Kenntnis der normalen Darmbakterien des Säuglings. *Jahrb f. Kinderheilkunde*, t. **II**, p. 38-55, 1900.

([1753]) Morooka N., Nakano N., Nakazawa S. et Tsunoda H. — On the chemical properties of fusarenon and the related compound obtained from toxic metabolites of *Fusarium nivale*. *J. Agric. Chem. Soc. Jap.*, t. **XLV**, p. 151-155, 1971.

([1754]) Morooka N., Nakano N. et Uchida N. — Biochemical and histopathological studies on the tumor-developing livers of mice fed on the diets containing luteoskyrin. *Jap. J. Med. Sci. Biol.*, t. **XIX**, p. 293-303, 1966.

([1755]) Morooka N. et Tatsuno T. — Toxic substances (fusarenon and nivalenol) produced by *Fusarium niveale. in* Herzberg M., *Toxic Micro-organisms*, p. 114-119, 1970.

([1756]) Morquer R., Lombard C., Berthelon M. et Lacoste L. — Pouvoir pathogène des Mucorales dans le règne animal. Une nouvelle mycose chez les Bovidés et les Porcins. *C.R. Acad. Sci.*, t. **CCLX**, p. 6173-6176, 1965.

([1757]) Morquer R., Lombard C., Berthelon M. et Lacoste L. — Pathogénie de quelques Mucorales pour les animaux. Une nouvelle mucormycose chez les Bovidés. *Bull. Soc. Mycol. Fr.*, t. **LXXXI**, p. 421-449, 1965.

([1758]) Morquer R., Redon P. et Roquebert-Hubert F. — Action de quelques champignons toxiques des fourrages sur le métabolisme des animaux. *Bull. Soc. Mycol. Fr.*, t. **LXXXVII**, p. 101-120, 1971.

([1759]) Mortimer P. H. — The experimental intoxication of sheep with sporidesmin, a metabolic product of *Pithomyces chartarum*. III. *Res. Vet. Sci.*, t. **III**, p. 269, 1962.

([1760]) Mortimer P. H. — The experimental intoxication of sheep with sporidesmin, a metabolic product of *Pithomyces chartarum*. IV. *Res Vet. Sci.*, t. **IV**, p. 166-185, 1963.

([1761]) Mortimer P. H. et Collins B. S. — The in vitro toxicity of the sporidesmins and related compounds to tissue-culture cells. *Res. Vet. Sci.*, t. **IX**, p. 136-142, 1968.

([1762]) Mortimer P. H. et Stanbridge T. A. — Changes in biliary secretion following sporidesmin poisoning in sheeps. *J. Comp. Pathol.*, t. **LXXXIX**, p. 267-275, 1969.

([1763]) Mortimer P. H. et Taylor A. — The experimental intoxication of sheep with sporidesmin, a metabolic product of *Pithomyces chartarum*. I. Clinical observations and findings at *post-mortem* examinations. *Research Vet. Sc.*, t. **III**, p. 147-160, 1962.

([1764]) Morton R. A. et Earlam W. T. — Absorption spectra in relation to quinones-1,4-naphthoquinone, anthraquinone and their derivatives. *J. Chem. Soc.*, p. 159, 1941.

([1765]) Moruzi C., Moldoveanu O., Balliu S. et Boteanu S. — Studii aspura activitatii antibiotice a unor tulpini de *Aspergillus clavatus* Desmazières. *Comunle Bot.*, t. **IX**, p. 67-76, 1969.

([1766]) Moseliani D. V. — Stakhibotriotoksikoz loshadei. *Veterinariya*; t. **XVII**, p. 42-44, 1940.

([1767]) Moss M. O. — Mycotoxins. *Int. Biodet. Bull.*, t. V, p. 141-150, 1969.

([1768]) Moss M. O. — The rubratoxins, toxic metabolites of *Penicillium rubrum* Stoll. *in* Ciegler A. et al., *Micrcbial Toxins*, t. **VI**, p. 381-407, 1971.

([1769]) Moss M. O. et Hill I. W. — Strain variation in the production of rubratoxins by *Penicillium rubrum* Stoll. *Mycopathol. Mycol. Appl.*, t. **XL**, p. 81-88, 1970.

([1770]) Moss M. O., Robinson F. V. et Wood A. B. — Rubratoxin B, a toxic metabolite of *Penicillium rubrum. Chem. Ind*,. p. 587-588, 1968.

([1771]) Moss M. O., Robinson F. V., Wood A. B. et Morrison A. — Observations on the structure of the toxins from *Penicillium rubrum. Chem. Ind.*, p. 755-757, 1967.

([1772]) Moss M. O., Robinson F. V., Wood A. B., Paisley H. M. et Feeney J. — Rubratoxin B, a proposed structure for a bis-anhydride from *Penicillium rubrum* Stoll. *Nature, G.B.*, t. **CCXX**, p. 767-770, 1968.

([1773]) Moss M. O., Wood A. B. et Robinson F. V. — The structure of rubratoxin A, a toxic metabolite of *Penicillium rubrum. Tetrahedron Lett.*, n° 5, p. 367-370, 1969.

([1774]) Moubasher A. H., Elnaghy M. A. et Abdel Hafez S. I. — Studies on the fungus flora of three grains in Egypt. *Mycopathol. Mycol. Appl.*, t. **XLVII**, p. 261-274, 1927.

([1775]) Moule G. R., Braden A. W. H. et Lamond D. R. — The significance of oestrogens in pasture plants in relation to animal production. *Anim. Breeding Abstr.*, t. **XXXI**, p. 139-157, 1965.

([1776]) Moule Y. et Frayssinet C. — Effect of aflatoxin on transcription in liver cell. *Nature, G.B.*, t. **CCXVIII**, p. 93-95, 1968.

([1777]) Moule Y. et Frayssinet C. — Action de deux hépatocarcinogènes sur l'activité RNA-polymérase des noyaux isolés du foie. *Bull. Soc. Chim. Biol.*, t. **LI**, p. 1544, 1969.

(1778) MÜCKE W. et KIERMEIER F. — Sprühreagens zum Nachweis von Aflatoxinen. *Z. Lebensm.-Untersuch.-Forsch.*, t. **CXLVI**, p. 329-331, 1971.

(1779) MULINGE S. K. et APINIS A. E. — Occurence of thermophilous fungi in stored moist barley grain. *Trans. Brit. Mycol. Soc.*, t. **LIII**, p. 361-370, 1969.

(1780) MULINGE S. K. et CHESTERS C. G. C. — Ecology of fungi associated with moist stored barley grain. *Ann. Appl. Biol.*, t. **LXV**, p. 277-284, 1970.

(1781) MULINGE S. K. et CHESTERS C. G. C. — Methods of isolating the microflora of moulding high moisture barley in partially sealed silos. *Ann. Appl. Biol.*, t. **LXV**, p. 285-292, 1970.

(1782) MULL R. P., TOWNLEY R. W. et SCHOLZ C. R. — Production of gliotoxin and a second active isolate by *Penicillium obscurum* Biourge. *Amer. Chem. Soc. Jour.*, t. **LXVII**, p. 1626-1627, 1945.

(1783) MÜLLER G. — Toxinbildende Schimmelpize in Lebensmitteln pflanzlicher Herkunft. *Lebensmittel Ind.*, t. **XVIII**, p. 289-294, 1971.

(1784) MULLER K. O. — Einige einfache Versuche zum Nachweis von Phytoalexinen. *Phytopathol. Z.*, t. **XXVII**, p. 237-254, 1956.

(1785) MULLER R. D., CARLSON C. W., SEMENIUK G. et HARSHFIELD G. S. — The response of chicks, ducklings, goslings, pheasants and poults to graded levels of aflatoxins. *Poult Sci.*, t. **XLIX**, p. 1346-1350, 1970.

(1786) MUNDKUR B. B. — Some preliminary feeding experiments with scabby barley. *Phytopathology*, t. **XXIV**, p. 1237, 1934.

(1787) MUNDKUR B. B. et COCHRAN R. L. — Some feeding tests with scabby barley. *Phytopathology*, t. **XX**, p. 132, 1930.

(1788) MURAKAMI H., OUWAKI K. et TAKASE S. — An aflatoxin strain, ATCC 15517. *J. Gen. Appl. Microbiol.*, t. **XII**, p. 195-206, 1966.

(1789) MURAKAMI M., SAGAWA H. et TAKASE S. — Non-productivity of aflatoxin by Japanese industrial strains of the *Aspergillus*. III. Common characteristics of the aflatoxin-producing strains. *J. Gen. Appl. Microbiol.*, t. **XIV**, p. 251-262, 1968.

(1790) MURAKAMI H. et SUZUKI M. — Mycological differences between the producer and non-producer of aflatoxin of *Aspergillus*. *in* HERZBERG M., *Toxic Micro-organisms*, p. 198-201, 1970.

(1791) MURAKAMI H., TAKASE S. et ISHI T. — Non productivity of aflatoxin by Japanese industrial strains of *Aspergillus*. I. Production of fluorescent substances in agar slant and shaking cultures. *J. Gen. Appl. Microbiol.*, t. **XIII**, p. 323-334, 1967.

(1792) MURAKAMI H., TAKASE S. et KUWABARA K. — Non productivity of aflatoxin by Japanese industrial strains of *Aspergillus*. II. Production of fluorescent substances in rice koji, and their identification by absorption spectrum. *J. Gen. Appl. Microbiol.*, t. **XIV**, p. 97-110, 1968.

(1793) MURASHKINSKII N. E. — Materialy po izucheniiu fuzarioza khlebov. I. Vidy roda *Fusarium* na khlebakh v Sibiri. *Trudy Sibir. Selskokhoz. Akad.*, 1934.

(1794) MUTH O. H. et HAAG J. R. — Disease of Oregon Cattle associated with hypomagnesemia and hypocalcemia. *N. Am. Vet.*, t. **XXVI**, p. 216, 1945.

(1795) MUTHU M., SRINIVASAN K. S. et MAJUMDER S. K. — Serial fumigation of processed and packaged foods with airwashed residual methyl bromide from fumigation chambers. *Food Technol.*, t. **XV**, p. 295, 1961.

(1796) NABNEY J., BURBAGE M. B., ALLCROFT R. et LEWIS C. — Metabolism of aflatoxin in sheep; excretion pattern in the lactating ewe. *Indian J. Chem.*, t. **IV**, p. 11-17, 1966.

(1797) NABNEY J. et NESBITT B. F. — Determination of the aflatoxins. *Nature, G.B.*, t. **CCIII**, p. 862, 1964.

(1798) NABNEY J. et NESBITT B. F. — A spectrophotometric method for determining the aflatoxins. *Analyst*, t. **XC**, p. 155-160, 1965.

([1799]) Nagai J., Hayashi M. et Mizobe K. — Citrinin. *Fukuoka Igaku Zassi*, t. **XLVIII**, p. 311-319, 1957.

([1800]) Nagarsekar C. L. — Aflatoxins. *Bombay Technol.*, t. **XVII**, p. 70-75, 1967.

([1801]) Nagel D. W., Steyn P. S. et Scott D. B. — Citreoviridin, a toxic metabolite of *Penicillium* species. *C.R. 5e Congr. ISHAM*, p. 216-217, Paris, 1971.

([1802]) Nagornyi I. S., Possoginskii M. M., Govzdov A. V. et Rybka N. V. — (Stachybotryotoxicosis in cattle). *Veterinariya*, t. **XXXVII**, p. 69, 1960.

([1803]) Nagy R. — Control of fungi in bakery plants. *Baker's Digest*, t. **XXII**, p. 47-48, 1948.

([1804]) Naim M. S. et El-Esawy A. A. — Antagonism between *Rhizoctonia solani* causing damping-off and selected rhizosphere microflora of some Egyptian cotton varieties. *Mycopathol. Appl.*, t. **XXVII**, p. 169-174, 1965.

([1805]) Naito H. — Studies on mold microflora in stored rice grain, with special reference to osmotic pressure. *Ann. Phytopathol. Soc. Japan.* t. **XVIII**, p. 41-45, 1953.

([1806]) Naji A. F. — Bronchopulmonary aspergillosis. Report of two new cases; review of literature and suggestion for classification. *Arch. Pathol.*, t. **LXVIII**, p. 282-291, 1959.

([1807]) Narasimhan M. J. — On the presence of aflatoxins in copra. *Hindustan Antibiot. Bull.*, t. **XI**, p. 104-105, 1968.

([1808]) Narasimhan M. J. — Sterochemical pharmacology of the « estrogenic » syndrome caused by the mycotoxin zearalenone. *Hindustan Antibiot. Bull.*, t. **XVI**, p. 1-3, 1971.

([1809]) Nartey F. — Aflatoxins of *Aspergillus flavus* grown on cassava. *Physiologia Plantarum*, t. **XIX**, p. 818-822, 1966.

([1810]) Natarajan C. P. et al. — Studies on the storage of coffee beans. *Food Sci. (Mysore)*, t. X, p. 315-321, 1961.

([1811]) Natori S., Sakaki S., Kurata H., Udagawa S., Ichinoe M., Saito M. et Umeda M. — Chemical and cytotoxicity survey on the production of ochratoxins and penicillic acid by *Aspergillus ochraceus* Wilhelm. *Chem. Pharm. Bull.*, t. **XVIII**, p. 2259-2268, 1970.

([1812]) Natori S., Sakaki S., Kurata H., Udagawa S., Ichinoe M., Saito M., Umeda M. et Ohtsubo K. — Production of rubratoxin B by *Penicillium purpurogenum* Stoll. *Appl. Microbiol.*, t. **XIX**, p. 613-617, 1970.

([1813]) Naumann K. — Mineralstoffversorgung von Boden-Pflanze-Tier. *Landwirtsch. Z. der Nord. Rheinprovinz*, 28 avr. 1962.

([1814]) Naumann K. et Barth K. — Chemische Untersuchungen in Weidetetaniegebiet des Niederrheins. *Landw. Forsch.*, t. **XII**, p. 186-195, 1959.

([1815]) Nauta W. T., Oösterhuis H. K., Van Der Linden A. C., Van Duyn P. et Dienske J. W. — On the structure of expansine, a bactericidal and fungicidal substance in *Penicillium expansum*. *Rec. Trav. Chim.*, t. **LXV**, p. 865-976, 1946.

([1816]) Navellier E. — Deux techniques modernes de préparation des aliments : surgélation et lyophilisation. *Fruits*, t. **XXIII**, p. 183-184, 1968.

([1817]) Neelakantan S., Pocker A. et Raistrick H. — Studies in the biochemistry of microorganisms. 101. *Biochem. J.*, t. **LXVI**, p. 234-237, 1957.

([1818]) Neill J. C. — The endophyte of Rye-grass (*Lolium perenne*). *New Zealand Journ. Sci, Tech., A*, t. **XXI**, p. 280, 1940.

([1819]) Neill J. C. — The endophyte of *Lolium* and *Festuca*. *New Zealand Journ. Sci. Tech., A*, t. **XXIII**, p. 185, 1941.

([1820]) Neill J. C. et Hyde E. O. C. — Blind-seed disease of rye-grass. *New Zealand Journ. Sci. Tech., A*, t. **XX**, p. 281-301, 1939 ; t. **XXIV**, p. 65-71, 1942.

([1821]) Nel W., Kempff P. G. et Pitout M. J., The metabolism and some metabolic effects of sterigmatocystin. *in* Purchase I. F. H., *Mycotoxins in human health*, p. 11-18, 1971.

([1822]) Nelson G. H., Christensen C. M. et Mirocha C. J. — Feeds, fungi and animal health. *Minn. Sci.*, t. **XXIII**, p. 12-13, 1966.

([1823]) NELSON R. R., MIROCHA C. J., HUISINGH D. et TIJERINA-MENCHACA A. — Effects of F-2, an estrogenic metabolite from *Fusarium*, on sexual reproduction of certain Ascomycetes. *Phytopathology*, t. **LVIII**, p. 1061-1062, 1968.

([1824]) NEMETH I. et JUHASZ S. — Effect of aflatoxin on serum protein fractions of day old ducklings. *Acta Vet. Acad. Sci. Hung.*, t. **XVIII**, p. 95-105, 1968.

([1825]) NESBITT B., O'KELLY J., SARGEANT K. et SHERIDAN A. — Toxic metabolites of *Aspergillus flavus. Nature*, t. **CXCV**, p. 1062-1063, 1962.

([1826]) NESHEIM S. — Mycotoxins : studies of the rapid procedure for aflatoxins in peanuts, peanut meal and peanut butter. *J. Ass. Off. Agric. Chemists*, t. **XLVII**, p. 1010-1017, 1964.

([1827]) NESHEIM S. — Note on aflatoxin analysis in peanuts and peanut products. *J. Ass. Off. Agr. Chem.*, t. **XLVII**, p. 586, 1964.

([1828]) NESHEIM S. — Note on ochratoxins (*Aspergillus ochraceus*). *Ass. Off. Anal. Chem. J.*, t. **L**, p. 370-371, 1967.

([1829]) NESHEIM S. — Conditions and techniques for thin layer chromatography of aflatoxins. *J. Amer. Oil Chem. Soc.*, t. **XLVI**, p. 335-338, 1969.

([1830]) NESHEIM S. — Isolation and purification of ochratoxin A and B and preparation of their methyl and ethyl esters. *J. Ass. Off. Anal. Chem.*, t. **LII**, p. 975-979, 1969.

([1831]) NESHEIM S., BANES D., STOLOFF L. et CAMPBELL A. D. — Note on aflatoxin analysis in peanuts and products. *J. of the A.O.A.C.*, t. **XLVII**, p. 586, 1964.

([1832]) NESTERIN M. F. et VISSARIONOVA V. Y. — Aflatoksiny (Obzor). *Priklad Biokhim. Mikrobiol.*, t. **V**, p. 121-126, 1969.

([1833]) NESTEROV V. S. — *Clinical symptoms of Septic Angina.* Voronezh, 1948.

([1834]) NEWBERNE P. M. — Carcinogenicity of aflatoxin-contaminated peanut meals. *Mycotoxins in Foodstuffs*, p. 187-208, 1965.

([1835]) NEWBERNE P. M., CARLTON W. W. et WOGAN G. N. — Hepatomas in rats and hepatorenal injury in ducklings fed peanut meal or *Aspergillus flavus* extract. *Pathologia Veterinaria*, t. **I**, p. 105-132, 1964.

([1836]) NEWBERNE P. M., HUNT C. E. et WOGAN G. N. — Neoplasms in the rat associated with administration of urethan and aflatoxin. *Experimental Molecular Pathology*, t. **VI**, p. 285-299, 1967.

([1837]) NEWBERNE P. M. et ROGERS A. E. — Carcinoma thymidine uptake and mitosis in the livers of rats exposed to aflatoxin carcino. *N.Z. Med. J.*, p. 8-17, 1968.

([1838]) NEWBERNE P. M. et ROGERS A. E. — Aflatoxin carcinogenesis in rats : dietary effects. *in* Purchase I. F. H., Mycotoxins in human health, p. 195-208, 1971.

([1839]) NEWBERNE P. M., ROGERS A. E. et WOGAN G. N. — Hepatorenal lesions in rats fed a low lipotrope diet and exposed to aflatoxin. *J. Nutrition*, t. **XCIV**, p. 331-342, 1968.

([1840]) NEWBERNE P. M., RUSSO R. et WOGAN G. N. — Acute toxicity of aflatoxin B_1 in the dog. *Pathol. Veter.*, *Suisse*, t. **III**, p. 331-340, 1966.

([1841]) NEWBERNE P. M. et WILLIAMS G. — Inhibition of aflatoxin carcinogenesis by diethylstilbestrol in male rates. *Arch Environ. Health*, t. **XIX**, p. 489-498, 1969.

([1842]) NEWBERNE P. M. et WOGAN G. N. — Effect of cirrhosis and aflatoxin on liver tumor induction. *Feder. Proc.*, *U.S.A.*, t. **XXIV**, p. 431, 1965.

([1843]) NEWBERNE P. M., WOGAN G. N., CARLTON W. W. et ABDEL KADER M. M. — Histopathological lesions in ducklings caused by *Aspergillus flavus* cultures, culture extracts and crystalline aflatoxins. *Tox. Appl. Pharmacol.*, t. **VI**, p. 542-556, 1964.

([1844]) NEWBERNE P. M., WOGAN G. N. et HALL A. — Effects of dietary modifications on response of the duckling to aflatoxin. *J. Nutr.*, t. **XC**, p. 123-130, 1966.

([1845]) NEWTON B. A. — Mechanisms of antibiotic action. *Ann. Review of Microbiology*, t. **XIX**, p. 209-240, 1965.

([1846]) NEZVAL J., BÖSENBERG H. et LINZEL U. — (Investigations on the effect of aflatoxin on bacteria). *Arch. Hyg.*, t. **CLIV**, p. 143-147, 1970.

([1847]) N'GABALA A. Z. et ZAMBETTAKIS C. — *Aspergillus flavus* Link. *Fiche de phytopathologie tropicale*, n° 22 (*Rev. de Mycol.*) 7 p., 1970.

([1848]) NICCOLINI P — Über einem hypotensorischen Wirkstoff von *Ustilago Maydis. Arch. Ital. Sci. Farmacol.*, t. **XI**, p. 137-152, 1942.

([1849]) NICHOLAS D. J. D. — Utilization of inorganic nitrogen compounds and amino acids by fungi, in AINSWORTH G. C. et SUSSMAN A. S. *The fungi* I, p. 349-376, Academic Press, New York et Londres, 1965.

([1850]) NICKELL L. G. et FINLAY A. C. — Antibiotics and their effects on plant growth. *J. Agr. Food Chem.*, t. **II**, p. 178-182, 1954.

([1851]) NICOT J. — Quelques moisissures des substances alimentaires. *Catalogue des Collections vivantes, Herbiers et Documents. Mus. Nat. Hist. Nat.*, t. **VI**, p. 24-29, 1953.

([1852]) NICOT J. — Remarques sur la taxinomie des *Penicillium* et genres voisins. *Ann. Sc. Nat., Bot.*, 12e sér., t. **VI**, p. 595-610, 1966.

([1853]) NICOT J. — Sur l'identité de l'organisme producteur de l'acide fusidique, antibiotique antistaphylococcique. *C. R. Acad. Sci.*, D, t. **CCLVII**, p. 290-292, 1968.

([1854]) NICOU R., GOARIN P. et DELASSUS M. — Traitements des semences d'arachide au Sénégal. *C.R. Congrès Protection Cultures Tropicales, Marseille, mars, 1965*, p. 421-423, 1965.

([1855]) NIELSEN H. E. et HASSELAGER E. — Muggent korn til tidligt fravaennede smagrise. *Bilag. Landok. Forsogslav. efterarsmode* (*arbog*) 1965, p. 91-95, 1965.

([1856]) NIELSEN J. — Trimethylammonium compounds in *Tilletia* ssp. *Canad. J. Bot.*, t. **XLI**, fasc. 3, p. 335-339, 1963.

([1857]) NIELSON R. S. — *Sport Fishery Res. Progr. Circ.*, t. **CI**, p. 6-9, 1960.

([1858]) NIGRELLI R. F. et JAKOWSKA S. — Spontaneous neoplasms in fishes. IX. Hepatomas in Rainbow Trout, *Salmo gairdnerii. Proc. Am. Assoc. Cancer Res.*, t. **II**, p. 38, 1955.

([1859]) NIGRELLI R. F. et JAKOWSKA S. — Fatty degeneration, regenerative hyperplasia and neoplasia in the liver of Rainbow Trout (*Salmo gairdnerii*). *Zoologica*, t. **XLVI**, p. 49-61, 1961.

([1860]) NIKOL'S'KA O. O. — The second All-Union conference on mycotoxicoses of man and agricultural animals. *J. Microbiol. Kiev*, t. **XXIX**, p. 64-66, 1962.

([1861]) NINARD B. — Modalités d'emploi des antiseptiques et désinfectants en hygiène individuelle. *Colloque Antiseptique et Désinfectants*, Inst. Pasteur, mai 1964.

([1862]) NINARD B. et HINTERMANN J. — Pathologie comparée et expérimentale des tumeurs du foie. *Bull. Inst. Hyg. Maroc*, t. **V**, p. 49, 1945.

([1863]) NISHIMURA S. — Observations on the fusaric acid production of the genus *Fusarium. Ann. Phytopath. Soc. Japan*, t. **XXII**, p. 274-275, 1957.

([1864]) NISHIMURA S. — Pathochemical studies on watermelon wilt (Part II). Observations on the fusaric acid, production of the fungi of the genus *Fusarium. Ann. Phytopath. Soc. Japan*, t. **XXIII**, p. 210-214, 1958.

([1865]) NOGUEIRA D. M. et STRUFALDI B. — Aflatoxinas ; problema bioquimico. *Rev. Fac. Farm. Bioquim. Univ. Sao Paulo*, t. **VI**, p. 5-15, 1968.

([1866]) NOLAN A. F. et HULL F. E. — Grass tetany in cattle. *Am. J. Vet. Res.*, t. **II**, p. 41, 1941.

([1867]) NORMAND M. L. — Les perspectives offertes par la lyophilisation dans l'évolution économique du marché des fruits et des légumes. *Aspects théoriques et industriels de la lyophilisation*, p. 593-601, 1964.

([1868]) NORSTADT F. A. et MCCALLA T. M. — Phytotoxic substance from a species of *Penicillium. Science*, t. **CXL**, p. 410-411, 1963..

([1869]) NORSTADT F. A. et MCCALLA T. M. — Microbially induced phytotoxicity in stubble mulched soil. *Soil Sci. Soc. Am. Proc.*, t. **XXXII**, p. 241-245, 1968.

([1870]) NORSTADT F. A. et MCCALLA T. M. — Patulin production by *Penicillium urticae* Bainier in batch culture. *Appl. Microbiol.*, t. **XVII**, p. 193-196, 1969.

([1871]) NORTON D. C. — Factors in the development of blue damage of Spanish peanuts. *Phytopathology*, t. **XLIV**, p. 300-302, 1954.

([1872]) NORTON D. C., MENON S. K. et FLANGAS A. L. — Fungi associated with unblemished spanish peanuts in Texas. *Plant Dis. Reptr.*, t. **XL**, p. 374-376, 1956.

([1873]) NOSE K. et ISHIBASHI K. — Absence of aflatoxins in takadiastase. *Ann. Rep., Sankyo Res. Lab.*, t. **XIX**, p. 74-80, 1967.

([1874]) ODELIEN M. — (La fumure peut-elle être la cause d'hypomagnésiémie et de tétanie chez les bovins ?). *Landbruksh. Instit. Jordkultur., Bull*, n° 48, 19 p., 1960.

([1875]) OETTLE A. G. — Cancer in Africa, especialy in regions South of the Sahara. *J. Nat. Cancer Inst.*, t. **XXXIII**, p. 383-439, 1964.

([1876]) OETTLE A. G. — The aetiology of primary carcinoma of the liver in Africa : a critical appraisal of previous ideas with an ouline of the mycotoxin hypothesis. *S. Afr. Med. J.*, t. **XXXIX**, p. 817-825, 1965.

([1877]) OGLOBLIN A. et JAUCH C. — Reacciones patologicas de los acridios atacados por *Aspergillus parasiticus. Rev. Arg. Agron.*, t. **X**, p. 256-267, 1943.

([1878]) OHARA I. — Classification of the *Aspergillus tamarii-oryzae* group. The production of kojic acid from various compounds as an aid in identification. *Res. Bull. Fac. Agr., Gifu Univ.*, n° 1, p. 71-85, 1951. *Suppl. J. Agr. Chem. Soc. Japan*, t. **XXVI**, p. 547-551, 1952.

([1879]) OHARA I. — Classification of the *Aspergillus tamarii-oryzae* group. Diagnosis of the series, species and subspecies. *Res. Bull. Gifu Imp. Coll. Agr.*, t. **XXVIII**, p. 75-85, 1953.

([1880]) OHMORI Y., ISONO C. et UCHIDA H. — On toxicity of Thailand yellowsis rice and Islandia yellowsis rice. *Jap. J. Pharmacol.*, t. **L**, p. 246, 1954.

([1881]) OHTSUBO K., YAMADA M. A. et SAITO M. — Inhibitory effect of nivalenol, a toxic metabolite of *Fusarium nivale*, on the growth cycle and bipolymer synthesis of HeLa cells. *Jap. J. Med. Sci. Biol.*, t. **XXI**, p. 185-194, 1968.

([1882]) OKSAMITNYI N. K. et VLASOV A. T. — O sluchaye massovogo otravlenniia svenoi golovnyevime gribami. *Veterinariya*, t. **XXXV**, p. 83, 1958.

([1883]) OKUBO K. et ISODA M. — Studies on the essential nature of intoxication by the products of *Fusarium nivale*. I. Histopathological examination of experimental acute intoxication by crude products of *Fusarium nivale*. *Bull. Nippon Vet. Zootechn. Coll.*, t. **XVI**, p. 22-42, 1967.

([1884]) OKUBO Y., URAKAMI N., HAYAMA T., SETO Y., MIURA T., KANO Y., MOTOYOSHI S., YAMAMOTO S., ISHIDA K., IIZUKA H. et IIDA M. — Note on some cases of malt rootlet poisoning in dairy cattle. *Jap. J. Vet. Sci.*, t. **XVII**, p. 144-151, 1955.

([1885]) OKUDA S., IWASAKI S., SAIR M. I., MACHIDA Y., INOUE A. et TSUDA K. — Stereochemistry of helvolic acid. *Tetrahedron L.*, n° 24, p. 2295-2302, 1967.

([1886]) OLAFSON J. H., CHRISTENSEN C. M. et GEDDES W. F. — Grain storage studies. XV. Influence of moisure content, commercial grade, and maturity on the respiration and chemical deterioration of corn. *Cereal Chem.*, t. **XXXI**, p. 333-340, 1954.

([1887]) OLAFSON P. — Hyperkeratosis (X disease) of cattle. *Cornell Vet.*, t. **XXXVII**, p. 279, 1947.

([1888]) OLAFSON P. et MC ENTEE K. — The experimental production of hyperkeratosis (X disease) by feeding a processed concentrate. *Cornell Vet.*, t. **XLI**, p. 107-109, 1951.

([1889]) OLDHAM L. S., OEHME F. W. et KELLEY D. C. — Production of aflatoxin in prepackaged luncheon meat and cheese at refrigerator temperatures. *J. Milk Food Technol.*, t. **XXXIV**, p. 349-351, 1971.

([1890]) OLIFSON L. E. — Toxins isolated from overwintered cereals and their chemical nature. *Monitor, Orenburg Sect. U.S.S.R., D.J. Mendeleyev Chem. Soc.*, t. **VII**, p. 21-35, 1957.

([1891]) OLIFSON L. E. — Chemical action of some fungi on overwintered cereals. *Monitor, Orenburg Sect. U.S.S.R., D.J. Mendeleyev Chem. Soc.*, t. **VII**, p. 37-45, 1957.

([1892]) OLLIVER M. et RENDLE T. — A new problem in fruit preservation. Studies on *Byssochlamys fulva* and its effects on the tissues of processed fruit. *J. Soc. Chem. Ind.*, t. **LIII**, p. 166-172, 1934.

([1893]) OLLIVER M. et SMITH G. — *Byssochlamys fulva* n. sp. *Jour. Bot.* (*London*), t. **LXXII**, p. 196-197, 1933.

([1894]) ONEGOV A. P. — On toxicity of some Hyphomycetes on clover hay. *Veterinariya*, t. **XXVI**, p. 41-43, 1949.

([1895]) OOSTERHUIS H. K. — Antibiotica uit Schimmels. *Mededel. Lab. Physiol. Chem. Univ.*, t. **X**, 116 p., 1945.

([1896]) OPEL M. — Patulinnachweis und -bestimmung mit dem Benzidinreagenz. *Naturwissenschaften*, t. **XLIV**, p. 306, 1957.

([1897]) OPITZ E. — Beiträge zur Frage der Durchgängigkeit von Darm und Nieren für Bakterien. *Zeitsch. f. Hyg. u. Inf.*, t. **XXIX**, p. 505-552, 1898.

([1898]) OSER B. L. — Regulatory aspects of control of mycotoxins in foods and feeds. *in* GOLDBLATT L. A., *Aflatoxin*, Academic Press, p. 393-400, 1969.

([1899]) OSIYEMI F. O., BABABUNMI E. A. et BASSIR O. — Studies on *Aspergillus flavus* Link and its metabolites. I. Production of aflatoxin in cultures of *Aspergillus flavus. West Afr. J. Biol. Appl. Chem.*, t. **IX**, p. 31-34, 1967.

([1900]) OTTO M. — Ueber die Giftwirkung einiger Stämme von *Aspergillus fumigatus* und *Penicillium glaucum* nebst einigen Bemerkungen über Pellagra. *Zeitschr. f. Klin. Med.*, t. **LIX**, p. 322-339, 1906.

([1901]) OUDEMANS C. A. J. A. — Contribution à la flore mycologique des Pays-Bas. XI. *Ned. Kruid. Archief*, 2e sér., t. **IV**, p. 502-562, 1886.

([1902]) OVCHINNIKOV J. A., KIRJUSHKIN A. A. et SHEMJAKIN M. M. — (Synthèse complète des sporidesmolides III et IV). *Zh. Obshch. Khim.*, t. **XXXVI**, p. 620-627, 1966.

([1903]) OWEN S. P. et JOHNSON M. J. — The effect of temperature change on the production of penicillin by *Penicillium chrysogenum* W 49-133. *Appl. Microbiol.*, t. **III**, p. 375-379, 1955.

([1904]) OWENS R. G., WELTY R. E. et LUCAS G. B. — Gas chromatographic analysis of the mycotoxins kojic acid, terreic acid, and terrein. *Anal. Biochem.*, t. **XXXV**, p. 249-258, 1970.

([1905]) OXFORD A. E. — Antibacterial substances from moulds. III. Some observations on the bacteriostatic powers of the mould products citrinin and penicillic acid. *Chem. and Indus.*, t. **LXI**, p. 48-51, 1942.

([1906]) OXFORD A. E. et RAISTRICK H. — Antibacterial substance from moulds. Part 4. Spinulosin and fumigatin, metabolic products of *Penicillium spinulosum* Thom. and *Aspergillus fumigatus* Fresenius. *Chem. and Ind.*, t. **LXI**, p. 128-129, 1942.

([1907]) OXFORD A. E., RAISTRICK H. et SIMONART P. — Studies in the biochemistry of micro-organisms. 60. Griseofulvin, $C_{17}H_{17}O_6Cl$, a metabolic product of *Penicillium griseofulvum* Dierckx. *Bioch. J.*, t. **XXXIII**, p. 240, 1939.

([1908]) OZEGOVIC L. et GRIN E. I. — Haemolytische Eigenschaft der Dermatophyten. *Mykosen*, t. **X**, p. 325-330, 1967.

([1909]) PAGET G. E. — Mitotic activity in the human liver. *J. Path. Bact.*, t. **LXVII**, p. 401-406, 1954.

([1910]) PAGET G. E. et WALPOLE A. L. — Some cytological effects of griseofulvin. *Nature*, t. **CLXXXII**, p. 1320-1321, 1958.

([1911]) PALYUSIK M. — Experimental stachybotryotoxicosis of young chicks. *Sabouraudia, J. Int. Soc. Hum. Anim. Mycol.*, t. **VIII**, p. 4-8, 1970.

([1912]) PALYUSIK M. — Biological test for the toxic substance of *Stachybotrys alternans Acta Vet. Acad. Sci. Hung.*, t. **XX**, p. 57-67, 1970.

([1913]) PALYUSIK M. — Experimental swine fusariotoxicosis (vulvovaginitis) induced with *Fusarium graminearum. C.R. 5e Congr. ISHAM*, p. 222-223, Paris 1971.

([1914]) PALYUSIK M., SZEP I. et SZOKE F. — Data on susceptibility to mycotoxins of day-old goslings. *Acta Vet. Acad. Sci. Hung.*, t. **XVIII**, p. 363-372, 1968.

([1915]) PAMMEL L. H. — *A manual of poisonous plants.* The Torch Press. Cedar Rapids, Iowa, 1911.

([1916]) PAMMEL L. H. — Darnel Poisonous. *Am. J. Vet. Med.*, t. **XV**, p. 491, 1920.

([1917]) PAMMEL L. H. — Probably a cryptogam poisoning due to mold on the corn. *Vet. Med.*, t. **XXIII**, p. 29, 1928.

([1918]) PANASENKO V. T. — Species of *Monilia* and *Torula* from food products. *Mycologia*, t. **LVI**, fasc. 6, p. 805-808, 1964.

([1919]) PAPAVIZAS G. C. et CHRISTENSEN C. M. — Grain storage studies. XXVI. Fungus invasion and deterioration of wheat stored at low temperatures and moisture contents of 15 and 18 per cent. *Cereal Chem.*, t. **XXXV**, p. 27-34, 1958.

([1920]) PARKE D. V. — *The biochemistry of foreign compounds.* 274 p., Pergamon, New York, 1968.

([1921]) PARKER W. A. et MELNICK D. — Absence of aflatoxin from refined vegetable oils. *J. Amer. Oil Chem. Soc.*, t. **XLIII**, p. 635-638, 1966.

([1922]) PARKIN J. C. — Deterioration of stored grains by fungi. *Rhod. Sci. News*, t. **V**, p. 15-17, 1971.

([1923]) PARPIA H. A. B. — *Report of the work on aflatoxin carried out at Central Food Technological Research Institute, Mysore. India.* 5 p., 1965.

([1924]) PARRISH F. W., WILEY B. J., SIMMONS E. G. et LONG L. — A survey of some species of *Aspergillus* and *Penicillium* for production of aflatoxins and kojic acid. *Tech. Rep. U.S. Army Mat. Comm., Microbiol.* Ser., n° 20, 21 p., 1965.

([1925]) PARRISH F. W., WILEY B. J., SIMMONS E. G. et LONG L. — Production of aflatoxins and kojic acid by species of *Aspergillus* and *Penicillium. Appl. Microbiol.*, t. **XIV**, p. 139, 1966.

([1926]) PASSMORE F. R. — Depreciation of prepared copra due to molds and insects. *Bull. Imper. Inst.*, t. **XXIX**, p. 171-180, 1931.

([1927]) PASSMORE F. R. — A survey of damage by insects and molds to West African cacao before storage in Europe; season 1930-31. *Bull. Imp. Inst. Lond.*, t. **XXX**, p. 296-305, 1932.

([1928]) PATERSON J. S., CROOK J. C., SHAND A., LEWIS G. et ALLCROFT R. — Groundnut toxicity as the cause of exudative hepatitis (oedema disease) of guinea pig. *Vet. Rec.*, t. **LXXIV**, p. 639-640, 1962.

([1929]) PATIL V. D. et SHINDE P. A. — Studies on seed-borne fungi of groundnut. *Nagpur Agr. Coll. Magazine*, t. **XXXVIII**, p. 35-41, 1964.

([1930]) PATTEE H. E. et DICKENS J. W. — A study of the relationship between aflatoxin and damage in North Carolina peanuts. *Rept. to Peanut Improvement Working Group. Meeting, Washington, 27-28 avr.*, 10 p., 1965.

([1931]) PATTEE H. E. et SESSOMS S. L. — Relation between *Aspergillus flavus* growth fat acidity and aflatoxin content in peanuts. *J. Amer. Oil Chem. Soc.*, t. **XXXIX**, p. 61-63, 1967.

([1932]) PATTEE H. E., SESSOMS S. L. et DICKENS J. W. — Influence of biologically modified atmospheres on aflatoxin production by *Aspergillus flavus* growing on peanut kernels. *Oléagineux*, t. **XXI**, p. 747-748, 1966.

([1933]) PATTERSON D. S. P. et ALLCROFT R. — Metabolism of aflatoxin in susceptible and resistant animal species. *Food Cosmet. Toxicol.*, t. **VIII**, p. 43-53, 1970.

([1934]) PATTERSON D. S. P. et ROBERTS B. A. — The formation of aflatoxins B_2 and G_2 and their degradation products during the in vitro detoxification of aflatoxin by livers of certain avian and mammalian species. *Food Cosmet. Toxicol.*, t. **VIII**, p. 257-538, 1970.

([1935]) PATTERSON D. S. P. et ROBERTS B. A. — Differences in the effect of phenobarbital treatment on the *in vitro* metabolism of aflatoxin and aniline by duck and rat liver. *Biochem. Pharmacol.*, t. **XX**, p. 3377-3383, 1971.

(1936) PATTERSON D. S. P., SWEASEY D., HERBERT N. C. et CARNAGHAN R. B. A. — Comparative biological and biochemical studies in hybrid chicks: I. The development of electrophoretic patterns of normal serum proteins. *Brit. Poult. Sci.*, t. **VIII**, p. 273-278, 1967.

(1937) PAULIN A. — La conservation frigorifique du raisin de table en emballage de polyéthylène en présence d'une émission d'anhydride sulfureux. *Coll. Inst. du Froid, Commissions* 4-5, 10 p., Bologne, juin 1966.

(1938) PAULMAN V. C. — Poisoning from burned, sweet-clover hay. *Vet. Med.*, t. **XVIII**, p. 8, 1923.

(1939) PAYET M., CROS J., QUENUM C., SANKALE M. et MOULANIER M. — Deux observations d'enfants ayant consommé de façon prolongée des farines souillées par *Aspergillus flavus. Presse Méd.*, t. **LXXIV**, p. 649-651, 1966.

(1940) PEARSON L. — A preliminary report upon forage-poisoning of horses (so-called cerebro-spinal meningitis). *J. Comp. Med. and Vet. Arch.*, t. **XXI**, p. 654, 1900.

(1941) PECKHAM J. C., DOUPNIK B. et JONES O. H. — Acute toxicity of ochratoxins A and B in chicks. *Appl. Microbiol.*, t. **XXI**, p. 492-494, 1971.

(1942) PEDERSEN H. — The Danish grain quality Commitee and the scope of its work. *Proc. Symp. Agric. Coll. Norway*, 1967, 7 p. ronéot.

(1943) PEERS F. G. — Aflatoxin: a summary of recent work. *Trop. Sci.*, t. **IX**, p. 186-203, 1967.

(1944) PELHATE J. — Moisissures dangereuses dans l'alimentation animale. *C.R. Acad. Agr. Fr.*, t. **LII**, p. 850-855, 1966.

(1945) PELHATE J. — Inventaire de la mycoflore des blés de conservation. *Bull. Soc. Mycol. Fr.*, t. **LXXXIV**, p. 127-143, 1968.

(1946) PELHATE J. — Contribution à l'étude écologique des moisissures des blés de conservation. *Thèse Doctorat Etat, Fac Sci. Brest,* 66 p., 19 juin 1968.

(1947) PELHATE J. — Evolution de la mycoflore des blés en cours de conservation. (Comité de Technologie Agricole D.G.R.S.T., 12 avr. 1967). *Ind. Aliment. Agric.*, t. **LXXXV**, p. 769-773, 1968.

(1948) PELHATE J. — Etude expérimentale des interactions de moisissures caractéristiques des grains. *Rev. de Mycol.*, t. **XXXIII**, p. 1-28, 1968.

(1949) PELHATE J. — Recherche des besoins en eau chez quelques moisissures des grains *Mycopathol. Mycol. Appl.*, t. **XXXVI**, p. 117-128, 1968.

(1950) PELHATE J. — Etat thermodynamique de l'eau dans les grains de blé sains ou altérés. *Ann. Technol. Agric.*, t. **XVII**, p. 227-236, 1968.

(1951) PELHATE J. — Analyses mycologiques systématiques d'aliments suspects. *Symposium INSERM Mycotoxines et Alimentation, Le Vésinet,* 24 oct. 1969.

(1952) PELHATE J. — Résistance du caryopse de blé à l'envahissement par les moisissures. *Bull. Soc. Bot. Fr.*, Mém. 1968, t. **CXV**, p. 82-91, 1969.

(1953) PELHATE J. — Longévité des espèces et maintien de la mycoflore des grains. *Phytopathol. Z.*, t. **LXIV**, p. 6-20, 1969.

(1954) PELHATE J. — Analyses mycologiques systématiques d'aliments suspects. *Cah. Nutr. Diét.* t. V, p. 75-77, 1970.

1955) PELHATE J. — Moisissures des maïs-grains en cours de conservation. Leur inhibition par l'acide propionique. *Bull. Soc. Mycol. Fr.*, t. **LXXXIX**, p. 53-65, 1973.

(1956) PENNINGTON E. S. — A study of clinical sensivity to air-borne molds. *Journ. Allergy*, t. **XII**, p. 388-402, 1941.

(1957) PEPPER E. H. — The microflora of barley kernels ; their isolation, characterization, etiology, and effects on barley, malt products. *Ph. D. Thesis, Dept. Pl. Pathol., Michigan State Univ., East Lansing*, 1960.

(1958) PERCIVAL J. C. — Photosensivity diseases in New Zealand. XVII. The association of *Sporidesmium bakeri* with facial eczema. *New Zealand J. Agr. Research*, t. **II**, p. 1041-1056, 1959.

([1959]) PERCIVAL J. C. et THORNTON R. H. — Relationship between the presence of fungal spores and a test for hepatotoxic grass. *Nature*, t. **CLXXXII**, p. 1095-1096, 1958.

([1960]) PÉRESSE M. — Résultats inédits, 1966.

([1961]) PERLMAN D., GIRFFRE N. A. JACKSON P. N. et GIARDINELLO F. E. — Effects of antibiotics on multiplication of L-cells in suspension culture. *Proc. Soc. Exp. Biol. Med.*, t. **CII**, p. 290-292, 1959.

([1962]) PERONE V. B., SCHEEL L. D. et MEITUS R. J. — A bioassay for the quantitation of cutaneous reactions associated with pinkrot celery. *J. Invest. Dermatol.*, U.S.A., t. **XLII**, fasc. 3, p. 267-271, 1964.

([1963]) PERRIN D. D. — Photosensivity diseases in New Zealand. X. The guinea pig as an experimental animal in the investigation of facial eczema. *New Zealand J. Sci. Technol.*, sér. A, t. **XXXVII**, p. 669-679, 1957.

([1964]) PERRIN D. D. — Photosensivity diseases in New Zealand. XV. A chemical procedure for the detection of facial eczema toxicity in pasture. *New Zealand J. Agr. Research*, t. **II**, p. 266-273, 1959.

([1965]) PERRIN D. R. et BOTTOMLEY W. — Studies on phytoalexins V. The structure of pisatin from *Pisum sativum* L. *J. Am. Chem. Soc.*, t. **LXXXIV**, p. 1919-1922, 1962.

([1966]) PETERS A. — Studien über den Zusammenhang zwischen Mängeln in der Winterstallfütterung hochleistender Milchkühe und der Grastetanie. *Schriftenreihe der Landwirtschaftlischen Fakultät, Kiel*, n⁰ 24, 62 p., 1960.

([1967]) PETERS A. T. — Cornstalk disease. *Nebraska Agr. Exp. Stn.*, Bull. n⁰ 52, 1898.

([1968]) PETERS J. A. — Mechanism of early sporidesmin intoxication in sheep. *Nature*, t. **CC**, p. 286, 1963.

([1969]) PETERS J. A. — Effect of sporidesmin on lipid metabolism in rabbits. *Nature*, t. **CCX**, n⁰ 5036, p. 601-603, 1966.

([1970]) PETERS J. A. et SMITH L. M. — The composition of liver lipids of sheep and the effect of early sporidesmin poisoning. *Biochem. J.*, t. **XII**, p. 379-385, 1964.

([1971]) PETERS R. H. et SUMNER H. H. — Spectra of anthraquinone derivatives. *J. Chem. Soc.*, p. 2101-2110, 1953.

([1972]) PETERSON R. E. et CIEGLER A. — Separation of aflatoxins by two-dimensional thin-layer chromatography. *J. Chromatogr.*, t. **XXXI**, p. 250-151, 1967.

([1973]) PETERSON R. et CIEGLER A. — Note on a water-based aflatoxin standard. *J. Ass. Off. Anal. Chem.*, t. **L**, p. 1201-1202, 1967.

([1974]) PETERSON R. E., CIEGLER A. et HALL H. H. — Densitometric measurement of aflatoxin (*Aspergillus flavus*). *J. Chromatogr.*, t. **XXVII**, p. 304-307, 1967.

([1975]) PETIT J. P. — Procédé chromatographique rapide pour l'étude de la fluorescence des aflatoxines. *Rev. Elevage Méd. Vétér. Pays Trop.*, t. **XIX**, p. 87-96, 1966.

([1976]) PETIT J. P., RIVIÈRE R., PERREAU P. et PAGOT J. — Recherches sur l'aflatoxine. Revue des travaux effectués pendant le premier semestre 1964 dans les laboratoires centraux de l'I.E.M.V.T. *Rev. Elevage Méd. Vétér. Pays Trop.*, t. **XVII**, p. 239-253, 1964.

([1977]) PETTERSSON G. — Biosynthesis of fumigatin. *Acta Chem. Scand.*, t. **XVII**, p. 1323-1329, 1963.

([1978]) PETTERSSON G. — Toluquinones from *Aspergillus fumigatus*. *Acta Chem. Scand.*, t. **XVII**, p. 1771-1776, 1963.

([1979]) PETTIT R. E. et TABER R. A. — Factors influencing aflatoxin accumulation in peanut kernels and the associated mycoflora. *Appl. Microbiol.*, t. **XVI**, p. 1230-1234, 1968.

([1980]) PETTIT R. E. et TABER R. A. — Fungal invasion of peanut kernels as influenced by harvesting and handling procedure. *Phytopathology*, t. **LX**, p. 1307, 1970.

([1981]) PETTIT R. E., TABER R. A. et SCHROEDER H. W. — Aflatoxin levels in fresh-dug peanuts. *Phytopathology*, t. **LX**, p. 586, 1970.

([1982]) PETTY M. A. et QUIGLEY G. D. — The microflora of wheat feeds as related to incidence of blue comb in chickens. *Poultry Sci.*, t. **XXVI**, p. 7-13, 1947.

([1983]) PEYRONEL B. — Caractérisation des mycocénoses de climats et de milieux divers et nouvelle méthode pour les représenter graphiquement. 6e *Congr. Int. Sci. Sol, Paris.* Rapports C, p. 45-49, 1956.

([1984]) PFEIFER V. F. et GRAVES R. R. — Microbiology of wheat and flour : reduction of microbial population during milling. *Rep. 4th Nat. Conf. Wheat Util. Res. Boise, Idaho,* p. 60-64, U. S. Dept. Agr. ARS, 74-35, 1965.

([1985]) PFEIFER V. F. et VOJNOVICH C. — Reducing the microbial population of wheat flour. *Bull. Ass. Oper. Millers,* p. 3022-3024, 1968.

([1986]) PFEIFER V. F., VOJNOVICH C. et GRAVES R. R. — Wheat and flour microbial population can be reduced during milling process. *Amer. Miller Processor,* t. **XCIV**, p. 15-17, 1966.

([1987]) PHILLIPS C. R. et KAYES. — Sterilizing action of gazeous ethylene oxide. *Am. Journn. Hyg.,* t. **L**, p. 270-306, 1949.

([1988]) PHILIP J. M. — Eurotox symposium on natural products. Aflatoxin-mammalian toxicity. *Food Cosmet. Toxiol.,* t. **II**, p. 674-675, 1964.

([1989]) PILLAI N. C. et SRINAVASAN K. S. — The amino acid metabolism of *Aspergillus flavus. J. Gen. Microbiol.,* t. **XIV**, p. 248-255, 1956.

([1990]) PIONNAT J. C. — Etude des altérations fongiques des grains d'orge en cours de conservation. *Ann. Epiphyties,* t. **XXII**, p. 203-214, 1966.

([1991]) PITOUT M. J. — The effect of ochratoxin A on glycogen storage in the rat liver. *Toxicol. Appl. Pharmacol,* t. **XIII**, p. 299-306, 1968.

([1992]) PITOUT M. J. — The hydrolysis of ochratoxin A by some proteolytic enzymes. *Biochem. Pharmacol.,* t. **XVIII**, p. 485-491, 1969.

([1993]) PITOUT M. J. — Biochemical studies on ochratoxin A. *in* PURCHASE I. F. H., *Mycotoxins in human health,* p. 53-64, 1971.

([1994]) PITOUT M. J., MCGEE H. A. et SCHABORT J. C. — The effect of aflatoxin B_1, aflatoxin B_2, and sterigmatocystin on nuclear deoxyribonucleases from rat and mouse livers. *in* PURCHASE I. F. H., *Mycotoxins in human health,* p. 31-46, 1971.

([1995]) PLATEL A., UENO Y. et FROMAGEOT P. — Sur l'origine du groupe méthyle en 3 et 3'de la lutéoskyrine. *Bull. Soc. Chim. Biol.,* t. **XLIX**, p. 1892-1984, 1967.

([1996]) PLATONOW N. — Effect of prolonged feeding of toxic groundnut meal in mice. *Vet. Rec.,* t. **LXXVI**, p. 589-590, 1964.

([1997]) PLATT B. S., STEWART R. J. C. et GUPTA R. — The chick embryo as a test organism for toxic substances in food. *Proc. Nutr. Soc. Engl. Scot.,* t. **XXI**, p. 30-31, 1962.

([1998]). PLOQUIN J. — Le bore dans l'alimentation. *Bull. Soc. Scient. Hyg. Alim.,* t. **LX**, p. 70-113, 1967.

([1999]) PLOTNIKOV N. P., SILAEV A. B., MAKSIMOVA R. A. et SEMENOV N. M. — Antibiotik iz *Aspergillus fumigatus* tsitotoksicheskim i protivovirusnym deistviem. *Antibiotiki,* t. **XIII**, p. 316-319, 1968.

([2000]) PLUM N. — Verschiedene Hyphomyceten-Arten als Ursache sporadischer Falle von Abortus beim Rinde. *Acta Pathol. Microbiol. Scand.,* t. **IX**, p. 150-157, 1932.

([2001]) POD'YAPOL'SKAYA O. P. — K voprosu o vliyanii ioniziruyushchikh izluchenii na mikroorganizmy zerna. *Sobshcheniya i referaty VNIIZ,* n⁰ 5, p. 1, 1957.

([2002]) POHLAND A. E. et ALLEN R. — Analysis and chemical confirmation of patulin in grains. *Journ. Ass. Off. Anal. Chem.,* p. 686-687, 1970.

([2003]) POHLAND A. E., CUSHAC M. E. et ANDRELLOS P. J. — Aflatoxin B_1 hemiacetal. *J. Ass. Off. Anal. Chem.,* t. **LI**, p. 907-910, 1968.

([2004]) POHLAND A. E., SANDERS K. et THORPE C. — Paper n⁰73 presented at the *82nd Annual Meeting of the Assoc. Offic. Anal. Chemists,* 15 oct. 1968.

([2005]) POISSON J. — Problèmes de la recherche des substances étrangères des grains, farines et produits de transformation. Recherche de la contamination bactériologique. *Mises au point de Chimie analytique,* t. **XV**, p. 127-148, 1967.

([2006]) Poisson J., Cahagnier B. et Guilbot A. — Sur la radiosensibilité des conidiospores des moisissures dominantes du maïs. Incidence sur la radurisation des grains. *Mycopathol. Mycol. Appl.*, t. **XLV**, p. 193-209, 1971.

([2007]) Poisson J. et Guilbot A. — Conditions de stockage et durée de conservation des grains. *La Meunerie francaise*, n⁰ 163, 11 p., 1963.

([2008]) Pokrovsky A. A., Kravchenko L. V. et Tutelyan V. A. — Investigation of the activity of lysosomal enzymes under the action of aflatoxin and mitomycin *C. Biokhimiya*, t. **XXXVI**, p. 690-696, 1971.

([2009]) Pokrovsky A. A., Kravchenko L. V. et Tutelyan V. A. — Effect of aflatoxin on rat liver lysosomes. *Toxicon*, t. **X**, p. 25-30, 1972.

([2010]) Pokrovsky A. A. et Nesterin M. F. — Vyyavlenie flyuorestsiruyushchikh veshchestv v fermentnykh preparatakh gribkovogo proikhozhdeniya, prednaznachennykh dlya primeneniya v pischchevoi promyshlennosti. *Vop. Pitan*, t. **XXVIII**, p. 49-53, 1969.

([2011]) Poliakov A. A. — Desinfektsiia pri stakhibotriotoksikoze. *Rukovodstvo po Veterinarnoi Dezinfeksii, Ogiz-Selskhozgiz, Moscou*, p. 154-158, 1948.

([2012]) Pollock A. — Production of citrinin by five species of *Penicillium. Nature*, t. **CLX**, p. 331, 1947.

([2013]) Pomeranz Y. — Proteolytic and starch-liquefying properties of mold cultures. *Bull. Research Council, Israel*, t. **VI**, C. p. 53-58, 1957.

([2014]) Pomeranz Y. — The pantothenic acid content of moldy wheat and flour. *Bull. Research Council, Israel*, t. **VI** C, p. 72, 1957.

([2015]) Pomeranz Y. — Formation of toxic compounds in storage-damaged foods and feedstuffs. *Cereal Science Today*, t. **IX**, fasc. 4, p. 93-96, 1964.

([2016]) Pomeranz Y., Halton P. et Peers F. G. — The effects on flour dough and bread quality of molds grown in wheat and those added to flour in the form of specific cultures. *Cereal Chem.*, t. **XXXIII**, p. 157-169, 1956.

([2017]) Pong R. S. et Wogan G. N. — Time course of alterations of rat liver polysome profiles induced by aflatoxin B_1. *Biochem. Pharmacol.*, t. **XVIII**, p. 2357-2361, 1969.

([2018]) Pong R. S. et Wogan G. N. — Time course and dose response characteristics of aflatoxin B_1 effects on rat liver RNA polymerase and ultrastructure. *Cancer Res.*, t. **XXX**, p. 294-304, 1970.

([2019]) Pons W. A. — Collaborative study on the determination of aflatoxins in cottonseed products. *J. Ass Off. Anal. Chem.*, t. **LII**, p. 61-72, 1969.

([2020]) Pons W. A., Cucullu A. F., Lee L. S., Robertson J. A., Franz A. O. et Goldblatt L. A. — Determination of aflatoxins in agricultural products : use of aqueous acetone for extraction. *J. Ass. Off. Anal. Chem.*, t. **XLIX**, p. 554-562, 1966.

([2021]) Pons W. A., Cucullu A. F., Franz A. O. et Goldblatt L. A. — Improved objective fluorodensitometric determination of aflatoxins in cottonseed products. *J. Amer. Oil Chem. Soc.*, t. **XLV**, p. 694-699, 1968.

([2022]) Pons W. A. et Eaves P. H. — Aqueous acetone extraction of cottonseed. *J. Amer. Oil Chem. Soc.*, t. **XLIV**, p. 460-464, 1967.

([2023]) Pons W. A. et Goldblatt L. A. — The determination of aflatoxins in cottonseed products. *J. Amer. Oil Chemists Soc.*, t. **XLII**, fasc. 6, p. 471-475, 1965.

([2024]) Pons W. A. et Goldblatt L. A. — Instrumental evaluation of aflatoxin resolution in TLC plates. *J. Ass. Off. Anal. Chem.*, t. **LI**, p. 1194-1197, 1968.

([2025]) Pons W. A. et Goldblatt L. A. — Physicochemical assay of aflatoxins. in Goldblatt L. A., *Aflatoxin*, Academic Press, p. 77-105, 1969.

([2026]) Popescu V. — Influenta mediului de cultura a supra ciupercii *Fusarium graminearum* Schw. agent patogen al fuzariozei unor plante furajere si a fuzariotoxicozei animalelor de ferma. *Int. Agron. Dr Petru Groza Cluj Lucr. Stiint, Ser. Agr.*, t. **XXI**, p. 249-267, 1965.

([2027]) Popper H. et Schaffner F. — Liver : Structure and Function. Mc Graw-Hill Publ. Co, Blakiston Division, New York, 1957.

(2028) POPPER K. et NURY F. — DEPC sterilization of cold storage rooms. *Ashrae J.*, p. 80-81, août 1966.

(2029) PORTER D. M. et WRIGHT F. S. — Proliferation of *Aspergillus flavus* in artificially infested windrow-dried peanut fruit in Virginia. *Phytopathology*, t. **LXI**, p. 1194-1197, 1971.

(2030) PORTMAN R. S., PLOWMAN K. M. et CAMPBELL T. C. — Aflatoxin metabolism by liver microsomal preparations of two different species. *Biochem. Biophys. Res. Commun.*, t. **XXXIII**, p. 711-715, 1968.

(2031) POSTERNAK T. — *Helv. Chim. Acta*, t. **XXI**, p. 1326, 1938.

(2032) PREBUDA H. J. — Observations on the hemorrhagic condition. *Feed Age*, t. **V**, p. 40-41, 1955.

(2033) PRENTICE N., DICKSON A. D. et DICKSON J. G. — Production of emetic material by species of *Fusarium*. *Nature*, t. **CLXXXIV**, p. 1319, 1954.

(2034) PRÉVOT A. — Dosage et élimination de l'aflatoxine aux U.S.A. *Rev. fr. Corps Gras*, t. **XVIII**, p. 390-393, 1971.

(2035) PRILLIEUX E. et DELACROIX G. — Travaux du laboratoire de Pathologie végétale. *Bull. Soc. Myc. Fr.*, t. **VIII**, p. 22-23, 1892.

(2036) PRINCE A. E. — Fungi isolated from Peanuts collected in South Carolina in 1944. *Plant Dis. Reptr.*, t. **XXIX**, p. 367-368, 1945.

(2037) PRINGLE R. B. et BRAUN A. C. — The isolation of the toxin of *Helminthosporium victoriae*. *Phytopathology*, t. **XLVII**, p. 369-371, 1957.

(2038) PRINGLE R. B. et SCHEFFER R. P. — Purification of the selective toxin of *Periconia circinata*. *Phytopathology*, t. **LIII**, p. 785-787, 1963.

(2039) PRINGLE R. B. et SCHEFFER R. P. — Host specific plant toxins. *Ann. Rev. Phytopathology*, t. **II**, p. 133-156, 1964.

(2040) PROHASZKA L. et JUHASZ S. — Persistent anisotropy in the livers of ducklings with aflatoxin poisoning. *Avian Dis.*, t. **XI**, p. 130-136, 1967.

(2041) PUETZER B., NIELD C. H. et BARRY R. H. — The synthesis of a clavacin isomer. *Science*, t. **CI**, p. 307-308, 1945.

(2042) PUETZER B., NIELD C. M. et BARRY R. M. — The synthesis of a clavacin isomer and related compounds. *J. Amer. Chem. Soc.*, t. **LXVII**, p. 832-837, 1945.

(2043) PULLAR E. M. et LEREW W. M.—Vulvovaginitis in swine. *Austral Vet. J.*, t. **XIII**, p. 28, 1937.

(2044) PULSS G. — Hypomagnesaemie und Weidetetanie. *Naturwissenschaften*, t. **XLVIII**, p. 59-60, 1961.

(2045) PULLS G. — Atmungshemmungen durch einen unbekannten Faktor in Gras von Tetanieweiden, nach Untersuchungen an lebenden Ratten sowie an Leberschnitten und Helezellen. *Naturwissenschaften*, t. **XLVIII**, p. 224-225, 1961.

(2046) PURCHASE I. F. H. — Aflatoxin in milk. *S. Afr. Med. J.*, t. **XL**, p. 774, 1966.

(2047) PURCHASE I. F. H. — Acute toxicity of aflatoxins M_1 and M_2 in one-day-old ducklings. *Food Cosmet. Toxicol.*, t. **V**, p. 339-342, 1967.

(2048) PURCHASE I. F. H. — *Mycotoxins in human health. Proc. Sympos.* 2-4 sept. 1970, Pretoria, 306 p., MacMillan edit., 1971.

(2049) PURCHASE I. F. H. et GONCALVES T. — Preliminary results from food analyses in the Inhambane area. *in* PURCHASE I. F. H., *Mycotoxins in human health*, p. 263-269, 1971.

(2050) PURCHASE I. F. H. et JOUBERT H. J. B. — Biological screening as a laboratory aid in determining cancer aetiology. *in* PURCHASE I. F. H., *Mycotoxins in human health*, p. 291-297, 1971.

(2051) PURCHASE I. F. H. et NEL W. — Recent advances in research on ochratoxin. Part 1 Toxicological aspects. *Biochemistry of some foodborne microbial toxins. Symp. Amer. Chem. Soc. New York, sept.* 1966, p. 153-156, M. I. T. Press, 1967.

([2052]) PURCHASE I. F. H. et STEYN M. — Estimation of aflatoxin M in milk. *J. Ass. Off. Anal. Chem.*, t. **L**, p. 363-366, 1967.

([2053]) PURCHASE I. F. H. et STEYN M. — The production of aflatoxin M on various substrates. *Mycopathol. Mycol. Appl.*, t. **XXXV**, p. 239-244, 1968.

([2054]) PURCHASE I. F. H. et STEYN M. — Aflatoxin metabolism. *in* PURCHASE I. F. H., *Mycotoxins in human health*, p. 47-51, 1971.

([2055]) PURCHASE I. F. H., STEYN M. et PRETORIUS H. E. — The production of aflatoxin M on various substrates. *Mycopathol. Mycol. Appl.*, t. **XXXV**, p. 239-244, 1968.

([2056]) PURCHASE I. F. H. et THERON J. J. — The acute toxicity of ochratoxin A to rats. *Food Cosmet. Toxicol.*, t. **VI**, p. 479-483, 1968.

([2057]) PURCHASE I. F. H. et VAN DER WATT J. J. — Acute toxicity of sterigmatocystin to rats. *Food Cosmet. Toxicol.*, t. **VII**, p. 135-139, 1969.

([2058]) PURCHASE I. F. H. et VAN DER WATT J. J. — Carcinogenicity of sterigmatocystin. *Food Cosmet. Toxicol.*, t. **VIII**, p. 289-295, 1970.

([2059]) PURCHASE I. F. H. et VAN DER WATT J. J. — The acute and chronic toxicity of sterigmatocystin. *in* PURCHASE I. F. H., *Mycotoxins in human health*, p. 209-213, 1971.

([2060]) PURCHASE I. F. H. et VORSTER L. J. — Aflatoxin in commercial milk samples *S. Afr. Med. J.*, t. **XLII**, p. 219, 1968.

([2061]) PURCHIO A. et CAMPOS R. — Molluscicidal activity of aflatoxin. *Rev. Inst. Med. Trop. Sao Paulo*, t. **XII**, p. 236-238, 1970.

([2062]) PURVES L. R. — Surveys for alpha-feto-protein among Bantu goldminers. *in* PURCHASE I. F. H., *Mycotoxins in human health*, p. 71-73, 1971.

([2063]) PUSEY D. F. G. et ROBERTS J. C. — Mycological chemistry. XIII. Averufin, a red pigment from *Aspergillus versicolor*. *J. Chem. Soc.*, p. 3542-3547, 1963.

([2064]) QASEM S. A. et CHRISTENSEN C. M. — Influence of moisture content, temperature and time on the deterioration of stored corn by fungi. *Phytopathology*, t. **XLVIII**, p. 544-549, 1958.

([2065]) QASEM S. A. et CHRISTENSEN C. M. — Influence of various factors on the deterioration of stored corn by fungi. *Phytopathology*, t. **L**, p. 703-709, 1960.

([2066]) QUENUM C., MARTINEAUD M. et CROS J. — Hépatite nécrosante provoquée par un régime contenant de l'aflatoxine chez un *Phacochoerus aethiopicus*. *C. R. Soc. Biol.*, t. **CLXI**, p. 1767-1772, 1967.

([2067]) QUEVAUVILLER A. et BITNET P. — Modalités de l'emploi des antiseptiqes et des désinfectants en hygiène collective. *Colloque Antiseptiques et Désinfactans, Inst. Pasteur*, mai 1964.

([2068]) QUILICO A. et CARDANI C. — The diffusion of echinulin in molds of the group *Aspergillus glaucus*. *Atti Accad. Nazl. Lincei, Rend. Classe Sci. Fis. Mat. e Nat.*, t. IX, p. 220-228, 1950.

([2069]) QUILICO A. et CARDANI C. — Echinulin III. Chemical investigations of the *Aspergillus glaucus* group VI. *Gazz. Chim. Ital.*, t. **LXXXIII**, p. 155-178, 1953.

([2070]) QUILICO A., CARDANI C. et PIOZZI F. — Echinulin IV. Chemical investigations of the *Aspergillus glaucus* group VII. *Gazz. Chim. Ital.*, t. **LXXXIII**, p. 179-191, 1953.

([2071]) QUILICO A., CARDANI C. et PIOZZI F. — Chemical investigations of the *Aspergillus glaucus* group. XI. Echinulin. *Gazz. Chim. Ital.*, t. **LXXXV**, p. 3-33, 1955.

([2072]) QUILICO A., CARDANI C. et PIOZZI F. — Chemical investigations of the *Aspergillus glaucus* group. XII. Echinulin. VI. *Gazz. Ital.*, t. **LXXXVI**, p. 211-233, 1956.

([2073]) QUILICO A., PANIZZI L. et ROSNATI V. — Echinulin. *Proc. Intern. Congr. Pure and Appl. Chem., 11th Congr. London, 1947*, t. **II**, p. 259-266, 1950.

([2074]) RABIE C. J. — Influence of temperature on the production of aflatoxin by *Aspergillus flavus*. *South African Dept. Agr. and Tech. Serv. Symposium on mycotoxins in Foodstuffs (Agricultural Aspects). Pretoria, 25-26 févr.*, p. 18-29, 1965.

([2075]) RABIE C. J. — New toxic fungi and physiology of toxin production. *1st Int. Congr. Pl.. Pathol., Londres*, Abstr. p. 158, 1968.

([2076]) RABIE C. J., DE KLERCK W. A. et TERBLANCHE M. — Toxicity of *Aspergillus amstelodami* to poultry and rabbits. *South Afr. Journ. Agric. Sci.*, t. V, p. 341-346, 1964.

([2077]) RABIE C. J. et SMALLEY E. B. — Influence of temperature on the production of aflatoxin by *Aspergillus flavus. Symp. Mycotoxins Foodstuffs, Agr. Aspects, Pretoria*, p. 18-19, 1965.

([2078]) RABIE C. J. et TERBLANCHE M. — Influence of temperature on the toxicity of different isolates of *Aspergillus wentii* Wehm. *S. Afr. J. Agric. Sci.*, t. X, p. 263-266, 1967.

([2079]) RABIE C. J., TERBLANCHE M., SMIT J. D. et DE KLERK W. A. — Toxicity of *Aspergillus wentii* Wehmer. *S. Afr. T. Lansbouwet*, t. **VIII**, p. 875-879, 1965.

([2080]) RADELEFF R. D. — *Veterinary toxicology*, 2e éd. 352 p., Lea et Febiger édit., Philadelphia, 1970.

([2081]) RAILLO I. A. — *Gribi roda Fusarium.* 415 p., Moscou, 1950.

([2082]) RAINEY D. P., SMALLEY E. B., CRUMP M. H. et STRONG F. M. — Isolation of salivation factor from *Rhizoctonia leguminicola* on red clover hay. *Nature, G.B.*, t. **CCV**, p. 203-204, 1965.

([2083]) RAISTRICK H., BIRKINSHAW J. H., MICHAEL S. E., BRACKEN A., GYE W. E., HOPKINS W. A. et GREENWOOD M. — Patulin in the common cold. *Lancet*, t. **II**, p. 625-635, 1943.

([2084]) RAISTRICK H., ROBINSON R. et TODD A. R. — The chemistry of *Aspergillus* colouring matters. I. *Jour. Chem. Soc. London*, p. 80-88, 1937.

([2085]) RAISTRICK H. et RUDMAN P. — Studies in the biochemistry of micro-organisms. 97. *Biochem. J.*, t. **LXIII**, p. 395-405, 1956.

([2086]) RAISTRICK H. et SIMONART P — Studies in the biochemistry of microorganisms. 29.2 : 5-Dihydroxybenzoic acid (gentisic acid) a new product of the metabolism of glucose by *Penicillium griseofulvum* Dierckx. *Biochem. J.*, t. **XXVII**, p. 628-633, 1933.

([2087]) RAISTRICK H. et SMITH G. — Studies in the biochemistry of microorganisms. XXXV. The metabolic products of *Byssochlamys fulva* Olliver et Smith. *Biochem. J.*, t. **XXVII**, p. 1814-1819, 1933.

([2088]) RAISTRICK H. et SMITH G. — Biochemistry of microorganisms XLII. *Biochem. J.*, t. **XXIX**, p. 606-611, 1935.

([2089]) RAISTRICK H. et SMITH G. — Studies in the biochemistry of microorganisms LI. *Biochem. J.*, t. **XXX**, p. 1315-1322, 1936.

([2090]) RAISTRICK H. et SMITH G. — Antibacterial substances from moulds. I. Citrinin, a metabolic product of *Penicillium citrinum* Thom. *Chem. and Indus.*, t. **LX**, p. 828-830, 1941.

([2091]) RAISTRICK H. et STOSSL A. — Metabolites of *Penicillium atrovenetum* G. Smith : β-nitropropionic acid, a major metabolite. *Biochem. J.*, t. **LXVIII**, p. 647-653, 1958.

([2092]) RAJ H. G., SHANKARAN R. et VENKITASUBRAMANIAN T. A. — Metabolic effects of aflatoxin in one day old chicks. *Indian J. Biochem.*, t. **VII**, p. 55-56, 1970.

([2093]) RAJ H. G., SHANKARAN R., VISWANATHAN L. et VENKITASUBRAMANIAN T. A. — Phospholipid biosynthesis in *Aspergillus flavus. Abstr. Ann. meeting Soc. Biol. Chem. India, Hyderabad*, p. 48, 1969.

([2094]) RAMAUT J. L. — Biosynthèse de pigments par *Aspergillus versicolor* en fonction des conditions de milieu. *Rev. Fermentations et Ind. Alimentaires*, t. **XVII**, p. 77-81, 1962.

([2095]) RAMUT J. L. — Contribution à l'étude des produits du métabolisme d'*Aspergillus versicolor* (Vuil.) Tiraboschi. II. Fraction souble dans l'éther sulfurique. *Phytochemistry*, t. **I**, p. 259-262, 1962.

([2096]) RAMAUT J. L., VANDERHOVEN C. et REMACLE J. — Etude de la production d'aflatoxine B_1 en fonction des étapes du développement morphologique d'*Aspergillus flavus* Link. *Rev. Ferment. Ind. Alim.*, t. **XXV**, p. 184-188, 1970.

([2097]) RANCE M. J. et ROBERTS J. C. — Total synthesis of racemic dihydro-o-methyl sterigmatocystin. *Tetrahedron L.*, n° 4, p. 277-278, 1969.

([2098]) RANDERATH K. — *Chromatographie sur couches minces.* (Traduit de l'allemand par NGUYEN-DANG-TAM). 294 p., GAUTHIER-VILLARS edit., Paris, 1964.

([2099]) RAO A. S. — Fungal populations in the rhizosphere of peanut (*Arachis hypogea* L.). *Plant Soil*, t. **XVII**, p. 260-266, 1962.

([2100]) RAO K. R. et THIRUMALACHAR M. J. — Control of black rot cabbage with citrinin. *Hindustan Antibiotics Bull.*, t. **II**, p. 126-127, 1960.

([2101]) RAO K. S. — Aflatoxin B_1 induced inhibition of liver protein synthesis in vivo and its role in fatty liver. *Biochem. Pharmacol.*, t. **XX**, p. 2825-2831, 1971.

([2102]) RAO K. S. et GEHRING P. J. — Acute toxicity of aflatoxin B_1 in monkeys. *Toxicol. Appl. Pharmacol.*, t. **XIX**, p. 169-175, 1971.

([2103]) RAO K. S., MADHAVAN T. V. et TULPULE P. G. — Incidence of toxigenic strains of *Aspergillus flavus* affecting groundnut crop in certain coastal districts of India. *Indian J. Med. Res.*, t. **LIII**, p. 1196-1202, 1965.

([2104]) RAO K. S. et TULPULE P. G. — Varietal differences of groundnut in the production of aflatoxin. *Nature, G.B.*, t. **CCXIV**, p. 738, 1967.

([2105]) RAPER K. B. et FENNELL D. I. — The genus *Aspergillus.* 686 p. The Williams and Wilkins Co, Baltimore, 1965.

([2106]) RAPER K. B. et THOM C. — *A manual of the Penicillia.* 875 p., Williams et Wilkins Co, Baltimore, 1949.

([2107]) RAU E. M., KOENIG V. L. et TILDEN E. B. — Partial purification and characterization of the endotoxin from *Aspergillus fumigatus. Mycopathologia*, t. **XIV**, p. 347-358, 1961.

([2108]) RAYNAUD J. P. — Une épidémie d'hépatite cirrhose du porc sévissant à Madagascar. Etude des tests hépatiques chez le porc et utilisation de la vitesse de sédimentation pour un diagnostic précoce. *Rev. Elev. Méd. Vét. Pays Trop.*, t. **XVI**, p. 429-437, 1961.

([2109]) RAYNAUD P. — Une dystrophie hépatique toxique du porc à Madagascar. Etude clinique, lésions, reproduction expérimentale par ingestion de tourteau d'arachide. *Rev. Elev. Méd. Vét. Pays Trop.*, t. **XVI**, p. 23-32, 1963.

([2110]) RAYNER E. T. et DOLLEAR F. G. — Removal of aflatoxins from oilseed meals by extraction with aqueous isopropanol. *J. Amer. Oil Chem. Soc.*, t. **XLV**, p. 622-624, 1968 et *Proc. 5th Nat. Peanut Res. Conf. Virginia*, p. 118, 1969.

([2111]) RAYNER E. T., DOLLEAR F. G. et CODIFER L. P. — Extraction of aflatoxins from cottonseed and peanutmeals with ethanol. *J. Amer. Oil Chem. Soc.*, t. **XLVII**, p. 26, 1970.

([2112]) REBSTOCK M. C. — New metabolite of patulin producing Penicillia. *Arch. Biochem. Biophys.*, t. **CIV**, p. 156-159, 1964.

([2113]) RECONDO A. M. de — Inhibition de la synthèse du DNA par l'aflatoxine. Comparaison avec le mode d'action de l'actinomycine. *Colloque aflatoxines, Villejuif*, 7 mars 1966.

([2114]) RECONDO A. M. de — Mode d'action de l'aflatoxine B_1 sur la synthèse du DNA au cours de l'hypertrophie compensatrice du foie. *Rapport activité C.N.R.S oct. 1964-oct. 1965*, p. 352, 1966.

([2115]) RECONDO A. M. de, FRAYSSINET C., LAFARGE C. et LE BRETON E. — Inhibition de la synthèse du DNA par l'aflatoxine B_1 au cours de l'hypertrophie compensatrice du foie chez le rat. *C.R. Acad. Sci.*, t. **CCLXI**, p. 1409-1412, 1965.

([2116]) RECONDO A. M. de, FRAYSSINET C., LAFARGE C. et LE BRETON E. — Action de l'aflatoxine sur le métabolisme du DNA au cours de l'hypertrophie compensatrice du foie après hépatectomie partielle. *Biochim. Biophys. Acta, Pays-Bas*, t. **CXIX**, p. 322-330, 1966.

([2117]) REDDY T. V., VISWANATHAN L. et VENKITASUBRAMANIAN T. A. — Thin-layer chromatography of aflatoxins. *Anal. Biochem.*, t. **XXXVIII**, p. 568-570, 1970.

([2118]) REDMOND H. E. — Wheat poisoning in cattle. *Southw. Vet.*, t. III, p. 22, 1950.

([2119]) REES K. R. — Aflatoxin. *Gut., G.B.*, t. **VII**, p. 205-207, 1966.

([2120]) REES K. B. — The inhibition of protein synthesis and fatty change in drug induced liver injury. (2nd Int. Symp. on drugs affecting lipid metabolism, Milan, 1966) *Progr. Biochem. Pharmacol.*, t. **III**, p. 439-444, 1967.

([2121]) REESE E. T., SIU R. G. H. et LEVINSON H. S. — The biological degradation of soluble cellulose derivatives and its relationship to the mechanism of cellulose hydrolysis. *J. Bacteriol.*, t. **LIX**, p. 485-497, 1950.

([2122]) REHM H. J. et MEYER H. — Mykotoxinbildung in Fruchtsaften. *Die Industrielle Obst-u. Gemüseverwertung*, t. **LII**, p. 675-677, 1967.

([2123]) REHM H. J. et SCHMIDT I. — Mycotoxins in foodstuffs. III. Production of ochratoxins in different foodstuffs. *Zentralbl. Bakteriol.*, Abt. 2, t. **CXXIV**, p. 364-368, 1970.

([2124]) REILLY H. C., SCHATZ A. et WAKSMAN S. A. — Antifungal properties of antibiotic substances. *J. Bacteriol.*, t. **XLIX**, p. 585-594, 1945.

([2125]) REIO L. — A method for the paper chromatographic separation and identification of phenol derivatives, mould metabolites and related compounds. *J. Chromatogr.*, t. **I**, p. 338-373, 1958.

([2126]) REISS J. — Mycotoxine I. Mycotoxine von *Aspergillus*-Arten. *Z. allg. Mikrobiol.*, t. **VIII**, p. 303-330, 1968.

([2127]) REISS J. — Hemmung des Sprosswachstums von *Caralluma frerei* Rowl. durch Aflatoxin. *Planta*, t. **LXXXIX**, p. 369-371, 1969.

([2128]) REISS J. — Untersuchungen über die Phytotoxizität von *Penicillium expansum*, *Aspergillus niger* und *Rhizopus nigricans*. *Phytopathol. Z.*, t **LXIX**, p. 78-82, 1970.

([2129]) REISS J. — Untersuchungen über den Einfluss von Aflatoxin B_1 auf die Morphologie und die Cytochemisch fassbare Aktivität einiger Enzyme von *Mucor hiemalis* (Mucorales). *Mycopathol. Mycol. Appl.*, t. **XLII**, p. 225-231, 1970.

([2130]) REISS J. — Auftrennung der Aflatoxine B_1, B_2, G_1 und G_2 mit Hilfe dünnschicht-chromatographischer Horizontaltecknik (Circular-und Radialmethode). *Fresenius Z. Anal. Chem.*, t. **CCLI**, p. 306-307, 1970.

([2131]) REISS J. — Förderung der Aktivität von β-Indolylessigsäure durch Aflatoxin B_1. *Z. f. Pflzphysiol.*, t. **LXIV**, p. 260-262, 1971.

([2132]) REISS J. — Inhibition of fungal sporulation by aflatoxin. *Arch. Mikrobiol.*, t. **LXXVI** p. 219-222, 1971.

([2133]) REISS J. — Der Einfluss von Aflatoxin B_1 auf *Paramecium caudatum* und *Paramecium bursaria*. *Arch. Hyg. Bakteriol.*, t. **CLIV**, p. 533-536, 1971.

([2134]) REISS J. — Wachstum von *Aspergillus flavus* und Aflatoxinbildung in geschnittenem und verpacktem Wiezenvollkornbrot. *Z. Lebensm.-Unters.-Forsch.*, t. **CXLV**, p. 155-158, 1971.

([2135]) REISS J. — Förderung der Aktivität von β-Indolylessigsäure durch Aflatoxin B_1. *Z. Pflanzenphysiol.*, t. **LXIV**, p. 260-262, 1971.

([2136]) REISS J. — Hemmung der Keimung der Kresse (*Lepidium sativum*) durch Aflatoxin B_1 und Rubratoxin B. *Biochem. Physiol. Pflanz.*, t. **CLXII**, p. 363-367, 1971.

([2137]) REMACLE J. — Population micromycétique de quelques types forestiers du Plateau du Sart Tilman (Liège, Belgique). *Plant and Soil*, t. **XXIII**, p. 285-294, 1965.

([2138]) REMACLE J., VANDERHOVEN C. et RAMAUT J. — Recherche d'organismes inhibiteurs du développement d'*Aspergillus flavus* Link ou destructeurs de l'aflatoxine B_1. *Rev. Ferment. Ind. Aliment.*, t. **XXV**, p. 227-235, 1970.

([2139]) RENAUD R. — Les moisissures du Cacao marchand. *Bull. Centre Rech. Agron. Bingerville*, t. **VII**, p. 45-64, 1953.

([2140]) RENAUD R. — La qualité du Cacao. Les moisissures des fèves fermentées. *L'Agronomie tropicale*, fasc. 4-5, p. 145-165, 1954.

([2141]) RENAULT P., ARNAUD G. et PRESLE G. D. — Aspergillose nasale et sinusienne. *Mycol. Méd. Exp. Sci. Fr.*, p. 44-47, 1958.

(2142) RENON J. — *Etude sur l'aspergillose chez les animaux et chez l'homme.* 301 p., Paris, Masson édit., 1897.

(2143) RENON L — Deux cas familiaux de tuberculose aspergillaire simple chez des peigneurs de cheveux, *C.R. Soc. Biol.*, t. **XLVII**, p. 694-696, 1895.

(2144) REUSSER F. — Biosynthesis of antibiotic U-22, 324, a cyclic polypeptide. *J. Biol. Chem.*, t. **CCXLII**, p. 243-247, 1967.

(2145) REY L. — *Aspects théoriques et industriels de la lyophilisation.* 653 p., Hermann édit., Paris, 1964.

(2146) RHODES A. — *Progr. Ind. Microbiol.*, t. **IV**, p. 167, 1963.

(2147) RHODES A. et FLETCHER D. L. — *Principles of industrial microbiology.* 320 p., Pergamon Press, 1966.

(2148) RHODES A., MC GONAGLE M. P. et SOMMERFIELD G. A. — Biosynthesis of geodin and asterric acid. *Chem. et Ind.*, p. 601-612, 1962.

(2149) RICHARD J. L. — Toxigenic fungi associated with stored corn with special emphasis on the toxigenicity of *Trichothecium roseum. Diss. Abstr.*, Ser. B, t. **XXIX**, p. 3227-3228, 1969.

(2150) RICHARD J. L. et CYSEWSKI S. J. — Occurrence of aflatoxin producing strains of *Aspergillus flavus* Link in stored corn. *Mycopathol. Mycol. Appl.*, t. **XLIV**, p. 221-229, 1971.

(2151) RICHARD J. L., ENGSTROM G. W., PIER A. C. et TIFFANY L. M. — Toxigenicity of *Trichothecium roseum* Link : isolation and partial characterization of a toxic metabolite. *Mycopathol, Mycol. Appl.*, t. **XXXIX**, p. 231-240, 1969.

(2152) RICHARD J. L., PIER A. C. et TIFFANY L. M. — Biological effects of toxic products from *Trichothecium roseum* Link. *Mycopathol. Mycol. Appl.*, t. **XL**, p. 161-170, 1970.

(2153) RICHARD J. L., TIFFANY L. H. et PIER A. C. — Toxigenic fungi associated with stored corn. *Mycopathol. Mycol. Appl.*, t. **XXXVIII**, p. 313-326, 1969.

(2154) RICHIR C. — Expérimentation réalisée sur le rat et le singe. *Colloque Aflatoxines. Villejuif*, 7 mars 1966.

(2155) RICHIR C., CROS J. et DUPIN M. — Les problèmes des toxicoses fongiques. *Le Concours médical*, t. **LXXXVI**, p. 371, 1964.

(2156) RICHIR C., CROS J., TOURY J., QUENUM C., DUPIN M. et TRELLU M. — Lettre à l'éditeur à propos d'un hépatome métastasiant survenu chez le rat Whistar à un régime contenant de l'aflatoxine. *Nutritio et Dieta*, t. **VI**, p. 76, 1964.

(2157) RICHIR C., MARTINEAUD M., TOURY J., GIORGY R. et DUPIN M. — Sur les effets cancérigènes des régimes contenant des arachides contaminées par *Aspergillus flavus. C.R. Soc. Biol.*, t. **CLVIII**, p. 1375-1379, 1964.

(2158) RICHIR C., TOURY J., GIORGI R. et DUPIN M. — Analyse des lésions hépatiques du caneton traité par des extraits d'arachides contaminées par *Aspergillus flavus. Pathol. et Biol.*, t. **XII**, p. 980-987, 1964.

(2159) RIDDLE H. F. V., CHANNELL S., BLYTH W., WEIR D. M., LLOYD M., AMOS W. M. G. et GRANT I. W. B. — Allergic alveolitis in a maltworker. *Thorax*, t. **XXIII**, p. 271-280, 1968.

(2160) RIEMANN H. — *Food-borne infections and intoxications.* 698 p., Academic Press, New York, 1969.

(2161) RINDERKNECHT H., WARD J. L., BERGEL F. et MORRISON A. L. — Studies on antibiotics. *Biochem. J.*, t. **XLI**, p. 463-469, 1947.

(2162) ROBERT M., BARBIER M., LEDERER E., ROUX L., BIEMANN K. et VETTER W. — Two new natural phytotoxins : aspergillomarasmine A and B and their relation to lycomarasmine and its derivatives. *Bull. Soc. Chim. Fr.*, p. 187-188, 1962.

(2163) ROBERTS B. A. et ALLCROFT R. — A note on the semi-quantitative estimation of aflatoxin M_1 in liquid milk by thin layer chromatography. *Food Cosmet. Toxicol.*, t. **VI**, p. 339-340, 1968.

([2164]) ROBERTS D. W. — Toxins produced by insect-pathogenic fungi. *1st Int. Mycol. Congr. Exeter. Abstr.* p. 80; 1971.

([2165]) ROBERTS J. C., SHEPPARD A. H., KNIGHT J. A. et ROFFEY O. — Studies in mycological chemistry. Part XXII. Total synthesis of racemic aflatoxin B_2. *J. Chem. Soc.* (C), p. 22-24, 1968.

([2166]) ROBERTS J. C. et WARREN C. W. H. — Studies in mycological chemistry. IV. Purpurogenone, a metabolic product of *Penicillium purpurogenum* Stoll. *J. Chem. Soc.*, p. 2992-2998, 1955.

([2167]) ROBERTS J. C. et WOOLLVEN P. — Studies in mycological chemistry. XXIV. Synthesis of ochratoxin A, a metabolite of *Aspergillus ochraceus* Wilh. *J. Chem. Soc., Sect. C, Org. Chem.*, t. **II**, p. 278-281, 1970.

([2168]) ROBERTSON J. A., LEE L. S., CUCULLU A. F. et GOLDBLATT L. A. — Assay of aflatoxin in peanuts and peanut products using acetone-hexane-water for extraction. *J. Amer. Oil Chem. Soc.*, t. **XLII**, p. 467-471, 1965.

([2169]) ROBERTSON J. A. et PONS W. A. — Solid state fluorescence emission of aflatoxins on silica gel. *J. Ass. Off. Anal. Chem.*, t. **LI**, p. 1190-1192, 1968.

([2170]) ROBERTSON J. A., PONS W. A. et GOLBLATT L. A. — Preparation of aflatoxins and determination of their ultraviolet and fluorescent characteristics. *J. Agr. Food Chem.*, t. **XV**, p. 798-801, 1967.

([2171]) ROBERTSON J. A., PONS W. A. et GOLDBLATT L. A. — Stability of individual aflatoxin B_1, B_2, G_1 and G_2 standards in benzene and chloroform solutions. *J. Ass. Off. Anal. Chem.*, t. **LIII**, p. 299-302, 1-70.

([2172]) ROBERTSON J. A., TEUNISSON D. J. et BOUDREAUX G. J. — Isolation and structure of a biologically reduced aflatoxin B_1. *J. Agr. Food Chem.*, t. **XVIII**, p. 1090-1091, 1970.

([2173]) ROBERTSON O. H. — Accelerated development of testis after unilateral gonadectomy with observations on normal testis of Rainbow Trout. *Fish. Bull.*, t. **CXXVII**, p. 9-30, 1958.

([2174]) ROCHE B. H. et BOHSTEDT G. — Scabbed barley and oats and their effect on various classes of livestock. *Proc. Am. Soc. Anim. Prod.*, t. **XXIII** (1930), p. 219, 1931.

([2175]) ROCHE B. H., BOHSTEDT G. et DICKSON J. G. — Feeding scab-infected barley. *Phytopathology*, t. **XX**, p. 132, 1930.

([2176]) RODERICK L. M. — The pathology of sweet clover disease in cattle. *J. Am. Vet. Med. Assoc.*, t. **LXXIV**, p. 314, 1929.

([2177]) RODERICK L. M. — A problem in the coagulation of the blood. « Sweet clover disease of cattle ». *Am. J. Physiol.*, t. **XCVI**, p. 413, 1931.

([2178]) RODERICK L. M. et SCHALK A. F. — Studies on sweet clover disease. *North Dakota Agr. Exp. Stn., Bull.* 250, 1931.

([2179]) RODRICKS J. V. — Separation and purification of aflatoxins B_1, B_2, G_1 and G_2 and comparison of semi-synthetic aflatoxins B_2 and G_2. *J. Amer. Chem. Soc.*, t. **XLVI**, p. 149-151, 1969.

([2180]) RODRICKS J. V. — Note on adsorption of aflatoxin standards to glass. *J. Ass. Off. Anal. Chem.*, t. **LII**, p. 979-980, 1969.

([2181]) RODRICKS J. V., HENERY-LOGAN K. R., CAMPBELL A. D., STOLOFF L. et VERRETT M. J. — Isolation of a new toxin from cultures of *Aspergillus flavus*. *Nature, G.B.*, t. **CCXVII**, p. 668, 1968.

([2182]) RODRICKS J. V., LUSTIG E., CAMPBELL A. D., STOLOFF C. et HENERY-LOGAN K. R. — Aspertoxin, a hydroxy derivative of o-methyl-sterigmatocystin from aflatoxin producing culture of *Aspergillus flavus*. *Tetrahedron*, t. **XXV**, p. 2975-2978, 1968.

([2183]) ROEGNER F. R. — The FDA looks at aflatoxins. *38th Ann. Tech. Conf. Grain Elevator Proceeding Superintendents Assoc. Chicago*, Feb. 28, 1967.

([2184]) ROGERS A. E. et NEWBERNE P. M. — The effects of aflatoxin B_1 and dimethylsulfoxide on thymidine-^{3}H uptake and mitosis in rat liver. *Cancer Res.*, t. **XXVII**, p. 855-864, 1967.

([2185]) ROGERS A. E. et NEWBERNE P. M. — Early stages of aflatoxin carcinogenesis in normal and lipotrope deficient rats. *Fed. Proc.*, t. **XXVII**, p. 607, 1968.

([2186]) ROGERS A. E. et NEWBERNE P. M. — Aflatoxin B_1 carcinogenesis in lipotrope-deficient rats. *Cancer Res.*, t. **XXIX**, p. 1965-1972, 1969.

([2187]) ROGERS L. A., DAHLBERG A. O. et EVANS A. E. — The cause and control of «buttons» in sweet condensed milk. *J. Dairy Sci.*, t. **III**, p. 122-129, 1920.

([2188]) RONALDSON J. W., TAYLOR A., WHITE E. P. et ABRAHAM R. J. — Sporidesmins. Part I. Isolation and characterization of sporidesmin and sporidesmin B. *J. Chem. Soc.*, p. 3172-3180, 1963.

([2189]) RONDANELLI E. G., GORINI P., STROSSELTI E. et PICORADI D. — Inhibition of the metaphase effect of patulin by thiols. Experimental research *in vivo* and *in vitro*. *Haematologia*, Pavia, t. **XLII**, p. 1427-1440, 1957.

([2190]) RONDELET J. — Action des rayons ultra-violets sur quelques champignons pathogènes. *Thèse Médecine*. Lyon, 1957.

([2191]) RONK S. E. et CARRICK C. W. — Feeding moldy corn to young chickens. *Poultry Sci.*, t. **X**, p. 236-244, 1931.

([2192]) ROOK J. A. F. et BALCH C. C. — Magnesium metabolism in the dairy cow. III. The intake and excretion of calcium, phosphorus, sodium, potassium, water and dry matter in relation to the development of hypomagnesaemia. *J. Agric. Sci.*, t. **LIX**, p. 103-108, 1962.

([2193]) ROOK J. A. F. et WOOD M. — Mineral composition of herbage in relation to the development of hypomagnesaemia in grazing cattle. *J. Sci. Food. Agric.*, t. **XI**, p. 137-142, 1960.

([2194]) ROSENFELD I. et BEATH O. A. — Pathology of Selenium poisoning. *Wyoming Agr. Expt. Stn. Bull.*, n° 275, 1946.

([2195]) ROSS V. C. et LEGATOR M. S. — Alternation in thymidine kinase activity by mitotic inhibitors. *In vitro*, t. **II**, p. 141, 1966.

([2196]) ROSSI P. et BRUYERE A. — Brucellose bovine et antibiothérapie. *Bull. Acad. Vet. Fr.*, t. **XXIII**, p. 443-449, 1950.

([2197]) ROY A. K. — Effects of aflatoxin B_1 on polysomal profiles and RNA synthesis in rat liver. *Biochim. Biophys. Acta*, t. **CLXIX**, p. 206-211, 1968.

([2198]) ROZKHOVA A., PAKCHUKH et al. — Isolation, purification and antibacterial properties of cephalosporin R. *Antibiotiki*, t. **I**, p. 9, 1960.

([2199]) RUBIN B. A. — The trypanocidal effect of antibiotic lactones and their analogs. *Yale J. Biol. and Med.*, t. **XX**, p. 233-271, 1947-48.

([2200]) RUBIN B. A. et GIARMAN N. J. — The therapy of experimental influenza in mice with antibiotic lactones and related compounds. *Yale J. Biol. and Med.*, t. **XIX**, p. 1017-1022, 1947.

([2201]) RUCKER R. R., YASUTAKE W. T. et WOLF H. — Trout hepatoma — a preliminary report. *Prog. Fish-Cult.*, t. **XXIII**, p. 3-7, 1961.

([2202]) RUDOLPH E. D. — The effect of some physiological and environmental factors on sclerotial Aspergilli. *Am. J. Bot.*, t. **XLIX**, p. 71-78, 1962.

([2203]) RUSSELL G. R. — Detection and estimation of a facial eczema toxin, sporidesmin. *Nature*, t. **CLXXXVI**, p. 788-789, 1960.

([2204]) RUSSELL G. R. — Detection of sporidesmin in pasture extracts. *Nature*, t. **CXCIII**, p. 354-356, 1962.

([2205]) RUTQVIST L. — Haemolytic, toxic and proteolytic effects of filtrates from *Aspergillus fumigatus*. *Proc. Symp. Agric. Coll. Norway*, 1967, 1 p. ronéot.

([2206]) RUTQVIST L. et PERSSON P. A. — Studies on *Aspergillus fumigatus*, experimental mycotoxicosis in mice, chicks and pigs with the appearance, in pigs, of perirenal edema. *Acta Veterinaria Scand.*, t. **VII**, p. 21-34, 1966.

([2207]) RYBAKOVA S. G. — β-Naftol v borbie c plesnivmi gribami porazkayuotshimi razlichioi klyloi na bumagii i kartonii. *Mikrobiologiya*, t. **XXIV**, p. 608-610, 1955.

([2208]) SACCARDO P. A. — Spegazzini C. Nova addenda ad mycologiam venetam. *Michelia*, t. **V**, p. 478, 1879.

([2209]) SAEZ H. — Quelques cas d'aspergillose aviaire observés au parc zoologique de Paris. Le parasite et l'hôte. *Ann. parasitol. humaine et comparée*, t. **LXIX** p. 89-99, 1961.

([2210]) SAEZ H. — Un champignon souvent isolé dans des prélèvements d'origine animale : l'*Aspergillus candidus* Link. *Bull. Mus. Nat. Hist., Nat.*, t. **XXXIII**, p. 341-345, 1961.

([2211]) SAFE S. et TAYLOR A. — Sporidesmins. X. Synthesis of polysuphides by reaction of dihydrogen disulphide with disulphides and thiols. *J. Chem. Soc. Sect. C. Org. Chem.*, t. **III**, p. 432-435, 1970.

([2212]) SAINT-LÈBE L. et BERGER G. — Conservation par irradiation des denrées alimentaires. *B.I.S.T., C.E.A.*, n° 141, p. 77-84, 1969.

([2213]) SAINT-LÈBE L., GUILBOT A. et PELEGRIN P. — Etude d'un traitement combiné pour le conditionnement du maïs récolté humide : irradiation gamma et séchage par ventilation contrôlée. *C.E.N. Cadarache*, 16 p., 1968.

([2214]) SAITO M. — Liver cirrhosis induced by metabolites of *Penicillium islandicum* Sopp. *Acta Pathol. Jap.*, t. **IX**, p. 785-790, 1959.

([2215]) SAITO M., ENOMOTO M., UMEDA M., OHTSUBO K., ISHIKO T., YAMAMOTO S. et TOYOKAWA H. — Field survey of mycotoxin-producing fungi contaminating human foodstuffs in Japan, with epidemiological background. *in* Purchase I.F.H., *Mycotoxins in human health*, p. 179-183, 1971.

([2216]) SAITO M., OHTSUBO K., UMEDA M., ENOMOTO M., KURATA H., UDAGAWA S., SAKABE F. et ICHINOE M. — Screening tests using He La cells and mice for detection of mycotoxin-producing fungi isolated from foodstuffs. *Japan. J. Exper. Med.*, t. **XLI**, p. 1-20, 1971.

([2217]) SAITO M. et OKUBO K. — Studies on the target injuries in experimental animals with the mycotoxins of *Fusarium nivale. in* Herzberg M., *Toxic Micro-organisms*, p. 82-93, 1970.

([2218]) SAITO M. et TATSUNO T. — Toxins of *Fusarium nivale. in* Kadis S. et al., Microbial Toxins, t. VII, p. 293-316, 1971.

([2219]) SAITO M., UMEDA M., OHTSUBO K., KURATA H., UDAGAWA S. et NATORI S. — Studies on the detection of carcinogens in natural products. I. Toxic effects of fungi isolated from foodstuffs. *Proc. Jap. Cancer Assoc., 27th Ann. Meet. Tokyo*, p. 59, 1968.

([2220]) SAKABE N., GOTO T. et HIRATA Y. — The structure of citreoviridin, a toxic compound produced by *Penicillium citreoviride* on rice. *Tetrah. Lett.*, n° 27, p. 1825-1830, 1964.

([2221]) SAKAGUCHI O. et YOKOTA K. — Studies on biological characteristics of extracted fractions of several filamentous fungi, especially *Aspergillus fumigatus* toxin. *Jap. J. Bacteriol.*, t. **XXIV**, p. 15-21, 1969.

([2222]) SAKAI F. — Experimental studies on rice yellowsis caused by *Penicillium citrinum* Thom and toxicity especially kidney damaging effect of citrinin pigment produced by the fungus. *Jap. J. Pharmonicol.*, t. **LI**, p. 431-442, 1955.

([2223]) SAKAI F. et URAGUCHI K. — Studies by long term feeding experiments with rats on development of chronic poisoning by toxic substance from yellowsis rice. VII. Pharmacological studies on toxicity of yellowsis rice. *Nisshin Igaku*, t. **XLII**, p. 609, 1955.

([2224]) SAKAKI J. — Toxicological studies on moldy rice. Report I. *J. Tokyo Med. Soc.*, t. **V**, p. 1097, 1891.

([2225]) SAKSENA S. — Ecological factors governing the distribution of soil microfungi in some forest soils of Sagar. *J. Indian Botan. Soc.*, t. **XXXIV**, p. 262-298, 1955.

([2226]) SALIKOV M. I. — Kekologii gribka *Stachybotrys alternans* vinovnika stakhibotriotoksikoza loshadei. *Sovet. Botan.*, t. **VI**, p. 53-56, 1940.

([2227]) SALMON W. D. et NEWBERNE P. M. — Occurrence of hepatomas in rats fed diets containing peanut meal as a major source of protein. *Cancer Res.*, t. **XXIII**. p. 571-576, 1963.

([2228]) SALVIN S. B. — Endotoxin in pathogenic fungi. *J. Immunol.*, t. **LXIX**, p. 89-99, 1952.

([2229]) SAMSONOV P. F. et SAMSONOV A. P. — Mikotoksikozy respiratornye. *Sporovye rasteniya srdnei Azii i Kazakhstana*, p. 123-127, 1965.

([2230]) SANDERS A. G. — Effect of some antibiotics on pathogenic fungi. *Lancet*, t. **CCL**, p. 44-46, 1946.

(2231) SANDERS T. H., DAVIS N. D. et DIENER U. L. — Effect of carbon dioxide, temperature, and relative humidity on production of aflatoxin in peanuts. *J. Amer. Oil Chem. Soc.*, t. **XLV**, p. 683-685, 1968.

(2232) SANDOS J., CLARE N. T. et WHITE E. P. — Photosensitivity diseases in New Zealand. XVI. Improved procedure for the beaker test for facial-eczema toxicity. *New Zealand J. Agr. Research*, t. **II**, p. 623-626, 1959.

(2233) SAPONARO A. et MADALUNI A. L. — Ricerche sulla microflora del grano conservato in magazzino. *Bol. Staz. Patol. Vegetale*, sér. 3, t. **XVII**, p. 247-266, 1960.

(2234) SARGEANT K., ALLCROFT R. et CARNAGHAN R. B. A. — Groundnut toxicity. *Vet. Rec.*, t. **LXXIII**, p. 865, 1961.

(2235) SARGEANT K. et CARNAGHAN R. B. A. — Groundnut toxicity in poultry : experimental and chemical aspects. *Brit. Vet. J.*, t. **CXIX**, p. 178-184, 1963.

(2236) SARGEANT K., CARNAGHAN R. B. A. et ALLCROFT R. — Toxic products in groundnuts : chemistry and origin. *Chemistry and Industry*, t. **II**, p. 53-55, 1963.

(2237) SARGEANT K., O'KELLY J., CARNAGHAN R. B. A. et ALLCROFT R. — The assay of a toxic principle in Brazilian groundnut meal. *Vet. Rec.*, t. **LXXIII**, p. 1219-1223, 1961.

(2238) SARGEANT K., SHERIDAN A., O'KELLY J. et CARNAGHAN R. B. A. — Toxicity associated with certain samples of groundnuts. *Nature*, t. **CXCII**, p. 1096-1097, 1961.

(2239) SARKISOV A. Kh. — Etologiia septicheskoi anginy. *Doklad na Komissi po Septicheskoi Angine Ychenogo Meditsinskikh Soveta Narkomsdrava SSR*, 1946.

(2240) SARKISOV A. Kh. — Veterinaria mikologiia. *Veterinariya*, t. **XXIV**, p. 25-27, 1947.

(2241) SARKISOV A. Kh. — *Mikotokochisikozi* (*Mycotoxicoses*). 216 p., Gosudarstvennoe Izdatelstvo Selskohoziastwennor Literatury. Moscou, 1954.

(2242) SARKISOV A. K. — Mycotoxicoses (Etiopathogenesis and control measures). *C. R. 5e Congr. ISHAM*, p. 224-225, Paris, 6 juil. 1971.

(2243) SARKISOV A. Kh. et KVASHNINA E. S. — (Nouvelles propriétés toxico-biologiques du *Fusarium sporotrichioides*.) *Doklady Akad.* Nauk S.S.S.R., t. **LXIII**, p. 77-79. 1948.

(2244) SARKISOV A. Kh. et ORSANSKAYA V. N. — Laboratonaia diagnostika toksichneskogo shtammi griba *Stachybotrys alternans*. *Veterinariya*, t, **XXI**, p. 38-40, 1944.

(2245) SARTORY A. et SARTORY R. — Sur la présence d'*Aspergillus fumigatus* (type Fresenius) dans une farine avariée. *J. Pharm. Alsace-Lorraine*, t. **LIII**, p. 58-59, 1926.

(2246) SASAKI M., ASAO Y. et YOKOTSUKA T. — Studies on the compounds produced by molds. IV et V. Isolation of non fluorescent pyrazine compounds. *Nippon Nogei Kagaku Kaishi*, t. **XLII**, p. 346-355, 1968.

(2247) SASAKI Y. — A study of molds in butter. *J. Fac. Agric. Hokkaido Univ.*, t. **XLIX**, p. 121-249, 1950.

(2248) SATO M. et TATSUNO T. — Chemical studies on chlorine containing peptide. One of the toxic metabolites of *P. islandicum* Sopp. I. Structure and synthesis of dehydrochlorinated-peptide amide. *Chem. Pharm. Bull.*, t. **XVI**, p. 2182-2190, 1968.

(2249) SAVAGE A. — Nitrate poisoning from sugar beet tops. *Canad. J. Comp. Med. and Vet. Sci.*, t. **XIII**, p. 9, 1949.

(2250) SAVEL'YEV V. F. — (Sur les caractères de la mycoflore des cobs de Maïs en Ukraine méridionale.) *J. Microbiol* (*Ukraine*), t. **XXIV**, p. 39-44, 1962.

(2251) SAVEL'YEV V. F. — Sklad ta korotka kharakteristika hribiv-zbudnykiv khvorob kachaniv Kukurudzy pri zroshenni na privdni Ukrayiny. *Vikorist. Zroshuv. zem. Kiev Urozhai*, p. 244-247, 1965.

(2252) SAVIN M. P. — Otravleniye svenoi plesnevim gorochom. *Veterinariya*, t. **XXXVI**, p. 73, 1959.

(2253) SCAIFE J. F. — Aflatoxin B_1 : cytotoxic mode of action evaluated by mammalian cell cultures. *FEBS Lett.*, t. **XII**, p. 143-147, 1971.

([2254]) SCARPELLI D. G. — Ultrastructural and biochemical observations on trout hepatoma. *Res. Rept.*, t. **LXX**, p. 60-71, 1967.

([2255]) SCARPELLI D. G., GREIDER M. M. et FRAJOLA W. J. — Observations on hepatic-cell hyperplasia, adenoma and hepatoma of rainbow trout (*Salmo gairdnerii*). *Cancer Res.*, t. **XXIII**, p. 848, 1963.

([2256]) SCHABORT J. C. et PITOUT M. J. — The effect of aflatoxins on pancreatic deoxyribonuclease. *in* PURCHASE I. F. H., *Mycotoxins in human health*, p. 19-29, 1971.

([2257]) SCHALK A. F. — Cattle disease resulting from eating damaged or spoiled sweet clover hay or silage. *North Dakota Agr. Exp. Stn.*, Circ. 27, 1926.

([2258]) SCHARFF J. W. et CATANEI A. — Champignons inférieurs isolés de l'humus obtenu à Alger par la méthode d'Indore. *Arch. Inst. Pasteur Alger*, t. **XXII**, p. 162-165, 1944.

([2259]) SCHEFFER R. P., NELSON R. R. et ULLSTRUP A. J. — Inheritance of toxin production and pathogenicity in *Cochliobolus carbonum* and *Cochliobolus victoriae*. *Phytopathology*, t. **LVII**, p. 1288-1291, 1967.

([2260]) SCHEFFER R. P. et PRINGLE R. B. — Respiratory effects of the selective toxin of *Helminthosporium victoriae*. *Phytopathology*, t. **LIII**, p. 465-468, 1963.

([2261]) SCHENK R. U. — Development of the peanut fruit. *Univ. Georgia Agr. Exp. Sta. Tech. Bull.*, N. S., n° 22, 53 p., 1961.

([2262]) SCHINDLER A. F. et EISENBERG W. V. — Growth and production of aflatoxins by *Aspergillus flavus* on red pepper (*Capsicum frutescens* L.) *J. Ass. Off. Anal. Chem.*, t. **LI**, p. 911-912, 1968.

([2263]) SCHINDLER A. F. et HARDY D. B. — Periodic production of ochratoxin A by an isolate of *Aspergillus ochraceus*. *Phytopathology*, t. **LX**, p. 587, 1970.

([2264]) SCHINDLER A. F. et NESHEIM S. — Effect of moisture and incubation time on ochratoxin A production by an isolate of *Aspergillus ochraceus*. *J. Ass. Off. Anal. Chem.*, t. **LIII** p. 89-91, 1970.

([2265]) SCHINDLER A. F., PALMER J. G. et EISENBERG W. V. — Aflatoxin production by *Aspergillus flavus* as related to various temperatures. *Appl. Microbiol.*, t. **XV**, p.1006-1009, 1967.

([2266]) SCHLEG M. C. et KNIGHT S. G. — The hydroxylation of progesterone by conidia from *Aspergillus ochraceus*. *Mycologia*, t. **LIV**, p. 317-319, 1962.

([2267]) SCHMID R. — Mycotoxine. *Naturwiss. Rundsch.*, t. **XXI**, p. 253-254, 1968.

([2268]) SCHMIDT E. L. — Nitrate formation by a soil fungus. *Science*, t. **CXIX**, p. 187-189, 1954.

([2269]) SCHMIDT E. L. — Cultural conditions influencing nitrate formation by *Aspergillus flavus*. *J. Bacteriol.*, t. **LXXIX**, p. 553-557, 1960.

([2270]) SCHMIDT E. L. — Nitrate formation by *Aspergillus flavus* in pure and mixed culture in natural environments. *Trans. Intern. Congr. Soil Sci. 7th Congr.*, *Madison* ; *Wisconsin*, t. **II**, p. 600-607, 1960.

([2271]) SCHMIDT I. et REHM H. F. — Mykotoxine in Lebensmitteln. I. Bestimmung von Byssochlaminsäure. *Z. Lebensmittel Untersuch.-Forsch.*, t. **CXXXIX**, p. 20-22, 1968. II. (Production of byssochlamic acid in fats and fat-containing foods), *Id.*, t. **CXLI**, p. 313-317, 1970.

([2272]) SCHMIDT I. et REHM H. J. — Mykotoxine in Lebensmitteln. IV. Wirkung von Byssochlaminsäure, Aflatoxin B_1 und Luteoskyrin auf einige Enzyme. *Zentralbl. Bakteriol.*, Abt. II, t. **CXXV**, p. 520-524, 1970.

([2273]) SCHMIDT L. et BANKOLE R. O. — Detection of *Aspergillus flavus* in soil by immunofluorescent staining *Science*, t. **CXXXVI**, p. 776-777, 1962.

([2274]) SCHMITZ H. — Ueber den Nachweis von Dehydrasen am Asitestumor der Maus mit der Rhunberg-Methode. *Z. Krebsforsch.*, t. **LVI**, p. 596-600, 1950.

([2275]) SCHNATHORST W. C. et HALINSKY P. M. — Severity, prevalence, and ecology of cotton boll rots as related to temperature. *Phytopathology*, t. **L**, p. 653-654, 1960.

([2276]) SCHOENTAL R. — Liver changes and primary liver tumours in rats given toxic guinea pig diet. *Brit. J. C.*, t. **XV**, p. 812-815, 1961.

([2277]) SCHOENTAL R. — Aflatoxin B and G. *Ann. Rev. Pharmacology*, t. **VII**, p. 343, 1967.

([2278]) SCHOENTAL R. — Hepatotoxic activity of retrorsine, senkirkine and hydroxysenkirkine in newborn rats, and the role of epoxides in carcinogenesis by pyrrolizidine alkaloids and aflatoxins, *Nature, G. B.*, t. **CCXVII**, p. 401-402, 1970.

([2279]) SCHOENTAL R., HEAD M. A. et PEACOCK P. R. — *Senecio* alkaloids : Primary liver tumours in rats as a result of treatment with 1) a mixture of alkaloids from *S. jacobaea* Linn ; 2) retrorsine ; 3) isatidine. *Brit. J. Cancer*, t. **VIII**, p. 458, 1954.

([2280]) SCHOENTAL R. et MAGEE P. N. — Chronic liver changes in rats after a single dose of lasiocarpine, a pyrrolizidine alkaloid. *J. Path. Bacteriol.*, t. **LXXIV**, p. 305, 1957.

([2281]) SCHOENTAL R. et MAGEE P. N. — Further observations on the subacute and chronic liver changes in rats after a single dose of various pyrrolizidine alkaloids. *J. Path. Bacteriol.*, t. **LXXVIII**, p. 471-482, 1959.

([2282]) SCHOENTAL R. et WHITE A. F. — Aflatoxins and albinism in plants. *Nature*, t. **CCV**, p. 57-58, 1965.

([2283]) SCHOFIELD F. W. — A brief account of a disease in cattle simulating hemorrhagic septicaemia due to feeding sweet clover. *Canad. Vet. Rec.*, t. **III**, p. 74, 1922.

([2284]) SCHOFIELD F. W. — Damaged sweet clover : the cause of a new disease in cattle simulating hemorrhagic septicemia and blackleg. *J. Am. Vet. Med. Assoc.*, t. **LXIV**, p. 553-575, 1924.

([2285]) SCHOOP G. et KLETTE H. — Oestrogenwirkung pilzbefallenen Roggens auf Schweine. *Fortpflanzung, Zuchthygiene und Haustierbesammung*, t. **V**, p. 37-40, 1955.

([2286]) SCHOTT A. — Berechtigen experimentelle oder klinische Erfahrungen zu der Annahme, dass pathogene oder nicht pathogene Bakterien die Wand des gesunden Magendarmkanals durchwandern können ? *Centralbl. f. Bakt.* Abt. I, t. **XXIX**, p. 239-255 ; 290-297, 1901.

([2287]) SCHROEDER H. W. — Orange stain, a storage disease of rice caused by *Penicillium puberulum. Phytopathology*, t. **LIII**, p. 843-845, 1963.

([2288]) SCHROEDER H. W. — Fungus deterioration of rice : effects of fungus infection on free amino acids and reducing sugars in white and parboiled rice. *Cereal Chem.*, t. **XLII**, p. 539-545, 1965.

([2289]) SCHROEDER H. W. — Factors affecting aflatoxin production *in vitro* by *Aspergillus flavus-oryzae* spp. Rept. to *Peanut Improvement Working Group Meeting, Washington, 27-28 avr.*, 7 p., 1965.

([2290]) SCHROEDER H. W. — Effect of corn steep liquor on mycelial growth and aflatoxin production in *Aspergillus parasiticus. Appl. Microbiol.*, t. **XIV**, p. 381-385, 1966.

([2291]) SCHROEDER H. W. — Factors influencing the development of aflatoxins in some field crops. *1st Int. Congr. Pl. Pathol.*, Londres, juil. 1968.

([2292]) SCHROEDER H. W. — Metabolite of *Macrophomna phaseoli* that can confuse thin-layer chromatographic identification of aflatoxin B_2. *Appl. Microbiol.*, t. **XVI**, p. 946-947, 1968.

([2293]) SCHROEDER H. W. — Aflatoxins in rice in the United States. *in* HERZBERG M. *Toxic Micro-organisms*, p. 56-60, 1970.

([2294]) SCHROEDER H. W. et ASHWORTH L. J. — Aflatoxins in Spanish peanuts in relation to pod and kernel condition. *Phytopathology*, t. **LV**, p. 464-465, 1965.

([2295]) SCHROEDER H. W. et ASHWORTH L. J. — Aflatoxins : some factors affecting production and location of toxins in *Aspergillus flavus-oryzae. J. Stored Prod. Res.*, t. **I**, p. 267-271, 1966.

([2296]) SCHROEDER H. W., BOLLER R. A. et HEIN H. — Reduction in aflatoxin contamination of rice by milling procedures. *Cereal Chem.*, t. **XLV**, p. 574-580, 1968.

([2297]) SCHROEDER H. W. et HEIN H. — Aflatoxins : production of the toxins *in vitro* in relation to temperatures. *Appl. Microbiol.*, t. **XV**, p. 441-445, 1967.

([2298]) SCHROEDER H. W. et HEIN H. — Effect of diurnal temperature cycles on the production of aflatoxin. *Appl. Microbiol.*, t. **XVI**, p. 988-990 ; 1968.

([2299]) SCHROEDER H. W. et SORENSON J. W. — Mold development in rough rice as affected by aeration during storage. *Rice J.*, t. **LXIV**, p. 8-10, 12, 21-23, 1961.

([2300]) SCHROEDER H. W. et VERRETT M. J. — Production of aflatoxin by *Aspergillus wentii* Wehmer. *Can. J. Microbiol.*, t. **XV**, p. 895-898, 1969.

([2301]) SCHULTZ H. W. et LEE J. S. — Food preservation by irradiation. Present status. *Food Tech.*, t. **XX**, p. 38-43, 1966.

([2302]) SCHULTZ J. et MOTZ R. — Toxinbildende *Aspergillus flavus* — Stämme in Futtermitteln. *Arch. exper. Veterinärmed.*, t. **XXI**, p. 129-140, 1967.

([2303]) SCHULTZ J., MOTZ R. et SCHÄFER M. — Zur Toxizität von *Aspergillus flavus* haltigen Malzkeimen. *Mh. Vet. Med.*, t. **XXI**, p. 458-461, 1966.

([2304]) SCHULZ A. — Neuere Untersuchungsergebnisse über die Klumpenbildung bei Mehl, Getreide. *Mehl u. Brot*, t. **III**, p. 149-151, 1949.

([2305]) SCHUMAIER G., DEVOLT H. M., LAFFER N. C. et CREEK R. D. — Stachybotryotoxicosis of chicks. *Poultry Sci.*, t. **XLII**, p. 70-74, 1963.

([2306]) SCHUMAIER G., PENDA B., DEVOLT H. M., LAFFER C. N. et CREEK R. D. — Haemorrhagic lesions in chickens resembling naturally occurring « haemorrhagic syndrome » produced experimentally by mycotoxins. *Poult. Sci.*, t. **XL**, p. 1132-1134, 1961.

([2307]) SCHWARTE L. H., BIESTER H. E. et MURRAY C. — A disease of horses caused by feeding moldy corn. *J. Am. Vet. Med. Assoc.*, t. **XC**, p. 76-85, 1937.

([2308]) SCHWEITZER A. — Pharmacological studies on the effects of clavatin. *Exp. med. Surg.*, t. **IV**, p. 289-305, 1946.

([2309]) SCOLARI C. — Contributo alla conoscenza degli adenocarcinomi epatici della trota iridea. *Atti Soc. Ital. Sci. Vet.*, t. **VII**, p. 599-605, 1953.

([2310]) SCOLARI C. — Su di una epizoogia delle trote iridee d'allevamento la « lipoidosi epatica ». *Clin. Veter.*, t. **LXXVII**, p. 4, 1954.

([2311]) SCOPPA P., MARAFANTE E. et RODIGHERO L. — Eliminazione biliare dell'aflatossina B_1 da parte del ratto. *Boll. Soc. Ital. Biol. Sper.*, t. **XLVII**, p. 475-477, 1971.

([2312]) SCOTT A. I., SUTHERLAND S. A., YOUNG D. W. GUGLIELMETTI L., ARIGONI D. et SIM G. A. — The structure and absolute configuration of rosololactone and related diterpenoid lactones. *Proc. Chem. Soc.*, p. 19-21, 1964.

([2313]) SCOTT A. I. et YALPANI M. — A mass-spectrometric study of biosynthesis : conversion of deutero-m-cresol into patulin. *Chem. Commun.*, t. XVIII, p. 945-946, 1967.

([2314]) SCOTT A. I., YOUNG D. W., HUTCHINSON S. A. et BHALLA N. S. — Diterpenoids. V. Isorosenolic acid, a new diterpenoid constituent of *Trichothecium roseum*. *Tetrah. Lett.*, p. 849-854, 1964.

([2315]) SCOTT B. de. — Toxigenic fungi isolated from cereal and legume products. *Mycopathol. Mycol. Appl.*, t. **XXV**, p. 213-222, 1965.

([2316]) SCOTT P. M. — Note on analysis of aflatoxins in green coffee. *J. Ass. Off. Anal. Chem.*, t. **LI**, p. 609, 1968.

([2317]) SCOTT P. M. — Analysis of cocoa beans for aflatoxins. *J. Ass. Off. Anal. Chem.*, t. **LII**, p. 72-74, 1969.

([2318]) SCOTT P. M. — The analysis of foods for aflatoxins and other fungitoxins : a review. *Canad. Inst. Food Technol. J.*, t. **II**, p. 173-177, 1969.

([2319]) SCOTT P. M. et HAND T. B. — Method for the detection and estimation of ochratoxin A in some cereal products. *Ass. Off. Anal. Chem. J.*, t. **L**, p. 366-370, 1967.

([2320]) SCOTT P. M., KENNEDY B. et VAN WALBEEK W. — Simplified procedure for the purification of ochratoxin A from extracts of *Penicillium viridicatum*. *J. Ass. Off. Anal. Chem.*, t. **LIV**, p. 1445-1447, 1971.

([2321]) SCOTT P. M. et LAWRENCE J. W. — Stemphone, a biologically active yellow pigment produced by *Stemphylium sarcinaeforme* (Cav.) Wiltshire. *Can. J. Microbiol.*, t. **XIV**, p. 1015-1016, 1968.

([2322]) SCOTT P. M. et PRZYBYLSKI W. — Collaborative study of a method for the analysis of cocoa beans for aflatoxins. *J. Ass. Off. Anal. Chem.*, t. **LIV**, p. 540-544, 1971.

([2323]) SCOTT P. M. et SOMERS E. — Stability of patulin and penicillic acid in fruit juices and flour. *J. Agr. Food Chem.*, t. **XVI**, p. 483-485, 1968.

([2324]) SCOTT P. M., VAN WALBEEK W. et FORGACS J. — Formation of aflatoxins by *Aspergillus ostianus.* Wehmer. *Appl. Microbiol.*, t. **XV**, p. 945, 1967.

([2325]) SCOTT P. M., VAN WALBEEK W., HARWIG J. et FENNELL D. I. — Occurrence of a mycotoxin, ochratoxin A, in wheat and isolation of ochratoxin A and citrinin producing strains of *Penicillium viridicatum. Can. J. Plant Sci.*, t. **L**, p. 583-585, 1970.

([2326]) SCOTT W. J. — Water relations of food spoilage microorganisms. *Adv. Food Res.*, t. **VII**, p. 83-127, 1957.

([2327]) SCURTI J. C., CANTINI G. et COLOMB B. — Ricerche sul miceti di foraggi conservati e sulla loro tossicita. *Allionia*, t. **XIV**, p. 113-121, 1968.

([2328]) SEARCY J. W., DAVIS N. D. et DIENER V. L. — Biosynthesis of ochratoxin *A. Appl. Microbiol.*, t. **XVIII**, p. 622-627, 1969.

([2329]) SEARLE C. E. — EXPERIMENTS on the carcinogenicity and reactivity of β-propiolactone. *Brit. J. Cancer*, t. **XV**, p. 804, 1961.

([2330]) SEEKLES L. — Intestinal autointoxication in cattle in relation to nutrition tetany (hypomagnesaemic tetany). *Proc. 16th Internat. Vet. Congr.*, t. **II**, p. 83-84, Madrid, 1959.

([2331]) SEEMÜLLER E. — Untersuchungen über die morphologische und biologische Differenzierung in der *Fusarium*-Sektion *Sporotrichiella. Mitt. Biol. Bundesanst. f. Land-u. Forstwissch. Berlin-Dahlem*, fasc. 127, 93 p., 1968.

([2332]) SELLSCHOP J. P. F. — Field observations on conditions conducive to the contamination of groundnuts with the mould *Aspergillus flavus* Link ex Fries. *South African Dept. Agr. Tech. Serv. Symposium on Mycotoxins in Foodstuffs (Agricultural Aspects). Pretoria, 25-26 févr.*, p. 47-52, 1965.

([2333]) SEMENIUK G. — Storage of cereal grains and their products. *Microflora. Amer. Assoc. Cereal Chemists, St-Paul, Minn.*, p. 77-151, 1954.

([2334]) SEMENIUK G. et BARRE H. J. — Pathology and mycology of corn. *Rept. Iowa State Coll. Agr. Expt. Stn., 1942-1943*, Part II, p. 52-57, 1944.

([2335]) SEMENIUK G. et GILMAN J. C. — Relation of molds to the deterioration of corn in storage. A Review. *Iowa Acad. Sci.*, t. **LI**, p. 256-280, 1944.

([2336]) SEMENIUK G., HARSHFIELD G. S., CARLSON C. W., HESSELTINE C. W. et KWOLEK W. F. — Occurrence of mycotoxins in *Aspergillus in* Herzberg M., *Toxic Microorganisms*, p. 185-190, 1970.

([2337]) SEMENIUK G., HARSHFIELD G. S., CARLSON C. W., HESSELTINE C. W. et KWOLEK W. F. — Mycotoxins in *Aspergillus. Mycopathol. Mycol. Appl.*, t. **XLIII**, p. 137-152, 1971.

([2338]) SEMENIUK G., NAGEL C. M. et GILMAN J. C. — Observations on mold development and on deterioration in stored yellow dent shelled corn. *Iowa State Coll. Agr. Expt. Stn.*, Bull. n° 349, p. 253-284, 1947.

([2339]) SENSER F. — Vorkommen und Bestimmung toxinbildender Schimmelpilze der Gruppe *Aspergillus flavus. Dtsch. Lebensm.-Rdsch.*, t. **LXIII**, p. 140-144, 1967.

([2340]) SENSER F. — Toxinbildende Schimmelpilze in Lebensmitteln. *Gordian*, t. **LXIX**, p. 159-163, 1969.

([2341]) SENSER F., REHM H. J. et RAUTENBERG E. — Zur Kenntnis fruchtsaftverderbender Mikroorganismen. II. Schimmelpilzarten in verschiedenen Fruchtsäften *Zbl. Bakt.*, II. Abt., t. **CXXI**, p. 737-746, 1967.

([2342]) SENTEIN P. — Altérations du fuseau mitotique et fragmentation des chromosomes par l'action de la patuline sur l'oeuf d'Urodèles en segmentation. *C.R. Soc. Biol.*, t. **CXLIX**, p. 1621-1622, 1955.

([2343]) SERCK-HANSSEN A. — Aflatoxin-induced fatal hepatitis ? A care report from Uganda. *Arch. Environ. Health*, t. **XX**, p. 729-731, 1970.

([2344]) SERRES H. — La toxicité de certains tourteaux d'arachide à Madagascar. *Bull. Madagascar*, n° 209, p. 575-580, 1963.

([2345]) SEXTON W. A. — *Chemical Constitution and Biological Activity.* 3e éd., 517 p. ,Spon édit. Londres, 1963.

([2346]) SHANK R. C. — Dietary aflatoxin loads and the incidence of human hepatocellular carcinoma in Thailand. *in* Purchase I. F. H., *Mycotoxins in human health*, p. 245-262, 1971.

([2347]) SHANK R. C. et WOGAN G. N. — Distribution and excretion of ^{14}C labeled aflatoxin B_1 in the rat. *Feder. Proc. U.S.A.*, t. **XXIV**, p. 627, 1965.

([2348]) SHANK R. C. et WOGAN G. N. — Acute effects of aflatoxin B_1 on liver composition and metabolism in the rat and duckling. *Toxicol. Appl. Pharmacol.*, t. **IX**, p. 468-476, 1966.

([2349]) SHANK R. C. et al. — Dietary aflatoxins and human liver cancer. I. Toxigenic moulds in foods and foodstuffs of tropical South-East Asia. *Food Cosmet. Toxicol.*, t. **X**, p. 51-60, 1972. II. Aflatoxins in market foods and foodstuffs of Thailand and Hong Kong. *Id.*, t. **X**, p. 61-69, 1972. III. Field survey of rural Thai families for ingested aflatoxins. *Id.*, t. **X**, p. 71-84, 1972.

([2350]) SHANKARAN P., SHANKARAN R., RAJ H. G. et VENKITASUBRAMANIAN T. A. — Biochemical changes in liver due to aflatoxin. *Brit. J. Exp. Pathol.*, t. **LI**, p. 487-491, 1970.

([2351]) SHANNON J. S. — Mass spectrometry. V. Sporidesmin and sporidesmin B. *Tetrah. Lett.*, n° 13, p. 801-806, 1963.

([2352]) SHANTHA T., SREENIVASAMURTHY V. et PARPIA H. A. B. — Urinary metabolites of ^{14}C aflatoxin in some laboratory animals. *J. Food Sci. Technol.*, t. **VII**, p. 135-138, 1970.

([2353]) SHAW F. W. et WARTHEN H. J. — Aspergillosis of bone. *Southern Med. J.*, t. **XXIX**, p. 1070-1071, 1936.

([2354]) SHAW M. K. — Die Desinfektion der Luft von Kühlräumen mittels Milchsäure *Fleischwirtschaft*, t. **XLIV**, p. 52, 1964.

([2355]) SHEMYAKIN M. M. et KOKHLOV A. S. — *Chemistry of Antibiotics.* Goskhimizdat, 1953.

([2356]) SHIBATA S. et KITAGAWA I. — Metabolic products of fungi. X. The structure of rubroskyrin and its relation to the structure of luteoskyrin. *Pharm. Bull. Tokyo*, t. **IV**, p. 309-313, 1956.

([2357]) SHIBATA S. et KITAGAWA I. — Metabolic products of fungi. XVI. The structure of rubroskyrin and luteoskyrin. *Pharm. Bull. Tokyo*, t. **VIII**, p. 884-888, 1960.

([2358]) SHIBATA S., NATORI S. et UDAGANA S. — *List of fungal products.* 170 p. Univ. Tokyo Press, Tokyo, 1964.

([2359]) SHIBKO S. I., ARNOLD D. L., MORNINGSTAR J. et FRIEDMAN L. — Studies on the effect of aflatoxin B_1 on the development of the chick embryo. *Proc. Soc. Exp. Biol. Med.*, t. **CXXVII**, p. 835-839, 1968.

([2360]) SHIGEURA N. T. et GORDON C. N. — Hadacidin, a new inhibitor of purine biosynthesis. *J. Biol. Chem.*, t. **CCXXXVII**, p. 1932-1936, 1962.

([2361]) SHIGEURA N. T. et GORDON C. N. — Mechanism of action of hadacidin *J. Biol. Chem.*, t. **CCXXXVII**, p. 1937-1940, 1962.

([2362]) SHIH C. N. et MARTH E. H. — Improved procedures for measurement of aflatoxins with thin layer chromatography and fluorimetry. *J. Milk Food Technol.* t. **XXXII**, p. 213-217, 1969.

([2363]) SHIH C. N. et MARTH E. H. — A procedure for rapid recovery of aflatoxins from cheese and other foods. *J. Milk Food Technol.*, t. **XXXIV**, p. 119-123, 1971.

([2364]) SHILO Y. M. — Toxic action of *Aspergillus flavus* in the animal body. *Sbornik Trudov Kharkovskogo Inst.*, t. **XIX**, p. 63-73, 1940.

([2365]) SHIMADA K. et MATSUSHIMA K. I. — Studies on the production of protease inhibitor by molds ; examination of the ability of molds to produce protease inhibitor. *Nippon Nogei Kagaku Kaishi*, t. **XLII**, p. 325-329, 1968.

([2366]) SHIMKIN M. B. et KRAYBILL H. F. — (Non-toxicité de l'huile d'arachide) *J. Am. Med. Assoc.*, t. **CLXXXIV**, p. 57, 1963.

([2367]) SHIMODA C. — An antibiotic substance, oryzacidin, against saké-putrefying bacteria produced by *Aspergillus oryzae*. I. Isolation of oryzacidin. *J. Agr. Chem. Soc. Japan*, t. **XXV**, p. 254-260, 1951.

([2368]) SHIMODA C. et NISHIWAKI Y. — An antibiotic substance, oryzacidin, produced by *Aspergillus oryzae*. IV. Assay of oryzacidin by the cup method. *J. Agr. Chem. Soc. Japan*, t. **XXVIII**, p. 832-837, 1954.

([2369]) SHIMODA C., SHIMADA S. et SUGITA O. — An antibiotic substance, oryzacidin, produced by *Aspergillus oryzae*. II. The mutation of *Aspergillus oryzae* by irradiation with ultraviolet rays. *J. Agr. Chem. Soc. Japan*, t. **XXVI**, p. 645-647, 1952.

([2370]) SHIPP H. L. — Food poisoning caused by biological agencies other than bacteria. *Scientific and technical Surveys, Brit. Food Manufact. Ind. Res. Ass.*, t. **XLIV**, p. 17-23, 1966.

([2371]) SHISHIYAMA J., FUKUTOMI M. et AKAI S. — Effect of some fungicides on the synthesis of chlorophylls, desoxyribonucleic acid and ribonucleic acid in onion leaves. *Phytopathology*, t. **LV**, fasc. 8, p. 844-847, 1965.

([2372]) SHOENTAL R. — Aflatoxins. *Ann. Rev. Pharmacol.*, t. **VII**, p. 343-356, 1967.

([2373]) SHOJI J. et SHIBATA S. — The structure of erythroskyrine, a nitrogen-containing coloring matter of *Penicillium islandicum*. *Chem. and Ind.*, p. 419-421, 1964.

([2374]) SHORTRIDGE E. H. — Hypomagnesaemia (Grass staggers) in beef cattle. *N.Z.J. Agric.*, déc. 1960.

([2375]) SHOTWELL O., HESSELTINE C. W., BURMEISTER H. R., KWOLEK W. F., SHANNON G. M. et HALL H. H. — Survey of cereal grains and soybeans for the presence of aflatoxin : I. Wheat, grain, sorghum and oats. *Cereal Chem.*, t. **XLVI**, p. 446-454, 1969. II. Corn and Soybeans. *Id.*, t. **XLVI**, p. 454-463, 1969.

([2376]) SHOTWELL O. L., HESSELTINE C. W. et GOULDEN M. L. — Note on the natural occurrence of ochratoxin A. *J. Ass. Off. Agr. Chem.*, t. **LII**, p. 81-83, 1969.

([2377]) SHOTWELL O. L., HESSELTINE C. W., GOULDEN M. L. et VANDEGRAFT E. E. —Survey of corn for aflatoxin zearalenone and ochratoxin. *Cereal Chem.*, t. **XLVII**, p. 700-707, 1970.

([2378]) SHOTWELL O. L., HESSELTINE C. W., STUBBLEFIELD R. D. et SORENSON W. G. — Production of aflatoxin on rice. *Appl. Microbiol.*, t. **XIV**, p. 425-428, 1966.

([2379]) SHOTWELL O. L., HESSELTINE C. W., VANDEGRAFT E. E. et GOULDEN M. L. — Survey of corn from different regions for aflatoxin, ochratoxin, and zearalenone. *Cereal Sci. Today*, t. **XVI**, p. 266-273, 1971.

([2380]) SHOTWELL O. L., SHANNON G. M., GOULDEN M. L., MILBURN M. S. et HALL H. H. — Factors in oats that could be mistaken for aflatoxin. *Cereal Chem.*, t. **XLV**, p. 236-241, 1968.

([2381]) SHUI-CHIN CHEN et FRIEDMAN L. — Aflatoxin determination in seed meal. *J. Ass. Off. Anal. Chemists*, t. **XLIX**, p. 28-33, 1966.

([2382]) SIDRANSKY H. et FRIEDMAN L. — Pulmonary aspergillosis associated with cortisone and antibiotic administration : human and experimental studies. *Am. J. Pathol.*, t. **XXXIV**, p. 585-586, 1958.

([2383]) SIDRANSKY H. et VERNEY E. — Experimental aspergillosis. *Lab. Invest.*, t. **XI**, p. 1172-1183, 1962.

([2384]) SIGG H. P., MAULI R., FLURY E. et HAUSER D. — The constitution of diacetoxyscirpenol. *Helv. Chim. Acta*, t. **XLVIII**, p. 962-988, 1965.

([2385]) SILAEV A. B., BEKKER Z. E., MAKSIMOVA R. A., GUS'KOVA T. M., KARNAUKHOVA M. V. et ZELENEVA R. N. — (La trichothécine, ses conditions de biosynthèse et d'extraction). *Sel'skokhozjajstrv. Biol., S.S.S.R.*, t. I, p. 627-631, 1966.

([2386]) SILLER W. G. et OSTLER D. C. — The histopathology of an enterohepatic syndrome of turkey poults. *Vet. Rec.*, t. **LXXII**, p. 134-138, 1961.

([2387]) SILOW S. A. — The potential contribution of atomic energy to development in Agriculture and related Industries. *Int. J. Appl. Radiation and Isotopes*, t. **III**, p. 257-280, 1958.

([2388]) SIMENSEN M. G. — Hypomagnesaemia and grass tetany. *Nord. Vet. Med.*, t. **IX**, p. 305-321, 1957.

([2389]) SIMONART P. et LITHOUWER R. — Formation de patuline par *Penicillium griseofulvum* Dierckx. *Ztrbl. f. Bact.*, Abt. II, t. **CX**, p. 107, 1956.

([2390]) SIMS F. H. et CROOKSHANK H. H. — Wheat pasture poisoning in cattle. *Texas Agr. Exp. Stn.*, Prog. Rep. 1739, 1954.

([2391]) SIMS F. H. et CROOKSHANK H. H. — Wheat pasture poisoning. *Texas Agr. Exp. Stn.*, *Bull.* 842, 1956.

([2392]) SIMS F. H. et CROOKSHANK H. H. — Wheat pasture poisoning in cows. *Southwest. Vet.*, t. **X**, p. 277, 1957.

([2393]) SINCLAIR D. P. — *Pithomyces chartarum* spores on pasture and their relation to facial eczema in sheep. *New Zealand J. Agr. Res.*, t. **IV**, p. 492-503, 1961.

([2394]) SINCLAIR D. P. et HOWE M. W. — Effect of thiabendazole on *Pithomyces chartarum* (Berk. et Curt.) M. B. Ellis. *N.Z.J. Agr. Res.*, t. **XI**, p. 59-62, 1968.

([2395]) SINGER R. H., GRAINGER R. B. et BAKER F. H. — Investigations on a complicated grass tetany syndrome in ruminants of Kentucky. I. *Kentucky Agr. Exp. Stn.*, Bull. 658, 1958.

([2396]) SINGH J. — Mechanisms of antifungal action of patulin. *Ph. D. Thesis Univ. Illinois*, Urbana, 1966.

([2397]) SINGH J. — Patulin. *in* GOTTLIEB D. et SHAW P. D., *Antibiotics* I, p. 621-630, 1967.

([2398]) SINGH K., SEGHAL S. M. et VEZINA C. — Steroid hydroxylation and carbohydrate utilization by conidia of *Aspergillus ochraceus*. *Intern. Congr. Microbiol. 8th Congr. Montreal*, p. 71, 1962.

([2399]) SINGH-VERMA S. B. — Ueber den Einsatz der Propionsäure zur Konservierung von industriell hergestellten Mischfuttermitteln sowie von feuchten Getreide und Mais. *Zentralbl. Bakteriol.*, Abt. II, t. **CXXV**, p. 100-111, 1970.

([2400]) SINGLETON V. L., BOHONOS N. et ULLSTRUP A. J. — Decumbin a new compound from a species of *Penicillium*. *Nature*, t. **CLXXXI**, p. 1072-1073, 1958.

([2401]) SINHA R. H. et WALLACC M. A. H. — Ecology of a fungus induced hot spot in stored grain. *Canad. J. Plant Sci.*, t. **XLV**, p. 48-59, 1965.

([2402]) SINKOVIC B. — « Disease » of turkey poults. *Vet. Rec.*, t. **LXXIII**, p. 86, 1961.

([2403]) SINNHUBER R. D. — Aflatoxin in cottonseed meal and liver cancer in rainbow trout. *U.S. Fish Wildlife Sew. Res. Rep.*, t. **LXX**, p. 48-55, 1967.

([2404]) SINNHUBER R. O., LEE D. J., WALES J. H. et AYRES J. L. — Dietary factors and hepatoma in rainbow trout (*Salmo gairdneri*) II. Cocarcinogenesis by cyclopropenoid fatty acids and the effects of gossypol and altered lipids on aflatoxin-induced liver cancer. *J. Nat. Cancer Inst.*, t. **XLI**, p. 1293-1301, 1968.

([2405]) SINNHUBER R. O., LEE D J., WALES J. H., LANDERS M. K. et KEYL A. C. — Aflatoxin M_1 a potent liver carcinogen for rainbow trout. *Fed. Proc.*, t. **XXIX**, p. 568, 1970.

([2406]) SINNHUBER R. O., WALES J. H., AYRES J. L., ENGEBRECHT R. H. et AMEND D. L. — Dietary factors and hepatoma in rainbow trout (*Salmo gairdneri*) I. Aflatoxins in vegetable protein feedstuffs. *J. Nat. Cancer Inst.*, t. **XLI**, p. 711-718, 1968.

([2407]) SINNHUBER R. O., WALES J. H., ENGEBRECHT R. H., AMEND D. F., KRAY W. D. et AYRES J. L. — Aflatoxins in cottonseed meal and hepatoma in rainbow trout. *Feder. Proc. U.S.A.*, t. **XXIV**, p. 627, 1965.

([2408]) SIPPEL W. L. — Mold intoxication of livestock. *Iowa Vet.*, t. **XXVIII**, p. 15-17, 42-43, 1957.

([2409]) SIPPEL W. L., BURNSIDE J. E. et ATWOOD M. B. — A disease of swine and cattle caused by eating moldy corn. *Proc. 90th Ann. Meeting, Am. Vet. Med. Assn., Toronto, Canada*, p. 174-181, 1953.

(2410) SIRIEZ H. — La nature et ses poisons : toxines dangereuses et substances cancérigènes secrétées par des moisissures. *Phytoma*, N° 1830, p. 46-47, 1966.

(2411) SJOLLEMA B. — Nutritional and metabolic disorders in cattle. *Nutrition Abstr. Rev.*, t. **I**, p. 621-632, 1932.

(2412) SJOLLEMA B. et SEEKLES L. — Ueber Störungen des mineralen Regulationsmechanismus bei Krankheiten des Rindes (Ein Beitrag zur Tetaniefrage). *Biochem. Ztschr.*, t. **CCXXIX**, p. 358-380, 1930.

(2413) SLATER T. F., STRAEULI U. D. et SAWYER B. C. — Changes in liver nucleotide concentrations in experimental liver injury. *Biochem. J.*, t. **XCIII**, p. 260-270, 1964.

(2414) SLIFKIN M., MERKOW L. P., PARDO M., EPSTEIN S. M., LEIGHTON J. et FARBER E. — Growth in vitro of cells from hyperplastic nodules of liver induced by 2-fluorenylacetamide or aflatoxin B_1. *Science*, t. **CLXVII**, p. 285-287, 1970.

(2415) SLOWATIZKY I., MAYER A. M. et POLJAKOFF-MAYBER A. — The effect of aflatoxin on greening of etiolated leaves. *Isr. J. Bot.*, t. **XVIII**, p. 31-36, 1969.

(2416) SMALLEY E. B. — Mycotoxicosis associated with moldy corn. *1st Congr. Pl. Pathol.*, Londres, 1968.

(2417) SMALLEY E. B., MARASAS W. F. O., STRONG F. M., BAMBURG J. R., NICHOLS R. E. et KOSURI N. R. — Mycotoxicoses associated with moldy corn. *in* HERZBERG M., *Toxic Micro-organisms*, p. 163-173, 1970.

(2418) SMITH G. — The pigments of the lower fungi (moulds). *Comm. 8e Congr. Int. Bot. Paris*, Sect. 18, p. 83-89, 1954.

(2419) SMITH G. — Some new species of moulds and some new British records. *Trans. Brit. Mycol. Soc.*, t. **XLIV**, p. 42-50, 1960.

(2420) SMITH J. D., LEES F. T. et CRAWLEY W. E. — Facial eczema on long and short herbage. *New Zealand J. Agric. Res.*, t. **VI**, p. 518-525, 1963.

(2421) SMITH J. D., LEES F. T. et CRAWLEY W. E. — Weather conditions, spore counts, and facial eczema in test sheep. *New Zealand J. Agr. Res.*, t. **VIII**, p. 63-87, 1965.

(2422) SMITH J. H. — Anticoagulant and anticomplementary activity of aflatoxin B_1 in rats. *Clin. Res.*, t. **XVI**, p. 314k 1968.

(2423) SMITH J. K. et LOURIA D. B. — Biochemical effects of fungal toxins. *Bull. N.Y. Acad Med.*, t. **XLIV**, p. 1144, 1968.

(2424) SMITH J. W. et HAMILTON P. B. — Aflatoxicosis in the broiler chicken. *Poult. Sci.*, t. **XLIX**, p. 207-215, 1970.

(2425) SMITH J. M. B. — Animal mycoses in New Zealand. *Mycopathol. Mycol. Appl.*, t. **XXXIV**, p. 323-336, 1968.

(2426) SMITH K. M. — « Disease » of turkey poults. *Vet. Rec.*, t. **LXXII**, p. 652, 1960.

(2427) SMITH R. H. — The influence of toxins of *Aspergillus flavus* on the incorporation of ^{14}C leucine into proteins. *Biochem. J.*, t. **LXXXVIII**, p. 50-51, 1963.

(2428) SMITH R. H. — The inhibition of amino acid activation in liver and *Escherichia coli* preparations by aflatoxin *in vitro*. *Biochem. J.*, t. **XCV**, p. 43-44, 1965.

(2429) SMITH R. H. et MC KERNAN N. — Hepatotoxic action of chromatographically separated fractions of *Aspergillus flavus* extracts. *Nature*, t. **CXCV**, p. 1301-1303, 1962.

(2430) SMITH W. K. et BRINK R. A. — Relation of bitterness to the toxic principle in sweet-clover. *J. Agr. Res.*, t. **LVI**, p. 145, 1938.

(2431) SNIESZKO S. F. — *New York Fish and Game J.*, t. **VIII**, 145-149, 1961.

(2432) SNOW D. — Mould deterioration of stored feeding stuffs. *Fertiliser, Feeding stuffs, Farm Supplies J.*, 1944.

(2433) SNOW D. — Mould deterioration of feeding stuffs in relation to humidity of storage. Part.III. The isolation of mould species from feeding stuffs stored at different humidities. *Ann. Appl. Biol.*, t. **XXXII**, p. 40-44, 1945.

([2434]) SNOW D. — The germination of mould spores at controlled humidities. *Ann. Appl. Biol.*, t. **XXXVI**, p. 1-13, 1949.

([2435]) SNOW D., CRICHTON M. M. G. et WRIGHT N. C. — Mould deterioration of feeding stuffs in relation to humidity of storage. I. The growth of moulds at low humidities. II. The water uptake of feeding stuffs at different humidities. *Ann. Appl. Biol.*, t. **XXXI**, p. 102-110, 111-116, 1944.

([2436]) SNOW D. et WATTS P. S. — The effect of sulphanilamide and other bacteriostatic drugs on the growth of moulds. *Ann. Appl. Biol.*, t. **XXXII**, p. 102-112, 1945.

([2437]) SNYDER W. C. et HANSEN H. N. — The species concept in *Fusarium. Amer. J. Bot.*, t. **XXVII**, p. 64-67, 1940.

([2438]) SNYDER W. C. et HANSEN H. N. — The species concept in *Fusarium* with reference to section *Martiella. Amer. J. Bot.*, t. **XXVIII**, p. 738-742, 1941.

([2439]) SNYDER W. C. et HANSEN H. N. — The species concept in *Fusarium* with reference to *Discolor* and other sections. *Amer. J. Bot.*, t. **XXXII**, p. 657-666, 1945.

([2440]) SNYDER W. C. et TOUSSOUN T. A. — Current status of taxonomy in *Fusarium* species and their perfect stages. *Phytopathology*, t. **LV**, p. 833-837, 1965.

([2441]) SOBERS E. K. et DOUPNIK B. — Pathogenicity and animal toxicity of single spore isolates of *Alternaria longipes. Phytopathology*. t. **LVII**, p. 1068, 1968.

([2442]) SOENEN M. et PINGUAIR R. — A bacteriological study of flour and its importance in milling and baking. *C. R. 5e Congr. Int. Tech. Chim. Ind. Agr.*, Budapest, t. **II**, p. 189-208, 1937.

([2443]) SOLOMON G., JENSEN R. et TANNER H. — Hepatic changes in rainbow trout (*Salmo gairdneri*) fed diets containing peanut, cottonseed, and soybean meals. *Amer. J. Veter. Res.*, t. **XXVI**, fasc. 112, p. 764-770, 1965.

([2444]) SOMMER N. F., FORTLAGE R. J., BUCKLEY P. M. et MAXIE E. C. — Radiation-heat synergism for inactivation of marked disease fungi of stone fruits. *Phytopathology*, t. **LVII**, p. 428-433, 1967.

([2445]) SORGER-DOMENIGG H., CUENDET L. S., CHRISTENSEN C. M. et GEDDES W. F. — Grain storage studies XVII. Effect of mold growth during temporary exposure of wheat to high moisture contents upon the development of germ damage and other indices of deterioration during subsequent storage. *Cer. Chem.*, t. **XXXII**, p. 270-284, 1955.

([2446]) SORENSON W. G., HESSELTINE C. W. et SHOTWELL O. L. — Effect of temperature on production of aflatoxin on rice by *Aspergillus flavus. Mycopathol. Mycol. Appl.*, t. **XXXIII**, p. 49-55, 1967.

([2447]) SOSEDOV N. I. et VAKAR A. B. — Effect of ionizing radiations on the biochemical and breadmaking properties of wheat, in Kretovitch, Biochemistry of grain and of breadmaking. *Acad. Sci. U.S.S.R.*, Israel Program for Scientific Translation, p. 131-153, 1965.

([2448]) SOUVERAIN R. — L'application des méthodes d'analyse des caractères organoleptiques. *Méthodes subjectives et objectives d'appréciation des caracteres organoleptiques des denrées alimentaires*, p. 575-606, édit. C.N.R.S., 1966.

([2449]) SOZZI T. et SHEPHERD D. — The factors influencing the development of undesirable moulds during the maturation of cheese. *Milchwissenschaft*, t. **XXVI**, p. 280-282, 1971.

([2450]) SPECTOR W. C. — Handbook of Toxicology. II. Antibiotics. Saunders édit., Philadelphia, 1957.

([2451]) SPEERS G. M., MERONUK R. A., BARNES D. M. et MIROCHA C. J. — Effect of feeding *Fusarium roseum* f. sp. *graminearum* contaminated corn and the mycotoxin F-2 on the growing chick and laying hen. *Poult. Sci.*, t. **L**, p. 627-633, 1971.

([2452]) SPENCER E. R. — Decay of Brazil nuts. *Botan. Gaz.*, t. **LXXII**, p. 265-292, 1921.

([2453]) SPENSLEY P. C. — L'aflatoxine, principe actif de la maladie « X » du dindon. *Endeavour*, t. **XXII**, p. 75-79, 1963.

([2454]) SPENSLEY P. C. — *Nature, G. B.*, t. **CXCVII**, p. 31, 1963.

([2455]) SPENSLEY P. C. et TOWNSEND R. J. — Toxic fungal products. *Inst. Biol. Symp.*, t. **XVI**, p. 103-121, 1967.

([2456]) SPESIVTSEVA N. A. — Mycoses et mycotoxicoses des animaux (en russe). *Gosud. izdatel'stvo sel'shokhoz. literatury*, 453 p., Moscou, 1960.

([2457]) SPICHER G. — Schimmelpilze in der Luft von Backstuben und Brotlagerräumen. *Brot u. Gebäck*, t. **XIX**, p. 148-153, 1965.

([2458]) SPICHER G. — Mykotoxine in Getreide und Getreide-produkten. *Brot u. Gebäck*, t. **XXII**, p. 237-241, 1968.

([2459]) SPICHER G. — Studien zur Frage des Vorkommens von Aflatoxinbildnern und Aflatoxin auf Brot. *Brot u. Gebäck*, t. **XXIII**, p. 149-153, 1969.

([2460]) SPICHER G. — Untersuchungen über das Vorkommen von Aflatoxin im Brot. *Zentralbl. f. Bakt. Parasitenbade, Infektionskrkh. u. Hygiene*, t. **CXXIX**, p. 697-706, 1970.

([2461]) SPICHER G. et STEPHAN H. — Auswirkungen einer Propionsäure-Behandlung von Getreide auf eine Mikroflora und das Backtechnische Verhalten des Mehles. *Getreide u. Mehl*, t. **XX**, 1970.

([2462]) SPILSBURY J. F. et WILKINSON S. — The isolation of festuclavine and two new clavine alkaloids from *Aspergillus fumigatus* Fres. *J. Chem. Soc.*, p. 2085-2091, 1961.

([2463]) SPINER D. R. et HOFFMAN R. K. — Method for disinfecting large enclosures with β-propiolactone vapor. *Appl. Microb.*, t. **VII**, p. 152-155, 1960.

([2464]) SPLENDIANI F. et LUCISANO A. — Ricerca di aflatossine in insaccati di produzione nazionale. *Zooprofilassi*, t. **XXVI**, p. 343-347, 1971.

([2465]) SPORN A. — Actiunea porumbului infestat cu *Fusarium moniliforme* (*Gibberella fujikuroi*) asupra organismului animal. *Igiena, Romin.*, t. **XIV**, fasc. 3, p. 171-172, 1965.

([2466]) SPORN M. B., DINGMAN C. W., PHELPS H. L. et WOGAN G. N. — Aflatoxin B_1 : binding to DNA *in vitro* and alteration of RNA metabolism *in vivo*. *Science*, t. **CLI**, p. 1539-1541, 1966.

([2467]) SPRENGER R. D. et RUOFF P. M. — The chemistry of citrinin. I. The synthesis of 2, 4-dimethoxy-3-ethyl benzoic acid and 2,6-dimethoxy-3-methyl-benzoic acid. *Jour. Organic Chem.*, t. **XI**, p. 189-193, 1946.

([2468]) SPURGIN M. M. — Suspected occurrence of *Byssochlamys fulva* in Queensland-grown canned strawberries. *Queensland J. Agric. Sci.*, t. **XXI**, p. 247-250, 1964.

([2469]) SREENIVASAMURTHY V., JAYARAMAN A. et PARPIA H. A. B. — A rapid paper chromatographic method for the detection of aflatoxin in groundnuts. *Indian J. Technol.*, t. **II**, p. 415-416, 1964.

([2470]) SREENIVASAMURTHY V., JAYARAMAN A. et PARPIA H. A. B. — Aflatoxin in Indian Peanuts : analysis and extraction. *Mycotoxins in Foodstuffs*, p. 251-260, 1965.

([2471]) SREENIVASAMURTHY V., PARPIA H. A. B., SRIKANTA S. et SHANKAR MURTI A. — Detoxification of aflatoxin in peanut meal by hydrogen peroxide. *J. Ass. Off. Anal. Chem.*, t. **L**, p. 350-354, 1967.

([2472]) STACCHINI A. et MANZONE A. M. — Determinazione spettrofluorimetrica de l'aflatossina B_1 nei panelli di arachide. *Boll. Lab. Chim. Prov.*, t. **XVI.**, p. 619-632, 1965.

([2473]) STACK M. et RODRICKS J. V. — Method for analysis and chemical confirmation of. sterigmatocystin. *J. Ass. Off. Anal. Chem.*, t. **LIV**, p. 86-90, 1971.

([2474]) STAHMAN M. A., HUEBNER C. F. et LINK K. P. — Studies on the hemorrhagic sweet clover disease. V. *J. Biol. Chem.*, t. **CXXXVIII**, p. 513, 1941.

([2475]) STALKER A. L. et MC LEAN D. L. — *J. Anim. Tech. Ass.*, t. **VIII**, p. 18, 1957.

([2476]) STAMATOVIC S., LJESEVOC Z. et DJURICKOVIC S. — On a fungal alimentary intoxication in swine (vulvovaginitis suum). *Vet. Glas.*, t. **XVII**, p. 507-510, 1963.

([2477]) STANKUSHEV K. H. — (Champignons du genre *Fusarium* se développant sur l'orge et l'avoine entreposées pour l'alimentation du bétail.) *Veter. Med. Nauki, Balg.*, t. **II**, p. 989-995, 1965.

([2478]) STANKUSHEV K. H. et MATEEV M. — Opiti za termichno obczvredyavane na porazena ot mikroskopicheski g'bichki tsarevitsa. *Vet. Med. Nauki*, t. **IV**, p. 47-52, 1967.

([2479]) STANLEY N. F. — The biological activity of a substance resembling gliotoxin produced by a strain of *Aspergillus fumigatus. Australian J. Exp. Biol. Med. Sci.*, t. **XXII**, p. 133-138, 1946.

([2480]) STANSLY P. G. et ANANENKO N. H. — Candidulin : an antibiotic from *Aspergillus candidus. Arch. Biochem.*, t. **XXIII**, p. 256-261.

([2481]) STARON T. — Recherches sur les structures azotées favorables au développement de quelques microorganismes. *Phytiatrie-Phytopharmacie*, t. **XIV**, p. 67-71, 1965.

([2482]) STARON T. et ALLARD C. — Propriétés antifongiques du 2-(4 thiazolyl) benzimidazole ou thiabendazole. *Phytiatrie-Phytopharmacie*, t. **XIII**, p. 163-168, 1964.

([2483]) STARON T., ALLARD C., NGUYEN DAT XUONG, CHAMBRE M. M., GRABOWSKI H. et KOLLMANN A. — Isolement de trois substances toxiques à partir des jus de culture et du mycélium d'un *Aspergillus ochraceus. Phytiatrie-Phytopharmacie*, t. **XIV**, p. 73-79, 1965.

([2484]) STEINEGGER E. et LEUPI H. — Influence of plant substances on root growth of *Allium cepa* and germination of *Lepidium sativum. Pharm. Acta Helv.*, t. **XXXI**, p. 45-51, 1956.

([2485]) STEINKRAUS K. H., YAP BWEE HWA, VAN BUREN S. P., PROVVIDENTI M. I. et HAND D. B. — Studies on tempeh, an Indonesian fermented soybean food. *Food Research*, t. **XXV**, p. 777, 1960.

([2486]) STEPHENSON E. L. — « Disease » of turkey poults. *Vet. Rec.*, t. **LXXII**, p. 956, 1960.

([2487]) STEVENS A. J., SAUNDERS C. N., SPENCE J. B. et NEWNHAM A. G. — Investigations into « disease » of turkey poults. *Vet. Rec.*, t. **LXXII**, p. 627-628, 1960.

([2488]) STEVENSON D. E. — Toxicity associated with certain batches of groundnuts. *Brit. Veter. J.*, t. **CXVIII**, p. 531-532, 1962.

([2489]) STEWARD G. F. — Mycotoxicoses... syndrome of the future. *Food Technol.*, t. **XX**, p. 715, 1966.

([2490]) STEYN D. G. — Fungi in relation to health in man and animal. *Onderstepoort J. Vet. Sci.*, t. **I**, p. 183, 1933.

([2491]) STEYN D. G. — The toxicology of plants in South Africa. *S. Afr. Agric.*, Ser. n⁰ 13. Central News Agency Ltd, 1934.

([2492]) STEYN D. G. — Poisoning of human beings by weeds contained in cereals (Bread poisoning). *Farming in S. Africa*, t. **IX**, p. 45, 1934.

([2493]) STEYN M. — Rapid method for the isolation of pure aflatoxins. *Ass. Off. Anal. Chem.*, t. **LIII**, p. 619-621, 1970.

([2494]) STEYN P. S. — The separation and detection of several mycotoxins by thin-layer chromatography. *J. Chromatogr.*, t. **XLV**, p. 473-475, 1969.

([2495]) STEYN P. S. — The isolation, structure and absolute configuration of secalonic acid D, the toxic metabolite of *Penicillium oxalicum, Tetrahedron*, t. **XXVI**, p. 51-57, 1970.

([2496]) STEYN P. S. — Ochratoxin and other dihydroisocoumarins. *in* Ciegler A. et al., Microbial Toxins, t. **VI**, p. 179-205, 1971.

([2497]) STEYN P. S. et HOLZAPFEL C. W. — *J. South Afr. Chem. Inst.*, t. **XX**, p. 186-189, 1967.

([2498]) STICKINGS C. E. et MAHMOODIAN A. — Metabolites of *Penicillium frequentans* and their significance for the biosynthesis of sulochrin. *Chem. and Ind.*, p. 1718-1719, 1962.

([2499]) STILL P. E., MACKLIN A. W., RIBELIN W. E. et SMALLEY E. B. — Relationship of ochratoxin to foetal death in laboratory and domestic animals. *Nature. G. B.*, t. **CCXXXIV**, p. 563-564, 1971.

([2500]) STITT F. — *Fundamental Aspects of the dehydration of foodstuffs.* Macmillan Co., 1958.

([2501]) STOBB M., BALDWIN R. S., TUITE J., ANDREWS F. N. et GILLETTE K. G. — Isolation of an anabolic, uterotrophic compound from corn infected with *Gibberella zeae. Nature*, t. 196, p. 1318, 1962.

([2502]) STÖCK B. L. — Case report : Generalised granulomatous lesions in chickens and wild ducks caused by *Aspergillus* species. *Avian diseases*, t. **V**, p. 89-93, 1961.

([2503]) STOLK A. C. et SAMSON R. A. — Studies on *Talaromyces* and related genera. I. *Hamigera* gen. nov. and *Byssochlamys*. *Persoonia*, t. **VI**, p. 341-357, 1971.

([2504]) STOLOFF L. — Collaborative study of a method for the identification of aflatoxin B_1 by derivative formation. *Ass. Off. Anal. Chem. J.*, t. **L**, p. 354-360, 1967.

([2505]) STOLOFF L., BECKWITH A. C. et CUSHMAC M. E. — TLC spotting solvent for aflatoxins. *J. Ass. Off. Anal. Chem.*, t. **LI**, p. 65-67, 1968.

([2506]) STOLOFF L., CAMPBELL A. D., BECKWITH A. C., NESHEIM J., WINBUSH S. et FORDHAM O. M. — Sample preparation for aflatoxin assay : the nature of the problem and approaches to a solution. *J. Amer. Oil Chem. Soc.*, t. **XLVI**, p. 678-684, 1969.

([2507]) STOLOFF L., GRAFF A. et RICH H. — Rapid procedure for aflatoxin analysis in cottonseed products. *J. Ass. Off. Anal. Chem.*, t. **XLIX**, p. 740-743, 1966.

([2508]) STOLOFF L., NESHEIM S., YIN L., RODRICKS J. V., STACK M. et CAMPBELL A. D. — A multimycotoxin detection method for aflatoxins, ochratoxins, zearalenone, sterigmatocystin and patulin. *J. Ass. Off. Anal. Chem.*, t. **LIV**, p. 91-97, 1971.

([2509]) STOLOFF L. et TRAGER W. — Recommended decontamination procedures for aflatoxin. *J. Ass. Off. Agric. Chemists*, t. **XLVIII**, fasc. 3, p. 681-682, 1965.

([2510]) STOUT G. H., DREYER D. L. et JENSEN L. H. — Structure of rubrofusarin. *Chem. and Ind.*, p. 289-290, 1961.

([2511]) STREET J. P. — Losses in the fat of corn meal due to the action of moulds. *N. J. Agr. Exp. Stn. Rep.*, t. **XXIII**, p. 125-129, 1903 et *Bull.* n° 160, p. 72-76, 1903.

([2512]) STREET J. P. — Changes in the composition of corn meal due to the action of moulds. *N. J. Agr. Exp. Stn. Rep.*, t. **XXIV**, p. 123-147. 1904.

([2513]) STREZLECK S. et KOGAN L. — Note on thin layer chromatography of aflatoxin. *J. Ass. Off. Anal. Chemists*, t. **XLIX**, p. 33, 1966.

([2514]) STRUFALDI B. — Influencia da aflatoxina B_1 sobre a atividade metabolica da fraçao mitocondrial de figado de rato (*Rattus norvegicus albinus*). *Thèse Fac. Pharm. Biochim. Univ. Sao Paulo*, 57 p., 1969.

([2515]) STUBBLEFIELD R. D., SHANNON G. M. et SHOTWELL O. L. — Aflatoxins : improved resolution by thin layer chromatography. *J. Ass. Off. Anal. Chem.*, t. **LII**, p. 669-672, 1969.

([2516]) STUBBLEFIELD R. D., SHANNON G. M. et SHOTWELL O. L. — Aflatoxins M_1 and M_2 : preparation and purification. *J. Amer. Oil Chem. Soc.*, t. **XLII**, p. 389-392, 1970.

([2517]) STUBBLEFIELD R. D., SHOTWELL O. L., HESSELTINE C. W., SMITH M. L. et HALL H. H. — Production of aflatoxin on wheat and oats : measurement with a recording densitometer. *Appl. Microbiol.*, t. **XV**, p. 186-190, 1967.

([2518]) STUBBLEFIELD R. D., SHOTWELL O. L. et SHANNON G. M. — Aflatoxins B_1, B_2, G_1 and G_2 : separation and purification. *J. Amer. Oil Chem. Soc.*, t. **XLV**, p. 686-688, 1967.

([2519]) SU H. J. — Some cultural studies on *Ustilago esculenta*. *Spec. Publ. Coll. Agric. Taiwan Univ.*, fasc. 10, p. 139-160, 1961.

([2520]) SULLMAN S. F., ARMSTRONG S. J., ZUCKERMAN A. J. et REES K. R. — Further studies on the toxicity of the aflatoxins on human cell cultures. *Brit. J. Exp. Pathol.*, t. **LI**, p. 314-316, 1970.

([2521]) SUND J. M. — Nitrate poisoning can kill your cows. *Hoard's Dairyman*, t. **CV**, p. 453, 1960.

([2522]) SURYANARA YANA RAO K. et TULPULE P. G. — Variated differences of groundnut in the production of aflatoxin. *Nature. G. B.*, t. **CCXIV**, p. 738-739, 1967.

([2523]) SUSSMAN A. S. — Studies of an insect mycosis III. Histopathology of an aspergillosis of *Platysamia cecropia* L. *Ann. Entomol. Soc. Am.*, t. **XLV**, p. 233-245, 1952.

([2524]) SUSSMAN A. S. — Studies of an insect mycosis. IV. The physiology of the host parasite relationship of *Platysamia cecropia* and *Aspergillus flavus*. *Mycologia*, t. **XLIV**, p. 493-506, 1952.

([2525]) SUSSMAN A. S. et HALVORSON H. O. — *Spores. Their dormancy and germination.* 354 p., Harper et Row édit., 1966.

([2526]) SUSUKI T., TAKEDA M. et TANABE H. — A new mycotoxin produced by *Aspergillus clavatus*. *Chem. Pharm. Bull.*, t. **XIX**, p. 1786-1788, 1971.

([2527]) SUTIC M., AYRES J. C. et KOEHLER P. E. — Identification and aflatoxin production of molds isolated from country cured hams. *Appl. Microbiol.*, t. **XXIII**, p. 656-658, 1972.

([2528]) SVINHUFVUD V. E. — Untersuchungen über die bodenmikrobiologischen Unterschiede der Cajander'schen Waldtypen. *Acta For. Fenn.*, t. **XLIV**, 67 p., 1937.

([2529]) SVOBODA D., GRADY H. J. et HIGGINSON J. — Aflatoxin B_1 injury in rat and monkey liver. *Amer. J. Pathol.*, t. **XLIX**, p. 1023-1051, 1966.

([2530]) SVOBODA D. J., REDDY J. K. et LIU C. — Multinucleate giant cells in livers of marmosets given aflatoxin B_1. *Arch. Pathol.*, t. **XCI**, p. 452-454, 1971.

([2531]) SWAN J. B. et JAMIESON N. O. — Studies on metabolic disorders in dairy cows. *N. Z. J. Sci. Technol.*, t. **XXXVIII**, p. 137-151 ; 316-325 ; 362-382, 1956.

([2532]) SWARBRICK O. — Disease of turkey poults. *Vet. Rec.*, t. **LXXII**, p. 671, 1960.

([2533]) SYKES G. — *Disinfection and sterilization. Theory and practice.* 2e éd., 486 p., Londres, E. et F. N. Spon Ltd, 1965.

([2534]) SYNGE R. L. M. et WHITE E. P. — Sporidesmin : a substance from *Sporidesmium bakeri* causing lesions characteristic of facial eczema. *Chem. and Ind.*, t. **XLIX**, p. 1546-1547, 1959.

([2535]) SYNGE R. L. M. et WHITE E. P. — Photosensitivity diseases in New Zealand. XXIII. Isolation of sporidesmin, a substance causing lesions characteristic of facial eczema, from *Sporidesmium bakeri* Syd. *New Zealand J. Agric. Res.*, t. **III**, p. 907-921, 1960.

([2536]) SYSK D. B., CARLTON W. W. et CURTIN T. M. — Experimental aflatoxicosis in young swine. *Amer. J. Vet. Res.*, t. **XXIX**, p. 1591-1602, 1968.

([2537]) TABAK H. H. et COOK W. B. — Growth and metabolism of fungi in an atmosphere of nitrogen. *Mycologia*, t. **LX**, p. 115-140, 1968.

([2538]) TABER R. A., PETTIT R. E., TABER W. A. et DOLLAMITE J. W. — Isolation of *Pithomyces chartarum* in Texas. *Mycologia*, t. **LX**, p. 727-730, 1968.

([2539]) TABER R. A. et SCHROEDER H. W. — Aflatoxin-producing potential of the *Aspergillus flavus-oryzae* group from peanuts. *Appl. Microbiol.*, t. **XV**, p. 140-144, 1967.

([2540]) TADASHI A., MORI Z., MAJIMA R. et URITANI I. — Effects of aflatoxins on metabolic changes in plant tissue in response to injury. *1st Int. Congr. Pl. Pathol.*, Londres, 1968.

([2541]) TAGA M., TURBURI A., CIUPERCESCU V. et DANALACHE D. — La présence de fongi aflatoxigènes dans les produits fourragers et leur possibilité de synthétiser des aflatoxines dans les substrats fourragers. *Arch. Vet.* (Bucuresti), t. **IV**, p. 225-231, 1968.

([2542]) TAINSH A. R. — The gross national waste of food grain. *Proc. Symp. Agric. Coll. Norway*, 1967, 4 p. ronéot.

([2543]) TAKEDA S., OGASAWARA K., OHARA H. et ONISHI T. — Studies on the food poisoning caused by the *Fusarium* scab disease of wheat and barley. Part 4. Animal test for toxic substances obtained from deteriorated wheat grain. *Rep. Hokkaido Inst. Public Health*, t. **IV**, p. 22-28, 1953.

([2544]) TANABE H. et SUZUKI T. — A new mycotoxin produced by *Aspergillus clavatus*. *in* HERZBERG M., *Toxic Micro-organisms*, p. 127-128, 1970.

([2545]) TANAKA H. et TAMURA T. — Constitution of rubrofusarin. *Tetrahedron Lett.*, p. 151-155, 1961.

([2546]) TANDON R. N. — Physiological studies on some pathogenic fungi. 80 p. Scientific Res. Committee, Allahabad, India, 1961.

([2547]) TANDON R. N. et CHAUMAN R. P. S. — A note on the effect of temperature on the wheat mold caused by *Aspergillus flavus* and *Aspergillus tamarii*. *Sci. and Cult.*, t. **XX**, p. 503-504, 1955.

([2548]) TANNENBAUM S. W. and BASSET E. W. — The biosynthesis of Patulin. I. Related aromatic substances from *Penicillium patulum*, strain 2159. *A. Biochim. Biophys. Acta*, t. XXVIII, p. 21-31, 1958 et t. **XL**, p. 3, 1960.

([2549]) TANNENBAUM S. W. — Biosynthesis of aromatic compounds and their congeners in cell-free extracts from Fungi Imperfecti. *Develop. Ind. Microbiol.*, t. **VIII**, p. 88-95, 1967.

([2550]) TANNER F. W. — *The Microbiology of Foods.* 2_e éd, 720 p., Garrard Press, Champaign, III., 1954.

([2551]) TARR S. A. J. — Control of seedbed losses of groundnuts by seed treatment. *Ann. Appl. Biol.*, t. **XLVI**, p. 178-185, 1958.

([2552]) TATARENKO E. — The influence of light on the development of mold fungi. *Microbiology, Moscou*, t. **XXIII**, p. 29-33, 1964.

([2553]) TATSUNO T. — Recherches biochimiques sur les substances toxiques du riz jauni par *Penicillium islandicum* Sopp. *Prod. et Prob. Pharm.*, t. **XVIII**, p. 180-186, 1963.

([2554]) TATSUNO T. — The chronic toxicity of naturally-occurring substances. Metabolites of *Penicillium islandicum* Sopp. *Food Cosmet. Toxicol.*, t. **II**, p. 678-680, 1964.

([2555]) TATSUNO T. — Biochemical findings of *Fusarium* substances, *Fusarium* toxicosis, toxicological studies of *Fusarium* substances. *Seikagaku*, t. **XLI**, p. 153-171, 1969.

([2556]) TATSUNO T., MORITA Y., TSUNODA H. et UMEDA M. — Recherches toxicologiques des substances métaboliques du *Fusarium nivale.* VII. La troisième substance métabolique de *F. nivale*, le diacétate de nivalénol. *Chem. Pharm. Bull.*, t. **XVIII**, p. 1485-1487, 1970.

([2557]) TATSUNO T., MOROOKA N., SAITO M., ENOMOTO M., UMEDA M., OKUBO K. et TSUNODA H. — Toxicological studies on the reddish wheat by *Fusarium graminearum* genera (Report 1). *Folia Pharmacol. Jap.*, t. **LXII**, p. 26-27, 1966.

([2558]) TATSUNO T., SAITO M., ENOMOTO M. et TSUNODA H. — Nivalenol, a toxic principe of *Fusarium nivale. Chem. Pharm. Bull.* (Tokyo), t. **XVI**, p. 2519-2520, 1968.

([2559]) TAUCHMANN F. et LEISTNER L. — Detoxifizierung von Aflatoxinen in wässeriger Lösung. *Fleischwirtschaft*, t. **XLIX**, p. 1640-1641, 1969.

([2560]) TAUCHMANN F. et LEISTNER L. — Aflatoxinbildung in Rohwurst und Reis in Abhängigkeit von Zeit und Temperatur. *Fleischwirtschaft*, t. **LI**, p. 77-79, 1971.

([2561]) TAYLOR A. — The chemistry and biochemistry of sporidesmins and other 2,5-epidithia-3,6-dioxopiperazines. Biochemistry of some Foodborne Microbial Toxins. *Symp. Amer. Chem. Soc. New York, sept. 1966*, p. 69-118, M.I.T. Press, 1967.

([2562]) TAYLOR A. — The toxicology of sporidesmins and other epipolythiadioxopiperazines. *in* KADIS S. et AL., *Microbial toxins*, t. **VII**, p. 337-376, 1971.

([2563]) TEASE S. C. — Progress of food freeze-drying in the U.S.A. *Researches and development in Freeze Drying*, p. 581-591, 1964.

([2564]) TEITELL L. — Effects of relative humidity on viability of conidia of Aspergilli. *Am. J. Bot.*, t. **XLV**, p. 748-753, 1958.

([2565]) TENG J. I. et HANZAS P. C. — Note on a new developer for thin layer chromatography of aflatoxins on silica gel plates. *J. Ass. Off. Anal. Chem.*, t. **LII**, p. 83-84, 1969.

([2566]) TENG J. I. et HANZAS P. C. — Determination of aflatoxins in moldy sugarbeet pulp. *J. Amer. Soc. Sugar Beet Technol.*, t. **XV**, p. 438-443, 1969.

([2567]) TERAO K. et MIYAKI K. — The effect of aflatoxin on chick-embryo liver cells. *Exper. Cell. Ress.*, t. **XLVIII**, p. 151-155, 1967.

([2568]) TERAO K. et MIYAKI K. — Different susceptibilities of chick-embryo liver cells in vitro to aflatoxin, actinomycin D and mitomycin C. *Z. Krebsforsch.*, t. **LXXI**, p. 199-207, 1968.

([2569]) TERAO K. et MIYAKI K. — New bioassay method of aflatoxins using the cultured liver cells of chicken embryo. *J. Food Hyg. Soc. Jap.*, t. **IX**, p. 285-288, 1968.

([2570]) TERAO K. et MIYAKI K. — The effect of aflatoxin on chick-embryo liver cells in vitro. *in* HERZBERG M., *Toxic Micro-organisms*, p. 67-71, 1970.

([2571]) TERBLANCHE M. et RABIE C. J. — A toxic fungus (*Sclerotium rolfsii*) isolated from groundnuts. *S. Afr. J. Agric. Sci.*, t. **X**, p. 253-262, 1967.

(2572) TERBLANCHE M., VAN RENSBURG J., ADELAAR T. F., NAUDE T. W. et SMIT J. D. — A preliminary note on the toxicity of *Aspergillus flavus* Link as grown in pure culture. *J. S. Afr. Vet. Med. Ass.*, t. **XXXIV**, p. 645-648, 1963.

(2573) TERRACINI B. — Rischi oncogeni da alimentazione. *Tumori*, t. **LIV**, p. 93-98, 1968.

(2574) TERTZAKIAN G., HASKINS R. H., SLATER G. P. et NESBITT L. R. — The structure of cephalochromin. *Proc. Chem. Soc.*, p. 195-196, 1964.

(2575) TERVET L. W. — The influence of fungi on storage, on seed viability and seedling vigor of soybeans. *Phytopathology*, t. **XXXV**, p. 3-15, 1945.

(2576) TETER N. C. et ROANE C. W. — Molds impose limitations in grain drying. *Agr. Eng.*, t. **XXXIX**, p. 24-27, 1958.

(2577) TETERNIKOVA-BABAYAN D. N. et ABRAMYAN D. G. — Rezul'taty izucheniya vozdeistviya nekotorykh gribov rizosfery na seyantsy Pomidora. *Biol. Zh. Armen.* t. **XIX**, p. 5-13, 1966.

(2578) TEUNISSON D. J. et ROBERTSON J. A. — Degradation of pure aflatoxins by *Tetrahymena pyriformis. Appl. Microbiol.*, t. **XV**, p. 1099-1103, 1967.

(2579) T'HART M. L. — Some problems of intensive grassland farming in the Netherlands. *7th internat. Grassland Congress*, p. 70-79, Nouvelle-Zélande, 1956.

(2580) THATCHER F. S. et CLARK D. S. — Microorganisms in foods. Their significance and methods of enumeration. 234 p., *Univ. Toronto Press*, 1968.

(2581) THATCHER F. S., COUTU C. et STEVENS F. — The sanitation of Canadian flour mills and its relationship to the microbial content of flour. *Cereal Chem.*, t. **XXX**, p. 71-102, 1953.

(2582) THEILER A. — Die Diplodiosis der Rinder und Schafe in Südafrika. *Dtsch., Tierärztl. Wschr.*, t. **XXXV**, p. 395-399, 1927.

(2583) THEODOSSIADES G. — Le caneton, réactif biologique pour le dépistage de la toxicité des tourteaux d'arachide. *Rev. Elev. Méd. Vét. Pays Trop.*, t. **XVI**, p. 229-236, 1963.

(2584) THERON J. J. — Acute liver injury in ducklings as a result of aflatoxin poisoning. *Lab. Invest., U. S. A.*, t. **XIX**, p. 1586-1603, 1965.

(2585) THERON J. J., LIEBENBERG N. et JOUBERT H. J. B. — Acute liver injury in ducklings as a result of aflatoxin and ochratoxin poisoning. *Symposium on mycotoxins, Pretoria*, 1965.

(2586) THERON J. J., VAN DER MERWE R. J., LIEBENBERG N., JOUBERT H. J. B. et NEL W.— Acute liver injury in ducklings and rats as a result of ochratoxin poisoning. *J. Pathol. Bacteriol.*, t. **XCI**, p. 521-529, 1966.

(2587) THIBAUT M. — *Les maladies par allergie mycosique*, 233 p., Arnette édit., Paris, 1963.

(2588) THIER H. P. — On aflatoxin. *Deutsch. Lebensmittel-Rundschau*, t. **LXII**, p. 118-120, 1966.

(2589) THOM C. — *The Penicillia.* The Williams and Wilkins Co, 643 p., Baltimore, 1930 (Nelle éd. : RAPER K. B. et THOM C. — A manual of the Penicillia. 875 p., Baltimore, 1949.)

(2590) THOM C. et AYERS S. M. — Effect of pasteurization on mold spores. *J. Agr. Res.*, t. **VI**, p. 153-166, 1916.

(2591) THOM C. et LEFEVRE E. — Flora of corn meal. *J. Agr. Res.*, t. **XXII**, p. 179-188, 1921 (*Abstr. Bacteriol.*, t. V, p. 10-11, 1921).

(2592) THOM C. et RAPER K. B. — *A manual of the Aspergilli.* 373 p. Williams et Wilkins Co, Baltimore, 1945.

(2593) THOMAS R. — *Biogenesis of antibiotic substances.* p. 160-161, Academic Press, 1965.

(2594) THOMKE S. — Influence of various factors on the mould flora of newly harvested feed grain, stored under mould-promoting conditions. *Proc. Symp. Agric. Coll. Norway*, 1967, 3 p. ronéot.

(2595) THOMPSON A. — Soil-plant animal relationships and hypomagnesaemia. *Conf. on Hypomagnesaemia*, p. 75-87, Londres, 1960.

(2596) THORNTON R. H. et ROSS D. J. — The isolation and cultivation of some fungi from soils and pastures associated with facial eczema disease of sheep. *New Zealand J. Agr. Research*, t. **II**, p. 1002-1016, 1959.

(2597) THORNTON R. H., SHIRLEY G. et SALISBURY R. M. — A nephrotoxin from *Aspergillus fumigatus* and its possible relationship with New Zealand mucosal disease-like syndrome in cattle. *N. Z. J. Agr. Res.*, t. **XI**, p. 1-14, 1968.

(2598) THORNTON R. H. et SINCLAIR D. P. — *Sporidesmium bakeri* and facial eczema of sheep in the field. *Nature*, t. **CLXXXIV**, p. 1327-1328, 1959.

(2599) THORNTON R. H. et SINCLAIR D. P. — Some observations on the occurrence of *Sporidesmium bakeri* Syd. and facial eczema disease in the field. *New Zealand J. Agr. Research*, t. **III**, p. 300-313, 1960.

(2600) THORNTON R. H. et TAYLOR W. B. — Antifungal activity of fatty acids to *Pithomyces chartarum* (Berk. et Curt.) M. B. Ellis in field trials. *New Zealand J. Agric. Res.*, t. **VI**, p. 329-342, 1963.

(2601) THOUKIS G., BOUTHILET R. J., UEDA M. et CAPUTIA. — Fate of DEPC in wine. *Am. J. Enol. Vit.*, t. **XIII**, p. 105-113, 1962.

(2062) TICKTIN H. E. et ZIMMERMAN H. J. — Hepatic dysfunction and jaundice in patients receiving triacetyloleandomycin. *New Engl. J. Med.*, t. 267, p. 964-968, 1962.

(2603) TIEMSTRA P. J. — A study of the variability associated with sampling peanuts for aflatoxin. *J. Amer. Oil Chem. Soc.*, t. **XLVI**, p. 667-672, 1969.

(2604) TILDEN E. B., FREEMAN S. et LOMBARD L. — Further studies of the *Aspergillus* endotoxins. *Mycopathol. Mycol. Appl.*, t. **XX**, p. 253-271, 1963.

(2605) TILDEN E. B., HATTON E. H., FREEMAN S., WILLIAMSON W. M. et KÖNIG V. L. — Preparation and properties of endotoxin of *Aspergillus flavus*. *Mycopathol. Mycol. Appl.*, t. **XIV**, p. 325-335, 1961.

(2606) TIMONIN M. I. — Another mould with antibacterial ability. *Science*, **XLVI**, p. 494, 1942.

(2607) TIMONIN M. I. — Activity of patulin against *Ustilago tritici* (Pers.) Jen. *Canad. J. Agr. Sci.*, t. **XXVI**, p. 358-360, 1946.

(2608) TIMONIN M. I. et ROUATT J. W. — Bacteriostatic activity of citrinin *in vitro*. *Canad. Pub. Health Jour.*, t. **XXXV**, p. 396-406, 1944.

(2609) TIMONIN M. I. et ROUATT J. W. — Production of citrinin by *Aspergillus* sp. of the *candidus* group. *Canad. Pub. Health Jour.*, t. **XXXV**, p. 80-88, 1944.

(2610) TODD G. C., SHALKOP W. T., DOOLEY K. L. et WISEMAN H. G. — Effects of ration modifications on aflatoxicosis in the rat. *Amer. J. Vet. Res.*, t. **XXIX**, p. 1855-1861, 1968.

(2611) TOGUKAWA Y. et EMOTO Y. — Ueber einen kurz nach der letzten Feuerbrunst entwickelten Schimmelpilz. *Jap. Journ. Bot.*, t. **II**, p. 175-188, 1924.

(2612) TOMOV A. — (*Aspergillus clavatus* Desm., cause d'intoxication de vaches ayant reçu du germe de malt dans leur ration.) *Veter. Med. Nauki, Bulg.*, t. **II**, p. 997-1003, 1965.

(2613) TOTH L., TAUCHMANN F. et LEISTNER L. — Fluorometrische Bestätigung des Nachweises von Aflatoxinen in Routineuntersuchungen. *Fleischwirtschaft*, t. **L**, p. 349-350, 1970.

(2614) TOTH L., TAUCHMANN F. et LEISTNER L. — Quantitativer Nachweis der Aflatoxine durch photofluorometrische Bestimmung direkt von Dünnschichtplatten. *Fleischwirtschaft*, t. **L**, p. 1235-1237, 1970.

(2615) TOUMANOFF C. — Actions des Champignons entomophytes sur la pyrale du Maïs (*Pyrausta nubilalis* Hübner). *Ann. parasitol. humaine et comparée*, t. **XI**, p. 129-143, 1933.

(2616) TOURY J. — L'aflatoxine de l'arachide, aspect chimique et bromatologique. *Méd. Afr. noire, Sénégal*, t. **XII**, p. 459-462, 1965.

(2617) TOURY J. — Incidence des conditions de production des arachides sur leur contamination par l'aflatoxine. *Colloque aflatoxines, Villejuif*, 7 mars 1966.

(2618) TOURY J., DUPIN H., CROS J., RICHIR C. — Contaminations alimentaires par *Aspergillus flavus*. *Rev. Gén. Pat. et Biol.*, p. 346-351, 1963.

([2619]) TOURY J. et GIORGI R. — Aflatoxine et fluorescence. *Ann. Nutrit. Aliment. Fr.*, t. **XX**, p. 111-118, 1966.

([2620]) TOURY J., GIORGI R., MONJOUR G., QUENUM C., DELASSUS M., GOARIN P. et VALETTE R. — Contaminants fongiques de produits vivriers africains. *C. R. Soc. Biol.*, t. **CLXII**, p. 1462-1463. 1968.

([2621]) TOURY J., MARTINEAUD M. et RICHIR C. — Considérations sur l'aflatoxine et autres mycotoxines. Résultats préliminaires de l'aflatoxine dans l'expérimentation sur le rat et le singe. *Méd. Afr. noire, Sénégal*, t. **XII**, p. 463-465, 1965.

([2622]) TOWNSEND R. J. — Toxic moulds and their metabolites. *Int. Biodet. Bull.*, t. **III**, p. 47-58, 1967.

([2623]) TOWNSEND R. J., MOSS M. O. et PECK H. M. — Isolation and characterisation of hepatotoxins from *Penicillium rubrum. J. Pharm. Pharmacol.*, t. **XVIII**, p. 471, 1966.

([2624]) TOWNSLEY P. M. et LEE E. G. H. — Response of fertilized eggs of the mollusk *Banksia setacea* to aflatoxin. *Ass. Off. Anal. Chem. J.*, t. **L**, p. 361-363, 1967.

([2625]) TRAGER W. et STOLOFF L. — Possible reactions for aflatoxin detoxification. *J. Food Chem.*, t. **XV**, p. 679-681, 1967.

([2626]) TRAGER W. T., STOLOFF L. et CAMPBELL A. D. — A comparison of assay procedures for aflatoxin in peanut products. *J. Ass. Off. Agric. Chemists*, t. **XLVII**, p. 993-1001, 1964.

([2627]) TRAXLER P. et TAMM C. — Die Struktur des Antibioticums Roridine H. Verrucarine und Roridine. 20. Mitt. *Helv. Chim. Acta* t. **LIII**, p. 1846-1869, 1970.

([2628]) TRENK H. L., BUTZ B. E. et FUN SUN CHU. — Production of ochratoxins in different cereal products by *Aspergillus ochraceus. Appl. Microbiol.*, t. **XXI**, p. 1032-1035, 1971.

([2629]) TRENK H. L. et CHU F. S. — Improved detection of ochratoxin A on thin layer plates. *J. Assoc. Off. Anal. Chem.*, t. **LIV**, p. 1307-1309, 1971.

([2630]) TRENK H. L. et HARTMAN P. A. — Effects of moisture content and temperature on aflatoxin production in corn. *Appl. Microbiol.* t. **XIX**, p. 781-784, 1970.

([2631]) TRESSLER D. K. et EVERS C. F. — *The freezing preservation of foods.* The Avi Publish. Co, Connecticut, 1957.

([2632]) TRIQUE B. — Recherches sur le développement de l'*Aspergillus versicolor* (Vuill.) Tiraboschi et du *Penicillium cyclopium* Westl., moisissures des blés de conservation ; étude particulière de certaines hydrolases du compartiment extra-cellulaire. *Thèse 3e cycle. Fac. Sci. Brest*, 20 juin 1969, 110 p.

([2633]) TRIQUE B. — Croissance et sporulation de l'*Aspergillus versicolor* (Vuil.) Tiraboschi et du *Penicillium cyclopium* Westl. en fonction des sources de carbone et d'azote. *Bull. Soc. Bot. Fr., Mém.* 1968, t. **CXV**, p. 101-109, 1969.

([2634]) TRIQUE B. — Croissance et sporulation de l'*Aspergillus versicolor* (Vuill.) Tiraboschi et du *Penicillium cyclopium* Westl. en culture stationnaire ou agitée. *Rev. de Mycol.* t. **XXXIV**, p. 365-375, 1970.

([2635]) TROLLER J. A. et OLSEN R. A. — Derivatives of sorbic acid as food preservatives. *J. Food Sci.*, t. **XXXII**, p. 228-236, 1967.

([2636]) TROPICAL PRODUCTS INSTITUTE. — Recommended procedures for the detection and estimation of aflatoxin B_1 in groundnuts and groundnut materials. T. P. I. Report n° G-13, Mai 9165.

([2637]) TRUELOVE V., DAVIS D. E. et THOMPSON O. C. — The effect of aflatoxin B_1 on protein synthesis by cucumber cotyledon discs. *Can. J. Bot.* t. **XLVIII**, p. 485-491, 1970.

([2638]) TRUM B. F. — Grass intoxication and tetany. *U. S. Army Vet. Bull.*, t. **XXXVI**, p. 110, 1942.

([2639]) TSCHERNIAK W. S. — Zur Lehre den Broncho- und Pneumonomykosen beim Pferde. *Arch. Wiss. und Prakt. Tierkeilk.*, t. **LVII**, p. 417-444, 1928.

([2640]) TSUBAKI K. — *Penicillium* isolated from toxic ensilage. *Trans. Mycol. Soc. Japan*, t. **I**, p. 6-7, 1956.

([2641]) Tsunoda H. — Studies on a poisonous substance produced on the cereals by a *Penicillium* Sopp under the storage. *Jap. J. Nutrition*, t. VIII, p. 185-199, 1951 ; t. **IX**, p. 1-6, 1952.

([2642]) Tsunoda H. — Study on damage of stored rice, caused by microorganisms. III On yellowsis rice from Thailand. *Food Res. Inst. Report 8. Trans. Jap. Phytopathol. Soc.*, t. **XIII**, p. 3, 1953.

([2643]) Tsunoda H. — Researches for the microorganisms which deteriorate the stored cereals. XXVI. Classification of genus *Penicillium* which is parasitic on the rice. *Rep. Food. Res. inst.*, t. **XV**, p. 98-100, 1961.

([2644]) Tsunoda H. — *Penicillium brunneum. Shokuhin Eiseigaku Zasshi*, t. **III**, p. 347-351, 1962.

([2645]) Tsunoda H. — Microorganisms of spoiled rice in storage. *Food. Its Science and Technology*, Suppl. p. 161, mai 1963.

([2646]) Tsunoda H. — Studies on damage of stored grains by microorganisms. Report 29. On *Penicillium brunneum*, a fungus parasitic to rice. *Food Res. Inst. Report 17*, p. 238, 1963.

([2647]) Tsunoda H. — Microorganisms which deteriorate stored cereals and grains. *in* Herzberg M., *Toxic Microorganisms*, p. 143-162, 1970.

([2648]) Tsunoda H. et Jaruki Y. — On the cultivation of *Penicillium toxicarium Rep. Food Res. Inst.*, t. **II**, p. 157-164, 1949.

([2649]) Tsunoda H., Toyazaki N., Morooka N., Nakano N., Yoshiyama H., Okubo K. et Isoda M. — Researches on the microorganisms which deteriorate the stored cereals and grains (part 34). Detection of injurious strains and properties of their toxic substance of scab *Fusarium* blight grown on the wheat. *Rep. Food Res. Inst.*, t. **XXIII**, p. 89-116, 1968.

([2650]) Tsunoda H., Tsuruta O. et Matsunami S. — Researches on the microorganisms which deteriorate the stored cereals. XVI. Studies on rice parasitic molds, *Giberella* and *Fusarium*. Results on animal test. *Rep. Food. Res. inst.* t. **XIII**, p. 26-28, 1958.

([2651]) Tsunoda H., Tsuruta O. et Takahashi M. — Researches for the microorganism which deteriorate the stored cereals. XIX. Parasites of imported rice (3). *Rep. Inst.*, t. **XIII**, p. 38-42, 1959.

([2652]) Tsuruta O. — A study on the parasitism and control of micro-organisms affecting stored rice with specific reference to family *Aspergillaceae. J. Medical Soc. Toho Univ.*, t. **IX**, p. 1-16, 1962.

([2653]) Tuite J. F. et Christensen C. M. — Fungi important in storage of barley. *Phytopathology*, t. **XLII**, p. 476, 1952.

([2654]) Tuite J. F. et Christensen C. M. — Grain storage studies. XVI. Influence of storage conditions upon the fungus flora of barley seed. *Cereal Chem.*, t. **XXXII**, p. 1-11, 1955.

([2655]) Tuite J. F. et Christensen C. M. — Grain storage studies. XXIV. Moisture content of wheat seed in relation to invasion of the seed by species of the *Aspergillus glaucus* group, and effect of invasion upon germination of the seed. *Phytopathology*, t. **XLVII**, p. 323-327, 1957.

([2656]) Tulpule P. G. — Aflatoxicosis. *Indian J. Med. Res.*, t. **LVII**, p. 102-114, 1969.

([2657]) Tulpule P. G., Madhavan T. V. et Gopalan C. — Effect of feeding aflatoxin to young monkeys. *Lancet*, t. **I**, p. 962-963, 1964.

([2658]) Tung T. C. et Ling K. H. — Study on aflatoxin of foodstuffs in Taiwan. *J. vitaminol.* (*Osaka*), t. **XIV** (suppl.), p. 48-52, 1968.

([2659]) Turesson G. — The presence and significance of moulds in the alimentary canal of man and higher animals. *Svensk Botanisk Tidskrift*, t. **X**, fasc. 1, p. 1-27, 1916.

([2660]) Turley H. — The microscopy of flour. *Baking Technol.*, t. **I**, p. 327-329, 1922.

([2661]) Tyler V. E. — Poisonous Mushrooms. *Progress in Chemical Toxicology*, t. **I**, p. 339-384, 1963.

([2662]) Udagawa S. — Taxonomic studies of fungi on stored rice grains. III. Penicillium group (Penicillia and related genera). *J. Agr. Sci.* Tokyo, t. **V**, p. 5-21, 1959.

(2663) UDAGAWA S. I., ICHINOE M. et KURATA H. — Occurrence and distribution of mycotoxin producers in Japanese foods. *in* HERZBERG M., *Toxin Micro-organism*, p. 174-184, 1970.

(2664) UDALL R. H. — Low blood magnesium and associated tetany occurring in cattle in the winter. *Cornell Vet.*, t. **XXXVII**, p. 314, 1947.

(2665) UENO Y. — Inhibition of protein synthesis in animal cells by nivalenol and related metabolites ; toxic principales of rice infested with *Fusarium nivale. in* HERZBERG M., *Toxic micro-organisms*, p. 76-79, 1970.

(2666) UENO Y. — Production of citreoviridin, a neurotoxic mycotoxin of *Penicillium citreoviride* Biourge. *in* PURCHASE I. F. H., *Mycotoxins in human health*, p. 115-132, 1971.

(2667) UENO Y. — Toxicological and biological properties of Fusarenon-X, a cytotoxic mycotoxin of *Fusarium nivale* Fn-2B. *in* PURCHASE I. F. H., *Mycotoxin in human health*, p. 163-178, 1971.

(2668) UENO Y. et ENOMOTO M. — *Japan J. Exp. Med.*, 1971.

(2669) UENO Y. et FUKUSHIMA K. — Inhibition of protein and DNA synthesis in Ehrlich ascites tumour by nivalenol, a toxic principle of *Fusarium nivale* growing rice. *Experientia*, t. **XXIV**, p. 1032-1033, 1968.

(2670) UENO Y. et HOSOYA M. — Bioassay of toxic principles of rice infested with *Fusarium nivale* employing rabbit reticulocytes. *in* HERZBERG M., *Toxic Micro-organisms*, p. 80-81, 1970.

(2671) UENO Y. et ISHIKAWA I. — Production of luteoskyrin, a hepatotoxic pigment, by *Penicillium islandicum* Sopp. *Appl. Microbiol.*, t. **XVIII**, p. 406-409, 1969.

(2672) UENO Y., ISHIKAWA Y., SAITO-AMAKAI K. et TSUNODA H. — Environmental factors influencing the production of fusarenon-X, a cytotoxic mycotoxin of *Fusarium nivale* Fn 2B. *Chem. Pharm. Bull.*, t. **XVII**, p. 304-312, 1970.

(2673) UENO Y., PLATEL A. et FROMAGEOT P. — Interaction entre pigments et acides nucléiques. II. Interaction *in vitro* entre la lutéoksyrine et le DNA de thymus de veau. *Biochim. Biophys. Acta*, t. **CXXXIV**, p. 27-36, 1967.

(2674) UENO Y., SAITO K. et TSUNODA H. — Isolation of toxic principles from the culture filtrate of *Fusarium nivale. in* HERZBERG M., *Toxic Micro-organisms*, p. 120, 1970.

(2675) UENO Y., UENO I. et MIZUMOTO K. — The mode of binding of luteoskyrin, a hepatotoxic pigment of *Penicillium islandicum* Sopp to deoxyribonucleic acid. *Jap. J. Exp. Med.*, t. **XXXVIII**, p. 47-64, 1968.

(2676) UENO Y., UENO I., ITO K. et TATSUNO t. — Impairments of RNA synthesis in Ehrlich ascites tumor by luteoskyrin, a hepatotoxic pigment of *Penicillium islandicum* Sopp. *Experientia*, t. **XXIII**, p.1001-1002, 1967.

(2677) UENO Y., UENO I., MIZUMOTO K. et TATSUNO T. — The binding of luteoskyrin, a hepatotoxic pigment of *Penicillium islandicum* Sopp. to deoxyribonucleohistone. *J. Biochem.*, t. **LXIII**, p. 395-397, 1968.

(2678) UENO Y., UENO I., TATSUNO T. et URAGUCHI K. — Les effets de la lutéoskyrine, substance toxique du *Penicillium islandicum* Sopp, sur le gonflement des mitochondries. *Japan J. Exp. Med.*, t. **XXXIV**, p. 197-209, 1964.

(2679) UKAI T., YAMAMOTO Y. et YAMAMOTO T. — Studies on the poisonous substance from a strain of *Penicillium* (Hori-Yamamoto strain). II. Culture method of Hori-Yamamoto strain and chemical structure of its poisonous substance. *J. Pharmacol. Soc. Japan*, t. **LXXIV**, p. 450-454, 1954.

(2680) ULLOA-SOSA M. et SCHROEDER H. W. — Note aflatoxin decomposition in the process of making tortillas from corn. *Cereal Chem.* t. **XLVI**, p. 397-400, 1969.

(2681) ULREY D. G. — Detoxifying tung meal. U.S. Patent nr 2641542, 1953.

(2682) ULRICH R. — *Conservation par le froid des denrées d'origine végétale.* 330 p., 1954.

(2683) ULRICH R. — Quelques aspects des réactions physiologiques des fruits aux traitements frigorifiques. *Fruits.* t. **XII**, p. 139-145, 1957.

([2684]) UMEDA M. — Cytotoxic effects of the mycotoxins of *Penicillium islandicum* Sopp, luteoskyrin and chlorine-containing peptide on Chang's liver cells and HeLa cells. *Acta Path. Japan*, t. **XIV**, p. 373-394, 1964.

([2685]) UNGAR H. et JOFFE A. Z. — Acute liver lesions resulting from percutaneous absorption of aflatoxins. *Pathol. Microbiol.*, t. **XXXIII**, p. 65-76, 1969.

([2686]) UNUMA TAO, MORRIS H. P. et BUSCH H. — Comparative studies of the nucleoli of Morris hepatomas, embryonic liver and aflatoxin B_1. *Cancer Res.*, t. **XXVII**, p. 221-233, 1967.

([2687]) URAGUCHI K. — Existence of toxic substance in the moldy rice. *Nisshin Igaku*, t. **XXXIV**, p. 155-161, 1947.

([2688]) URAGUCHI K. — The present situation in studies on toxicity of yellowsis rice. *J. Jap. Med. Assoc.*, t. **XXXII**, p. 507, 1954.

([2689]) URAGUCHI K. — Citreoviridin. *in* CIEGLER A. et AL. , *Microbial Toxins*, t. **VI**, p. 299-380, 1971.

([2690]) URAGUCHI K., SAITO M., NOGUCHI Y., TAKAHASHI K., ENOMOTO M. et TATSUNO T. — Chronic toxicity and carcinogenicity in mice of the purified mycotoxins luteoskyrin and cyclochlorotin. *Food Cosmet. Toxicol.*, t. X, p. 193-207, 1972.

([2691]) URAGUCHI R., SAKAI Y., TATSUNO T. et WAKAMATSU H. — Toxicological approach to metabolites of *Fusarium roseum* and other « red » molds growing on rice, wheat and other cereal grains I. *Folia Pharmacol. Jap.*, t. **LIV**, p. 127, 1958.

([2692]) URAGUCHI K., SAKAI F., TSUKIOKA M., NOGUCHI Y., TATSUNO T., SAITO M., ENOMOTO M., SHIKATA T. et MIYAKE M. — Acute and chronic toxicity in mice and rats of the fungus mat of *Penicillium islandicum* Sopp added to diet. *Jap. J. Exp. Med.*, t. **XXXI**, p. 435-461, 1961.

([2693]) URAGUCHI K., TATSUNO T., SAKAI F., TSUKIOKA M., SAKAI Y., YONEMITSU O., ITO H., MIYAKE M., SAITO M., ENOMOTO M. et MIYAKE M. — Toxicological approach to the metabolites of *Penicillium islandicum* Sopp growing on the yellowed rice. *Jap. J. Exp. Med.*, t. **XXXI**, p. 1-18, 1961.

([2694]) URAGUCHI K., TATSUNO T., SAKAI F., TSUKIOKA M., SAKAI Y., YONEMITSU O., ITO H., MIYAKE M., SAITO M., ENOMOTO M., SHIKATA T. et ISHIKO T. — Isolation of two toxic agents, luteoskyrin and chlorine-containing peptide from the metabolites of *Penicillium islandicum* Sopp, with some properties thereof. *Jap. J. Exptl. Med.*, t. **XXXI**, p. 19-46, 1961.

([2695]) URITANI I., ASAHI T., MAJIMA R. et MORI Z. — The biochemical effects of aflatoxins and other toxic compounds related to parasitic fungi on the metabolism of plant tissues. *in* Herzberg M., *Toxic Micro-organisms*, p. 107-113, 1970.

([2696]) URRY W. H., WEHRMEISTER H. L., HODGE E. B. et HIDY P. H. — The structure of zearalenone. *Tetrahedron L.*, n° 27, p. 3109-3114, 1966.

([2697]) URSINY J. et GROCH L. — Prispevok k aktualnej problematike mykotocksikjoz hospodarskych zvierat v CSSR a jej profylaxie. *Veterinarni Medicina*, t. **XII**, p. 311-320, 1967.

([2698]) U.S. DEPARTMENT OF AGRICULTURE. — Agricultural Marketing Service. — Methods for determining moisture content as specified in the official grain standards of the United States and in the United States standards for beans, peas, lentils, and rice. *Service and Regulatory Announcements*, n° 147, 3 p., 1959.

([2699]) UZZAN A. — Caractères organoleptiques des corps gras alimentaires : causes, évolution, appréciation. *Méthodes subjectives et objectives d'appréciation des caractères organoleptiques des denrées alimentaires*, Coll. C.N.R.S., A311-A334, 1965.

([2700]) VANDERHAEGHE H., VAN DIJCK P. et DE SOMER P. — Identity of ramycin with fusidic acid. *Nature, G.B.*, t. **CCV**, p. 710-711, 1965.

([2701]) VANDERHOVEN C., REMACLE J. et RAMAUT J. L. — Recherche d'un rapport éventuel entre la morphologie de diverses souches d'*Aspergillus flavus* Link et leur production d'aflatoxines. *Rev. Ferment. Ind. Aliment.*, t. **XXV**, p. 179-183, 1970.

([2702]) VAN DER LAAN P. A. — Emploi de substances antibiotiques comme fongicides dans la lutte contre *Cercospora nicotianae* attaquant le tabac. *Tijdschr. Plantenziekten*, t **LIII**, p. 180-187, 1945.

([2703]) VAN DER LINDE J. A., FRENS A. M. et VAN ESCH G. J. — Experiments with cows fed groundnut meal aflatoxin. *Mycotoxins in Foodstuffs*, p. 247-249, 1965.

([2704]) VAN DER LINDE J. A., FRENS A. M., DE IONGH H. et VLENS R. O. — Onderzoek van melk afkomstig van koeien govoed met aflatoxinehoudend grondnoten meel. *Tijdschr. v. Diergeneesk.*, t. **LXXXIX**, p. 1082-1088, 1964.

([2705]) VAN DER MERWE K. J., FOURIE L. et DE SCOTT B. — On the structure of the aflatoxins. *Chem. Ind.*, p. 1660-1661, 1963.

([2706]) VAN DER MERWE R. J., STEYN P. S. et FOURIE L. — Mycotoxins II. The constitution of ochratoxins A, B and C, metabolites of *Aspergillus ochraceus* Wilh. *J. Chem. Soc.*, p. 7083-7088, 1965.

([2707]) VAN DER MERWE K. J., STEYN P. S., FOURIE L., SCOTT DE B. et THERON J. J. — Ochratoxin A, a toxic metabolite produced by *Aspergillus ochraceus* Wilh. *Nature*, t. **CCV**, p. 1112-1113, 1965.

([2708]) VAN DER WALT J. P. — Toxigenic fungi isolated from cereal and legume products. *South African Dept. Agr. Tech. Serv. Symposium on Mycotoxins in Foodstuffs Human Nutritional Aspects Pretoria, 25-26 févr.*, p. 1-14, 1965.

([2709]) VAN DER WALT J. P. — A review of progress in mycotoxin research by the South Africa CSIR. *1st Int. Congr. Pl. Pathol., Londres*, 1968.

([2710]) VAN DER WATT J. J. et PURCHASE I. F. H. — Subacute toxicity of sterigmatocystin to rats. *S. Afr. Med. J.*, t. **XLIV**, p. 59-160, 1970.

([2711]) VAN DER WATT et PURCHASE I. F. H. — The acute toxicity of retrorsine, aflatoxin and sterigmatocystin in vervet monkeys. *Brit. J. Exp. Pathol.*, t. **LI**, p. 183-190, 1970.

([2712]) VAN DER ZIJDEN A. S. M., KOELENSMID W. A. A. B., BOLDINGH J., BARRETT C. B., ORD W. O. et PHILP J. — Isolation in crystalline form of a toxin responsible for turkey X disease. *Nature*, t. **CXCV**, p. 1060-1062, 1962.

([2713]) VAN DIJCK P. J. — Occurrence of ramycin in strains of *Mortierella ramanniana. Trans. Brit. Mycol. Soc.*, t. **LIII**, p. 142-143, 1969.

([2714]) VAN DIJCK P. et DE SOMER P. — Ramycin : a new antibiotic. *J. Gen. Microbiol.*, t. **XVIII**, p. 377-381, 1958.

([2715]) VAN DORP D. A., VAN DER ZIJDEN A. S. M., BEERTHUIS R. F., SPARRE BOOM S., ORD W. O., DE IONGH H. et KENNING R. — Dihydro-aflatoxin B, a metabolite of *Aspergillus flavus. Rec. Trav. Chim. des Pays-Bas*, T. **LXXXII**, p. 587-592, 1963.

([2716]) VAN EIJK G. W. — Isolation and identification of orsellinic acid and penicillic acid produced by *Penicillium fenelliae* Stolk. *Antonie van Leeuwenhoek*, t. **XXXV**, p. 497-504, 1969.

([2717]) VAN RENSBURG S. J., PURCHASE I. F. H. et VAN DER WATT J. J. — Hepatic and renal pathology induced in mice by feeding fungal cultures. *in* PURCHASE I.F.H., *Mycotoxins in human health*, p. 153-161, 1971.

([2718]) VAN SOEST T. C. et PEERDEMAN A. F. — The crystal structures of aflatoxin B_1. I. Structure of the chloroform solvate of aflatoxin B_1 and the absolute configuration of aflatoxin B_1. *Acta crystallogr.*, Sect. B, t. **XXVI**, p. 1940-1947, 1970. II. The structure of an orthorhombic and a monoclinic modification. *Id.*, t. **XXVI**, p. 1947-1955, 1970.

([2719]) VAN SOEST T. C. et PEERDEMAN A. F. — The crystal structure of aflatoxin B2. *Acta crystallogr., Sect. B*, t. **XXVI**, p. 1959-1963, 1970.

([2720]) VAN VEEN A. G. — Bongkrek acid, a new antibiotic. *Documenta Neerlandica Indonesica Morbis Tropicus*, t. **II**, p. 185, 1950.

([2721]) VAN VEEN A. G., GRAHAM D. C. W. et STEINKRAUS K. H. — Fermented peanut presscake. *Cereal Sci. Today*, t. **XII**, p. 96-99, 1968.

([2722]) VAN VEEN A. G. et MERTENS W. K. — Die Giftstoffe der sogenannten Bongkrek — Vergiftungen in Java. *Rec. Trav. Chim,*, t. **LIII**, p. 257, 1934.

([2723]) VAN VEEN A. G. et MERTENS W. K. — Das Toxoflavin, der gelbe Giftstoff der Bongkrek. *Rec. Trav. Chim.*, t. **LIII**, p. 398, 1934.

([2724]) Van Veen A. G. et Schaefer G. — The influence of the tempeh fungus on the soybean. *Documenta Neerlandica Indonesica Morbis Tropicus*, t. **II**, p. 270, 1950.

([2725]) Van Walbeek W., Clademenos T. et Thatcher F. S. — Influence of refrigeration on aflatoxin production by strains of *Aspergillus flavus*. *Can. J. Microbiol.*, t. **XV**, p. 629-632, 1969.

([2726]) Van Walbeek W., Scott P. M., Harwig J. et Lawrence J. W. — *Penicillium viridicatum* Westling : a new source of ochratoxin A. *Can. J. Microbiol.*, t. **XV**, p. 1281-1285, 1969.

([2727]) Van Walbeek W., Scott P. M. et Thatcher F. S. — Mycotoxins from food-borne fungi. *Canad. J. Microbiol.*, t. **XIV**, p. 131-137, 1968.

([2728]) Van Warmelo K. T. — The fungus flora of stock feeds in South Africa. *Onderstepoort Vet. Res.*, t. **XXXIV**, p. 439-405, 1967.

([2729]) Van Warmelo K. T. et Marasas W. F. O. — *Phomopsis leptostromiformis* : the causal fungus of lupinosis, a mycotoxicosis, in sheep. *Mycologia*, t. **LXIV**, p. 316-324, 1972.

([2730]) Van Warmelo K. T., Marasas W. F. O., Adelaar T. F., Kellerman T. S., Van Rensburg I. B. J. et Minne J. A. — Experimental evidence that lupinosis in sheep is a mycotoxicosis caused by the fungus *Phomopsis leptostromiformis* (Kühn) Bubak. *J. South Afr. Veter. Med. Assoc.*, t. **XLI**, p. 235-247, 1970. in Purchase I.F.H., *Mycotoxins in human health*, p. 185-193, 1971.

([2731]) Van Warmelo K. T., Van der Westhuizen G. C. A. et Minne J. A. — The production of aflatoxins in naturally infected high quality maize. *Tech. Commun.* n° 71, *Dept. Agr. Tech. Serv. Pretoria*, 5 p., 1968.

([2732]) Van Zytveld W. A., Kelley D. C. et Dennis S. M. — Aflatoxicosis : the presence of aflatoxins or their metabolites in livers and skeletal muscles of chickens. *Poult. Sci.*, t. **XLIX**, p. 1350-1356, 1970.

([2733]) Varham S. D. et Yadava I. S. — Biochemistry of aflatoxins. A review. *J. Nutr. Diet.*, t. **V**, p. 87-89, 1968.

([2734]) Veen W. A. G. — Changes in the fatty acid pattern of the liver of young piglets under the influence of aflatoxin B_1. *Acta Physiol. Pharmacol. Neerl.*, t. **XIV**, p. 448-459, 1967.

([2735]) Velan M. et Reynaud J. — Sur le destin de l'aflatoxine au cours du raffinage par distillation de l'huile d'arachide. *Rev. Fr. Corps Gras*, t. **XIV**, p. 305-310, 1967.

([2736]) Velasco J. — A simplified procedure for the determination of aflatoxin B_1 in cottonseed meals. *J. Amer. Oil Chem. Soc.*, t. **XLVI**, p. 105-107, 1969.

([2737]) Velasco J. — Determination of aflatoxin in cottonseed by ferric hydroxide gel cleanup. *J. Ass. Off. Anal. Chem.*, t. **LIII**, p. 611-616, 1970.

([2738]) Verona O. — Studio microbiologico di un terreno torboso. *Arch. Mikrobiol.*, t. **V**, p. 328-337, 1934.

([2739]) Verona O. et Picci G. — *Microbiologia degli alimenti : alterazioni, deterioramenti, techniche di conservazione*. 586 p., U.T.E.T., Torino, 1968.

([2740]) Verrett M. J., Marliac J. P. et McLaughlin J. — Use of the chicken embryo in the assay of aflatoxin toxicity. *J. Ass. Off. Agric. Chemists*, t. **XLVII**, p. 1003-1006, 1964.

([2741]) Vertinskii K. I. — Stakhibotriotoksikoz loshadei. *Veterinarya*, t. **XVII**, p. 61-68, 1940.

([2742]) Vertinskii K. I. — Toksicheskaia dispepsia i dizenteriia prosyat. *Veterinariya*, t. **XXXIII**, p. 14-22, 1956.

([2743]) Vertinskii K. I., Dzhilavyan K. A. et Koroleva V. P. — *Byull. vses. Inst. eksp. Vet. Mosk.*, p. 86-90, 1967.

([2744]) Vey A. — Expériences avec l'aflatoxine B sur les insectes. *Symposium INSERM Mycotoxines et Alimentation*, Le Vésinet, 24 oct. 1969 ; *Cah. Nutr. Diét.*, t. **V**, p. 73-74, 1970.

([2745]) Vezina C., Sehgal S. N. et Singh K. — Transformation of steroids by spores of micro-organisms. I. Hydroxylation of progesterone by conidia of *Aspergillus ochraceus*. *Appl. Microbiol.*, t. **XI**, p. 50-57, 1963.

([2746]) VIDAL P. — L'irradiation des produits agricoles et alimentaires en France. *Bull. Inf. Sci. et Tech. C.E.A.*, n° 99, p. 43-55, 1965.

([2747]) VIDHYASEKARAN P., MUTHUSWAMY G. et SUBRAMANIAN C. L. — Role of seed-borne microflora in paddy seed spoilage. I. Production of hydrolytic enzymes. *Indian Phytopath.*, t. **XIX**, p. 333-341, 1966.

([2748]) VIENNOT-BOURGIN G. — *Les Champignons parasites des plantes cultivées.* 2 vol., 1851 p., Masson édit. Paris, 1949.

([2749]) VIENNOT-BOURGIN G. et BRUN J. — Les pourritures des pommes et des poires sur le marché français. *Ann. Ec. Nat. Agric. Grignon,* sér. 3, t. **IV**, p. 181-215, 1944.

([2750]) VILLA-TREVINO S. et LEAVER D. D. — Effects of the hepatoxic agents retrorsine and aflatoxin B_1 on hepatic protein synthesis in the rat. *Biochem. J.*, t. **CIX**, p. 87-94, 1968.

([2751]) VOISIN A. — *Tétanie d'herbe.* 296 p., La Maison Rustique, 1963.

([2752]) VOJNOVICH C. et PFEIFER C. F. — Reducing the microbial population of flour by warm storage. *Northwest. Miller,* t. **CCLXXIII**, p. 12-14, 1966.

([2753]) VOJNOVICH C. et PFEIFER V. F. — Reducing the microbial population of flour during milling. *Cereal Sci. Today,* t. **XII**, p. 54-55, 58-60, 1967.

([2754]) VOJNOVICH C., PFEIFFER V. F. et GRIFFIN E. L. — Reducing the microbial count of flour. *Cereal Sci. Today,* t. **XI**, p. 16-18 et 31, 1966.

([2755]) VOLLMAR H. — Versuche ueber die Beeinflussung des Wachstums von Gewebe durch Patulin. *Z. Hyg. Infektionskrkh.*, t. **CXXVII**, p. 316-321, 1947.

([2756]) VORSTER L. J. — Etudes sur la détoxification des arachides contaminées par l'aflatoxine et destinées à l'huilerie. *Rev. Fr. Corps Gras,* t. **XIII**, p. 7-12, 1966.

([2757]) VORSTER L. J. et PURCHASE I. F. H. — A method for the determination of sterigmatocystin in grains and oilseeds. *Analyst,* t. **XCIII**, p. 694-696, 1968.

([2758]) VYSHELESSKI S. N. — Spetsialna Epizootologiia. *Ogiz-Selskhozgiz, Moscou,* p. 374-382, 1948.

([2759]) WAGNER L., DREWS J. — The effect of aflatoxin B_1 on RNA synthesis and breakdown in normal and regenerating rat liver. *Eur. J. Cancer,* t. **VI**, p. 465-476, 1970.

([2760]) WAID J. S. — Influence of oxygen upon growth and respiratory behaviour of fungi decomposing rye-grass roots. *Trans. Brit. Mycol. Soc.*, t. **XLV**, p. 479-487, 1962.

([2761]) WAISS A. C., WILEY M., BLACK D. R. et LUNDIN R. E. — 3-Hydroxy-6,7-dimethoxydifuroxanthone. A new metabolite from *Aspergillus flavus. Tetrahedron L.*, t. **XXVIII**, p. 3207-3210, 1968.

([2762]) WAKSMAN S. A. et BUGIE E. — Action of antibiotic substances upon *Ceratostomella ulmi. Proc. Soc. Exp. Biol. Med.*, t. **LIV**, p. 79-82, 1943.

([2763]) WAKSMAN S. A., HORNING E. S. et SPENCER E. L. — The production of two antibacterial substances, fumigacin and clavacin by *Aspergillus fumigatus* and *Aspergillus clavatus. Science*, t. **XCVI**, p. 202-203, 1942.

([2764]) WAKSMAN S. A., HORNING E. S. et SPENCER E. L. — Two antagonist fungi, *Aspergillus fumigatus* and *Aspergillus clavatus*, and their antibiotic substances. *J. Bact.*, t. **XLV**, p. 233-248, 1943.

([2765]) WALES P. et SOMERS E. — Susceptibility of aflatoxin-producing strains of *Aspergillus flavus* to a range of fungicides. *Can. J. Plant Sci.*, t. **XLVII**, p. 377-379, 1968.

([2766]) WALKER A. E. et JOHNSON H. C. — Convulsive factor in commercial penicillin. *Arch. Surg.*, t. **L**, p. 69-73, 1965.

([2767]) WALLACE H. A. H. et SINHA R. H. — Fungi associated with hot spots in farm stored grain. *Can J. Plant Sci.*, t. **XLII**, p. 130-141, 1962.

([2768]) WALLACH P. et VERLOT J. B. — Le développement de la consommation des produits surgelés. *Rev. Gén. Froid,* t. **LVII**, p. 1335-1341, 1966.

([2769]) WALLEN V. R. et SKOLKO A. J. — Activity of antibiotics against *Ascochyta pisi. Can. J. Bot.*, t. **XXIX**, p. 316-323, 1951.

([2770]) WALSH T. et DONOHOE O. — Magnesium deficiency in some crop plants in relation to the level of potassium nutrition. *J. Agric. Sci.*, t. **XXXV**, p. 254-263, 1945.

([2771]) WALTKING A. E. — Collaborative study of three methods for determination of aflatoxin in peanuts and peanut products. *J. Ass. Off. Anal. Chem.*, t. **LIII**, p. 104-113, 1970.

([2772]) WALTKING A. E. — Fate of aflatoxin during roasting and storage of contaminated peanut products. *J. Ass. Off. Anal. Chem.*, t. **LIV**, p. 533-539, 1971.

([2773]) WALTKING A. E., BLEFFERT G. et KIERNAN M. — An improved rapid physicochemical assay method for aflatoxin in peanuts and peanut products. *J. Amer. Oil Chem. Soc.*, t. **XLV**, p. 880-884, 1968.

([2774]) WANG C. Y. et PARKS P. F. — Synthesis and toxicology of 5,7-dibenzyloxycyclopentenone (2,3-c) coumarin. A model compound of aflatoxin B_1. *J. Med. Chem.*, t. **XIV**, p. 447-448, 1971.

([2775]) WANG F. H. — The effect of clavacin on root growth. *Botan. Bull. Acad. Sinica*, t. **II**, p. 265-269, 1948.

([2776]) WANNOP C. C. — Disease of turkey poults. *Vet. Rec.*, t. **LXXII**, p. 671-672, 1960.

([2777]) WANNOP C. C. — Turkey « X » disease. *Vet. Rec.*, t. **LXXIII**, p. 310-311, 1961.

([2778]) WANNOP C. C. — The histopathology of Turkey«X»disease in Great Britain. *Avian Diseases*, t. **V**, p. 371-381, 1961.

([2779]) WARCUP J. H. — Studies on the occurrence and activity of fungi in wheat field soil. *Trans. Brit. Mycol. Soc.*, t. **XL**, p. 237-259, 1957.

([2780]) WARD F. S. — Deterioration of copra caused by bacteria and molds. *Sci. Ser. Dept. Agr.*, n° 20, p. 95-108, 1937.

([2781]) WARD H. S. et DIENER U. L. — Biochemical changes in shelled peanuts caused by storage fungi. I. Effects of *Aspergillus tamarii*, four species of *A. glaucus* group, and *Penicillium citrinum*. *Phytopathology*, t. **LI**, p. 244-250, 1961.

([2782]) WASHKO F. V. et MUSHETT C. W. — Some observations on the pathology of the hemorrhagic conditions of chickens. *Proc. 92nd Ann. Meeting Am. Vet. Med. Assoc., Minneapolis, Minn.*, p. 360-363, 1955.

([2783]) WATSON S. A. et YAHL K. R. — Survey of aflatoxins in commercial suplies of corn and grain sorghum used for wet-milling. *Cereal Sci. Today*, t. **XVI**, p. 153-155, 1971.

([2784]) WATT J. M. et BREYER-BRANDWIJK M. G. — The Medicinal and Poisonous Plants of Southern and East Africa. 1457 p., Livingstone édit. Londres, 1962.

([2785]) WEAVER E. A. et KENNEY H. E. — Microbial transformation reactions. U.S. Patent n° 2 989 439, 20 juin 1961.

([2786]) WEBB B. D., THIERS H. D. et RICHARDSON L. R. — Studies in feed spoilage. Inhibition of mold growth by gamma radiation. *Appl. Microbiol.*, t. **VII**, p. 329-333, 1959.

([2787]) WEBSTER J. — Culture studies on *Hypocrea* and *Trichoderma*. I. Comparison of perfect and imperfect states of *H. gelatinosa*, *H. rufa* and *Hypocrea* sp. *Trans. Brit. Mycol. Soc.*, t. **XLVII**, p. 75-96, 1964.

([2788]) WEBSTER J. et NORMA LOMAS. — Does *Trichoderma viride* produce glitoxin and viridin ? *Trans. Brit. Mycol. Soc.*, t. **XLVII**, p. 535-540, 1964.

([2789]) WEHMER F. C. et RABIE C. J. — The microorganisms in nuts and dried fruits. *Phytophylactia*, t. **II**, p. 165-170, 1970.

([2790]) WEI C. — Miscellaneous pharmacologic actions of citrinin. *Jour. Lab. and Clin. Med.*, t. **XXXI**, p. 72-78, 1946.

([2791]) WEIGMANN et WOLFF A. — Über einige zum Rübengeschmak der Butter beitragende Mycelpilze. *Centralbl. f. Bakt.*, Abt. II, t. **XXII**, p. 657-671, 1908-1909.

([2792]) WEINDLING R. — *Trichoderma lignorum* as a parasite of other soil fungi. *Phytopathology*, t. **XXII**, p. 837-845, 1932.

([2793]) WEINDLING R. — Studies on a lethal principle effective in the parasitic action of *Trichoderma lignorum* on *Rhizoctonia solani* and other soil fungi. *Phytopathology*, t. **XXIV**, p. 1153-1179, 1934.

([2794]) WEINDLING R. — Isolation of toxic substances from the culture filtrates of *Trichoderma* and *Gliocladium. Phytopathology*, t. **XXVII**, p. 1175-1177, 1937.

([2795]) WEINDLING R. — Experimental consideration of the mold toxins of *Gliocladium* and *Trichoderma. Phytopathology*, t. **XXXI**, p. 991-1003, 1941.

([2796]) WEINDLING R. et EMERSON O. H. — The isolation of a toxic substance from the culture filtrate of *Trichoderma. Phytopathology*, t. **XXVI**, p. 1068-1070, 1936.

([2797]) WEISER H. H. — *Practical Food Microbiology and Technology.* 345 p., The Avi Publishing Co, Westport, Conn., 1962.

([2798]) WEISER S. — Der Nährstoffgehalt von brandsporenghaltigen und brandsporenfreier Koppereistauben. *Fortschr. der Landwirtsch.*, t. **I**, p. 169-171, 1926.

([2799]) WELLER W. A., JOSEPH P. J. et HORA J. F. — Deep mycotic involvement of the right maxillary and ethmoid sinuses, the orbit and adjacent structures; case report evaluating the use of Mycostatin locally and amphotericin B (Fungizone) intravenously against *Aspergillus flavus. Laryngoscope*, t. **LXX**, p. 999-1016, 1960.

([2800]) WELTY R. E. et COOPER W. E. — Prevalence and development of storage fungi in peanut *Arachis hypogaea* seed. *Mycopathol. Mycol. Appl.*, t. **XXXV**, p. 290-296, 1968.

([2801]) WERCH S. C., OESTER Y. T. et FRIEDEMANN T. E. — Kojic acid — a convulsant. *Science*, t. **CXXVI**, p. 450-451, 1957.

([2802]) WERNER W. — Ueber den Mineralstoffgehalt in jungen Weidefutter unter besonderer Berücksichtigung des K : (Ca + Mg) — Verhältnisses. *Landw. Forschung*, t. **XII**, p. 133-139, 1959.

([2803]) WETZEL R. et ELKSNITIS W. — Die Stachybotryotoxikose der Pferde. *Kolonialtierarzt*, t. **VIII**, p. 319-320, 1943.

([2804]) WHEELER B. — An examination of cereals and feeds for aflatoxins. *Irish J. Agric. Res.*, t. **VIII**, p. 172-174, 1969.

([2805]) WHEELER H. et LUKE H. H. — Microbial toxins in plant disease. *Ann. Rev. Microbiol.* t. **XVII**, p. 223-242, 1963.

([2806]) WHITAKER T. B. et WISER E. H. — Theoretical investigations into the accuracy of sampling shelled peanuts for aflatoxin. *J. Amer. Oil Chem. Soc.*, t. **XLVI**, p. 377-379, 1969.

([2807]) WHITE E. C. — Bactericidal filtrates from a mold culture. *Science*, t. **XCII**, p. 127, 1940.

([2808]) WHITE E. C. et HILL J. H. — Antibacterial filtrates from a strain of *Aspergillus flavus. J. Bacteriol.*, t. **XLIII**, p. 12, 1942.

([2809]) WHITE E. C. et HILL J. H. — Studies on antibacterial products formed by molds. I. Aspergillic acid, a product of a strain of *Aspergillus flavus. J. Bacteriol.* t. **XLV**, p. 433-444, 1943.

([2810]) WHITE E. P. — Photosensitivity diseases in New Zealand. XII. Concentration of the facial eczema poison. *New Zealand J. Agr. Research*, t. **I**, p. 433-446, 1958.

([2811]) WHITE E. P. — Chemical extraction and fractionation of the toxin. *Symposium on facial eczema Research. Proc. New Zealand Soc. Animal Production*, t. **XIX**, p. 64-68, 1959.

([2812]) WHITE M. F. — A note on the status of *Aspergillus giganteus. Trans. Brit. Mycol. Soc.*, t. **XLVI**, fasc. 4, p. 482-484, 1963.

([2813]) WHITEHAIR C. K., YOUNG H. C., GIBSON M. E. et SHORT G. E. — A nervous disturbance in cattle caused by a toxic substance associated with nature Bermuda grass. *Oklahoma Agr. Exp. Stn. Misc. Publ.* n° 22, p. 57, 1951.

([2814]) WHITEHEAD M. D. et THIRUMALACHAR M. J. — Effect of aureofungin spray on the control of fungal diseases and *Aspergillus flavus* infestation in peanuts. *Hindustan Antibiot. Bull.*, t. **XII**, p. 79-80, 1971.

([2815]) WHITTAKER R. H. et FEENY P. P. — Allelochemics : chemical interactions between species. *Science*, t. **CLXXI**, p. 757-770, 1971.

([2816]) WHITTEN M. E. — Screening cottonseed for aflatoxin. *J. Amer. Oil Chem. Soc.*, t. **XLVI**, p. 39-40, 1968.

([2817]) WHITTEN M. E. — Aflatoxins in cottonseed hulls. *J. Amer. Oil Chem. Soc.*, t. **XLVII**, p. 5-6, 1970.

([2818]) WICTOR C. E. — Molybdenum poisoning. *Los Angeles County Livestock Dept. Ann. Rept.* 1951-1952, p. 22, 1952.

([2819]) WIEDER R. — Pulmonary tumors in strain A mice given injections of aflatoxin B_1. *J. Nat. Cancer Inst.*, t. **XL**, p. 1195-1197, 1968.

([2820]) WIESNER B. P. — Bacterial effects of *Aspergillus clavatus. Nature, G.B.*, t. **CXLIX**, p. 356-357, 1942.

([2821]) WILDMAN J. D., STOLOFF L. et JACOBS R. — Aflatoxin production by a potent *Aspergillus flavus* Link isolate. *Biotechnol. Bioeng.*, t. **IX**, p. 429-437, 1967.

([2822]) WILEY J. R. — « Disease » of turkey poults. *Vet. Rec.*, t. **LXXII**, p. 786-787, 1960.

([2823]) WILEY M. — Note on analysis of aflatoxins. *J. Ass. Off. Anal. Chem.*, t. **XLIX**, p. 1223-1224, 1966.

([2824]) WILEY M. et WAISS A. C. — An improved separation of aflatoxin. *J. Amer. Oil Chem. Soc.*, t. **XLV**, p. 870-871, 1968.

([2825]) WILEY M., WAISS A. C. et BENNETT N. — Reaction of aflatoxin B_1 with acetic acid — thionylchloride. *J. Ass. Off. Anal. Chem.*, t. **LII**, p. 75-76, 1969.

([2826]) WILKINSON S. et SALISBURY J. F. — Gliotoxin from *Aspergillus chevalieri* (Mangin) Thom et Church. *Nature*, Lond., t. **CCVI**, fasc. 4984, p. 619, 1965.

([2827]) WILLIAMS L. E. et SCHMITTHENNER A. F. — Effect of growing crops and crop residues on soil fungi and seedling blights. *Phytopathology*, t. **L**, p. 22-25, 1960.

([2828]) WILSON B. J. — Toxic substances formed by filamentous fungi growing on feedstuffs. *Mycotoxins in Foodstuffs*, p. 147-149, M.I.T. Press, 1965.

([2829]) WILSON B. J. — Toxins other than aflatoxins produced by *Aspergillus flavus. Bacteriol. Rev.*, t. **XXX**, p. 478-484, 1966.

([2830]) WILSON B. J. — Recently dicovered metabolites with unusual toxic manifestations. *in* PURCHASE I.F.H., *Mycotoxins in human health*, p. 223-229, 1971.

([2831]) WILSON B. J. — Miscellaneous *Aspergillus* toxins. *in* Ciegler A. et al., Microbial Toxins, t. **VI**, p. 208-295, 1971.

([2832]) WILSON B. J. — Miscellaneous *Penicillium* toxins. *in* Ciegler A. et al., Microbial Toxins, t. **VI**, p. 460-521, 1971.

([2833]) WILSON B. J., CAMPBELL T. C., HAYES A. W. et HANLIN R. T. — Investigation of reported aflatoxin production by fungi outside the *Aspergillus flavus* group. *Appl. Microbiol.*, t. **XVI**, p. 819-821, 1968.

([2834]) WILSON B. J. et HAYES A. W. — Mycotoxins : a new food problem. *J. Tenn. Acad. Sci.*, t. **XLIII**, p. 42, 1968.

([2835]) WILSON B. J., HARRES T. M. et HAYES A. W. — Mycotoxin from *Penicillium puberulum. J. Bacteriol.*, t. **XCIII**, p. 1737-1738, 1967.

([2836]) WILSON B. J., TEER P. A., BARNEY G. H. et BLOOD F. R. — Relationship of aflatoxin to epizootics of toxic hepatitis among animals in Southern United States. *Amer. J. Vet. Res.*, t. **XXVIII**, p. 1217-1230, 1967.

([2837]) WILSON B. J. et WILSON C. H. — Oxalate formation in moldy feedstuffs as a possible factor in livestock toxic disease. *Am. J. Vet. Research*, t. **XXI**, p. 261-269, 1961.

([2838]) WILSON B. J. et WILSON C. H. — Extraction and preliminary characterization of a hepatoxic substance from cultures of *Penicillium rubrum. J. Bacteriol.*, t. **LXXXIV**, p. 283-290, 1962.

([2839]) WILSON B. J. et WILSON C. H. — Toxin from *Aspergillus flavus*. Production on food materials of a substance causing tremors in mice. *Science*, t. **CXLIV**, p. 177-178, 1964.

([2840]) WILSON B. J., WILSON C. M. et HAYES A. W. — Tremorgenic toxin from *Penicillium cyclopium* grown on food materials. *Nature, G.B.*, t. **CCXX**, p. 77-78, 1968.

([2841]) Wilson B. J., Yang D. T. C. et Boyd M. R. — Toxicity of mould-damaged sweet potatoes (*Ipomoea batatas*). *Nature, G.B.*, t. **CCXXVII**, p. 521-522, 1970.

([2842]) Wilson C. — A survey of fungi associated with peg and seed rots of peanuts in Southern Alabama. Abstr. in *Phytopathology*, t. **XXXVII**, p. 24, 1947.

([2843]) Wilson C. — Concealed damage of peanuts in Alabama. *Phytopathology*, t. **XXXVII**, p. 657-668, 1947.

([2844]) Wilson M., Noble M. et Gray E. — The blind seed disease of rye-grass and its causal fungus. *Trans. Royal Soc. Edinburgh*, t. **LXI**, p. 327-340, 1945.

([2845]) Wilson M., Noble M. et Gray E. — *Gloeotinia* — a new genus of the Sclerotiniaceae. *Trans. Brit. Mycol. Soc.*, t. **XXXVII**, p. 29-32, 1954.

([2846]) Wilson R. K. — An attempt to induce hypomagnesaemia in wethers by feeding high levels of urea. *Vet. Rec.*, t. **LXXV**, p. 698, 1963.

([2847]) Wink W. A. et Sears G. R. — Instrumentation studies. 52. Equilibrium relative humidities above saturated salt solutions at various temperatures. *Tappi*, t. **XXXIII**, p. 96A-99A, 1950.

([2848]) Winstead J. A. et Suhadolnik R. I. -- Biosynthesis of gliotoxin. II. Further studies on the incorporation of carbon-14 and tritium-labeled precursors. *J. Am. Chem. Soc.*, t. **LXXXII**, p. 1644-1646, 1960.

([2849]) Wiseley D. V. et Falk H. L. — *J. Amer. Med. Ass.*, t. **CLXXIII**, p. 1161, 1960.

([2850]) Wiseman H. G., Jacobson W. C. et Harmeyer W. C. — Note on removal of pigments from chloroform extracts of aflatoxin cultures with copper carbonate (*Aspergillus flavus*). *J. Ass. Off. Anal. Chem.*, t. **L**, p. 981-983, 1967.

([2851]) Withers. — Wastage and disease incidence in dairy herds. *Vet. Rec.*, t. **LXVII**, p. 605-612, 1955.

([2852]) Wogan G. N. — Report on Symposium on mycotoxins in foodstuffs and résumé of current research activities in the U.S. relating to mycotoxins. WHO/FAO/UNICEF. *Protein Advisory Group News Bull.* N° 4, p. 60-72, 1964.

([2853]) Wogan G. N. — Experimental toxicity and carcinogenicity of aflatoxins. *Mycotoxins in Foodstuffs*, p. 163-173, 1965.

([2854]) Wogan G. N. — Mycotoxin contamination of foodstuffs. Symposium of the Division of Agricultural and Food Chemistry at the 150th Meet. of the Amer. Chem. Soc. on world protein resources, sept. 1965. *Adv. Chem. Ser.*, n° 57, p. 195-215, 1966.

([2855]) Wogan G. N. — Chemical nature and biological effects of aflatoxins. *Bacteriol. Rev.*, t. **XXX**, p. 460-470, 1966.

([2856]) Wogan G. N. — Metabolism and biochemical effects of aflatoxins. *in* Goldblatt L. A., *Aflatoxin*, Academic Press, p. 151-186, 1969.

([2857]) Wogan G. N. — Effects of aflatoxins on in vivo nucleic acid metabolism in rats. *in* Purchase I. F. H. , *Mycotoxins in human health*, p. 1-10, 1971.

([2858]) Wogan G. N., Edwards G. S. et Newberne P. M. — Acute and chronic toxicity of rubratoxin B. *Toxicol. Appl. Pharmacol.*, t. **XIX**, p. 712-720, 1971.

([2859]) Wogan G. N., Edwards G. S. et Shank R. C. — Excretion and tissue distribution of radioactivity from (*Aspergillus flavus*) aflatoxin B_1 — ^{14}C in rats. *Cancer Res.*, t. **XXVII**, p. 1729-1736, 1967.

([2860]) Wogan G. N. et Friedman M. A. — Effects of aflatoxin B_1 on tryptophan pyrrolase induction in rat liver. *Feder. Proc.* U.S.A., t. **XXIV**, p. 627, 1965.

([2861]) Wogan G. N. et Newberne P. M. — Dose-response characteristics of aflatoxin B carcinogenesis in the rat. *Cancer Res.*, t. **XXVII**, p. 2370-2376, 1967.

([2862]) Wolff E. Von. — *Farm foods, or the rational feeding of farm animals.* Gurney et Jackson édit., Londres, 1895.

([2863]) Wolf F. T. — Relation of various fungi to otomycosis. *Arch. Otolaryngol.*, t. **XLVI**, p. 361-374, 1947.

([2864]) WOLF H. et JACKSON E. W. — Hepatomas in Rainbow Trout. Descriptive and experimental epidemiology. *Science*, t. **CXLII**, p. 676-678, 1963.

([2865]) WOLLENWEBER H. W. et REINKING O. A. — *Die Fusarien.* 335 p., Paul Parey, Berlin, 1935.

([2866]) WOODS C. T. et WILLIAMS M. — Certain cancers traced to moldy grain. *Sunshine State Agr. Res. Rep.*, t. **XIII**, p. 11, 1968.

([2867]) WOOD E. M. et LARSON C. P. — Hepatic carcinoma in Rainbow Trout. *Arch. Path.*, t. **LXXI**, p. 471, 1961.

([2868]) WOOD E. M. et YASUTAKE W. T. — Ceroid in fish. *Am. J. Path.*, t. **XXXII**, p. 591-603, 1956.

([2869]) WOOD E. M., YASUTAKE W. T., WOODALL A. N. et HALVER J. E. — The nutrition of salmonoid fishes. *J. Nutr.*, t. **LXI**, p. 465-488, 1957.

([2870]) WOODHAM A. A. et DAWSON R. — The effect of heat on defatted groundnut flour. *Proc. Nutr. Soc. Engl. Scot.*, t. **XXV**, p. 8-9, 1966.

([2871]) WOODWARD R. B. et SINGH G. — The structure of patulin. *J. Amer. Chem. Soc.*, t. **LXXI**, p. 758-759, 1949.

([2872]) WOODWARD R. B. et SINGH G. — The synthesis of patulin. *J. Amer. Chem. Soc.*, t. **LXXII**, p. 1428, 1950.

([2873]) WOOLEY D. W., BERGER J., PETERSON W. H. et STEENBOCK H. — Toxicity of *Aspergillus sydowii* and its correction. *J. Nutr.*, t. **XVI**, p. 465- 1938.

([2874]) WORKER N. A. — Biological assays for toxicity. *Symposium on facial eczema Research. Proc. New Zealand Soc. Animal Production*, t. **XIX**, p. 59-63, 1959.

([2875]) WORKER N. A. — A hepatotoxin causing liver damage in facial eczema in sheep. *Nature*, t. **CLXXXV**, p. 909-910, 1960.

([2876]) WORONIN M. — Über das « Taumelgetreide » in Süd-Ussurien. *Botan. Z.*, t. **XLIX**, p. 81-93, 1891.

([2877]) WRAGG J. B., ROSS V. C. et LEGATOR M. S. — Effect of aflatoxin B_1 on the deoxyribonucleic acid polymerase of *Escherichia coli* (32274). *Proc. Soc. Exp. Biol. Med.*, t. **CXXV**, p. 1052-1055, 1967.

([2878]) WRIGHT C. E. — Blind seed disease of ryegrass. *Euphytica*, t. **XVI**, p. 122-130, 1967.

([2879]) WRIGHT D. E. et FORRESTER I. T. — Some biochemical effects of sporidesmin. *Canad. J. Biochem.*, t. **XLIII**, p. 881-888, 1965.

([2880]) WRIGHT D. E. — Sporidesmin, aflatoxin and other mycotoxins. *Australas. J. Pharm. Sci., Suppl.*, t. **LXVI**, p. 72, 1968.

([2881]) WRIGHT D. E. — Phytotoxicity of sporidesmin. *New Zeal. J. Agric. Res.*, t. **XII**, p. 271-274, 1969.

([2882]) WRIGHT J. M. — Phytotoxic effects of some antibiotics. *Ann. Bot.*, t. **XV**, p. 493-499, 1951.

([2883]) WRIGHT J. M. — The production of antibiotics in soil. II. Production of griseofulvin by *Penicillium nigricans. Ann. Appl. Biol.*, t. **XLIII**, p. 288-296, 1955.

([2884]) WU C. M., KOEHLER P. E. et AYRES J. C. — Isolation and identification of xanthotoxin (8-methoxy psoralen) and bergapten (5-methoxypsoralen) from celery infected with *Sclerotinia sclerotiorum. Appl. Microbiol.*, t. **XXIII**, p. 852-856, 1972.

([2885]) WUNDER W. et OTTO H. — Aflatoxine als Ursache des Leberkrebs der Forelle. *Naturwissenschaften*, t. **LVI**, p. 352-355, 1969.

([2886]) WYLLIE J. — A method for the rapid production of citrinin. *Canad. Pub. Health Jour.*, t. **XXXVI**, p. 477-483, 1945.

([2887]) X. — L'acide propionique, agent de conservation du maïs-grain humide. *Actualités Céréales*, p. 9, 17 avr. 1970.

([2888]) X. — Microbiological problems in food preservation by irradiation. *I.A.E.A. Publ.*, 148 p., 1966.

(2889) X. — Preservation of fruit and vegetables by radiation. *I.A.E.A. Publ.*; 152 p., 1966.

(2890) X. — Searching for resistant crops. *New Scientist*, t. **XXXIV**, p. 649, 1967.

(2891) X. — *Storage of cereal grains and their products.* A.A.C.C., St Paul, Minnesota, 1954.

(2892) X. — International rules for seed testing. *Int. Seed Testing Ass. Proc.*, t. **XXIV**, p. 475-584, 1959.

(2893) X. — Rep. of the Secretary to the Federal Ministry of Agriculture in Rhodesia and Nyasaland for the year ended 20th Sept. 1961. *Vet. Bull.*, t.**XXXIII**, p. 215, 1961.

(2894) X. — *The interdep. Working Party on groundnut toxicity* Research. D.S.I.R., Londres 1962.

(2895) X. — *Unicef meeting on groundnut toxicity problems.* London, Tropical Products Institute, 28-29 oct. 1963, 50 p., 1963.

(2896) X. — Symposium européen sur les maladies des poissons. *Bull. Off. Inter. Epiz.*, t. **LIX**, 1963.

(2897) X. — Toxicité et pouvoir cancérigène des moisissures des oléagineux et des céréales. *D.G.R.S.T. Les actions concertées. Rapport d'activité 1963*, p. 262-263, 1964.

(2898) X. — Report by a commercial Research Group. *Vet. Rec.*, t. **LXXVI**, p. 498, 1964.

(2899) X. — Aflatoxine. *Les Actions Concertées, 1964*; *D.G.R.S.T.*, p. 377-379, 1965.

(2900) X. — Aflatoxin bibliography 1960-1967. *Rep. Trop. Prod. Inst.*, 50 p., 1968.

(2901) Yabuta T. — The constitution of kojic acid, a γ pyrone derivative formed by *Aspergillus oryzae* from carbohydrates. *J. Chem. Soc.*, t. **CXXV**, p. 575, 1924.

(2902) Yadgiri B., Reddy V., Tulpule P. G., Srikantia S. G. et Gopalan C. — Aflatoxin and Indian childhood cirrhosis. *Amer. J. Clin. Nutr.*, t. **XXIII**, p. 94-98, 1970.

(2903) Yahl C. R., Watson S. A., Smith R. J. et Barabolok R. — Laboratory wet-milling of corn containing high levels of aflatoxin and a survey of commercial wet-milling products. *Cereal Chem.*, t. **XLVIII**, p. 385-391, 1971.

(2904) Yamaguchi S. et Kiruchi M. — Notes on the production of organic acids and mycelial pigments during growth of *Penicillium islandicum. The Botanical Magazine, Tokyo*, t. **LXXVII**, fasc. 908, p. 49-53, 1964.

(2905) Yamamoto T. — Studies on the poison producing mold isolated from dry malt. I. Distribution, isolation, cultivation, and formation of the toxic substance. II. Toxicity. *J. Pharmacol. Soc. Japan*, t. **LXXIV**, p. 797 et 810-812, 1954.

(2906) Yamamoto T. — Studies on the poison-producing mold isolated from dry malt. VI. Microdetection and determination of patulin. *J. Pharm. Soc. Japan*, t. **LXXVI**, p. 1375-1381, 1956.

(2907) Yamamoto Y., Nitta K. et Jinbo A. — Studies on the metabolic products of a strain of *Aspergillus fumigatus* (DM 413). III. Biosynthesis of toluquinones. *Chemical Pharmaceutical Bull.*, t. XV, p. 427-431, 1967.

(2908) Yamano T., Kishino K., Yamatodani S. et Abe M. — Investigation ot ergot alkaloids found in cultures of *Aspergillus fumigatus. Takeda Kenkyusho Nempo*, t. **XXI**, p. 95-101, 1962.

(2909) Yamazaki M. — Isolation of *Aspergillus ochraceus* producing ochratoxins from Japanese rice. *in* Purchase I. F. H., *Mycotoxins in human health*, p. 107-114, 1971.

(2910) Yamazaki M., Suzuki S. et Miyaki K. — Tremorgenic toxins from *Aspergillus fumigatus* Fres. *Chem. Pharm. Bull.*, t. **XIX**, p. 1739-1740, 1971.

(2911) Yamazoe S., Kanoh T. et Nakamura I. — Effects of poisons of *Penicillium islandicum* Sopp on the metabolism in isolated liver from animals. II. *Gunma J. Med. Sci. Jap.*, t. **XII**, fasc. 4, p. 265-268, 1963.

(2912) Yanagita T. — Biochemical aspects of germination of *Aspergillus niger* conidiospores. *Arch. Mikrobiol.*, t. **XXVI**, p. 329-344, 1957.

(2913) Yates A. R. — *Byssochlamys nivea*, a food spoilage mould of unusual nature. *Rep. Sch. Agric. Univ., Nott., 1964*, p. 99-102, 1965.

(2914) YATES S. G. — Toxicity of tall fescue forage : a review. *Econ. Botany*, t. **XVI**, p. 295-303, 1962.

(2915) YATES S. G. — Paper chromatography of alkaloids of tall fescue hay. *J. Chromatogr.*, t. **XII**, p. 423-426, 1963.

(2916) YATES S. G. — Toxin-producing fungi from fescue pasture. *in* Kadis S. et al., Microbial Toxins, t. **VII**, p. 191-206, 1971.

(2917) YATES S. G. et TOOKEY H. L. — Festucine, an alkaloid from tall fescue (*Festuca arundinacea* Schreb.) : Chemistry of the functional groups. *Austral. J. Chem.*, t. **XVIII**, p. 53-60, 1965.

(2918) YATES S. G., TOOKEY H. L., ELLIS J. J. et BURKHARDT H. J. — Toxic butenolide produced by *Fusarium nivale* (Fries) Cesati isolated from tall fescue (*Festuca arundinacea* Schreb.). *Tetrahedron L.*, n° 7, p. 621-625, 1967.

(2919) YATES S. G., TOOKEY H. L. et ELLIS J. J. — Correlation of toxicity of *Fusarium* isolates to known toxins in tall fescue pasture. *Appl. Microbiol.*, t. **XIX**, p. 103-105, 1970.

(2920) YATES S. G., TOOKEY H. L., ELLIS J. J. et BURKHARDT H. J. — Mycotoxins produced by *Fusarium nivale* isolated from tall fescue (*Festuca arundinacea* Schreb.). *Phytochemistry*, t **VII**, p. 139-146, 1968.

(2921) YATES S. G., TOOKEY H. L., ELLIS J. J., TALLENT W. H. et WOLF F. — Mycotoxins as a possible cause of fescue toxicity. *J. Agr. Food Chem.*, t. **XVII**, p. 437-442, 1969.

(2922) YESAIR J. — The action of disinfectants on mold. *Ph. D. dissertation, Univ. Chicago*, 1928.

(2923) YIN L. — Note on acetonitrile as an extracting solvent for aflatoxins. *J. Ass. Off. Anal. Chem.*, t. **LII**, p. 880, 1969.

(2924) YOKOTSUKA T., ASAO Y., SASAKI M. et OSHITA K. — Pyrazine compounds produced by molds. *in* Herzberg M., *Toxic Microorganisms*, p. 133-142, 1970.

(2925) YOKOTSUKA T., KIKUCHI T., SASAKI M. et OSHITA K. — Aflatoxin G like compounds with green fluorescence produced by Japanese industrial molds. *Nippon Nogei Kagaku Kaishi*, t. **XLII**, p. 581-585, 1968.

(2926) YOKOTSUKA T., SASAKI M., KIKUCHI T., ASAO Y. et NOBUHARA A. — Fluorescent compounds, different from aflatoxins, produced by Japanese industrial molds. Abstr. A-016, *152nd Meeting, Amer. Chem. Soc., New York*, sept. 1966, p. 131-152, M.I.T. Press. 1967.

(2927) YOKOTSUKA T., SASAKI M., KIKUCHI T., ASAO Y. et NOBUHARA A. — Studies on the compounds produced by moulds. I. Fluorescent compounds produced by Japanese industrial moulds. *Nippon Nogei Kagaku Kaishi*, t. **XLI**, p. 32-38, 1967.

(2928) YOSHIKURA H. et AIBARA K. — Synchronized culture as an assay system of aflatoxin. *in* Herzberg M. , *Toxic Micro-organisms*, p. 19-22, 1970.

(2929) ZAPLETAL O., KRIZ H. et ZIMA S. — Qualitative differentiation of aflatoxins and Kurasane in feeding mixtures for poultry by thin-layer chromatography. *Acta Vet.*, t. **XXXVIII**, p. 163-170, 1969.

(2930) ZAROOGIAN G. E. et CURTIS R. W. — Isolation and identification of assymetrin. *Plant and Cell. Physiol., Tokyo*, t. **V**, p. 291-296, 1964.

(2931) ZELENIN B. N. et PAVLOVA G. L. — Organolepticheskie i fisikokhimicheskie izmeneiya pishchevykh produktov pri ikh konservirovanii γ-luchami. Tezisy doklada na Vsesoyuznoi nauchno-teknicheskoi konferentsii po primeneniyu radioktivnykh i stabil'nykh izotopov i izluchenii v narodnom khozyaistve i nauke ; p. 188, Moscou, 1957.

(2932) ZELENY L. — Chemical, physical, and nutrional changes during storage. *in* Anderson J. A. et Alcock A. W., *Storage of cereal grains and their products*, chap. II. *Amer. Ass. Cereal Chem.*, St Paul, Minn., 1954.

(2933) ZELENY L. — Ways to test seeds for moisture. *U.S. Dept. Agr. Yearbook*, p. 443-447, 1961.

(2934) ZINEHENKO A. V. — Zabolevanie swenoi vizvannoie toksicheskemi gribami. *Veterinariya*, t. **XXXVI**, p. 37, 1959.

([2935]) ZIPFEL. — Vergiftungsversuche mit *Penicillium glaucum*. *Zeitschr. f. Veterinärkunde*, t. **VI**, p. 57, 1894.

([2936]) ZUCKERMAN A. J., REES K. R., INMAN D. et PETTS V. — Site of action of aflatoxin on human liver cells in culture. *Nature*, *G.B.*, t. **CCXIV**, p. 814-815, 1967.

([2937]) ZUCKERMAN A. J., REES K. R., INMAN D. R. et ROBB I. A. — The effects of aflatoxins on human embryo liver cells in culture. *Brit. J. Exp. Pathol.*, t. **XLIX**, p. 33-39, 1968.

([2938]) ZUCKERMAN A. J., TSIQUAYE K. N. et FULTON F. — Tissue culture of human embryo liver cells and the cytotoxicity of aflatoxin B. *Brit. J. Exp. Pathol.*, t. **XLVIII**, p. 20-26, 1966.

REFERENCES TO CHAPTER 13

BOOKS PUBLISHED SINCE 1973

(2939) Purchase, I. F. H. (ed) – *Mycotoxins,* Elsevier, 1974.

(2940) Jemmal, M. (ed) – *Mycotoxins in foodstuffs,* Pergamon, 1977.

(2941) Rodricks J. V. (ed) –*Mycotoxins in human and animal health,* Pathotox, 1977.

(2942) Wyllie, T. D., and Morehouse, L. G. – *Myxotoxic fungi, mycotoxins, mycotoxicoses, an encyclopaedic handbook.* Marcel Dekker Inc., 1977.

RECENT PUBLICATIONS

(2943) Adamson, R. H., Correa, P., and Dalgard, D. W. – Occurrence of a primary liver carcinoma in a rhesus monkey fed aflatoxin B_1. *J. Natl. Cancer Inst.,* **56**, p. 549–553, 1973.

(2944) Adamson, R. H., Correa, P., Sieber, S. M., Mcintire, K. R., and Dalgaard, D. W. – Carcinogenicity of aflatoxin B_1 in Rhesus monkeys: Two additional cases of primary liver cancer. *J. Natl. Cancer Inst.,* **57**, p. 67–71, 1976.

(2945) Ames, B. N., Durston, W. E., Yamasaki, E., and Lee, F. D. – Carcinogens as mutagens: A simple test system combining liver homogenate for activation and bacteria for detection. *Proc. Natl. Acad. Sci. US,* **70**, p. 2281–2285, 1973.

(2946) Anderegg, R. J., Biemann, K., Buchi, G., and Cushman, M. – Malformin C, a new metabolite of *Aspergillus niger. J. Amer. Chem. Soc.,* **98**, p. 3365–3370, 1976.

(2947) Andersen, R., Buchi, G., Kobbe, B., and Demain, A. L. – Secalonic acids D and F are toxic metabolites of *Aspergillus aculeatus. J. Org. Chem.,* **42**, p. 352–353, 1977.

(2948) Ashoor, S. H., and Chu, F. S. – Interaction of aflatoxin B_{2a} with amino acids and proteins. *Biochem. Pharmacol.,* **24**, p. 1799–1805, 1975.

(2949) Austwick, P. K. C. – Balkan nephropathy. *Proc. roy. Soc. Med.,* **68**, p. 219–221, 1975.

(2950) Barnes, J. M. – *In* Wolstenholme, G. E. W. and Knight, J. *The Balkan nephropathy.* Ciba Study Group **30**, Churchill, London, 1967.

(2951) Barnes, J. M., Carter, R. L., Peristianis, G. C., Austwick, P. K. C., Flynn, F. V., and Aldridge, W. N. – Balkan (endemic) nephropathy and a toxin producing strain of *Penicillium verrucosum* var. *cyclopium*: an experimental model in rats. *The Lancet* p. 671–675, 1977.

(2952) Berger, Y., and Ramout, J. – Isolement et structure de la deoxyaverufinone et de la dehydroaverufine: deux metabolites de l'*Aspergillus versicolor* (Vuillemin) Tiraboschi. *Bull. Soc. Chim. Belg.,* **85**, p. 161–166, 1976.

(2953) Berget, Y., and Jadot, J. Biosynthese de l'averufine et de la versicolorine B: deu metabolites de l'*Aspergillus versicolor. Bull. Soc. Chim. Belg.,* **85**, p. 271–276, 1976.

(2954) Bergot, B. J., Stanley, W. L., and Masri, M. S. – Reaction of coumarin with aqueous ammonia. Implication in detoxification of aflatoxin. *J. Agric. Food Chem.,* **25**, p. 965–966, 1977.

(2955) Bernard C., and Dumas, P. – Action de la rubratoxine B sur la respiration des mitochondries hepatiques de rat. *Mycopathologia,* **55**, p. 53–55, 1975.

(2956) Brekhe, O. L., Sinnhuber, R. O., Peplinski, A. J., Wales, J. H., Putnam, G. B., Lee, D. J., and Ciegler, A. – Aflatoxin in corn: Ammonia inactivation and bioassay with rainbow trout. *Appl. and Environ. Microbiol.*, **34**, p. 34–37, 1977.

(2957) Buchi, G., Kitaura, Y., Yaun, S. S., Wright, H. E., Clardy, J., Demain, A. L., Glinsuka, T., Hunt, W., and Wogan, G. N. – Structure of cytochalasin E, a toxic metabolite of *Aspergillus clavatus. J. Amer. Chem. Soc.,* **95**, p. 5423–5425, 1973.

(2958) Buchi, G., Luk, K. C., Kobbe, B., and Townsend, J. M. – Four new mycotoxins from *Aspergillus clavatus* related to tryptoquivaline. *J. Org. Chem.,* **42**, p. 244–246, 1977.

(2959) Burka, L. T., Kuhnert, L., Wilson, B. J., and Harris, T. M. – Biogenesis of lung-toxic furans produced during microbial infection of sweet potatoes (*Ipomoea batatas*). *J. Amer. Chem. Soc.*, **99**, p. 2302–2305, 1977.

(2960) Burroughs, L. F. – Stability of patulin to SO_2 and to yeast fermentations. *J.A.O.A.C.*, **60**, p. 100–103, 1977.

(2961) Carlton, W. W., Stack, M. E., and Eppley, R. M. – Hepatic alterations produced in mice by xanthomegnin and viomellein, metabolites of *Penicillium viridicatum. Toxicol. Appl. Pharmacol.,* **38**, p. 455–459, 1976.

(2962) Carter, S. B. – Effects of cytochalasins on mammalian cells. *Nature, London,* **213**, p. 261–264, 1967.

(2963) Chang, F. C., and Chu, F. S. – The fate of ochratoxin A in rats. *Fd. Cosmet. Toxicol.,* **15**, p. 199–204, 1977.

(2964) Chedid, A., Bundeally, A. E., and Mendenhall, C. L. – Inhibition of hepatocarcinogenesis by adrenocorticotropin in aflatoxin B_1-treated rats. *J. Natl. Cancer Inst.*, **58**, p. 339–349, 1977.

(2965) Chexal, K. K., Springer, J. P., Clardy, J., Cole, R. J., Kirksey, J. W., Dorner, J. W., Cutler, H. G., and Strawter, B. J. – Austin, a novel polyisoprenoid mycotoxin from *Aspergillus ustus. J. Amer. Chem. Soc.*, **98**, p. 6748–6750, 1976.

(2966) Chu, F. S., Hsia, M.- T. S., and Sun, P. S. – Preparation and characterization of aflatoxin B_1-1-(*o*-carboxymethyl) oxime. *J.A.O.A.C.*, **60**, p. 791–794, 1977.

(2967) Chu, F. S., and Ueno, I. – Production of antibody against aflatoxin B_1. *Appl. Environ. Microbiol.,* **33**, p. 1125–1128, 1977.

(2968) Clardy, J., Springer, J. P., Buchi, G., Masao, K., and Wightman, R. – Tryptoquivaline and tryptoquivalone, two tremorgenic metabolites of *Aspergillus clavatus. J. Amer. Chem. Soc.,* **97**, 663–665, 1975.

(2969) Cole, R. J., Dorner, J. W., Lansden, J. A., Cox, R. H., Pape, C., Cunfer, B., Nicholson, S. S., and Bedell, D. M. – Paspalum staggers: isolation and identification of tremorgenic metabolites from sclerotia of *Claviceps paspali. J. Agric. Fd. Chem.,* **25**, p. 1197–1201, 1977.

(2970) Cole, R. J., Kirksey, J. W., Dorner, J. W., Wilson, D. M., Johnson, J. C., Johnson, A. N., Bedell, D. M., Springer, J. P., Chexal, K. K., Clardy, J. C., and Cox, R. M. – Mycotoxins produced by *Aspergillus fumigatus* species isolated from moulded silage. *J. Agric. Fd. Chem.,* **25**, p. 826–830, 1977.

(2971) Cole, R. J., Kirksey, J. W., and Wells, J. M. – A new tremorgenic metabolite from *Penicillium paxilli. Can. J. Microbiol.,* **20**, p. 1159–1162, 1974.

(2972) Cox, R. H., Churchill, F., Cole, R. J., and Dorner, J. W. – Carbon-13 n.m.r. studies of the structure and biosynthesis of versiconal acetate. *J. Amer. Chem. Soc.*, **99**, p. 3159–3166, 1977.

(2973) Cox, R. H., and Cole, R. J. – Carbon-13 nuclear magnetic resonance studies of fungal metabolites, aflatoxins and sterigmatocystins. *J. Org. Chem.*, **42**, p. 112–114, 1977.

(2974) Curtis, R. W. – Root curvatures induced by culture filtrates of *Aspergillus niger. Science,* **128**, p. 661–662, 1958.

(2975) Demain, A. L., Hunt, N. A., Malik, V., Kobbe, B., Hawkins, M., Matsuo, K., and Wogan, G. N. – Improved procedure for production of cytochalasin E and tremorgenic mycotoxins by *Aspergillus clavatus. Appl. Environ. Microbiol.,* **31**, p. 138–140, 1976.

(2976) Desaiah, D., Hayes, A. W., and Ko, I. K. – Effect of rubratoxin B on adenosine triphosphatase activities in the mouse. *Toxicol. Appl. Pharmacol.,* **39**, p. 71–79, 1977.

(2977) Diebold, G. J., and Zare, R. N. – Laser fluorimetry subpicogram detection of aflatoxins using high pressure liquid chromatography. *Science,* **196**, p. 1439–1441, 1977.

(2978) Dimitrov, M. – *In* Puchlev, A. *Endemic nephritis in Bulgaria.* Sofia, p. 201–207, 1960.

(2979) Eickmann, N., Clardy, J., Cole, R. J., and Kirksey, J. W. – The structure of fumitremorgin A. *Tetrahedron letters,* p. 1051–1054, 1975.

(2980) Elling, F., and Krogh, P. – Fungal toxins and Balkan (endemic) nephropathy. *The Lancet,* p. 1213, 1977.

(2981) Emeh, C. O., and Marth, E. H. – Growth and synthesis of rubratoxin by *Penicillium rubrum* in a chemically defined medium fortified with organic acids and intermediates of the tricarboxylic acid cycle. *Mycopathologia,* **59**, p. 137–142, 1976.

(2982) Engel, G. – Untersuchen zur bildung von mykotoxinen und deren quantitative analyse. *J. Chromatog.,* **130**, p. 293–297, 1977.

(2983) Engstrom, G. W., Richard, J. L., and Cysewski, S. J. – High pressure liquid chromatographic method for detection and resolution of rubratoxins, aflatoxin, and other mycotoxins. *J. Agric. Fd. Chem.,* **25**, p. 833–836, 1977.

(2984) Enomoto, M., and Saito, M. – Carcinogens produced by fungi. *Annual Review of Microbiology,* **26**, p. 279–312, 1972.

(2985) Eppley, R. M., Mazzola, E. P., Highet, R. J., and Bailey, W. J. – Structure of satratoxin H, a metabolite of *Stachybotrys atra.* Application of proton and carbon-13 nuclear magnetic resonance. *J. Org. Chem.*, **42**, p. 240–243, 1977.

(2986) Essigmann, J. M., Croy, R. G., Nadzan, A. M., Busby, W. F., Reinhold, V. N., Buchi, G., and Wogan, G. N. – Structural identification of the major DNA adduct formed by aflatoxin B_1 *in vitro. Proc. Natl. Acad. Sci. US*, **74**, p. 1870–1874, 1977.

(2987) Evans, M. A., and Harbison, R. D. – Prenatal toxicity of rubratoxin B and its hydrogenated analogue. *Toxicol. Appl. Pharmacol.,* **39**, p. 13–22, 1977.

(2988) Fayos, J., Lokensgard, D., Clardy, J., Cole, R. J., and Kirksey, J. W. – Structure of verruculogen, a tremor producing peroxide from *Penicillium verruculosum. J. Amer. Chem. Soc.*, **96**, p. 6785–6787, 1974.

(2989) Fitzell, D. L., Singh, R., Hsieh, D. P. H., and Motell, E. L. – Nuclear magnetic resonance identification of versiconal hemiacetal acetate as an intermediate in aflatoxin biosynthesis. *J. Agric. Fd. Chem.,* **25**, p. 1193–1197, 1977.

(2990) Gallagher, R. T., Clardy, J., and Wilson, B. J. – Aflatrem, a tremorgenic toxin from *Aspergillus flavus. Tetrahedron letters,* in press.

(2991) Gallagher, R. T., and Latch, G. C. M. – Production of the tremorgenic mycotoxins verruculogen and fumitremorgin B by *Penicillium piscarium* Westling. *Appl. Environ. Microbiol.*, **33**, p. 730–731, 1977.

(2992) Gallagher, R. T., Mccabe, T., Hirotsu, K., Clardy, J., Nicholson, J., and Wilson, B. J. – Aflavinine, a novel indole mevalonate metabolite from tremorgen-producing *Aspergillus flavus* species. *Tetrahedron letters,* in press.

(2993) Garner, R. C., Miller, E. C., and Miller, J. A. – Liver microsomal metabolism of aflatoxin B_1 to a reactive derivative toxic to *Salmonella typhimurium* TA 1530. *Cancer Res.,* **32,** p. 2058–2066, 1972.

(2994) Garner, R. C., and Wright, C. M. – Induction of mutations in DNA repair deficient bacteria by a liver microsomal metabolite of aflatoxin B_1. *Brit. J. Cancer,* **28,** p. 544–551, 1973.

(2995) Glinsukon, T., Shank, R. C., Wogan, G. N., and Newberne, P. M. – Acute and subacute toxicity of cytochalasin E in the rat. *Toxicol. Appl. Pharmacol.,* **32,** p. 135–146, 1975.

(2996) Gopolan, C., Tulpule, P. G., and Krishnamurthi, D. – Induction of hepatic carcinoma with aflatoxin in the Rhesus monkey. *Fd. Cosmet. Toxicol.,* **10,** p. 519–521, 1972.

(2997) Gorst-Allman, C. P., Pachler, K. G. R., Steyn, P. S., and Scott, D. B. – Carbon-13 nuclear magnetic resonance assignments of some fungal C_{20} anthraquinones; their biosynthesis in relation to that of aflatoxin B_1. *J. Chem. Soc. Perkin Trans. I,* p. 2181–2188, 1977.

(2998) Gorst-Allman, C. P., Pachler, K. G. R., Steyn, P. S., Wessels, P. L., and Scott, D. B. – Biosynthesis of averufin in *Aspergillus parasiticus* from $[^{13}C]$ acetate. *J. Chem. Soc. Chem. Commun.,* p. 916–917, 1976.

(2999) Gurtoo, H. L., and Motycka, L. – Effect of sex difference on the *in vitro* and *in vivo* metabolism of aflatoxin B_1 by the rat. *Cancer Res.,* **36,** p. 4663–4671, 1976.

(3000) Hacking, A. – Personal communication of material from the Annual Report of the East Midland Region of ADAS, Ministry of Agriculture, Fisheries and Food, 1977.

(3001) Hacking, A. and Harrison, J. – Mycotoxins in animal feeds. *In* Skinner, F. A. and Carr, J. G. *Microbiology in agriculture, fisheries and food,* SAB Symposium, **4,** p. 243–250, 1976.

(3002) Hamilton, P. B. – Interrelationships of mycotoxins with nutrition. *Fed. Proc.,* **36,** p. 1899–1902, 1977.

(3003) Hayes, A. W. – Action of rubratoxin B on mouse liver mitochondria. *Toxicology,* **6,** p. 253–261, 1976.

(3004) Hayes, A. W., and Mccain, H. W. – A procedure for the extraction and estimation of rubratoxin B from corn. *Fd. Cosmet. Toxicol.,* **13,** p. 221–229, 1975.

(3005) Heathcote, J. G., Dutton, M. F., and Hibbert, J. R. – Biosynthesis of aflatoxins. Part II. *Chemistry and Industry,* p. 270–272, 1976.

(3006) Hesseltine, C. W. – Natural occurrence of mycotoxins in cereals. *Mycopath. Mycol. Applicata,* **53,** p. 141–153, 1974.

(3007) Holder, C. L., Nony, C. R., and Bowman, M. C. – Trace analysis of zearalenone and/or zearalenol in animal chow by HPLC and GLC. *J.A.O.A.C.,* **60,** p. 272–278, 1977.

(3008) Hsia, M. T. S., and Chu, F. S. – Reduction of aflatoxin B_1 with zinc borohydride: an efficient preparation of aflatoxicol. *Experientia,* **33,** p. 1132–1133, 1977.

(3009) Hsieh, D. P. H., Lin, M. T., Yao, R. C., and Singh, R. – Biosynthesis of aflatoxin. Conversion of norsolorinic acid and other hypothetical intermediates into aflatoxin B_1. *J. Agric. Fd. Chem.,* **24,** p. 1170–1174, 1976.

(3010) Hsieh, D. P. H., Seiber, J. N., Reece, C. A. Fitzell, D. L., Yang, S. L., Dalezios, J. I., La Mar, G. N., Budd, D. L., and Mottell, E. – ^{13}C-nuclear magnetic resonance spectra of aflatoxin B_1 derived from acetate. *Tetrahedron,* **31,** p. 661–663, 1975.

(3011) Jarvis, B. – Mycotoxins in food – their occurrence and significance. *Intern. J. Environmental Studies,* **8**, p. 187–194, 1975.

(3012) Jarvis, B. – Mycotoxins in food. *In* Skinner, F. A., and Carr, J. G., *Microbiology in agriculture, fisheries and food, S.A.B. Symposium 4*, p. 251–267, 1976.

(3013) Jarvis, B. and Burke, C. S. – Practical and legislative aspects of the chemical preservation of food. *In* Skinner, F. A., and Hugo, W. B., *Inhibition and inactivation of vegetative microbes. SAB Symposium 5*, p. 345–367, 1976.

(3014) Josefsson, B. G. E., and Moller, T. E. – Screening method for the detection of aflatoxin, ochratoxin, patulin, sterigmatocystin, and zearalenone in cereals. *J.A.O.A.C.,* **60**, p. 1369–1371, 1977.

(3015) Judah, D. J., Legg, R. F., and Neal, G. E. – Development of resistance to cytotoxicity during aflatoxin carcinogenesis. *Nature,* **265**, p. 343–345, 1977.

(3016) Kellerman, T. S., Pienaar, J. G., Van der Westhuizen, G. C. A., Anderson, L. A. P., and Naude, T. W. – A highly fatal tremorgenic mycotoxicosis of cattle caused by *Aspergillus clavatus. Onderstepoort J. Vet. Res.*, **43**, p. 147–154, 1977.

(3017) Kobbe, B., Cushman, M., Wogan, G. N., and Demain, A. L. – Production and antibacterial activity of malformin C, a toxic metabolite of *Aspergillus niger. Appl. Environ. Microbiol.*, **33**, p. 996–997, 1977.

(3018) Kramer, R. K., Davis, N. D., and Diener, U. L. – Byssotoxin A, a secondary metabolite of *Byssochlamys fulva. Appl. Environ. Microbiol.,* **31**, p. 249–253, 1976.

(3019) Krogh, P. – Mycotoxic nephropathy, *in* Purchase, I. F. H., ed., Mycotoxins p. 419–428, 1974.

(3020) Krogh, P., Elling, F., Gyrd-Hansen, N., Hald, B., Larsen, A. R. Lillehoj, E. B., Madsen, A., Mortensen, H. P., and Ravnskov, U. – Experimental porcine nephropathy. Changes of renal function and structure perorally induced by crystalline ochratoxin A. *Acta pathol. microbiol. scand. Sec. A.,* **84**, p. 429–434, 1976.

(3021) Krogh, P., Elling, F., Hald, B., Larsen, A. E., Lillehoj, E. B., Madsen, A., and Mortensen, H. P. – Time-dependent disappearance of ochratoxin A residues in tissues of bacon pigs. *Toxicology,* **6**, p. 235–242, 1976.

(3022) Langone, J. J., and Van Vunakis, H. – Aflatoxin B_1, specific antibodies and their use in radioimmunoassay. *J. Natl. Cancer Inst.,* **56**, p. 591–595, 1976.

(3023) Lansden, J. A. – A clean up procedure for high pressure liquid chromatography of aflatoxins in agricultural commodities. *J. Agric. Fd. Chem.,* **25**, p. 969–971, 1977.

(3024) Lawellin, D. W., Grant, D. W., and Joyce, B. K. – Aflatoxin localization by the enzyme-linked immunocytochemical technique. *Appl. Environ. Microbiol.,* **34**, p. 88–93, 1977.

(3025) Leaich, L. L., and Papa, K. E. – Aflatoxins in mutants of *Aspergillus flavus. Mycopathol. Mycol. Appl.,* **52**, p. 223–229, 1974.

(3026) Liao, L.-L., Grollman, A. P., and Horowitz, S. B. – Mechanisms of action of the 12, 13-epoxytrichothecene, anguidine, an inhibitor of protein synthesis. *Biochim. Biophys. Acta,* **454**, p. 273–284, 1976.

(3027) Malaiyandi, M., Vesonder, R. F., and Ciegler, A. – Large-scale production, purification, and a study of some spectral properties of penitrem A. *J. Environ. Sci. Health. Part B.,* **B11**, p. 139–164, 1976.

(3028) Mantle, P. G., Mortimer, P. H., and White, E. P. – Mycotoxic tremorgens of *Claviceps paspali* and *Penicillium cyclopium*: a comparative study of effects on sheep and cattle in relation to natural staggers syndromes. *Res. Vet. Sci.,* **24**, p. 49–56, 1977.

(3029) Martin, C. N., and Garner, R. C. – Aflatoxin B-oxide generated by chemical or enzymic oxidation of aflatoxin B_1 causes guanine substitution in nucleic acids. *Nature,* **265**, p. 863–865, 1977.

(3030) Moreau, C. – Les mycotoxines dans les produits laitiers. *Revue Générale des Questions Laitières (Tome L VI) No. 555–556,* p. 286–303, 1976.

(3031) Moreau, S., and Caean, M. – Eremofortin C. A new metabolite obtained from *Penicillium roqueforti* cultures and from the biotransformation of PR toxin, *J. Org. Chem.*, 42, p. 2632–2634, 1977.

(3032) Moreau, S., Gaudemer, A., Lablanche-Combier, A., and Biguet, J. – Metabolites de *Penicillium roqueforti*: PR toxin et metabolites associes. *Tetrahedron letters*, p. 833–834, 1976.

(3033) Moss, M. O. – Aflatoxin and related mycotoxins. *In* Harborne, J. B., *Phytochemical ecology,* Phytochemical Society Symposium, **8**, p. 125–144.

(3034) Moss, M. O. – Role of mycotoxins in disease of man. *Int. J. Environ. Studies,* **8**, p. 165–170, 1975.

(3035) Moss, M. O., Robinson, F. V., and Wood, A. B. – Rubratoxins. *J. Chem. Soc.,* Section C, p. 619–624, 1971.

(3036) Moule, Y., and Darracq, N. – Effect of aflatoxin B_1 on transcription and translation in the liver of mice. *Biochimie (Paris),* **58**, p. 1011–1013, 1976.

(3037) Moule, Y., Jemmali, M., and Rousseau, N. – Mechanism of the inhibition of transcription by PR toxin, a mycotoxin from *Penicillium roqueforti. Chem. Biol. Interactions,* **14**, p. 207–216, 1976.

(3038) Moule, Y., Moreau, S., and Bousquet, J. F. – Relationships between the chemical structure and the biological properties of some eremophilane compounds related to PR toxin. *Chem. Biol. Interactions,* **17**, p. 185–192, 1977.

(3039) Pachler, K. G. R., Steyn, P. S., Vleggaar, R., Wessels, P. L., and Scott, D. B. – Carbon-13 nuclear magnetic resonance assignments and biosynthesis of aflatoxin B_1 and sterigmatocystin. *J. Chem. Soc. Perkin I,* p. 1182–1189, 1976.

(3040) Papa, K. E. – The parasexual cycle in *Aspergillus flavus. Mycologia,* **65**, p. 1201–1205, 1973.

(3041) Papa, K. E. – Linkage groups in *Aspergillus flavus. Mycologia,* **68**, p. 159–165, 1976.

(3042) Papa, K. E. – Genetic analysis of a mutant of *Aspergillus flavus* producing more aflatoxin B_2 than B_1. *Mycologia,* **69**, p. 556–562, 1977.

(3043) Papa, K. E. – Mutant of *Aspergillus flavus* producing more aflatoxin B_2 than B_1. *Appl. Environ. Microbiol.*, **33**, p. 206, 1977.

(3044) Patterson, D. S. P. – Metabolism as a factor in determing the toxic action of the aflatoxins in different animal species. *Fd. Cosmet. Toxicol.,* **11**, p. 287–294, 1973.

(3045) Patterson, D. S. P., and Roberts, B. A. – Aflatoxin metabolism in duck-liver homogenates: the relative importance of reversible cyclopentenone reduction and hemiacetal formation. *Fd. Cosmet. Toxicol.,* **10**, p. 501–512, 1972.

(3046) Pons, W. A., and Franz, A. O. – High performance liquid chromatography of aflatoxins in cottonseed products. *J.A.O.A.C.,* **60**, p. 89–95, 1977.

(3047) Preston, R. S., Haynes, J. R., and Campbell, T. C. – The effect of protein deficiency on the *in vivo* binding of aflatoxin B_1 to rat liver macromolecules. *Life Sciences,* **19**, p. 1191–1198, 1976.

(3048) Rajkovic, I. A. – The effect of some nutritional parameters on rubratoxin biosynthesis by *Penicillium rubrum.* M. Phil. Thesis, University of Surrey, 1977.

(3049) Reye, R. D. K., Morgan, G. and Boral, J. – Encephalopathy and fatty degeneration of viscera, a disease entity in childhood. *Lancet*, **2**, p. 749–752.

(3050) Richard, J. L., Thurston, J. R., and Graham, C. K. – Changes in complement activity, serum proteins, and prothrombin time in guinea-pigs fed rubratoxins alone or in combination with aflatoxin. *American J. Vet. Res.*, **35**, p. 957–959, 1974.

(3051) Riche, C., Paccard-Billy, C., Devys, M., Gaudemer, A., Barbier, M., and Bousquet, J. F. – Structure cristalline et moleculaire de la phomenone, phytotoxine produite par le champignon *Phoma exigua* var. *inoxydabilis*. *Tetrahedron letters*, p. 2765–2766, 1974.

(3052) Robberts, B. A., and Patterson, D. S. P. – The multimycotoxin screening method and the use of confirmatory tests in the detection of mycotoxins in animal feedstuffs. 2nd Meeting on Mycotoxins in animal disease. Patterson, D. S. P., Pepin, G. A., and Shreeve, B. J., eds., p. 40–45, 1976.

(3053) Rodricks, J. V., and Eppley, R. M. – *Stachybotrys* and stachybotryotoxicosis. *In* Purchase, I. F. H., *Mycotoxins*, p. 181–197, 1974.

(3054) Roebuck, B. D., and Wogan, G. N. – Species comparison of *in vitro* metabolism of aflatoxin B_1. *Cancer Res.*, **37**, p. 1649–1656, 1977.

(3055) Salhab, A. S., Abramson, F. P., Geelhoed, G. W., and Edwards, G. S. – Aflatoxicol M_1, a new metabolite of aflatoxicol. *Xenobiotica*, **7**, p. 401–408, 1977.

(3056) Salhab, A. S., and Edwards, G. S. – Comparative *in vitro* metabolism of aflatoxicol by liver preparation from animals and humans. *Cancer Res.*, **37**, p. 1016–1021, 1977.

(3057) Sarasin, A., Goze, A., Devoret, R., and Moule, Y. – Induced reactivation of u.v.-damaged phage λ in *E. coli* K 12 host cells treated with aflatoxin B_1 metabolites. *Mutation Res.*, **42**, p. 205–215, 1977.

(3058) Sarasin, A. R., Smith, C. A., and Hanawalt, P. C. – Repair of DNA in human cells after treatment with activated aflatoxin B_1. *Cancer Res.*, **37**, p. 1786–1793, 1977.

(3059) Schroeder, H. W., and Carlton, W. W. – Accumulation of only aflatoxin B_2 by a strain of *Aspergillus flavus*. *Appl. Microbiol.*, **25**, p. 146–148, 1973.

(3060) Scott, P. M., Merrien, M.-A., and Polansky, J. – Roquefortine and isofumigaclavine A, metabolites from *Penicillium roqueforti*. *Experientia*, **32**, p. 140–142, 1976.

(3061) Senn, P. – Zur biosynthese von Rubratoxin B. Ph.D. Thesis, University of Basel, 1976.

(3062) Seto, H., Cary, L. W., and Tanabe, M. – Utilization of the ^{13}C–^{13}C coupling in structural and biosynthetic studies. V. The ^{13}C n.m.r. spectrum of sterigmatocystin. *Tetrahedron letters*, p. 4491–4494, 1974.

(3063) Shank, R. C., Gordon, J. E., and Wogan, G. N. – Dietary aflatoxins and human liver cancer. III. Field survey of rural Thai families for ingested aflatoxins. *Fd. Cosmet. Toxicol.*, **10**, p. 71–84, 1972.

(3064) Shank, R. C., Bourgeois, C. H., Keschamras, N., and Chandavimol, P. – Aflatoxins in autopsy specimens from Thai children with an acute disease of unknown aetiology. *Fd. Cosmet. Toxicol.*, **9**, p. 501–507, 1971.

(3065) Shotwell, O. L., and Gorden, M. L. – Aflatoxin, comparison of analysis of corn by various methods. *J.A.O.A.C.*, **60**, p. 83–88, 1977.

(3066) Shotwell, O. L., Goulden, M. L., Bennett, G. A., Plattner, R. D., and Hesseltine, C. W. – Survey of 1975 wheat and soybeans for aflatoxin zearalenone and ochratoxin. *J.A.O.A.C.*, **60**, p. 778–783, 1977.

(3067) Singh, R., and Hsieh, D. P. H. – Aflatoxin biosynthetic pathway: elucidation by using blocked mutants of *Aspergillus parasiticus*. *Arch. Biochm. Biophys.*, **178**, p. 285–292, 1977.

(3068) Smith, S. J., Deen, K. C., Caldhoun, W. J., and Beittenmiller, H. F. – Effects of dietary feeding of aflatoxin B_1 on ribosomal RNA metabolism in rat liver. *Cancer Res.*, **37**, p. 2226–2231, 1977.

(3069) Sphon, J. A., Dreiffuss, P. A., and Schulten, H.-R. – Field desorption mass spectrometry of mycotoxins and mycotoxin mixtures, and its application as a screening technique for foodstuffs. *J.A.O.A.C.*, **60**, p. 73–82, 1977.

(3070) Stack, M. E., Eppley, R. M., Dreifuss, P. A., and Pohland, A. E. – Isolation and identification of xanthomegnin, viomellein, rubrosulphin and viopurpurin as metabolites of *Penicillium viridicatum. Appl. Environ. Microbiol.*, **33**, p. 351–355, 1977.

(3071) Steyn, P. S. – Austamide, a new toxic metabolite from *Aspergillus ustus. Tetrahedron letters,* p. 3331–3334, 1971.

(3072) Steyn, P. S., and Vleggaar, R. – Austocystins. Six novel dihydrofuro (3′2′:4 5) furo (3, 2–6) xanthenones from *Aspergillus ustus. J. Chem. Soc. Perkin Trans. I,* p. 2250–2256, 1974.

(3073) Steyn, P. S., and Vleggaar, R. – The structure of dihydrodeoxy 8-epi austdiol and the absolute configuration of the azaphilones. *J. Chem. Soc. Perkin Trans. I,* p. 204–206, 1976.

(3074) Stinson, E. E., Huhtanen, C. N., Zell, T. E., Schwartz, D. P., and Osman, S. F. – Determination of patulin in apple juice products as the 2:4-dinitrophenylhydrazone derivative. *J. Agric. Fd. Chem.*, **25**, p. 1220–1222, 1977.

(3075) Stoloff, L., and Dalrymple, B. – Aflatoxin and zearalenone occurrence in dry-milled corn products. *J.A.O.A.C.*, **60**, p. 579–582, 1977.

(3076) Swensen, D. H., Lin, J. K., Miller, E. C., and Miller, J. A. – Aflatoxin B_1- – 2,3-epoxide as a probable intermediate in the covalent binding of aflatoxins B_1 and B_2 to rat liver DNA and ribosomal RNA *in vivo. Cancer Res.*, **37**, p. 172–181, 1977.

(3077) Tamura, S., Kuyama, S., Kodaira, Y., and Higashikawa, S. – The structure of destruxin B, a toxic metabolite of *Oospora destructor. Agric. Biol. Chem.*, **28**, p. 137–138, 1964.

(3078) Tanabe M., Uramoto, M., Hamasaki, T., and Cary. L. – Biosynthetic studies with ^{13}C: a novel Favorsky rearrangement route to the fungal metabolite aspyrone. *Heterocycles,* **5**, p. 355–364, 1976.

(3079) Takahashi, D. M. – Reversed-phase high performance liquid chromatographic analytical system for aflatoxins in wines with fluorescence detection. *J. Chromatog.*, **131**, p. 147–156, 1977.

(3080) Tilak, T. B. – Induction of cholangiocarcinoma following treatment of a Rhesus monkey with aflatoxin. *Fd. Cosmet. Toxicol.*, **13**, p. 247–249, 1975.

(3081) Tilak, T. B. G., Nagarajan, V., and Tulpule, P. G. – Resistance of cold-exposed rats to aflatoxin. *Experientia,* **32**, p. 1179–1180.

(3082) Vleggaar, R., Steyn, P. S., and Nagel, D. W. – Constitution and absolute configuration of austdiol, the main toxic metabolite from *Aspergillus ustus. J. Chem. Soc. Perkin Trans. I,* p. 45–49, 1974.

(3083) Wallace, E. M., Pathoe, S. V., Mirocha, C. J., Robison, T. S., and Fenton, S. W. – Synthesis of radiolabelled T-2 toxin. *J. Agric. Fd. Chem.*, **25**, p. 836–838, 1977.

(3084) Watson, S. A., and Hayes, A. W. – Evaluation of possible sites of action of rubratoxin B – induced polyribosomal disaggregation in mouse liver. *J. Toxicol. Environmental Health,* **2**, p. 639–650, 1977.

(3085) Wei, R.-D., Schnoes, H. K., Hart, P. A., and Strong, F. M. – The structure of PR toxin, a mycotoxin from *Penicillium roqueforti. Tetrahedron,* **31**, 109–114, 1975.

(3086) Wei, R.-D., Still, P. E., Smalley, E. R., Schnoes, H. K., and Strong, F. M. – Isolation and characterization of a mycotoxin from *Penicillium roqueforti*. *Appl. Environ. Microbiol.,* **25**, p. 111–114, 1973.

(3087) Yamazaki, M., Fujimoto, H., and Kawasaki, T. – The structure of a tremorgenic metabolite from *Aspergillus fumigatus* Fres.: fumitremorgin A. *Tetrahedron letters,* p. 241–244, 1975.

(3088) Yoshizawa, T., Morooka, N., Sawada, Y., and Udagawa, S. I. – Tremorgenic mycotoxins from *Penicillium paraherquei* Abe ex G. Smith. *Appl. Environ. Microbiol.*, **32**, p. 441–442, 1976.

(3089) Zwicker, G. M., Carlton, W. W., and Tuite, J. – Prolonged administration of *Penicillium viridicatum* to mice: Preliminary report of carcinogenicity. *Fd. Cosmet. Toxicol.,* **11**, p. 989–994, 1973.

TAXONOMIC INDEX

*indicates illustration

SUBJECT INDEX

*indicates formula